any event on the timeline could be placed in more than one category. Some scholars claim that globalization has been happening for a long time, and others say it has only been happening for a relatively short time. As you read through this timeline, think about how current events happening in your social world are part of this larger process of global change.

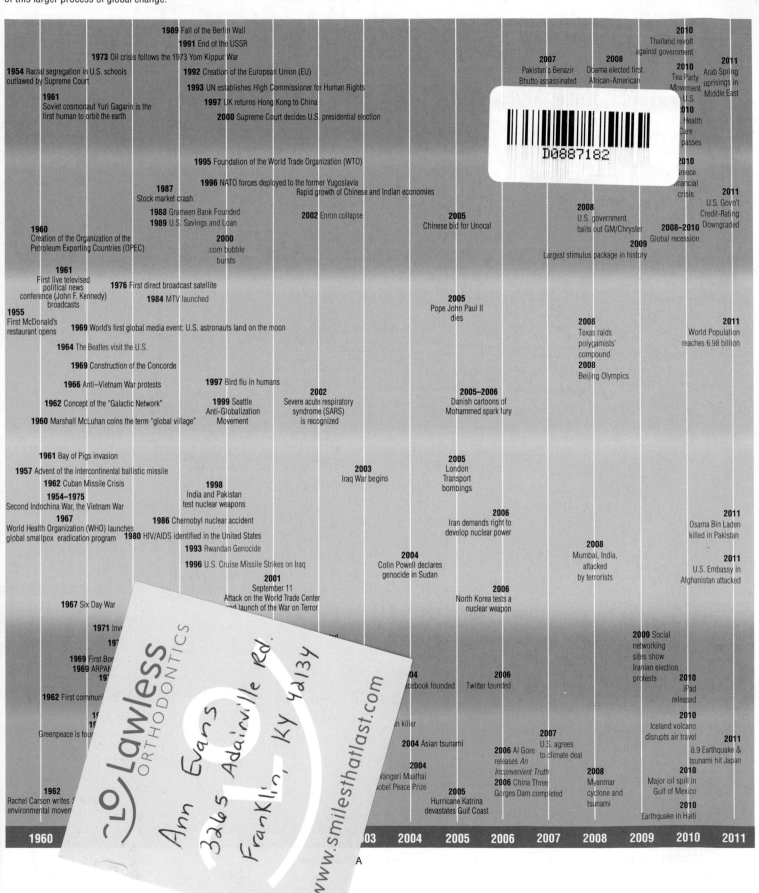

SECOND EDITION

OUR SOCIAL WORLD

Condensed Version

Introduction to Sociology

Jeanne H. Ballantine | Keith A. Roberts

To the Student

Improve Your Grade With These Student Study Tools!

Our praised student study site at **www.sagepub.com/oswcondensed2e** features

- Updated and revised audio episodes of National Public Radio's *This American Life*
- New video links from YouTube, PBS's *Frontline*, and TED
- Updated chapter quizzes, flashcards, and internet exercises
- Updated "Learning From SAGE Journal Articles" features for each chapter with related discussion questions
- New recommended readings
- and much more!

Chapter-opening visual representations of the **Social World Model**—streamlined for clarity and simplicity—display the micro, meso, and macro levels for that chapter's topic, with "**Think About It**" questions to focus your thinking on key ideas.

This feature provides active learning opportunities to encourage you to further engage with the subject matter.

"**Thinking Sociologically**" features throughout the text challenge your understanding of core concepts and ask you to apply the material to your own life.

Color bars at the end of each chapter pose important questions for you to consider and provide you with a brief look ahead to what will be covered in the next chapter.

Chapter-ending review material includes "**What Have We Learned?**" boxes features to ensure mastery of each chapter's core material.

Thinking Sociologically 🌐

Think about the meaning, belonging, and structural system of a religion with which you are familiar. Illustrate how the elements influence and are influenced by the social world from individual to national and global systems.

Think About It	
Self and Inner Circle	How can sociology help me understand my own life and my sense of self?
Local Community	How can sociology help me to be a more effective employee and citizen in my community?
National Institutions; Complex Organizations; Ethnic Groups	How do sociologists help us understand and even improve our lives in families, classrooms, and health care offices?
National Society	How do national loyalty and national policies affect my life?
Global Community	How might global events affect my life?

The next issue, then, is how we gather data that inform how we understand and influence the social world. When we say we know something about society, how is it that we know? What is considered evidence in sociology, and what lens (theory) do we use to interpret the data? These are the central issues of the next chapter.

Contributing to Our Social World: What Can We Do?

At the end of this and all subsequent chapters, you will find suggestions for work, service learning, internships, and volunteering that encourage you to apply the concepts, principles, and ideas discussed in the chapter in practical contexts.

At the Local Level:

At the Regional and National Levels:

- *The American Sociological Association (ASA)* is the leading professional organization of sociologists in the United States. It has several programs and initiatives of special interest to students. Visit the ASA website at www.asanet .org and click on the "Teaching & Learning/Students: Undergraduate" link at the top of the page. Read the

What Have We Learned?

We live in a complex social world with many layers of interaction. If we really want to understand our own lives, we need to comprehend the various levels of analysis that affect our lives and the dynamic connections between those levels. Moreover, as citizens of democracies, we need to understand how to influence our social environments, from city councils, school boards, and state legislatures to congressional, presidential, and other organizations with major policy makers. To do so wisely, we need both objective lenses for viewing this complex social world and accurate, valid information (facts) about the society. As the science of society, sociology can provide empirical data and

lives and personal troubles are shaped by historical and structural events outside of our everyday lives. It also prods us to see how we can influence our society. (See pp. 9–10.)

- Sociology is a social science and, therefore, uses the tools of the sciences to establish credible evidence to understand our social world. As a science, sociology is scientific and objective rather than value laden. (See pp. 10–13.)

- Sociology has pragmatic value and benefits, including those (See pp. 13–17.)

Excerpt From Review in *Teaching Sociology*

Student Praise for *Our Social World*

To the Instructor

A New Intro Text for a New Generation of Students . . .
Incredibly Successful in Its First Edition and Now Even Better!

Written by two award-winning sociology instructors who are passionate about excellence in teaching, this textbook has been adopted at a variety of community colleges, four-year colleges, and comprehensive research universities, and these schools have used it in traditional as well as online Introduction to Sociology courses. Some of the many schools that have adopted the book include

- American River College
- Boise State University
- Buffalo State College
- Cape Cod Community College
- Capital Community College
- Capital University
- Chaffey College
- Cleveland State University
- College of Staten Island, The City University of New York
- Drexel University
- Hanover College
- Hofstra University
- Indiana University

- Indiana University of Pennsylvania
- Ithaca College
- La Salle University
- Le Moyne College
- Long Island University
- Messiah College
- Minot State University
- Monroe County Community College
- New Mexico State University
- Ohio State University
- Pennsylvania State University
- Quinebaug Valley Community College
- St. Cloud State University

- San Francisco State University
- Shippensburg University
- Sinclair Community College
- The State University of New York
- Tennessee State University
- Towson University
- University of Central Oklahoma
- University of Maryland
- University of Nebraska
- University of Richmond
- University of Tennessee
- University of the Ozarks
- Villa Julie College
- Wright State University

. . . . and many, many more! We would like to say THANK YOU to our loyal adopters who chose *Our Social World* in its previous editions, making it such an overwhelming success. We hope you find the second edition of *Our Social World, Condensed Version* even more exciting.

Instructor Praise for *Our Social World*

"Unlike most textbooks that I have read, the breadth and depth of coverage . . . is very impressive. This text forces the students deep into the topics covered and challenges them to see the interconnectedness of them."
—Keith Kerr, Texas A&M University

"I love the global emphasis, the applied material, and the emphasis on solutions to social problems."
—Gina Carreno, Florida Atlantic University

"So often students ask, 'What can I do with sociology?' Having this applied information interspersed in the text allows them to get answers to that question over and over again."

"Finally, a text that brings sociology to life! . . . This is so well written that I'm not sure the students will even realize they are learning theory!"
—Martha Shockey, St. Louis University

"This is an excellent textbook it has definitely made life easier, and the ancillary material is extremely helpful."
—Jamie M. Dolan, Carroll College

New to This Edition!

- A brand-new feature, "**Engaging Sociology**," appears in every chapter. Each Engaging Sociology exercise encourages active learning, providing students with an opportunity to use sociology to solve a problem or analyze some aspect of society.
- The **Social World Model** has been simplified for greater clarity and increased ease of use.
- "**Think About It**" questions have moved to the first page of each chapter, giving students a framework for understanding what they are about to read.
- Twenty-three new featured essays and five new "**The Applied Sociologist at Work**" boxes have been added.
- Chapters 2 (**Examining the Social World: How Do We Know?**), 3 (**Society and Culture: Hardware and Software of Our Social World**), and 13 (**Politics: Penetrating Power**) have been reorganized.
- A new section opener in Part IV clearly defines and introduces institutions—a concept that introductory students typically find difficult to understand.
- Dozens of new and updated tables have been added, as have more than 300 new references, more than 120 new photos, and two new Photo Essays.

Global Community

Society

National Organizations, Institutions, and Ethnic Subcultures

Local Organizations and Community

Me (and My Network of Close Friends)

Micro: Networks in organizations—alumni, civic groups

Meso: Ethnic organizations, political parties, religious denominations

Macro: Connections between citizens of a nation

Macro: Global networks; United Nations, international courts; transnational corporations

Photo Essay
Houses as Part of a Society's Material Culture

Homes are good examples of material culture. Their construction is influenced by local materials, but also ideas of what a home is. Homes shape the context in which family members interact. Indeed, in some cases, homes become status symbols that are far larger than the family needs, but the family is making a prestige statement about their socioeconomic standing.

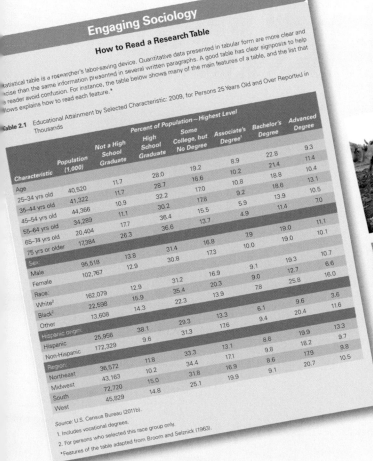

Engaging Sociology
How to Read a Research Table

A statistical table is a researcher's labor-saving device. Quantitative data presented in tabular form are more clear and concise than the same information presented in several written paragraphs. A good table has clear signposts to help the reader avoid confusion. For instance, the table below shows many of the main features of a table, and the list that follows explains how to read each feature.*

Table 2.1 Educational Attainment by Selected Characteristic: 2009, for Persons 25 Years Old and Over Reported in Thousands

Characteristic	Population (1,000)	Not a High School Graduate	High School Graduate	Some College, but No Degree	Associate's Degree[1]	Bachelor's Degree	Advanced Degree
Age					8.9	22.8	9.3
25–34 yrs old	40,520	11.7	28.0	19.2	10.2	21.4	11.4
35–44 yrs old	41,322	11.7	28.7	16.6	10.8	18.8	10.4
45–54 yrs old	44,366	10.9	32.2	17.0	9.2	18.6	13.1
55–64 yrs old	34,289	11.1	30.2	17.8	5.9	13.9	10.5
65–74 yrs old	20,404	17.7	36.4	15.5	4.9	11.4	7.0
75 yrs or older	17,384	26.3	36.6	13.7	7.9	19.0	11.1
Sex:				16.8	10.0	19.0	10.1
Male	95,518	13.8	31.4	17.3		19.3	10.7
Female	102,767	12.9	30.8		9.1	12.7	6.6
Race:			31.2	16.9	9.0	25.8	16.0
White[2]	162,079	12.9	35.4	20.3	7.8	9.6	3.6
Black[2]	22,598	15.9	22.3	13.9	6.1	20.4	11.6
Other	13,608	14.3	29.3	13.3	9.4	19.9	13.3
Hispanic origin:		38.1	31.3	17.6	8.6	18.2	9.7
Hispanic	25,956	9.6		13.1	9.8	17.9	9.8
Non-Hispanic	172,329		33.3	17.1	8.6	20.7	10.5
Region:		11.8	34.4	16.9	9.1		
Northeast	36,572	10.2	31.8	19.9			
Midwest	43,163	15.0	25.1				
South	72,720	14.8					
West	45,829						

Source: U.S. Census Bureau (2011b).

1. Includes vocational degrees.

2. For persons who selected this race group only. From Broom and Selznick (1963).

*Features of the table adapted from Broom and Selznick (1963).

The invention of the plow was essential for agricultural societies to develop, and in the early period of agriculture, plows were pushed by people and then pulled by animals. The harnessing of energy was taken to another level when gasoline engines could pull the plow and cultivate thousands of acres. This represents the beginning of industrialization. Modern machinery such as this harvester has pushed farming into the new level of productivity.

This Japanese schoolgirl might think that eating with a fork or spoon is quite strange. She has been well socialized to know that polite eating involves competent use of chopsticks.

Thinking Sociologically

Think about the meaning, belonging, and structural systems of a religion with which you are familiar. Illustrate how these elements influence and are influenced by the social world, from individual to national and global systems.

Finally, here is a text that engages students. *Our Social World* uses a unique, dynamic approach to focus on developing sociological skills of analysis rather than simply emphasizing memorization of basic ideas.

The text is both personal and global. It introduces sociology clearly, with the updated Social World Model providing a clear and cohesive framework that lends integration and clarity to the course. With the theme of globalization intertwined throughout the text and an emphasis on deep learning and the applied side of sociology, *Our Social World*, Third Edition, inspires both critical thinking and community participation.

Teach Students the Basics of Sociology via the Newly Simplified, Visually Compelling "Social World Model"

This model, which provides the organizing framework for the text and visually introduces each chapter, illustrates the level of analysis of each topic (from micro to meso to macro), and addresses how each topic is related.

Engage Students . . . With Engaging Sociology

Active learning exercises keep sociology fun, drawing students into an analysis of a relevant table, the application of a population pyramid to the business world, or an interactive survey on the differences in social and cultural capital for first generation students. This element replaces the last edition's "How Do We Know?" feature, although you will still see your favorite "How Do We Know" examples reprised within "Sociology in Our Social World."

Introduce a Global Perspective, Asking Students to Be Citizens of the World

This text uniquely weaves a truly global perspective into each part of the book, challenging students to think of themselves as global citizens rather than as local citizens looking out at others in the world. "Sociology Around the World" features introduce fascinating historical or global examples of key concepts.

Encourage Deep Learning and Critical Thinking with Uniquely Effective Pedagogy

"Thinking Sociologically" questions, "What Have We Learned?" end-of-chapter summaries, and "Sociology in Our Social World" features encourage critical thought, challenging students to reflect on how the material is relevant and applicable to their lives.

Inspire an Active Engagement in Sociology Using Motivating Chapter Features

"Contributing to Our Social World: What Can We Do?" suggestions include work and volunteer opportunities in which students can apply their newfound sociological knowledge right away. "The Applied Sociologist at Work" boxes—five of which are new to this edition—introduce students to a variety of people with sociological degrees, illuminating post-degree career options.

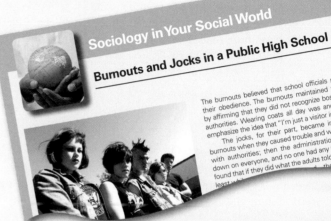

Sociology Around the World
Tunisian Village Meets the Modern World

This is a story of change as macro-level innovations enter a small traditional village. It illustrates how the social units of the social world model and the three levels of analysis enter into sociological analysis. As you read, try to identify both the units and levels of analysis being discussed and the impact of globalization on a community that cannot know what these changes will bring.

The workday began at dawn as usual in the small fishing village on the coast of Tunisia, North Africa. Men prepared their nets and boats for the day, while women prepared breakfast and dressed the young children for school. About 10 A.M., it began—the event that would [change] this picturesque village forever. Bulldozers [came] followed by trench diggers and cement [...]

beautiful village; others viewed the changes as destroying a lifestyle that was all they and generations before them had known.

Today, the village is dwarfed by the huge hotel, and the locals are looked on as quaint curiosities by the European tourists. Fishing has become a secondary source of employment to working in the hotel and casino or selling local crafts and trinkets to souvenir-seeking visitors. Many women are now employed outside the home by the hotel, creating new family structures as grandparents, unemployed men, and other relations take over child-rearing responsibilities.

To understand the changes in this one small village [...] change, a soci-

Sociology in Your Social World
Burnouts and Jocks in a Public High School

The burnouts believed that school officials should [...] their obedience. The burnouts maintained their di[...] by affirming that they did not recognize bossy adu[...] authorities. Wearing coats all day was another w[...] emphasize the idea that "I'm just a visitor in this so[...]

The jocks, for their part, became irritated [...] burnouts when they caused trouble and were bel[...] with authorities; then the administration wou[...] down on everyone, and no one had any freedom[...] found that if they did what the adults told them t[...] least [...] they got a [...]

What Have We Learned?

Theories serve as lenses to help us make sense of the data that we gather with various research strategies. However, the data themselves can be used to test the theories, so there is an ongoing reciprocal relationship between theory (the lens for making sense of the data) and the research (the evidence used to test the theories) The most important ideas in this chapter are what sociology considers evidence and how sociology operates as a science.

Key Points:

[...]derstand society have existed for at le[...]

(through secondary sources or through content analysi[...] (See pp. 37–40.)

- Use of multiple methods—triangulation—incr[...] confidence in the findings. (See p. 39.)
- Scientific confidence in results also requires repre[...] tive samples, usually drawn randomly. (See pp. [...]
- Sociology has three strong traditions: One str[...] objective nature of science; another emphas[...] uniquely human qualities, such as humans [...]

Contributing to Our Social World: What Can We Do?

At the Local Level:

- Local service organizations, found in every community, work to provide for unmet needs of community members: housing, legal aid, medical care, elder care, and so on. United Way works with most local service organizations. Volunteer to work with the organization in its needs assessment research and learn more about the sociological principles and research methods that are used.

At the Meso Level:

Applied and Clinical Sociol[...]

At the National and Global Levels:

- The U.S. Bureau of the Census is best known for it[...] nial (occurring every 10 years) enumeration of [...] lation, but its work continues each year as i[...] special reports, population estimates, and re[...] lications (including *Current Population Report*[...] Census Bureau's website at www.census.gov [...] the valuable and extensive amount of quan[...] and other information available, or visi[...] Census Bureau office to discuss volunteer [...]

Ancillaries

Compatible With Blackboard and Other Course Management Systems

To facilitate the use of the Second Edition of *Our Social World, Condensed Version* with Blackboard and other course management platforms, SAGE provide the following teaching and learning ancillaries at **www.sagepub.com/oswcondensed2e**:

A password-protected **Instructor Teaching Site**

An open-access **Student Study Site**

Instructor Teaching Site

The Instructor Teaching Site features a variety of popular and effective teaching aids, including

Chapter Outlines: Carefully crafted outlines follow the structure of each chapter, providing an essential reference and teaching tool.

Chapter Exercises and Activities: These include lively and stimulating ideas for use both in and out of class to reinforce active learning. The activities apply to individual or group projects.

Course Syllabi: Sample syllabi—for semester, quarter, and online classes—provide suggested models for creating the syllabus for your course.

Web Resources: These links to relevant websites direct both instructors and students to additional resources for further research on important chapter topics.

Photographic Essay Projects: Unique assignments encourage students to observe and evaluate social issues in creative ways.

SAGE Journal Articles: A "Learning From SAGE Journal Articles" feature provides access to recent, relevant full-text articles from SAGE's leading research journals. Each article supports and expands on the concepts presented in the chapter. Also provided are discussion questions to focus and guide student interpretation.

Video Resources: Carefully selected, web-based video resources feature relevant interviews, lectures, personal stories, inquiries, and other content for use in independent or classroom-based explorations of key topics. Discussion questions are provided to guide interpretation of material. The site also includes new video links for YouTube and for PBS's *Frontline*.

Test Bank (Word): This Word test bank offers a diverse set of multiple-choice, true/false, short-answer, and essay test questions and answers for every chapter to aid instructors in assessing students' progress and understanding.

Test Bank (Diploma): This electronic test bank using Diploma software is available for use with a PC or Mac. The test bank offers a diverse set of multiple-choice, true/false, short-answer, and essay test questions and answers for every chapter to aid instructors in assessing students' progress and understanding.

PowerPoint Slides: Chapter-specific slide presentations offer assistance with lecture and review preparation by highlighting essential content, features, and artwork from the book.

Teaching Tips: Teaching Tips provide suggestions and resources for using the Social World Model for traditional and online Introduction to Sociology courses.

Additional Resource: Adopters will receive free online access to Bryant's *21st Century Sociology: A Reference Handbook* (SAGE, ©2007), to help make teaching globally even more impactful. Contact Customer Care at 1-800-818-SAGE (7243) for more information on this extraordinary teaching aid.

Student Study Site

To further enhance students' understanding of and interest in the course material, we have created a Student Study Site, which includes

Podcasts and Audio Clips: Each chapter includes links to podcasts, which cover important topics and are designed to supplement key points within the text. The site also includes updated and revised audio episodes of National Public Radio's *This American Life*.

Video Resources: Carefully selected, web-based video resources feature relevant interviews, lectures, personal stories, inquiries, and other content for use in independent or classroom-based explorations of key topics. Discussion questions are provided to guide interpretation of material. The site includes new video links for YouTube and for PBS's *Frontline*.

Web Quizzes: Self-quizzes allow students to independently assess their progress in learning course material.

Web Exercises and Activities: These links direct both instructors and students to useful and current web resources, as well as creative activities to extend and reinforce learning.

Flashcards: This study tool reinforces student understanding of key terms and concepts that have been outlined in the chapters.

SAGE Journal Articles: A "Learning From SAGE Journal Articles" feature provides access to recent, relevant full-text articles from SAGE's leading research journals. Each article supports and expands on the concepts presented in the chapter. Discussion questions focus and guide student interpretation.

Recommended Readings: Interesting and relevant supplements provide a jumping-off point for course assignments, papers, research, group work, and class discussion.

and much more!

SAGE Teaching Innovations and Professional Development Awards Fund

Largely inspired by the authors of *Our Social World*, SAGE has created this awards fund to help graduate students and pretenure faculty attend the annual American Sociological Association preconference, hosted by the Section on Teaching and Learning in Sociology, with grants of $500 per recipient. In its first year, there were 13 award recipients. Since that time, there have been 23 additional authors who have joined the program as cosponsors, and in the five years of the program, 95 recipients have benefitted from this award. In 2008, the ASA's Section on Teaching and Learning in Sociology awarded SAGE with a glass plaque to honor the publisher and its participating authors and editors—including the authors of *Our Social World*—for their commitment to excellence in teaching.

About the Authors

Jeanne H. Ballantine (far right) is University Professor of Sociology at Wright State University, a state university of about 17,000 students in Ohio.

She has also taught at several four-year colleges, including an "alternative" college and a traditionally black college, and at international programs in universities abroad. Jeanne has been teaching introductory sociology for more than 30 years with a mission to introduce the uninitiated to the field and to help students see the usefulness and value in sociology. Jeanne has been active in the teaching movement, shaping curriculum, writing and presenting research on teaching, and offering workshops and consulting in regional, national, and international forums. She is a Fulbright Senior Scholar and serves as a Departmental Resources Group consultant and evaluator.

Jeanne has written several textbooks, all with the goal of reaching the student audience. As the original director of the Center for Teaching and Learning at Wright State University, she scoured the literature on student learning and served as a mentor to teachers in a wide variety of disciplines. Local, regional, and national organizations have honored her for her teaching and for her contributions to helping others become effective teachers. In 1986, the American Sociological Association's Section on Undergraduate Education (now called the Section on Teaching and Learning in Sociology) recognized her with the Hans O. Mauksch Award for Distinguished Contributions to Undergraduate Sociology. In 2004, she was honored by the American Sociological Association with its Distinguished Contributions to Teaching Award.

Keith A. Roberts (top left) is Professor of Sociology at Hanover College, a private liberal arts college of about 1,100 students in Indiana. He has been

teaching introductory sociology for more than 35 years with a passion for active learning strategies and a focus on "deep learning" by students that transforms the way students see the world. Prior to teaching at Hanover, he taught at a two-year regional campus of a large university. Between them, these authors have taught many types of students at different types of schools.

Keith has been active in the teaching movement, writing on teaching and serving as a consultant to sociology departments across the country in his capacity as a member of the ASA Departmental Resources Group. He is the coauthor of a very popular textbook in the sociology of religion (with David Yamane), has coauthored a book on writing in the undergraduate curriculum, and annually runs workshops for high school sociology teachers. He has been honored for his teaching and teaching-related work at local, state, regional, and national levels. The American Sociological Association's Section on Teaching and Learning awarded him the Hans O. Mauksch Award for Distinguished Contributions to Undergraduate Sociology in 2000. In 2010 he was awarded the American Sociological Association's Distinguished Contributions to Teaching Award.

2004 AMERICAN SOCIOLOGICAL ASSOCIATION

DISTINGUISHED CONTRIBUTIONS TO TEACHING AWARD

2010 AMERICAN SOCIOLOGICAL ASSOCIATION

DISTINGUISHED CONTRIBUTIONS TO TEACHING AWARD

Visit www.sagepub.com for valuable
Intro to Sociology supplemental texts
1-800-818-SAGE (7243)
www.sagepub.com

The world is condensing, and it will be a different social place for our children and grandchildren.

It is in that spirit that we dedicate this book to our respective grandchildren:

Hannah Ballantine, Caleb Ballantine, Kai Jolly-Ballantine, Ayla Jolly-Ballantine, Zainakai Blair-Roberts, and others "on the way to becoming." May the condensed global society they inhabit be an increasingly humane one.

OUR SOCIAL WORLD
Condensed Version | SECOND EDITION

Jeanne H. Ballantine | Keith A. Roberts
Wright State University | *Hanover College*

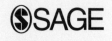

SAGE

Los Angeles | London | New Delhi
Singapore | Washington DC

Los Angeles | London | New Delhi
Singapore | Washington DC

FOR INFORMATION:

SAGE Publications, Inc.
2455 Teller Road
Thousand Oaks, California 91320
E-mail: order@sagepub.com

SAGE Publications Ltd.
1 Oliver's Yard
55 City Road
London, EC1Y 1SP
United Kingdom

SAGE Publications India Pvt. Ltd.
B 1/I 1 Mohan Cooperative Industrial Area
Mathura Road, New Delhi 110 044
India

SAGE Publications Asia-Pacific Pte. Ltd.
33 Pekin Street #02-01
Far East Square
Singapore 048763

Acquisitions Editor: Dave Repetto
Associate Editor: Maggie Stanley
Editorial Assistant: Lydia Balian
Production Editor: Eric Garner
Copy Editor: Melinda Masson
Typesetter: C&M Digitals (P) Ltd.
Proofreader: Theresa Kay
Indexer: Diggs Publication Services
Cover Designer: Edgar Abarca
Marketing Manager: Erica Deluca
Permissions Editor: Karen Ehrmann

Copyright © 2012 by SAGE Publications, Inc.

Printed in Canada

Library of Congress Cataloging-in-Publication Data

Ballantine, Jeanne H.
Our social world : condensed version / Jeanne H. Ballantine, Keith A. Roberts.—2nd ed.

p. cm.
Includes bibliographical references and index.

ISBN 978-1-4129-8727-1 (pbk.)

1. Sociology. 2. Sociology—Cross-cultural studies. I. Roberts, Keith A. II. Title.

HM586.B352 2012
301—dc23   2011031322

This book is printed on acid-free paper.

12 13 14 15 10 9 8 7 6 5 4 3

Brief Contents

Detailed Contents

Chapter 14 • The Process of Change: Can We Make a Difference? 438

Preface

To Our Readers

This book asks you to think outside the box. Why? The best way to become a more interesting person, to grow beyond the old familiar thoughts and behaviors, and to make life exciting is to explore new ways to view things. The world in which we live is intensely personal and individual in nature, with much of our social interaction occurring in intimate circles of friends and family. Our most intense emotions and most meaningful links to others are at this "micro" level of social life.

However, these intimate micro-level links in our lives are influenced by larger social structures and global trends. At the start of the second decade of the twenty-first century, technological advances make it possible to connect with the farthest corners of the world. Multinational corporations cross national boundaries, form new economic and political unions, and change job opportunities of people everywhere. Some groups embrace the changes, while others try to protect their members from the rapid changes that threaten to disrupt their traditional lives. Even our most personal relationships or what we eat tonight may be shaped by events on the other side of the continent or the globe. From the news headlines to family and peer interactions, we confront sociological issues daily. The task of this book and of your instructor is to help you see world events and your personal lives from a sociological perspective. Unless you learn how to look at our social world with an analytical lens, many of its most intriguing features will be missed.

The social world you face in the job markets of the twenty-first century is influenced by changes and forces that are easy to miss. Like the wind, which can do damage even if the air is unseen, social structures are themselves so taken for granted that it is easy to miss them. However, their effects can be readily identified. Sociology provides new perspectives, helping students understand their families, their work lives, their leisure, and their place in a diverse and changing world.

A few of you will probably become sociology majors. Others will find the subject matter of this course relevant to your personal and professional lives. Some of the reasons the authors of this book and your own professor chose to study sociology many years ago are the factors that inspire undergraduates today to choose a major in sociology: learning about the social world from a new perspective; working with people and groups; developing knowledge, inquiry, and interpersonal skills; and learning about social life, from small groups to global social systems. As the broadest of the social sciences, sociology has a never-ending array of fascinating subjects to study. This book touches only the surface of what the field has to offer and the exciting things you can do with this knowledge. These same considerations motivated us and gave us direction in writing this introductory book.

Where This Book Is Headed

A well-constructed course, like an effective essay, needs to be organized around a central question, one that spawns other subsidiary questions and intrigues the participants. The problem with introductory courses in many disciplines is that there is no central question and thus no coherence to the course. They have more of a flavor-of-the-week approach (a different topic each week), with no attempt at integration. We have tried to correct that problem in this text.

The Social World Model

For you to understand sociology as an integrated whole rather than a set of separate chapters in a book, we have organized the chapters in this book around the *social world model:* a conceptual model that demonstrates the relationships between individuals (micro level); organizations, institutions, and subcultures (meso level); and national societies and global structures (macro level). At the beginning of each chapter, a visual diagram of the model will illustrate this idea as it relates to the topic of that chapter, including how issues related to the topic have implications at various levels of analysis in the social world. For example, socialization is explored in Chapter 4 as a part of the whole social world, influencing and being affected by other parts of society. No aspect of society exists in a vacuum. On the other hand, this model does not assume that everyone always gets along or that relationships are always harmonious or supportive. Sometimes, different parts of the society are in competition for resources, and intense conflict and hostility may be generated.

This micro- to macro-level analysis is a central concept in the discipline of sociology. Many instructors seek first and foremost to help students develop a *sociological imagination,* an ability to see the complex links between various levels of the social system. This is a key goal of this book. Within a few months, you may not remember all of the specific concepts or terms that have been introduced, but if the way you see the world has been transformed during this course, a key element of deep learning has been accomplished. Learning to see things from alternative perspectives is a precondition for critical thinking. This entire book attempts to help you recognize connections between your personal experiences and problems and larger social forces of society. You will be learning to take a new perspective on the social world in which you live.

A key element of that social world is diversity. We live in societies in which there are people who differ in a host of ways: ethnicity, socioeconomic status, religious background, political persuasion, gender, sexual orientation, and so forth. Diversity is a blessing in many ways to a society because the most productive and creative organizations and societies are those that are highly diverse. This is the case because people with different backgrounds solve problems in very different ways. When people with such divergences come together, the outcome of their problem solving can create new solutions to vexing problems. However, diversity often creates challenges as well. Misunderstanding and "we" versus "they" thinking can divide people. These issues will be explored throughout this book. We now live in a global village, and in this book, you will learn something about how people on the other side of the village live and view the world.

We hope you enjoy the book and get as enthralled with sociology as we are. It genuinely is a fascinating field of study.

Jeanne H. Ballantine
Wright State University

Keith A. Roberts
Hanover College

Instructors

How to Make This Book Work for You

pecial features woven throughout each chapter support the theme of the book. These will help students comprehend and apply the material and make the material more understandable and interesting. These features are also designed to facilitate deep learning, to help students move beyond rote memorization and increase their ability to analyze and evaluate information.

For students to understand both the comparative global theme and sociology as an integrated whole rather than as a set of separate chapters in a book, we have organized the chapters in this book around the *social world model:* a conceptual model that demonstrates the relationships between individuals (micro level); organizations, institutions, and subcultures (meso level); and national societies and global structures (macro level). At the beginning of each chapter, a visual diagram of the model will illustrate this idea as it relates to the topic of that chapter, including how issues related to the topic have implications at various levels of analysis in the social world.

"Think About It"

So that students can become curious, active readers, we have posed questions as part of the introductory figure that we hope are relevant to everyday life but that are also tied to the micro–meso–macro levels of analysis that serve as the theme of the book. The purpose is to transform students from passive readers who run their eyes across the words into curious, active readers who read to answer a question and to be reflective. Active or deep reading is key to comprehension and retention of reading material. Instructors can also use this feature to encourage students to think critically about the implications of what they have read. Instructors might want to ask students to write a paragraph about one of these questions before coming to class each day. These questions might also provide the basis for in-class discussions.

Students should be encouraged to start each chapter by reading and thinking about these questions, looking at the topics in the chapter, and asking some questions of their own. This will mean they are more likely to stay focused, remember the material long-term, and be able to apply it to their own lives.

A Global Perspective and the Social World Model

We are part of an ever-shrinking world, a global village. What happens in distant countries not only is news the same day but also affects relatives living in other countries, the cost of goods, work and travel possibilities, and the balance of power in the world. Instead of simply including cross-cultural examples of strange and different peoples, this book incorporates a global perspective throughout. This is done so students can see how others live different but rewarding lives and so they can see the connections between others' lives and their own. Students will need to think about and relate to the world globally in future roles as workers, travelers, and global citizens. Our analysis illustrates the interconnections of the world's societies and their political and economic systems and demonstrates that what happens in one part of the world affects others. For instance, if a major company in the area moves much of its operations to another country with cheaper labor, jobs are lost, and the local economy is hurt.

This approach attempts to instill interest, understanding, and respect for different groups of people and their lifestyles. Race, class, and gender are an integral part of understanding the diverse social world, and these features of social life have global implications. The comparative global theme is carried throughout the book in headings and written text, in examples, and in boxes and selection of photos. As students read this book, they should continually think about how the experiences in their private world are influenced by and may influence events at other levels: the community, organizations and institutions, the nation, and the world.

Opening Vignettes

Chapters typically open with an illustration relevant to the chapter content. For instance, in Chapter 2, "Examining the Social World," the case of Hector, a Brazilian teenager living in poverty in a favela, is used to illustrate research methods and theory. In Chapter 4, "Socialization," the case of an immigrant child attending school in his new society and facing a new learning environment is discussed. Chapter 8 begins with an actual account of a girl who is among the 27 million slaves in the modern world. These vignettes are meant to interest students in the upcoming subject matter by helping them relate to a personalized story. In several cases, including Hector's story, the vignettes serve as illustrations throughout the chapters.

"Thinking Sociologically" Questions

Following major topics, students will find questions that ask them to think critically and apply the material just read to some aspect of their lives or the social world. The purpose of this feature is to encourage students to apply the ideas and concepts in the text to their lives and to develop critical thinking skills. These questions can be the basis for in-class discussions and can be assigned as questions to start interesting conversations with friends and families to learn how the topics relate to students' own lives. Note that some of these questions have a miniature icon—a small version of the chapter-opening model—signifying that these questions reinforce the theme of micro, meso, and macro levels of social influence.

Key Concepts, Examples, and Writing Style

Key terms that are defined and illustrated within the running narrative appear in **bold**. Other terms that are defined but are of less significance are italicized. The text is rich in examples that bring sociological concepts to life for student readers. Each

chapter has been student tested for readability. Both students and reviewers describe the writing style as reader friendly, often fascinating, and accessible but not watered down.

Special Features

Although there are numerous examples throughout the book, featured inserts provide more in-depth illustrations of the usefulness of the sociological perspective to understand world situations or events with direct relevance to a student's life. There are four kinds of special features. "Sociology in Your Social World" features focus on a sociological issue or story, often with policy implications. "Sociology Around the World" features take readers to another part of the globe to explore how things are different from (or how they are the same as) what they might experience in their own lives. "The Applied Sociologist at Work" features appear in many chapters and examine profiles of contemporary sociologists who are working in the field. This helps students grasp what sociologists can actually do with sociology. The most innovative new feature in this edition is called "Engaging Sociology"—and the double entendre is intentional. We want students to think of sociology as engaging and fun, and these features are designed to engage—to draw students into active analysis of a table, application of a population pyramid to the business world, or taking a survey to understand why differences in social and cultural capital make first-generation students feel alienated on a college campus.

Technology and Society

Nearly every chapter examines issues of technology that are relevant to that chapter. We have especially sought out materials that have to do with the Internet and with communications technology.

Social Policy and Becoming an Active Citizen

Some chapters include discussion of social policy issues: an effort to address the concerns about public sociology and the relevance of sociological findings to current social debates. Further, because students sometimes feel helpless to know what to do about social issues that concern them at macro and meso levels, we have concluded every chapter with a few ideas about how they might become involved as active citizens, even as undergraduate students. Suggestions in the "Contributing to Our Social World: What Can We Do?" sections may be assigned as service learning or term projects or simply used as suggestions for ways students can get involved on their own time.

Summary Sections

Each chapter ends with a brief "What Have We Learned?" summary to ensure mastery of the chapter's core material.

A Little (Teaching) Help From Our Friends

Whether the instructor is new to teaching or an experienced professor, there are some valuable ideas that can help invigorate and energize the classroom. The substantial literature on teaching methodology tells us that student involvement is key to the learning process. Built into this book are discussion questions and projects that students can report on in class. In addition, there are a number of suggestions in the supplements and teaching aids for active learning in large or small classes.

Instructor Resources Site

The Instructor Resources Site (**www.sagepub.com/osw condensed2e**) contains a number of helpful teaching aids, from goals and objectives for chapters to classroom lecture ideas, active learning projects, collaborative learning suggestions, and options for evaluating students. Suggestions for the use of visual materials—videos, transparencies, and multimedia—are also included.

Test Bank

Compatible with both PC and Mac computers, the computerized test bank is available on a CD separate from the Instructor Resources Site. The test bank allows for easy question sorting and exam creation. It includes multiple-choice, true/false, short-answer, and essay questions with the page number(s) or section of the book from which the question

is taken. In keeping with the deep-learning thrust of this book, however, the test questions will have more emphasis on application skills than on rote memorization—the latter a too-common characteristic of test banks.

PowerPoint Slides

Because visuals are an important addition to classroom lectures, recognizing the varying learning styles of students, PowerPoint slides that include lecture outlines and relevant tables, maps, diagrams, pictures, and short quotes are included in the Instructor Resources Site for instructors to use in the classroom.

Student Study Site

To further enhance students' understanding of and interest in the material, we have created a student website to accompany the text. This website includes the following:

- **Flash cards** that allow for easy reviewing of key terms and concepts;
- **Self-quizzes** that can be used to check students' understanding of the material or can be sent in to the professor for a grade;
- **Web exercises** that direct students to various sites on the web and ask them to apply their knowledge to a particular topic;
- *This American Life* radio segments that illustrate each chapter's concepts;
- **"Learning From Journal Articles"** features that include original research from SAGE journal articles and teach students how to read and analyze a journal article;
- A list of **recommend websites** that students can explore for research or their own edification;
- Information on how to create **photo essays**;
- and much more!

Visit **www.sagepub.com/oswcondensed2e**.

What Is Different About This *Condensed* Edition?

The title of this book, like the subtitles of each chapter, plays on double entendre. The world has indeed condensed—becoming more accessible and interconnected, with communication and transportation making possible contacts with and travel to the most remote parts of the globe. That reality comes through in this book. However, this book is also a condensed version of our longer introductory book, *Our Social World*. We did surveys of sociologists to find out what was essential and what was dispensable. The one thing that was not dispensable was the thematic nature of the book—that we absolutely not turn this into another "core concepts" book that focuses on memorization of concepts rather than transformative learning through perspective taking. We were committed to making this a book that avoids the typical cross between a dictionary and an encyclopedia. Like the more comprehensive book, this is a coherent essay on the sociological imagination, understood globally. Moreover, many "brief" books are edited and reduced by developmental editors rather than by the authors. That is not true of this volume. The authors used a fine scalpel to refine use of language, cut some boxed features, reduced the size and number of photos, and kept an eye on both quality of the text and cost to students of the final product.

There are more than two dozen new boxed features in this book, including the innovative "Engaging Sociology" features that are described above. Faculty members who have test-driven the "Engaging Sociology" features in class have found them highly engaging to students and deeply enhancing to the learning-teaching process.

Although Chapter 13 still has more focus on politics than economics (with economics discussed in many other chapters—such as Chapter 7 on socioeconomic inequality), the discussion of economics was expanded and more fully integrated into the chapter. We have also made hundreds of other changes—adding new information, updating data, and in some cases changing topics within chapters. For example, this edition has more than 400 entirely new references and 120 new photos.

The core elements of the book—with the unifying theme and the social world model at the beginning of every chapter—have not changed. For greater simplicity in understanding the social world model, the "Think About It" questions were moved to the first page of the chapter rather than placed on the opening page.

Finally, although we have been told that the writing was extraordinarily readable, we have tried to simplify sentence structure in a number of places. In short, we have tried to respond to what we heard from all of you—both students and instructors (and, yes, we *do* hear from students)—to make this book more engaging and more accessible.

A Personal Note to the Instructor

We probably share many of the same reasons for choosing sociology as our careers. Our students also share these reasons for finding sociology a fascinating and useful subject: learning about the social world from a new perspective; working with people and groups; developing a range of knowledge, inquiry, and interpersonal skills; and learning the broad and interesting subject matter of sociology, from small groups to societies. In this book, we try to share our own enthusiasm for the subject with students. The following explains what we believe to be unique features of this book and some of our goals and methods for sharing sociology. We hope you share our ideas and find this book helps you meet your teaching and learning goals.

What Is Distinctive About This Book?

What is truly distinctive about this book? This is a text that tries to break the mold of the typical textbook synthesis, the cross between an encyclopedia and a dictionary. *Our Social World, Condensed* is a unique course text that is **a coherent essay on the sociological imagination—understood globally**. We attempt to radically change the feel of the introductory book by emphasizing coherence, an integrating theme, and current knowledge about learning and teaching, as we present traditional content. Instructors will not have to throw out the well-honed syllabus and begin from scratch, but they can refocus each unit so it stresses understanding of micro-level personal troubles within the macro-level public issues framework. Indeed, in this book, we make clear that the public issues must be understood as global in nature.

Here is a text that engages students. *They* say so! From class testing, we know that the writing style, the structure of chapters and sections, the "Thinking Sociologically" features, the wealth of examples, and other pedagogical aids help students stay focused, think about the material, and apply it to their lives. It neither bores them nor insults their intelligence. It focuses on deep learning rather than memorization. It develops sociological skills of analysis rather than emphasizing memorization of vocabulary. Key concepts and terms are introduced but only in the service of a larger focus on the sociological imagination. The text is both personal and global. It speaks to sociology as a science as well as addressing applied aspects of sociology. It has a theme that provides integration of topics as it introduces the discipline. This text is an analytical essay, not a disconnected encyclopedia.

As one of our reviewers noted,

> Unlike most textbooks I have read, the breadth and depth of coverage in this one is very impressive. It challenges the student with college-level reading. Too many textbooks seem to write on a high school level and give only passing treatment to most of the topics, writing in nugget-sized blocks. More than a single definition and a few sentences of support, the text forces the student deep into the topics covered and challenges them to see interconnections.

Normally, the global-perspective angle within textbooks, which seemed to grow in popularity in the mid-to-late 1990s, was implemented by using brief and exotic examples to show differences between societies—a purely comparative approach rather than a globalization treatment. They gave, and still give to a large extent, a token nod to diversity. This textbook, however, forces the student to take a broader look at similarities and differences in social institutions around the world and at structures and processes operating in all cultures and societies.

So our focus in this book is on deep learning, especially expansion of students' ability to role-take or "perspective-take." Deep learning goes beyond the content of concepts and terms and cultivates the habits of thinking that allow one to think critically. Being able to see things from the perspective of others is essential to doing sociology, but it is also indispensible to seeing weaknesses in various theories or recognizing blind spots in a point of view. Using the sociological imagination is one dimension of role-taking because it requires a step back from the typical micro-level understanding of life's events and fosters a new comprehension of how meso- and macro-level forces—even global ones—can shape the individual's life. Enhancement of role-taking ability is at the core of this book because it is a *prerequisite* for deep learning in sociology and it is the core competency needed to *do* sociology. One cannot do sociology unless one can see things from various places on the social landscape.

This may sound daunting for some student audiences, but we have found that instructors at every kind of institution have had great success with the book because of the writing style and teaching tools used throughout. We have made some strategic decisions based on these principles of learning and teaching. We have focused much of the book on higher-order thinking skills rather than memorization and regurgitation. We want students to learn to think sociologically: to apply, analyze, synthesize, evaluate, and comprehend the interconnections of the world through a globally informed sociological imagination. However, we think it is also essential to do this with an understanding of how students learn.

Many introductory-level books offer several theories and then provide a critique of the theory. The idea is to teach critical thinking. We have purposefully refrained from extensive critique of theory (although some does occur) for several reasons. First, providing critique to beginning-level students does not really teach critical thinking. It trains them to memorize someone else's critique. Furthermore, it simply confuses many of them, leaving students with the feeling that sociology is really just contradictory ideas and the discipline really does not have anything firm to offer. Teaching critical thinking needs to be done in stages, and it needs to take into account the building steps that occur before effective critique is possible. That is why we focus on the concept of deep learning. We are working toward building the foundations that are necessary for sophisticated critical thought at upper levels in the curriculum.

Therefore, in this beginning-level text, we have attempted to focus on a central higher-order or deep-learning skill—synthesis. Undergraduate students need to grasp this before they can fully engage in evaluation. Deep learning involves understanding of complexity, and some aspects of complexity need to be taught at advanced levels. While students at the introductory level are often capable of synthesis, complex evaluation requires some foundational skills. Thus, we offer contrasting theories in this text, and rather than telling what is wrong with each one, we encourage students through "Thinking Sociologically" features to analyze the use of each and to focus on honing synthesis and comparison skills.

Finally, research tells us that learning becomes embedded in memory and becomes long-lasting only if it is related to something that learners already know. If they memorize terms but have no unifying framework to which they can attach those ideas, the memory will not last until the end of the course, let alone until the next higher-level course. In this text, each chapter is tied to the social world model that is core to sociological thinking. At the end of a course using this book, we believe that students will be able to explain coherently what sociology is and construct an effective essay about what they have learned from the course as a whole. Learning to develop and defend a thesis, with supporting logic and evidence, is another component of deep learning. A text that is mostly a dictionary does not enhance that kind of cognitive skill.

Organization and Coverage

Reminiscent of some packaged international tours, in which "it is Day 7, so this must be Paris," many introductory courses seem to operate on the principle that it is Week 4 so this must be deviance week. Students do not sense any integration, and at the end of the course, they have trouble remembering specific topics. This book is different. A major goal of the book is to show the integration between topics in sociology and between parts of the social world. The idea is for students to grasp the concept of the interrelated world. A change in one part of the social world affects all others, sometimes in ways that are mutually supportive and sometimes in ways that create intense conflict.

Although the topics are familiar, the textbook is organized around levels of analysis, explained through the social world model. This perspective leads naturally to a comparative approach and discussions of diversity and inequality.

Each chapter represents a part of the social world structure (society, organizations and groups, and institutions) or a process in the social world (socialization, stratification, and change). Chapter order and links between chapters clarify this idea. Part I (Chapters 1 and 2) introduces the student to the sociological perspective and tools of the sociologist: theory and methods. Part II (Chapters 3–6) examines "Social Structure, Processes, and Control," exploring especially processes such as socialization, interaction, and networks. Part III (Chapters 7–9) covers the core issue of inequality in society, with emphasis on class, race or ethnicity, and gender. Part IV (Chapters 10–12) turns to the structural dimensions of society, as represented in institutions. Unlike most section openers, which are limited to one page, the section opener on institutions is more substantial, and we recommend that you include the section opener in your assignments regardless of which institutions you cover. It defines institutions, examines their contribution to society at the meso level, and explains the strong ties between them. Rather than trying to be all-encompassing, we examine family, education and religion, and politics and economics to help students understand how structures affect their lives. We do not cover sports, science, mass media, health, or military in separate chapters, but aspects of emerging institutions are woven into many chapters. It was a painful decision to cut our coverage of material from the comprehensive book, but attention to length and cost required hard choices. Part V (Chapters 13 and 14) turns to social dynamics: how societies change. Population patterns, urbanization and environmental issues, social movements, technology, and other aspects of change are included.

As instructors and authors, we value books that provide students with a well-rounded overview of approaches to the field. Therefore, this book takes an eclectic theoretical approach, drawing on the best insights of various theories and stressing that multiple perspectives enrich our understanding. We give attention to most major theoretical perspectives in sociology: structural-functionalist and conflict theories at the meso and macro levels of analysis and symbolic interaction and rational choice theories at the micro and meso levels of analysis. Feminist, postmodern, and ecological theories are discussed where relevant to specific topics. Each of these is integrated into the broad social world model, which stresses development of a sociological imagination.

The book includes 14 chapters plus additional online materials, written to fit into a semester or quarter system. It allows instructors to use the chapters in order or to alter the order, because each chapter is tied into others through the social world model. We strongly recommend that Chapter 1 be used early in the course because it introduces the integrating model and explains the theme. Otherwise, the book has been designed for flexible use. Instructors may also want to supplement the core book with other materials, such as those suggested in the Instructors' Resources on CD-ROM. While covering all of the key topics in introductory sociology, the cost and size of a "condensed" book allows for this flexibility. Indeed, for a colorful introductory-level text, the cost of this book is remarkably low—roughly a third of the cost of some other popular introductory texts.

A Unique Program Supporting Teaching of Sociology

There is one more way in which *Our Social World* has been unique among introductory sociology textbooks. In 2007, the authors teamed with SAGE to start a new program to benefit the entire discipline. Using royalties from *Our Social World* and *Our Social World, Condensed,* there is a new award program called the SAGE Teaching Innovations and Professional Development Awards Fund. It is designed to prepare a new generation of scholars within the teaching movement in sociology. People in their early career stages (graduate students, assistant professors, newly minted PhDs) can be reimbursed $500 each for expenses entailed while attending the daylong American Sociological Association (ASA) Section on Teaching and Learning in Sociology's preconference workshop. The workshop is the day before ASA meetings. In 2007, 13 people received this award and benefited from an extraordinary workshop on learning and teaching. Subsequently joined by 23 other SAGE authors and by the publisher, a total of 95 people have been beneficiaries by 2011. We are pleased to have had a hand in initiating and continuing to support this program.

We hope you find this book engaging. If you have questions or comments, please contact us.

Jeanne H. Ballantine
Wright State University
jeanne.ballantine@gmail.com

Keith A. Roberts
Hanover College
robertsk@hanover.edu

Acknowledgments

Knowledge is improved through careful, systematic, and constructive criticism. The same is true of all writing. This book is of much greater quality because we had such outstanding critics and reviewers. We, therefore, wish to honor and recognize the outstanding scholars who served in this capacity. These scholars are listed on this page and the next.

We also had people who served in a variety of other capacities: drafting language for us for special features, doing library and Internet research to find the most recent facts and figures, and reading or critiquing early manuscripts. Contributors include Khanh Nguyen, Kate Ballantine, Jessica Hoover, Kelly Joyce, Justin Roberts, Kent Roberts, Susan Schultheis, and Vanessa M. Simpson. Authors of short sections within the book include Jeremy Castle, Leslie Elrod, Peter Dreier, James Faulkner, Melanie Hughes, Wendy Ng, Amy J. Orr, and Elise Roberts.

Both of us are experienced authors, and we have worked with some excellent people at other publishing houses. However, the team at SAGE Publications was truly exceptional in support, thoroughness, and commitment to this project. Our planning meetings have been fun, intelligent, and provocative. Ben Penner, former SAGE acquisitions editor, signed us for this project; David Repetto, now the acquisitions editor, has continued to provide excellent support; and Jerry Westby has been a sustaining member of the team as executive editor. Other folks who have meant so much to the quality production of this book include Maggie Stanley, associate editor who has taken care of so very many details; Lydia Balian, the new editorial assistant; Eric Garner, our production manager; Melinda Masson, copy editor extraordinaire and in reality a research assistant; John O'Neill, permissions editor; Erica DeLuca, senior marketing manager; Claudia Hoffman, managing editor; Scott Hooper, manufacturing manager; Ravi Balasuriya, art director; Steven Martin, vice president—production; Michele Sordi, vice president and editorial director—books acquisitions; Helen Salmon, director—books marketing; David Horwitz, vice president—sales; Tom Taylor, vice president—marketing and sales; and Blaise Simqu, president and chief executive officer. We have become friends and colleagues with the staff at SAGE Publications. They are all greatly appreciated.

Thanks to the following reviewers:

Sabrina Alimahomed
University of California at Riverside

Richard Ball
Ferris State University

Fred Beck
Illinois State University

David L. Briscoe
University of Arkansas at Little Rock

Jamie M. Dolan
Carroll College (MT)

Obi N. I. Ebbe
The University of Tennessee at Chattanooga

Lance Erickson
Brigham Young University

Stephanie Funk
Hanover College

Loyd R. Ganey, Jr.
Western International University

Mary Grigsby
University of Missouri at Columbia

Chris Hausmann
University of Notre Dame

Todd A. Hechtman
Eastern Washington University

Keith Kerr
Blinn College

Elaine Leeder
Sonoma State University

Jason J. Leiker
Utah State University

Stephen Lilley
Sacred Heart University

David A. Lopez
California State University at Northridge

Ali Akbar Mahdi
Ohio Wesleyan University

Gerardo Marti
Davidson College

Laura McCloud
The Ohio State University

Meeta Mehrotra
Roanoke College

Melinda S. Miceli
University of Hartford

Leah A. Moore
University of Central Florida

Katy Pinto
California State University at Dominguez Hills

R. Marlene Powell
University of North Carolina at Pembroke

Suzanne Prescott
Central New Mexico Community College

Olga Rowe
Oregon State University

Paulina Ruf
Lenoir-Rhyne University

Sarah Samblanet
Kent State University

Martha L. Shockey-Eckles
Saint Louis University

Toni Sims
University of Louisiana at Lafayette

Terry L. Smith
Harding University

Frank S. Stanford
Blinn College

Tracey Steele
Wright State University

Rachel Stehle
Cuyahoga Community College

Amy Stone
Trinity University

John Stone
Boston University

Stephen Sweet
Ithaca College

Ruth Thompson-Miller
Texas A & M University

Tim Ulrich
Seattle Pacific University

Thomas L. Van Valey
Western Michigan University

Connie Veldink
Everett Community College

Chaim I. Waxman
Rutgers University

Debra Welkley
California State University at Sacramento

Debra Wetcher-Hendricks
Moravian College

Deborah J. White
Collin County Community College

Jake B. Wilson
University of California at Riverside

Laurie Winder
Western Washington University

Robert Wonser
College of the Canyons

Luis Zanartu
Sacramento City College

John Zipp
University of Akron

PART I

Understanding Our Social World

The Scientific Study of Society

Why are we studying what seems to be common sense? Why would anyone want to study how friends and family get along, how groups work, and where a society fits into the global system? What can we learn from scientifically studying our everyday lives? What exactly does it mean to see the world sociologically? Can sociology make our lives any better as the study of biology or chemistry does through new medications?

Studying sociology takes us on a trip to a deeper level of understanding of ourselves and our social world. The first chapter of this book helps answer two questions: What is sociology, and why study it? The second chapter addresses how sociology began and how sociologists know what they know. Like your sociology professor, this book will argue that sociology is valuable because it gives us new perspectives on our personal and professional lives and because sociological insights and skills can help all of us make the world a better place.

When sociologists make a statement about the social world, how do they know it is true? What perspective or lens might sociologists employ to make sense of their information? For example, when sociologists find that education does not treat all children equally, what can be done about it? What evidence would be considered reliable, valid, dependable, and persuasive to support this statement? By the time you finish reading the first two chapters, you should have an initial sense of what sociology is, how it can help you understand your social world, why the field is worth taking your time to explore, and how sociologists know what they know. We invite you to take a seat and come on a trip through the fascinating field of sociology, our social world.

Sociology

A Unique Way to View the World

Sociology involves a transformation in the way one sees the world—learning to recognize the complex connections of our intimate personal lives with small groups, with large organizations and institutions, and with national and global structures and events.

Our Social World Model

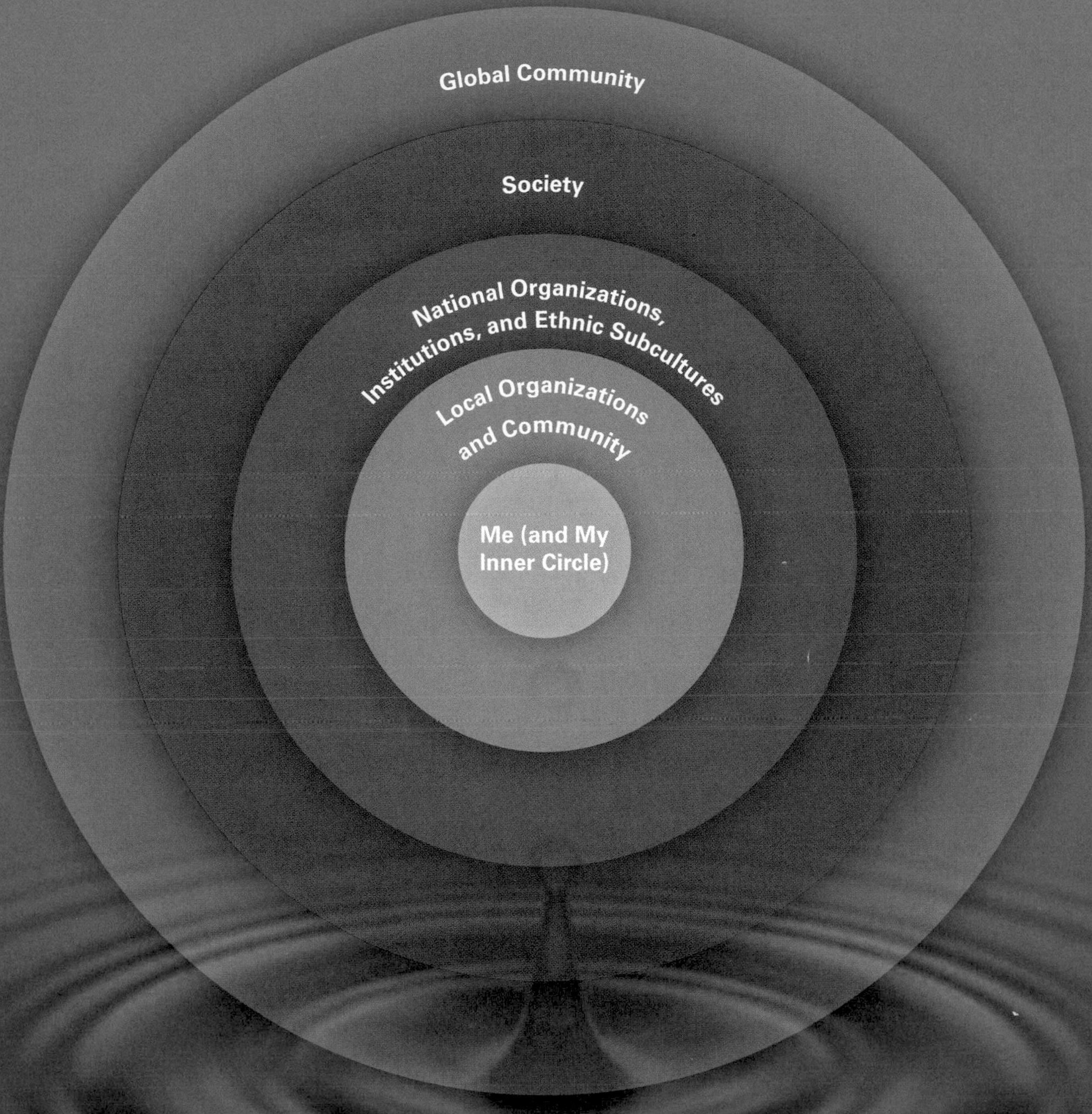

Global Community

Society

National Organizations, Institutions, and Ethnic Subcultures

Local Organizations and Community

Me (and My Inner Circle)

This model expresses a core idea carried throughout the book—the way in which your own life is embedded in, is shaped by, and influences your family, community, society, and world. It is a critically important reality that can make you a more effective person and a more knowledgeable citizen.

Think About It	
Self and Inner Circle	How can sociology help me understand my own life and my sense of self?
Local Community	How can sociology help me to be a more effective employee and citizen in my community?
National Institutions; Complex Organizations; Ethnic Groups	How do sociologists help us understand and even improve our lives in families, classrooms, and health care offices?
National Society	How do national loyalty and national policies affect my life?
Global Community	How might global events affect my life?

"It may win the prize for the strangest place to get a back massage," but according to a recent scientific article, twins do a good deal of it (Weaver 2010). Scientists studied the movement of five pairs of twin fetuses using ultrasonography, a technique that visualizes internal body structures. By the fourth month of gestation, twin fetuses begin reaching for their "womb-mates," and by 18 weeks, they spend more time touching their neighbors than themselves or the walls of the uterus. Fetuses that have single-womb occupancy tend to touch the walls of the uterus a good deal to make contact with the mother. Nearly 30% of the movement of twins was directed toward their companions. Movements toward the partner, such as stroking the back of the head, are more sustained and more precise than movements toward themselves, such as touching their own mouths or other facial features. As the authors

Within hours after their birth in October 2010, Jackson and Audrey Pietrykowski became highly fussy if the nurses tried to put them in separate bassinets. At one point shortly after birth, both babies were put in a warmer, and Jackson cried until he found Audrey, proceeding to intertwine his arms and legs with hers. Twins, like all humans, are hardwired to be social and in relationships with others.

put it, they're "wired to be social" (Castiello et al. 2010). In short, humans are innately social creatures.

Strange as it may seem, the social world is not merely something that exists outside of us. As the twins illustrate, the social world is also something we carry inside of us. We are part of it, we reflect on it, and we are influenced by it, even when we are alone. The patterns of the social world engulf us in ways both subtle and obvious, with profound implications for how we create order and meaning in our lives.

Sometimes it takes a dramatic and shocking event for us to realize just how deeply embedded we are in a social world that we take for granted. "It couldn't happen in the United States," read typical world newspaper accounts. "This is something you see in Bosnia, Kosovo, the Middle East, Central Africa, and other war-torn areas. . . . It's hard to imagine this happening in the economic center of the United States." Yet on September 11, 2001, shortly after 9 A.M., a commercial airliner crashed into a New York City skyscraper, followed a short while later by another pummeling into the paired tower, causing this mighty symbol of financial wealth—the World Trade Center—to collapse. After the dust settled and the rescue crews finished their gruesome work, nearly 3,000 people were dead or unaccounted for. The world as we knew it changed forever that day. This event taught U.S. citizens how integrally connected they are with the international community.

Following the events of September 11, the United States launched its highly publicized War on Terror, and many terrorist strongholds and training camps were destroyed. Still, troubling questions remain unanswered. Why did this extremist act occur? How can such actions be deterred in the future? How do the survivors recover from such a horrific event? Why was this event so completely disorienting to Americans and to the world community? These terrorist acts horrified people because they were unpredicted and unexpected in a normally predictable world. They violated the rules that foster our connections to one another. They also brought attention to the discontent and disconnectedness that lie under the surface in many societies—discontent that expressed itself in hateful violence. That discontent and hostility is likely to continue until the root causes are addressed.

Terrorist acts represent a rejection of modern civil society (Smith 1994). The terrorists themselves see their acts as

These signs were put up right after the September 11 attacks on the World Trade Center by people looking for missing loved ones. The experience of New Yorkers was alarm, fear, grief, and confusion—precisely the emotions that the terrorists sought to create. Terrorism disrupts normal social life and daily routines and undermines security. It provides an effective tool for those with no power.

justifiable, but few outside their inner circle can sympathize with their behavior. When terrorist acts occur, we struggle to fit such events into our mental picture of a just, safe, comfortable, and predictable social world. The events of September 11 forced U.S. citizens to realize that, although they may see a great diversity among themselves, people in other parts of the world view them as all the same. U.S. citizens may also be despised for what they represent, as perceived by others. In other words, terrorists view U.S. citizens as intimately connected. For many U.S. citizens, their sense of loyalty to the nation was deeply stirred by the events of 9/11. Patriotism abounded. So, in fact, the nation's people became more connected as a reaction to an act against the United States.

Most of the time, we live with social patterns that we take for granted as routine, ordinary, and expected. These social patterns, or social facts, characterize social groups. The social expectations are external to each individual (unlike motivations or drives), but they still guide (or constrain) our behaviors and thoughts. Without shared expectations between humans about proper patterns of behavior, life would be chaotic. Connections require some basic rules of interaction, and these rules create routine and safe normality in everyday interaction. For the people in and around the World Trade Center, the social rules governing everyday life broke down that awful day. How could anyone live in society if there were no rules?

This first chapter examines the social ties that make up our social world, as well as sociology's focus on those ties. We will learn what sociology is and why it is valuable to study it; how sociologists view the social world and what they do; how studying sociology can help us in our everyday life; and how the social world model is used to present the topics we will study throughout this book.

What Is Sociology?

Whether we are in a coffee shop, in a classroom, in a dining hall, at a party, in our residence hall, at work, or within our home, we interact with other people. Such interactions are the foundation of social life, and they are the subject of interest to sociologists. According to the American Sociological Association (2002),

> **Sociology** is the scientific study of social life, social change, and the social causes and consequences of human behavior. Since all human behavior is social, the subject matter of sociology ranges from the intimate family to the hostile mob; from organized crime to religious cults; from the divisions of race, gender, and social class to the shared beliefs of a common culture; and from the sociology of work to the sociology of sports. (p. 1)

As we shall see, sociology is relevant and applicable to our lives in many ways. Sociologists conduct scientific research on social relationships and problems that range from tiny groups of two people to national societies and global social networks.

Unlike the discipline of psychology, which focuses on attributes, motivations, and behaviors of individuals, sociology tends to focus on group patterns. Whereas a psychologist might try to explain behavior by examining the personality traits of individuals, a sociologist would examine the position of different people within the group and how positions influence what people do. Sociologists seek to analyze and explain why people interact with others and belong to groups, how groups work, who has power and who does not, how decisions are made, and how groups deal with conflict and change. From the early beginnings of their discipline (discussed in Chapter 2), sociologists have asked questions about the rules that govern group behavior; about the causes of social problems, such as child abuse, crime, and poverty; and about why nations declare war and kill each other's citizens.

Two-person interactions—*dyads*—are the smallest units sociologists study. Examples of dyads include roommates discussing their classes, a professor and student going over an assignment, a husband and wife negotiating their budget, and two children playing. Next in size are small groups consisting of three or more interacting people—a family, a neighborhood or peer group, a classroom, a work group, or a street gang. Then come increasingly larger groups—organizations such as sports or scouting clubs, neighborhood associations, and local religious congregations. Among the largest groups contained within nations are ethnic groups and national organizations, including economic, educational, religious, health, and political systems. Nations themselves are still larger and can sometimes involve hundreds of millions of people. In the past several decades, social scientists have also pointed to globalization, the process by which the entire world is becoming a single interdependent entity. Of particular interest to sociologists are how these various groups are organized, how they function, why they conflict, and how they influence one another.

Thinking Sociologically

Identify several dyads, small groups, and large organizations to which you belong. Did you choose to belong, or were you born into membership in the group? How does each group influence decisions you make?

Underlying Ideas in Sociology

All sciences rest on certain fundamental ideas or principles. The idea that one action can cause something else is a core idea in all science—for example, that heavy drinking before driving might cause an automobile accident. Sociology is based on several principles that sociologists take for granted about the social world. These ideas about humans and social life are supported by considerable evidence, but they are no longer matters of debate or controversy—they are assumed at this point to be true. Understanding these core principles helps us see how sociologists approach the study of people in groups.

People are social by nature. This means that humans seek contact with other humans, interact with each other, and influence and are influenced by the behaviors of one another. Furthermore, humans need groups to survive. Although a few individuals may become socially isolated

An athletic team teaches members to interact, cooperate, develop awareness of the power of others, and deal with conflict. Here, children experience ordered interaction in the competitive environment of a football game. What values, skills, attitudes, and assumptions about life and social interaction do you think these young boys are learning?

as adults, they could not have reached adulthood without sustained interactions with others. The central point here is that we become who we are because other people and groups constantly influence us.

People live much of their lives belonging to social groups. It is in social groups that we interact, learn to share goals and to cooperate, develop identities, obtain power, and have conflicts. Our individual beliefs and behaviors, our experiences, our observations, and the problems we face are derived from connections to our social groups.

Interaction between the individual and the group is a two-way process in which each influences the other. Individuals can influence the shape and direction of groups; groups provide the rules and the expected behaviors for individuals.

Recurrent social patterns, ordered behavior, shared expectations, and common understandings among people characterize groups. A degree of continuity and recurrent behavior is present in human interactions, whether in small groups, large organizations, or society.

The processes of conflict and change are natural and inevitable features of groups and societies. No group can remain stagnant and hope to perpetuate itself. To survive, groups must adapt to changes in the social and physical environment. Rapid change often comes at a price. It can lead to conflict within a society—between traditional and new ideas and between groups that have vested interests in particular ways of doing things. Rapid change can give rise to protest activities; changing in a controversial direction or failing to change fast enough can spark conflict, including revolution. The collapse of Soviet domination of Eastern Europe and the violence of citizens against what some saw as a corrupt election in Kenya illustrate the demand for change that can spring from citizens' discontent with corrupt or authoritarian rule.

As you read this book, keep in mind these basic ideas that form the foundation of sociological analysis: People are social; they live and carry out activities largely in groups; interaction influences both individual and group behavior; people share common behavior patterns and expectations; and processes such as change and conflict are always present. In several important ways, sociological understandings differ from our everyday views of the social world and provide new lenses for looking at our social world.

Sociology Versus Common Sense

Human tragedy can result from inaccurate commonsense beliefs. For example, both the Nazi genocide and the existence of slavery have their roots in false beliefs about

racial superiority. Consider for a moment some events that have captured media attention, and ask yourself questions about these events: Why do some families remain poor generation after generation? Are kids from certain kinds of neighborhoods more likely to get into trouble with the law than kids from other neighborhoods? Why do political, religious, and ethnic conflicts exist in the Congo, Rwanda, Sudan, the Middle East, and other areas? Our answers to such questions reflect our beliefs and assumptions about the social world. These assumptions often are based on our experiences, our judgments about what our friends and family believe, what we have read or viewed on television, and common stereotypes, which are rigid beliefs, often untested and unfounded, about a group or a category of people.

Common sense refers to ideas that are so completely taken for granted that they have never been seriously questioned and seem to be sensible to any reasonable person. Commonsense interpretations based on personal experience are an important means of processing information and deciding on a course of action. Although all of us

An East German border guard shakes hands with a West German woman through a hole in the Berlin Wall. Although their governments were hostile to one another, the people themselves often had very different sentiments toward those on the other side of the divide.

hold such ideas and assumptions, that does not mean they are accurate. Sociologists assume human behavior can be studied scientifically; they use scientific methods to test the accuracy of commonsense beliefs and ideas about human behavior and the social world. Would our commonsense notions about the social world be reinforced or rejected if examined with scientifically gathered information? Many commonsense notions are actually contradictory:

Birds of a feather flock together	Opposites attract
Absence makes the heart grow fonder	Out of sight, out of mind
Look before you leap	He who hesitates is lost
You can't teach an old dog new tricks	It's never too late to learn
Above all to thine own self be true	When in Rome, do as the Romans do
Variety is the spice of life	Never change horses in midstream
Two heads are better than one	If you want something done right, do it yourself
You can't tell a book by its cover	The clothes make the man
Haste makes waste	Strike while the iron is hot
There's no place like home	The grass is always greener on the other side

These are examples of maxims that people use as "absolute" guides to live by. They become substitutes for real analysis of situations. The fact is that all of them are accurate *at some times, in some places, about some things.* Sociological thinking and analysis are about studying the conditions in which they hold and do not hold (Eitzen, Zinn, and Smith 2009).

The difference between common sense and sociology is that sociologists test their beliefs by gathering information and analyzing the evidence in a planned, objective, systematic, and replicable (repeatable) scientific way. Indeed, they set up studies to see if they can disprove what they think is true. This is the way science is done. Consider the following examples of commonsense beliefs about the social world and some research findings about these beliefs.

Thinking Sociologically

Do you know any other commonsense sayings that contradict one another? You may also want to take the common sense quiz online at **www.sagepub.com/oswcondensed2e**.

Commonsense Beliefs and Social Science Findings

Belief: Most of the differences in the behaviors of women and men are based on "human nature"; men and women are just plain different from each other. Research shows that biological factors certainly play a part in the behaviors of men and women, but the culture (beliefs, values, rules, and way of life) that people learn as they grow up determines how biological tendencies are played out. A unique example illustrates this: In the Wodaabe tribe in Africa, women do most of the heavy work while men adorn themselves with makeup, sip tea, and gossip (Beckwith 1983). Variations in behavior of men and women around the world are so great that it is impossible to attribute behavior to biology or human nature alone.

Wodaabe society in Niger in sub-Saharan Africa illustrates that our notions of masculinity and femininity—which common sense tells us are innate and universal—are actually socially defined, variable, and learned. Wodaabe men are known for their heavy use of makeup to be attractive to women.

Belief: As developing countries modernize, the lives of their female citizens improve. This is generally false. In fact, the status of women in many developed and developing countries is getting worse. Women make up roughly 51% of the world's approximately 6.8 billion people and account for two thirds of the world's hours-at-work. However, in no country for which data are available do they earn what men earn, and sometimes, the figures show women earning less than 50% of men's earnings for similar work. Women hold many unpaid jobs in agriculture, and they own only 1% of the world's property. Furthermore, of the world's 1 billion illiterate adults, two thirds are women (World Factbook 2009). Only 77% of the world's women over age 15 can read and write compared to 87% of men. Illiteracy rates for women in South Asia, sub-Saharan Africa, and the Middle East are highest in the world, implying lack of access to education. These are only a few examples of the continuing poor status of women in many countries (Institute for Statistics 2006a; World Factbook 2009; youthxchange 2007).

Belief: Given high divorce rates in the United States and Canada, marriages are in serious trouble. Although the divorce rate in North America is high, the rate of marriage is also one of the highest in the world (Coontz 2005). If the fear-of-commitment hypothesis were true, it is unlikely the marriage rate would be so high. Moreover, even those who have been divorced tend to remarry. Despite all the talk about decline and despite genuine concern about high levels of marital failure, Americans now spend more years of their lives in marriage than at any other time in history. Divorce appears to be seen as rejection of a particular partnership rather than as rejection of marriage itself (Coontz 2005; Wallerstein and Blakeslee 1996). The divorce rate reached a peak in the United States in 1982 and has declined modestly since that time (Newman and Grauerholz 2002).

As these examples illustrate, many of our commonsense beliefs are challenged by social scientific evidence. On examination, the social world is often more complex than our commonsense understanding of events, which is based on limited evidence. Throughout history, there are examples of beliefs that seemed obvious at one time but have been shown to be mistaken through scientific study. Social scientific research may also confirm some common notions about the social world. For example, the unemployment rate among African Americans in the United States is higher than that of most other groups; women with similar education and jobs earn less income than men with the same education and jobs; excessive consumption of alcohol is associated with high levels of domestic violence; people tend to marry others who are of a similar social class. The point is that the discipline of sociology provides a method to assess the accuracy of our commonsense assumptions about the social world.

Literacy is a major issue for societies around the globe. These Chinese children are learning to read, but in many developing countries, boys have more access to formal education than girls have. The commonsense notion is that most children in the world, boys and girls, have equal access to education, yet many children, especially girls, do not gain literacy.

To improve lives of individuals in societies around the world, decision makers must rely on an accurate understanding of the society. Accurate information gleaned from sociological research can be the basis for more rational and just social policies—policies that better meet the needs of all groups in the social world. The sociological perspective, discussed below, helps us gain reliable understanding.

Thinking Sociologically

Think of a commonsense belief that you disagree with. Why did you develop this belief?

The Sociological Perspective and the Sociological Imagination

What happens in the social world affects our individual lives. If we are unemployed or lack funds for our college education, we may say this is a personal problem when often broader social issues are at the root of our situation. The sociological perspective holds that we can best understand our personal experiences and problems by examining their broader social context—by looking at the big picture.

As sociologist C. Wright Mills (1959) explains, individual problems or private troubles are rooted in social or public issues, what is happening in the social world outside of one's personal control. This relationship between individual experiences and public issues is the **sociological imagination**. For Mills, many personal experiences can and should be interpreted in the context of large-scale forces in the wider society.

Consider, for example, the personal trauma caused by being laid off from a job. The unemployed person often experiences feelings of inadequacy or lack of worth. This, in turn, may produce stress in a marriage or even result in divorce. These conditions not only are deeply troubling to the person most directly affected but also are related to wider political and economic forces in society. The unemployment may be due to corporate downsizing or to a corporation taking operations to another country where labor costs are cheaper and where there are fewer environmental regulations on companies. People may blame themselves or each other for personal troubles such as unemployment or a failed marriage, believing that they did not try hard enough. Often, they do not see the connection between their private lives and larger economic forces beyond their control. They fail to recognize the public issues that create private troubles.

Families also experience stress as partners have, over time, assumed increasing responsibility for their mate's and their children's emotional and physical needs. Until the second half of the twentieth century, the community and the extended family unit—aunts, uncles, grandparents, and cousins—assumed more of that burden. Extended families continue to exist in countries where children settle near their parents, but in modern urban societies, both the sense of community and the connection to the extended family are greatly diminished. There are fewer intimate ties to call on for help and support. Divorce is a very personal condition for those affected, but it can be understood far more clearly when considered in conjunction with the broader social context of economics, urbanization, changing gender roles, lack of external support, and legislated family policies.

As we learn about sociology, we will come to understand how social forces shape individual lives, and this will help us understand aspects of everyday life we take for granted. In this book, we will investigate how group life influences our behaviors and interactions and why some individuals follow the rules of society and others do not. A major goal is to help us incorporate the sociological perspective into our way of looking at the social world and our place in it. Indeed, the notion of sociological imagination—connecting events from the global and national level to the personal and intimate level of our own lives—is the core organizing theme of this book.

Thinking Sociologically

How does poverty, a war, or a recession cause personal troubles for someone you know? Why is trying to explain the causes of these personal troubles by examining only the personal characteristics of those affected not adequate?

Some sociologists study issues and problems and present their results for others to use. Others become involved in solving the very problems they study. The "Sociology in Your Social World" feature on the next page provides an extension on the sociological imagination, illustrating how some use their sociological knowledge to become involved in their communities or the larger world; these students of sociology advocate an active role in bringing about change.

Questions Sociologists Ask—and Don't Ask

Sociologists ask questions about human behavior in social groups and organizations—questions that can be studied scientifically. Sociologists, like other scientists, cannot answer certain questions—philosophical questions about the existence of God, the meaning of life, the ethical implications of stem cell research, or the morality of physician-assisted suicide. What sociologists do ask, however, is this: What effect does holding certain ideas or adhering to certain ethical standards have on the behavior and attitudes of people? For example, are people more likely to obey rules if they believe that there are consequences for their actions in an afterlife? What are the consequences—positive and negative—of allowing suicide for terminally ill patients who are in pain? Although sociologists may study philosophical or religious beliefs held by groups, they do not make judgments about what beliefs are right or wrong or about moral issues involving philosophy, religion, values, or opinion. They focus on issues that can be studied objectively and scientifically, rather than those that are judgmental or value based.

Applied sociologists, those who carry out research to help organizations solve problems, agree that the research itself should be as objective as possible. After the research is completed, the applied sociologists might use the research findings to explore policy implications and make recommendations for change.

Consider the following examples of questions sociologists might ask:

- Who gets an abortion, why do they do so, and how does society as a whole view abortion? These are matters of fact that a social scientist can explore. However, sociologists avoid making

Sociology in Your Social World

How Will You Spend the Twenty-First Century?

By Peter Dreier

Today, Americans enjoy more rights, better working conditions, better living conditions, and more protection from disease in childhood and old age than anyone could have imagined 100 years ago. . . . But that doesn't let you off the hook! There are still many problems and much work to do. Like all agents for social change, . . . social reformers [such as] Martin Luther King, Jr., [a sociology major] understood the basic point of sociology—that is, to look for the connections between people's everyday personal problems and the larger trends in society. Things that we experience as personal matters—a woman facing domestic violence, or a low-wage worker who cannot afford housing, or middle-class people stuck in daily traffic jams—are really about how our institutions function. Sociologists hold a mirror up to our society and help us see our society objectively. One way to do this is by comparing our own society to others. This sometimes makes us uncomfortable—because we take so much about our society for granted. Conditions that we may consider "normal" may be considered serious problems by other societies. For example, if we compare the United States to other advanced industrial countries such as Canada, Germany, France, Sweden, Australia, Holland, and Belgium, we find some troubling things:

- The United States has the highest per capita income among those countries. At the same time, the United States has, by far, the widest gap between the rich and the poor.
- Almost 30% of American workers work full-time, year-round, for poverty-level wages.
- The United States has the highest overall rate of poverty. More than 33 million Americans live in poverty.
- More than 12 million of these Americans are children. In fact, 1 out of 6 American children is poor. They live in slums and trailer parks, eat cold cereal for dinner, share a bed or a cot with their siblings and sometimes with their parents, and are often one disaster away from becoming homeless.
- Only 3 out of 5 children eligible for the Head Start program are enrolled because of the lack of funding.
- The United States has the highest infant mortality rate among the major industrial nations.
- The United States is the only one of these nations without universal health insurance. More than 43 million Americans—including 11 million children—have no health insurance.
- Americans spend more hours stuck in traffic jams than people of any of these other countries. This leads to more pollution, more auto accidents, and less time spent with families.
- Finally, the United States has a much higher proportion of our citizens in prison than any of these societies. . . .

. . . What would you like your grandchildren to think about how you spent the twenty-first century? . . . No matter what career you pursue, you have choices about how you will live your lives. As citizens, you can sit on the sidelines and merely be involved in your society, or you can decide to become really committed to making this a better world.

Today, there are hundreds of thousands of patriotic Americans committed to making our country live up to its ideals. . . . They are asking the same questions that earlier generations of active citizens asked: Why can't our society do a better job of providing equal opportunity, a clean environment, and a decent education for all? They know there are many barriers and obstacles to change, but they want to figure out how to overcome these barriers and to help build a better society.

So ask yourselves: What are some of the things that we take for granted today that need to be changed? What are some ideas for changing things that today might seem "outrageous" but that—25 or 50 or 100 years from now—will be considered common sense?

. . . A record number of college students today are involved in a wide variety of "community service" activities—such as mentoring young kids in school, volunteering in a homeless shelter, or working in an AIDS hospice. As a result of this student activism, more than 100 colleges and universities have adopted "anti-sweatshop" codes of conduct for the manufacturers of clothing that bear the names and logos of their institutions.

Positive change is possible, but it is not inevitable. . . . I am optimistic that your generation will follow a lifelong commitment to positive change.

I know you will not be among those who simply "see things the way they are and ask: why?" Instead, you will "dream things that never were and ask: why not?" [Robert Kennedy].

ethical judgments about whether abortion is right or wrong. Such a judgment is a question of values, not one that can be answered through scientific analysis. The question about the morality of abortion is very important to many people, but it is based on philosophical or theological rationale, not on sociological findings.

- Who is most beautiful? Cultural standards of beauty impact individual popularity and social interaction, and this issue interests some social scientists. However, the sociologist would not judge which individuals are more or less attractive. Such questions are matters of aesthetics, a field of philosophy and art.

- What are the circumstances around individuals becoming drunk and drunken behavior? This question is often tied more to social environment than to alcohol itself. Note that a person might be very intoxicated at a fraternity party but behave differently at a wedding reception, where the expectations for behavior are very different. The researcher does not make judgments about whether use of alcohol is good or bad or right or wrong and avoids—as much as possible—opinions regarding responsibility or irresponsibility. The sociologist does, however, observe variations in the use of alcohol in social situations and resulting behaviors. An applied sociologist researching alcohol use on campus for a college or for a national fraternity may, following the research, offer advice based on that research about how to reduce the number of alcohol-related deaths or sexual assault incidents on college campuses (Sweet 2001).

Sociologists learn techniques to avoid letting their values influence their research designs, data gathering, and analysis. Still, complete objectivity is difficult at best, and what one chooses to study may be influenced by one's interests and concerns about injustice in society. The fact that sociologists know they will be held accountable by other scientists for the objectivity of their research is a major factor in encouraging them to be objective when they do their research.

What is acceptable or unacceptable drinking behavior varies according to the social setting. Binge drinking, losing consciousness, vomiting, or engaging in sexual acts while drunk may be a source of storytelling at a college party but be offensive at a wedding reception. Sociologists study different social settings and how the norms of acceptability vary in each, but they do not make judgments about those behaviors.

Thinking Sociologically

From the information you have just read, what are some questions sociologists might ask about divorce or cohabitation or same-sex unions? What are some questions sociologists *would not* ask about these topics, at least while in their roles as researchers?

The Social Sciences: A Comparison

Not so long ago, our views of people and social relationships were based on stereotypes, intuition, superstitions, supernatural explanations, and traditions passed on from one generation to the next. Natural sciences first used the scientific method, a model later adopted by social sciences. Social scientists, including anthropologists, psychologists, economists, cultural geographers, historians, and political scientists, apply the scientific method to study social relationships, to correct misleading and harmful misconceptions about human behaviors, and to guide policy decisions. Consider the following examples of specific studies a social scientist might conduct. These are followed by a brief description of the focus of sociology as a social science.

One anthropological study focused on garbage, studying what people discard to understand their patterns of life. Anthropology is closely related to sociology. In fact, the two areas have common historical roots. *Anthropology* is the study of humanity in its broadest context. There are four subfields within anthropology: physical anthropology (which is related to biology), archaeology, linguistics, and cultural anthropology (sometimes called *ethnology*). This last field has the most in common with sociology. Cultural anthropology focuses on the culture, or way of life, of the society being studied and uses methods appropriate to understanding culture.

After wiring research subjects to a machine that measures their physiological reaction to a violent film clip, a psychologist asks them questions about what they were feeling. *Psychology* is the study of individual behavior and mental processes (e.g., sensation, perception, memory, and thought processes). It differs from sociology in that it focuses on individuals, rather than on groups, institutions, and societies as sociology does. Although there are different branches of psychology, most psychologists are concerned with what motivates individual behavior, personality attributes, attitudes, beliefs, and perceptions. Psychologists also explore abnormal behavior, the mental disorders of individuals, and stages of normal human development (Wallerstein 1996; Wallerstein and Blakeslee 2004).

A political scientist studies opinion poll results to predict who will win the next election, how various groups of people are likely to vote, or how elected officials will vote on proposed legislation. *Political science* is concerned with government systems and power—how they work, how they are organized, forms of government, relations between governments, who holds power and how they obtain it, how power is used, and who is politically active. Political science overlaps with sociology, particularly in the study of political theory and the nature and the uses of power.

An economist studies the banking system and market trends, trying to determine what will remedy the global recession. *Economists* analyze economic conditions and explore how people organize, produce, and distribute material goods. They are interested in supply and demand, inflation and taxes, prices and manufacturing output, labor organization, employment levels, and comparisons of industrial and nonindustrial nations.

Psychology as a discipline tends to focus on individuals, including such fields as sensation, perception, memory, and thought processes. In this study, the researcher is using some equipment and a computer to measure how the eye and the brain work together to help create depth perception.

What all of these social sciences—sociology, anthropology, psychology, economics, political science, cultural geography, and history—have in common is that they study aspects of human behavioral and social life. Social sciences share many common topics, methods, concepts, research findings, and theories, but each has a different focus or perspective on the social world. Each of these social sciences relates to topics studied by sociologists, but sociologists focus on human interaction, groups, and social structure, providing the broadest overview of the social world.

Thinking Sociologically

Consider other issues such as the condition of poverty in developing countries or homelessness in North America. What question(s) might different social sciences ask about these problems?

Why Study Sociology . . . and What Do Sociologists Do?

Did you ever wonder why some families are close and others are estranged? Why some work groups are very productive while others are not? Why some people are rich and others remain impoverished? Why some people engage in criminal behaviors and others conform rigidly to rules? Although sociologists do not have all the answers to such questions, they do have the perspective and methods to search for a deeper understanding of these and other patterns of human interaction.

Two ingredients are essential to the study of our social world: a keen ability to observe what is happening in the social world and a desire to find answers to the question of why it is happening. The value of sociology is that it affords us a unique perspective from which to examine the social world, and it provides the methods to study systematically important questions about human interaction, group behavior, and social structure. The practical significance of the sociological perspective is that it

- fosters greater self-awareness, which can lead to opportunities to improve one's life;
- encourages a more complete understanding of social situations by looking beyond individual explanations to include group analyses of behavior;
- helps people understand and evaluate problems by enabling them to view the world systematically and objectively rather than in strictly emotional or personal terms;
- cultivates an understanding of the many diverse cultural perspectives and how cultural differences are related to behavioral patterns;
- provides a means to assess the impact of social policies;
- reveals the complexities of social life and provides methods of inquiry to study them; and
- provides useful skills in interpersonal relations, critical thinking, data collection and analysis, problem solving, and decision making.

This unique perspective has practical value as we carry out our roles as workers, friends, family members, and citizens. For example, an employee who has studied sociology may better understand how to work with groups and how the structure of the workplace affects individual behavior, how to approach problem solving, and how to collect and analyze data. Likewise, a schoolteacher trained in sociology may have a better understanding of classroom management, student motivation, causes of poor student learning that have roots outside the school, and other variables that shape the professional life of teachers and the academic success of students. Consider the example in "Sociology in Your Social World," which explores how high school groups such as "jocks" and "burnouts" behave and why each clique's behavior might be quite logical in certain circumstances. *Burnouts and Jocks in a Public High School* explores a social environment very familiar to most of us, the social cliques in a high school.

What Sociologists Do

Sociologists are employed in a variety of settings. Although students may first encounter them as teachers and researchers in higher education, sociologists also hold nonacademic, applied sociology jobs in social agencies, government, and business. Table 1.1 illustrates that a significant portion of sociologists work in business, government, and social service agencies (American Sociological Association 2006; Dotzler and Koppel 1999).

Table 1.1   Where Sociologists Are Employed

Places of Employment	Percentage Employed
College or university	75.5
Government (all positions)	7.1
Private, for-profit business	6.2
Not-for-profit public service organizations	7.6
Self-employed	0.4

Source: American Sociological Association (2006).

College graduates who seek employment immediately after college (without other graduate work) are most likely to find their first jobs in social services, administrative assistantships, or some sort of management position. The areas of first jobs of sociology majors are indicated in Figure 1.1 on page 16. With a master's or a doctorate degree, graduates usually become college teachers, researchers, clinicians, and consultants.

Consider your professor. The duties of professors vary depending on the type of institution and the level of courses offered. Classroom time fills only a portion of the professor's working days. Other activities include preparing for classes, preparing and grading exams and assignments, advising students, serving on committees, keeping abreast of new research in the field, and conducting research studies and having them published. This "publish or perish" task is deemed the most important activity for faculty in some major universities.

In businesses, applied sociologists use their knowledge and research skills in human resources or to address organizational needs or problems. In government jobs, they provide data such as population projections for education and health care planning. In social service agencies, such as police departments, they help address deviant behavior, and in health agencies, they may be concerned with doctor-patient interactions. Applied sociology is an important aspect of the field; you will find featured inserts in some chapters discussing the work of an applied sociologist.

Sociology in Your Social World

Burnouts and Jocks in a Public High School

High schools are big organizations made up of smaller friendship networks and cliques; a careful examination can give us insight into the tensions that exist as the groups struggle for resources and power in the school.

Sociologist Penelope Eckert (1989) focused on two categories of students that exist in many high schools in North America: "burnouts" and "jocks." The burnouts defied authorities, smoked in the restrooms, refused to use their lockers, made a public display of not eating in the school cafeteria, and wore their jackets all day. Their open and public defiance of authority infuriated the jocks—the college prep students who participated in choir, band, student council, and athletics and who held class offices. The burnouts were disgusted with the jocks. In their view, by constantly sucking up to the authorities, the jocks received special privileges and, by playing the goody-two-shoes role, made life much more difficult for the burnouts.

Despite their animosity toward one another, the goal of both groups was to gain more autonomy from the adult authorities who constantly bossed students around. As the burnouts saw things, if the jocks would have even a slight bit of backbone and stand up for the dignity of students as adults, life would be better for everyone.

The burnouts believed that school officials should earn their obedience. The burnouts maintained their dignity by affirming that they did not recognize bossy adults as authorities. Wearing coats all day was another way to emphasize the idea that "I'm just a visitor in this school."

The jocks, for their part, became irritated at the burnouts when they caused trouble and were belligerent with authorities; then the administration would crack down on everyone, and no one had any freedom. Jocks found that if they did what the adults told them to do—at least while the adults were around—they got a lot more freedom. When the burnouts got defiant, however, the principal got mad and removed everyone's privileges.

Sociologist Eckert (1989) found that the behavior of both groups was quite logical for their circumstances and ambitions. Expending energy as a class officer or participating in extracurricular activities is a rational behavior for college preparatory students because those leadership roles help students get into their college of choice.

However, those activities do not help students get a better job in a factory in town. In fact, hanging out at the bowling alley makes far more sense. For the burnouts, having friendship networks and acquaintances in the right places is more important to achieving their goals than a class office listed on their résumé.

Eckert's (1989) method of gathering information was effective in showing how the internal dynamics of schools—conflicts between student groups—were influenced by outside factors such as working- and upper-middle-class status. Recent research upholds Eckert's findings on the importance cliques play in shaping school behavior. Like Eckert, Bonnie Barber, Jacquelynne Eccles, and Margaret Stone (2001) followed various friendship cliques starting in 10th grade in a Michigan high school. The jocks in their study were the most integrated to mainstream society in adult life. The burnouts (or criminals, as they are labeled in Barber's research) were most likely to have been arrested or incarcerated, showing that the propensity to defy authority figures may carry on into adult life.

These studies show that sociological analysis can help us understand some ways that connections between groups—regardless of whether they are in conflict or harmony—shape the perceptions, attitudes, and behaviors of people living in this complex social world.

focus on writing, speaking, analytical skills—especially when faced with complex problems, comprehension of other cultures and of diversity within the United States, ability to work effectively in diverse teams, and ability to gather and interpret quantitative information. As Table 1.3 indicates, employers want more of these kinds of skills from college graduates.

The following skills and competencies are part of most sociological training:

1. Communication skills (listening, verbal and written communication, working with peers, and effective interaction in group situations)

2. Analytical and research skillso

3. Computer and technical literacy (basic understanding of computer hardware and software programs)

4. Flexibility, adaptability, and multitasking (ability to set priorities, manage multiple tasks, adapt to changing situations, and handle pressure)

5. Interpersonal skills (working with coworkers)

6. Effective leadership skills (ability to take charge and make decisions)

7. Sensitivity to diversity in the workplace and with clients

8. Organizing thoughts and information and planning effectively (ability to design, plan, organize, and implement projects and to be self-motivated)

9. Ability to conceptualize and solve problems and be creative (working toward meeting the organization's goals)

10. Working with others (ability to work toward a common goal)

11. Personal values (honesty, flexibility, work ethic, dependability, loyalty, positive attitude, professionalism, self-confidence, willingness to learn) (Hansen and Hansen 2003)

These competencies reflect skills stressed in the sociology curriculum: an ability to understand and work with others, research and computer skills, planning and organizing skills, oral and written communication skills, and critical thinking skills (WorldWideLearn 2007).

We now have a general idea of what sociology is and what sociologists do. It should be apparent that sociology is a broad field of interest; sociologists study all aspects of human social behavior. The next section of this chapter shows how the parts of the social world that sociologists study relate to each other, and it outlines the model you will follow as you continue to learn about sociology.

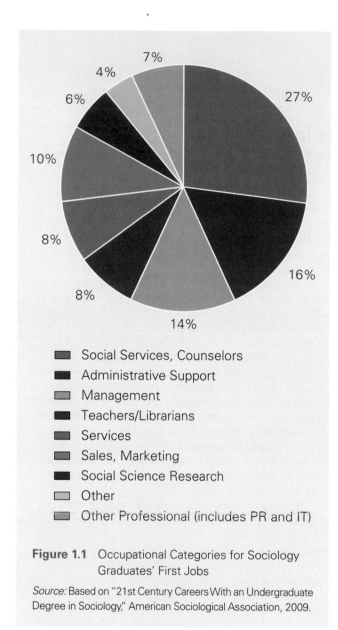

Figure 1.1   Occupational Categories for Sociology Graduates' First Jobs

Source: Based on "21st Century Careers With an Undergraduate Degree in Sociology," American Sociological Association, 2009.

These examples will provide a picture of what one can do with a sociology degree. In addition, at the end of some chapters you will find a section discussing policy examples and implications related to that chapter topic. Table 1.2 provides some ideas of career paths for graduates with a degree in sociology.

What Employers Want and What Sociology Majors Bring to a Career

Sociologists and other social scientists have studied what job skills and competencies employers seek in new employees, in addition to subject matter expertise. They tend to

Table 1.2 What Can You Do With a Sociology Degree?

Business or Management	Human Services	Education
Market researcher	Social worker	Teacher
Sales manager	Criminologist	Academic research
Customer relations	Gerontologist	Administration
Manufacturing representative	Hospital administrator	School counselor
Banking or loan officer	Charities administrator	Policy analyst
Data processor	Community advocate or organizer	College professor
Attorney		Dean of student life
Research	**Government**	**Public Relations**
Population analyst	Policy advisor or administrator	Publisher
Surveyor	Labor relations	Mass communications
Market researcher	Legislator	Advertising
Economic analyst	Census worker	Writer or commentator
Public opinion pollster	International agency representative	Journalist
Interviewer	City planning officer	
Policy researcher	Prison administrator	
Telecommunications researcher	Law enforcement	
	FBI agent	
	Customs agent	

Note: Surveys of college alumni with undergraduate majors in sociology indicate that this field of study prepares people for a broad range of occupations. Notice that some of these jobs require graduate or professional training. For further information, contact your department chair or the American Sociological Association in Washington, DC, for a copy of *Careers in Sociology,* 6th edition (2002).

Source: American Sociological Association (2006).

Table 1.3 Percentage of Employers Who Want Colleges to "Place More Emphasis" on Essential Learning Outcomes

Knowledge of Human Culture	
• Global Issues	72%
• The role of the United States in the world	60%
• Cultural values and traditions—U.S. and global	53%
Intellectual and Practical Skills	
• Teamwork skills in diverse groups	76%
• Critical thinking and analytic reasoning	73%
• Written and oral communication	73%
• Information literacy	70%
• Complex problem solving	64%
• Quantitative reasoning	60%
Personal and Social Responsibility	
• Intercultural competence (teamwork in diverse groups)	76%
• Intercultural knowledge (global issues)	72%

Source: American Sociological Association (2009).

Thinking Sociologically

What are some advantages of mayors, legislators, police chiefs, or government officials making decisions based on information gathered and verified by sociological research rather than on their own intuition or assumptions?

The Social World Model

Think about the different groups you depend on and interact with on a daily basis. You wake up to greet members of your family or your roommate. You go to a larger group—a class—that exists within an even larger organization—the college or university. Understanding sociology and comprehending the approach of this book requires a grasp of **levels of analysis**, social groups from the smallest to the largest. It may be relatively easy to picture small groups, such as a family, a sports team, or a sorority or fraternity. It is more difficult to visualize large groups such as corporations—the

These men carry the supplies for a new school to be built in their local community—Korphe, Pakistan. The trek of more than 20 miles up mountainous terrain was difficult, but their commitment to neighbors and children of the community made it worthwhile. The project was a local one (micro level), but it also was made possible by an international organization—Central Asia Institute—founded as a charitable organization by Greg Mortenson of Montana.

Gap, Abercrombie & Fitch, Eddie Bauer, General Motors Corporation, or Starbucks—or organizations such as local or state governments. The largest groups include nations or international organizations, such as the sprawling networks of the United Nations or the World Trade Organization. Groups of various sizes shape our lives. Sociological analysis requires that we understand these groups at various levels of analysis.

The **social world model** helps us picture the levels of analysis in our social surroundings as an interconnected series of small groups, organizations, institutions, and societies. Sometimes, these groups are connected by mutual support and cooperation, but sometimes, there are conflicts and power struggles over access to resources. What we are asking you to do here and throughout this book is to develop a sociological imagination—the basic lens used by sociologists. Picture the social world as a linked system made up of increasingly larger circles. To understand the units or parts in each circle of the social world model, look at the social world model shown on this page:

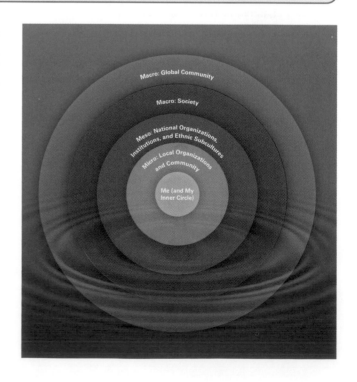

This social world model will be used throughout our book to illustrate how each topic fits into the big picture, our social world. No social unit of our social world can stand alone. All units affect each other, either because they serve needs of other units in the system or because of intense conflict and tension affecting different units. The social world is organized into two parts—structures and processes. Now, let us take a trip through our social world.

Social Structures

The social world model is made up of a number of **social units**, interconnected parts of the social world. These range from interaction in dyads and small groups to negotiating between warring societies. All these units combine into a system to form the **social structure** that holds societies together and brings order to our lives by regulating the way the units work in combination. Compare this to a picture of our body's skeleton, which governs how our limbs are attached to the torso and how they move.

Sometimes the interconnections in the social structure conflict, however, due to divergent beliefs or self-interests of units. For example, a religion that teaches that it is wrong to have blood transfusions may conflict with the health care system regarding how to save the life of a child. Business executives want to produce products at the lowest possible cost, but this may mean paying workers low wages and causing damage to the environment. All levels of analysis are linked. Some links are supportive; others are in conflict.

Social institutions are found in every society—family, education, religion, politics, economics, science, sports, and health care. They provide the rules, roles, and relationships to meet human needs and guide human behavior. They are the parts of the social structure in societies through which organized social activities take place, and they provide the setting for activities essential to human and societal survival. For example, we cannot survive without an economic institution to provide guidelines and a structure for meeting our basic needs of food, shelter, and clothing. Likewise, we would never make it to adulthood as functioning members of society without the family, the most basic of all institutions. Like the system of organs that make up our bodies—heart, lungs, kidneys, bladder—all social institutions are interrelated. Just as a change in one part of the body affects all others, a change in one institution affects the others.

The **national society**, one of the largest social units in our model, includes a population of people, usually living within a specified geographic area, who are connected by common ideas and are subject to a particular political authority. It also features a social structure with groups and institutions. Although a national society is one of the largest social units, it is still a subsystem of the interdependent global system. France, Kenya, Brazil, and Laos are all national societies on separate continents, but they are linked as part of the global system of nations. In addition to having relatively permanent geographic and political boundaries, national societies also have one or more languages and a way of life. In most cases, national societies involve countries or large regions where the inhabitants share a common identity as members. In certain other instances, such as contemporary Great Britain, a single national society may include several groups of people who consider themselves distinct nationalities (Welsh, English, Scottish, and Irish within the United Kingdom). Such multicultural societies may or may not be harmonious.

Thinking Sociologically

Think about how a major conflict or change in your family (micro level) might affect your education, economic situation, or health care. How might change in one national institution such as health care affect change in another institution (such as the family or the economy)?

Social Processes

Picture **social processes** as the actions taken by people in social units. Processes keep the social world working, much as the beating heart keeps the body working. Consider the processes of socialization and stratification. The process of socialization teaches individuals how to become productive members of society. It takes place through actions of families, educational systems, religious organizations, and other social units. Socialization is essential for the continuation of any society. Similarly, our social positions in society are the result of stratification, the process of layering people into social strata based on such factors as birth, income, occupation, and education.

Sociologists generally do not judge these social processes as good or bad. Rather, sociologists try to identify and explain processes that take place within social units. Picture these processes as overlying and penetrating our whole social world, from small groups to societies. Social units would be lifeless without the action brought about by social processes, just as body parts would be lifeless without the processes of electrical impulses shooting from the brain to each organ or the oxygen transmitted by blood coursing through our arteries to sustain each organ.

The Environment

Surrounding each social unit is an **environment**. It includes everything that influences the social unit, such as its physical and organizational surroundings and technological

This refugee mother and child from Mozambique represent the smallest social unit, a dyad. In this case, they are trying to survive with help from larger groups such as the United Nations.

to teach the children to read. A religious group may also be affected by other religious bodies, competing with one another for potential members from the community. These religious groups may work cooperatively—organizing a summer program for children or jointly sponsoring a holyday celebration—or they may define one another as evil, each trying to stigmatize the other. Moreover, one local religious group may be composed primarily of professional and business people, and another group mostly of laboring people. The religious groups may experience conflict in part because they each serve different socioeconomic constituencies.

The point is that to understand a social unit or the human body, we must consider the structure and processes within the unit, as well as the interaction with the surrounding environment. No matter what social unit the sociologist studies, the unit cannot be understood without considering the interaction of that unit with its unique environment.

Perfect relationships or complete harmony between the social units is unusual. Social units are often motivated by self-interests and self-preservation, with the result that they compete with other groups and units for resources (time, money, skills, energy of members). Therefore, social units within the society are often in conflict. Whether groups are in conflict or mutually supportive does not change their interrelatedness; units are interdependent. The nature of that interdependence is likely to change over time and can be studied using the scientific method.

Studying the Social World: Levels of Analysis

Picture for a moment your sociology class as a social unit in your social world. Students (individuals) make up the class, the class (a small group) is offered by the sociology department, the sociology department (a large group) is part of the college or university, the university (an organization) is located in a community and follows the practices approved by the social institution (education) of which it is a part, and education is an institution located within a nation. The practices the university follows are determined by a larger accrediting unit that provides guidelines and oversight for institutions. The national society, represented by the national government, is shaped by global events—technological and economic competition between nations, natural disasters, global warming, wars, and terrorist attacks. Such events influence national policies and goals, including policies for the educational system. Thus, global tensions and conflicts may shape the curriculum that the individual experiences in the sociology classroom.

Each of these social units—from the smallest (the individual student) to the largest (society and the global system)—is referred to as a level of analysis (see Table 1.4).

innovations. Each unit has its own environment to which it must adjust, just as each individual has a unique social world, including family, friends, and other social units that make up the immediate environment. Some parts of the environment are more important to the social unit than others. Your local church, synagogue, or mosque is located in a community environment. That religious organization may seem autonomous and independent, but it depends on its national organization for guidelines and support, the local police force to protect the building from vandalism, and the local economy to provide jobs to members so that the members, in turn, can support the organization. If the religious education program is going to train children to understand the scriptures, the local schools are needed

Table 1.4 The Structure of Society and Levels of Analysis

	Level	Parts of Education
Micro-level analysis	Interpersonal	Sociology class; study group cramming for an exam
	Local organizations	University; sociology department
Meso-level analysis	Organizations and institutions	State boards of education; National Education Association
	Ethnic groups within a nation	Islamic madrassas or Jewish yeshiva school systems
Macro-level analysis	Nations	Policy and laws governing education
	Global community	World literacy programs

These levels are illustrated in the social world model at the beginning of each chapter, and relation to that chapter's content is shown through examples in the model.

 MICRO-LEVEL ANALYSIS

Sometimes, sociologists ask questions about face-to-face interactions in dyads or small groups. A focus on individual or small-group interaction entails **micro-level analysis**. Micro-level analysis is important because face-to-face interaction forms the basic foundation of all social groups and organizations to which we belong, from families to corporations to societies. We are members of many groups at the micro level.

To illustrate micro-level analysis, consider the problem of spousal abuse. Why does a person remain in an abusive relationship, knowing that each year thousands of people are killed by their lovers or mates and millions more are severely and repeatedly battered? To answer this, several possible micro-level explanations can be considered. One view is that the abusive partner has convinced this person that she is powerless in the relationship or that she "deserves" the abuse. Therefore, she gives up in despair of ever being able to alter the situation. The abuse is viewed as part of the interaction—of action and reaction—and the partners come to see abuse as what comprises "normal" interaction.

Another explanation for remaining in the abusive relationship is that the person may have been brought up in a family situation where battering was an everyday part of life. However unpleasant and unnatural this may seem to outsiders, it may be seen by the abuser or by the abused as a "normal" and acceptable part of intimate relationships.

Another possibility is that an abused woman may fear that her children will be harmed or that she will be harshly judged by her family or church if she "abandons" her mate. She may have few resources to make leaving the abusive situation possible. To study each of these possible explanations involves analysis at the micro level because each focuses on interpersonal interaction factors rather than on society-wide trends or forces. Meso-level concerns lead to quite different explanations for abuse.

 MESO-LEVEL ANALYSIS

Analysis of intermediate-size social units, called **meso-level analysis**, involves looking at units smaller than the nation but larger than the local community or even the region. This level includes national institutions (e.g., the economy of a country, the national educational system, or the political system within a country); nationwide organizations (e.g., a political party, a soccer league, or a national women's rights organization); nationwide corporations (e.g., Ford Motor Company or IBM); and ethnic groups that have an identity as a group (e.g., Jews, Mexican Americans, or the Lakota Sioux in the United States). Organizations, institutions, and ethnic communities are smaller than the nation or global social forces, but they are still beyond the everyday personal

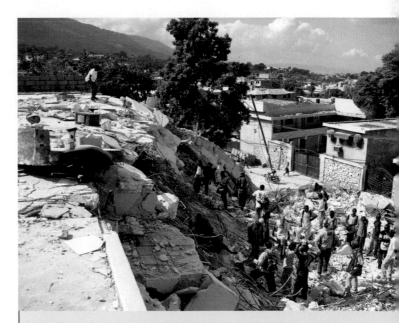

This photo depicts the damage following the catastrophic earthquake that hit Haiti on January 12, 2010. This event not only changed the lives of people in Haiti—one of the poorest countries in the world—but had ripple effects on economic exchange, relief efforts around the globe, and international trade. Those, in turn, can affect the cost of various products such as the foods you put on your table.

experience and control of individuals, unless those individuals organize to collectively change these structures. They are intermediate in the sense of being too large to know everyone in the group, but they are not nation-states at the macro level. Consider the fact that it is easier to bring about change in a state in the United States, a province in Canada, or a prefecture in Japan than the national bureaucracies of countries.

In discussing micro-level analysis, we used the example of domestic violence. We must be careful not to "blame the victim"—in this case, the abused person—for getting into an abusive relationship and for failing to act in ways that stop the abuse. To avoid blaming victims for their own suffering, many social scientists look for broader explanations of spousal abuse, such as the social conditions at the meso level of society that cause the problem (Straus and Gelles 1990). When a pattern of behavior in society occurs with increasing frequency, it cannot be understood solely from the point of view of individual cases or micro-level causes. For instance, sociological findings show that fluctuations in spousal or child abuse are related to levels of unemployment. Frustration resulting in abuse erupts within families when poor economic conditions make it nearly impossible for people to find stable and reliable means of supporting themselves and their families. Economic issues must be addressed if violence in the home is to be lessened.

 MACRO-LEVEL ANALYSIS

Studying the largest social units in the social world, called **macro-level analysis**, involves looking at entire nations, global forces, and international social trends. Macro-level analysis is essential to our understanding of how larger social forces, such as global events, shape our everyday lives. A natural disaster such as the 2005 tsunami in Indonesia, the heat waves of summer 2006 in Europe, the floods in the United States in the summer of 2008 and winter of 2009, or frequent earthquakes around the world may change the foods we are able to put on our family dinner table because much of our cuisine is now imported from other parts of the world. Map 1.1 shows some of the most deadly natural disasters of the past few years. Likewise, a political conflict on the other side of the planet can lead to war, which means that a member of your family may be called up to active duty and sent into harm's way more than 7,000 miles from your home. Each member of the family may experience individual stress, have trouble concentrating, and feel ill with worry. The entire globe has become an interdependent social unit. If we are to prosper and thrive in the twenty-first century, we need to understand connections that go beyond our local communities.

Even patterns such as domestic violence, considered as micro- and meso-level issues above, can be examined at the macro level. A study of 95 societies around the world found that violence against women (especially rape) occurs at very different rates in different societies, with some societies being completely free of rape (Benderly 1982) and others having a "culture of rape." The most consistent predictor of violence against women was a macho conception of masculine roles and personality. Societies that did not define masculinity in terms of dominance and control were virtually free of rape. Some sociologists believe that the same pattern holds for domestic violence: A society or subgroup within society that teaches males that the finest expression of their masculinity is physical strength and domination is very likely to have battered women (Burn 2005). The point is that understanding individual human behavior often requires investigation of larger societal beliefs that support that behavior. Worldwide patterns may tell us something about a social problem and offer new lenses for understanding variables that contribute to a problem. Try the "Engaging Sociology" activity on page 24 to test your understanding of levels of analysis and the sociological imagination.

Thinking Sociologically

What factors influenced you to take this sociology class? Micro-level factors might include your advisor, your schedule, and a previous interest in sociology. At the meso and macro levels, what other factors influenced you?

Distinctions between each level of analysis are not sharply delineated. The micro level shades into the meso level, and the lines between the meso level and the macro level are blurry. Still, it is clear that in some micro-level social units, you know everyone, or at least every member of the social unit is only two degrees of relatedness away. Every person in the social unit knows someone whom you also know. We also all participate in meso-level social units that are smaller than the nation but can be huge. Millions of people may belong to the same religious denomination or the same political party. We have connections with those people, and our lives are affected by people we do not even know. Consider political activities in the United States and other countries that take place on the Internet. In political campaigns, millions of individuals join organizations such as MoveOn .org and TrueMajority, participate in dialogues online, and contribute money to political organizations. People living thousands of miles from one another united financially and in spirit to support Obama-Biden or McCain-Palin in the 2008 U.S. election. Thus, the meso level is different from the micro level, but both influence us. The macro

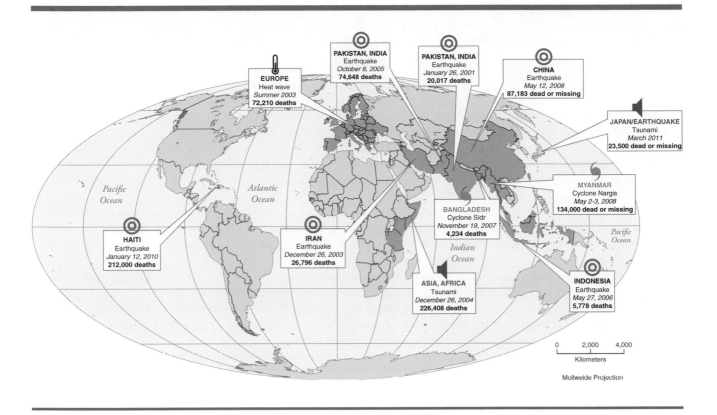

Map 1.1 The Deadliest Natural Disasters From January 2001 to June 2011

Source: Centre for Research on the Epidemiology of Disasters 2011. Map by Anna Versluis.

level is even more removed from the individual, but its impact can change our lives.

The social world model presented in the chapter opening illustrates the interplay of micro-, meso-, and macro-level forces, and Figure 1.2 illustrates that this micro-to-macro model should be seen as a continuum. In "Sociology Around the World" on page 25 we examine a village in Tunisia to see how macro-level forces influence a meso-level local community and individual micro-level lives.

Thinking Sociologically

Place the groups to which you belong in a hierarchy from micro to meso to macro levels. Note how each social unit and its subunits exist within a larger unit until you reach the level of the entire global community.

Micro social units   Meso social units   Macro social units

Figure 1.2 The Micro-to-Macro Continuum

Engaging Sociology

Micro-Meso-Macro

Look at the list of various groups and other social units below. Identify which group would belong in each level—(1) micro, (2) meso, or (3) macro. The definitions should help you make your decisions, but keep in mind that not all social units fall clearly into one level. Answers are found online at www.sagepub.com/oswcondensed2e.

Micro-level groups: Small, local community social units in which everyone knows everyone or knows someone whom you also know.

Meso-level groups: Social units of intermediate size, usually large enough that many members may never have heard the names of many other members and may have little access to the leaders, yet not so large as to seem distant or the leaders unapproachable. If you do not know the leaders yourself, you probably know someone who is friends with them.

Macro-level groups: Large social units, usually quite bureaucratic, which operate at a national or a global level. Most members are unlikely to know or have communicated with the leaders personally or know someone who knows them. The "business" of these groups is of international import and implication. Some research indicates that every person on the planet is within seven degrees of relatedness to every other human being. A macro-level system is one in which *most* of the members are within at least five degrees of relatedness to each other—that is, you know someone who knows someone who knows someone who knows someone who knows the person in question.

1. Micro social units

_____ Your nuclear family

_____ The United Nations

_____ A local chapter of the Lions Club or the Rotary Club

_____ Your high school baseball team

_____ India

_____ NATO (North Atlantic Treaty Organization)

_____ The First Baptist Church in Muncie, Indiana

_____ World Bank

_____ A family reunion

_____ Google, Inc. (international)

_____ The Department of Education for the Commonwealth of Kentucky

_____ The show choir in your local high school

_____ African Canadians

2. Meso social units

_____ The Dineh (Navajo) people

_____ Canada

_____ The Republican Party in the United States

_____ The World Court

_____ A fraternity at your college

_____ International Monetary Fund (IMF)

_____ The Ministry of Education for Spain

_____ The Roman Catholic Church (with its headquarters at the Vatican in Rome)

_____ Australia

_____ The Chi Omega National Sorority

_____ Boy Scout Troop #3 in Marion, Ohio

_____ Al-Qaeda (an international alliance of terrorist organizations)

_____ The provincial government for the Canadian province of Ontario

_____ The United States of America

3. Macro social units

Most of these fall into clear categories, but some are "on the line," and one could legitimately place them in more than one group. See how your authors rate these at **www.sagepub.com/oswcondensed2e.** There is a second exercise online that asks you to apply these categories to identify some connections.

Sociology Around the World

Tunisian Village Meets the Modern World

This is a story of change as macro-level innovations enter a small traditional village. It illustrates how the social units of the social world model and the three levels of analysis enter into sociological analysis. As you read, try to identify both the units and levels of analysis being discussed and the impact of globalization on a community that cannot know what these changes will bring.

The workday began at dawn as usual in the small fishing village on the coast of Tunisia, North Africa. Men prepared their nets and boats for the day, while women prepared breakfast and dressed the young children for school. About 10 A.M., it began—the event that would change this picturesque village forever. Bulldozers arrived first, followed by trench diggers and cement mixers, to begin their overhaul of the village.

Villagers had suspected something was afoot when important-looking officials arrived two months earlier with foreign businessmen, followed by two teams of surveyors. Without their approval, the government had sold land that the village had held communally for generations to the foreigners so they could build a multimillion-dollar hotel and casino. When concerned citizens asked what was happening in their village, they were assured that their way of life would not change. The contractor from the capital city of Tunis said they would still have access to the beach and ocean for fishing. He also promised them many benefits from the hotel project—jobs, help from the government to improve roads and housing, and a higher standard of living.

The contractor had set up camp in a trailer on the beach, and word soon got around that he would be hiring some men for higher hourly wages than they could make in a day or even month of fishing. Rivalries soon developed between friends over who should apply for the limited number of jobs.

As the bulldozers moved in, residents had mixed opinions about the changes taking place in their village and their lives. Some saw the changes as exciting opportunities for new jobs and recognition of their beautiful village; others viewed the changes as destroying a lifestyle that was all they and generations before them had known.

Today, the village is dwarfed by the huge hotel, and the locals are looked on as quaint curiosities by the European tourists. Fishing has become a secondary source of employment to working in the hotel and casino or selling local crafts and trinkets to souvenir-seeking visitors. Many women are now employed outside the home by the hotel, creating new family structures as grandparents, unemployed men, and other relations take over child-rearing responsibilities.

To understand the changes in this one small village and other communities facing similar change, a sociologist uses the sociological imagination. This involves understanding the global, political, and economic trends that are affecting this village and its inhabitants (macro-level analysis). It requires comprehension of transformation of social institutions within the nation (meso-level analysis). Finally, sociological investigation explores how change impacts the individual Tunisian villagers (micro-level analysis).

To sociologically analyze the process of change, it is important to understand what is going on in this situation. The government officials and the international business representatives negotiated a lucrative deal to benefit both Tunisia and the business corporation. The community and its powerless residents presented few obstacles to the project from the point of view of the government, and in fact, government officials reasoned that villagers could benefit from new jobs. However, economic and family roles of the villagers—how they earned a living and how they raised their children—changed dramatically with the disruption to their traditional ways. The process of change began with the demand of people far from Tunisia for vacation spots in the sun. Ultimately, this process reached the village's local environment, profoundly affecting the village and everyone in it. For this Tunisian village, the old ways are gone forever.

The Social World Model and This Book

Throughout this book, the social world model will be used as the framework for understanding the social units, social processes, and surrounding environment. Each social unit and process is taken out, examined, and returned to its place in the interconnected social world model so that you can comprehend the whole social world and its parts, like putting a puzzle together. Look for the model at the beginning of every chapter. You can also expect the micro-, meso-, and macro-level dimensions of issues to be explored throughout the text.

The social world engulfs each of us from the moment of our birth until we die. Throughout our lives, each of us is part of a set of social relationships that provide guidelines for how we interact with others and how we see ourselves. This does not mean that human behavior is strictly determined by our links to the social world. Humans are more than mere puppets whose behavior is programmed by social structure. It does mean, however, that influence between the individual and the larger social world is reciprocal. We are influenced by and we have influence on our social environment. The social world is a human creation, and we can and

do change that which we create. It acts on us, and we act on it. In this sense, social units are not static but are constantly emerging and changing in the course of human interaction.

The difficulty for most of us is that we are so caught up in our daily concerns that we fail to see and understand the social forces that are at work in our personal environments. What we need are the conceptual and methodological tools to help us gain a more complete and accurate perspective on the social world. The concepts, theories, methods, and levels of analysis employed by sociologists are the very tools that will help give us that perspective. To use an analogy, each different lens of a camera gives the photographer a unique view of the world. Wide-angle lenses, close-up lenses, telephoto lenses, and special filters each serve a purpose in creating a distinctive picture or frame of the world. No one lens will provide the complete picture. Yet, the combination of images produced by various lenses allows us to examine in detail aspects of the world we might ordinarily overlook. That is what the sociological perspective gives us: a unique set of tools to see the social world with more penetrating clarity. In seeing the social world from a sociological perspective, we are better able to use that knowledge constructively, and we are better able to understand who we are as social beings. Practice the levels of analysis in the following "Engaging Sociology."

Building and staffing of this resort in Tunisia—which is patronized by affluent people from other continents (global)—changed the economy, the culture, the social structure (meso level), and individual lives (micro level) in the local community.

Engaging Sociology

Micro-Meso-Macro: An Application Exercise

Imagine that there has been a major economic downturn (recession) in your local community. Identify four possible events at each level (micro, meso, and macro) that could contribute to the economic troubles in your town.

The micro (local community) level:

1. _____

2. _____

3. _____

4. _____

The meso (intermediate—state, organizational, or ethnic subculture) level:

1. _____

2. _____

3. _____

4. _____

The macro (national/global) level:

1. _____

2. _____

3. _____

4. _____

The next issue, then, is how we gather data that inform how we understand and influence the social world. When we say we know something about society, how is it that we know? What is considered evidence in sociology, and what lens (theory) do we use to interpret the data? These are the central issues of the next chapter.

What Have We Learned?

We live in a complex social world with many layers of interaction. If we really want to understand our own lives, we need to comprehend the various levels of analysis that affect our lives and the dynamic connections between those levels. Moreover, as citizens of democracies, we need to understand how to influence our social environments, from city councils, school boards, and state legislatures to congressional, presidential, and other organizations with major policy makers. To do so wisely, we need both objective lenses for viewing this complex social world and accurate, valid information (facts) about the society. As the science of society, sociology can provide both tested empirical data and a broad, analytical perspective.

Key Points:

- Humans are, at their very core, social animals—more akin to pack or herd animals than to individualistic cats. (See pp. 5–7.)

- A core concept in sociology is the sociological imagination. It requires that we see how our individual lives and personal troubles are shaped by historical and structural events outside of our everyday lives. It also prods us to see how we can influence our society. (See pp. 9–10.)

- Sociology is a social science and, therefore, uses the tools of the sciences to establish credible evidence to understand our social world. As a science, sociology is scientific and objective rather than value laden. (See pp. 10–13.)

- Sociology has pragmatic applications, including those that are essential for the job market. (See pp. 13–17.)

- Sociology focuses on social units or groups, on social structures such as institutions, on social processes that give a social unit its dynamic character, and on their environments. (See pp. 17–20.)

- The social world model is the organizing theme of this book. Using the sociological imagination, we can understand our social world best by clarifying the interconnections between micro, meso, and macro levels of the social system. Each chapter of this book will examine society at these three levels of analysis. (See pp. 20–27.)

Contributing to Our Social World: What Can We Do?

At the end of this and all subsequent chapters, you will find suggestions for work, service learning, internships, and volunteering that encourage you to apply the concepts, principles, and ideas discussed in the chapter in practical contexts.

At the Local Level:

- *Sociology departments' student organizations or clubs.* You can meet other students interested in sociology, get to know faculty members, and attend presentations by guest speakers. Visit a meeting and consider joining whether you are a sociology major or not. If no such organization exists, consider forming one with the help of a faculty member.

- *Undergraduate honors society, Alpha Kappa Delta (AKD).* Visit the AKD website at http://sites.google.com/site/alphakappadeltainternational and learn more about it and what it takes to join or form a chapter.

- *Volunteer opportunities.* This sociology course will give you ideas of many volunteer opportunities in which you may want to become involved. This is rewarding and good experience for future jobs.

At the Regional and National Levels:

- *The American Sociological Association (ASA)* is the leading professional organization of sociologists in the United States. It has several programs and initiatives of special interest to students. Visit the ASA website at www.asanet.org and click on the "Teaching & Learning/Students: Undergraduate" link at the top of the page. Read the items and follow the links to additional material on the advantages of studying sociology. Most sociologists also participate in a major state or regional association. These groups are especially student friendly and feature publications and sessions at their annual meetings specifically for undergraduates. The organizations and website addresses are listed by the ASA, with direct links to their home pages at http://www2.asanet.org/governance/aligned.html.

Visit **www.sagepub.com/oswcondensed2e** for online activities, sample tests, and other helpful information. Select "Chapter 1: Sociology" for chapter-specific activities.

CHAPTER 2

Examining the Social World

How Do We Know?

Science is about knowing through scientific research. Pictured here are scientists: archaeologists, a sociologist, a geologist, and a biologist. Sociology is a social science because of the way we gather scientific evidence.

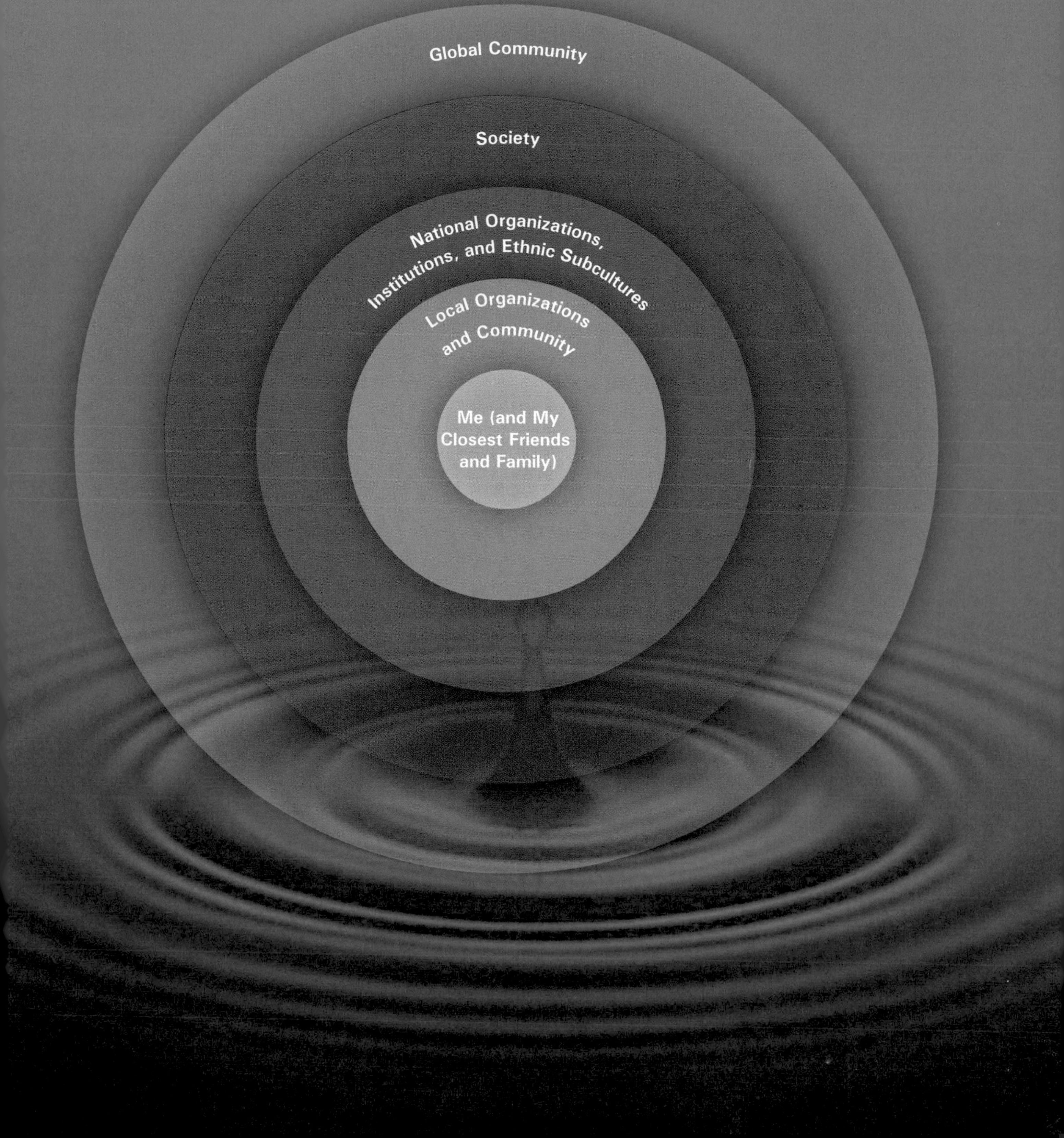

Global Community

Society

National Organizations,
Institutions, and Ethnic Subcultures

Local Organizations
and Community

Me (and My
Closest Friends
and Family)

Think About It

Me (and My Closest Friends/Family)	When you say that you know something, how do you know it?
Local Community	When you are trying to convince neighbors or people in your community to accept your opinion, why is evidence important?
National Institutions; Complex Organizations; Ethnic Groups	How do sociologists gather dependable data about families, educational institutions, or ethnic groups?
National Society	What does it mean to study national societal patterns scientifically?
Global Community	How can theories about global interactions help us understand our own lives at the micro level?

Let us travel to the Southern Hemisphere to meet a teenage boy, Hector. He is a 16-year-old living in a favela (slum) on the outskirts of São Paulo, Brazil. He is a polite, bright boy, but his chances of getting an education and a steady job in his world are limited. Like millions of other children around the world, he comes from a poor rural farm family that migrated to the urban area in search of a better life. However, his family ended up in a crowded slum with only a shared spigot for water and one string of electric lights along the dirt road going up the hill on which they live. The sanitary conditions in his community are appalling—open sewers, no garbage collection, contagious diseases. His family is relatively fortunate, for they have cement walls and wood flooring, although no bathroom, running water, or electricity. Many adjacent dwellings are little more than cardboard walls with corrugated metal roofs and dirt floors.

Hector wanted to stay in school but was forced to drop out to help support his family. Since leaving school, he has picked up odd jobs—deliveries, trash pickup, janitorial work, gardening—to help pay the few centavos for the family's dwelling and to buy food to support his parents and six siblings. Even when he was in school, Hector's experience was discouraging. He was not a bad student, and some teachers encouraged him to continue, but other students from the city teased the favela kids and made them feel unwelcome because they were poor and sometimes dirty. Most of his friends dropped out before he did. Hector often missed school because of other obligations—opportunities for part-time work, a sick relative, or a younger sibling who needed care. The immediate need to put food on the table outweighed the long-term value of staying in school. What is the bottom line for Hector and millions like him? Because of his limited education and work skills, obligations to his family, and limited opportunities, he most likely will continue to live in poverty. Sociologists are interested in what influences the social world of children like Hector: family,

Slum dwellers of São Paulo, Brazil. Hector lives in a neighborhood with shelters made of available materials such as boxes, no electricity or running water, and poor sanitation.

friends, school, community, and the place of one's nation in the global political and economic systems. Because of many previous studies of poverty and theories as to why it occurs, sociologists have some understanding of Hector's problems.

To understand how sociologists study poverty and many other social issues, we consider the theories and methods they use to do their work. Understanding the *how* helps us see that sociology is more than guesswork or opinion. Rather, it involves the use of scientific methods based on a systematic process for expanding knowledge of the social world. Some of the ideas and terms in this chapter may be new to you, but they are important as they lay the foundation for future chapters.

Whatever our area of study or job interests, we are likely to find ourselves asking sociological questions. Consider some examples: How do we select our mates? Why do we feel so strongly about our preferred sports team? Why is there binge drinking on college campuses? Do sexually explicit videos and magazines reinforce sexist stereotypes or encourage sexual violence? Do tough laws and longer prison sentences deter people from engaging in criminal conduct? How do we develop our religious and political beliefs? How are the Internet and other technologies affecting everyday life for people around the world?

This chapter will introduce you to the basic tools used to plan studies and gather dependable information on topics of interest. It will help you understand how sociology approaches research questions. To this end, we will consider ideas underlying science, social (empirical) research and social theory, how sociologists study the social world, the development of sociology as a discipline, the relationship between sociological theories and research methods, major social theories, ethical issues, and practical applications and uses of sociological knowledge. We will start with some background on the origins of sociology to provide a better understanding of how sociology developed.

Ideas Underlying Science

Throughout most of human history, people came to "know" the world by traditions being passed down from one generation to the next. Things were so because authoritative people in the culture said they were so. Often, there was reliance on magical or religious explanations of the forces in nature, and these explanations became part of tradition. With advances in the natural sciences, observations of cause-and-effect processes became more systematic and controlled. As the way of knowing about the world shifted, tradition and magic as primary means to understand the world were challenged. It was only a little more than 200 years ago that people

Lightning has been understood as a form of electricity rather than a message from an angry god only since 1752, thanks to an experiment by Benjamin Franklin.

thought lightning storms were a sign of an angry god, not electricity caused by meteorological forces.

The scientific approach is based on several core ideas: First, there is a real social world that can be studied scientifically. Second, there is a certain order to the world with identifiable patterns that results from a series of causes and effects. The world is not a collection of unrelated random events but rather is a collection of events that are causally related and patterned. Third, the way to gain knowledge of the world is to subject it to empirical testing. **Empirical knowledge** means that the facts have been objectively (without opinion or bias) gathered and carefully measured and that what is being measured is the same for all people who observe it.

For knowledge to be scientific, it must come from phenomena that can be observed and measured. Phenomena that cannot be subject to measurement are not within the realm of scientific inquiry. The existence of God, the devil, heaven, hell, and the soul cannot be observed and measured and therefore cannot be examined scientifically.

Finally, science is rooted in **objectivity**, using methods that limit the impact of the researcher's personal opinions or biases on the study being planned, data collection, and analysis of evidence about the social world. **Evidence** refers to facts and information that are confirmed through systematic testing using the five senses and sometimes enhanced with research tools. Moreover, scientists are obliged not to distort their research findings so as to promote a particular point of view. Scientific research is judged first on whether it passes the test of being conducted objectively, without bias.

One way to evaluate evidence is to see whether the research does or does not support what the researcher thinks is true. The failure to disprove one's ideas offers evidence to support the findings. For example, if we wish to study whether gender is a major factor in a person being altruistic (helpful to others), then we must plan the research study so that we can either support or disprove our hypothesis (our educated guess or prediction).

A failure to meet these standards—empirical knowledge, objectivity, and scientific evidence—means that a study is not scientific. Someone's ideas can seem plausible and logical but still not be supported by the facts. This is why evidence is so important. Sociology is concerned with having accurate evidence, and it is important to know what is or is not considered accurate evidence. You have probably seen an episode of the *CSI: Crime Scene Investigation* series on television. The shows in this series depict the importance of careful collection of data and commitment to objective analysis. Sociologists deal with different issues, but the same sort of concern for accuracy in gathering data guides their work.

Sociologists have come a long way from the early explanations of the social world that were based on the moralizing of the eighteenth- and nineteenth-century social philosophers and the guesswork of social planners. Part of the excitement of sociology today comes from the challenge to improve the scientific procedures applied to the study of humans and social behavior and to base policy on carefully collected data.

Thinking Sociologically

If you wanted to learn something about the morale of one of your university's athletic teams or of a Greek sorority or fraternity, which of the following would produce better results: (a) interviewing those who have quit the organization, or (b) interviewing a wide range of campus athletes or Greek organization members? Why? Why might one method end up with bias in the results and not give the full story?

Empirical Research and Social Theory

We all have beliefs about how our social world works, and we take these beliefs for granted. Social researchers use the scientific method to examine these views of society. They assume that (a) there are predictable social relations in the world, (b) social situations will recur in certain patterns, and (c) social situations have causes that can be understood. If the social world were totally chaotic, little would be gained by scientific study.

Just as individuals develop preferences for different religious or political beliefs that guide their lives, sociologists develop preferences for different explanations of the social world, called social theories. The main difference

between individual beliefs and social theories is that the latter are subject to ongoing, systematic empirical testing.

Theories are statements or explanations of how two or more facts about the social world are related to each other. Theories try to explain social interactions, behaviors, and problems. A good theory should allow the scientist to make predictions about the social world. Different theories are useful at each level of analysis in the social world—micro, meso, or macro; which theory a sociologist uses to study the social world depends on the level of analysis to be studied, as illustrated in the model on this page.

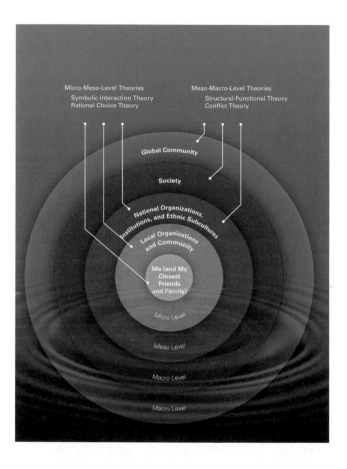

Each of the major theoretical perspectives discussed in this chapter—symbolic interaction, rational choice, structural-functionalist, and conflict theory—gives a perspective on the way the social world works. To study Hector's life in Brazil, researchers might focus on the micro-level interactions between Hector and his family members, peers, teachers, and employers as factors that contribute to his situation. For example, why does Hector find some activities (e.g., working) more realistic or immediately rewarding than others (e.g., attending school)? A meso-level focus might examine organizations and institutions—such as the business world, the schools, and the religious communities in Brazil—to see how they shape the forces that affect Hector's life. Alternatively, the focus might be on macro-level analysis—the class structure (rich to poor) of the society and the global forces such as trade relations between Brazil and other countries that influence opportunities for the Brazilians who live in poverty.

Sociologists focus on group or societal explanations to understand issues such as the poverty that haunts Hector's life. Different research methods to collect data are appropriate at different levels of analysis and depend on the theory being used and the question being researched.

Scientists, including sociologists, often use theories to predict why things happen and under what conditions they are likely to happen. Explanations about the relationships between social variables need to be tested by collecting data. This is where research methods—the procedures one uses to gather data—are relevant. Theory and research are used together and are mutually dependent. The facts (data) must be carefully gathered and are then used to assess the accuracy of the theory. If a theory is not supported by the data, it must be reformulated or discarded. Theory and research are used together and are mutually dependent. We turn first to a discussion of how we gather data.

How Sociologists Study the Social World

How do we know? Sociologists design and refine scientific tools for studying sociological questions. Presented here is a skeleton of the research process that sociologists and other social scientists spend years studying and perfecting. If you study more sociology, you will learn more about the research process. It is exciting to collect accurate information and learn answers to important, meaningful questions that can make a difference in people's lives.

A researcher follows a number of logically related steps.

I.   Planning a Research Study

   Step 1:   Define a topic or problem that can be investigated scientifically.

   Step 2:   Review existing relevant research studies and theory to refine the topic and define variables.

   Step 3:   Formulate hypotheses or research questions and operationalize variables.

II. Designing the Research Method and Collecting the Data

   Step 4:   Design the research method that specifies how the data will be gathered.

   Step 5:   Select a sample of people or groups from the population to study.

   Step 6:   Collect the data using appropriate research methods.

III. Making Sense of the Data

   Step 7:   Analyze the data, figuring out exactly what the study says about the research question(s).

   Step 8:   Draw conclusions and present the final report, including suggestions for future research.

Each of these steps is very important, whether you become a professional sociologist or do a study for a local organization or your workplace. To introduce the research process, we provide a quick summary: planning a research study; designing the research method and collecting data; making sense of the data; and drawing conclusions.

Planning a Research Study

Planning a research study involves four initial steps: defining the topic or problem clearly, reviewing existing research to find out what is already known about the topic, formulating hypotheses, and operationalizing definitions of variables.

Step 1, the most important step, is to define a topic or problem that can be investigated scientifically. Without a clear problem, the research will go nowhere. Using poverty as an example of a broad topic area, sociologists could focus on a number of issues—for example, rates of poverty in different countries, poverty and levels of education, race or gender and poverty, or the relationship between poverty and drug or alcohol abuse. These broad topic areas are a good start, but they must be more specific and defined before they become suitable topics for systematic research. One must clarify specifically what one wants to know about the topic.

Specific research topics, ones that can be measured and tested, are usually posed in the form of questions: Why is it that, in some countries, large segments of the population live in poverty? What causes some students—especially those in poor families—to drop out of school? Why are there disproportionate numbers of women and people of color living in poverty? Are people who are poor more vulnerable to excessive drug or alcohol use? The research question must be asked in a precise way. Otherwise, it cannot be tested empirically. Developing good research questions is extremely important in sound scientific research.

Step 2 in planning the research is reviewing existing relevant research studies and theory to determine what other researchers have already learned about the topic. One needs to know how the previous research was done, how terms were defined, and the strengths and limitations

Gathering data—through interviews, direct observations of behavior, experiments, and other methods—is part of the science of sociological investigation.

by collecting data to test hypothesized relationships between variables. The process of determining how to measure concepts is called *operationalization*.

To operationalize variables, researchers link social concepts such as poverty to specific measurement indicators. Using the hypothesis above, concepts such as dropouts and poverty can be measured by determining the number of times they occur. Dropouts, for instance, might be defined by the number of days of school missed in a designated period of time according to school records. Poverty could be defined as having an annual income that is less than half of the average income for a similar size of family in Brazil. Measurement of poverty could also include family assets such as ownership of property or other tangible goods such as cattle, automobiles, and indoor plumbing, as well as access to medical care, education, transportation, and other services available to citizens. It is important to be clear, precise, and consistent about how one measures poverty. This is essential so that those reading the study are clear on how the study was done and those doing a follow-up study can critique and improve on the study. Background research—including examination of previous studies of poverty—can provide guidelines or examples for how to operationalize variables.

Thinking Sociologically

Pick a topic of interest to you. Now write a research question on your topic. What are your variables, and how could you operationalize them?

Being Clear About Causality

Underlying this discussion about defining the problem is the ability to plan a study that accurately measures the predicted relationship between concepts that are stated in our hypothesis. The following key research terms are important in understanding how two variables (concepts that vary in frequency and can be measured) are related:

of that research. Social scientists can then link their study to existing findings to build the knowledge base. This step usually involves combing through scholarly journals and books on the topic. In the case of Hector, researchers would need to identify what previous studies have said about why impoverished young people drop out of school.

In *Step 3*, based on the review of the literature on the topic, social scientists often formulate **hypotheses**—reasonable, educated guesses about how variables are related to each other, including causal relationships. These are speculations not yet with supporting data but based on the knowledge researchers have gained on the topic. A hypothesis predicts the relationship between two or more variables and ways in which variables are related to each other (e.g., poverty and education). An example of a hypothesis to study Hector's situation might be as follows: "Poverty is a major cause of favela teenagers dropping out of school because they need to earn money for the family." Another hypothesis could predict the opposite—that dropping out of school leads to poverty. Again, *a hypothesis provides a statement to be tested*.

Researchers then identify the key concepts, or ideas, in the hypothesis (e.g., poverty and dropping out of school). These concepts can be measured by collecting facts or data. Key concepts such as poverty and dropping out of school can be measured and vary in degree. For example, poverty levels vary from high to low, and this is referred to as a variable because of this variation. **Variables** are concepts (ideas) that can vary in frequency of occurrence from one time, place, or person to another. Examples include levels of poverty, percentage of people living in poverty, or number of years of formal education. Variables can then be measured

- **Correlation** refers to a relationship between variables such as poverty and low levels of educational attainment in which change in one variable is associated with change in another. The hypothesis above predicts that poverty and teenagers dropping out of school are related and vary together.

- **Cause-and-effect relationships** help establish the relationship between two variables. Once we have determined that there is probably a correlation—that both variables, poverty and dropping out of school, occur in the same situation—we need to take the next step of figuring out how they are related or which comes first by testing for a causal relationship between variables. The **independent variable** is hypothesized to cause the change; in terms of time sequence, it comes first and affects the **dependent variable**. If we hypothesize that poverty causes Hector and others to drop out of school, *poverty* is the independent variable in this hypothesis, and *dropping out of school* is the dependent variable, dependent on the poverty. In determining cause and effect, the independent variable must always precede the dependent variable in time sequence if we are to say one variable causes another.

- **Spurious relationships** occur when there is no causal relationship between the independent and dependent variables but they vary together, often due to a third variable working on both of them. For example, if the quantity of ice cream consumed is highest during those weeks of the year when the most swimming pool drownings occur, these two events are correlated. However, eating ice cream does not cause the increase in pool deaths. Indeed, hot weather may cause more people both to purchase ice cream and to go swimming, thus resulting in more chance of drowning incidents. This is a spurious relationship. **Controls** are used by researchers to eliminate all spurious variables, leaving only those related to the hypothesis. Using controls helps ensure the relationship is not spurious.

Designing the Research Method and Collecting the Data

After the researcher has carefully planned the study, *Step 4* is to select appropriate data collection methods. These depend on the levels of analysis of the research question (micro, meso, and macro) the researcher is asking. If researchers want to answer a macro-level research question, such as the effect of poverty on school success nationwide, they are likely to focus on large-scale social and economic data sources. To learn about micro-level issues such as the influence of peers on an individual's decision to drop out of school, researchers will focus on small-group interactions. Figure 2.1 illustrates different levels of analysis.

Different data collection techniques will be appropriate for research questions and hypotheses at different levels of analysis. The method selected is one of the most important decisions in research because the quality of the data collected is directly related to the value of the study findings. Accuracy in planning can mean the difference between a fruitful scientific study and data that have little meaning.

The primary methods used to collect data for research studies include surveys (both interviews and questionnaires), observation studies, controlled experiments, and use of existing information. Our discussion cannot go into detail about how researchers use these techniques, but a brief explanation gives an idea of why they use these primary methods.

The **survey method** is used when sociologists want to gather information directly from a number of people regarding how they think or feel or what they do. Two forms of surveys are common: the interview and the questionnaire. Both involve a series of questions asked of respondents.

Interviews are conducted by talking directly with people and asking questions in person or by telephone. Questionnaires are written questions to which respondents reply in writing. In both cases, questions may be open-ended, allowing the respondent to say or write whatever comes to mind, or closed-ended, requiring the respondent to choose from a set of possible answers.

Micro Level	
Individual	Hector
Small group	Hector's family and close friends
Local community	The *favela;* Hector's local school, church, neighborhood organizations
Meso Level	
Organizations	Brazilian corporations, Catholic Church, and local school system in Brazil
Institutions	Family; education; political, economic, and health systems in the region or nation of Brazil
Ethnic subcultures	Native peoples, African-Brazilians
Macro Level	
National society	Social policies, trends, and programs in Brazil
Global community of nations	Status of Brazil in global economy; trade relations with other countries; programs of international organizations or corporations

Figure 2.1 The Social World Model and Levels of Analysis

One example of a survey is the census, done in the United States and many other countries every 10 years. Sometimes it is difficult to gather accurate data on the entire population. In this situation, the census worker is counting homeless people in Penn Station in New York City.

One method to study school dropouts is to survey teenagers about what caused them to leave school. The researcher would have to evaluate whether an interview or a questionnaire would provide the best information. Interviews are more time consuming and labor intensive than questionnaires but are better for gathering in-depth information. However, if the researcher wants information from a large number of teenagers, questionnaires are often more practical and less costly.

Thinking Sociologically

How would you word questions in a survey about the effects of peer influence on dropping out of school? How could you find out whether your questions are good ones?

Observation studies (also called field methods) involve systematic, planned observation and recording of behavior or interaction in natural settings, and can take several forms. In *nonparticipant observation,* the researcher is not involved in group activities but observing as an outsider. For instance, the researcher may observe or videotape a classroom that Hector and his friends attend to study the group dynamics without their awareness of the researcher's presence.

Participant observation occurs when the researcher actually participates in the activities of the group being studied. The researcher, for instance, might hang out with Hector and his friends to observe group processes as they occur

naturally. Participation is particularly useful when studying illegal or deviant activities, where it might be impossible to gain scientifically useful information any other way. The results from observation studies are referred to as *qualitative* research. Qualitative research provides valuable information that is rich in detail and describes the context of the situation, such as the dynamics of the favela or Hector's classroom. The research evolves in response to what the researcher learns as the research progresses. Unlike research driven by hypotheses to be tested, this method allows the researcher to respond to new ideas that come up during the research. Quantitative research, on the other hand, provides hard data such as percentages or other numbers that answer the researcher's questions and can be illustrated in neat tables (Creswell 2009).

In **controlled experiments**, all variables except the one being studied are held constant. In this way, researchers can manipulate the variable being studied to determine its effects. A controlled experiment usually requires an **experimental group**, in which people are exposed to the variable being studied to test its effect, and a **control group**, in which the subjects are not exposed to the variable the experimenter wants to test. The control group provides a baseline to which the experimental group can be compared. Controlled experiments are powerful because they are the most accurate test of cause and effect.

Despite the advantages of controlled experiments, there are some drawbacks. First, many sociological questions cannot be studied in controlled settings because they deal

Experiments are especially effective at controlling all of the variables to be able to know which outcomes result from the independent variable. Of course, they are not as effective as observation studies in terms of being real-life situations, and they are more subject to researcher effects—that is, respondents being influenced by the researcher or the research setting.

with meso- and macro-level organizations and social forces and cannot be placed in a controlled situation like the lab pictured above. For example, Hector's situation in the favela cannot be studied in a laboratory setting. Second, the mere fact of being in a laboratory setting and knowing one is in an experiment may affect the research results—another form of research effects. Third, because of ethical constraints, social scientists cannot do studies that might harm people. Thus, there are many variables that social scientists cannot introduce in the laboratory.

Existing sources refer to data that already exist but are being employed in a new way or analyzed to understand a new relationship. There are two important approaches to using existing sources: secondary analysis and content analysis. **Secondary analysis** uses existing data, information that has already been collected by other researchers in other studies. Often, large data-collecting organizations such as the United Nations or a country's census bureau, a country's national education department, or a private research organization will make data available for use by researchers. Consider the question of the dropout rate in Brazil. Ministries or departments of education often collect such data. Likewise, if we want to compare modern dropout rates with those of an earlier time, we may find data from previous decades to be invaluable.

Content analysis involves systematic categorizing and recording of information from written or recorded sources. With content analysis (a common method in historical research and study of organizations), sociologists can gather the data they need from printed materials (books, magazines, newspapers, laws, letters), videos, archived radio broadcasts, or even artworks. A researcher studying variations in levels of concern about child abuse could do a content analysis of popular magazines to see how many pages or stories were devoted to child abuse in 1970, 1980, 1990, 2000, and 2010. Content analysis has the advantage of being relatively inexpensive and easy to do. It is also unobtrusive, meaning that the researcher does not influence the subjects being investigated by having direct contact. Furthermore, using materials in historical sequence can be effective in recognizing patterns over time. However, accuracy of the study relies on what content is available.

Thinking Sociologically

What methods would be appropriate to collect data on your research question?

Triangulation, or multiple methods of social research, combines two or more methods of data collection to enhance the accuracy of the findings. To study Hector's situation, a research study could use macro-level quantitative data on poverty and on educational statistics in Brazil and micro-level qualitative interviews with Hector and his peers to determine their goals and their attitudes toward education. Thus, if all findings point to the same conclusion, the researcher can feel much more confident about the study results. Data collection techniques—survey, observation study, controlled experimentation, and analysis of existing sources—represent the dominant methods used to collect data for sociological research.

Step 5 involves selecting a sample. It would be impossible to survey the reasons for poverty or for dropping out of school for every teenager in Brazil. Researchers must select a representative group to study that will help them to understand the larger group. A part of the research design is determining how to make sure the study includes people who are typical of the total group. When the research study involves a survey or field observation, sociologists need to decide who will be observed or questioned to provide a group representative of the whole. This involves careful selection of a **sample**, a small and manageable group of people to study, systematically chosen to accurately represent the characteristics of the entire group (or population) being studied.

Researchers use many types of samples. A common one, the *representative sample,* attempts to accurately reflect the group being studied so that the sample results can be generalized or applied to the larger population. In the case of Hector's favela, a sample for a study could be drawn from all 13- to 16-year-olds in his region or city in Brazil.

Social scientists are not the only professionals who use triangulation. Journalists also consult a variety of sources including social scientists to put together news broadcasts.

The most common form of representative sample is the *random sample*. People from every walk of life and every group within the population have an equal chance of being selected for the study. By observing or talking with this smaller group selected from the total population under study, the researcher can get an accurate picture of the total population and have confidence that the findings apply to the larger group. Developing an effective sampling technique is often a complex process, but it is important to have a sample that represents the group being studied.

Step 6 involves collecting the data. Now that the appropriate research method for collecting data and the sample to study have been selected, the researcher's next step is to collect the data and then analyze those data and see what they say about the research questions and hypotheses. *Step 7* is to analyze the data and evaluate the relationship of variables to each other.

Analysis: Making Sense of the Data

Now that we have, say, 500 questionnaires or a notebook full of field observation notes, what do we do with them? Social researchers use multiple techniques to analyze data. First, the variables being tested must be clear: If dropout rates are high among Hector's friends, was their school achievement low? What other factors are present? This analysis can become quite complex, but the purpose is the same—to determine the relationship between the variables being examined. Second, sociologists determine the most effective tools to analyze relationships between variables. Sociology is not guesswork or speculation. It involves careful, objective analysis of specific data.

Discussion and criticism are an important part of science and make possible more accurate findings, new ideas and interpretations, and more sophisticated methodological approaches to problems. Interpretation of data involves judgment and opinion, often based on the theory guiding the research. As such, it can be challenged by other researchers, who might interpret the results differently or challenge the methods used to collect and analyze the data. Sociologists grow through these critiques, as does the field of sociology itself. Because every research study should be replicable—capable of being repeated—enough information must be given to ensure that another researcher could repeat the study and compare results.

Step 8 involves discussion of results and drawing conclusions for the analysis. A report is developed, outlining the research project and analysis of the data collected. The final section of the report presents a discussion of results, draws conclusions as to whether or not the hypotheses were supported, interprets the results, and makes recommendations from the researcher's point of view, if appropriate. As part of the presentation and discussion of results, the report may contain tables or figures presenting summaries of data that are useful in understanding the data. The next "Engaging Sociology," *How to Read a Research Table,* provides useful tips on reading research tables found in journal articles and newspapers.

Given that social science research focuses on humans and humans are changeable, it is difficult to say that any single study has definitively proven the hypothesis or answered a research question with finality. Indeed, the word *prove* is never used to describe the interpretation of findings in the social and behavioral sciences. Rather, social scientists say that findings tend to support or reject hypotheses.

Science and Uncertainty

So, why bother doing social science research if findings can be challenged, nothing seems absolute, and conclusions seem so uncertain? We can base social policy and our understanding of society on information that is as close to the truth as possible even if we can never claim absolute truth. Systematic, scientific research brings us closer to the reality of the social world than guesswork and opinions. Having supportive findings from numerous studies, not just one, builds a stronger case to support a theory and conclusions. The search for truth is ongoing—a kind of mission in life for those of us who are scholars. Furthermore, as we get closer to accurate understanding of society, our social policies can be based on the most accurate knowledge available.

What makes a discipline scientific is not the subject matter. It is how we conduct our research and what we consider valid evidence. The core features of a science are (a) commitment to empirically validated evidence, facts, and information that is confirmed through systematic processes of testing using the five senses; (b) a focus on being convinced by the evidence rather than by our preconceived ideas; (c) absolute integrity and objectivity in reporting and in conducting research; and (d) continual openness to having our findings reexamined and new interpretations proposed.

It is the fourth feature that causes us to be open to criticism and alternative interpretations. To have credible findings, we always consider the possibility that we have overlooked alternative explanations of the data and alternative ways to view the problem. This is one of the hardest principles to grasp, but it is one of the most important in reaching the truth. Science—including social science—is not facts to be memorized. Science is a process that is made possible by a social exchange of ideas, a clash of opinions, and a continual search for truth. Knowledge in the sciences is created by vigorous debate. Rather than just memorizing concepts in this book to take a test, we hope you will engage in the creation of knowledge by entering into these debates.

Engaging Sociology

How to Read a Research Table

A statistical table is a researcher's labor-saving device. Quantitative data presented in tabular form are more clear and concise than the same information presented in several written paragraphs. A good table has clear signposts to help the reader avoid confusion. For instance, the table below shows many of the main features of a table, and the list that follows explains how to read each feature.*

Table 2.1 Educational Attainment by Selected Characteristic: 2009, for Persons 25 Years Old and Over Reported in Thousands

| Characteristic | Population (1,000) | Percent of Population—Highest Level | | | | | |
		Not a High School Graduate	High School Graduate	Some College, but No Degree	Associate's Degree[1]	Bachelor's Degree	Advanced Degree
Age							
25–34 yrs old	40,520	11.7	28.0	19.2	8.9	22.8	9.3
35–44 yrs old	41,322	11.7	28.7	16.6	10.2	21.4	11.4
45–54 yrs old	44,366	10.9	32.2	17.0	10.8	18.8	10.4
55–64 yrs old	34,289	11.1	30.2	17.8	9.2	18.6	13.1
65–74 yrs old	20,404	17.7	36.4	15.5	5.9	13.9	10.5
75 yrs or older	17,384	26.3	36.6	13.7	4.9	11.4	7.0
Sex:							
Male	95,518	13.8	31.4	16.8	7.9	19.0	11.1
Female	102,767	12.9	30.8	17.3	10.0	19.0	10.1
Race:							
White[2]	162,079	12.9	31.2	16.9	9.1	19.3	10.7
Black[2]	22,598	15.9	35.4	20.3	9.0	12.7	6.6
Other	13,608	14.3	22.3	13.9	7.8	25.8	16.0
Hispanic origin:							
Hispanic	25,956	38.1	29.3	13.3	6.1	9.6	3.6
Non-Hispanic	172,329	9.6	31.3	17.6	9.4	20.4	11.6
Region:							
Northeast	36,572	11.8	33.3	13.1	8.6	19.9	13.3
Midwest	43,163	10.2	34.4	17.1	9.8	18.2	9.7
South	72,720	15.0	31.8	16.9	8.6	17.9	9.8
West	45,829	14.8	25.1	19.9	9.1	20.7	10.5

Source: U.S. Census Bureau (2011b).

1. Includes vocational degrees.

2. For persons who selected this race group only.

*Features of the table adapted from Broom and Selznick (1963).

(Continued)

(Continued)

TITLE: The title provides information on the major topic and variables in the table.

"Educational Attainment by Selected Characteristic: 2009"

HEADNOTE (or Subtitle): Many tables will have a headnote or subtitle under the title, giving information relevant to understanding the table or units in the table.

For this table, the reader is informed that this includes all persons over the age of 25, and they will be reported in thousands.

HEADINGS AND STUBS: Tables generally have one or two levels of headings under the title and headnotes. These instruct the reader about what is in the columns below.

In this table, the headings indicate the level of education achieved so the reader can identify the percentage with a specified level of education.

The table also has a stub: the far-left column. This lists the items that are being compared according to the categories found in the headings. In this case, the stub indicates population of various characteristics: age, sex, race, Hispanic origin, and region.

MARGINAL TABS: In examining the numbers in the table, try working from the outside in. The marginals, the figures at the margins of the table, often provide summary information.

In this table, the first column of numbers is headed "Population (1,000)," indicating (by thousands) the total number of people in each category who were part of the database. The columns to the right indicate—by percentages—the level of educational attainment for each category.

CELLS: To make more detailed comparisons, examine specific cells in the body of the table. These are the boxes that hold the numbers or percentages.

In this table, the cells contain data on age, sex, racial/ethnic (white, black, Hispanic), and regional differences in education.

UNITS: Units refer to how the data are reported. It could be in percentages, in number per 100 or 1,000, or in other units.

In this table, the data are reported first in raw number in thousands and then in percentages.

FACTS FROM THE TABLE: After reviewing all of the above information, the reader is ready to make some interpretations about what the data mean.

In this table, the reader might note that young adults are more likely to have a college education than older citizens.

Likewise, African Americans and Hispanics are less likely than whites to have college or graduate degrees, but the rate for "Other" is even higher than that for whites, probably because it includes Asian Americans. People in the Northeast and West have the highest levels of education. What other interesting patterns do you see?

FOOTNOTES: Some tables have footnotes, usually indicating something unusual about the data or where to find more complete data.

In this table, two footnotes are provided so the reader does not make mistakes in interpretation.

SOURCE: The source note, found under the table, points out the origin of the data. It is usually identified by the label "source."

Ethical Issues in Social Research

Sociologists and other scientists are bound by ethical codes of conduct governing research. This is to protect any human subjects used in research from being harmed by the research. The following standards are a part of the American Sociological Association's code of ethics:

- How will the findings from the research be used? Will they hurt individuals, communities, or nations if they get into the hands of "enemies"? Can they be used to aggravate hostilities? In whose interest is the research being carried out?
- How can the researcher protect the privacy and identities of respondents and informants? Is the risk

to subjects justified by an anticipated outcome of improved social conditions?

- Is there informed consent among the people being studied? This issue relates to the issue of deception by social scientists. If the public learns that social scientists have actively deceived them, what will be the consequences for trusting researchers at a later time when a different study is being conducted?

- Will there be any harm to the participants—including injury to self-esteem or feelings of guilt and self-doubt?

- How much invasion of privacy is legitimate in the name of research? How much disclosure of confidential information, even in disguised form, is acceptable?

Codes of ethics provide general guidelines. In addition, most universities and organizations where research is common have Human Subjects Review Boards whose function is to ensure that human participants are not harmed by the research.

Thinking Sociologically

What might be some ethical problems in the research project you are thinking about as you read this chapter? What, for example, would be the ethical issues of studying a setting or situation but not informing those people involved that you are studying them?

The Development of Sociology as a Social Science

Throughout recorded history, humans have been curious about how and why people form groups. That should not be surprising because the groups we belong to are so central to human existence and to a sense of satisfaction in life.

Early Sociological Thought

Religion influenced the way individuals thought about the world and their social relationships. Christianity dominated European thought systems during the Middle Ages, from the end of the Roman Empire to the 1500s, while Islamic beliefs ruled in much of the Middle East and parts of Africa. North African Islamic scholar Ibn

Khaldun (1332–1406) is generally considered the first social scientist as he was probably the first to suggest a systematic approach to explain the social world. Influenced by Aristotle's work, Khaldun wrote about the processes of social and political change; the rise and fall of societies, cities, and economic life; and the feelings of solidarity that held tribal groups together during his day, a time of great conflict and wars. He discussed the importance of individuals identifying with their groups and subordinating their own interests to their societies and kinship groups. He felt that this subordination made societies possible and allowed for the rule of monarchs, seen as the natural ruling structure at his time (Alatas 2006; Hozien n.d.).

Sociology has its modern roots as a scientific discipline in the ideas of nineteenth-century social, political, and religious philosophers. Mostly European, they laid the groundwork for the scientific study of society. Until the nineteenth century, social philosophers provided the primary approach to understanding society, one that invariably had a strong moral tone. Their opinions were derived from abstract reflection about how the social world works. Often, they advocated forms of government that they believed would be just and good, denouncing those they considered inhumane or evil. Even today, people still debate the ideal communities proposed by Plato, Aristotle, Machiavelli, Thomas More, and other philosophers from the past.

Conditions Leading to Modern Sociology

Several conditions in the nineteenth century gave rise to the emergence of sociology: First, Europeans wanted to learn more about the people in their new colonies. Second, they sought to understand the social revolutions taking place and changes brought about by the Industrial Revolution. Finally, advances in the natural sciences demonstrated the value of the scientific method, and some wished to apply this scientific method to understand the social world.

The social backdrop for the earliest European sociologists was the Industrial Revolution (which began around the middle of the 1700s) and the French Revolution (1789–1799). No one had clear, systematic explanations for why the old social structure, which had lasted since the early Middle Ages, was collapsing or why cities were exploding with migrants from rural areas. French society was in turmoil, and new rules of justice were taking hold. Churches were being subordinated to the state, equal rights under the law were established for citizens, and democratic rule emerged. These dramatic changes marked the end of the traditional monarchy and the beginning of a new social order. Leaders in France relied on the social philosophies of the time to react to the problems that surrounded them, philosophies not informed by facts about social life.

It was in this setting that the scientific study of society emerged. Two social thinkers, Henri Saint-Simon (1760–1825) and Auguste Comte (1798–1857), decried the lack of systematic data collection or objective analysis in social thought. These Frenchmen are considered the first to suggest that a science of society could help people understand and perhaps control the rapid changes and unsettling revolutions taking place. Comte officially coined the term *sociology* in 1838. Just as the natural sciences provided basic facts about the physical world, so, too, there was a need to gather scientific knowledge about the social world. Only then could leaders systematically apply this scientific knowledge rather than use religious or philosophical speculation to improve social conditions.

Comte asked two basic questions: What holds society together and gives rise to a stable order rather than anarchy, and why is there change in society? Comte conceptualized society as divided into two parts. *Social statics* referred to aspects of society that give rise to order, stability, and harmony. *Social dynamics* referred to change and evolution in society over time. Simply stated, Comte was concerned with what contemporary sociologists and the social world model in this book refer to as *structure* (social statics) and *process* (social dynamics). By understanding these aspects of the social world, Comte felt leaders could strengthen society and could respond appropriately to change. His optimistic belief was that sociology would be the "queen of sciences," guiding leaders to construct a better social order.

Although he helped found a new discipline, Comte had some ideas that are not above reproach. He maintained strong philosophical biases toward order and stability, even though the outcomes of that order might reinforce inequality and oppression. Nevertheless, the two main contributions

Industrialization had a number of positive outcomes, including the expansion of prosperity to a larger class of people, but it also had some high costs in exploitation of workers and slums in places like London.

of Comte stand the test of time: The social world can and should be studied scientifically, and the knowledge gained should be used to improve the human condition.

Massive social and economic changes in the eighteenth and nineteenth centuries brought about restructuring and sometimes the demise of political monarchies, aristocracies, and feudal lords. Scenes of urban squalor were common in Great Britain and other industrializing European nations. Machines replaced both agricultural workers and cottage (home) industries because they produced an abundance of goods faster, better, and cheaper. Peasants were pushed off the land by the new technology and migrated to urban areas to find work at the same time that a powerful new social class of capitalists was emerging. Industrialization brought the need for a new skilled class of laborers, putting new demands on an education system that had served only the elite. Families now depended on wages from their labor in the industrial sector to stay alive.

These changes stimulated other social scientists to study society and its problems. Writings of Émile Durkheim, Karl Marx, Harriet Martineau, Max Weber, W. E. B. Du Bois, and many other early sociologists set the stage for development of sociological theories. Accompanying the development of sociological theory was the utilization of the scientific method—systematic gathering and recording of reliable and accurate data to test ideas. First let us examine several scholarly traditions that emerged during the development of sociology.

Three Sociological Traditions

First we have the situation of Hector living in poverty (research problem). Then we have possible explanations to help us understand his situation (theories). Finally we have methods of collecting information (data) to study his situation. Sociologists put these three parts together to carry out research. Here we discuss three long-standing traditions in sociology that speak to the relationship between social scientific (empirical) research and social theory (Buechler 2008). All three traditions have been around for as long as the discipline has existed. The founders of sociology tended to combine them, especially the first and third traditions. So from its beginning, sociology focused on the relationship between micro-, meso-, and macro-level processes.

One tradition, strongly influenced by the ideas of Auguste Comte, Émile Durkheim, and other early sociologists, stresses *scientific sociology*. The thrust is on objectivity and on patterning the discipline after the natural sciences. For these scholars, the net result of sociological work is pure, unbiased analysis for the purposes of scientific understanding. As they see it, others will be left to decide how to use the facts. Sociology is to stick to fact finding and testing hypotheses.

A second tradition is called *humanistic sociology*: "When scientific sociologists do science, they emphasize the word *science,* linking it to other sciences. When humanistic sociologists do social science, they emphasize the word *social,* separating it from other sciences" (Buechler 2008:326). This tradition puts much more emphasis on the unique capacity of humans to create meaning, and it recognizes the subjective nature of meaning. However, when one is studying meaning, it is virtually impossible to produce objective quantitative data that are often the standard of the natural sciences. Humanistic sociologists argue that studying humans is qualitatively different from research in the natural sciences, and that focusing only on the standards of objectivity causes researchers to miss some of the most interesting and important dimensions of human social behavior. Studies should be relevant to humans by improving society and the quality of human life, they assert.

A third tradition sees the role of the discipline as improving society. This often takes two forms. One approach is to help us think more critically about the society in which we live, with special focus on issues of social justice. *Critical sociology* was the focus of Karl Marx, and many in this tradition call for rather drastic changes in the society to promote fairness.

Another strand of "social improvement" tradition is applied sociology. **Applied sociology** is concerned with pragmatic ways to improve the society, sometimes with major reorganization of the society and sometimes with modest policy proposals that will make modern organizations operate more smoothly. Applied sociologists generally work within the current system to fine-tune it rather than overhaul it. In any case, practical applications of the field and concern for justice for women and other "second-class citizens" have been a central focus of the field. Whether the concern is justice or compassion and humane treatment of all people, a common concern is with practical uses of sociology in formulating social policy.

Each of the three traditions plays a part in sociological research. Which is most relevant to a study depends on the purpose of the study. These traditions focus on how the discipline itself is understood. All three will be given attention in this book, for they are all part of mainstream sociology today. From these traditions stem the theories—modes of interpreting data used by all three traditions—that we use to guide modern research.

Sociology's Major Theoretical Perspectives

Let us return to our opening example of Hector. Sociologists have several perspectives that can help us understand the poverty Hector experiences in São Paulo, Brazil. Hector's interactions with family, peers, schoolteachers, employers, and religious leaders can be studied at the micro level of analysis stressing the small groups and neighborhood of which he is a part.

By contrast, meso-level analysis focuses on institutions, large organizations, and ethnic communities. For instance, forces or trends in Brazilian institutions affect the health and vitality of the economy. Likewise, migration from rural areas into the big city of São Paulo, where there are few employment opportunities for those with low skills, would result in meso-level problems for study.

Macro-level analysis considers the larger social context—national and global—within which Hector and his family live. From this perspective, Hector's position in society is part of a total system in which Brazilian people in poverty constitute a reserve labor force, available to work in unskilled jobs as needed. International trade, or a change in the Brazilian economy, could have an influence on Hector's job prospects quite independent of his individual motivation to work. Macro-level theories would consider questions related to the policies of Brazil as a nation and the position of Brazil in the world system.

Sociologists draw on major theoretical perspectives at each level of analysis to guide their research and to help them understand social interactions, organizations, structures, and processes. A **theoretical perspective** is a basic view of society that guides sociologists' ideas and research. Theoretical perspectives are the broadest theories in sociology, providing overall approaches to understanding the social world and social problems. These perspectives help sociologists develop explanations of social patterns and see relationships between patterns.

Recall the description of the social world model presented in Chapter 1. It stresses the levels of analysis—smaller units existing within larger social systems. Some theories are especially effective in understanding micro-level interactions, and others illuminate macro-level structures, although the distinctions are not absolute. Either type of theory—those most useful at the micro or macro levels—can be used at the meso level, depending on the research question being asked. To illustrate how a social problem is approached differently by four major theoretical perspectives, we will further delve into our examination of Hector's circumstances (Ashley and Orenstein 2009; Turner 2003). Keep in mind that sociologists use many theories, some of which will be introduced where relevant in future chapters.

Micro- to Meso-Level Theories: Symbolic Interaction and Rational Choice

Symbolic interaction theory is concerned with how people create shared meanings regarding symbols and events and then interact on the basis of those meanings. For example,

Hector interacts with his family and friends in the favela in ways that he has learned are necessary for survival there. In Hector's world of poverty, the informal interactions of the street and the more formal interactions of the school carry different sets of meanings that guide Hector's behavior.

Symbolic interaction theory assumes that groups form around interacting individuals. Through these interactions, people learn to share common understandings and to learn what to expect from others. They make use of symbols such as language, or verbal and nonverbal communication (words and gestures), to interpret interactions with others. Symbolic communication (e.g., language) helps people construct a meaningful world. This implies that humans are not merely passive agents responding to their environments. Instead, they are actively engaged in creating their own meaningful social world based on their constructions and interpretations of the social world. In some cases, people may passively accept the social definitions and interpretations of others. This saves them the effort of making sense of the stimuli around them and of figuring out interpretations (Blumer 1969; Fine 1990). Still, more than any other theory in the social sciences, symbolic interaction theory stresses *human agency*—the active role of individuals in creating their social environment.

George Herbert Mead (1863–1931) is prominently identified with the symbolic interaction perspective through

his influential work, *Mind, Self, and Society* (Mead [1934] 1962). Mead linked the human mind, the social self, and the structure of society to the process of interaction. He believed that humans have the ability to decide how to act based on their perceptions of the social world, or what sociologists refer to as their *definition of the situation*. Mead explored the mental processes associated with how humans define situations. He placed special emphasis on human interpretations of gestures and symbols (including language) and the meanings we attach to our actions. He also examined how we learn our social roles in society such as mother, teacher, and friend and how we learn to carry out these roles. Indeed, as we will see in Chapter 4, he insisted that our notion of who we are—our self—emerges from social experience and interaction with others. These ideas of how we construct our individual social worlds and maintain some control over them have come from the Chicago School of symbolic interaction theory.

Another symbolic interaction approach—the Iowa School (named after scholars from the University of Iowa)—makes a more explicit link to the meso level of the social system. The Iowa School sees individual identities connected to roles and positions within organizations. If we hold several positions—honors student, club president, daughter, sister, student, athlete, thespian, middle-class person—those positions form a relatively stable core self. We will interpret new situations in light of our social positions, some of which are very important and anchor how we see the social world. Once a core self is established, it guides and shapes the way we interact with people in many situations—even new social settings (Kuhn 1964). Thus, if you are president of an organization and have the responsibility for overseeing the organization, part of your self-esteem, your view of responsible citizenship, and your attitude toward life will be shaped by that position. Thus, the Iowa School of symbolic interaction places a bit less emphasis on individual choice, but it does a better job of recognizing the link between the micro, meso, and macro levels of society (Carrothers and Benson 2003; Stryker 1980).

The following principles summarize the modern symbolic interaction perspective and how individual action results in groups, institutions, and societies (Ritzer 2004b):

- Humans are endowed with the capacity for thought, shaped by social interaction.
- In social interaction, people learn the meanings and symbols that allow them to exercise thought and participate in human action.
- People modify the meanings and symbols as they struggle to make sense of their situations and the events they experience.
- Interpreting the situation involves seeing things from more than one perspective, examining possible courses of action, assessing the advantages

Micro-level interactions occur in classrooms every day, both among peers and between teacher and students.

and disadvantages of each, and then choosing one course of action.

- These patterns of action and interaction make up the interactive relations that we call groups, institutions, and societies.
- Our positions or memberships in these groups, organizations, and societies may profoundly influence the way we define our selves and may lead to fairly stable patterns of interpretation of our experiences and of social life.

To summarize, the modern symbolic interaction perspective emphasizes the process each individual goes through in creating and altering his or her social reality and identity within a social setting. Without a system of shared symbols, humans could not coordinate their actions with one another, and hence society as we understand it would not be possible.

Theorists from the symbolic interaction perspective have made significant contributions to understanding the development of social identities and interactions as the basis for groups, organizations, and societies.

By contrast, **rational choice theory** (also referred to as exchange theory) assumes that human behavior involves choices and that people chart a course of action based on rational decisions and self-interest. People act in ways that maximize their rewards and minimize their costs as they make their choices. Where the balance lies determines our behavior. Women may decide to stay in abusive relationships if they think the costs of leaving the relationship are higher than the costs of staying. In considering a divorce, for example, do the rewards of escaping a conflicted, abusive marriage outweigh the costs of daily economic and emotional stresses related to going it alone?

Rational choice has its roots in several disciplines—economics, behavioral psychology, anthropology, and philosophy (Cook, O'Brien, and Kollock 1990). Social behavior is seen as an exchange activity—a transaction in which resources are given and received (Blau 1964; Homans 1974). Every interaction involves an exchange of something valued—money, time, material goods, attention, sex, allegiance. People stay in relationships because they get something from the exchange, and they leave relationships that cost them without providing adequate benefits. Simply stated, people are more likely to act if they see some reward or success coming from their behavior. The implication is that self-interest for the individual is the guiding element in human interaction.

Applying the rational choice theory to Hector's situation, we could examine how he evaluates his life options. What are the costs and rewards for staying in school versus those associated with dropping out and earning money from temporary jobs? For Hector, the immediate benefit of earning a meager income to put food on his family's table

comes at a cost—the loss of a more hopeful and prosperous future. When hungry, the reward of being paid for odd jobs may outweigh the cost of leaving school and losing long-term opportunities.

Rational choice theorists see human conduct as self-centered. Even charitable, unselfish, or altruistic behaviors are seen as self-centered, with benefits of making people feel good about themselves or of possible reward in the afterlife. These benefits are said to outweigh the cost of helping others.

Some theorists (Denzin 1992; Fiske 1991) have attempted a synthesis between the interaction level of analysis and macro-level theories. For instance, friendships, employee-employer contracts, relations in families, and consumer-salesperson transactions can all be represented as exchange relationships. By combining the millions of exchange relationships, these researchers contend that the interactions make up the larger patterns in society. Therefore, we can also study macro-level patterns in the social world (Fiske 1991).

Thinking Sociologically

How can symbolic interaction and rational choice perspectives help explain dating behavior? For example, how might a hookup mean something different to females as opposed to males? How would each micro theory answer this question?

Meso- and Macro-Level Theories: Structural-Functionalist, Conflict, and Feminist

Meso- and macro-level theories consider larger units in the social world such as organizations, institutions (e.g., education, politics, or economics), societies, or global systems. For example, Hector lives in a modernizing country, Brazil, which is struggling to raise the standard of living of its people and compete in the world market. The decisions that are made by his government at a national and an international level affect his life in a variety of ways. As Brazil industrializes, the nature of jobs and modes of communication change. Local village cultures modify as the entire nation gains more uniformity of values, beliefs, and norms. Similarly, resources such as access to clean water in Tanzania may be allotted at the local level, but local communities need national and sometimes international support, as illustrated in the photo of tribal elders on the next page. We can begin to understand how the process of modernization influences Hector, this village in Tanzania, and other people around the globe by looking at two major macro-level approaches: structural-functionalist and conflict perspectives.

PART II

Social Structure, Processes, and Control

Picture a house. First there is the wood frame and then the walls and roof. This provides the framework or structure. Within that structure, activities or processes take place—electricity allows for turning on lights and appliances, water is available to wash in and drink, and people carry out these processes. If something goes wrong in the house, we take steps to control the damage and repair it.

Our social world is constructed in a similar way: Social structure is the framework of society with its organizations, and social processes are the dynamic activities of the society. This section begins with a discussion of the structure of society, followed by the processes of culture and socialization through which individuals are taught cultural rules—how to function and live effectively within their society. Although socialization takes place primarily at the micro level, we will explore its implications at the meso and macro levels as well.

If we break the social structure into parts, like the wood frame, walls, and roof of a house, it is the groups and organizations that make up the structure. To work smoothly, these organizations depend on people's loyalty so that they do what society and its groups need to survive. However, smooth functioning does not always happen. Things break down. This means those in control of societies try to control disruptions and deviant individuals in order to maintain control and function smoothly.

As we explore the next chapters, we will continue to examine social life at the micro, meso, and macro levels, for each of us as an individual is profoundly shaped by social processes and structures at larger and more abstract levels, all the way to the global level.

Society and Culture

Hardware and Software of Our Social World

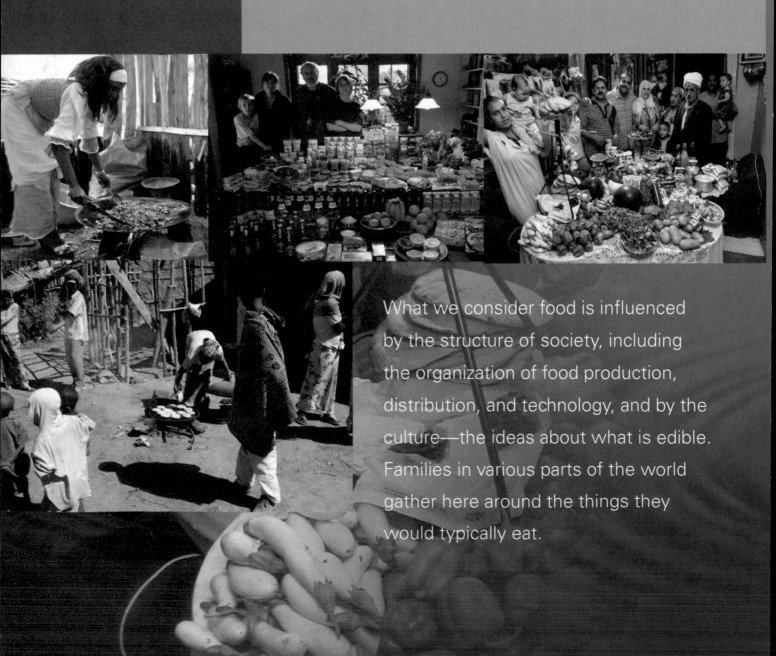

What we consider food is influenced by the structure of society, including the organization of food production, distribution, and technology, and by the culture—the ideas about what is edible. Families in various parts of the world gather here around the things they would typically eat.

Global Community

Society

National Organizations,
Institutions, and Ethnic Subcultures

Local Organizations
and Community

Me (and
My Close
Associates)

Micro: Community Microculture: Your family; a local boy scout troop; a college sorority chapter, a high school soccer team—and their microcultures

Meso: Large bureaucratic corporations; ethnic groups—and their subcultures

Macro: The social structure of a nation —and that nation's culture

Macro: Multinational organizations such as United Nations and World Health Organization —and global culture

Think About It	
Self and Inner Circle	Could you be human without culture?
Local Community	How do microcultures—the values and beliefs in your fraternity, choir, or athletic team—affect you?
National Institutions; Complex Organizations; Ethnic Groups	How do subcultures and counter cultures shape the character of the nation and influence your own life?
National Society	How do the nation's social structures and culture impact you, and how can you influence the national structures and culture?
Global Community	Why do people live so differently in various parts of the world, and how can those differences be relevant to your own country?

What do people around the world eat? To answer this question, researchers asked families to buy food supplies for a week. They then took pictures of families with their weekly diets laid out. The differences in these foods and what each family paid to eat give us an insight into differences in one aspect of cultures around the world.

Mrs. Ukita, the mom in the Ukita family, rises early to prepare a breakfast of miso soup and a raw egg on rice. The father and two daughters eat quickly and rush out to catch their early morning trains to work and school in Kodaira City, Japan. The mother cares for the house, does the shopping, and prepares a typical evening meal of fish, vegetables, and rice for the family.

The Ahmed family lives in a large apartment building in Cairo, Egypt. The 12 members of the extended family include the women who shop for and cook the food—peppers, greens, potatoes, squash, tomatoes, garlic, spices, and rice along with pita bread and often fish or meat.

In the Breidjing refugee camp in Chad, many Sudanese refugees eat what relief agencies can get to them—and that food source is not always reliable. Typical for the Aboubakar family, a mother and five children, is rice or another grain, oil for cooking, occasionally some root plants or squash that keep longer than fresh fruits and vegetables, dried legumes, and a few spices. The girls and women go into the desert to fetch firewood for cooking and to get water from whatever source has water at the time. This is a dangerous trip as they may be attacked and raped or even killed outside of the camps.

The Walker family from Norfolk, Virginia, grabs dinner at a fast-food restaurant on their way to basketball practice and an evening meeting. Because of their busy schedules and individual activities, they cannot always find time to cook and eat together.

Although most diets include some form of grain and starch, locally available fruits and vegetables, and perhaps meat or fish, broad variations in food consumption exist even within one society. Yet all of these differences have

something in common: Each represents a society that has a unique culture that includes what people eat. Food is only one aspect of our way of life and what is necessary for survival. Ask yourself why you sleep on a bed, brush your teeth, or listen to music with friends. Our way of life is called culture.

Culture consists of ideas and "things" that are passed on from one generation to the next in a society—the knowledge, beliefs, values, rules or laws, language, customs, symbols, and material products (e.g., food, houses, and transportation) that help meet human needs. Culture provides guidelines for living. Learning our culture puts our social world in an understandable framework, providing a tool kit with a variety of "tools" we can use to help us construct the meaning of our world (Bruner 1996; Nagel 1994).

A **society** consists of individuals who live together in a specific geographic area, who interact more with each other than they do with outsiders, and who cooperate for the attainment of common goals. Each society includes key institutions—such as family, education, religion, politics, economics, and health—that meet basic human needs. Members of a society share a common culture over time. The way people think and behave in any society is largely prescribed by its culture, which is reshaped, learned, and transmitted from generation to generation. All activities in the society, whether educating young members, preparing and eating dinner, selecting leaders for the group, finding a mate, or negotiating with other societies, are guided by cultural rules and expectations. In each society, culture provides the social rules for how individuals carry out necessary tasks, just as software provides the rules for computer applications.

While culture provides the "software" for the way people live, society represents the "hardware"—the structure that gives organization and stability to group life. Society—organized groups of people—and culture—their way of life—are interdependent. The two are not the same thing, but they cannot exist without each other, just as computer hardware and software are each useless without the other.

Two of the photos on the opening page of this chapter show families from Egypt and Germany gathered around a typical week's food supply. Members of the Aboubakar family of Sudan gather here in front of their tent with a week's worth of food. Note the difference.

Increasingly in the United States, where both parents work, dinner is a fast-food takeout or ready-to-cook packaged meal—a quick dash for nourishment in front of the TV rather than a communal event.

This chapter explores the ideas of society and culture and discusses how they relate to each other. We will look at what society is and how it is organized, how it influences and is influenced by culture, what culture is, how and why culture develops, the components that make up culture, and cultural theories. After reading this chapter we will have a better idea of how we as individuals fit into society and learn our culture.

Society: The Hardware

Families, groups of friends, neighborhoods or communities, and workplace and school groups all provide the structure for culture, just as the computer hardware provides the framework within which the software operates. Society provides the framework for culture.

All societies have geographical boundaries or borders and individuals who live together in families and communities and who share a culture. The structures that make up society include the interdependent positions we hold (parents, workers), the groups to which we belong (family, work group, clubs), and the institutions in which we participate. This "hardware" (structure) of our social world provides the framework for "software" (culture) to function.

Societies differ because they exist in different places with unique resources. Societies change over time with new technology and leadership. Although human societies have become more complex over time, especially in recent human history, people have been hunters and gatherers for 99% of human existence. A few groups remain hunter-gatherers today.

Evolution of Societies

The Saharan desert life for the Tuareg tribe is pretty much as it has been for centuries. In simple traditional societies, individuals are assigned to comparatively few social positions or statuses. Today, however, few societies are isolated from global impact. Even the Tuareg are called on to escort adventurous tourists through the desert for a currency unknown to them and unneeded until recently.

In such traditional societies, men teach their sons everything they need to know, for all men do much the same jobs, depending on where they live—hunting, fishing, or farming and protecting the community from danger. Likewise, girls learn their jobs from their mothers—such as child care, fetching water, food preparation, farming, weaving, and perhaps house building. By contrast, more complex societies, such as industrial or "modern" societies, have thousands of interdependent job statuses based on complex divisions of labor with designated tasks. An interesting question is how traditional societies such as the Tuareg change into new types of societies such as the United States or Canada.

Émile Durkheim ([1893] 1947), an early French sociologist, pictured a continuum between simple and complex societies. He described simple premodern societies as held together by **mechanical solidarity**—the glue that holds society together through shared beliefs, values, and traditions. Furthermore, the division of labor is based largely on male/female distinctions and age groupings. Members of premodern societies tend to think the same way on important matters, and everyone fulfills his or her expected social positions. This provides the glue that holds the society together. The entire society may involve only a few hundred people, with no meso-level institutions, organizations, and

The Tuareg live a simple traditional life in the Saharan desert in the Sahel region of Niger (Africa), and their social structure has few social positions except those defined by gender.

subcultures. Prior to the emergence of nation-states, there was no macro level either—only tribal groupings.

According to Durkheim ([1893] 1947), as societies transform, they become more complex through increasingly complex divisions of labor and changes in the ways people carry out necessary tasks for survival. **Organic solidarity** is found in societies in which social cohesion (glue) is based on division of labor, with each member playing a highly specialized role in the society and each person being dependent on others due to interrelated, interdependent tasks. Prior to the factory system, for example, individual cobblers made shoes to order. With the Industrial Revolution, factories took over the process, with many individuals carrying out interdependent tasks. The division of labor is critical because it leads to new forms of social cohesion based on interdependence, not emotional ties. Durkheim called this organic solidarity because he likened the interdependence of people in different social positions to the interdependence of an animal's internal organs; if any one organ ceases to work, other organs may malfunction, and the animal may die. This is also true of a society. Changes from mechanical (traditional) to organic (modern) society also involve harnessing new forms of energy and finding more efficient ways to use them (Nolan and Lenski 2008). For example, the use of steam engines and coal for fuel triggered the Industrial Revolution, leading to the development of industrial societies.

Also, as macro-level societies changed, they developed more levels, and the meso level—institutions and large bureaucratic organizations—became more influential. Still,

as recently as 200 years ago, even large societies had little global interdependence, and life for the typical citizen was influenced mostly by events at the micro and meso levels. Keep in mind that it was just over 200 years ago when American colonies selected representatives to form a loosely federated government that came to be called the United States.

As you read about each of the following types of societies, from the simplest to most complex, notice the presence of these variables: division of labor, interdependence of people's positions, increasingly advanced technologies, and new forms and uses of energy. Although none of these variables appears *sufficient* to trigger evolution to a new type of society, they may all be *necessary* for a transition to occur.

According to Durkheim ([1893] 1947), then, in traditional societies with mechanical solidarity, interpersonal interaction and community life at the micro level are the most important aspects of social life. Societies have developed as a result of changes toward more organic solidarity. As societies become more complex, meso- and macro-level institutions become more important and have more profound impacts on the lives of individuals.

Hunter-Gatherer Societies

In the Kalahari Desert of Southwest Africa live hunter-gatherers known as the !Kung. (The ! is pronounced with a click of the tongue.) The !Kung live a nomadic life, moving from one place to another as food supplies become available or are used up. As a result, they carry few personal possessions and live in temporary huts, settling around water holes for a few months at a time. Settlements are small, rarely more than 20 to 50 people, for food supplies are not great enough to support large, permanent populations (Lee 1984). !Kung women gather edible plants, while !Kung men hunt. Beyond division of labor by gender and age, however, there are few differences in roles or status.

Life is organized around kinship ties and reciprocity—that is, helping each other—for the well-being of the whole community. When a large animal is killed, people gather from a wide area to share in the bounty, and great care is taken to ensure that the meat is distributed fairly. The !Kung are a typical **hunter-gatherer society** in which people rely directly on plants and animals in their habitat to live. People make their clothing, shelter, and tools from available materials and through trade with other nearby groups. People migrate seasonally to new food sources. Population size remains small as the number of births and deaths is balanced.

From the beginning of human experience until recently, hunting and gathering (or foraging) were the sole means of sustaining life. Other types of societies emerged only recently (Nolan and Lenski 2008), and today, only a handful of societies still rely on hunting and gathering. The hunter-gatherer lifestyle is becoming extinct. For example, much of the wild game on which the !Kung subsisted has

A mother in Côte d'Ivoire (West Africa), carrying her load on her head, returns to the village with her daughter after gathering wood. Carrying wood and water is typically women's work. In this hunting-gathering society, the primary social cohesion is mechanical solidarity.

been overhunted or is now protected on game preserves. South Africa and Botswana have attempted to settle such groups on reservations, and contact with modern governments has forever changed the !Kung way of life and that of other hunting and gathering societies.

Herding and Horticultural Societies

The Masai inhabit the grasslands of Kenya and Tanzania, depending on their herds for survival. Their cattle and goats provide meat, milk, blood (which they drink), and hides. A seminomadic herding society, the Masai move camp to find grazing land for their animals and set up semipermanent shelters for the few months they will remain in one area. Settlements consist of huts constructed in a circle with a perimeter fence surrounding the compound. At the more permanent settlements, the Masai grow short-term crops to supplement their diet. The Masai are under some pressure to become agriculturalists, however, as the government of Kenya—a relatively new macro-level influence in their lives—now restricts their territory and herding practices that encroach on wild animal refuges. Tourism also greatly influences the Masai's economy.

Herding societies have food-producing strategies based on keeping herds of domesticated animals, whose care is the central focus of their activities. Domesticating animals has replaced hunting them. In addition to providing food and other products, cattle, sheep, goats, pigs, horses, and camels represent forms of wealth that result in higher status.

Horticultural societies keep domesticated animals but focus on primitive agriculture or gardening. Horticulturalists use digging sticks and wooden hoes to cultivate tree crops, such as date palms or bananas, and to plant and maintain produce in garden plots, such as yams, beans, taro, squash, or corn. This is more efficient than gathering wild vegetables and fruits.

Both herding and horticultural societies differ from hunter-gatherer societies in that they make their living by cultivating food and have some control over its production (Ward and Edelstein 2009). The ability to control food sources was a major turning point in human history. Societies became more settled and built surpluses of food, which led to increases in population size. A community could contain as many as 3,000 individuals. More people, surplus food, and greater accumulation of possessions encouraged the development of private property and created new status differences between individuals and families. Forms of social inequality became even more pronounced in agricultural societies.

The end of the horticultural stage saw advances in irrigation systems, fertilization of land, and crop rotation. The Neolithic Revolution, occurring as early as 14,500 years ago in Egypt and then evolving in Melanesia, the Middle East, and parts of Asia, was characterized by more permanent settlements, land ownership, human modification of the natural environment, higher population density (cities), changes in diets to more vegetables and cereals, and power

Masai men in Kenya herd their cattle, leading them to water or better grazing. The strategy of domesticating cattle rather than hunting game has been a survival strategy for the Masai, but it also affects the culture in many other ways.

hierarchies. However, the technological breakthrough that moved many societies into the agricultural stage was the plow, introduced more than 6,000 years ago. It marked the beginning of the Agricultural Revolution and brought massive changes in social structures to many societies.

Agricultural Societies

Pedro and Lydia Ramirez, their four young children, and Lydia's parents live as an extended family in a small farming village in Nicaragua. They rise early, and while Pedro heads for the fields to do some work before breakfast, Lydia prepares his breakfast and lunch and sees that their eldest son is up and ready to go to school, while Lydia's mother looks after the younger children. After school, the boy also helps in the fields. Most of the land in the area is owned by a large company that grows coffee, but the Ramirezes are fortunate to have a small garden plot of their own where they grow some vegetables for themselves. At harvest time, all hands help, including the young children. The family receives cash for the crops they have grown, minus rent for the land. The land is plowed with the help of strong animals such as horses and oxen, and fertilizers are used. Little irrigation is attempted, although the garden plots may be watered by hand in the dry season.

The Ramirez family's way of life is typical of agrarian or agricultural farming. Like horticulturalists, **agricultural societies** rely primarily on raising crops for food, but agriculture is more efficient than horticulture. Technological advances such as the plow, irrigation, use of animals, and fertilization allow for intensive and continuous cultivation of the same land, thus permitting permanent settlements and greater food surpluses. Through time, the size of population centers increased to as much as a million or more.

As surpluses accumulated, land in some societies became concentrated in the hands of a few individuals. Wealthy landowners built armies and expanded their empires and could control the labor sources and acquire serfs or slaves. Thus, the feudal system was born. Serfs (the peasant class) were forced to work the land for their survival. Food surpluses also allowed some individuals to leave the land and to trade goods or services in exchange for food. For the first time, social inequality became extensive enough that we could refer to social classes. At this point, religion, political power, a standing army, and other meso-level institutions and organizations came to be independent of the family. The meso level became well established.

As technology advanced, goods were manufactured in cities. Peasants moved from farming communities where land could not support the population to rapidly growing urban areas where the demand for labor was great. It was not until the mid-1700s in England that the next major transformation of society took place, resulting largely from technological advances and harnessing of energy.

The invention of the plow was essential for agricultural societies to develop, and in the early period of agriculture, plows were pushed by people and then pulled by animals. The harnessing of energy was taken to another level when gasoline engines could pull the plow and cultivate thousands of acres. This represents the beginning of industrialization. Modern machinery such as this harvester has pushed farming into the new level of productivity.

Industrial Societies

The Industrial Revolution brought the harnessing of steam power and gasoline engines, permitting machines to replace human and animal power; a tractor can plow far more land in a week than a horse, and an electric pump can irrigate more acres than an ox-driven pump. As a result of the new technologies, raw mineral products such as ores, raw plant products such as rubber, and raw animal products such as hides could be transformed into mass-produced consumer goods. The Industrial Revolution brought enormous changes in products and social structures.

Industrial societies rely primarily on mechanized production for subsistence, resulting in greater division of labor based on expertise. Economic resources were distributed more widely among individuals in industrial societies, but inequities between owners and laborers persisted. Wage earning gradually replaced slavery and serfdom, and highly skilled workers earned higher wages, leading to the rise of a middle class. Farm workers moved from rural areas to cities to find work in factories, which produced consumer goods such as cars and washing machines. Cities came to be populated by millions of people.

Family and kinship patterns at the micro level also changed. Agricultural societies need large, land-based extended family units to do the work of farming, but industrial societies need individuals with specific skills, ability to move to where the jobs are, and smaller families to support. Family roles have changed. For example, children are an asset in agrarian societies and begin work at an early age. In an industrial society, however, children become a liability because they contribute less to the finances of the family and compete with adults for jobs.

Meso- and macro-level dimensions of social life expand in industrializing societies and become more influential in the lives of individuals. National institutions and multinational organizations develop. Today, for example, global organizations such as the World Bank, the World Court, the United Nations, and the World Health Organization address social problems and sometimes even make decisions that change policies within a nation. Medical organizations such as Doctors Without Borders work cross-nationally; corporations such as Nike and Gap become multinational; and voluntary associations such as Amnesty International lobby for human rights around the globe.

Perhaps the most notable characteristic of the industrial age is the rapid rate of change compared to other stages of societal development. The beginning of industrialization in Europe was gradual, based on years of population movement, urbanization, technological development, and other factors in modernization. Today, however, societal change occurs so rapidly that societies at all levels of development are being drawn together into a new age—the postindustrial era.

Postindustrial or Information Societies

The difference between India and the United States is "night and day"—literally. As Keith and Jeanne finished chapters for this book, they were sent to India in the evening and returned typeset by morning, a feat made possible by the time difference. Efficiency of overnight electronic delivery and cost of production have led many publishing companies to turn to businesses halfway around the world for much of the book production process. As India and other developing countries increase their trained, skilled labor force, they are being called on by national and multinational companies to carry out global manufacturing processes. India has some of the world's best technical

This Buddhist monk uses modern technology, including a laptop that can connect him with colleagues on the other side of the globe. In a postindustrial or information society, even rather traditional positions can be affected.

training institutes and the most modern **technology**—that is, knowledge and tools used to extend human abilities. Although many people in India live in poverty, a relatively new middle class is rapidly emerging in major business centers around the country.

After World War II, starting in the 1950s, the transition from industrial to **postindustrial society** began in the United States, Western Europe (especially Germany), and Japan. This shift was characterized by movement from human labor to automated production and from a predominance of manufacturing jobs to a growth in service jobs, such as computer operators, bankers, scientists, teachers, public relations workers, stockbrokers, and salespeople. More than two thirds of all jobs in the United States now reside in organizations that produce and transmit information, thus the reference to an "information age." Daniel Bell (1973, 1989) describes this transformation of work, information, and communication as "the third technological revolution" after industrialization based on steam (what he calls the first technological revolution) and the invention of electricity (the second technological revolution). According to Bell, the third technological revolution was the development of the computer, which has led to this postindustrial or information age.

As in the other types of society, postindustrial societies are undergoing significant structural changes. For example, postindustrial societies require workers with high levels of technical and professional education, such as those in India. Those without technical education are less likely to find rewarding employment in the technological revolution (Tapscott 1998). This results in new class lines being drawn, based in part on skills and education in new technologies. Inequalities in access to education and technology will create or perpetuate social inequalities. For example, one fifth of adults in the United States do not use the Internet, and 18% have no Internet access. That is 20 million households. One half of these nonusers are age 65 or older, and 56% have only a high school education (Sachoff 2008). However, among teens and young adults from 12 to 29, 93% go online, compared to 74% of adults 18 or over. Only 38% of those 65 or over use the Internet (Pew Internet and American Life Project 2010).

Note the decline in use with the end of the work years and among earlier generations, who had less—or no—access to computers when they were growing up. The number of older generations online has been growing steadily, and they are engaging in more varied activities such as shopping, looking for health information, and making travel reservations online (Jones and Fox 2009). Female and male computer users are about equal, at 75% and 73%, respectively. Among English-speaking Hispanics, 58% use computers, compared to 77% of whites and 64% of blacks. Computer usage grows as household incomes and education become greater, with more than 90% of households above $50,000 using computers (Pew Internet and American Life Project 2008).

Postindustrial societies rely on new sources of power, such as atomic, wind, and solar energy, and new uses of computer automation, such as computer-controlled robots, which eliminate the need for human labor other than highly skilled technicians. In an age of global climate change, there is also a lot of interest in technologies that reduce pollution, such as hybrid buses. The core issue in a postindustrial society is this: Control of information and ability to develop technologies or provide services rather than control of money or capital become most important.

Values of twenty-first-century postindustrial societies favor scientific and creative approaches to problem solving, research, and development, along with attitudes that support the globalization of world economies. Satellites, cell phones, fiber optics, and especially the Internet are further transforming postindustrial societies of the Information Age.

The relationship between creativity, local cultural climate, and economic prosperity is the topic of Richard Florida's research. "The Applied Sociologist at Work" explores this fascinating research and explains why it is important to policy makers in local communities. As his research makes clear, the organization of the society and the means of providing the necessities of life have a profound impact on values, beliefs, lifestyle, and other aspects of culture.

Thinking Sociologically

What are likely to be the growth areas for jobs in your society? What competencies and skills will be essential in the future for you to find employment and be successful on the job?

What will the future bring? Among the many ideas for the future, technological advances dominate the field. Predictions include use of cell phones, connecting the poorest corners of the globe with the rest of the world. One billion mobile phone users are predicted for China by 2020, with 80% of the population having phones. Alternative energy sources from wind and solar power to hydrogen will replace gasoline. One million hydrogen-fueled cars are predicted for the United States. Rechargeable batteries that will run for 40 hours without recharging will run most home appliances by 2030. Those who are paralyzed will find help from brain-computer interfaces, giving them ability to control their environments. Many advances will occur in space travel; a hotel may open on the moon by 2025, with transport to the moon. Medical advances will result in stem-cell breakthroughs that can develop into various types of body tissue (Future for All 2011; News of Future 2007). These are just a few of the many predictions that will affect societies and alter some human interactions. The point is that rapid change is inevitable, and the future looks exciting.

In much of this book, we focus on complex, multilevel societies, for this is the type of system in which most of us reading this book now live. Much of this book also focuses on social interaction and social structures, including interpersonal networking, the growth of bureaucratic structures, social inequality within the structure, and the core institutions necessary to meet the needs of individuals and society. In short, "hardware"—society—is the focus of many subsequent chapters. The remainder of this chapter will focus primarily on the "software" dimension (culture) and how we study it.

Culture: The Software

I sleep on a bed. Perhaps you sleep on a tatami mat. I brush my teeth with a toothbrush and toothpaste. Perhaps you chew on a special stick to clean your teeth. I speak English, along with 325 million English speakers for whom it is their first language or native tongue. You may speak one of the many other languages in the world, as do over 1 billion Chinese speakers (mostly Mandarin);

The Applied Sociologist at Work— Richard Florida

Creativity, Community, and Applied Sociology

Like the transformations of societies from the hunter-gatherer to the horticultural stage or from the agricultural to the industrial stage, our own current transformation seems to have created a good deal of "cultural wobble" in society. How does one identify the elements or the defining features of a new age while the transformation is still in process? This was one of the questions that intrigued sociologist Richard Florida, who studied U.S. communities.

Dr. Florida (2002) combined several methods. First, he traveled around the country to communities that were especially prosperous and seemed to be on the cutting edge of change in U.S. society. In these communities, he did both individual interviews and focus-group interviews. Focus-group interviews are semistructured group interviews with seven or eight people in which their views and ideas can be generated by asking open-ended questions. Professor Florida recorded the discussion and analyzed the transcript of the discussion. The collected data helped him identify the factors that caused people to choose a place to live. His informants discussed quality of life and the way they make decisions. As certain themes and patterns emerged, he tested the ideas by comparing statistical data for regions that were vibrant, had growing economies, and seemed to be integrated into the emerging information economy. He used another method to compare communities and regions of the country—analyzing already existing archival data collected by various U.S. government agencies, especially the U.S. Bureau of Labor Statistics and the U.S. Census Bureau.

Florida argues that the economy of the twenty-first century is largely driven by creativity, and creative people often decide where to live based on certain features of the society. Currently, more than one third of the jobs in the United States—and almost all of the extremely well-paid professional positions—require creative thinking. These include not just the creative arts but scientific research, computer and mathematical occupations, educational and library science positions, and many media, legal, and managerial careers. People in this "creative class" are given an enormous amount of autonomy; in their work, they have problems to solve and freedom to figure out how to do so. Florida found that modern businesses flourish when they hire highly creative people. Thus, growing businesses tend to seek out places where creative people locate.

Through his research, Florida identified regions and urban areas that are especially attractive to the creative class. Florida found that creative people thrive on diversity—ethnic, gender, religious, and otherwise—for when creative people are around others who think differently, it tends to spawn new avenues of thinking and problem solving. Tolerance of difference and even the enjoyment of individual idiosyncrasies are a hallmark of thriving communities. Interestingly, Florida is now very much in demand as a consultant to mayors and urban-planning teams, and his books have become required reading for city council members. Some elected officials have decided that fostering an environment conducive to creativity that attracts creative people leads to prosperity because business will follow. Key elements for creative communities are local music and art festivals, the presence of organic food grocery stores, legislation that encourages interesting mom-and-pop stores (and keeps out large "box stores" that crush such small and unique endeavors), encouragement of quaint locally owned bookstores and distinctive coffee shops, provisions for bike and walking paths throughout the town, and ordinances that establish an environment of tolerance for people who are "different."

Note: Richard Florida heads the Martin Prosperity Institute at the Rotman School of Management at the University of Toronto. He also runs the private Creative Class Group. He earned his bachelor's degree from Rutgers University and his doctorate in urban planning from Columbia University.

474 million Hindustani speakers in just India, and many others around the world; and 700 million Spanish speakers in 23 countries in which Spanish is the official language (New York Times Almanac 2011). I like meat and veggies. Perhaps you like tofu and grasshoppers. I wear jeans and a T-shirt. Perhaps you wear a sari or burqa. In the United States, proper greetings include a handshake, a wave, or saying "hello" or "hi." The greeting ceremony in Japan includes bowing, with the depth of the bow defined by the relative status of each individual. To know the proper bowing behavior, Japanese businesspeople who are strangers exchange business cards on introduction. Their titles, as printed on their cards, disclose each person's status and thus provide clues as to how deeply each should bow. The proper greeting behavior in many European countries calls for men as well as women to kiss acquaintances on both cheeks.

The point of these examples is to show that the culture—the ideas and "things" that are passed on from one generation to the next in a society, including the knowledge, beliefs, values, rules and laws, language, customs, symbols, and material products—varies greatly as we travel the globe. Each social unit of cooperating and interdependent people, whether at the micro, meso, or macro level, develops a unique way of life. This culture provides guidelines for the actions and interaction of individuals and groups within the society. The cultural guidelines that people follow when they greet another person are examples.

As you can see, the sociological definition of culture refers to far more than "high culture"—such as fine art, classical music, opera, literature, ballet, or theater—and also far more than "popular culture"—such as reality TV, professional wrestling, YouTube, and other mass entertainment. In fact, much of pop culture has been shaped by technology, as is illustrated in the next "Engaging Sociology" feature. The sociological definition of culture includes both high culture and pop culture and has a much broader meaning besides.

Several characteristics make human culture unique, shared by no other animals: Every human shares a culture with others. There may be different views within and between cultures about what rules and behaviors are most important, but no one could survive without culture, for without culture, there would be no guidelines or rules of behavior. Societies would be chaotic masses of individuals. Culture provides the routines, patterns, and expectations for carrying out daily rituals and interactions. Within a society, the process of learning how to act is called socialization (discussed in detail in Chapter 4). From birth, we learn the patterns of behavior approved in our society.

Culture evolves over time and is adaptive. What is normal, proper, and good behavior in hunter-gatherer societies, where cooperation and communal loyalty are critical to the hunt, differs from appropriate behavior in an information age, where individualism and competition may be encouraged and enhance one's position and well-being.

The creation of culture is ongoing and cumulative. Individuals and societies continually build on existing culture to adapt to new challenges and opportunities. The behaviors, values, and institutions that seem natural to you are actually shaped by your culture. Culture is so much a part of life that you may not even notice behaviors that outsiders find unusual or even abhorrent. You may not think about touching someone on the head, putting a baby in a crib, or picking up food with your left hand, but in some other cultures, such acts may be defined as inappropriate or morally wrong.

The transmission of culture is the feature that most separates humans from other animals. Some societies of higher primates have shared cultures but do not systematically enculturate (teach a way of life to) the next generation. Primate cultures focus on behaviors relating to food getting, use of territory, and social status. Human cultures have significantly more content and are mediated by language. Humans are the only mammals with cultures that enable them to adapt to and even modify their environments so they can survive on the equator, in the Arctic, or even beyond the planet.

This Japanese schoolgirl might think that eating with a fork or spoon is quite strange. She has been well socialized to know that polite eating involves competent use of chopsticks.

Ethnocentrism and Cultural Relativity

As scientists, sociologists must rely on careful use of the scientific method to understand behavior. The scientific method calls for objectivity, the practice of considering observed behavior independently of one's own beliefs and values. The study of social behavior thus requires both sensitivity to a wide variety of human social patterns and a perspective that reduces bias. This is more difficult than it sounds because sociologists themselves are products of society and culture. All of us are raised in a particular culture that we view as normal or natural. Yet, not every culture views the same things as "normal."

The Arapesh of New Guinea, a traditional and stable society, encourage premarital sex. It is a way for a girl to prove her fertility, which makes her more attractive as a potential marriage partner (Mead [1935] 1963). Any babies born out of wedlock are simply absorbed into the girl's extended family. The baby's care, support, belonging, and lineage are not major issues for the Arapesh. The babies are simply accepted and welcomed as new members of the mother's family because the structure is able to absorb them. For the Arapesh, sexual behavior outside of marriage is not a moral issue.

From studies of 154 societies documented in the *Human Relations Area Files* (Ford 1970), a source of comparative information on many traditional societies around the world, scientists have found that about 42% of the 154 included societies encourage premarital sex, whereas 29% forbid such behavior and punish those who disobey this rule. The remainder fall in between. As you can see, social values, beliefs, and behaviors can vary dramatically from one society to the next. Differences can be threatening and

Engaging Sociology

Pop Culture and Technology

How surprising to think that digital telephones, high-speed lines for computers, digitized print media, and the World Wide Web have all occurred within about the past half-century. Vinyl records, dial telephones, and VHS videos have been surpassed by CDs, high-speed cell phones, and DVDs. Slim laptops and handheld computers have replaced bulky desktops. The following timeline shows the rapid advances of the Internet and World Wide Web in recent years; the point of this timeline is to illustrate the rapid advance of technology and the place it holds in our lives and in conveying pop culture.

INTERNET AND WORLD WIDE WEB TIMELINE

1844: The telegraph constitutes a data network forerunner.

1866: Transoceanic telegraph service begins.

1876: The telephone is introduced.

1915: The first transcontinental phone call is made.

1946: ENIAC, the first general-purpose computer, is developed for military purposes.

1951: UNIVAC becomes the first civilian computer.

1962: The first communications satellite and the first digital phone networks are introduced.

1965: A highly usable computer language, BASIC, is developed.

1969: The U.S. Department of Defense launches ARPANET, the first communications network.

1971: Microprocessors are developed, making possible personal computer technology.

1972: The first video game, *Pong*, is introduced. E-mail is developed on ARPANET for military use.

1975: The first personal computer, Altair, is launched.

1977: The first fiber-optic network is created.

1978: Cellular phone service begins.

1980s: Hypertext is developed in the mid-1980s, making possible the eventual creation of the World Wide Web.

1982: The National Science Foundation sponsors a high-speed communications network, leading to the Internet.

1984: Apple Macintosh introduces the first PC with graphics.

1989: A new company called AOL (America Online)—the first successful Internet service provider—is formed.

1991: The Internet opens to commercial uses, HTML is developed, and the Web is finally launched.

1993: The first point-and-click Web browser, Mosaic, is introduced.

1994: The first Internet cafés open. Jeff Bezos launches Amazon.com.

1995: Digital cellular phones are introduced. The first online auction house, eBay, is launched.

1996: Google makes its debut.

2001: Instant messaging services expand to allow exchanges between different service providers.

2002: Broadband technology is developed in South Korea.

Mid-2000s: The Internet converges with previous media (radio, television, etc.) to produce online versions of all media forms. It also becomes a continuing source of new forms of communication, including websites such as Myspace, Facebook, and YouTube.

The continuing rapid advances in technology have paralleled development of shared pop culture in the United States and around the world, culture that is accessible to everyone, not just the elite. Music groups from Africa, Asia, the Middle East, and Europe have gained audiences in the United States and vice versa. New musical groups can become instant success stories through YouTube. Streaming allows for people around the world to listen to broadcasts on the Internet. The spread of pop culture—music, mass art, fashion—has been made possible by advances in technology.

We know that "there is no turning back the clock once an innovation is introduced that makes communication more rapid, cheap, efficient, and broadly accessible" (Danesi 2008:21).

(Continued)

(Continued)

Engaging with Sociology:

1. Identify four of the historic innovations that particularly interest you. What are some ways your life is different today because of those four introductions to our popular culture?

2. Identify three positive ways these rapid changes are likely to affect less developed parts of the world.

3. Identify three negative ways these rapid changes are likely to affect less developed parts of the world.

4. Are technology and the spread of pop culture bringing the world closer together, or are these innovations more likely to cause tensions as governments attempt to protect their citizens from e-fraud, pornography, political uses of technology (including revolutions), and change of the local culture?

even offensive because most people judge others according to their own perspectives, experiences, and values.

The tendency to view one's own group and its cultural expectations as right, proper, and superior to others is called **ethnocentrism**. If you were brought up in a society that forbids premarital or extramarital sex, for instance, you might judge the Arapesh—or many Americans—to be immoral. In a few Muslim societies, people who violate this taboo may be severely punished, or even executed, because premarital sex is seen as an offense against the faith and the family and as a weakening of social bonds. In turn, the Arapesh would find rules of abstinence to be strange and even wrong.

Societies instill some degree of ethnocentrism in their members because ethnocentric beliefs hold groups together and help members feel that they belong to the group. Ethnocentrism promotes loyalty, unity, high morale, and conformity to the rules of society. Fighting for one's country, for instance, requires some degree of belief in the rightness of

one's own society and its causes. Ethnocentric attitudes also help protect societies from rapid, disintegrating change. If most people in a society did not believe in the rules and values of their own culture, the result could be widespread dissent, deviance, or crime.

Ethnocentrism can take many forms (see, for example, how you react to Map 3.1). Unfortunately, ethnocentrism often leads to misunderstandings between people of different cultures. In addition, the same ethnocentric attitudes that strengthen ties between people may also encourage hostility, racism, war, and even genocide against others—even others within the society—who are different. Virtually all societies tend to "demonize" their adversary—in movies, the news, and political speeches—while the conflict is most intense. Dehumanizing another group with labels makes it easier to torture or kill its members or to perform acts of discrimination and brutality. We see this in current Iraqi conflicts in which both sides in the conflict feel hatred for the other combatants. However, as we become a part of a global social world, it becomes increasingly important to accept those who are "different." Bigotry and attitudes of superiority do not enhance cross-national cooperation and trade—which is what the increasing movement toward a global village and globalization entail.

U.S. foreign relations also illustrate how ethnocentrism can produce hostility. Many U.S. citizens are surprised to learn that the United States—great democracy, world power, and disseminator of food, medicine, and technological assistance to developing nations—is despised in many countries. One cause is the political dominance of the United States and the threat it poses to other people's way of life (Hertsgaard 2003). U.S. citizens are regarded by some as thinking ethnocentrically and only about their own welfare as their country exploits weaker nations. U.S. tourists are often seen as loudmouthed ignoramuses whose ethnocentric attitudes prevent them from seeing value in other cultures or from learning other languages.

Anti-U.S. demonstrations in South America, the Middle East, and Asia have brought this reality to life through television. Indeed, politicians in several Latin American and European countries have run for office on platforms aimed at reducing U.S. influence. Thus, U.S. ethnocentrism

In 1923, the Hollywood Association started a campaign to expel the Japanese from their community. Signs like these were prominent throughout Hollywood and other California communities. They illustrate ethnocentrism.

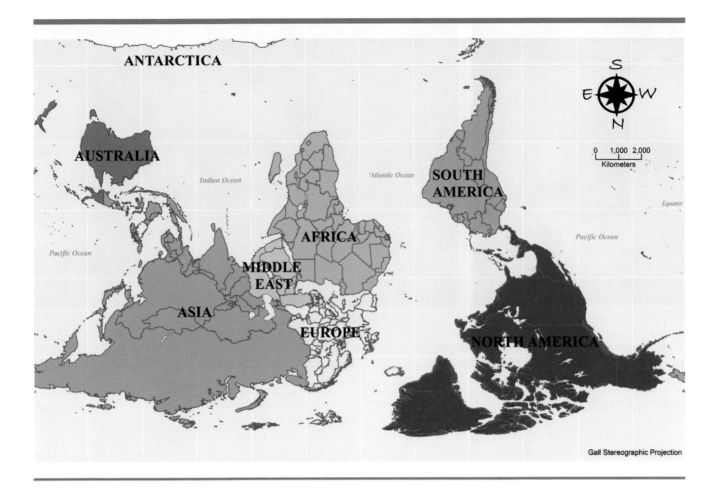

Map 3.1 "Southside Up" Global Map

Source: Map by Anna Versluis.

Note: This map illustrates geographic ethnocentrism. Americans tend to assume it is natural that the north should always be "on top." The fact that this map of the world is upside-down, where south is "up," seems incorrect or disturbing, yet it illustrates an ethnocentric view of the world. Most people think of their countries or regions as occupying a more central and larger part of the world.

may foster anti-American ethnocentrism by people from other countries. Note that even referring to citizens of the United States as "Americans"—as though people from Canada, Mexico, and South America do not really count as Americans—is seen as ethnocentric by many people from these other countries. *America* and the *United States* are not the same thing, but many people in the United States, including some presidents, fail to make the distinction, much to the dismay of other North and South Americans.

Not all ethnocentrism is hostile; some is just reactions we have to the strange ways of other cultures. An example is making judgments about what is proper food to eat and what is just not edible. While food is necessary for survival, there are widespread cultural differences in what people eat. Some people in New Guinea tribes savor grasshoppers; Europeans and Russians relish raw fish eggs (caviar); Eskimo children find seal eyeballs a treat; some Indonesians eat

dog; and some Nigerians prize termites. Whether it is from another time period or another society, variations in food can be shocking to us.

In contrast to ethnocentrism, **cultural relativism** is a view that requires setting aside cultural and personal beliefs and prejudices to understand another group or society through the eyes of a member of that group. Cultural relativism requires that we shrug our shoulders and admit, "Well, they are getting vitamins, proteins, and other nutrients, and it seems to work for them." Instead of judging cultural practices and social behavior as good or bad according to one's own cultural practices, the goal is to be impartial in learning the purposes and consequences of practices and behaviors in the group under study. That does not mean we have to accept others' practices—just try to understand why they exist. Just as we may have preferences for certain software programs to do word processing, we can recognize that other software

In the United States, dogs are family members and are highly cherished, with dog care becoming a growth industry. In Nanking, China, dogs are valued more as a culinary delicacy and a good source of protein. If you feel a twinge of disgust, that is part of the ethnocentrism we may experience as our perspectives conflict with those elsewhere in the world.

programs are quite good, may have some features that are better than the one we use, and are ingeniously designed.

Yet, being tolerant and understanding is not always easy, and even for the most careful observer, the subtleties of other cultures can be elusive. The idea of being "on time," which is so much a part of the cultures of the United States, Canada, Japan, and parts of Europe, is a rather bizarre concept in many societies. Among many Native American people, such as the Dineh (Apache and Navajo), it is ludicrous for people to let a timepiece that one wears on one's arm govern the way one lives life. The Dineh orientation to time—that one should do things according to a natural rhythm of the body and not according to an artificial ticking mechanism—is difficult for many North Americans to grasp. Misunderstandings occur when those of European heritage think "Indians are always late" and jump to an erroneous conclusion that "Indians" are undependable. Native Americans, on the other hand, think whites are neurotic about letting some instrument control them (Basso 1979; Farrer 1996; Hall 1959, 1983).

Cultural relativism does not require that social scientists accept or agree with all the beliefs and behaviors of the societies or groups they study. Certain behaviors, such as infanticide, cannibalism, slavery, female genital mutilation (removal of part of the female anatomy), forced marriage, or genocide, may be regarded as unacceptable by almost all social scientists. Yet, it is still important to try to understand those practices in the social and cultural contexts in which they occur. Many social scientists take strong stands against violations of human rights, environmental destruction, and other social policies. They base their judgments on the concept of universal human rights and the potential for harm

to individuals. Yet, note that most Western democracies are adaptations to systems that thrive on values of extreme individualism, differentiation, and competition that are consistent with very complex society. Other societies do not always think our ideas of "universal rights" are universal. They are based on ideas of individualism that some other cultures do not use as the basis of their values, with the needs of communities having higher priority than individual rights.

Thinking Sociologically

Small, tightly knit societies with no meso or macro level often stress cooperation, conformity, and personal sacrifice for the sake of the community. Complex societies with established meso- and macro-level linkages are more frequently individualistic, stressing personal uniqueness, achievement, individual creativity, and critical thinking. Why might this be so?

Things and Thoughts: The Components of Culture

Things and thoughts—these make up much of our culture. From our things—**material culture**—to our thoughts or beliefs—**nonmaterial culture**—culture provides the guidelines for our lives.

Material Culture: The Artifacts of Life

Material culture includes all the objects we can see or touch, all the artifacts of a group of people—their grindstones for grinding cassava root, microwave ovens for cooking, bricks of mud or clay for building shelters, hides or woven cloth for making clothing, books or computers for conveying information, tools for reshaping their environments, vessels for carrying and sharing food, and weapons for dominating and subduing others. Material culture includes anything you can touch that is made by humans.

Some material culture is of local, micro-level origin. The kinds of materials with which homes are constructed and the materials used for clothing often reflect the geography and resources of the local area. Likewise, types of

Multinational corporations know no national barriers. Thousands of products you use every day were likely made or assembled on the other side of the globe. Industrialization in the less developed countries is also changing life for them, as well as affecting jobs and consumption patterns in North America.

jewelry, pottery, musical instruments, or clothing reflect tastes that emerge at the micro and meso levels of family, community, and ethnic subculture. At a more macro level, national and international corporations interested in making profits work hard to establish trends in fashion and style that may cross continents and oceans.

Material culture in many ways drives the globalization process. Many of our clothes are now made in Asian or Central American countries. Our shoes may well have been produced in the Philippines. The oil used to make plastic water bottles and devices in our kitchens likely came from the Middle East. Even food is imported year-round from around the planet. That romantic diamond engagement ring—a symbol that represents the most intimate tie—may well be imported from a South African mine using low-paid or even slave labor. Our cars are assembled from parts produced on nearly every continent. Moreover, we spend many hours in front of a piece of material culture—our computers—surfing the World Wide Web. Material culture is not just for local homebodies anymore.

Thinking Sociologically

Think of examples of material culture that you use daily: stove, automobile, cell phone, refrigerator, clock, money, and so forth. How do these material objects influence your way of life and the way you interact with others? How would your behavior be different if these material objects, say watches or cars, did not exist as is the case in many cultures?

Nonmaterial Culture: Beliefs, Values, Rules, and Language

Nonmaterial culture refers to invisible and intangible parts of culture; it is of equal or even greater importance than material culture for it involves the society's rules of behavior, ideas, and beliefs that shape how people interact with others and with their environment. Although we cannot touch the nonmaterial components of our culture, they pervade our life and are instrumental in determining how we think, feel, and behave. Nonmaterial culture is complex, comprising four main elements: values, beliefs, norms or rules, and language. These elements are expressed in two main forms: an ideal culture and a real culture.

Ideal culture consists of practices, beliefs, and values that are regarded as most desirable and are consciously taught to children. Not everyone, however, follows the approved cultural patterns. **Real culture** refers to the way things in society are actually done. For the most part, we hardly question these practices and beliefs that we see

Photo Essay

Houses as Part of a Society's Material Culture

Homes are good examples of material culture. Their construction is influenced by local materials, but also ideas of what a home is. Homes shape the context in which family members interact. Indeed, in some cases, homes become status symbols that are far larger than the family needs, but the family is making a prestige statement about their socioeconomic standing.

around us. Rather like animals that have always lived in a rain forest and cannot imagine a treeless desert, we often fail to notice how our culture helps us make sense of things. Because of this, we do not recognize the gap between what we tell ourselves about ourselves and what we actually do.

For example, the ideal in many societies is to ban extramarital sex. Sex outside marriage can raise questions of paternity (who is the father) and inheritance (who is in line to inherit wealth), besides leading to spousal jealousy and family conflict. In the United States, about one fourth of all married men and one in seven married women report that they have had at least one extramarital affair (National Opinion Research Center 2010; Newman and Grauerholz 2002). While in ideal culture we claim that we believe in marital fidelity and this is at the core of family morality, actual behavior or real culture is frequently different.

Values are shared judgments about what is desirable or undesirable, right or wrong, good or bad. They express the basic ideals of any society and are therefore abstract. In industrial and postindustrial societies, for instance, a good education is highly valued. Gunnar Myrdal (1964), a Swedish sociologist and observer of U.S. culture, referred to the U.S. value system as the "American creed." Values become a creed when they are so much a part of the way of life that they acquire the power of absolute conviction. We tend to take our core values for granted, including freedom, equality, individualism, democracy, free enterprise, efficiency, progress, achievement, and material comfort (Macionis 2010; Williams 1970).

At the macro level, conflicts may arise between groups in society because of differing value systems. For example, there are major differences between the values of the dominant culture and those of various native or "first-nations" peoples (Lake 1990; Sharp 1991). Consider the story in the next "Sociology Around the World" about Rigoberta Menchú Tum and the experiences of Native American populations living in Guatemala.

Thinking Sociologically

The experiences of Rigoberta Menchu Tum, described in the next "Sociology Around the World", illustrate clashes between powerful and powerless groups. What are other ways that different cultural values cause problems between Native Americans and dominant groups?

The conflict in values between Native Americans and the national cultures of Canada and the United States has had serious consequences. Cooperation is a cultural value that has been passed on through generations of Native

Americans because group survival depends on group cooperation in the hunt, in war, and in daily life. The value of cooperation can place native children at a disadvantage in North American schools that emphasize competition. Native American and Canadian First Nation children experience more success in classrooms that stress cooperation and sociability over competition and individuality (Lake 1990; Mehan 1992).

Another Native American value is the appreciation of and respect for nature. Conservation of resources and protection of the natural environment—Mother Earth—have always been important because of the people's dependence on nature for survival. Today, we witness disputes between native tribes and governments of Canada, the United States, Mexico, Guatemala, Brazil, and other countries over raw resources found on native reservations. While many other North and South Americans also value cooperation and respect for the environment, these values do not govern decision making in most communities (Brown 2001; Marger 2006). The values honored by governments and corporations are those held by the people with power, prestige, and wealth.

Beliefs are more specific ideas we hold about life, about the way society works, and about where we fit into the world. They are expressed as specific statements that we hold to be true. Beliefs come from traditions established over time, religious teachings, experiences people have had, and lessons given by parents and teachers or other individuals in authority. Beliefs influence the choices we make. Many Hindus, for example, believe that they will be rewarded in their next incarnation for fulfilling behavioral expectations of their own social caste. In the next life, good people will be born to a higher social status. By contrast, some Christians believe that their fate in the afterlife depends on whether they believe in certain ideas—for instance, that Jesus Christ is their personal savior. Beliefs tend to be based on values, which are broader and more abstract notions of something desirable. A value might be that the environment is worth preserving, but a belief might be that humans have caused global warming. Another value might be eternal life, but a belief might be that this occurs through reincarnation and eventually nirvana.

Norms are rules of behavior shared by members of a society and rooted in the value system. Norms range from religious warnings such as "thou shalt not kill" to the expectation that young people will complete their high school education. Sometimes, the origins of particular norms are quite clear. Few people wonder, for instance, why there is a norm to stop and look both ways at a stop sign. Other norms, such as the rule in many societies that women should wear skirts but men should not, have been passed on through the generations and become unconsciously accepted patterns and a part of tradition in some societies. Sometimes we may not know how norms originated or even be aware of norms until they are violated.

Sociology Around the World

Life and Death in a Guatemalan Village

In her four decades of life, Rigoberta Menchú Tum experienced the closeness of family and cooperation in village life. These values are very important in Chimel, the Guatemalan hamlet where she lives. She also experienced great pain and suffering with the loss of her family and community. A Quiche Indian, Menchú became famous throughout the world in 1992 when she received the Nobel Peace Prize for her work to improve the conditions for Indian peoples.

Guatemalans of Spanish origin hold the power and have used Indians almost as slaves. Some of the natives were cut off from food, water, and other necessities, but people in the hamlet helped support each other and taught children survival techniques. Most people had no schooling. Menchú's work life in the sugarcane fields began at age 5. At 14, she traveled to the city to work as a domestic servant. While there, she learned Spanish, which helped her be more effective in defending the rights of the indigenous population in Guatemala. Her political coming of age occurred at age 16 when she witnessed her brother's assassination by a group trying to expel her people from their native lands.

Her father started a group to fight repression of the indigenous and poor, and at 20, Menchú joined the movement, *Comité de Unidad Campesina* (CUC), which the government claimed was communist inspired. Her father was murdered during a military assault, and her mother was tortured and killed. Menchú moved to Mexico with many other exiles to continue the nonviolent fight for rights and democracy.

The values of the native population represented by Menchú focus on respect for and a profound spiritual relationship with the environment, equality of all people, freedom from economic oppression, the dignity of her culture, and the benefits of cooperation over competition. The landowners tended to stress freedom of people to pursue their individual self-interests (even if inequality resulted), the value of competition, and the right to own property and to do whatever one desired to exploit that property for economic gain. Individual property rights were thought to be more important than preservation of indigenous cultures. Economic growth and profits were held in higher regard than religious connectedness to the earth.

The values of the native population and landowners are in conflict. Only time will tell if the work of Indian activists such as Rigoberta Menchú Tum and her family will make a difference in the lives of this indigenous population.

Norms are generally classified into three categories—folkways, mores, and laws—based largely on how important the norms are in the society and people's response to the breach of those norms. *Folkways* are customs or desirable behaviors, but they are not strictly enforced: Some examples are responding politely when introduced to someone, not scratching your genitals in public, or using proper table manners. Violation of these norms causes people to think you are weird or even uncouth but not necessarily immoral or criminal.

Mores are norms that most members observe because they have great moral significance in a society. Conforming to mores is a matter of right and wrong, and violations of many mores are treated very seriously. The person who deviates from mores is considered immoral or bordering on criminal. Being honest, not cheating on exams, and being faithful in a marriage are all mores. Table 3.1 provides examples of folkways and mores.

Taboos are the strongest form of mores. They concern actions considered unthinkable or unspeakable in the culture. For example, most societies have taboos that forbid incest (sexual relations with a close relative) and prohibit eating a human corpse. Taboos are most common in societies that do not have centralized governments to enforce laws and maintain jails.

Taboos and other moral codes may be of the utmost importance to a group, yet behaviors that are taboo in one situation may be acceptable at another time and place. The **incest taboo**, a prohibition against sex with a close relative, is an example found in all cultures, yet application of the

Table 3.1   Norms and Violations of Norms

Folkways: Conventional Polite Behaviors
Violations viewed as "weird":
Swearing in house of worship
Wearing blue jeans to the prom
Using poor table manners
Picking one's nose in public
Mores: Morally Significant Behaviors
Violations viewed as "immoral":
Lying or being unfaithful to a spouse
Buying cigarettes or liquor for young teens
Having sex with a professor as a way to increase one's grade
Parking in handicap spaces when one is in good physical condition

incest taboo varies greatly across cultures (Brown 1991). In medieval Europe, if a man and a woman were within seven degrees of relatedness and wanted to marry, the marriage could be denied by the priest as incestuous. (Your first cousin is a third degree of relatedness from you.) On the other hand, the Balinese permit twins to marry because it is believed they have already been intimately bonded together in the womb (Leslie and Korman 1989). In some African and Native American societies, one cannot marry a sibling but might be expected to marry a first cousin. As Table 3.2 illustrates, the definition of what is and what is not incest varies even from state to state in the United States.

Laws are norms that have been formally encoded by those holding political power in society, such as laws against stealing property or killing another person. The violator of a law is likely to be perceived not just as a weird or immoral person but as a criminal who deserves formal punishment. Many mores are passed into law, and some folkways are also made into law. Formal punishments are imposed. Spitting on the sidewalk is not a behavior that has high levels of moral contempt, yet it results in fines in some cities. Furthermore, behaviors may be folkways in one situation and mores in another. For example, nudity or various stages of near nudity may be only mildly questionable in some social settings (the beach or certain fraternity parties) but would be quite offensive in others (a four-star restaurant or a house of worship) and against the law in others, incurring a penalty.

Sanctions reinforce norms through rewards and penalties. **Formal sanctions** to enforce the most important norms are implemented by official action. Fines for parking illegally, lowered grades on an assignment for plagiarism, and expulsion for bringing drugs or weapons to school are formal negative sanctions your school might impose. Honors and awards are formal positive sanctions. **Informal sanctions** are unofficial rewards or punishments. A private word of praise by your professor after class about how well you did on your exam would be an informal positive sanction;

Table 3.2   Incest Taboos in the United States: States That Allow First-Cousin Marriage

Alabama	Connecticut	Maryland	Rhode Island
Alaska	District of Columbia	Massachusetts	South Carolina
California	Florida	New Jersey	Tennessee
Colorado	Georgia	New Mexico	Vermont
Hawaii	New York	North Carolina	Virginia

States that allow only under certain conditions: The following states also allow first-cousin marriage but only under certain conditions such as marriage after a *certain age or inability to bear children:* Arizona, Illinois, Indiana, Utah, and Wisconsin.

Marriage of half cousins: Kansas, Maine, Montana, Nebraska, Nevada, and Oklahoma.

Marriage of adopted cousins: Louisiana, Mississippi, Oregon, and West Virginia.

States that do not allow first-cousin marriage: All other U.S. states disallow marriages to first cousins within the state: Arkansas, Delaware, Idaho, Iowa, Kentucky, Michigan, Minnesota, Missouri, New Hampshire, North Dakota, Ohio, Pennsylvania, South Dakota, Texas, Washington, and Wyoming.

Historically, in the United States, incest laws forbid in-law marriages far more than first-cousin marriages.

Source: "Cousin Marriage Laws in the U.S." (2008).

gossip or ostracism by other students because of clothes you are wearing would be an informal negative sanction. Sanctions vary with the importance of the norm and can range from a parent frowning at a child who misbehaves to a prison term or death sentence. Similarly, when we obey norms, we are rewarded, sometimes with simple acts such as jokes and pats on the back that indicate solidarity with others. Most often, adherence to norms is ingrained so deeply that our reward is simply "fitting in." Your reward for polite social behavior—for asking about someone's interests and activities or for not being overly pushy in conversation—is friendship and fitting into the group. Folkways and many mores are enforced through informal sanctions, yet sometimes penalties for deviant behavior can be severe.

Norms concerning sexual behaviors are often very strong and carry powerful sanctions, sometimes even imposed by national governments. A woman who becomes pregnant outside of marriage in societies that strongly condemn nonmarital sex is likely to be ostracized and her child labeled illegitimate. Such children may be stigmatized for life, living as outcasts in poverty on the streets, begging for food (Lawson 1991; Nguyen 2005). This was the case in Vietnam during the U.S.-Vietnam War, when many biracial children from U.S. soldiers were ostracized

Nudity may be considered a violation of law, or mores, or folkways, or it may simply be accepted as normal, as in the case of this nude beach.

by society. Biracial children are still suffering discrimination in Vietnam (DiversityInc 2008; Nguyen 2005).

When a society faces change, especially from war, rapid urbanization, industrialization, and modernization, traditional norms that have worked for the society for centuries are challenged. In the past few decades, examples have been seen in many Islamic countries in which modernization has met with a resurgence of religious fundamentalism. A case in point is Iran where rapid modernization and social changes in the post–World War II era were met by a backlash from religious fundamentalists. Radio music and drinking of alcoholic beverages were banned, and women were required to again wear the veil. Afghanistan and other Muslim nations are currently struggling with the conflicts between traditional religious and cultural values and those related to pressures from Western nations for modernization.

Communication is often mediated and enhanced through nonverbal indicators such as tone of voice, inflection, facial expressions, or other gestures that communicate emotion. In e-mail, the words are just words without context. To establish norms, many electronic mailing lists now have rules for polite communication to avoid "flaming" someone—insulting him or her with insensitive words by a faceless person. The development of emoticons (such as smileys) has allowed for adding combinations of characters that represent the emotional context of the message. Norms of Internet communication are still emerging, and you probably have experienced times when messages have been misunderstood because the norms of communication are ambiguous.

Thinking Sociologically

One of the problems of Internet communication has been that many norms of civil discourse are ignored and new norms emerge. What do you see as the current rules for Internet and Twitter communication? When dealing with conflict, how do these norms differ from verbal norms of communication?

Language is the foundation of every culture. The minidrama between infant and adult is played out every day around the world as millions of infants learn the language of the adults who care for them. In the process, they acquire an essential part of culture. Although many animals can communicate with a limited repertoire of sounds, the ability to speak a language is unique to humans. The human infant is capable of making roughly 1,000 sounds, but any given society considers only about 40 or 50 of these to be language sounds. The baby soon learns that some of those sounds elicit enthusiastic responses from the adult caretakers, reinforcing their use. At about one year of age, most infants begin to pronounce recognizable words in the language of their culture (Oller 2006).

Transport a baby from France to the Arapesh tribe in New Guinea and another baby from New Guinea to France, and each will learn to speak the language and adhere to the culture in which it is brought up. Language, in other words, is learned. Through the use of language, members of culture can pass on essential knowledge to children and can share ideas with other members of their society. Work can be organized; the society can build on its experiences and plan its future.

Language takes three primary forms: spoken, written, and nonverbal. Spoken language allows individuals to produce a set of sounds that symbolize an object or idea. That sound combination is learned by all who share a culture, and it generally holds similar meaning for all members. Written language enables humans to store ideas for future generations, accelerating the accumulation of ideas on which to build. It also makes possible communication over distances. Members of a society learn to read these shared symbols, some of which are displayed in Figure 3.1.

Nonverbal language consists of gestures, facial expressions, and body postures. This mode of communication may carry as much as 90% of the meaning of the message (Samovar and Porter 2003). Every culture uses nonverbal language to communicate, and just like verbal language, those cues may differ widely among cultures. For instance, an A-OK gesture or a hand wave that is positive in one culture may have a negative or even obscene meaning in another.

The power to communicate nonverbally is illustrated in American Sign Language, designed for the hearing-challenged and the mute. Complex ideas can be transmitted without vocalizing a word. Indeed, one can argue that the deaf have a distinctive culture of their own rooted in large part in the unique language that serves them. In addition, technology has aided communication among the hearing-impaired through text messaging.

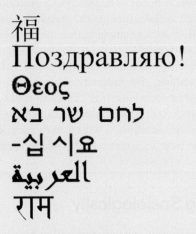

Figure 3.1   Societies Use Various Symbols to Communicate Their Written Language

Misunderstandings can occur between ethnic groups because of cultural differences in communication. The Apache Indians in the American Southwest tell "white-man jokes"—ridiculing white people because they engage in such grossly unacceptable actions as calling one another by name even when they do not know each other intimately, asking about one another's health, complimenting one another's clothing, and greeting others who walk into a room. Because so many white U.S. citizens do not understand that these behaviors are viewed as offensive, they may violate expectations and create difficulties for themselves in dealing with Apaches. They may act in ways that are considered ill-mannered and rude (Basso 1979).

Misinterpretation of nonverbal signals can also occur between males and females in Western societies. Consider the differences in meaning of a nod of the head. When

Language development is extremely important to becoming fully human, and it happens very rapidly from about the first year of life. Still, babies learn to communicate in a variety of other ways. Note how these infants communicate emotions nonverbally. What does each of them seem to be communicating?

women nod their heads in response to a person who is talking, they often are encouraging the speaker to continue, signaling that they are listening and that they want the speaker to carry on with clarification and explanation. It does not signal agreement. When men nod their heads when another person is speaking, they typically assume the message is "I agree with what you are saying." This can lead to awkward, confusing, and even embarrassing miscues when men and women talk to one another, with a man mistakenly confident that the woman agrees with his ideas (Stringer 2006).

Thinking Sociologically

Explain a situation of miscommunication between you and a member of the other sex. What happened?

Language also plays a critical role in perception and in thought organization. The *linguistic relativity theory* (Sapir 1929, 1949; Whorf 1956) posits that the people who speak a specific language make interpretations of their reality—they notice certain things and may fail to notice certain other things. "A person's 'picture of the universe' or 'view of the world' differs as a function of the particular language or languages that person knows" (Kodish 2003:384). Children in each different culture will learn about the world within the framework provided by their language. Scientists continue to debate the extent to which language can influence thought, but most agree that while language may contribute to certain ways of thinking, it does not totally determine human thinking (Casasanto 2008; Gumperz and Levinson 1996; Levinson 2000). Although this theory is controversial and aspects have been misinterpreted, the idea of language, culture, consciousness, and behavior affecting each other is influential when studying the role of language.

Consider the role of certain words in a language. In the English language, people tend to associate certain colors with certain qualities in a way that may add to the problem of racist attitudes. In *Webster's Encyclopedic Unabridged Dictionary of the English Language* (1989:153), the definition of the word *black* includes "dismal," "boding ill," "hostile," "harmful," "inexcusable," "without goodness," "evil," "wicked," "disgrace," and "without moral light." The word *white* (*Webster's* 1989:1628), on the other hand, is defined as "honest," "dependable," "morally pure," "innocent," and "without malice." If the linguistic relativity thesis is correct, it is more than coincidence that bad things are associated with the "black sheep" of the family, the "blacklist," or "Black Tuesday" (when the U.S. stock market dropped dramatically and crashed in 1929).

The societies that have negative images and language for black and positive images for white are the same societies that associate negative qualities with people of darker skin. The use of *white* as a synonym for "good" or "innocent"—as in reference to a "white wedding dress," "white noise machine," or "white lie"—may contribute to a cultural climate that devalues people with darker shades of skin. In essence, the language may influence our perception of color in a manner that contributes to racism.

Language evolves in specific settings. Within your own college, there are probably terms used to refer to course sequences, majors, or student organizations that people at other universities would find bewildering. At the Massachusetts Institute of Technology, for example, students often respond to the question "What is your major?" with the number from the catalog: "I'm majoring in 23." That would be a truly bizarre response at some other campuses.

White and black as colors have symbolic meaning—with phrases like "blackballed from the club" or "black sheep of the family" indicating negative judgment associated with blackness. Research shows that teams wearing black are called for more fouls than teams wearing white, which raises questions about how pervasive this association is in our perception.

Thinking Sociologically

The words *bachelor* and *spinster* are supposedly synonymous terms, referring to unmarried adult males and females, respectively. Generate a list of adjectives that describe each of these words and that you frequently hear associated with them (e.g., *eligible, swinging, old, unattractive*). Are the associated words positive or negative in each case? How are these related to the position of the unmarried in societies?

Neither culture nor society stands alone; they are interdependent. In the next section we examine the connections between society and culture at each level of analysis: micro, meso, and macro.

Society, Culture, and Our Social World

Whether their people are eating termite eggs, fish eggs, or chicken eggs, societies always have a culture, and culture is always linked to a society. Culture provides guidelines for actions that take place at each level of society, from the global system to the individual family. The social world model at the beginning of the chapter, with its concentric circles, represents the micro to macro levels of society. Smaller social units such as schools operate within larger social units such as the community, which is also part of a region of the country. What takes place in each of these units is determined by the culture. There is a social unit—a structural "hardware"—and a culture or "software" at each level.

Microcultures: Micro-Level Analysis

Micro-level analysis focuses on social interactions in small groups. To apply this idea to culture, we look at microcultures. Groups and organizations such as a Girl Scout troop, a local chapter of the Rotary club, or six participants in a Facebook group involve a small number of people. These organizations influence only a portion of members' lives. When the culture affects only a small segment of one's life, affecting a portion of one's week or influencing a limited time period in one's life, it is called a **microculture** (Gordon 1970). Other classic examples from sociology include a

street gang, a college sorority, a business office, or a summer camp group.

Hospitals are organizations for the care and treatment of the sick and injured, providing centralized medical knowledge and technology for treatment of patients. They are social units with microcultures where people in different-colored uniforms scurry around carrying out their designated tasks as part of the organizational culture. Hospital workers interact among themselves to attain goals of patient care. They have a common in-group vocabulary, a shared set of values, a hierarchy of positions with roles and behaviors for each position, and a guiding system of regulations for the organization—all of which shape interactions during the hours that each member works in the hospital. Yet, the hospital culture may have little relevance to the rest of the employees' everyday lives. Microcultures may survive over time, with individuals coming in and going out from the group, but in a complex society no one lives his or her entire life within a microculture. The values, rules, and specialized language used by the hospital staff continue as one shift ends and other medical personnel enter and sustain that microculture.

Every organization, club, and association has an organizational microculture—its own set of rules and expectations. Organizational microcultures may last for many years, but some microcultures exist for a limited period of time or for a special purpose. A summer camp microculture may develop but exists only for that summer. The following summer, a very different culture may evolve because of new counselors and campers. A girls' softball team may develop its own cheers, jokes, insider slang, and values regarding competition or what it means to be a good sport, but next year, the girls may be realigned into different teams, and the transitory culture of the previous year may change. In contrast to microcultures, subcultures continue across a person's life span.

Subcultures and Countercultures: Meso-Level Analysis

A **subculture** is the way of life of a group that is smaller than the nation but, unlike a microculture, large enough to support people throughout their life span. A subculture is unique to that group yet at the same time shares the culture of the dominant society (Arnold 1970; Gordon 1970). Many ethnic groups within the larger society have their own subculture with their own sets of conventions and expectations. For example, picture a person who is African Canadian, Chinese Canadian, or Hispanic Canadian, living within an ethnic community that provides food, worship, and many other resources. Despite unique cultural traits, that person and the group are still good Canadian citizens, living within the national laws and way of life. It is just that their life has guidelines from the subculture in addition to the dominant culture of the society.

Because the social unit plays a more continuous role in the life span of members through institutions and structures, we analyze subcultures at the meso level. (Table 3.3 illustrates the connection between the social unit at each level and the type of culture at that level.) One can be born, work, marry, and die in that social unit. Members maintain a feeling of "we" (those belonging to the group) versus "they" (those outside the group). Members also maintain a belief in the rightness of their customs, rituals, religious practices, dress, food, or whatever else distinguishes them as part of a subculture.

Note that many categories into which we group people are not subcultures. For example, redheads, left-handed people, individuals who read *Wired* magazine, people who are single, and visitors to Chicago do not make up subcultures because they do not interact as social units or share a common way of life. A motorcycle gang, a college fraternity, and a summer camp are also not subcultures because microcultures affect only a segment of one's life (Gordon 1970; Yablonski 1959). A subculture, by contrast, pervasively influences a person's daily life. You may participate in many different microcultures in a single day (the choir, a Greek organization, an athletic team, a classroom), but you are likely to live your entire life within the same subculture. Consider some examples:

In the United States, subcultures include ethnic groups, such as Mexican Americans and Korean Americans, and social class groups, including the exclusive subculture of the elite upper class on the East and West coasts of the United States. The superwealthy have networks, exclusive clubs, and the Social Register, which lists names and phone numbers of the elite so they can maintain contact with one another. They have a culture of opulence that differs from middle-class culture, and this culture is part of their experience throughout their lives.

Another example is Hasidic Jews, who adhere to the same laws as other Americans but follow additional rules specific to their religion. Clothing and hairstyles follow strict rules; men wear beards and temple locks (*payos*), and married women wear wigs. Their religious holidays are different from those of the dominant Christian culture. Hasidic Jews observe dietary restrictions such as avoiding pork and shellfish, and they observe the Sabbath from sunset Friday to sunset Saturday. Hasidic Jews are members of a subculture in the larger society of which they are citizens. In today's world, this is a global subculture maintained through websites of, by, and for Hasidic Jews. Again, one can live one's entire life under the influence of the values and rules of the subculture, and this is certainly true for Hasidic Jews.

A give-and-take exists between subcultures and the dominant culture, with each contributing to and influencing the other. Hispanic Americans have brought many foods to American cuisine, for example, including tacos, burritos, and salsa.

Subcultural practices can be the source of tensions with the dominant group, which has the power to determine cultural expectations in society. A small group of Mormons in the United States believes in and practices polygamy. Having more than one wife (polygyny) violates federal and state laws, but some hold to the old teachings of founder Joseph Smith and cling to this practice as sanctioned by God.

When conflict with the larger culture becomes serious and laws of the dominant society are violated, a different type of culture emerges. A **counterculture** is a group with expectations and values that contrast sharply with the dominant values of a particular society (Yinger 1960).

One type of counterculture is represented by the Old Order Amish of Pennsylvania and Ohio. The Amish drive horse-drawn buggies and seldom use electricity or modern machines. They reject many mainstream notions of success and replace them with their own work values and goals. Conflicts between federal and state laws and Amish religious beliefs have produced compromises by the Amish on issues of educating children, using farm machinery, and transportation. The Old Order Amish prefer to educate their children in their own communities, insisting that their children not go beyond an eighth-grade education in the public school curriculum. They also do not use automobiles or conventional tractors. The Amish are pacifists and will not serve as soldiers in the national military.

Table 3.3   Social Units of Society and Levels of Culture

Social Unit (the people who interact and feel they belong)	Culture (the way of life of that social unit)
Dyads; small groups; local community	Microculture
Ethnic community or social class community	Subculture
National society	Culture of a nation
Global system	Global culture

Other types of countercultures seek to withdraw from society or to operate outside its economic and legal systems or even to bring about the downfall of the larger society. Examples are survivalist groups such as racist militia and skinheads, who reject the principles of democratic pluralism. Countercultures of all types have existed throughout history. The "old believers" of seventeenth-century Russia committed group suicide rather than submit to the authority of the czar of Russia on matters of faith and lifestyle (Crummey 1970). There are now Russian old believer communities in Oregon, northern Alberta, and the Kenai Peninsula of Alaska. Some of their villages are so isolated that they are virtually inaccessible by car, and visitors are greeted with "no trespassing" signs.

Some countercultures continue over time and can sustain members throughout their life cycle—such as the Amish. They are like subcultures in that they operate at the meso level, but they strongly reject the mainstream culture. However, most countercultural groups, such as punk rock groups or violent and deviant teenage gangs, are short-lived or are relevant to people only at a certain age—operating only at the micro level.

Members of countercultures do not necessarily reject all of the dominant culture, and in some cases, parts of their culture may eventually come to be accepted by the dominant culture. During the Vietnam War, for instance, some antiwar protesters focused their lives on protesting the U.S. involvement in Southeast Asia (and related political and economic issues). By the early 1970s, opposition to the war had become widespread in society, so antiwar protesters were no longer outside of the mainstream. They were less likely to be labeled as anti-American for their beliefs. Following the war, many counterculture antiwar hippies became active in the mainstream culture and developed conventional careers in the society. Thus, ideas, protest songs, emblems, longer hair for men, and the peace symbol from the 1960s were absorbed into the larger culture.

Countercultures are not necessarily bad for society. According to the conflict perspective, which was introduced in Chapter 2, the existence of counterculture groups is clear evidence that there are contradictions or tensions within a society that need to be addressed. Countercultures often challenge unfair treatment of groups in society that do not hold power and sometimes develop into social organizations or protest groups. Extremist religious and political groups, whether Christian, Islamic, Hindu, or other, may best be understood as countercultures against Western or global influences that they perceive as threatening to their way of life. Figure 3.2 illustrates the types of cultures in the social world and the relationship of countercultures to their national culture. Countercultures, as depicted, view themselves and are viewed by others as "fringe" groups—partial outsiders within a nation.

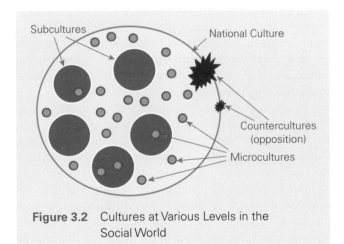

Figure 3.2   Cultures at Various Levels in the Social World

A young Orthodox Jewish boy prepares to pray according to Jewish law by wrapping the leather strap of his tefillin around his arm and a tallit (prayer shawl) around his shoulders. He is part of a subgroup of the larger society, the Jewish faith community. This community will affect his values throughout life—from infancy to death. It is a subculture.

Thinking Sociologically

Describe a counterculture group whose goals are at odds with those of the dominant culture. Do you see evidence that the group is influencing behavioral expectations and values in the larger society? What effect, if any, does it have on your life?

National and Global Cultures: Macro-Level Analysis

Canada is a national society, geographically bounded by the mainland United States to the south, the Pacific Ocean and Alaska to the west, the Atlantic Ocean to the east, and the Arctic to the north. The government in Ottawa passes laws that regulate activities in all provinces (which are similar to states or prefectures), and each province passes its own laws. These geographic boundaries and political structures make up the national society of Canada.

National Society and Culture

The **national society** includes a population of people usually living within a specified geographic area, who are connected by common ideas, are subject to a particular political authority, and cooperate for the attainment of common goals. Within the nation, there may be smaller groups, such as ethnic, regional, or tribal subcultures made up of people who identify closely with each other. Along with subcultures, most nations have a **national culture** of common values and beliefs that tie citizens together. The national culture affects the everyday lives of all its citizens to some extent. Within some countries of Africa and the Middle East that became self-governing nations during the twentieth century, local ethnic or religious loyalties are much stronger than any sense of national culture. Subcultural differences divide many nations. Consider, for example, the loyalties of Shiites, Sunnis, and Kurds in Iraq, where the national culture struggles for influence over its citizens through laws, traditions, and military force.

In colonial America, people thought of themselves as Virginians or Rhode Islanders rather than as U.S. citizens. Even during the "War Between the States" of the 1860s, the battalions were organized by states and often carried their state banners into battle. The fact that some Southern states still call it the War Between the States rather than the Civil War communicates the struggle over whether to recognize the nation or states as the primary social unit of loyalty and identity. People in the United States today are increasingly likely to think of themselves as U.S. citizens (rather than as Iowans or Floridians), yet the national culture determines only a few of the specific guidelines for everyday life. Still, the sense of nation has grown stronger in most industrialized societies over the past century, and primary identity is likely to be "United States" or "Canadian."

Global Society and Culture

Several centuries ago, it would have been impossible to discuss a global culture, but with expanding travel, economic interdependence of countries, international political linkages, global environmental concerns, and technology allowing for communication throughout the world, people now interact across continents in seconds. **Globalization** refers to the process by which the entire world is becoming a single interdependent entity—more uniform, more integrated, and more interdependent (Pieterse 2004; Robertson 1997; Stutz and Warf 2005). Globalization is not a product or an object but a process of increased connectedness and interdependency across the planet (Eitzen and Zinn 2006).

Western political and economic structures dominate in the development of this global society, largely as a result of the domination of Western (European and U.S.) worldviews and Western control over resources. For example, the very idea of governing a geographic region with a bureaucratic structure known as a nation-state is a fairly new notion. Formerly, many small bands and tribal groupings dominated areas of the globe. However, with globalization, nation-states now exist in every region of the world.

Global culture refers to behavioral standards, symbols, values, and material objects that have become common across the globe. For example, beliefs that monogamy is normal; that marriage should be based on romantic love; that people have a right to life, liberty, and the pursuit of happiness; that people should be free to choose their leaders; that women should have rights such as voting; that wildlife and fragile environments should be protected; and that everyone should have a cell phone and television set are spreading across the globe (Leslie and Korman 1989; Newman and Grauerholz 2002). During the twentieth century, the idea of the primacy of individual rights, civil liberties, and human rights spread around the world, creating conflicts with nations that traditionally lack democratic institutions and processes. Backlashes against these and other Western ideas also can be seen in acts of groups that have embraced terrorism (Eitzen and Zinn 2006; Misztal and Shupe 1998; Turner 1991a, 1991b).

Still, these trends are aspects of the emerging global culture. Even 100 years ago, notions of global culture would have seemed quite bizarre (Lechner and Boli 2005). However, in nations all over the globe, people who travel by plane know they must stand in line, purchase a ticket, negotiate airport security, and stay seated in the plane until they are told they can get up (Lechner and Boli 2005). Regardless of nationality, we know how to behave in any airport in the world. Nations are accepted as primary units of social control, and use of coercion is perfectly normal if it is done by the government. We compete in Olympic Games as citizens of nations, and the winners stand on the platform while their national anthems are played. Across the globe the idea of nations seems "normal," yet only a few centuries ago, the notion of nations would have seemed very strange (Lechner and Boli 2005).

Global culture probably has fairly minimal impact on the everyday interactions and lives of the average person, yet it affects nations, which in turn affect our lives. As the world community becomes more interdependent and addresses issues that can only be dealt with at the global level (e.g., global warming, pirates from Somalia and other countries, massive human rights violations as in the Libyan revolution or Sudan war, or international terrorism), the idea of a common "software" of beliefs, social rules, and common interests takes on importance. Common ideas for making decisions allow for shared solutions to conflicts that previously would have resulted in war and massive killing of people. Global culture at the macro level will increasingly be a reality in the third millennium.

However, global culture is not the only pattern that is new. Today, we are seeing a counterculture at the global level. Stateless terrorist networks reject the values of the World Court, the Geneva Convention, and other international systems designed to resolve disputes. Terrorists do not recognize the sovereignty of nations and do not acknowledge many values of respect for life or for civil discourse. This counterculture at the global level is a more serious threat than those at the micro and meso levels.

So social structure and culture operate at each of the levels of society. Various theories about culture help illuminate further the way culture operates in the society, and we turn to the main theories next.

Thinking Sociologically

Make a list of social units of which you are a part. Place each group into one of the following categories: microculture, subculture, national culture, or global culture. Consider which of them affects only a portion of your day or week (like your place of work) or only a very limited time in your entire life span. Consider which groups are smaller than the nation but will likely influence you over much of your life. To what cross-national (global) groups do you belong?

Cultural Theory at the Micro Level

When students first read about sociology or other social sciences, the underlying message may seem to be that humans are shaped by the larger society in which they live. External forces do shape us in many ways, but that is not the whole story, as we see when we examine the symbolic interaction approach to culture.

Symbolic Interaction Theory

How amazing it is that babies learn to share the ideas and meanings of complex cultures with others in those cultures. Symbolic interaction theory considers how we learn to share meanings of symbols, whether material or nonmaterial. Culture is about symbols, such as rings, flags, and words that stand for or represent something. A ring means love and commitment. A flag represents national identity and is intended to evoke patriotism and love of country. A phrase such as *middle class* conjures up images and expectations of what the phrase means, and we share this meaning with others with whom we interact. Together in our groups and societies, we define what is real, normal, and good.

Symbolic interaction theory maintains that our humanness comes from the impact we have on each other through these shared understandings of symbols that humans have created. When people create symbols, such as a new greeting ("give me five") or a symbolic shield for a fraternity or sorority, symbols come to have an existence and importance for a group. This is Step 1: The symbol is created. Who designed the Star of David and gave it meaning as a symbol of the Jewish people? Who initiated the sign of the cross for Catholics to use at the beginning of prayer? Who designed the fraternity's or sorority's shield? Who determined that an eagle should symbolize the United States? Most people do not know the answers to these questions, but they do know what the symbol stands for. They share with others the meaning of a particular object. This is Step 2: The symbol is objectified, assuming a reality independent of the creator. In fact, people may feel intense loyalty to the symbol itself. An entire history of a people may be recalled and a set of values rekindled when the symbol such as a national flag is displayed. This is Step 3: The group has internalized the symbol. This may be the case whether the symbol is part of material culture or a nonmaterial gesture. Members of a culture absorb the ideas or symbols of the larger culture—which were originally created by some individual or small group.

Symbolic interaction theory pictures humans as consciously and deliberately creating their personal and collective histories. The theory emphasizes the part language and gestures play in the shared symbols of individuals and the smooth operation of society. More than any other theory in the social sciences, symbolic interaction stresses the active decision-making role of individuals—the ability of individuals to do more than conform to the larger forces of the society.

Many of our definitions of what is "normal" in a society are shaped by what others define as "normal" or "good." This is sometimes called *the social construction of reality—* what is considered real is what our cultures define as real. One illustration of this is the notion of what is beautiful or ugly. In the late eighteenth and early nineteenth centuries, in Europe and the United States, beaches were considered eyesores since there was nothing there but crushed stone and dangerous water. A beach was not viewed as a place

In the late 1700s and early 1800s, these beach and mountain views would have been considered eyesores—too ugly to enjoy. The social construction of reality—the definition of what is beautiful in our culture—has changed dramatically over the past two centuries.

to relax in a beautiful environment. Likewise, when early travelers to the West encountered the Rocky Mountains, with soaring granite rising to snow-capped peaks, the idea was that these were incredibly ugly wounds in the earth's surface. The summits were anything but appealing, restful, and soothing. Still, over time, some individuals began to redefine these crests as breathtakingly beautiful. We now see both as beautiful, but the social construction of scenery has not always been so (Lofgren 1999, 2010).

This notion that individuals shape culture and that culture influences individuals is at the core of symbolic interaction theory. Other social theories tend to focus at the meso and macro levels.

Thinking Sociologically

Recall some of the local "insider" symbols that you used as a teenager—such as friendship bracelets. Some individual started each idea, and it spread rapidly from one school to another and one community to another. How are the three steps in the creation of symbols illustrated by a symbol you and your friends used?

Cultural Theories at the Meso and Macro Levels

How can we explain such diverse world practices as eating grubs and worshipping cows? Why have some societies allowed men to have four wives whereas others—such as the Shakers—prohibited sex between men and women entirely? Why do some groups worship their ancestors while others have many gods or believe in a single divine being? How can societies adapt to extremes of climate and geographical terrain—hot, cold, dry, wet, mountainous, flat? Humankind has evolved practices so diverse that it would be hard to find a practice that has not been adopted in some society at some time in history.

To explain cultural differences, we will examine two perspectives that have made important contributions to understanding culture at the meso and macro levels: structural-functional and conflict theories.

Structural-Functionalist Theory

Structural-functionalist theory seeks to explain why members of an ethnic subculture or a society engage in certain practices. To answer, structural-functionalists look at how those practices meet social needs or contribute to the survival or social solidarity of the group or society as a whole. A classic example is the reverence for cattle in India. The "sacred cow" is protected and treated with respect and is not slaughtered for food. The reasons relate to India's ancient development into an agricultural society that required sacrifices (Harris 1989). Cattle were needed to pull plows and to provide a source of milk and fuel as their dried dung was the only source. Cows gained religious significance because of their importance for the survival of early agricultural communities. They must, therefore, be protected from hungry people for the long-term survival of the group.

Functionalists view societies as composed of interdependent parts, each fulfilling certain necessary functions or purposes for the total society (Radcliffe-Brown 1935). Shared norms, values, and beliefs, for instance, serve the function of holding a social group together. At a global macro level, functionalists see the world moving in the direction of having a common culture, potentially reducing "we" versus "they" thinking and promoting unity across boundaries. Synthesis of cultures and even the loss of some cultures are viewed as a natural result of globalization.

Although most cultural practices serve positive functions for the maintenance and stability of the society, some practices, such as slavery or child abuse, may be dysfunctional for minority groups or individual members of society. Some critics argue that functional theory overemphasizes the need for consensus and integration among different parts of society, thus ignoring conflicts that may point to problems in societies (Dahrendorf 1959).

Conflict Theory

Whereas functionalists assume consensus because all people in society have learned the same cultural values, rules, and expectations, conflict theorists do not view culture as having this uniting effect. Conflict theorists describe societies as composed of groups—class, ethnic, religious, and political groups at the meso level—vying for power. Each group protects its own self-interests and struggles to make its own cultural ways dominant in the society. Instead of consensus, the dominant groups may impose their cultural beliefs on minorities and other subcultural groups, thus

Conflict theorists believe that society is composed of groups, each acting to meet its own self-interests, and those groups struggle to make their own cultural values supreme in the society. The recent conflict over immigration laws is illustrated by protesters who expressed their opinions in this "Day Without an Immigrant" march in Los Angeles.

laying the groundwork for conflict. Conflict theorists identify tension between meso and macro levels, whereas functionalists tend to focus on harmony and smooth integration between those levels.

Actually, conflict may contribute to a smoother-running society in the long run. The German sociologist Georg Simmel (1955) believed that some conflict could serve a positive purpose by alerting societal leaders to problem areas that need attention. This view is illustrated by the political changes taking place in several North African and Middle Eastern countries such as Tunisia, Egypt, Yemen, and Libya.

Conflict theorists argue that the people with privilege and power in society manipulate institutions such as religion and education. In this way, people learn the values, beliefs, and norms of the privileged group and foster beliefs that justify the dominant group's self-interests, power, and advantage. The needs of the privileged are likely to be met, and their status will be secured. For instance, schools that serve lower-class children usually teach obedience to authority, punctuality, and respect for superiors—behaviors that make for good laborers. The children of the affluent, meanwhile, are more likely to attend schools stressing divergent thinking, creativity, and leadership, attributes that prepare them to occupy the highly rewarded positions in their society. Conflict theorists point to this control of the education process by those with privilege as part of the overall pattern by which the society benefits the rich.

Conflict theory can also help us understand global dynamics. Poor nations feel that the global system protects the self-interests of the richest nations and that those rich nations impose their own culture, including their ideas about economics, politics, and religion, on the less affluent.

Some scholars believe there is great richness in local customs that is lost when homogenized by the cultural domination in the macro-level trends of globalization (Ritzer 2004a). Conflict theory is useful for analyzing relationships between societies (macro level) and between subcultures (meso level) within complex societies. It also helps illuminate tensions in a society when local (micro-level) cultural values clash with national (macro-level) trends.

Conflict theory is not as successful, however, in explaining simple, well-integrated societies in which change is slow to come about and cooperation is an organizing principle.

The Fit Between Hardware and Software

Computer software cannot work with incompatible machines. Some documents cannot be easily transferred to another piece of hardware, although sometimes a transfer can be accomplished with significant modification in the formatting of the document. The same is true with the hardware of society and the software of culture. For instance, having large extended families, typically valued in agricultural societies, does not work well in industrial and postindustrial societies. Other values such as rewarding people based on individual merit, emphasizing the idea that humans are motivated primarily by their own self-interests, or believing that change in cultures is inevitable and equals progress are not particularly compatible with the hardware (structure) of traditional horticultural or herding societies. Values and beliefs ("software") can be transferred to another type of society ("hardware"). However, there are limits to what can be transferred, and the change of "formatting" may mean the new beliefs are barely recognizable.

Attempts to transport U.S.-style "software" (culture)—individualism, capitalism, freedom of religion, and democracy—to other parts of the world illustrate that these ideas are not always successful in other settings. The hardware of other societies may be able to handle more than one type of software or set of beliefs, but there are limits to the adaptability. Thus, we should not be surprised when our ideas are transformed into something quite different when they are imported to another social system. If we are to understand the world in which we live and if we want to improve it, we must first fully understand these societies and cultures.

B ecause there is such variation between societies and cultures in what they see as normal, how do we ever adjust to our society's expectations? The answer is addressed in the next chapter. Human life is a lifelong process of socialization to social and cultural expectations.

What Have We Learned?

Individuals and small groups cannot live without the support of a larger society, the hardware of the social world. Without the software—culture—there could be no society for there would be no norms to guide our interactions with others in society. Humans are inherently social and learn their culture from others. Furthermore, as society has evolved into more complex and multileveled social systems, humans have learned to live in and negotiate conflicts between multiple cultures, including those at micro, meso, and macro levels. Life in an information age society demands adaptability to different sociocultural contexts and tolerance of different cultures and subcultures. This is a challenge to a species that has always had tendencies toward ethnocentrism.

Key Points:

- Society consists of individuals who live together in a specific geographic area, interact with each other more than with outsiders, cooperate to attain goals, and share a common culture. Each society has a culture, ideas and "things" that are passed on from one generation to the next in the society; the culture has both material and nonmaterial components. (See p. 60.)

- Societies evolve from very simple societies to more complex ones, from the simple hunter-gatherer society to the information societies of the postindustrial world. (See pp. 61–66.)

- The study of culture requires that we try to avoid ethnocentrism (judging other cultures by the standards of our culture), taking a stance of cultural relativity instead so that the culture can be understood from the standpoint of those inside it. (See pp. 68–72.)

- Just as social units exist at various levels of our social world, from small groups to global systems, cultures exist within different levels of the social system—microcultures, subcultures, national cultures, and global cultures. Some social units at the micro or meso level stand in opposition to the dominant national culture, and they are called countercultures. (See pp. 81–83.)

- Various theories offer different lenses for understanding culture. While symbolic interaction illuminates the way humans bring meaning to events (thus generating culture), the functionalist and conflict paradigms examine cultural harmony/seamless fit and conflict between cultures, respectively. (See pp. 85–88.)

- The metaphor of hardware (society's structure) and software (culture) describes their interdependent relationship, and as with computers, there must be some compatibility between the structure and the culture. If there is none, either the cultural elements that are transported into another society will be rejected, or the culture will be "reformatted" to fit the society. (See p. 88.)

Contributing to Our Social World: What Can We Do?

At the Local and National Level:

Most large (and many smaller) communities have organizations and clubs that focus on the interests of specific ethnic groups: Arab Americans, Chinese Americans, Italian Americans, Polish Americans, and so on.

- Select and contact one of these groups (of your own background or, even more interesting, of a background that differs from your own). Arrange to visit one of the group's meetings and learn about the activities in which members are involved.

- Many ethnically oriented organizations assist recent immigrants in dealing with adjustment to American life. Contact one of these groups and explore the possibility of volunteering or serving as an intern.

At the Global Level:

As the process of globalization accelerates, the cultures, languages, and basic rights of indigenous people are under increasing threat.

- *UN Permanent Forum on Indigenous Issues.* The United Nations has a program that is intended to assist the affected groups in facing these challenges. Visit the program's website at www.un.org/esa/socdev/unpfii, and contact the program about the possibility of volunteering.

- *Cultural Survival* at www.cs.org. Contact a nongovernmental organization engaged in action-oriented programs. Cultural Survival, for example, partners with indigenous peoples to promote respect for their right to self-determination; ensure their right to full and effective political, economic, and social participation in the country where they live; and enjoy their rights to their lands, resources, languages, and cultures.

Visit **www.sagepub.com/oswcondensed2e** for online activities, sample tests, and other helpful information. Select "Chapter 3: Society and Culture" for chapter-specific activities.

Socialization

Becoming Human and Humane

Whether at the micro, meso, or macro level, our close family and friends plus various organizations help us learn how to be human and humane in our society. Skills are taught, as are values such as loyalty and caregiving.

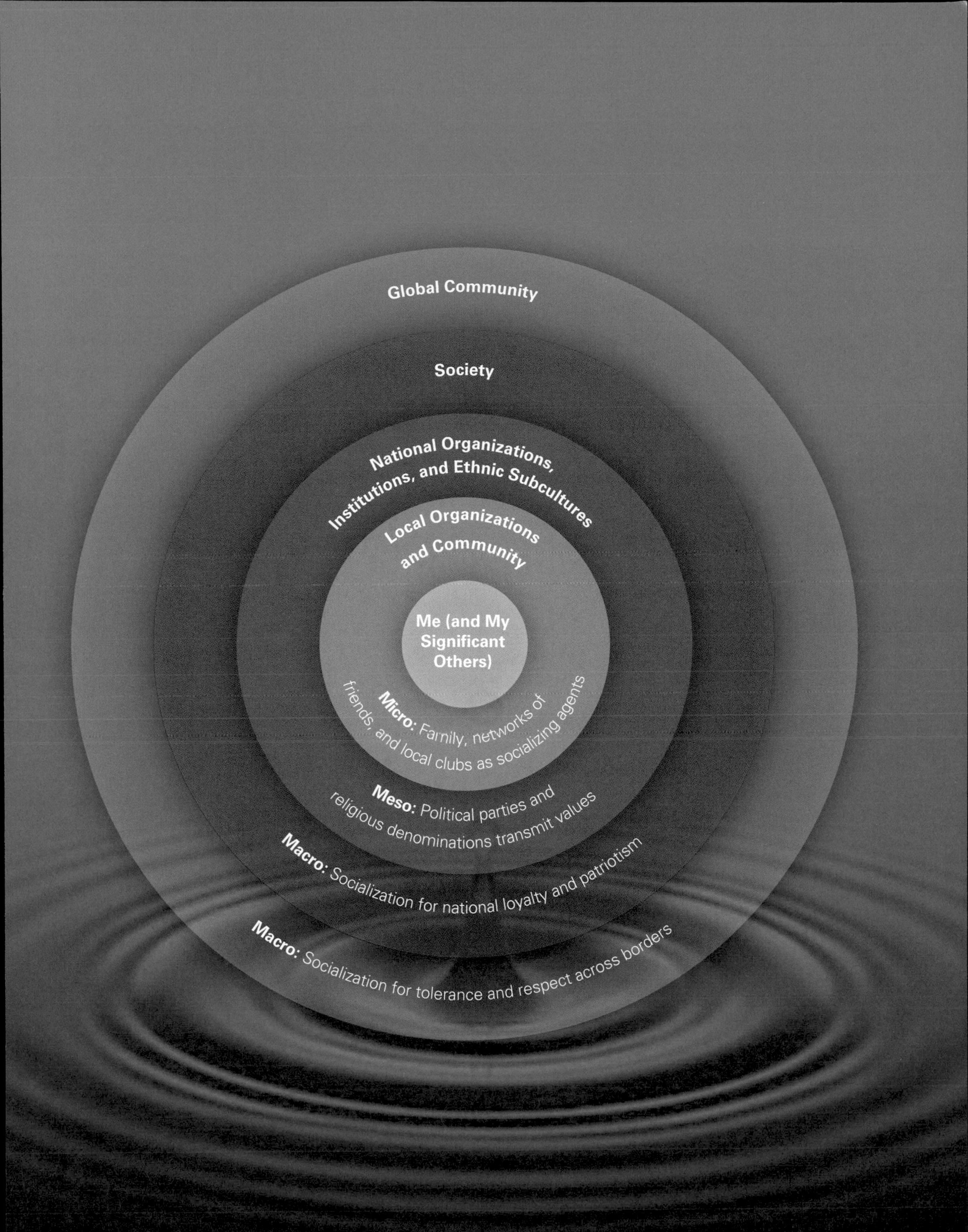

Global Community

Society

National Organizations,
Institutions, and Ethnic Subcultures

Local Organizations
and Community

Me (and My
Significant
Others)

Micro: Family, networks of
friends, and local clubs as socializing agents

Meso: Political parties and
religious denominations transmit values

Macro: Socialization for national loyalty and patriotism

Macro: Socialization for tolerance and respect across borders

Think About It	
Self and Inner Circle	What does it mean to have a "self"? How would you be different if you had been raised in complete isolation from other people?
Local Community	How have your local religious congregation and schools shaped who you are?
National Institutions; Complex Organizations; Ethnic Groups	How do various subcultures or organizations of which you are a member (your political party, your religious affiliation) influence your position in the social world?
National Society	What would you be like if you were raised in a different country? How does your sense of national identity influence the way you see things?
Global Community	How might globalization or other macro-level events impact you and your sense of self?

Ram, a first grader from India, had been attending school in Iowa for only a couple of weeks. The teacher was giving the first test. Ram did not know much about what a test meant, but he rather liked school, and the red-haired girl next to him, Elyse, had become a friend. He was catching on to reading a bit faster than she, but she was better at the number exercises. They often helped each other learn while the teacher was busy with a small group in the front of the class.

The teacher gave each child the test, and Ram saw that it had to do with numbers. He began to do what the teacher instructed the children to do with the worksheet, but after a while, he became confused. He leaned over to look at the page Elyse was working on. She hid her sheet from him, an unexpected response. The teacher looked up and asked what was going on. Elyse said that Ram was "cheating." Ram was not quite sure what that meant, but it did not sound good. The teacher's scolding of Ram left him baffled, confused, and entirely humiliated.

This incident was Ram's first lesson in the individualism and competitiveness that govern Western-style schools. He was being socialized into a new set of values. In his parents' culture, competitiveness was discouraged, and individualism was equated with selfishness and rejection of community. Athletic events were designed to end in a tie so that no one would feel rejected. Indeed, a well-socialized person would rather lose in a competition than cause someone else to feel bad because he or she lost.

Socialization is the lifelong process of learning to become a member of the social world, beginning at birth and continuing until death. It is a major part of what the family, education, religion, and other institutions do to prepare individuals to be members of their social world. Like Ram, each of us learns the values and beliefs of our culture. In Ram's case, he literally moved from one cultural group to another and had to adjust to more than one culture within his social world.

Have you ever interacted with a newborn human baby? Infants are interactive, ready to develop into members of the social world. As they cry, coo, or smile, they gradually learn that their behaviors elicit responses from other humans. This exchange of messages—this **interaction**—is the basic building block of socialization. Out of this process of interaction, a child learns its culture and becomes a member of society. This process of interaction shapes the infant into a human being with a social self—perceptions we have of who we are.

Three main elements provide the framework for socialization: human biological potential, culture, and individual experiences. Babies enter this world unsocialized, totally dependent on others to meet their needs, and completely lacking in social awareness and an understanding of the

Babies interact intensively with their parents, observing and absorbing everything around them and learning what kinds of sounds or actions elicit response from the adults. Socialization starts at the beginning of life.

rules of their society. Despite this complete vulnerability, they have the potential to learn the language, norms, values, and skills needed in their society. They gradually learn who they are and what is expected of them. Socialization is necessary not only for the survival of the individual but also for the survival of society and its groups. The process continues in various forms throughout our lives as we enter and exit various positions—from school to work to retirement to death.

In this chapter, we will explore the nature of socialization and how individuals become socialized. We consider why socialization is important. We also look at development of the self, socialization through the life cycle, who or what socializes us, macro-level issues in the socialization process, and a policy example illustrating socialization. First, we briefly examine an ongoing debate: Which is more influential in determining who we are—our genes (nature) or our socialization into the social world (nurture)?

Nature *Versus* Nurture—or *Both* Working Together?

What is it that most makes us who we are? Is it our biological makeup or the environment in which we are raised that guides our behavior and the development of our self? One side of the contemporary debate regarding nature versus nurture seeks to explain the development of the self and human social behaviors—violence, crime, academic performance, mate selection, economic success, gender roles, and other behaviors too numerous to mention here—by examining biological or genetic factors (Harris 2009; Winkler 1991). Sociologists call this sociobiology, and psychologists refer to it as evolutionary psychology. The theory claims that our human genetic makeup wires us for social behaviors (Wilson et al. 1978).

The idea is that we perpetuate our own biological family lines and the human species through various social behaviors. Human groups develop power structures, are territorial, and protect their kin. A mother ignoring her own safety to help a child, soldiers dying in battle for their comrades and countries, communities feeling hostility toward outsiders or foreigners, and neighbors defending property lines against intrusion by neighbors are all examples of behaviors that sociobiologists claim are rooted in genetic makeup of the species. Sociobiologists would say these behaviors continue because they result in an increased chance of survival of the species as a whole (Lerner 1992; Lumsden and Wilson 1981; Wilson 1980, 1987).

Most sociologists believe that sociobiology and evolutionary psychology explanations have flaws. If social behavior is genetically programmed, then it should manifest itself regardless of the culture in which humans are raised. Yet there are vast differences between cultures, especially in gender behaviors and traits. The range of differences would not occur if we were biologically hardwired to certain behaviors. In Chapter 1, for example, we saw that in some societies men wear makeup and are gossipy and vain, violating our stereotypes. The key is that what makes humans unique is not our biological heritage but our ability to learn the complex social arrangements of our culture.

Most sociologists recognize that individuals are influenced by biology, which limits the range of human responses and creates certain needs and drives, but they believe that nurture is far more important as the central force in shaping human social behavior through the socialization process. Many sociologists now consider the interplay of nature and nurture. Alice Rossi (1984) has argued that we need to build both biological and social theories—or biosocial theories—into explanations of social processes such as parenting. Just in the twenty-first century, a few sociologists are developing an approach called evolutionary sociology that takes seriously the way our genetic makeup—including a remarkable capacity for language—shapes our range of behaviors. However, it is also very clear from biological research that living organisms are often modified by their environments and the behaviors of others around them—with even genetic structure changing (Lopreato 2001; Machalek and Martin 2010). In short, biology influences human behavior, but interactive behavior can also modify biological traits. Indeed, nutritional history of grandparents can affect the metabolism of their grandchildren, and what the grandparents ate was largely shaped by cultural ideas about food (BBC's Science and Nature 2009; Freese, Powell, and Steelman 1999; Rossi 1984). The point is that socialization is key in the process of "becoming human and humane."

The Importance of Socialization

If you have lived on a farm, watched animals in the wild, or seen television nature shows, you probably have noticed that many animal young become independent shortly after birth. Horses are on their feet in a matter of hours, and the parents of turtles are long gone by the time the babies hatch from eggs. Many species in the animal kingdom do not require contact with adults to survive because their behaviors are inborn and instinctual. Generally speaking, the more intelligent the species, the longer the period of gestation and of nutritional and social dependence on the mother and family. Humans clearly take the longest time to socialize their young. Even among primates, human infants have the longest gestation and dependency period, generally six to eight years. Chimpanzees, very similar to humans in their DNA, take only 12 to 28 months. This extended dependency period for humans—what some have referred to as the *long childhood*—allows each human being time to learn the complexities of culture. This suggests that biology and social processes work together.

Normal human development involves learning to sit, crawl, stand, walk, think, talk, and participate in social interactions. Ideally, the long period of dependence allows children the opportunity to learn necessary skills, knowledge, and social roles through affectionate and tolerant interaction with people who care about them. Yet, what happens if children are deprived of adequate care or even human contact? The following section illustrates the importance of socialization by showing the effect of deprivation and isolation on normal socialization.

Isolated and Abused Children

Anna and Isabelle experienced extreme isolation in early childhood, cut off from other humans. In case studies comparing the two girls, Kingsley Davis (1947) found that even minimal human contact made some difference in their

Intense interaction by infants and their caregiver, usually a parent, occurs in all cultures and is essential to becoming a part of the society and to becoming fully human. This African father shares a tender moment with his son.

socialization. Both "illegitimate" girls were kept locked up by relatives who wanted to keep their existence a secret. Both were discovered at about age six and moved to institutions where they received intensive training. Yet, the cases were different in one significant respect: Prior to her discovery by those outside her immediate family, Anna experienced virtually no human contact. She saw other individuals only when they left food for her. Isabelle lived in a darkened room with her deaf-mute mother, who provided some human contact. Anna could not sit, walk, or talk and learned little in the special school in which she was placed. When she died from jaundice at age 11, she had learned the language and skills of a two- or three-year-old. Isabelle, on the other hand, progressed. She learned to talk and played with her peers. After two years, she reached an intellectual level approaching normal for her age but remained about two years behind her classmates in performance levels (Davis 1940, 1947).

Cases of children who come from war-torn countries (Povik 1994), live in orphanages, or are neglected or abused illustrate less extreme isolation. Although not totally isolated, these children also experience problems and disruptions in the socialization process. These neglected children's situations have been referred to as abusive, violent, and dead-end environments that are socially toxic because of their harmful developmental consequences for children.

What is the message? These cases illustrate the devastating effects of isolation, neglect, and abuse early in life on normal socialization. Humans need more from their environments than food and shelter. They need positive contact, a sense of belonging, affection, safety, and someone to teach them knowledge and skills. This is children's socialization into the world through which they develop a self. Before we examine the development of the self in depth, however, we consider the complexity of socialization in the multilayered social world.

Socialization and the Social World

Sociologists are interested in how individuals become members of their society and learn the culture to which they belong. Through the socialization process, individuals learn what is expected in their society. At the micro level, most parents teach children proper behaviors to be successful in life, and peers influence children to "fit in" and have fun. Individual development and behavior occur in social settings. Interaction theory, focusing on the micro level, forms the basis of this chapter, as you will see.

At the meso level, religious denominations espouse their versions of the Truth, and schools teach the knowledge and skills necessary for functioning in society. At the nation-wide macro level, television ads encourage viewers to buy products that will make them better and happier people. From interactions with our significant others to dealing with government bureaucracy, most activities are part of the socialization experience that teaches us how to function in our society. Keep in mind that socialization is a lifelong process. Even your grandparents are learning how to live at their stage of life.

The social world model at the start of the chapter illustrates the levels of analysis in the social world. The process of socialization takes place at each level, linking the parts. Small micro-level groups include families, **peer groups** (who are roughly equal in some status within the society, such as age or occupation), and voluntary groups such as civic clubs. Examples of meso-level institutions are educational and political systems, while an important macro-level unit is the federal government. All of these have a stake in how we are socialized because they all need trained and loyal group members to survive. Organizations need citizens who have been socialized to devote time, energy, and resources that these groups need to survive and meet their goals. Lack of adequate socialization increases the likelihood of individuals becoming misfits or social deviants.

Most perspectives on socialization focus on the micro level, as we shall see when we explore development of the self. Meso- and macro-level theories add to our understanding of how socialization prepares individuals for their roles in the larger social world. For example, structural-functionalist perspectives of socialization tend to see different levels of the social world operating to support each other. According to this perspective, education in many Western societies reinforces individualism and an achievement ethic. Families often organize holidays around patriotic themes, such as a national independence day, or around religious celebrations. These values are compatible with preparing individuals to support national political and economic systems.

Socialization can also be understood from the conflict perspective, with the linkages between various parts of the social world based on competition with or even direct opposition to another part. For example, demands from organizations for our resources (time, money, and energy devoted to the Little League, the Rotary club, and library associations) may leave nothing to give to our religious communities or even our families, setting up a conflict. Each organization and unit competes to gain our loyalty in order to claim some of our resources.

At the meso level, the purposes and values of organizations or institutions are sometimes in direct contrast with one another or are in conflict with the messages at other levels of the social system. Businesses and educational institutions try to socialize their workers and students to be serious, hardworking, sober, and conscientious, with lifestyles focused on the future. By contrast, many fraternal organizations and barroom microcultures favor lifestyles that celebrate drinking, sex, and living for the moment. This creates conflicting values in the socialization process.

Conflict can occur in the global community as well. For example, religious groups often socialize their members to identify with humanity as a whole ("the family of God"). However, in some cases, nations do not want their citizens socialized to identify with those beyond their borders. If religion teaches that all people are "brothers and sisters" and if religious people object to killing, the nation may have trouble mobilizing its people to arms when the leaders call for war.

Conflict theorists believe that those who have power and privilege use socialization to manipulate individuals in the social world to support the power structure and the self-interests of the elite. Although they may not realize it, most individuals have little power to control and decide their futures.

Each theoretical explanation has merit for explaining some situations. Whether we stress harmony in the socialization process or conflict rooted in power differences, the development of a sense of self through the process of socialization is an ongoing, lifelong process. Having considered the multiple levels of analysis and the issues that make socialization complicated, let us focus specifically on the micro level: Where does the sense of self originate?

Thinking Sociologically

Although the socialization process occurs primarily at the micro level, it is influenced by events at each level of analysis shown in the social world model. Give examples of family, community, subcultural, national, or global events that might have influenced how you were socialized or that might influence how you would socialize your child.

Development of the Self: Micro-Level Analysis

The main product of the socialization process is *the self*. Fundamentally, **self** refers to the perceptions we have of who we are. Throughout the socialization process, our self is derived from our perceptions of the way others are responding to us. The development of the self allows individuals to interact with other people and to learn to function at each level of the social world.

Humans are not born with a sense of self. It develops gradually, beginning in infancy and continuing throughout adulthood. Selfhood emerges through interaction with others. Individual biology, culture, and social experiences all play a part in shaping the self. The hereditary blueprint each person brings into the world provides broad biological outlines, including particular physical attributes, temperament, and a maturational schedule. However, nature is shaped by nurture. Each person is also born into a family that lives within a particular culture. This hereditary blueprint, in interaction with family and culture, helps create each unique person, different from any other person yet sharing the types of interactions by which the self is formed.

Most sociologists, although not all (Irvine 2004), believe that we humans are distinct from other animals in our ability to develop a self and to be aware of ourselves as individuals or objects. Consider how we refer to ourselves in the first person—*I* am hungry, *I* feel foolish, *I* am having fun, *I* am good at basketball. We have a conception of who we are, how we relate to others, and how we differ from and are separate from others in our abilities and limitations. We have an awareness of the characteristics, values, feelings, and attitudes that give us our unique sense of self (James [1890] 1934; Mead [1934] 1962).

Thinking Sociologically

Who are some of the people who have been most significant in shaping your *self*? How have their actions and responses helped shape your self-conception as musically talented, athletic, intelligent, kind, assertive, or any of the other hundreds of traits that make up your *self*?

Our sense of self is often shaped by how others see us and what is reflected back to us by the interactions of others. Cooley called this process, operating somewhat like a mirror, the "looking-glass self."

The Looking-Glass Self and Role-Taking

The theoretical tradition of symbolic interaction offers important insights into how individuals develop the self. Charles H. Cooley ([1909] 1983) believed the self is a social product, shaped by interactions with others from the time of birth. He likened interaction processes to looking in a mirror wherein each person reflects an image of the other.

> Each to each a looking-glass
> Reflects the other that doth pass.
>
> (Cooley [1909] 1983:184)

For Cooley ([1909] 1983), the **looking-glass self** is a reflective process based on our interpretations of the reactions of others. In this process, Cooley believed that there are three principal elements, shown in Figure 4.1. We experience feelings such as pride or shame based on this imagined judgment and respond based on our interpretation. Moreover, throughout this process, we actively try to manipulate other people's view of us to serve our needs and interests. This is one of the many ways we learn to be boys or girls—the image that is reflected back to us lets us know whether we have behaved in ways that are socially acceptable according to gender expectations. The issue of gender socialization in particular will be discussed in Chapter 9. Of course, this does not mean our interpretation of the other person's response is correct, but our interpretation does determine how we respond.

Our self is influenced by the many "others" with whom we interact, and each of our interpretations of their reactions feeds into our self-concept. Recall that the isolated children failed to develop this sense of self precisely because they lacked interaction with others.

Taking the looking-glass self idea a step further, George Herbert Mead ([1934] 1962) explained that individuals take others into account by imagining themselves in the position of that other, a process called **role-taking**. When children play mommy and daddy, doctor and patient, or firefighter, they are imagining themselves in another's shoes. Role-taking allows humans to view themselves from the standpoint of others. This requires mentally stepping out of our own experience to imagine how others experience and view the social world. Through role-taking, we begin to see who we are from the standpoint of others. In short, role-taking allows humans to view themselves as objects, as though they are looking at themselves from outside their bodies. For Mead, role-taking is prerequisite for the development of sense of self.

Mead ([1934] 1962) also argued that role-taking is possible because humans have a unique ability to use and respond to **symbols**. Symbols, discussed in Chapter 3 on culture, are human creations such as language and gestures

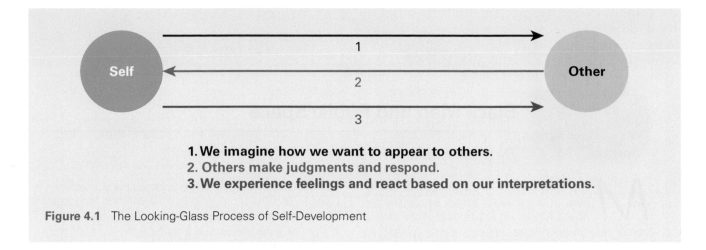

1. **We imagine how we want to appear to others.**
2. **Others make judgments and respond.**
3. **We experience feelings and react based on our interpretations.**

Figure 4.1 The Looking-Glass Process of Self-Development

that are used to represent objects or actions; they carry specific meaning for members of a culture. Symbols such as language allow us to give names to objects in the environment and to infuse those objects with meanings. Once the person learns to symbolically recognize objects in the environment, the self can be seen as one of those objects. In the most rudimentary sense, this starts with possessing a name that allows us to see our self as separate from other objects. Note that the connection of symbol and object is arbitrary, such as the name Michelle Obama and a specific human being. When we say that name, most listeners immediately think of the same person: First Lady of the United States, mother of two daughters, concerned with childhood obesity issues and poor diets. Using symbols is unique to humans. In the process of symbolic interaction, we take the actions of others and ourselves into account. Individuals may blame, encourage, praise, punish, or reward themselves. An example would be a basketball player missing the basket because the shot was poorly executed and thinking, "What did I do to miss that shot? I'm better than that!" A core idea is that the self has *agency*—it is an initiator of action and a maker of meaning, not just a passive responder to external forces.

The next "Sociology in Your Social World" on page 100 illustrates the looking-glass self process and the way role-taking affects selfhood for African American males.

Thinking Sociologically

First, read the essay on the following page. Brent Staples goes out of his way to reassure others that he is harmless. What might be some other responses to this experience of having others assume one is dangerous and untrustworthy? How might one's sense of self be influenced by these responses of others? How are the looking-glass self and role-taking at work in this scenario?

Parts of the Self

According to the symbolic interaction perspective, the self is composed of two distinct but related parts—dynamic parts in interplay with one another (Mead [1934] 1962). The most basic element of the self is what George Herbert Mead refers to as the **I**, the spontaneous, unpredictable, impulsive, and largely unorganized aspect of the self. These spontaneous, undirected impulses of the *I* initiate or give propulsion to behavior without considering the possible social consequences. We can see this at work in the "I want it now" behavior of a newborn baby or even a toddler. Cookie Monster on the children's television program *Sesame Street* illustrates the *I* in every child, gobbling cookies at every chance.

The *I* continues as part of the self throughout life, tempered by the social expectations that surround individuals. In stages, humans become increasingly influenced by interactions with others who instill society's rules. Children develop the ability to see the self as others see them and critique the behavior of the *I*. Mead ([1934] 1962) called this reflective capacity of the self the **Me**. The *Me* has learned the rules of society through socialization and interaction, and it controls the *I* and its desires. Just as the *I* initiates the act, the *Me* gives direction to the act. In a sense, the *Me* channels the impulses of the *I* in an acceptable manner according to societal rules and restraints yet meets the needs of the *I* as best it can. When we stop ourselves just before saying something and think to ourselves, "I'd better not say that," it is our *Me* monitoring and controlling the *I*. Notice that the *Me* requires the ability to take the role of the other, to anticipate the other's reaction.

Stages in the Development of the Self

The process of developing a social self occurs gradually and in stages. Mead ([1934] 1962) identified three critical stages: the imitation stage, the play stage, and the game stage, each

Sociology in Your Social World

Black Men and Public Space

By Brent Staples

Many stereotypes—rigid images of members of a particular group—surround the young African American male in the United States. How these images influence these young men and their social world is the subject of this feature. Think about the human cost of stereotypes and their effect on the socialization process as you read the following essay. If your sense of self is profoundly influenced by the ways others respond to you, how might the identity of a young African American boy be affected by public images of black males?

My first victim was a woman—white, well dressed, probably in her early 20s. I came upon her late one evening on a deserted street in Hyde Park, a relatively affluent neighborhood in an otherwise mean, impoverished section of Chicago. As I swung onto the avenue behind her . . . she cast back a worried glance. To her, the youngish black man—broad, six feet, two inches tall; with a beard and billowing hair; both hands shoved into the pockets of a bulky military jacket—seemed menacingly close. After a few more quick glimpses, she picked up her pace and was running in earnest. Within seconds, she disappeared into a cross street.

That was more than a decade ago. I was 22 years old, a graduate student newly arrived at the University of Chicago. It was in the echo of that terrified woman's footfalls that I first began to know the unwieldy inheritance I'd come into. . . . It was clear that she thought herself the quarry of a mugger, a rapist, or worse. Suffering a bout of insomnia, however, I was stalking sleep, not defenseless wayfarers. . . . I was surprised, embarrassed, and dismayed all at once. Her flight . . . made it clear that I was indistinguishable from the muggers who occasionally seeped into the area from the surrounding ghetto. That first encounter, and those that followed, signified that a vast, unnerving gulf lay between nighttime pedestrians—particularly women—and me. And I soon gathered that being perceived as dangerous is a hazard in itself. I only needed to turn a corner into a dicey situation; crowd some frightened, armed person in a foyer somewhere; or make an errant move after being pulled over by a policeman. Where fear and weapons meet—and they often do in urban America—there is always the possibility of death.

In that first year, my first away from my hometown, I was to become thoroughly familiar with the language of fear. At dark, shadowy intersections, I could cross in front of a car stopped at a traffic light and elicit the thunk, thunk, thunk, thunk of the driver—black, white, male, or female—hammering down the door locks. On less-traveled streets after dark, I grew accustomed to but never comfortable with people crossing to the other side of the street rather than pass me. Then there was the standard unpleasantness with policemen, doormen, bouncers, cabdrivers, and those whose business it is to screen out troublesome individuals before there is any nastiness.

After dark, on the warren-like streets of Brooklyn where I live, I often see women who fear the worst from me. They seem to have set their faces on neutral, and with their purse straps strung across their chests bandolier style, they forge ahead as though bracing themselves against being tackled. I understand, of course, that . . . women are particularly vulnerable to street violence and young black males are drastically overrepresented among the perpetrators of that violence. Yet these truths are no solace against the kind of alienation that comes of being ever the suspect. . . .

Over the years, I learned to smother the rage I felt at so often being taken for a criminal. Not to do so would surely have led to madness. I now take precautions to make myself less threatening. I move about with care, particularly late in the evening. I give a wide berth to nervous people on the subway platforms during the wee hours. . . . I have been calm and extremely congenial on those rare occasions when I've been pulled over by the police.

And on late-evening constitutionals, I employ what have proved to be excellent tension-reducing measures: I whistle melodies from Beethoven and Vivaldi and the more popular classical composers. Even steely New Yorkers hunching toward nighttime destinations seem to relax, and occasionally they even join in the tune. Virtually everybody seems to sense that a mugger wouldn't be warbling bright, sunny selections from Vivaldi's *Four Seasons*. It is my equivalent of the cowbell that hikers wear when they know they are in bear country.

of which requires the unique human ability to engage in role-taking. In the **imitation stage**, the child under three years old is preparing for role-taking by observing others and imitating their behaviors, sounds, and gestures.

The **play stage** involves a kind of play-acting in which the child is actually "playing at" a role. Listen to children who are three to five years old play together. You will notice that they spend most of their time telling each other what to do. One of them will say something like "You be the mommy, and I can be the daddy, and Julie, you be the dog. Now you say, 'Good morning, Dear,' and I'll say, 'How did you sleep?' and Julie, you scratch at the door like you want to go out." They will talk about their little skit for 15 minutes and then enact it, with the actual enactment taking perhaps one minute. Small children mimic or imitate role-taking based on what they have seen.

A child who is playing mommy or daddy with a doll is playing at *taking the role* of parent. The child is directing activity toward the doll in a manner imitative of how the parents direct activity toward the child. The child often does not know what to do when playing the role of a parent "going off to work" because children can play only roles they have seen or are familiar with. They do not know what the absent parent does at work. The point is that this "play" is actually extremely important "work" for children because they need to observe and imitate the relationships between roles to form the adult self (Handel, Cahill, and Elkin 2007).

Society and its rules are initially represented by **significant others**—parents, guardians, relatives, or siblings whose primary and sustained interactions with the child are especially influential. That is why much of the play stage involves role-taking based on these significant people in the child's life. The child does not yet understand the complex relations and multiple role players in the social world outside the immediate family. Children may have a sense of how Mommy or Daddy sees them, but children are not yet able to comprehend how they are seen by the larger social world. Lack of role-taking ability is apparent when children say inappropriate things such as "Why are you so fat?"

In the **game stage**, the child learns to take the role of multiple others concurrently. Have you ever watched a team of young children play T-ball (a pre–Little League baseball game in which the children hit the ball from an upright rubber device that holds the ball), or have you observed a soccer league made up of six-year-olds? If so, you have seen Mead's ([1934] 1962) point illustrated vividly. In soccer (or football), five- or six-year-old children will not play their positions despite constant urging and cajoling by coaches. They all run after the ball, with little sense of their interdependent positions. Likewise, a child in a game of T-ball may pick up a ball that has been hit, turn to the coach, and say, "Now what do I do with it?" Most still do not quite grasp throwing it to first base, and the first-base player may actually have left

By imitating roles she has seen, this child is learning both adult roles and empathy toward others. This kind of role enactment is an important prerequisite to the more complex interaction of playing a game with others.

the base to run for the ball. It can be hilarious for everyone except the coach, as a hit that goes seven feet turns into a home run because everyone is scrambling for the ball.

When the children enter the game stage at about age seven or eight, they will be developmentally able to play the roles of various positions and enjoy a complex game. Each child learns what is expected and the interdependence of roles because she or he is then able to respond to the expectations of several people simultaneously (Hewitt 2007; Meltzer 1978). This allows the individual to coordinate his or her activity with others.

In moving from the play stage to the game stage, children's worlds expand from family and day care to neighborhood playmates, school, and other organizations. This process gradually builds up a composite of societal

Very young children who play soccer do not understand the role requirements of games. They all—including the goalie—want to chase after the ball. Learning to play positions is a critical step in socialization, for it requires a higher level of role-taking than children can do at the play stage.

expectations—what Mead ([1934] 1962) refers to as the **generalized other**. The child learns to internalize the expectations of society—the generalized other—over and above the expectations of any "significant others." Behavior comes to be governed by abstract rules ("no running outside of the baseline" or "no touching the soccer ball with your hands unless you are the goalie") rather than guidance from and emotional ties to a "significant other." Children become capable of moving into new social arenas such as school, organized sports, and (eventually) the workplace to function with others in both routine and novel interactions. Individuals are active in shaping their social contexts, the self, and the choices they make about the future.

As we grow, we identify with new in-groups such as a neighborhood, a college sorority, or the military. We learn new ideas and expand our understanding. Some individuals ultimately come to think of themselves as part of the global human community. Thus, for many individuals, the social world expands through socialization. However, some individuals never develop this expanded worldview, remaining narrowly confined and drawing lines between themselves and others who are different. Such narrow boundaries often result in prejudice against others.

Two major approaches to symbolic interaction developed at different universities: the Universities of Chicago and Iowa. Much of the discussion thus far has focused on the Chicago School of symbolic interactionism. That perspective emphasizes the role of the *I* and focuses on the active agency of individuals in their own development. The Iowa School of symbolic interactionism places more emphasis on the *Me*—on the role of others and the

external social environment in shaping us (Carrothers and Benson 2003).

The Iowa School stresses that our identities are linked to our environments: institutions, organizations, and nations. Because of that, we have a vested interest in the stability of the society and the survival of those organizations that mean a lot to us—whether it is the college where we are a student, the Greek house that we join, the faith community with which we affiliate, or the nation of which we are a citizen. We will voluntarily give up our resources—time, energy, money, or even our lives—to preserve our beliefs, way of life, institutions, or nation. Thus, the self and its connection with meso-level structures such as organizations to which we belong are linked (Kuhn 1964; Stryker 1980, 2000; Stryker and Stratham 1985; Turner 2003).

Thinking Sociologically

Who are you? Write down 15 or 20 roles or attributes that describe who you are. How many of these items are characteristics associated with the *Me*—nouns such as *son, mother, student,* or *employee*? Which of the items are traits or attributes—adjectives such as *shy, sensitive, lonely, selfish,* or *vulnerable*? How do you think each of these was learned or incorporated into your conception of your *self*?

Socialization Throughout the Life Cycle

Markers in many societies point to rites of passage that mark movement from one stage to the next in the socialization process: birth; naming ceremonies or christenings; starting school at age five or six; rites of passage to mark puberty; obtaining a driver's license; becoming eligible for military draft; being able to vote and drink alcohol; retirement. Most social scientists emphasize the importance of *rites of passage*—celebrations or public recognitions when individuals shift from one status to another. The importance of this shift resides in how others come to perceive the individual differently, the different expectations that others hold for the person, and changes in how the person sees himself or herself.

Socialization is a lifelong process with many small and large passages. Infants begin the socialization process at birth. In childhood, one rite of passage is a child's first day at school—entrance into a meso-level institution. This turning point marks a child's entry into the larger world. The standards of performance are now defined by the child's teachers, peers, friends, and others outside the home. Adolescence is an important stage in Western industrial and postindustrial societies, but this stage is far from universal. Indeed, it is largely an invention of complex societies over the past two centuries, characterized by extensive periods of formal education and dependency on parents (Papalia, Olds, and Feldman 2006). Adolescence is, in a sense, a structurally produced mass identity crisis because Western societies lack clear rites of passage for adolescents. Teens come to view themselves as a separate and distinct group with their own culture, slang vocabulary, clothing styles, and opinions about appropriate sexual behavior and forms of recreation.

Most of our adult years are spent in work and home life, including marriage and parenting roles. It is not surprising, then, that graduation from one's final alma mater (whether it be high school, college, or graduate school), marriage, and acceptance of one's first full-time job are rites of passage into adulthood in modern societies. Even the retired and elderly members of society are constantly undergoing socialization and resocialization in the process of developing their sense of self. The type of society influences the socialization experience of the elderly and how they carry out their roles, as well as their status in society. Consider the changes that have taken place in the lifetimes of those born before 1945. They were born "before television, before polio shots, frozen foods, Xerox, plastic, contact lenses, Frisbees and the Pill . . . before credit cards, split atoms, laser beams and ball point pens; before pantyhose, dishwashers, clothes dryers, electric blankets, air conditioners . . . ," and many other familiar "necessities" of today (Grandpa Junior 2006). There are increasing numbers of people in the elder category. The average life expectancy in 1929 was only 57 years. Today, it is 78.1 years, with 50 countries having higher average life expectancies than the United States (Landau 2009).

Thinking Sociologically

What are the rites of passage from adolescence into adulthood in your country? Does each stage carry clear roles and responsibilities? If not, does this have consequences that create problems? Find someone who has grown up in a different culture and ask her or him about rites of passage from adolescence to adulthood. How are the patterns similar to or different from your own?

The elderly are vitally important to the ongoing group in more settled agricultural societies. They are the founts of wisdom and carry group knowledge, experiences, and traditions that are valued in societies where little change

Children learn many things in school, but one of the first is to master rules such as standing in lines.

takes place. In industrial and postindustrial countries, the number of elderly is growing rapidly as medical science keeps people alive longer, diets improve, and diseases are brought under control. Yet, in modern systems, social participation by the elderly often drops after retirement. Retirement is a rite of passage to a new status, like that of marriage or parenthood, for which there is little preparation. As a result, retired people sometimes feel a sense of uselessness when they abruptly lose their occupational status. Yet, retirees in Western societies generally have 20 or more years of life yet to live. Many retirees develop hobbies, enjoy sports, or have new jobs they can pursue.

Dying is the final stage of life (Kübler-Ross 1997). Death holds different meanings in different cultures: passing into another life, a time of judgment, a waiting for rebirth, or a void and nothingness. In some religious groups, people work hard or do good deeds because they believe they will be rewarded in an afterlife or with rebirth to a better status in the next life on earth. Thus, beliefs about the meaning of death can affect how people live their lives and how they cope with dying and death.

Each stage of the life cycle involves socialization into new roles in the social world. Many social scientists have studied these developmental stages and contributed insights

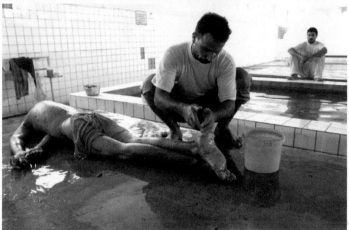

Death rituals differ depending on the culture and religion of the group. In India (top), this body is being cremated by the holy Ganges River to release the soul from earthly existence. The closest relative lights the funeral pyre. The middle photo shows the Muslim tradition of washing and wrapping the dead before burial in Najaf, Iraq. At the bottom, a U.S. Honor Guard carries a casket with the remains of U.S. Air Force personnel at Arlington National Cemetery.

into what happens at each stage (Clausen 1986; Erikson 1950; Freud [1923] 1960; Gilligan 1982; Handel et al. 2007; Kohlberg 1971; Papalia et al. 2006; Piaget 1989). Although examination of this topic is beyond the scope of this chapter, it is important to know that developmental theorists have detailed stages in the growth process.

Death ends the lifelong process of socialization, a process of learning social rules and roles and adjusting to them. When the individual is gone, society continues. New members are born, are socialized into the social world, pass through roles once held by others, and eventually give up those roles to younger members. Cultures provide guidelines for each new generation to follow, and except for the changes each generation brings to the society, the social world perpetuates itself and outlives the individuals who populate it.

The Process of Resocialization

If you have experienced life in the military, a boarding school, a convent, a mental facility, or a prison or had a major transition in your life such as a divorce or the death of a spouse or child, you have experienced resocialization. **Resocialization** is the process of shedding one or more positions and taking on others. It involves changing from established patterns learned earlier in life to new ones suitable to the newly acquired status (Goffman 1961). Resocialization may take place in a **total institution** in which a group of people is bureaucratically processed, physically isolated from the outside world, and scheduled for all activities. These include prisons, mental hospitals, monasteries, concentration camps, boarding schools, and military barracks. Bureaucratic regimentation and the manipulation of residents for the convenience of the staff are part of the routine (Goffman 1961).

We often associate resocialization with major changes in adult life—divorce, retirement, and widowhood. One must adjust to raising children alone, living alone, loneliness, and possible financial problems. One divorcée of three years told the authors: "There are many things to commend the single life, but I still have not adjusted to eating alone and cooking for myself. But worse than that are Sunday afternoons. That is the loneliest time."

Sometimes resocialization describes individuals' attempts to adjust to new statuses and roles, such as widowhood. In other cases, individuals are forced into resocialization to correct or reform behaviors that are defined as undesirable or deviant. Prison rehabilitation programs provide one example. However, research suggests the difficulty in resocializing prisoners is rooted in the nature of the prison environment itself. Prisons are often coercive and violent environments, which may not provide the social supports necessary for bringing about change in a person's attitudes and behaviors.

Although resocialization is the goal of self-help groups such as Alcoholics Anonymous, Gamblers Anonymous, Parents Anonymous, drug rehabilitation groups, and weight-loss groups, relapse is a common problem among participants. These groups aim to substitute new behaviors and norms for old undesirable ones, but the process of undoing socialization and achieving resocialization is difficult.

There are multiple individuals, groups, and institutions involved in the socialization process. These socialization forces are referred to as agents of socialization.

Agents of Socialization: The Micro-Meso Connection

Agents of socialization are the transmitters of culture— the people, organizations, and institutions that teach us how to thrive in our social world. Agents are the mechanism by which the self learns the values, beliefs, and behaviors of the culture. Agents of socialization help new members find their place, just as they prepare older members for new responsibilities in society. At the micro level, one's family, one's peer group, and local groups and organizations help people know what is expected of them. At the meso level, formal sources of learning— education, religion, politics, economics, and health—and other informal sources of learning such as the media and books are all agents that contribute to socialization. They transmit information to children and to adults throughout people's lives.

Thinking Sociologically

As you read this section, make a list of the socializing agents discussed in these pages. Indicate two or three central messages each agent of socialization tries to instill in people. Consider which agents are micro, meso, or macro agents. Are there different kinds of messages at each level? Do any of them conflict? If so, why, and what problems are caused?

In early childhood, the family acts as the primary agent of socialization, passing on messages about respect for property, authority, and neatness, for example (Handel et al. 2007). Peer groups are also important, especially during the teenage years. Some writers even argue that the peer group is most important in the socialization process

The primary socialization unit for young children is the family, but as they become teenagers, peers become increasingly important as a reference group, shaping their norms, values, and attitudes.

of children and teens (Aseltine 1995; Harris 2009). Each agent has its own functions or purposes and is important at different stages of the life cycle, but meso-level institutions play a more active role as one matures. For example, schools and religious bodies become more involved in socialization as children become six years old compared to when they were preschool age. The "Sociology in Your Social World" on page 100 discusses socialization in schools by exploring an important and widely cited research project—the issue of how schools reinforce notions of gender.

Lessons from one agent of socialization generally complement those of other agents. Parents work at home to support what school and religion teach. However, at times agents provide conflicting lessons. For example, family and faith communities often give teens messages that conflict with those of peer groups regarding sexual activity and drug use. This is an instance of mixed messages given by formal and informal agents.

For **formal agents**, socialization is the stated goal. Formal agents usually have some official or legal responsibility for instructing individuals. A primary goal of families is to teach children to speak and to learn proper behavior. In addition, schoolteachers educate by giving formal instruction, and religious training provides moral instruction. (These formal agents of socialization will be discussed in Chapters 10, 11, and 12.)

Informal agents do not have the express purpose of socialization, but they function as unofficial forces that shape values, beliefs, and behaviors. For example, the media, books, the Internet, and advertisements bring us

Sociology in Your Social World

Gender Socialization in American Public Schools

Pause for a moment as you pass a school yard, and observe the children at play. Children's behavior on the school playground translates into a powerful agent of gender socialization in a world that is very complex. Consider the evidence reported in the ethnographic study of Barrie Thorne, recounted in her award-winning book, *Gender Play* (1993).

Many people assume that gender differences are natural and that we are "born that way." By contrast, Thorne provides evidence that gender differences are social constructions, influenced by the setting, the players involved in the situation, and the control people have over the situation.

As an astute observer and researcher, Thorne suspected that girls and boys have complex relations that play out in the classroom and school yard. She chose the playground as the focus of her observation of the separate worlds of girls and boys. As her research strategy, she points out that "when adults seek to learn about and from children, the challenge is to take the closely familiar and to render it strange" (Thorne 1993:12). She is suggesting that we need to look at well-known everyday patterns from a fresh point of view.

Through systematic participant observation she found that children and adults play an active role in defining and shaping gender expectations through the collective practices of forming lines, choosing seats, teasing, gossiping, and participating in selected activities.

Thorne used two schools for her research and entered the world of the children, sometimes sitting apart on the playground taking notes, sometimes participating in their activities such as eating and talking in the lunchroom.

In each setting, she recorded her observations and experiences. For example, she noted what children call themselves and how they think of themselves.

She was intrigued by the reference to the *opposite sex*—a term that stresses difference and opposition rather than similarity and a sense of "we." Thorne was struck by the active meaning construction of children as they gained a notion of "normal" gender behavior. The real focus of her work is in taking seriously how children themselves make sense of sex differences—and how they sometimes ignore any difference as irrelevant to their activities.

Previous studies concluded that boys tend to interact in larger, more age-heterogeneous groups and in more rough-and-tumble play and physical fighting. Thorne also found that boys' play involves a much larger portion of the playground, and their play space was generally farther from the building, making them less subject to monitoring and sanctioning. In addition, boys would often run "sneak invasions" into the girls' space to take things belonging to the girls. Many boys felt they had a right to the geographical space that was occupied by females. Girls played close to the buildings in much smaller areas and rarely ventured into the boys' area.

Girls' play tended to be characterized by cooperation and turn-taking. They had more intense and exclusive friendships, which took shape around keeping and telling secrets, shifting alliances, and indirect ways of expressing disagreement. Instead of direct commands, girls more often used words like *let's* or *we gotta*. However, Thorne found that these notions of "separate girls' and boys' worlds" used in most previous studies miss the subtleties in the situation—race, class, and other factors.

Thorne's work alerts us to the complexity of the gender socialization process, helping us see the extent to which children are active agents who are creating their own definitions of social relations, not just short automatons who enact adult notions of what gender means.

continuous messages even though their primary purpose is not socialization but entertainment or selling products. Children watch countless advertisements on television, many with messages about what is good and fun to eat and how to be more attractive, more appealing, smarter, and a better person through the consumption of products. This bombardment is a particularly influential part of socialization at young ages.

Thinking Sociologically

Recall your own playground days or watch a sibling or child on the playground. What do your observations tell you about the role of play in gender socialization?

This distinction between formal, intentional socialization and informal socialization has important implications for the kinds of messages that are presented and for how such messages are received.

Micro-Level Socialization: Families as Formal Agents

One way in which families teach children what is right and wrong is through rewards and punishments, called *sanctions*. Children who steal cookies from the cookie jar may receive a verbal reprimand or a slap on the hand, be sent to their rooms, have "time out," or receive a beating, depending on differences in child-rearing practices. These are examples of negative sanctions. Conversely, children may be rewarded for good behavior with a smile, praise, a cookie, or a special event. These are examples of positive sanctions. The number and types of sanctions dispensed in the family shape the socialization process, including development of the self and the perceptions we have of who we are. Note that family influence varies from one culture to another.

In Japan, the mother is the key agent in the process of turning a newborn into a member of the group, passing on the strong group standards and expectations of family, neighbors, community, and society through the use of language with emotional meaning (Hendry 1987; Holloway 2001). The interaction of family and formal education in Japan is explored in more detail in the "Sociology Around the World" on page 108.

Thinking Sociologically

First, read the essay on the next page. How did agents of socialization influence who you are today, and how did your experience differ from that of Japanese children described in the essay?

In the United States, most parents value friendliness, cooperation, orientation toward achievement, social competence, responsibility, and independence as qualities their children should learn, in contrast to values of conformity and fitting into the group espoused in Japan. However, subcultural values and socialization practices may differ within the diverse groups in the U.S. population. Conceptions of what makes a "good person" or a "good citizen" and different goals of socialization bring about differences in the process of socialization around the world.

In addition, the number of children in a family and the placement of each child in the family structure can influence the unique socialization experience of the child. In large families, parents typically have less time with

Japanese fathers and their sons eat lunch during a festival. While mothers are the key agents of socialization in Japan, fathers also have a role, especially during special events in the life of the child.

each additional child. Where the child falls in the hierarchy of siblings can also influence the development of the self. In fact, birth order is a very strong predictor of social attitudes—perhaps more so than race, class, or gender—according to some studies (Benokraitis 2008; Freese et al. 1999), and firstborns are typically the highest achievers ("First Born Children" 2008; Paulhus, Trapnell, and Chen 1999). Younger children may be socialized by older siblings as much as by parents, and older siblings often serve as models that younger children want to emulate.

Meso-Level Socialization: Social Class

Our educational level, our occupation, the house we live in, what we choose to do in our leisure time, the foods we eat, and what we believe in terms of religion and politics are just a few aspects of our lives that are affected by socialization. Applying what we know from sociological research, the evidence strongly suggests that socialization varies by **social class**, or the wealth, power, and prestige rankings that individuals hold in society (Ellison, Bartkowski, and Segal 1996; Pearce 2009). Meso-level patterns of distribution of resources affect who we become. For example, upper-middle-class and middle-class parents in the United States usually have above-average education and managerial or professional jobs. They tend to pass on to their children skills and values necessary to succeed in this social class subculture. Autonomy, creativity, self-direction (the ability to make decisions and take initiative), responsibility, curiosity, and consideration of others are especially important for

Sociology Around the World

Socialization in Japan: The Family and Early Schooling

By Wendy Ng

Each of our families prepares us through the socialization process for the culture we are entering. How families carry out this process differs around the world, just as the cultures for which they are preparing their children differ. Here we consider meso-level family and early schooling in Japanese socialization.

The family is one of the most important socializing influences in Japan. The basis of the family unit in Japan is called the *ie* (pronounced ee-ay). Traditionally, it is made up of blood relatives who reside in the same household, as well as ancestors and descendents not yet born. Thus, family in Japan goes beyond those who belong to the immediate nuclear grouping and includes a broader array of individuals. Compared with the past, the modern *ie* in Japan relies more on the nuclear and living extended family and serves as the major reference group that socializes individuals within the family. Thus, family members within the *ie* are responsible for teaching individuals their family roles, values, and norms within the culture.

A unique feature of interdependence that is found within the family structure is that of *amae* (ah-may), which roughly translated means "passive love" but is often referred to as an emotional bond usually held between mother and child. Through this relationship, children are socialized to understand that they are an important part of the family, and they also learn that parents are to be respected and obeyed as the adults within the family. Although this appears hierarchical, the emotional bond of *amae* sets up a relationship of interdependency between child and parent for their lifetime. As children grow into adulthood, they will take care of their parents in the way that they were taken care of as children. This bond of loyalty between parent and child within the family structure is translated into other social structures outside the family. For example, in a business organization, there is a similar expectation of group loyalty.

In terms of early childhood socialization, Japanese children learn the distinction between two related yet distinct concepts: *uchi* and *soto*. Roughly translated these mean "inside" and "outside" and can apply to material distinctions of clean and unclean spaces. In behavior, this means taking one's shoes off outside the house because the inside is clean and the outside is unclean. In Japanese households, the bathroom has similar clean and unclean designations. The bath is "clean," and the toilet, used to dispose of bodily wastes, is "unclean."

Thus, one would never wear the same shoes or slippers in the toilet room as the bathroom because that would be mixing clean and unclean elements.

Within the family, immediate members are "insiders," and other people are "outsiders." Children learn that the family is a safe and secure environment where the emphasis is on harmony among the various family members. Interactions between individuals stress cooperation, and interpersonal disputes are avoided. If a disagreement happens, children are taught to apologize to one another. Reciprocity is yet another behavior that is emphasized within the family. Children are taught to put themselves in the role of the other person and to think of the consequences of their behavior before acting out. This type of role behavior suggests that harmony between and among family members is important and sets the foundation for the child's educational socialization.

Whereas the family serves as the central socializing force when children are very young, as they grow, the educational system continues to socialize children through group interaction and learning. When children enter kindergarten they become familiar with participating in a social group with peers. The Japanese kindergarten system emphasizes group equality among children and thus socializes children to be loyal to their classmates and group. Other children now form their new *uchi* or "inside" associations and friendships. The emphasis on group over individual identities is accomplished through wearing identical uniforms or smocks, having similar educational tools for all students, and having children take turns in different duties in the classroom. For example, the responsibility of passing out paper in the classroom or food at lunchtime is rotated among all the children in the classroom. Thus, cooperation and group participation become an important defining feature of kindergarten socialization.

At first glance, the emphasis on equality among individuals in the kindergarten classroom setting might seem to conflict with the emphasis on hierarchical authority present in much of Japanese society. In fact, the emphasis on group socializing helps encourage a sense of belonging and group identity that works well within hierarchical authority structures. By learning these behaviors at a young age, the children learn that they are individuals within a larger group and that their actions reflect not only on themselves but also on their family, their school, or whatever social group they belong to as adults.

middle-class success (Kohn 1989). If the child misbehaves, for example, middle-class parents typically analyze the child's reasons for misbehaving, and punishment is related to these reasons. Sanctions often involve instilling guilt and denying privileges.

Working-class parents tend to pass on to children their cultural values of respect for authority and conformity to rules, lessons that will be useful if the children also have blue-collar jobs (Kohn 1989). Immediate punishment with no questions asked if a rule is violated functions to prepare children for positions in which obedience to rules is important to success. They are expected to be neat, clean, well-mannered, honest, and obedient students (MacLeod 2008). Socialization experiences for boys and girls are often different, following traditional gender-role expectations. Moreover, these differences in behavior across social classes and parenting styles are apparent cross-culturally as well (Leung, Lau, and Lam 1998).

What conclusions can we draw from these studies? Members of each class are socializing their children to be successful in their social class and to meet expectations for adults of that class. Schools, like families, participate in this process. Although the extent to which schools create or limit opportunities for class mobility is debated, what is clear is that children's social class position on entering school has an effect on the socialization experiences they have in school (Ballantine and Hammack 2012). Families and schools socialize children to adapt to the settings in which they grow up and are likely to live.

Social class, however, is only one of many influencing agents. As we saw in *Black Men and Public Space* on page 100,

This father passes on a love for the piano to his young son. Because of the social class of this father, his son is likely to receive many messages about creativity, curiosity, and self-direction.

race and ethnicity are very important factors in socialization, and the influence of gender is enormous. Gender socialization will be discussed in more detail in Chapter 9, but we note here that race, class, and gender act as structural constraints on some members of the population. People who are not privileged receive different messages about who they are and how they should behave. Some of us receive messages about being privileged in certain respects and perceive constraining messages in others, and that is why it is important to recognize the interplay of the variables in people's lives.

Informal Agents of Socialization: Electronic Media

Television and computers are important informal agents of socialization. In developed countries, there is scarcely a home without a television set, and over 75% of homes have computers and Internet access (Web Site Optimization 2010).

Researchers have collected nearly five decades of information on how television has become a way of life in homes. By the time an average child in the United States reaches age 18, he or she will have spent more time watching television than participating in any other single activity besides sleeping. On average, children between ages 8 and 18 spend three hours a day watching television, one hour and 11 minutes watching videos or DVDs, one hour and 44 minutes with audio media, one hour using computers, and 49 minutes playing video games, with a total media exposure in a typical day of eight hours and 33 minutes. The "Engaging Sociology" feature on the next page shows total media exposure of children by several variables (Rideout, Foehr, and Roberts 2010). Examine this issue in more depth by answering the questions following the table.

The moguls of mass media—a meso-level social system—are able to influence socialization within the most intimate of environments. "Children [in the United States] use computers at very young ages—21% of children 2 years and younger, 58% of 3- to 4-year-olds, and 77% of 5- to 6-year-olds" (National Science Foundation 2005:1). Seventy-four percent of the U.S. population now uses the Internet (Internet World Statistics 2009).

A serious concern related to socialization centers on the messages children receive from television and computer games, along with the behavioral effects of these messages. There is ample evidence that children are affected in negative ways by excessive television viewing, especially television violence (National Science Foundation 2005), but a direct causal link between television viewing and behavior is difficult to establish. Researchers know, however, that parents who play an active role in helping children understand the content of television shows can have a powerful effect on mitigating television's negative impacts and enhancing the positive aspects of television shows. The television-viewing habits of parents—length of viewing time, types of shows watched, times of day—can also influence how their children respond to television.

Engaging Sociology

Media Exposure and Socialization

Examine Tables 4.1 and 4.2 and respond to the questions below.

Table 4.1 Total Media Exposure (Average Hours per Day)

Age	8 to 10 years old	7:51
	11 to 14 years old	11:53
	15 to 18 years old	11:23
Gender	Boys	11:12
	Girls	10:17
Race	White	8:36
	Black	12:59
	Hispanic	13:00
Parent education	High school or less	11:26
	Some college	11:30
	College graduate	10:00

Table 4.2 Average Amount of Time per Day Spent With Each Medium: 8- to 18-Year-Olds

Medium	1999	2009
Television	3:47	4:29
Music/audio	1:48	2:31
Computer	:27	1:29
Video games	:26	1:13
Print	:43	:38
Movies	:18	:25
Total media exposure	7:29	10:45
Multitasking proportion	16%	29%

Sociological Data Analysis:

1. Considering the data in Tables 4.1 and 4.2, how would you describe television-watching and other media-engaged patterns among different groups?

2. Are the trends in media exposure over the first decade of the twenty-first century a matter of concern? Why or why not?

3. How might media engagement affect other aspects of socialization of children?

4. What might be the social consequences of ethnic minorities (blacks and Hispanics) and those children whose parents do not have a college education having so much higher media exposure each day than whites and the more highly educated?

5. Do your conclusions cause any concerns about your society? Why or why not?

Source: Rideout, Foehr, and Roberts (2010).

Perhaps the most important aspect of television and computers is something that we do not fully understand but that has frightening potential. For the first time in human history, we have powerful agents of socialization in the home from a child's birth onward. Time spent attending to television or computer games means less time spent engaging in interaction with caregivers and peers. Intimate family bonds formed of affection and meaningful interaction are being altered by the dominant presence of electronic media in the home. In addition, those who control the flood of mass media messages received by children may have interests and concerns that are very much at odds with those of parents. A significant part of the informal socialization process occurs with the assistance of electronic equipment within the home that commands a significant portion of a child's time and attention.

With globalization, global knowledge and understanding also become important parts of school curricula and media coverage, and we move next to a discussion of some of the national and global processes that impact socialization.

Thinking Sociologically

What other agents of socialization in addition to family, social class, and electronic media are important in teaching us our roles, norms, values, and beliefs? What is the impact, for example, of friendship networks or peer groups?

Socialization and Macro-Level Issues

Sense of Self Versus the "Other": Diverse Global Societies

Immigration patterns and ethnic conflicts around the world have resulted in a fairly new phenomenon: transnationalism. **Transnationalism** is the process by which immigrants create multinational social relations that link together their original societies with their new locations. This means that an individual or a family has national loyalty to more than one country (Levitt 2001; Levitt and Waters 2006). Often, it occurs after migration of war refugees, when one's roots lie in the country of origin and many of one's close family members continue to live there. Consider children raised in war-torn countries. In the Palestinian territories, especially Gaza, and in Israeli settlements along the border, children grow up with fear

and hatred, major influences on their socialization. Some war refugees spend childhoods in refugee camps and may never return to their native countries.

For people experiencing transnationalism, there are conflicting messages about culturally appropriate behaviors and the obligations of loyalty to family and nation. However, one need not migrate to another country to experience global pressures. The Internet and cell phones have increasingly created a sense of connectedness to other parts of the world and an awareness of global interdependencies (Brier 2004; Roach 2004). Some commentators have even suggested that the Internet is a threat to the nation-state as it allows individuals to maintain traditions and loyalties to relatives and friends in more than one country (Drori 2006). Ideas of social justice or progress in many parts of the world are shaped not just by the government that rules the country but by international human rights organizations and ideas that are obtained from media that cross borders, such as the World Wide Web. In recent uprisings in some Middle Eastern and North African countries, social networking kept movement participants in touch with others in the uprisings and with outside media and supporters.

Access to international information and friendships across borders and boundaries are increasingly possible as more people have access to the Internet. Map 4.1, on Internet use around the world, illustrates variability of access but also how widespread this access is becoming. One interesting question is how access or lack of access will influence the strength of "we" versus "they" feelings.

At a time when people lived in isolated rural communities and did not interact with those unlike themselves, there was little price to pay for being bigoted or chauvinistic toward those who were different. However, we now live in a global village where we or our businesses will likely interact with very different people in a competitive environment. If we hold people in low regard because they are unlike us or because we think they are destined for hell because of their spiritual beliefs, there may be a high cost for this alienation toward those who are not like us. Among other problems, terrorism is fermented where people feel alienated. Therefore, diversity training and cultural sensitivity to those "others" have become an economic and political issue.

The reality is that children in the twenty-first century are being socialized to live in a globalized world. Increasingly, children around the world are learning multiple languages to enhance their ability to communicate with others. Some college campuses are requiring experiences abroad as part of the standard curriculum because faculty members and administrators feel that a global perspective is essential in our world today and part of a college education. Socialization to global sensitivity and tolerance of those who were once considered "alien" has become a core element of our day (Robertson 1992; Schaeffer 2003; Snarr and Snarr 2008).

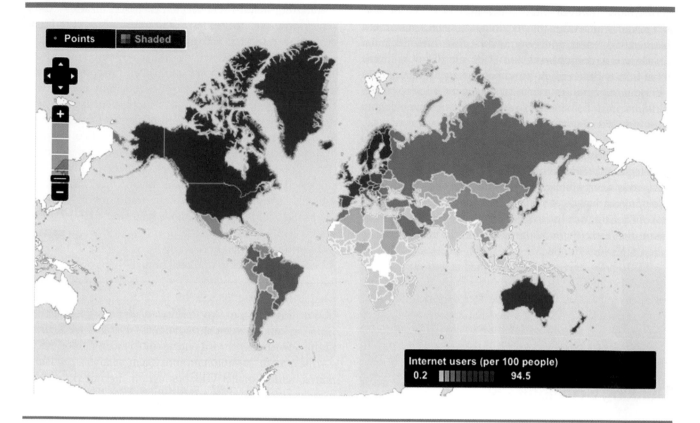

Map 4.1   Internet Users per 100 People in 2010

Source: World Bank 2011.

Sometimes, global events can cause a different turn: away from tolerance and toward defensive isolation. When 19 young men from Saudi Arabia and other Middle Eastern countries crashed planes into the World Trade Center in New York City and into the Pentagon in Washington, DC, the United States was shocked and became mobilized to defend itself and its borders. The messages within schools and from the government suddenly took a more patriotic turn. Then-U.S. Secretary of Defense Donald Rumsfeld stated that the problem with America is that citizens do not have a strong enough sense of "we" versus "they." So this event and other terrorist acts, clearly tragedies rooted in global political conflicts, can intensify boundaries between people and loyalty to the nation-state. Global forces are themselves complex and do not always result in more tolerance.

Indeed, the only thing we can predict with considerable certainty is that in this age of sharing a small planet, the socialization of our citizens will be influenced by events at the macro level, whether national or global.

Policy and Practice

Should preschoolers living in poverty be socialized in day care settings? Should adolescents work while going to school? Should new parents be required to take child-rearing classes? How should job-training programs be structured? How can communities use the talents and knowledge of retirees? Can the death process be made easier for the dying person and the family? Should we place emphasis in high school and college on in-group loyalty and patriotism or on developing a sense of global citizenship mobilized around common human issues? These are all policy questions—issues of how to establish governing principles that will enhance our common life.

These policy questions rely on an understanding of socialization—how we learn our beliefs and our positions in society. For example, making decisions about how to provide positive early childhood education experiences at a time when young children are learning the ways of their culture relies on understanding the socialization they receive at home and at school. The quality of child care we provide for young children will affect not only how effective our future workforce is but also whether children turn out to be productive citizens or a drain on society.

Some sociologists do research to try to help policy makers have accurate data and make good interpretations of the data so they can make wise decisions. Others are more activist, working in the field as applied sociologists trying to solve the problems through private foundations, consulting firms, or state agencies.

Now that we have some understanding of the process of socialization, we look next at the process of interaction and how individuals become members of small groups, networks, and large complex organizations.

What Have We Learned?

Human beings are not born to be noble savages or depraved beasts. As a species, we are remarkable in how many aspects of our lives are shaped by learning—by socialization. Human socialization is pervasive, extensive, and lifelong. We cannot understand what it means to be human without comprehending the impact of a specific culture on us, the influence of our close associates, and the complex interplay of pressures from micro, meso, and macro levels. Indeed, without social interaction, there would not even be a self. We humans are, in our most essential natures, social beings. The purpose of this chapter has been to open our eyes to the ways in which we become the individuals we are.

Key Points:

- Human beings must work with their biological makeup, but most of what makes us uniquely human has to do with things we learn from our culture and society—our socialization. Humans who are not socialized and live in isolation from others are tragic—barely human—creatures. (See pp. 94–97.)

- The self consists of the interaction of the *I*—the basic impulsive human with drives, needs, and feelings—and the *Me*—the reflected self one develops by role-taking to see how others might see one. (See pp. 99–102.)

- The self is profoundly shaped by others, but it also has agency—it is an initiator of action and the maker of meaning. (See pp. 99–102.)

- The self develops through stages, from mimicking others (the play stage) to more intellectually sophisticated abilities to role-take and to see how various roles fit together—the play and game stages. (See pp. 99–102.)

- Although the self is somewhat elastic in adjusting to different settings and circumstances, there is also a self that is often vested in meso-level organizations and institutions in which the self participates. (See pp. 107–109.)

- The self is modified as it moves through life stages, and some of those stages require major resocialization—shedding old roles and taking on new ones as one enters new statuses in life. (See pp. 104–105.)

- A number of agents of socialization are at work on each of us, communicating messages that are relevant at the micro, the meso, or the macro level of social life. At the meso level, for example, we may receive different messages about what it means to be a "good" person depending on our ethnic, religious, or social class subculture. (See pp. 105–106.)

- Some of these messages may be in conflict with each other, as when global messages about tolerance for those who are different conflict with a nation's desire to have absolute loyalty and a sense of superiority. (See pp. 111–112.)

Contributing to Our Social World: What Can We Do?

At the Local Level:

In every community, numerous opportunities exist for volunteer work in helping children from economically and otherwise disadvantaged backgrounds succeed in school.

- Tutor or mentor in the local schools. Contact an education faculty member for information.
- Volunteer in Head Start centers for poor preschool children. See the National Head Start Association website at www.nhsa.org.
- Help in a local boys' and girls' club that provides socialization experiences for children through their teens.
- Volunteer in care facilities and hospices for people who are ill or dying to help reduce loneliness and provide positive interaction. Check these opportunities for Academic Service Learning (ASL) credit in which course assignments include such community work under the supervision of the instructor. Find out about ASL programs on your campus.

At the National and Global Levels:

Literacy, one of the most vital components of socialization, remains an unmet need in many parts of the world, especially in the less-developed countries of Africa and Asia.

- World Education. Visit www.worlded.org to learn about this group's wide variety of projects and volunteer/work opportunities.
- CARE International and Save the Children provide funding for families to send children to school and to receive specialized training. Opportunities exist for fund-raising, internships, or eventually jobs with these organizations.

Visit **www.sagepub.com/oswcondensed2e** for online activities, sample tests, and other helpful information. Select "Chapter 4: Socialization" for chapter-specific activities.

CHAPTER 5

Interaction, Groups, and Organizations

Connections That Work

Human interaction results in connections—networks—that work to make life more fulfilling and that make our economic efforts more productive. These connections are critical in our social world—from small micro groups to large bureaucratic organizations.

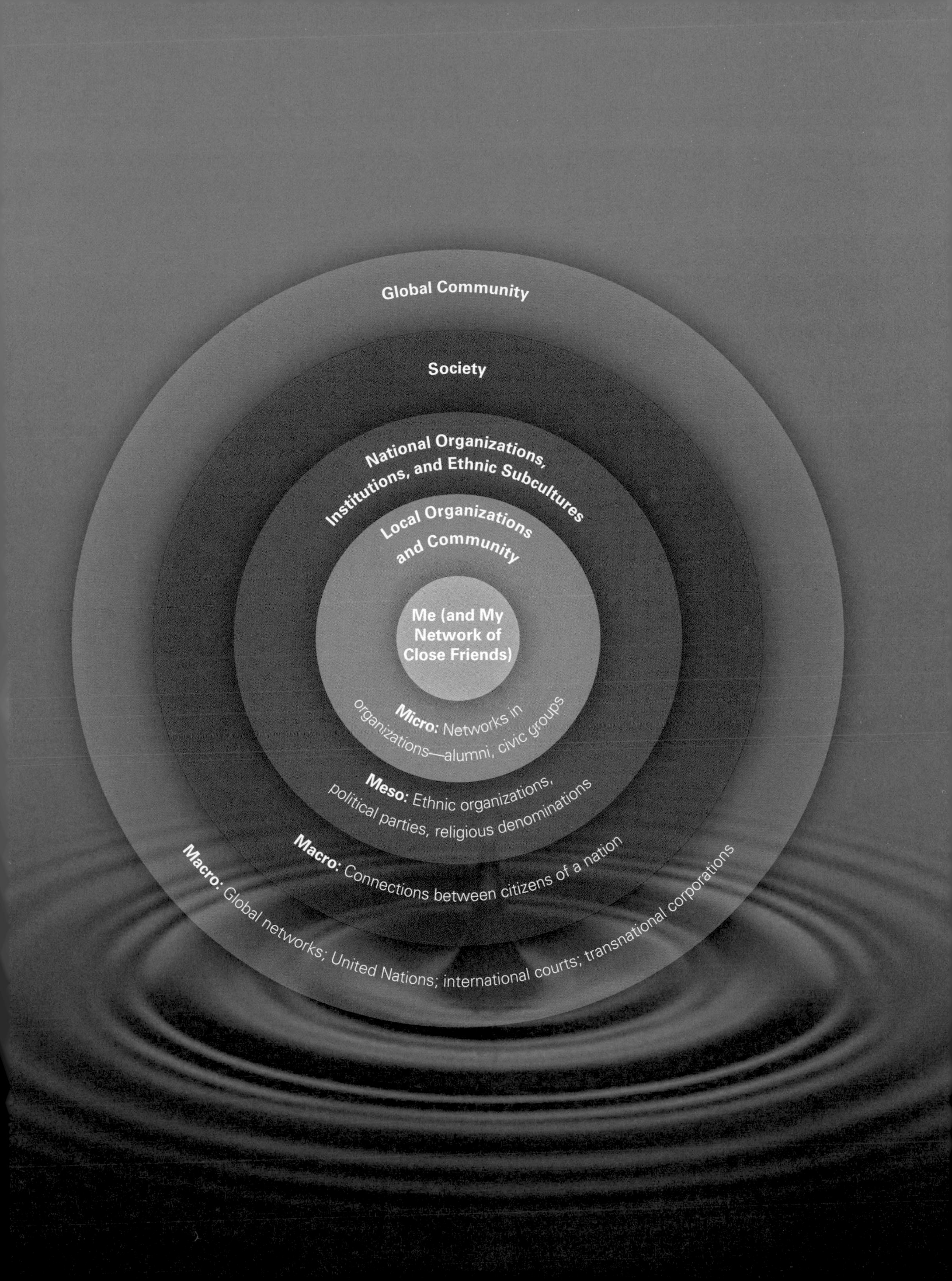

Global Community

Society

National Organizations,
Institutions, and Ethnic Subcultures

Local Organizations
and Community

Me (and My
Network of
Close Friends)

Micro: Networks in
organizations—alumni, civic groups

Meso: Ethnic organizations,
political parties, religious denominations

Macro: Connections between citizens of a nation

Macro: Global networks; United Nations; international courts; transnational corporations

Think About It	
Self and Inner Circle	Are you likely to meet your perfect mate over the Internet?
Local Community	How does interaction with others affect who you are and what you believe?
National Institutions; Complex Organizations; Ethnic Groups	What purpose does bureaucracy serve, and what are the alternatives?
National Society	How do national trends—such as the spread of fast-food chains and "box stores"—influence your quality of life?
Global Community	How do global networks across the globe affect you, your education, and your work?

Peaceful demonstrators, gathered to contest the 2009 election in Iran, were confronted with massive police forces. They were driven to disperse, many were beaten and arrested, and some were killed. Although the government imposed a news blackout and crackdown on communications, demonstrators used their cell phones, blogs, Twitter, and Facebook to send pictures and video footage around the world, documenting the events. Some have referred to this as the "Twitter Revolution." This cyber inspiration has spread to protesters in other countries such as Tunisia, Yemen, Egypt, Oman, Bahrain, and Libya. Cyberspace links people around the world in seconds in ways that few governments can stop (Stone and Cohen 2009).

No longer are paper and pencil the medium of academics. Most universities are now requiring students to have laptops or access to computers on campus. The colleges now provide Wi-Fi Internet access, thus expanding modes of teaching, learning, and communicating.

There is no need to wait for the mail or even talk on the phone. The information superhighway is opening new communication routes and allowing individuals with common interests to engage in networking. Yet only a few decades ago, we read about cyberspace in science fiction novels written by authors with a little science background and a lot of imagination. Indeed, the word *cyberspace* was coined in 1984 by science fiction writer William Gibson (Brasher 2004).

The implications of the rapidly expanding links in cyberspace are staggering. We cannot even anticipate some of them because change is so rapid. For entertainment, we can talk with friends on electronic mailing lists or with people we have "met" through cyber social groups. Some of these acquaintances have never left their own country, which is on the other side of the planet. All of this takes place in the comfort of our homes. Face-to-face communication may never occur except on webcams attached to our computers.

Universities now communicate with students and employees by computer. You may be able to register for class by "talking" to the computer. Computers track your registration and grades, and they may even write you letters about your status. They also monitor employee productivity. For doing certain types of research, library books are becoming secondary to the World Wide Web.

This chapter continues the discussion of how individuals fit into the social world, exploring the link between the individual and the social structure. Socialization prepares individuals to be part of the social world, and individuals interact with others to form groups and organizations. Interacting face-to-face and belonging to groups and organizations are the primary focus in the following pages. First, we consider how networks and connections link individuals and groups to different levels of analysis. Then, we focus on micro-level interactions, meso-level groups, and meso- and macro-level organizations and bureaucracies. Finally, we consider macro-level national and global networks.

Networks and Connections in Our Social World

Try imagining yourself at the center of a web, as in a spider's web. Attach the threads that spread from the center first to family members and close friends, next to peers, and then to friends of friends. Some thread connections are close and direct. Others are more distant but connect more and more people in an ever-expanding web. Now imagine trying to send a letter to someone you do not know. A researcher actually tried this experiment to discover how people are networked and how far removed citizens are from one another within the United States.

Perhaps you have heard it said that every American is only six steps (or degrees removed) from any other person in the country. This assertion is rooted in a study with evidence to support it. Stanley Milgram and his associates (Korte and Milgram 1970; Milgram 1967; Travers and Milgram 1969) studied social networks by selecting several target people in different cities. Then they identified "starting persons" in cities more than 1,000 miles away. Each starting person was given a booklet with instructions and the target person's name, address, and occupation, as well as a few other facts. The starting person was instructed to pass the booklet to someone he or she knew on a first-name basis who lived closer to or might have had more direct networks with the target person than the starting person had. Although many packages never arrived at their destination, one third did. The researchers were interested in how many steps were involved in the delivery of the packages that did arrive. The number of links in the chain to complete delivery ranged from two to ten, with most having

five to seven intermediaries. This is the source of the reference to "six degrees of separation." Clearly, networks are powerful linkages and create a truly small world.

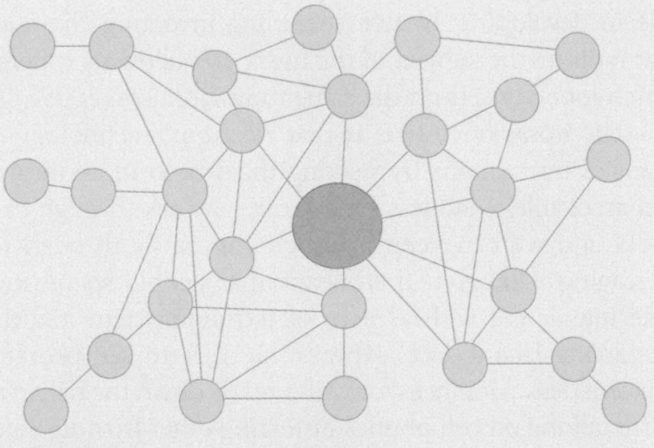

Our **social networks**, then, refer to individuals linked together by one or more social relationships connecting them to the larger society. We use our social network to get jobs or favors, often from people who are not very far removed from us in the web. Networks begin with micro-level contacts and exchanges between individuals in private interactions and expand to small groups, then to large (even global) organizations (Granovetter 2007; Tolbert and Hall 2009). The stronger one's networks, the more influential they can be in the person's life. However, types of networks differ. For example, women's networks for both getting jobs and promotions and succeeding in careers are broader (more extensive) than men's, but also weaker than men's in helping with promotions. This is because men's networks are more "instrumental"—that is, focused on the task at hand (Rothbard and Brett 2000). The web in Figure 5.1 illustrates that individuals are linked to other people, groups, organizations, and nations in the social world through networks.

Although network links can be casual and personal rather than based on official positions and channels, they place a person within the larger social structure, and it is from these networks that group ties emerge. People in networks talk to each other about common interests. This communication process creates linkages between clusters of people. For example, cyber networks on the Internet bring together people with common interests.

Networks at the Micro, Meso, and Macro Levels

At the most micro level, you develop close friends in college—bonds that may continue for the rest of your life. You introduce your friends from theater to your roommate's friends from the soccer team, and the network expands. These acquaintances from the soccer team may have useful information about which professors to avoid, how to make contacts to study abroad, and how to get a job in your field. Food cooperatives, self-help groups such as Alcoholics Anonymous and Weight Watchers, and computer-user groups are examples of networks that connect individuals with common interests (Powell 1990).

The golden circle represents an individual—perhaps you—and the blue dots represent your friends and acquaintances. Your network looks a bit like a web but is less complete in terms of every point connected to the adjacent point, since some of your friends do not know each other.

Figure 5.1   Networks: A Web of Connections

Thinking Sociologically

Map your social network web. What advantages do you get from your network? What economic benefits might your connections have for you? How have you tried to expand your network? What Internet sites help you network?

Network links create new types of organizational forms at the meso level, such as those in the opening example of cyberspace and the Internet. These networks cross societal, racial, ethnic, religious, and other lines that divide people. Networks also link groups at different levels of analysis. In fact, you are linked through networks to (a) local civic, sports, and religious organizations; (b) formal complex organizations such as a political party or national fraternity and ethnic or social class subcultures; (c) the nation of which you are a citizen and to which you have formal obligations (such as the requirement that you go to war as a draftee if the government so decides); and (d) global entities such as the United Nations, which use some of your taxes or donate resources to help impoverished people, tsunami victims, and earthquake survivors elsewhere in the world. These networks may open opportunities, but they also may create obligations that limit your freedom to make your own choices. As we move from the micro level to the larger meso and macro levels, interactions tend to become more formal. Formal organizations will be explored in the latter half of this chapter.

One of the most interesting developments at the beginning of the twenty-first century is the way Internet technology is influencing networking, linking individuals to people around the globe. Internet users have redefined networking through the creation of blogs, chat rooms, message boards, electronic mailing lists, newsgroups, and dozens of websites devoted to online networking ("Five Rules for Online Networking" 2005). Websites such as LinkedIn.com offer business and professional networking, whereas other sites such as Myspace.com, YouTube.com, and Friendster.com focus on personal and social networking. There is even an international journal, *Social Networks*, that publishes interdisciplinary studies about the structure and impact of social networks and sites.

Probably the most famous networking site is Facebook, and it became even more well known with the recent film about its founder, *The Social Network*. With members from all ages and places, Facebook truly does link the world. This social networking utility connects friends and friends of friends and people with common interests. There are more than 500 million active users around the world, and over 50% log on daily for 700 billion minutes per month. Users share blogs, news, web links, notes, photos, and more. Over 70% of users are outside of the home base, the United States, and Facebook is translated into 70 languages (Facebook.com 2011). Some analysts argue that Facebook is changing the way we interact and connect with people.

Thinking Sociologically

Compare the way you communicate with your friends and the way your parents or grandparents communicate. What are the differences, and the advantages and disadvantages, of each? Does one method result in closer relationships than another?

Websites dedicated to social networking focus on sharing mutually interesting information, finding and keeping up with friends, uploading photos and videos, finding activity partners, publishing notes, or establishing professional contacts. Facebook (Facebook.com 2011), which launched in February 2004 at Harvard University, is now used by university students around the world and has spread to high school students, adults, music groups, politicians, and businesses. Like many other social networking sites, it allows users to post a profile and pictures, offer testimonials about others, create and join groups, and link to the profiles of "friends"—often a mixture of friends, colleagues, relatives, and acquaintances of varying levels and from various periods of life. Users can view the profiles of anyone within four degrees of separation—friends of friends of friends of friends—and search the network for people with the same friends, location, hometown, occupation, schools attended, interests, hobbies, or taste in movies, books, or television shows. In many cases, users link to dozens of friends and can access thousands of profiles around the world without reaching beyond the friends-of-friends level of connectivity. In 2007 Facebook Platform was launched, allowing other software specialists to design new features such as games, music, and photo sharing. These new features have attracted more users to Facebook (Bahney 2009). One downside is that some critics believe that social networking sites may provide a means for predators such as pedophiles to solicit sex from minors (Bahney 2009). As a way of examining your own networks, try the exercise in the next "Engaging Sociology" feature.

An even more recent development is the rapidly increasing popularity of Twitter, a microblogging website that allows users to post 140-character messages ("tweets") using their personal computer or the text messaging function on their mobile phone. Interestingly, users have played a crucial role in developing Twitter, including inventing "hashtags" that indicate the subject of the tweet and allowing others to search for all tweets on the same topic (Johnson 2009).

The irony, of course, is that electronic technology has changed interactions by making them both more intimate and accessible—and less so. We can network through Facebook, and we can keep in touch constantly through text messaging. However, this means that people spend more time interacting with a piece of technology and less time interacting face-to-face. How ironic it is to see people at adjacent desks or offices, or with friends down the hall in the dorm talking on cell phones or texting other friends. Sometimes those in physical proximity are ignored. Likewise, distance learning courses are in one sense less intimate—the instructor and the student may never meet face-to-face. Yet through Internet contact, the students may have more access to the ideas and the personal counsel of a professor than they would if they were in a large lecture class. We are only beginning to understand the implications of this technology on human interaction and on interpersonal skills.

Engaging Sociology

Examining Your Social Networks

If you are on Facebook, go to your Facebook account and note the number of friends you have listed. Then look at them carefully to see if you can answer the following questions:

What is the age range of the friends on your list?

What is the gender composition of your list?

How many of each of the following racial or ethic groups are on your list?

___ African Americans

___ Whites of predominantly European heritage

___ Hispanics

___ Asians

___ Other (mixed)

What is the socioeconomic status of your friends?

___ Blue-collar (families where the primary wage earner[s] works for an hourly wage)

___ Middle class (families where the primary wage earner[s] earns a salary of less than $100,000 per year)

___ Professional (families where the primary wage earner[s] earns a salary of $100,000 to $500,000 per year)

___ Highly affluent corporate executives (families where the primary wage earner[s] earns a salary of $500,000 to $10 million per year)

___ Upper class (families where much of the wealth is inherited and annual income is multimillions)

What do you conclude about the diversity or homogeneity of your network of friends and acquaintances? You can carry this to the next degree by looking at friends' friends.

Thinking Sociologically

How have you or your friends used the Internet to expand your social or professional networks? How do Facebook and Twitter shape your interactions with friends?

The Process of Interaction: Connections at the Micro Level

Each morning as you rouse yourself and prepare for the challenges ahead, you consider what the day might bring, what activities and obligations are on your calendar, and whom you will talk to. As you lift your limp, listless body from a horizontal to an upright position and blood begins coursing through your veins, thoughts of the day's events begin to penetrate your semiconscious state. A cup of caffeine, cold water on the face, and a mouth-freshening brush bring you to the next stage of awareness. You evaluate what is in store for you, what roles you will play during the day, and with whom you are likely to interact.

Should you wear the ragged but comfortable jeans and T-shirt? No, there is that class trip to the courthouse today. Something a bit less casual is in order. Then, you are meeting with your English professor to discuss the last essay you wrote. What approach should you take? You could act insulted that she failed to think of you as a future J. K. Rowling. Maybe a meek, mild "please tell me what I did wrong; I tried so hard" approach would work. She seems a nice, sympathetic sort. After class, there is a group of students who chat in the hall. It would be nice to meet them. What strategy should you use? Try to enter the conversation? Tell

The same words, "let's have a drink," may have very different meanings in different social interaction contexts. Humans must learn not only the language but also how to interpret the messages.

a joke? Make small talk? Talk to the students individually so you can get to know each before engaging the whole group? Each of these responses is a strategy for interaction, and each might elicit different reactions.

The Elements of Social Interaction

"Let's have a drink!" Such a simple comment might have many different meanings. We could imagine two children playing together, men going to a bar after work, a couple of friends getting together to celebrate an event, fraternity brothers at a party, or a couple on a date. In all these cases, **social interaction** consists of two or more individuals purposefully relating to each other.

"Having a drink," like all interaction, involves action on the part of two or more individuals, is directed toward a goal that people hope to achieve, and takes place in a social context that includes cultural norms and rules governing the situation, the setting, and other factors shaping the way people perceive the circumstances. The action, goal, and social context help us interpret the meaning of statements such as "Let's have a drink."

The norms governing the particular social context tell us what is right and proper behavior. Recall from Chapter 3 that norms are rules that guide human interactions. People assume that others will share their interpretation of a situation. These shared assumptions about proper behavior provide the cues for your own behavior that become a part of your social self. You look for cues to proper

behavior and rehearse in your mind your actions and reactions. In the "let's have a drink" scenario, you assume that the purpose of the interaction is understood. What dress, mannerisms, speech, and actions you consider appropriate depend on expectations from your socialization and past experience in similar situations (Parsons 1951b), for in modern societies, a range of behaviors and responses is possible in any social situation.

Although most people assume that talking, or verbal communication, is the primary means of communication between individuals, words themselves are actually only a part of the message. In most contexts, they make up less than 35% of the emotional content of the message (Birdwhistell 1970). **Nonverbal communication**—interactions using facial expressions, the head, eye contact, body posture, gestures, touch, walk, status symbols, and personal space—make up the rest (Drafke and Kossen 2002). These important elements of communication are learned through socialization as we grow up. Although one may master another written and verbal language, it is much more difficult to learn the nonverbal language.

People who travel to a country other than their own often use gestures to be understood. Like spoken language, nonverbal gestures vary from culture to culture, as illustrated in the photos on the next page. Communicating with others in one's own language can be difficult enough. Add to this the complication of individuals with different languages, cultural expectations, and personalities using different nonverbal messages, and misunderstandings are likely. Nonverbal messages are the hardest part of another language to master because they are specific to a culture and learned through socialization.

Consider the following example: You are about to wrap up a major business deal. You are pleased with the results of your negotiations, so you give your hosts the thumb-and-finger A-OK sign. In Brazil, you have just grossly insulted your hosts—it is like giving them "the finger" in the United States. In Japan, you have asked for a small bribe. In the south of France, you have indicated the deal is worthless. Although your spoken Portuguese, Japanese, or French may have been splendid, your nonverbal language did not cut the deal. Intercultural understanding is more than being polite and knowing the language.

Consider another example of nonverbal language: personal space. Most people have experienced social situations, such as parties, where someone gets too close. One person backs away, the other moves in again, and the first backs away again—into a corner or a table with nowhere else to go. Perhaps the person approaching was aggressive or rude, but it is also possible that the person held different cultural norms or expectations in relation to personal space.

The amount of personal space an individual needs to be comfortable or proper varies with the cultural setting, gender, status, and social context of the interaction.

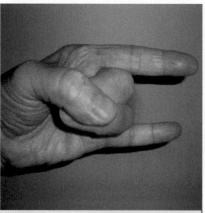

Gestures are symbolic forms of interaction. However, these gestures can have entirely different meanings in different cultures. A friendly gesture in one culture may be obscene in the next.

Individuals from Arab countries are comfortable at very close range. However, people from Scandinavia or the United States need a great deal of personal space. Consider the following four categories of social distance and social space based on a study of U.S. middle-class people. Each category applies to particular types of activity (Hall and Hall 1992):

1. *Intimate distance:* from zero distance (touching, embracing, kissing) to 18 inches. Children may play together in such close proximity, and adults and children may maintain this distance, but between adults, this intimate contact is reserved for private and affectionate relationships.

2. *Personal distance:* from 18 inches to 4 feet. This is the public distance for most friends and for informal interactions with acquaintances.

3. *Social distance:* from 4 feet to 12 feet. This is the distance for impersonal business relations, such as a job interview or class discussions between students and a professor. This distance implies a more formal interaction or a significant difference in the status of the two people.

4. *Public distance:* 12 feet and beyond. This is the distance most public figures use for addressing others, especially in formal settings and in situations in which the speaker has very high status.

Personal space also communicates one's position in relation to others. The higher one's position is, the more one is in control of her or his personal space. In social situations, individuals with higher positions spread out, prop their feet up, put their arms out, and use more sweeping gestures (Knapp and Hall 1997). Women and men differ with regard to personal space and other forms of nonverbal language. For instance, women are more sensitive to subtle cues such as status differences and the use of personal space (Henley, Hamilton, and Thorne 2000).

Sociologists study interactions, including verbal and nonverbal communication, to explain this very basic link between humans and the group. The following theoretical perspectives focus on the micro level of analysis in attempting to explain interactions.

A more formal setting calls for a distance of 4 to 12 feet. If the distance is quite large, the situation can feel very formal and intimidating. What is the message at this meeting?

When a famous or powerful person speaks in a public setting, listeners are expected to keep themselves at a distance of about 12 feet or more. President Barack Obama speaks here in a formal setting and the audience is at an appropriate public distance. It is not only the Secret Service that keeps people away; it is a sense of awed respect for the office of the President.

Thinking Sociologically

What are some complications that you or your friends have had in interactions involving cross-cultural contacts or male-female miscommunication? What might help clarify communication in these cases?

Theoretical Perspectives on the Interaction Process

How many people do you interact with each day, and what happens in each of these interactions? You probably have not given the question much thought or analysis, but that process is exactly what fascinates interaction theorists. Why do two people interact in the first place? What determines whether the interaction will continue or stop? How do two people know how to behave and what to say around each

other? What other processes are taking place as they "talk" to each other? Why do people interact differently with different people? What governs the way they make sense of messages and how they respond to them? These questions interest sociologists because they address the basic interaction processes that result in group formations that range in size from dyads (two people) to large organizations. The following section explores theories that attempt to provide explanations for interactions.

Rational Choice Theory

Rational choice or exchange theorists look at why relationships continue, considering the rewards and costs of interaction to the individual. They argue that choices we make are guided by reason. If the benefits of the interaction are high and if the costs are low, the interaction will be valued and sustained. Every interaction involves calculations of self-interest, expectations of reciprocity (a mutual exchange of favors), and decisions to act in ways that have current or eventual payoff for the individual (Smelser 1992).

Reciprocity is a key concept for rational choice theorists. The idea is that if a relationship is imbalanced over a period of time, it will become unsatisfying. As theorists from this perspective see human interaction, each person tends to keep a mental ledger of who "owes" whom. If I have done you a favor, you owe me one. If you have helped me in some way, I have an obligation to you. If I then fail to comply or even do something that hurts you, you will likely view it as a breach in the relationship and have negative feelings toward me. Moreover, if there is an imbalance in what we each bring to the relationship, one person may have more power in the relationship. In the study of families, scholars utilize the "principle of least interest," which says that the person with the least interest in the relationship has the most power. The person with the least interest is the person who brings more resources to the relationship and receives less. That person could easily leave. The person who offers less to the relationship or who has fewer assets (physical, financial, social, personal) is more dependent on the relationship. This person is likely to give in when there is a disagreement, so the person with less interest gets her or his way. Lack of reciprocity can be important for how relationships develop. It is this that interests rational choice theorists.

Sometimes a person may engage in a behavior where there is little likelihood of reciprocity from the other person—as in cases where the behavior is altruistic or self-giving. Rational choice theorists would argue that there is still a benefit. It might be enhanced feelings of self-worth, recognition from others, hope for a place in heaven, or just the expectation of indirect reciprocity. This later notion is that the person I help might not help me, but if I am in a similar situation, I would hope for and expect someone to come to my assistance (Gouldner 1960; Turner 2003).

Symbolic Interaction Theory

Symbolic interaction theory focuses instead on how individuals interpret situations, such as "let's have a drink," and how this, in turn, affects their actions. One variation on symbolic interaction theory—called dramaturgical analysis—explains aspects of interaction that involve manipulating how people perceive interactions. Dramaturgy theorists analyze life as a play or drama on a stage, with scripts and props and scenes to be played. The play we put on creates an impression for our audience. In everyday life, individuals learn new lines to add to their scripts through the socialization process, including influence from family, friends, films, and television. They perform these scripts for social audiences in order to maintain certain images, much like the actors in a play.

Every day in high schools, thousands of teenagers go on stage—in the classroom or the hallway with friends and peers and with adult authorities who may later be giving grades or writing letters of reference. The props these students use include their clothing, a backpack with paper and pen, and a smile or "cool" look. The set is the classroom, the cafeteria, and perhaps the athletic field. The script is shaped by the actors, with teachers establishing an authoritarian relationship, with classmates engaging in competition for grades, or with peers seeking social status among companions. The actors include hundreds of teens struggling with issues of identity, changing bodies, and attempts to avoid humiliation. Each individual works to assert and maintain an image through behavior, clothing, language, and friends.

As individuals perform according to society's script for the situation, they take into consideration how their actions will influence others. By carefully managing the impression they wish the acquaintance to receive—a process called impression management—people hope to create an impression that works to their advantage. In other words, the actor is trying to manipulate how others define the situation, especially as it relates to their opinion of the actor.

Most of the time, we engage in front-stage behavior, the behavior safest with casual acquaintances because it is scripted and acted for the public to create an impression. At home or with close friends with whom we are more intimate, we engage in backstage behavior, letting our feelings show and behaving in ways that might be unacceptable for other audiences (Goffman [1959] 2001, 1967). Each part or character an individual plays, as well as each audience, requires a different script. Dramaturgical analysis can be a useful approach to broadening our understanding of interactions.

Thinking Sociologically

What are some ways your life feels like a dramatic production—with a front-stage presentation and a backstage presentation?

Social Status: The Link to Groups

Our social status defines how, for example, we interact with others and how others react to us in a specific situation. We interact differently when in the daughter status with our parents, in a student status with our professor, or in friend status with our peers. Each individual holds many statuses, and this combination held by any individual is called a status set: daughter, mother, worker, teammate, student. Recall the network web you drew. Now add to that web your **social statuses**, positions you hold in the social world.

Each individual's unique status set is the product of family relationships and groups that the individual joins (a university or club) or into which she or he is born (an ethnic group or a gender). Many statuses change with each new stage of life, such as work, student status, or marital status. Statuses at each stage of life and the interactions that result from those statuses form each person's unique social world.

Statuses affect the type of interactions individuals have. In some interactions (as with classmates), people are equals. In other situations, individuals have interchanges with people who hold superior or inferior statuses. If you are promoted to supervisor, your interaction with subordinates will change. Consider the possible interactions shown in Figure 5.2, in which the first relationship is between equals and the others are between those with unequal statuses.

When individuals are in dominant or subordinate positions, power or deference affects their interactions. With

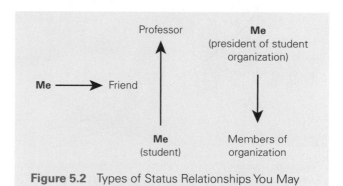

Figure 5.2 Types of Status Relationships You May Experience

a friend, these status relationships are constantly being negotiated and bargained: "I'll do what you want tonight, but tomorrow I choose." Studies of interaction between males and females find that in addition to gender, power and hierarchical relationships are important in determining interaction patterns. The more powerful person, such as one who has more wealth or privilege, can interrupt in a conversation with his or her partner and show less deference in the interaction (Kim et al. 2007; Reid and Ng 2006; Wood 2008).

People have no control over certain statuses they hold. These **ascribed statuses** are often assigned at birth and do not change during an individual's lifetime. Some examples are gender and race or ethnicity. Ascribed statuses are assigned to a person without regard for personal desires, talents, or choices. In some societies, one's caste, or the social position into which one is born (e.g., a slave), is an ascribed status because it is usually impossible to change.

Achieved status, on the other hand, is chosen or earned by decisions one makes and sometimes by personal ability. Attaining a higher education, for example, improves an individual's occupational opportunities and hence his or her achieved status. Being a guitarist in a band is an achieved status, and so is being a prisoner in jail, for both are earned positions based on the person's own decisions and actions.

At a particular time in life or under certain circumstances, one of an individual's statuses may become most important and take precedence over others. Sociologists call this a **master status**. Whether it is an occupation, parental status, or something else, it dominates and shapes much of an individual's life, activities, self-concept, and position in the community for a period of time. For a person who is very ill, for instance, that illness may occupy a master status needing constant attention from doctors, influencing social relationships, and determining what that person can do in family, work, or community activities.

Positions people carry out in different cultures depend on tasks important in that culture. Upper left: The two women from Ghana carry yams, a staple food, from the fields. Upper right: The market for these Asian women is on the water. Lower left: Men in China use traditional methods to keep the roads open. Lower right: Fishermen in India pull in a catch for their livelihood.

Thinking Sociologically

What are your statuses? Which ones are ascribed, and which are achieved statuses? Do you have a master status? How do these statuses affect the way you interact with others in your network of relationships?

The Relationship Between Status and Role

Every status (position) in your network includes certain behaviors and obligations as you carry out the expected behaviors, rights, and obligations of the status—referred to as **roles**. Roles are the dynamic, action part of statuses in a society. They define how each individual in an interaction is expected to act (Linton 1937). The role of a college student includes behaviors and obligations such as attending classes, studying, taking tests, writing papers, and interacting with professors and other students. Individuals enter most statuses with some knowledge of how to carry out the roles dictated by their culture. Through the process of socialization, individuals learn roles by observing others, watching television and films, reading, and being taught how to carry out the status. Both statuses (positions) and roles (behavioral obligations of the status) form the link with other people in the social world because they must be carried out in relationships with others. A father has certain obligations (or roles) toward his children and their mother. The position of father exists not on its own but in relationship to significant others who have reciprocal ties.

Your status of student requires certain behaviors and expectations, depending on whether you are interacting with a dean, a professor, an adviser, a classmate, or a prospective employer. This is because the role expectations of the status of student vary as one interacts with specific people in other statuses. Within a group, individuals may hold both formal and informal statuses. One illustration is the formal status of high school students, each of whom plays a number of informal roles in cliques that are not part of the formal school structure. They may be known as preps, goths, hicks, or druggies. Each of these roles takes place in a status relationship with others: teacher-student, peer-peer, coach-athlete. The connections between statuses and roles are illustrated in Table 5.1.

Our statuses connect us and make us integral parts of meso- and macro-level organizations. Sometimes, the link is through a status in a family group such as son or daughter, sometimes through an employer, and sometimes through our status as citizen of a nation. Social networks may be based on ascribed characteristics such as age, race, ethnicity, and gender or on achieved status such

Table 5.1 The Relationship Between Statuses and Roles

Status (position in structure)	Role (behavior, rights, obligations)
Student	*Formal*
	Study, attend class, turn in assignments
	Informal
	Be a jock, clown, cut-up; abuse alcohol on weekends
Parent	*Formal*
	Provide financial support, child care
	Informal
	Be a playmate, lead family activities
Employee	*Formal*
	Work responsibilities: punctuality, doing one's tasks
	Informal
	Befriend coworkers, join lunch group, represent company in bowling league

as education, occupation, or common interests (House 1994). These links, in turn, form the basis for social interactions and group structures (Hall 2002). However, at times, individuals cannot carry out their roles as others expect them to, creating role strain or conflict.

Thinking Sociologically

In your web, detail your network from close family and friends to distant acquaintances. Include your statuses and roles in each part of the network with family members, friends, and school and work acquaintances. What does your network tell you about yourself?

Role Strain and Role Conflict

Every status carries role expectations, the way the status is supposed to be carried out according to generally accepted societal or group norms. Most people have faced times in their lives when they simply could not carry out all the obligations of a status such as student—write two papers, study adequately for two exams, complete the portfolio for the studio art class, finish the reading assignments for five classes, and memorize lines for the oral interpretation class, all in the same

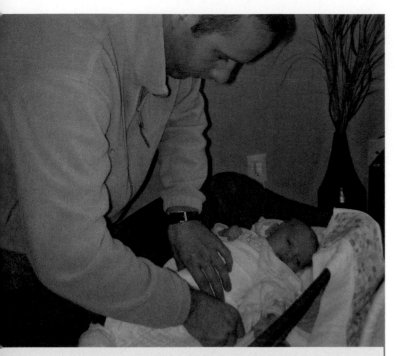

Parenting roles often need to be negotiated. In some traditional families in the past, the status had more explicit role expectations, and fathers were rarely expected to change diapers.

week. In these cases, individuals face **role strain**, tension between roles within one of the statuses. Role strain causes the individual to be pulled in many directions by various obligations of the single status, as in the example regarding the status of "student." Another such strain is often experienced by first-time fathers as they attempt to reconcile their role expectations of fathering with ideas held by their wives.

To resolve role strain, individuals cope in one of several ways: pass the problem off lightly (and thus not do well in classes), consider the dilemma humorous, become highly focused and pull a couple of all-nighters to get everything done, or become stressed, tense, fretful, and immobilized because of the strain. Most often, individuals set priorities based on their values and make decisions accordingly: "I'll work hard in the class for my major and let another one slide."

Role conflict differs from role strain in that conflict is between the roles of two or more statuses. The conflict can come from within an individual or be imposed from outside. College athletes face role conflicts from competing demands on their time (Adler and Adler 1991, 2004). They must complete their studies on time, attend practices and be prepared for games, perhaps attend meetings of a Greek house to which they belong, and get home for a little brother's birthday. Similarly, a student may be going to school, holding down a part-time job to help make ends meet, and raising a family. If the student's child gets sick, the status of parent comes into conflict with that of student and worker. In the case of role conflict, the person may choose—or be informed

by others—which status is the master status. Figure 5.3 illustrates the difference between role conflict and role strain.

Statuses and the accompanying roles come and go. You will not always be a student, and someday, you may be a parent and hold a professional job. Certainly, you will retire from your job. For instance, as people grow older, they disengage from some earlier statuses in groups and engage in new and different statuses and roles.

Our place within the social world is guaranteed, even obligatory, because of statuses we hold at each level of society—within small groups (family and peers), in larger groups and organizations (school and work organizations), in institutions (political parties or religious denominations), and ultimately as citizens of the society and the world (workers in global corporations). Each of these statuses connects us to a group setting.

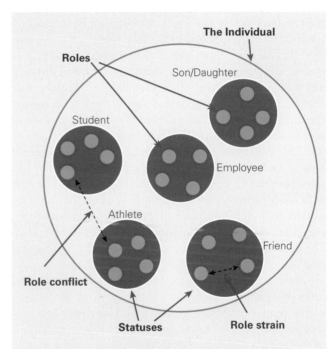

Figure 5.3   Role Strain and Role Conflict

Note: Each individual has many statuses: a status set. Each status has many roles: a role set. A conflict between two roles of the same status is a role strain. A conflict between the roles of two different statuses is a role conflict.

Thinking Sociologically

Using Figure 5.3, fill in the statuses you hold in your social world and roles you perform in these statuses. In your diagram, list three examples of role conflicts and three examples of role strains that you experience.

Groups in Our Social World: The Micro-Meso Connection

Groups refer to two or more people who interact with each other because of shared common interests, goals, experiences, and needs (Drafke and Kossen 2002). Few of us could survive without others. Most of us constantly interact with the people around us. We are born into a family group. Our socialization occurs in groups. We depend on the group for survival. Groups are necessary for protection, to obtain food, to manufacture goods, and to get jobs done. Groups meet our social needs for belonging and acceptance, support us throughout our lives, and place restrictions on us. Groups can be small, intimate environments—micro-level interactions such as a group of friends—or they can become quite large as they morph into meso-level organizations. In any case, it is through our group memberships that the micro and meso levels are linked.

The members of groups feel they belong to the group and are seen by others as thinking, feeling, and behaving with a common goal. Members consider each other's behavior and engage in structured interaction patterns. Groups have defined memberships and ways to take in new members. They also have rules that guide behavior of members. In this section, we look at several questions: What are groups, and how do they vary? How is interaction affected by small groups? What is the importance of groups for individuals?

Not all collections of individuals are groups, however. For instance, your family is a group, but people shopping at a mall or waiting for a bus are not a group because they do not interact or acknowledge shared common interests.

Groups form through a series of succeeding steps. Consider people forming a football team. The first step is initial interaction. If membership is rewarding and meets individuals' needs, the individuals will attempt to maintain the benefits the group provides (Mills 1984). A group of people interact to form this team. In the second step, a collective goal emerges. For example, team members work together to plan practice and game schedules, buy uniforms, and advertise the games. Groups establish their own goals and pursue them, trying to be free from external controls or constraints. In the third and final step, the group attempts to expand its collective goals by building on the former steps and by pursuing new goals. For example, the team may reach out to new players, to coaches, and to supporters for funding.

The Importance of Groups for the Individual

Groups are essential parts of human life (micro level) and of organizational structures (meso and macro levels). They

Even Buddhist nuns, who spend much of their lives devoted to private meditation, need the support and solidarity of a group.

establish our place in the social world, providing us with support and a sense of belonging. Few individuals can survive without groups. This becomes clear when we consider two problems: anomie and suicide.

Anomie and Suicide

Who commits suicide? Did you know that this is closely related to an individual's group affiliations? With the rapid changes and continued breakdown of institutional structures in Afghanistan as rival warlords vie for power over territory and in Iraq as religious groups vie for political and economic power, horrific problems abound. Civil disorder, conflicts for power, suicide bombings, murder of police officers, and looting are frequent occurrences. Social controls (police and military forces) are strained, and leaders struggle to cope. The result of this breakdown in norms is *anomie,* the state of normlessness; the rules for behavior in society break down under extreme stress from rapid social change or conflict (Merton 1968a). Consider the example of suicide.

Suicide seems like an individual act, committed because of personal problems. However, early sociologist Émile Durkheim took a unique approach. In his volume, *Suicide* (Durkheim [1897] 1964), he discussed the social factors contributing to suicide. Using existing statistical data to determine suicide rates in European populations, Durkheim looked at such variables as sex, age, religion, nationality, and the season in which suicide was committed. His findings were surprising to many, and they demonstrate that individual problems cannot be understood without also understanding the group context in which they occur.

Durkheim ([1897] 1964) found that Protestants committed suicide more often than Catholics, urban folks more often than people living in small communities, people in highly developed and complex societies more frequently than those in simple societies, and people who lived alone more than those situated in families. The key variable linking these findings was the degree to which an individual was integrated into the group—that is, the degree of social bond with others. During war, for instance, people generally felt a sense of common cause and belonging to their country. Thus, suicide rates were greater during peacetime because it offered less cause for feeling that bond.

Durkheim ([1897] 1964) describes three distinct types of suicide. Egoistic suicide occurs when the individual feels little social bond to the group or society and lacks ties such as family or friends that might prevent suicide. Egoistic suicide is a result of personal despair and involves the kind of motive most people associate with suicide.

Anomic suicide occurs when a society or one of its parts is in disorder or turmoil and lacks clear norms and guidelines for social behavior. This situation is likely during major social change or economic problems such as a severe depression.

Altruistic suicide differs from the others in that it involves such a strong bond and group obligation that the individual is willing to die for the group. Self-survival becomes less important than group survival (Durkheim [1897] 1964). Buddhist monks who burned themselves to protest the Vietnam War and terrorists who blow themselves up are examples. These are suicides rooted in extraordinarily high integration into a political or religious group.

Many sociologists have studied suicide, and while not all studies support Durkheim's original findings, they do support the general finding that suicide rates are strongly influenced by social and psychological factors that can operate at the meso or macro level. Suicide is not a purely individualistic decision (Hall 2002; Nolan, Triplett, and McDonough 2010; Pescosolido and Georgianna 1989). No individual is an island. The importance of groups and social influence from various levels in the system is an underlying theme of this text. Groups are essential to human life, but to understand them more fully, we must understand the various kinds of groups in which humans participate.

Types of Groups

Each of us belongs to several types of groups. Some groups provide intimacy and close relationships, whereas others do not. Some are required affiliations, and others are voluntary.

This photo is hard to look at, but reality sometimes is ugly and jarring in desperate places. This monk has set himself on fire in a public square, seeing this as one of the few ways he can make a dramatic statement against war. He was tightly integrated into his religious community and his people, and his own death seemed a worthy sacrifice if it would help bring attention to or end the suffering that the war brought to his nation. This illustrates altruistic suicide.

Some provide personal satisfaction, and others are obligatory or necessary for survival. The following discussion points out several types of groups and reasons individuals belong to them.

Primary groups are characterized by close contacts and lasting personal relationships—the most micro level. Your family members, best friends, school classmates, and close work associates are all of primary importance in your everyday life. Primary groups provide a sense of belonging and shared identity. Group members care about you, and you care about the other group members, creating a sense of loyalty. Approval and disapproval from the primary group influence the activities you choose to pursue. Belonging rather than accomplishing a task is the main reason for membership. The group is of intrinsic value—enjoyed for its own sake—rather than for some utilitarian value such as making money.

For individuals, primary groups provide an anchor point in society. You were born into a primary group—your family. You play a variety of roles in primary relationships—those of spouse, parent, child, sibling, relative, close friend, and so on. You meet with other members face-to-face or keep in touch on a regular basis and know a great deal about their lives. What makes them happy or angry? What are sensitive issues? In primary groups you share values, say what you think, let down your hair, dress as you like, and share your concerns and emotions, as well as your successes and failures (Goffman [1959] 2001, 1967). Charles H. Cooley ([1909] 1983), who first discussed the term *primary group,* saw these relationships as the source of close human feelings and emotions—love, cooperation, and concern.

Secondary groups are those with formal, impersonal, businesslike relationships. In the modern world, people cannot always live under the protective wing of primary group relationships. Secondary groups are usually large and task oriented because they have a specific purpose to achieve and focus on accomplishing a goal. As children grow, they move from the security and acceptance of primary groups—the home and neighborhood peer group—to the large school classroom, where each child is one of many students vying for the teacher's approval and competing for rewards. Similarly, the job world requires formal relations and procedures: applications, interviews, contracts, and so on. Employment is based on specific skills, training, and job knowledge, and there may be a trial period. In Western cultures, we assume that people should be hired not because of personal friendship or nepotism but rather for their competence to carry out the role expectations in the position.

Because each individual in secondary groups carries out a specialized task, communication between members is often specialized as well. Contacts with doctors, store clerks, and even professors are generally formal and impersonal parts of organizational life. Sometimes associations with secondary groups are long lasting, and sometimes they are of short duration—as in the courses you are taking this term. Secondary groups operate at the meso and macro levels of our social world, but they affect individuals at the micro level.

As societies modernize, they evolve from small towns and close, primary relationships to predominantly urban areas with more formal, secondary relationships. In the postindustrial world, as family members are scattered across countries and around the world, secondary relationships have come to play ever greater roles in people's lives. Large work organizations may provide day care, health clinics, financial planning, courses to upgrade skills, and sports leagues.

The small micro- and large macro-level groups often occur together. Behind most successful secondary groups are primary groups. These relationships help individuals feel a part of larger organizations, just as residents of large urban areas have small groups of neighborhood friends.

Megachurches began in the 1950s. They have more than 10,000 members, with the largest church in the world (in Korea) having 830,000 members. About half of these

These teens—a primary group of friends—enjoy one another's company during a cookout at the beach. Their interpersonal connections are valued for intrinsic reasons.

churches are nondenominational Protestant, and the rest are related to evangelical or Pentecostal groups. The focus of programming is on creating small support groups (primary groups) within the huge congregation ("O Come All Ye Faithful: Special Report on Religion and Public Life" 2007; Sargeant 2000; Thumma and Travis 2007). Table 5.2 summarizes some of the dimensions of primary and secondary groups.

Problems in primary groups can affect performance in secondary groups. Consider the problems of a student who has an argument with a significant other or roommate or experiences a failure of his or her family support system due to divorce or other problems. Self-concepts and social skills diminish during times of family stress and affect group relationships in other parts of one's life (Drafke and Kossen 2002; Parish and Parish 1991).

Reference groups are composed of members who act as role models and establish standards against which all members measure their conduct. Individuals look to reference groups to set guidelines for behavior and decision making. The term is often used to refer to models in one's chosen career field. Students in premed, nursing, computer science, business, or sociology programs watch the behavior patterns of those who have become successful professionals in their chosen career. When people make

Table 5.2 Primary and Secondary Group Characteristics

	Primary Group	*Secondary Group*
Quality of relationships	Personal orientation	Goal orientation
Duration of relationships	Usually long-term	Variable, often short-term
Breadth of activities	Broad, usually involving many activities	Narrow, usually involving few largely goal-directed activities
Subjective perception of relationships	As an end in itself (friendship, belonging)	As a means to an end (to accomplish a task, earn money)
Typical examples	Families, close friendships	Coworkers, political organizations

the transition from student to professional, they adopt clothing, time schedules, salary expectations, and other characteristics from reference groups. Professional organizations such as the American Bar Association or American Sociological Association set standards for behavior and achievements.

However, it is possible to be an attorney or an athlete and not aspire to be like others in the group if they are unethical or abuse substances such as steroids. Instead, a person might be shaped by the values of a church group or a political group with which he or she identifies. Not every group one belongs to is a reference group. It must provide a standard by which one evaluates his or her behavior for it to be a reference group. For example, ethnic groups provide some adolescents with strong reference group standards by which to judge themselves. The stronger the ethnic pride and identification, the more some teens may separate themselves from contact with members of other ethnic groups (Schaefer and Kunz 2008). This can be functional or dysfunctional for the teens, as shown in the next section on in-groups and out-groups.

An **in-group** is one to which an individual feels a sense of loyalty and belonging. It also may serve as a reference group and a primary group; these are different features of groups. An **out-group** is one to which an individual does not belong, but more than that, it is a group that is often in competition or in opposition to an in-group.

Membership in an in-group may be based on sex, race, ethnic group, social class, religion, political affiliation, the school one attends, an interest group such as the fraternity or sorority one joins, or the area where one lives. People tend to judge others according to their own in-group identity. Members of the in-group—for example, supporters of a high school team—often feel hostility toward or reject out-group members, boosters of the rival team. The perceived outside threat or hostility is often exaggerated, but it does help create the in-group members' feelings of solidarity.

Bloods gang members in Los Angeles use their in-group hand signal to identify one another.

Unfortunately, these feelings can result in prejudice and ethnocentrism, overlooking the individual differences of group members. Teen groups or gangs such as the Bloods and the Crips are examples of in-groups and out-groups in action, as are the ethnic and religious conflicts between Sunni and Shiite Muslims in Iraq. In each case, the group loyalty is enhanced by hostility toward the out-group, resulting in gang conflicts and war.

Thinking Sociologically

What are some examples of your own group affiliations: Primary groups? Secondary groups? Peer groups? Reference groups? In-groups and out-groups?

Organizations and Bureaucracies: The Meso-Macro Connection

Our days are filled with activities that involve us with complex organizations: from the doctor's appointment to college classes; from the political rally for the issue we are supporting to Sabbath worship in our church, temple, or mosque; from paying state sales tax for our toothpaste to buying a sandwich at a fast-food franchise. Figure 5.4 shows institutions of society, each made up of thousands of organizations and each following the cultural norms of our society. We have statuses and roles in each group, and these link us to networks and the larger social world.

How did these organizational forms develop? Let us consider briefly the transformations of organizations into their modern forms and the characteristics of meso-level organizations today.

The Evolution of Modern Organizations

Empires around the world have risen and fallen since the dawn of civilization. Some economic, political, and religious systems have flourished while others have withered.

We cannot understand our social world at any historical or modern time without comprehending the organizational structures and processes present at that time. Recall the discussion of types of societies, from hunter-gatherer to postindustrial, in Chapter 3. Each type of society entails different organizational structures, from early cities and feudal manors to craft guilds, heavy industries, and today's information- and web-based companies (Blau 1956; Nolan and Lenski 2008).

The development of modern organizations began with industrialization in the 1700s and became the dominant form of organization by the 1800s. What Max Weber called rationalization of social life—the attempt to maximize efficiency by creating rules and procedures focused solely on accomplishing goals—was thought to be the best way to run organizations (Weber 1947). People were expected to behave in purposeful, coordinated ways that advanced the organization. No longer were decisions made by tradition, custom, or the whim of a despot. Instead, trained leaders planned policies, tasks became more specialized, some manual jobs were taken over by machines, and standardization of products allowed for greater productivity, precision, and speed. Modern rational organizations are complex secondary groups deliberately formed to pursue and achieve certain goals. Sometimes they are called **formal organizations** because of the written charters, constitutions, bylaws, and procedures that govern them. The Red Cross, Ford Motor Company, the National Basketball Association (NBA), the Republican Party, and your university are all modern organizations.

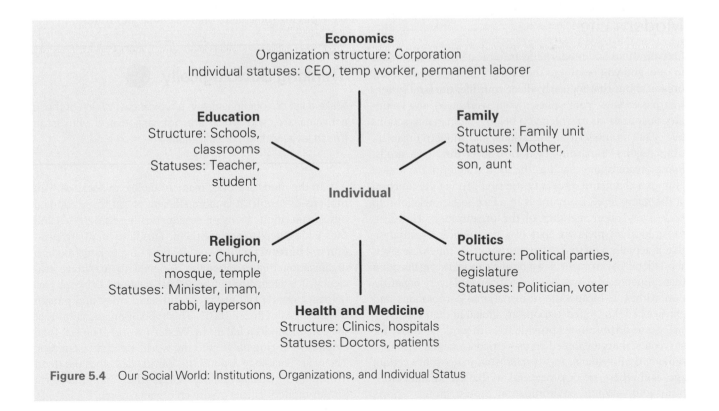

Economics
Organization structure: Corporation
Individual statuses: CEO, temp worker, permanent laborer

Education
Structure: Schools, classrooms
Statuses: Teacher, student

Family
Structure: Family unit
Statuses: Mother, son, aunt

Individual

Religion
Structure: Church, mosque, temple
Statuses: Minister, imam, rabbi, layperson

Politics
Structure: Political parties, legislature
Statuses: Politician, voter

Health and Medicine
Structure: Clinics, hospitals
Statuses: Doctors, patients

Figure 5.4   Our Social World: Institutions, Organizations, and Individual Status

As societies became dominated by large organizations, fewer people worked the family farm, owned a cottage industry in their homes, or were small shopkeepers. Today, some countries are organizational societies in which a majority of the members work in large organizations rather than being self-employed, and many others are moving in this direction.

Societies making the transition between traditional and modern organizational structures often have a mix of rural traditional agriculture and urban industrialization (Nolan and Lenski 2008). Bribery, corruption, and favoritism govern some nations as they move toward modern bureaucracies. In these countries, people in government jobs are promoted based on connections, not on their competence as assessed by formal training, examinations, or criteria derived from the needs of the position.

Postindustrial societies feature high dependence on technology and information sharing. Few people in such societies live and work on farms. Only 1% of U.S. citizens claim farming as their main occupation, and only 2% live on farms (U.S. Environmental Protection Agency 2009). About 1 in 5 citizens live in rural areas (U.S. Department of Agriculture 2008). This is compared to 15% in 1950 (Wright 2007) and 63.7% in 1850 (*Europa World Year Book* 2005). The number of U.S. farms decreased from 5,648,000 in 1950 to 2,090,000 in 2006, while the average size of farms grew from 213 acres to 445 acres as agribusiness bought up small family farms (Wright 2007).

Modern Organizations and Modern Life

Organizations and modern life are almost synonymous. Live in one, and you belong to the other. Think about the many organizations that regularly affect your life: the legal system that passes laws, your college, your workplace, and voluntary organizations to which you belong. Human interactions take place in modern organizations, and modern organizations require human interaction to meet their own needs. Some organizations, such as the local chapter of the Rotary club or a chapter of a sorority, are small enough to function at the micro level. Everyone is in a face-to-face relationship with every other member of the organization. However, those local chapters are part of a nationwide organization that is actually a meso-level organization. At the macro level, the federal government is a complex modern organization that influences the lives of every citizen and every organization within it. Meanwhile, transnational corporations and entities like the United Nations are global in their reach.

Some organizations provide us with work necessary for survival. Others are forced on us—prisons, mental hospitals, military draft systems, and even education up until a certain age. Still others are organizations we believe in and voluntarily join—scouts, environmental protest groups, sports

The hospital setting is an example of a modern organization that not only is governed by formal rules and impersonal relations but also increasingly involves extensive communication via computers.

leagues, and religious organizations (Etzioni 1975). Membership in voluntary organizations is higher in the United States than in many other countries (Johnson and Johnson 2006), yet in recent years, membership in many voluntary organizations has dropped. Some analysts argue that new types of affiliations and interactions, including the Internet, are replacing some older organizational affiliations.

Thinking Sociologically

Make a list of your activities in a typical day. Which of these activities are, and which are not, associated with large (meso-level) modern organizations?

In the modern world, many of us live in societies with millions of people. It is more efficient to get things done if our organizations focus on systematically setting goals and then taking steps to achieve them. This kind of an organization is a **bureaucracy**—a specific type of very large modern organization that is rationally designed to maximize efficiency. It is characterized by formal relations between participants, clearly laid-out procedures and rules, and pursuit of stated goals (Ritzer 2007). This type of organization is at the core of modern life in the Western world. The fast-food empires springing up around the world are vivid examples. The next "Sociology Around the World" describes the trend toward the *McDonaldization of society*—the pop-culture term for our modern efficiency-driven organizations.

Sociology Around the World

The McDonaldization of Society

McDonaldization has changed the way we live! The process of rationalization described by Max Weber—the attempt to reach maximum organizational efficiency—comes in a new modern version, expanded and streamlined, as exemplified by the fast-food restaurant business and the chain "box" stores found around the world. Efficient, rational, predictable sameness is sweeping the world—from diet centers such as Nutrisystem to 7-Eleven and from Wal-Mart to Gap clothing stores with their look-alike layouts. Most major world cities feature McDonald's or Kentucky Fried Chicken in the traditional main plazas or train stations for the flustered foreigners and curious native consumers.

The McDonaldization of society, as George Ritzer (2007) calls it, refers to several trends. First, *efficiency* is maximized by the sameness—same store plans, same mass-produced items, same procedures. Second is *predictability*, the knowledge that each hamburger or piece of chicken will be the same, leaving nothing to chance. Third, everything is *calculated* so that the organization can ensure that everything fits a standard—every burger is cooked the same number of seconds on each side. Fourth, there is *increased control* over employees and customers so there are fewer variables to consider—including substitution of technology for human labor as a way to ensure predictability and efficiency.

What is the result of this efficient, predictable, planned, automated new world? According to Ritzer (2007), the world is becoming more dehumanized, and the efficiency is taking over individual creativity and human interactions. The mom-and-pop grocery, bed and breakfasts, and local craft or clothing shops are rapidly becoming a thing of the past, giving way to clone-like businesses—the "McClones." This process of the McDonaldization of society, meaning principles of efficiency and rationalization exemplified by fast-food chains, is coming to dominate more and more sectors of our social world (Ritzer 1998, 2004, 2010). While there are aspects of this predictability that we all like, there is also a loss of uniqueness and local flavor that individual entrepreneurs bring to a community.

To try to re-create this "McCulture," Ritzer suggests, there is a movement toward "Starbuckization." Starbucks is unique because of its aesthetic contribution. Starbucks makes customers feel like they are purchasing a cultural product along with their coffee. This culture, however, is as controlled as any other McDonaldized endeavor, making Starbucks as much a McClone as its predecessors.

McDonaldization is so widespread and influential that many modern universities are McDonaldized. A large lecture hall is a very efficient way of teaching sociology or biology to many students at once. Similarly, distance learning, PowerPoint presentations, multiple-choice exams, and limited textbooks increase the predictability and standardization of course offerings. Grade point averages and credit hours completed are very calculated ways to view students. Moreover, universities control or standardize the situation by deciding what courses are offered, when they are offered, and what professors will teach them (Ritzer 2007).

Thinking Sociologically

When you shop or eat, is it at a fast-food chain or at individually owned stores? Can anything be done to protect the locally owned "mom and pop" businesses, or are they destined to be eliminated by the "McCompetition"?

Characteristics of Bureaucracy

To get a driver's license, pay school fees, or buy tickets for a popular concert or game, you may have to stand in a long line or fill out a form online. Some institutions have adopted telephone and online registration or ticket purchases. Finally, after waiting in line or online, you may discover that you have forgotten your Social Security number or cannot pay with a credit card. The rules and red tape can be irksome. Frustrations with bureaucracy are part of modern life, but the alternatives (e.g., bartering) are less efficient and sometimes more frustrating or bewildering to those who are used to bureaucratic efficiency. If you have been to a Caribbean, African, Asian, or Middle Eastern market, you know that the bartering system is used to settle on a mutually agreeable price. This system is more personal and involves intense interaction between the seller and the buyer, but it also takes more time, is less efficient, and can be frustrating to the uninitiated visitor accustomed to the relative efficiency and predictability of bureaucracy. As societies transition, they tend to adopt bureaucratic forms of organization. Most organizations in modern society—hospitals, schools, churches, government agencies, industries, banks, and even large clubs—are bureaucracies. Therefore, understanding these modern organizations is critical to understanding the modern social world in which we live.

At the beginning of the twentieth century, Max Weber (1864–1920) looked for reasons behind the massive changes taking place in societies that were causing the transition from traditional society to bureaucratic, capitalistic society. He wanted to understand why the rate of change was more rapid in some parts of Europe than others and why bureaucracy came to dominate forms of organization in some countries. Whereas traditional society looked to the past for guidance, bureaucratic industrial society required a new form of thinking and behavior, a change in attitude toward rationality. Weber observed that leadership in business and government was moving from traditional forms with powerful families and charismatic leaders toward more efficient and less personal bureaucracy.

Weber's (1947) term **ideal-type bureaucracy** refers to the dominant and essential characteristics of our modern organizations that are designed for rationality and efficiency. The concept describes not a good or perfect organization but merely an organization with a particular set of traits. Any one bureaucracy is unlikely to have all of the characteristics in the ideal type, but the degree of bureaucratization is measured by how closely an organization resembles the core characteristics of the ideal type. The following characteristics of Weber's ideal-type bureaucracy provide an example of how these characteristics relate to schools:

1. *Division of labor based on technical competence:* Administrators lead but do not teach, and instructors teach only in areas of their certification; staff are assigned positions for which credentials make them most qualified, and recruitment and promotion are governed by formal policies.

2. *Administrative hierarchy:* There is a specified chain of command and designated channels of communication, from school board to superintendent to principal to teacher.

3. *Formal rules and regulations:* Written procedures and rules—perhaps published in an administrative manual—spell out system-wide requirements, including discipline practices, testing procedures, curricula, sick days for teachers, penalties for student tardiness, field trip policies, and other matters.

4. *Impersonal relationships:* Formal relationships tend to prevail between teachers and students and between teachers and administrative staff (superintendents, principals, counselors); written records and formal communication provide a paper trail for all decisions.

5. *Emphasis on rationality and efficiency to reach goals:* Established processes are used, based on the best interests of the school. Efficiency is defined in terms of lowest overall cost to the organization in reaching a goal, not in terms of personal consequences.

6. *Provision of lifelong careers:* Employees may spend their entire careers working for the same organization, working their way up the hierarchy through promotions.

Although the list of characteristics makes bureaucracies sound formal and rigid, informal structures allow organizational members to deviate from rules both to meet goals of the bureaucracy more efficiently and to humanize an otherwise uncaring and sterile workplace. A teacher may help a student outside the formal class structure and curriculum, even though it is not part of her contract to do so. The *informal structure,* then, includes the unwritten norms and the interpersonal networks that people use within an organization to carry out roles. Likewise, although bylaws, constitutions, or contracts spell out the way things are supposed to be done, people often develop unwritten shortcuts to accomplish goals.

Informal norms are not always compatible with those of the formal organization. Consider the following example from a famous classical study. In the Western Electric plant near Chicago, the study found that new workers were quickly socialized to do "a fair day's work," and those who did more or less than the established norm—what the work group thought was fair—were considered "rate busters" or "chiselers" and experienced pressure from the group to conform. These informal mechanisms gave informal groups of workers a degree of power in the organization (Roethlisberger and Dickson 1939).

Thinking Sociologically

How closely does each of Weber's characteristics of ideal-type bureaucracy describe your college or your work setting? Is your college highly bureaucratized, with many rules and regulations? Are decisions based on efficiency and cost-effectiveness, educational quality, or both? To what extent is your work setting characterized by hierarchy and formal rules governing your work time? To what extent by informal relationships?

Individuals in Bureaucracies

Lynndie England, a young woman from a small town in West Virginia looking for opportunity and adventure, joined the military. Instead, she became a victim and a scapegoat. The bureaucracy was the U.S. military; the setting was the Abu Ghraib prison in Iraq; the group pressure was to conform and fit in under stressful conditions by "going along with the guards" who were abusing and humiliating prisoners (Zimbardo 2004). Caring and humane military officers may allow abuse of prisoners or order bombing strikes even though they know that some innocent people will suffer. In the military command in a war, a cold, impersonal cost-benefit analysis takes priority. In bureaucracies, self-preservation is the core value. Humans can be moral, altruistic, and self-sacrificing for the good of others, but nations and other extremely large organizations (e.g., corporations or government) are inherently driven by a cost-benefit ratio (Niebuhr 1932). As one moves toward larger and more impersonal bureaucracies, the nature of interaction often changes.

Professionals in bureaucracies include doctors, lawyers, engineers, professors, and others who have particular attributes, including competency in a field, high levels of autonomy on the job, strong commitment to their field, a service orientation and commitment to the needs of the client, and a sense of intrinsic satisfaction from the work (rather than motivation rooted mainly in external rewards, such as salary). Professionals also claim authority and control in their work, but they may face conflicting loyalties to their profession and to the bureaucratic organization in which they are employed (Hall 2002). A scientist hired by a tobacco company faced a dilemma when his research findings did not support the company position that nicotine is not addictive. His superiors wanted him to falsify his research, which would be a violation of professional ethics. Should he publicly challenge the organization? Several professional whistle-blowers have done so but lost their jobs as a result. Bureaucracy, some argue, can be the number-one enemy of professionalism because the goals may be in conflict.

The potential clash between professionals and bureaucracy raises key concerns as universities, hospitals, and other large organizations are governed increasingly by bureaucratic principles (Roberts and Donahue 2000). Alienation among professionals occurs when they are highly regulated rather than when they have some decision-making authority and are granted some autonomy (Tolbert and Hall 2009). For example, high-tech companies find that hierarchical structures undermine productivity, whereas factors such as intrinsic satisfaction, flexible hours, and relaxed work environments are central to creative productivity (Florida 2002; Friedman 2005; Molotch 2003).

Women and minorities in bureaucracies often face barriers that keep them from reaching high levels of management. The result is that individuals from these groups are found disproportionately in midlevel positions with little

Women in many countries have low-paid, dead-end jobs that make them feel uninvolved and unconnected. Such jobs result in alienation, but these women need the work and have little choice but to accept them.

authority and less pay than others with similar skills and credentials (Arulampalam, Booth, and Bryan 2005; Heilman and Chen 2003). When employees have little chance for promotion, they have less ambition and loyalty to the organization (Kanter 1977). Despite this problem, research indicates that women executives bring valuable alternative perspectives to organizational leadership. They share information readily, give employees greater autonomy, and stress interconnectedness between parts of the organization, resulting in a more democratic type of leadership style (Helgesen 1995; Kramer 2010). The more women in an organization's senior positions, the more likely newcomers are to find support.

The interaction of people who see things differently because of religious beliefs, ethnic backgrounds, and gender experiences increases productivity in many organizations. Having a wide range of perspectives can lead to better problem solving (Florida 2004; Molotch 2003), increased productivity, and reduction in the barriers to promotion that are dysfunctional for organizations.

Problems in Bureaucracies

Bureaucratic inefficiency and red tape are legendary. Yet, bureaucracies are likely to stay, for they are the most efficient form of modern organization yet devised. Nonetheless, several individual and organizational problems created by bureaucratic structures are important to understand.

Alienation, feeling uninvolved, uncommitted, unappreciated, and unconnected, occurs when workers experience routine, boring tasks or dead-end jobs with no possibility of advancement. Marx believed that alienation is a structural feature of capitalism with serious consequences: Workers lose their sense of purpose and seem to become dehumanized and objectified in their work, creating a product that they often do not see completed, that they feel no pride in, and for which they do not get the profits (Marx [1844] 1964).

Dissatisfaction comes from low pay and poor benefits; routine, repetitive, and fragmented tasks; lack of challenge and autonomy, leading to boredom; and poor working conditions. Workers who see possibilities for advancement put more energy into the organization, but those stuck in their positions are less involved and put more energy into activities outside the workplace (Kanter 1977).

Workers' productive behavior is not solely a result of rational factors such as pay and working conditions but is influenced by emotions, beliefs, and norms (Roethlisberger and Dickson 1939). To address these factors, some organizations have moved toward workplace democracy, employee participation in decision making, and employee ownership plans.

Thinking Sociologically

How might participation in decision making and increased autonomy for workers enhance commitment and productivity in your place of work or in your college? What might be some risks or downsides to such worker input and freedom?

Oligarchy, the concentration of power in the hands of a small group, is a common occurrence in organizations. In the early 1900s, Robert Michels, a French sociologist, wrote about the iron law of oligarchy, the idea that power becomes concentrated in the hands of a small group of self-perpetuating leaders in political, business, voluntary, or other organizations. Initially, organizational needs, more than the motivation for power, cause these few stable leaders to emerge. As organizations grow, a division of labor emerges so that only a few leaders have access to information, resources, and the overall picture. This, in turn, causes leaders who enjoy their elite positions of power to become entrenched (Michels [1911] 1967). Yet recent events in the Middle East and North Africa illustrate that concentrated authority is being challenged. Likewise, in the United States, Wall Street high rollers felt immune from interference in their insulated world, but the public demanded the reinstatement of regulations to limit their power and make them accountable to someone for their actions.

Goal displacement occurs when the original motives or goals of the organization are displaced by new secondary goals. Organizations form to meet specific goals. Religious organizations are founded to worship a deity and serve humanity on behalf of that deity; schools are founded to educate children; and social work agencies are organized to serve the needs of citizens who seem to have fallen between the cracks. Yet, over time, original goals may be met or become less important as other motivations and interests emerge (Merton 1968b; Whyte 1956).

Policy makers have explored other ways of organizing society. Recent approaches attempt to deal with problems discussed previously. Alternative organizational structures such as *employee-owned companies* are owned in part or in full by employees who are given or can buy shares. Profit sharing is generally part of the deal. If the company does well, all workers benefit. Studies show that this increases production and improves dedication, but the decision-making process is slower, and there is increased risk for the individual employees. However, the idea is that employees have a stake in their company and the job they do (Blasi, Kruse, and Bernstein 2003). *Democratic-collective organizations* rely on cooperation, place authority and decision making in the collective group, and use personal appeals to ensure that everyone participates in problem solving. The rules, hierarchy, and status distinctions are minimized, and hiring is often based on shared values and friendships (Deming 2000).

Although we may fantasize about escaping from the rat-race world or about isolating ourselves on an island, groups and bureaucracies are a part of modern life that few can escape.

Thinking Sociologically

In what areas of your college do you see goal displacement—that is, decisions being driven by goals other than the original purpose of the university (such as educating a new generation and expanding human knowledge)?

National and Global Networks: The Macro Level

In one sense, understanding people who are unlike us, networking with people who have different cultures, and making allowances for alternative ideas about society and human behavior have become core competencies in our globalizing social world. Increasingly, colleges have study-abroad programs, jobs open up for teaching English as a second language, and corporations seek employees who are multilingual and culturally competent in diverse settings.

One recent college graduate with a sociology degree found that she could use her sociology skills in leading groups of college-age students in international travel experiences. She explains this applied use of sociology in the next "Applied Sociologist at Work."

With modern communication and transportation systems and the ability to transfer ideas and money with a touch of the keyboard, global networks are superseding national boundaries. Multinational corporations now employ citizens from around the world and can make their own rules because there is no oversight body. National systems and international organizations, from the United Nations to the World Bank to multinational corporations, are typically governed through rational bureaucratic systems.

One complicating factor at the national and global levels is the reality that many languages are used to communicate. Although Spanish, Chinese, and Arabic are widely used in homes across the globe, English has become the language of commerce. This is because much of the political and economic power and control over the expansion of technology has been located in English-speaking countries.

Technology has some interesting implications for human interaction. E-mail, webpages, chat rooms, and blogs have made it possible for people around the world to talk with each other, exchange ideas, and even develop friendships. Although authoritarian governments and dictators in the world try, they can no longer keep the citizenry from knowing what people outside their country think. In China, as fast as officials censor information on the World Wide Web, computer experts find ways around the restrictions. Egyptian officials tried to block satellite communication and restrict access to the information highway during the January–February 2011 revolution. Despite these efforts by some nations to limit access, the process of change is occurring so rapidly that we cannot know what technology will mean for the global world, but it is almost certain that the processes of interaction across national lines will continue despite some efforts to curb communication. The World Wide Web was not invented until 1989, and e-mail has been available to most citizens—even in affluent countries—only since the early 1990s (Brasher 2004). Countries are still struggling with what this means for change.

Within nations, people with common interests, be it organic food or peace, can contact one another. Hate groups also set up webpages and mobilize others with their view of the world, and terrorists use the Internet to communicate with terrorist cells around the world. Clearly, ability to communicate around the globe is transforming the world and our nations in ways that affect local communities and private individuals. Indeed, the Internet has become a major outlet for sellers and a major source for consumers. A wide range of products can be obtained through Internet orders

from websites that have no geographical home base. In some cases, there are no actual warehouses or manufacturing plants, and there might not be a home office. Some businesses are global and exist in the Ethernet, not in any specific nation (Ritzer 2007). This reality makes one rethink national and global loyalties and boundaries.

Thinking Sociologically

How do you think global interaction will be transformed through Internet technology? How will individual connections at the micro level be affected by these transformations at the macro levels of our social world?

Policy Issues: Women and Globalization

Women in much of the world are viewed as second-class citizens, the most economically, politically, and socially marginalized people on the planet, caught in expectations of religion, patriarchy, and roles needed to sustain life (Schneider and Silverman 2006). To help their children survive, women do whatever their situation allows to make money: street-selling,

Around the world, two thirds of the poorest adults are women, even though women produce more than two thirds of food supplies in the world. In this photo an Indian woman transports goods to market (right).

The Applied Sociologist at Work— Elise Roberts

Using Sociology in International Travel and Intercultural Education

After graduating from college with a bachelor's degree in sociology, I left the country to backpack through Mexico and Central America. My studies helped me be more objective and aware as I experienced other societies. My international travel helped me examine my own societal assumptions and further understand how society creates so much of one's experience and view of the world. Eventually, I found a job leading groups of teenagers on alternative education trips abroad. I was excited to get the job, but I was soon to learn that leading groups of teenagers in other countries is actually very hard work.

What struck me on meeting my first group was that I had very few students who initially understood this sociological perspective that I took for granted and that was so helpful in dealing with others. At times, my students would make fun of the way things were done in other countries, calling them "weird" or "stupid." They would mock the local traditions, until we discussed comparable traditions in American culture. These students were not mean or unintelligent. In fact, they loved the places we were seeing and the people we were meeting. They just thought everything was factually, officially weird. They had been socialized to understand their own society's ways as "right" and "normal." They were fully absorbed in the U.S. society, and they had never questioned it before.

It was rewarding to apply concepts from my textbooks to the real world. My coleaders and I learned to have fun while encouraging our students to become more socially conscious and analytical about their travel experience. We sent the groups out on scavenger hunts, and they would inevitably come back proudly announcing what they had paid for a rickshaw ride—only to learn that they had paid 10 times the local price. We would use this experience to talk about the role of foreigners, the assumptions that the local population made due to our skin color, and the culture of bartering. They had to learn to understand "odd" gestures, like pointing with the lips or side-to-side nodding. We would use these experiences to discuss nonverbal communication and gestures that each culture takes for granted.

We would encourage our novice travelers to interact with the people around them, which helped them understand the struggles facing immigrants and non-English speakers in countries such as the United States. We would force them to have conversations while standing toe-to-toe with each other, and they would finish with backaches from leaning away from one another. We would not allow them to explain their behavior with "because it's creepy to stand so close together," even though this was the consensus. "Why do you feel uncomfortable?" we would ask. "Why is this weird?" The answer has to do with social constructions of what is "normal" in any society.

Of course, while traveling internationally, one is surrounded by various other sociological issues, such as different racial or ethnic conflicts, gender roles, or class hierarchies, and learning about these issues was a part of our program as well. Without realizing it, many group conversations and meetings began to remind me of some of my favorite Macalester College classes. "Study sociology!" I would say, plugging my major to the most interested students.

I have always thought that travel is an incredibly useful means not only to learn about the society and culture one is visiting but also to learn much about oneself and one's home society. For teenagers who otherwise might never step back to think about the role of being a foreigner or the traditions and social patterns they take for granted, it is even more important. Traveling abroad on my own and leading programs abroad were such extremely rich and rewarding experiences not only due to the cross-cultural exchanges and the intense personal examination that I saw in my students but also because it was fascinating and rewarding to be able to use my sociology degree every day on my job.

Note: Elise Roberts graduated from Macalester College with a major in sociology. Her recent travels took her through Central America, the South Pacific, and many parts of Asia. She competed a master's degree from Columbia University in international social work and is now a regional coordinator for Witness for Peace.

low-paying factory assembly line work, piecework (e.g., sewing clothing in their homes), prostitution and sex work, and domestic service. Women also produce 80% to 90% of the food crops in sub-Saharan Africa, and more than 50% in the world (Food and Agriculture Organization of the United Nations 2008). They run households and are often the main support for children in poor countries. Yet, due to the "feminization of poverty," two out of three poor adults are women (Enloe 2006; Robbins 2005). Causes of the many problems facing women are found at each level of analysis and will be discussed throughout this book.

One macro-level organization with policies to help raise the status of women through development is the United Nations. Each decade since the 1960s, this global organization has set forth plans to improve conditions for women. Early plans to help poor countries were driven by interests of capitalist countries in the first world. This left women out of the equation and planning. Women suffered greatly under some of these plans because their positions and responsibilities did not change. Conditions got worse as development money went to large corporations with the idea that profits would trickle down to the local level, an idea that did not materialize in most countries (Boulding and Dye 2002).

More recently, the United Nations has sponsored conferences on the status of women. In July 2010, the UN launched a new division, United Nations Women, headed by former Chilean president Michelle Bachelet. The United Nations Division for the Advancement of Women and United Nations Women initiates policies to help educate local women in health, nutrition, basic first aid, and business methods. Many projects have been spawned, including micro-lending organizations (Grameen Bank, SEWA, FINCA, CARE), allowing individuals and groups of women to borrow money to start cooperatives and other small-business ventures. These initiatives can have very beneficial effects if they are used wisely by policy makers who care about the people rather than just about their own benefits. In the next few chapters, we will learn about factors that cause some individuals and countries to be poorer than others.

Our networks play a major role in setting our norms and controlling our behaviors, usually resulting in our conformity to the social expectations of our network associates. This, of course, contributes to stability of the entire social system because deviation can threaten the existence of "normal" patterns. The next chapter explores deviance, including crime.

What Have We Learned?

Each of us has a network of people and groups that surround us. The scope of those networks has broadened with increased complexity and includes the global social world. Indeed, it is easy not to recognize how extensively our networks reach, even to the global level. Although some of our social experiences are informal (unstructured), we are also profoundly affected by another phenomenon of the past three centuries—highly structured bureaucracies. As a result of both, our intimate experiences of our personal lives are far more extensively linked to meso- and macro-level events and to people and places on the other side of the globe than was true for our parents' generation. If we hope to understand our lives, we must understand this broad context. Although it may have been possible to live without global connections and bureaucratic systems several centuries ago, these networks are intricately woven into our lifestyles and our economic systems today. The question is whether we control these networks or they control us.

Key Points:

- People in the modern world are connected through one acquaintance to another in a chain of links, referred to as networks. (See pp. 119–120.)

- Increasingly our electronic technology is creating networks that span the globe, but this same impersonal technology is used to enhance friendship networks and to meet a romantic life partner. (See pp. 119–121.)

- Interpersonal interactions at the micro level are affected by unspoken assumptions that are understood due to context, by nonverbal communication, and by the symbolism of space between people. (See pp. 121–125.)

- Many of our behaviors are shaped by the statuses (social positions) we hold and the roles (expectations associated with a status) we play. However, our multiple status occupancy can create role conflicts (between the roles of two statuses) and role strains (between the role expectations of a single status). (See pp. 125–129.)

- When the norms of behavior are unclear, we may experience anomie (normlessness), and this ambiguity compromises our sense of belonging and is linked to reasons for individuals performing one of the most personal of acts—suicide. (See pp. 129–130.)

- Various types of groups affect our behavior—from primary and secondary groups to peer groups and reference groups. (See pp. 130–132.)

- At the meso level, we find that formal organizations in the contemporary modern world are ruled by rational calculation of the organization's goals rather than by tradition or emotional ties. These modern formal organizations expand, are governed by impersonal formal rules, and stress efficiency and rational decision making. They have come to be called bureaucracies. (See pp. 133–134.)

- Bureaucracies often involve certain problems for individuals, and they may actually develop certain problems that make them inefficient or destructive. (See pp. 134–138.)

- While bureaucratization emerged at the meso level as the defining element of the modern world, it is also found at the national and global levels where people do not know each other on a face-to-face basis. (See pp. 137–138.)

- Some scholars think that this impersonal mode of organizing social life—so common in the West for several centuries now—is a critical factor in anti-American and anti-Western resistance movements. (See pp. 138–139.)

Contributing to Our Social World: What Can We Do?

At the Local Level:

- *Tutoring and other programs:* Most campuses have programs that are designed to help students who are struggling with their studies. If your campus has one or more of these, contact its office and arrange to observe and/or volunteer in its work. Helping students build their *social capital,* which includes their knowledge of ways to obtain help they need, can increase their chances of success.

At the Organizational or Institutional Level:

The social capital theory can also be applied to meso-level community organizations, including those that work to reduce conflict or improve intergroup relations.

- *The Anti-Defamation League, the Arab Anti-Defamation League, and the National Association for the Advancement of Colored People:* These organizations often use volunteers or interns and can provide you with the opportunity to learn about the extent to which social contacts and networks play a role in managing social conflict.

- *Microcredit organizations:* Organize or contribute to a microcredit organization (e.g., FINCA, www.finca.org) to help women in Global South countries.

Visit **www.sagepub.com/oswcondensed2e** for online activities, sample tests, and other helpful information. Select "Chapter 5: Interaction, Groups, and Organizations" for chapter-specific activities.

Deviance and Social Control

Sickos, Weirdos, Freaks, and Folks Like Us

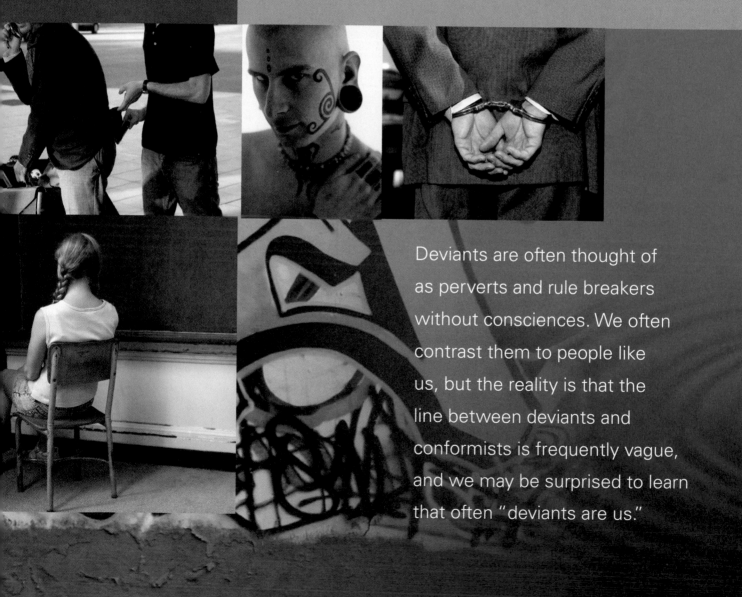

Deviants are often thought of as perverts and rule breakers without consciences. We often contrast them to people like us, but the reality is that the line between deviants and conformists is frequently vague, and we may be surprised to learn that often "deviants are us."

Global Community

Society

National Organizations,
Institutions, and Ethnic Subcultures

Local Organizations
and Community

Me (and My
Deviant
Friends)

Micro: Violations of local
ordinances: theft, burglary, local corruption

Meso: Violations of state laws;
crimes within and by corporations

Macro: Federal crimes (treason, tax fraud);
state crimes (domestic terrorism); Internet fraud

Macro: Global environmental destruction; international terrorism; human rights violations

Think About It	
Me (and My Minority Friends)	Are you deviant? Who says so?
Local Community	Why do some people in your community become deviant?
National Institutions; Complex Organizations; Ethnic Groups	What are the implications of organized crime or occupational crime in large bureaucratic organizations?
National Society	What are the costs—and the benefits—of deviance for the nation?
Global Community	How can a global perspective on crime enhance our understanding of international criminal activities?

Wafa Idris grew up as other girls do, but she was destined to make world headlines on January 27, 2002—as the first female suicide bomber in the Palestinian-Israeli conflict. She was the age of most university students reading this book, but she never had an opportunity to attend college. The story sounds simple: She was to take explosives across the Israeli border for the intended bomber, her brother. Instead, she blew up herself and an Israeli soldier. She was declared a martyr, a *sahida,* by the al-Aqsa Martyrs Brigade, which took credit for the attack. Her act was given approval by the group's political leadership, opening the way for other women to follow. Why did this happen? What motivated her to commit suicide and take another life in the process? Was she driven by ideology to participate in the Palestinian-Israeli struggle? Were there sociological factors that affected her decision? Most important, was she criminally deviant in carrying out this act, and according to whom?

Wafa grew up in Palestine. She was married at a young age but did not produce children. Because this leaves a woman with no role to play in Palestinian society, her husband divorced her and remarried. She had no future, for who would want a barren, divorced woman in a society that values women for their purity and their childbearing ability? She was a burden to her family. Her way out of an impossible and desperate situation was to commit suicide, bringing honor and wealth to her family and redeeming herself in the process. Other women who followed Wafa have similar stories: Most shared an inability to control their own lives in the patriarchal (male-controlled) society (Handwerk 2004; Victor 2003).

Ironically, terrorist cells in Palestine saw this as an opportunity to recruit other vulnerable women with nothing to lose. It is not a measure of equality that women, too, commit suicide bombings. This development results from a structure of inequality (Victor 2003). Our question is this: Are these women deviant criminal terrorists, mentally ill "crazies," powerless and invisible victims in a patriarchal society, or martyrs who should be honored for their acts? Who says so? Each of these views is held by commentators

on the situation. From this opening example, we can begin to see several complexities that arise when considering deviance, and some of these ideas may encourage new ways of looking at what is deviant. In this chapter, we will consider who is deviant, under what circumstances, and in whose eyes. We will find that most people around the world conform to social norms of their societies most of the time, and we will explore why some turn to deviance.

In this chapter, we discuss **deviance**—the violation of social norms—and the social control mechanisms that keep most people from becoming deviant. We also explore crime, the form of deviance on which formal penalties like fines, jail, and prison are imposed by societies' legal system. The content of this chapter may challenge some of your deeply held assumptions about human behavior and defy some conceptions about deviance. In fact, it may convince you that we are all deviant at some times and in some places. The self-test in the next "Engaging Sociology" illustrates this point. Try taking it to see whether you have committed a deviant act.

What Is Deviance?

Deviance refers to violation of the society's norms, which then evokes negative reactions from others. The definition is somewhat imprecise because of constantly changing ideas and laws about what acts are considered deviant. Some acts are deviant in most societies most of the time: murder, assault, robbery, and rape. Most societies impose severe penalties on these forms of deviance, and if the government imposes severe formal penalties, then the deviant actions become crimes. Occasionally, deviant acts may be overlooked or even viewed as understandable, as in the case of looting by citizens following Hurricane Katrina along the Gulf Coast in August 2005. Killing innocent people, looting, and burning houses during a civil or tribal war, which occurred in the aftermath of the 2008 presidential election

Engaging Sociology

Who Is Deviant?

Please jot down your answers to the following self-test questions. There is no need to share your responses with others.

Have you ever engaged in any of the following acts?

☐   1.   stolen anything, even if its value was under $10

☐   2.   used an illegal drug

☐   3.   misused a prescription drug

☐   4.   run away from home prior to age 18

☐   5.   used tobacco prior to age 18

☐   6.   drunk alcohol prior to age 21

☐   7.   engaged in a fist fight

☐   8.   carried a knife or gun

☐   9.   used a car without the owner's permission

☐   10.   driven a car after drinking alcohol

☐   11.   (boys) forced a girl to have sexual relations against her will

☐   12.   offered sex for money

☐   13.   damaged property worth more than $10

☐   14.   been truant from school

☐   15.   arrived home after your curfew

☐   16.   been disrespectful to someone in authority

☐   17.   accepted or transported property that you had reason to believe might be stolen

☐   18.   taken a towel from a hotel room after renting the room for a night

All of the above are delinquent acts (violations of legal standards), and most young people are guilty of at least one infraction. However, few teenagers are given the label of *delinquent*. If you answered yes to any of the preceding questions, you have committed a crime in the state of Ohio and in many other states. Your penalty or sanction for the infraction would range from a stiff fine to several years in prison—*if* you got caught!

Questions:

1. Do you think of yourself as being deviant? Why or why not?

2. Are deviants only those who get caught? For instance, if someone steals your car but avoids being caught, is he or she deviant?

in Kenya, also may seem reasonable to those committing the atrocities. Other acts of deviance are considered serious offenses in one society but tolerated in another. Examples include prostitution, premarital or extramarital sex, gambling, corruption, and bribery. Even within a single society, different groups may define deviance and conformity quite differently. The state legislature may officially define alcohol consumption by 19-year-olds as deviant, but on a Saturday night at the fraternity party, the 19-year-old "brother" who does not drink may be viewed as deviant by his peers. What do these cases tell us about what deviance is?

Deviance is socially constructed. This means that members of groups in societies define what is deviant. Consider the phenomenon of today's young people getting tattoos, studs, and rings anywhere on the body they can place one. Is this deviant? It depends on who is judging. Are tattoos, studs, and rings symbols of independence and rebellion, a "cool" and unique look? What if many people begin to adopt the behavior? Is it still a sign of independence, or does it become conformity to a group? When The Beatles started the long-hair rage in the 1960s, this was deviant behavior to many. Today, we pay no special attention to men with long hair.

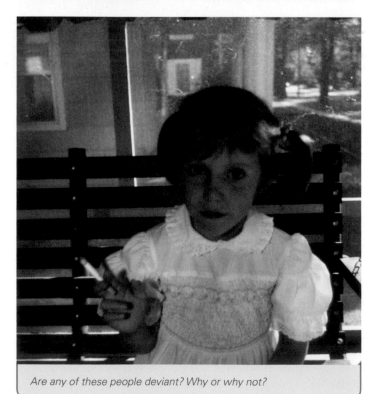

Are any of these people deviant? Why or why not?

Some acts are deviant at one time and place and not at others. Stem cell research has been viewed as unacceptable and a violation of U.S. law because of ethical concerns about using or destroying human cells. Yet, in many other countries, there are no moral restrictions, and scientists are proceeding with such research.

An individual's status or group may be defined as deviant. Some individuals have a higher likelihood of being labeled deviant because of the group into which they were born, such as a particular ethnic group, or because of a distinguishing mark or characteristic, such as a deformity. Others may escape being considered deviant because of their dominant status in society. The higher one's status, the less likely one is to be suspected of violating norms and characterized as "criminal."

Even the looting that happened following Hurricane Katrina was addressed differently when it was done by whites rather than African Americans (Huddy and Feldman 2006; Thompson 2009). (An image of a "looter" in New Orleans appears on this page.) The media showed photos of black "looters" who "stole food," but the same media described whites who "broke into grocery stores" in search of food as "resourceful." Likewise, gays and lesbians are often said to be deviant and accused of flaunting their sexuality. Heterosexuals are rarely accused of "flaunting" their sexuality, regardless of how overtly flirtatious or underdressed they are. So one's group membership or ascribed traits may make a difference in whether or not one is defined as deviant.

Deviance represents a breakdown in norms. However, according to structural-functionalist theory, it can be problematic or functional for society. Deviance serves vital functions by setting examples of what is considered unacceptable behavior, providing guidelines for behavior that is necessary to maintain the social order, and bonding people together through their common rejection of the deviant behavior. Deviance is also functional because it provides jobs for those who deal with deviants—police, judges, social workers, and so forth (Gans 2007). Furthermore, deviance can signal problems in society that need to be addressed and can therefore stimulate positive change. Sometimes, deviant individuals break the model of conventional thinking, thereby opening the society to new and creative paths of thinking. Scientists, inventors, and artists have often been rejected in their time but have been honored later for accomplishments that positively affected society. Vincent van Gogh, for example, lived in poverty and mental turmoil during his life, but he became recognized as a renowned painter after his death, with his paintings selling for millions of dollars.

Our tasks in this chapter are to understand what deviance is, what causes it, and where it fits in the social world. We look at theoretical perspectives that help explain deviance, how some deviant acts become crimes, and what policies might be effective in controlling or reducing crime.

Thinking Sociologically

What might be some negative consequences in a heterogeneous society if everyone conformed and no one ever deviated from social standards?

Misconceptions About Deviance

Many common beliefs about deviance are, in fact, misperceptions. Using scientifically collected data, the sociological perspective helps dispel false beliefs, as in the examples below.

Popular belief: Some acts are inherently deviant. Fact: Deviance is relative to time, place, and status of the individual. At some place or time, almost any behavior you can mention has been defined as deviant, just as most acts have been legal or even the typical behavior in other times and places. For example, homosexual liaisons were normal for men in ancient Greece and have been acceptable in various societies throughout time. Deviance is not inherent in certain behaviors but is defined by people and their governments (Erikson 1987). Some individuals are defined as deviant because they do not fit into the dominant system of values and norms. They may be seen as disruptive, a liability, or a threat to the system.

Famous individuals remembered in history books were often considered deviant in their time. In the Middle Ages, for example, anyone who questioned the concept of a flat Earth at the center of the universe (with a sun and stars circling it) was a deviant and a religious heretic. In the early seventeenth century, Galileo, following Copernicus's lead, wrote a treatise based on empirical observations that upheld the concept of Earth revolving around the sun. He was tried by the Inquisition, a Roman Catholic Church court, in Rome and condemned for heresy because of his theory.

Definitions of deviance also vary depending on the social situation or context in which the behavior occurs (Clinard and Meier 2004; McCaghy et al. 2006). If we take the same behavior and place it in a different social context, perceptions of whether the behavior is deviant may well change. In Greece, Spain, and other Mediterranean countries, the clothing norms on beaches are very different from those in most of North America. Topless sunbathing by women is not at all uncommon, even on beaches designated as family beaches. The norms vary, however, even within a few feet of the beach. Women will sunbathe topless, lying only 10 feet from the boardwalk where concessionaires sell beverages, snacks, and tourist items. If these women become thirsty, they cover up, walk the 15 feet to purchase a cola, and return to their beach blankets, where

Some sociologists point out that crime can be "functional" because it creates jobs for people like these above, and it unifies the society against the nonconformists.

they again remove their tops. To walk onto the boardwalk topless would be highly deviant.

Popular belief: Those who deviate are socially identified and recognized. Fact: Most of us deviate from some norm at some time, as you saw when completing the questionnaire on deviance. (See the preceding "Engaging Sociology.") Yet only about one third of all crime that is reported to the police in the United States ever leads to an arrest. This means that two thirds is never officially handled through the formal, legal structure, and the perpetrators escape being labeled deviant.

Popular belief: Deviants purposely and knowingly break the law. Fact: Although it is a popular notion that those who engage in deviant behavior make a conscious choice to do so, much deviance is driven by emotion, encouraged by friends, caused by disagreements over norms (as in the case of whether marijuana use should be criminalized), or a result of conditions in the immediate situation (as in cases of teens spontaneously engaging in a behavior in response to boredom or a struggle for prestige among peers) (McCaghy et al. 2006).

Popular belief: Deviance occurs because there is a dishonest, selfish element to human nature. Fact: While surveys show many of us believe propensity for deviance is inborn, most people who commit deviant acts do not attribute their own deviance to basic dishonesty or other negative personality factors. Most people believe they have clear understandings of what deviance is and that deviant individuals know when they are being deviant. Social scientists do not support the idea that deviance is inborn as the research reported in this chapter shows.

Medical marijuana patient Kay Mitchell, 82, of Sonoma joined more than 1,000 people protesting outside the California state capitol to continue state-approved and licensed medical marijuana dispensaries in California, which serve mostly terminally ill patients suffering from cancer, AIDS, and other ailments.

Thinking Sociologically

Think of examples in your life that illustrate the relative nature of deviance. For instance, are some of your behaviors deviant in one setting but not in another, or were they deviant when you were younger but not deviant now?

Crime: Deviance That Violates the Law

When the criminal justice system gets involved and formal penalties are imposed by society, we refer to deviance as *crime*—deviance that violates criminal law. Laws reflect opinions of what is considered right or wrong at a particular time and place in a society. Like all other norms, laws change over time, reflecting changing public opinion based on social conditions or specific events. Still, there are formal sanctions—punishments—that the government imposes for violation of laws, and the stigma associated with these deviant acts results in being identified as "criminal."

At the end of the 1920s, 42 of the 48 U.S. states had laws forbidding interracial marriage (Coontz 2005). Legislatures in half of these states removed those restrictions by the 1960s, but the rest of these laws became unconstitutional only after a U.S. Supreme Court ruling in 1968. Today, this legal bar on interracial marriage has been eliminated completely, illustrating that most laws change to reflect the times and sentiments of the majority of people. When members of society are in general agreement about the seriousness of deviant acts, these are referred to as **consensus crimes** (Brym and Lie 2007; Goodman and Brenner 2002; Hagan, Silva, and Simpson 1977). Predatory crimes (premeditated murder, forcible rape, and kidnapping for ransom) are consensus crimes that are considered wrong in and of themselves in most nations.

By contrast, **conflict crimes** occur when one group passes a law over which there is disagreement or that disadvantages another group (Hagan 1993, 1994). Examples include laws concerning public disorder, chemical (drug and alcohol) offenses, prostitution, gambling, property offenses, and political disenfranchisement (denying voting rights only to some citizens—females, particular ethnic groups, or those without property). Public opinion about the seriousness of these crimes is often divided, based on people's different social class, status, and interests. The severity of societal response also varies, with high disagreement over the harmfulness of conflict crimes. Consensus or conflict about whether the behavior is harmful has implications for the punishment or support experienced by the person who is accused.

Crimes are often thought to be the most threatening forms of deviance, but it is important to recognize that they

In the United States, up until 1968, this couple would have been violating the law in roughly half of the states, and their family would be "illegitimate." Interracial marriages were illegal until the Supreme Court decided otherwise. Change in definitions of what is illegal has been common in the past 50 years.

are still just one type of deviant behavior. We will discuss crime further at a later point in this chapter. Be aware that as we discuss theories in the next section, the theories explain a range of deviant acts, including (but not limited to) crimes.

What Causes Deviant Behavior? Theoretical Perspectives

Helena is a delinquent. Her father deserted the family when Helena was 10, and before that, he had abused Helena and her mother. Now, her mother has all she can cope with; she is just trying to survive financially and keep her three children in line. Helena gets little attention and little support or encouragement in her school activities. Her grades have fallen steadily. As a young teen, she sought attention from boys, and in the process, she became pregnant. Now, the only kids who have anything to do with her are others who have also been in trouble and labeled delinquent. Helena's schoolmates, teachers, and mother see her as a delinquent troublemaker, and it would be hard for Helena to change their views and her status.

How did this happen? Was Helena born with a biological propensity toward deviance? Does she have psychological problems? Is the problem in her social environment? Helena's situation is, of course, only one unique case. Sociologists cannot generalize from Helena to other cases, but

they do know from their studies that there are thousands of teens with problems like Helena's.

Throughout history, people have proposed explanations for why some members of society "turn bad"—from biological explanations of imbalances in hormones and claims of innate personality defects to social conditions within individual families or in the larger social structure. Biological and psychological approaches focus on personality disorders or abnormalities in the body or psyche of individuals, but they generally do not consider the social context in which deviance occurs.

Sociologists examine why certain acts are defined as deviant, why some people engage in deviant behavior, and how other people in the society react to deviance. Sociologists place emphasis on understanding the interactions, social structure, and social processes that lead to deviant behavior, rather than on individual characteristics. They consider the socialization process and interpersonal relationships, group and social class differences, cultural and subcultural norms, and power structures that influence individuals to conform to or deviate from societal expectations (Liska 1999). Theoretical explanations about why people are deviant are important because the interpretations influence social policy decisions about what to do with deviants.

This section explores several approaches to understanding deviance. Some theories explain particular types of crime (say, theft as opposed to sexual assault) better than others, and some illuminate micro-, meso-, or macro-level processes better than others. Taken together, these theories help us understand a wide range of deviant and criminal behaviors.

Micro-Level Explanations of Deviance: Rational Choice and Interactionist Perspectives

No one is born deviant. Individuals learn to be law-abiding citizens or to be deviant through the process of socialization as they develop their social relationships. *Social control theory* contends that most people are law-abiding citizens because the desire to fit into the group encourages conformity with the norms. Then why do some people become deviant and others follow the norms of society? A rational choice explanation focuses on cost-benefit analysis of one's behavioral choices. Social control theory suggests that deviance is to be expected unless a person has an investment in the existing social system. Another approach argues that deviance is learned, often through the process of *differential association* with others who commit deviant acts. Deviants are exposed to the opportunity to commit delinquent acts through their social relationships with peers and family members, and they come to be labeled deviant. *Labeling theory* points to society's response to unacceptable behaviors and the labeling of those who violate norms as deviant.

Rational Choice Theory

The basic idea behind *rational choice theory* is that when individuals make decisions, they calculate the costs and benefits to themselves. They consider the balance between pleasure and pain. Social control comes from shifting the balance toward more pain and fewer benefits for those who deviate from norms. However, some members of society find crime to be to their advantage within their situations and opportunities, the product of a conscious, rational, calculated decision made after weighing the costs and benefits of alternatives. Often they choose lives of crime after failure in school or work or after seeing others succeed in crime.

Rational choice theorists believe punishment—imposing high "costs" for criminal behavior, such as fines, imprisonment, or even the death penalty—is the way to dissuade criminals from choosing crime. When the cost outweighs the potential benefit and opportunities are restricted, it deters people from thinking crime is a "rational" choice (Earls and Reiss 1994; Winslow and Zhang 2008). Even just changing the perception of the cost-benefit balance can be important in lowering crime. Criminals make decisions based on the situational constraints and opportunities in their lives (Schmalleger 2009).

Social Control Theory

One of sociology's central concepts is *social control* (Gibbs 1989; Hagan 2007). Social control theory focuses on why most people conform most of the time and do not commit deviant acts. If human beings were truly free to do whatever

they wanted, they would likely commit more deviant acts. Yet, to live near others and with others requires individuals to control their behaviors based on social norms and sanctions—in short, social control.

A perpetual question in sociology is the following: How is order possible in the context of rapidly changing society? A very general answer is that social control results from norms that promote order and predictability in the social world. When many people fail to adhere to these norms or when the norms are unclear, the stability and continuance of the entire social system may be threatened, as is happening in some countries of the world such as Somalia.

Control theory contends that people are bonded to others by four powerful factors:

1. *Attachment* to other people who respect the values and rules of the society. Individuals do not want to be rejected by those to whom they are close or whom they admire.

2. *Commitment* to conventional activities (e.g., school and jobs) that they do not want to jeopardize.

3. *Involvement* in activities that keep them so busy with conventional roles and expectations that they do not have time for mischief.

4. *Belief* in the social rules of their culture, which they accept because of childhood socialization and indoctrination into conventional beliefs.

Should these variables be weakened, there is an increased possibility that the person could commit deviant acts (Hirschi [1969] 2002). In short, these factors increase the benefit of conformist behavior and the cost of deviance, so control theory is often considered one type of rational choice explanation.

Two primary factors shape our tendency to conform. The first is internal controls, those voices within us that tell us when a behavior is acceptable or unacceptable, right or wrong. The second is external controls, society's formal or informal controls against deviant behavior. Informal external controls include smiles, frowns, hugs, and ridicule from close acquaintances (Gottfredson and Hirschi 1990). Formal external controls come from the legal system through police, judges, juries, and social workers.

Rational choice theories hold that people weigh the possible negative consequences against the benefits, and if the benefits outweigh the costs, then violating official rules or expectations may be worth it. This would suggest that the cost must be increased to deter deviance. This little boy is weighing the costs and benefits of taking cookies from the cookie jar on the counter, and the benefits are looking pretty sweet!

Thinking Sociologically

Think of a time when you committed a deviant act or avoided doing so despite tempting opportunity. What factors influenced whether you conformed to societal norms or committed a deviant act?

Differential Association Theory

If someone offered you some heroin, what would determine whether you took it? First, would you define sticking a needle in your arm and injecting heroin as a good way to spend your afternoon? Second, do you typically hang around with others who engage in this type of behavior and define it as "the thing to do"? Third, would you know the routine—how to cook the heroin to extract the liquid—if you had never seen it done? You likely would not know the proper technique for how to prepare the drug or how to inject it. Why? Being a drug user depends on whether you have associated with drug users and whether your family and friends define drug use as acceptable or deviant.

Differential association theory refers to two processes that can result in individuals learning to engage in crime. First, association with others who share criminal values and commit crimes results in learning how to carry out a criminal act (Sutherland, Cressey, and Luckenbil 1992); second, learning in a particular social context—socialization into a counterculture—results in reinforcement of criminal behavior (Akers 1992, 1998; Lee, Akers, and Borg 2004). Differential association theory, then, is based on the idea of

Shooting drugs as a way of spending time and money is a way of life for some young people. Yet most teens would not know the technique for preparing and injecting illegal drugs, nor would they have learned from associates that this is a fun or acceptable way to spend one's time.

socialization, and since it involves accepting the definition of reality of one's friends or associates, it is sometimes seen as a specific application of symbolic interactionism.

Differential association theory focuses on the process of learning deviance from family, peers, fellow employees, political organizations, neighborhood groups such as gangs, and other groups in one's surroundings (Akers 1992; Akers et al. 1979; Sutherland et al. 1992). Helena, for example, came to be surrounded by people who made dropping out of school and other delinquent acts seem normal. If her close friends and siblings were sexually active as teens, her teen pregnancy might not be remarkable and might even be a source of some prestige with her group of peers.

According to differential association theory, the possibility of becoming deviant depends on four factors related to associating with a deviant group: the duration of time spent with the group, the intensity of interaction, the frequency of interaction, and the priority of the group in one's friendship networks (Sutherland et al. 1992). If deviant behavior exists in people's social circles and if they are exposed to deviance regularly and frequently (duration and intensity), especially if they are in close association with a group that accepts criminal behavior, they are more likely to learn deviant ways. Furthermore, individuals learn motives, drives, rationalizations, and attitudes, and they develop techniques that influence behavior and cause them to commit deviant acts.

Some theorists contend that lower-class life constitutes a distinctive subculture in which delinquent behavior patterns are transmitted through socialization. The values, beliefs, norms, and practices that have evolved in lower-class communities over time can often lead to violation of laws. These values and norms have been defined by those in power as deviant. Just as upper-class youth seem to be expected and destined to succeed, lower-class youth may learn other behaviors that those with privilege have defined as delinquent and criminal (Bettie 2003; Chambliss 1973). For instance, in a recent study, Bettie (2003) found that race, class, and gender intersect in important ways to increase the labeling of (discussed next) and decrease the opportunities for lower-class and minority high school girls. For some inner-city youth, the local norms are to be tough and disrespectful of authority, to live for today, to seek excitement, and to be "cool"—these are survival techniques. With time, these attitudes and behaviors become valued in and of themselves by their peer group.

Labeling Theory

Labeling theory is related to both the symbolic interaction perspective and the conflict theory perspective. Labels (e.g., "juvenile delinquent") are symbols that have meanings that affect an individual's self-concept. Those who are labeled are often the "have-nots" of society according to conflict theory. **Labeling theory** focuses on how people define deviance—what is or is not "normal"—a core issue in the symbolic interactionist paradigm. It points to society's response to unacceptable behaviors and labeling as deviant those who violate society's norms. Labels people carry affect their own and others' perceptions, resulting in conformity or deviance.

The basic social process of labeling someone is as follows: Members of society create deviance by defining certain behaviors as deviant—smoking pot, wearing long hair, holding hands in public, or whatever is seen as inappropriate at a particular time and place. They then react to the deviance by rejecting the miscreant or by imposing penalties.

Labeling theorists define two stages in the process of becoming a deviant. **Primary deviance** is a violation of a norm that may be an isolated act, such as a young teenager shoplifting something on a dare by friends. Most people commit acts of primary deviance. However, few of us are initially labeled deviant as a result of these primary acts.

Shoplifting is often done by young people who have not been arrested for anything—a form of primary deviance.

Remember how you marked the deviant behavior test that you took at the beginning of this chapter? If you engaged in deviant acts, you were probably not labeled deviant for the offense. If you were labeled, you would likely not be in college or taking this class.

If an individual continues to violate a norm and begins to take on a deviant identity, this is referred to as **secondary deviance**. Secondary deviance becomes publicly recognized, and the individual is identified as deviant, beginning a deviant career. If a teenager such as Helena in the opening example is caught, her act becomes known, perhaps publicized in the newspaper. She may spend time in a juvenile detention center, and parents of other teens may not want their children associating with her. Employers and store managers may refuse to hire her. Soon, there are few options open to her, and others expect her to be delinquent. The teen may continue the deviant acts and delinquent acquaintances, in part because few other options are available within the expectations of community members. Society's reaction, then, is what defines a deviant person (Lemert 1951, 1972).

The process of labeling individuals and behaviors takes place at each level of analysis, from individual to society. If community or societal norms define a behavior as deviant, individuals are likely to believe it is deviant. Sanctions for juvenile delinquents can have the effect of reinforcing the deviant behavior by (a) increasing alienation from the social world, (b) forcing increased interaction with deviant peers, and (c) motivating juvenile delinquents to positively value and identify with the deviant status (Kaplan and Johnson 1991).

The **self-fulfilling prophecy** occurs when a belief or a prediction becomes reality and affects one's actions. Individuals may come to see themselves as deviant because of harassment, ridicule, rejection by friends and family, and negative sanctions. James is eight years old and already sees himself as a failure because his parents, teachers, and peers tell him he is "dumb." In keeping with the idea of self-fulfilling prophecy, James accepts the label and acts accordingly. Unless someone—such as an insightful teacher—steps in to give him another image of himself, the label is unlikely to change. Again, labeling theory focuses on the micro level: the individual and the formation of the self.

Thinking Sociologically

What labels do you carry, and how do they affect your self-concept and behavior?

Another explanation of why certain individuals and groups are labeled as deviant has to do with their status and power in society. Those who are on the fringes, away from power and nonparticipants in the mainstream, are more likely to be labeled as deviants—the poor, minorities, members of new religious movements, or those who in some way do not fit into the dominant system. Because the powerful have the influence to define what is acceptable, they protect themselves from being defined as deviants. People from different subcultures, social classes, or religious groups may be accorded deviant labels.

A study by Chambliss (1973) illustrates the process of labeling in communities and groups and the relationship between interaction and conflict theories. Chambliss looked at the behavior of two groups of boys and at the reactions of community members to their behavior. The Saints, boys from "good" families, were some of the most delinquent boys at Hanibal High School. Although the Saints were constantly occupied with truancy, drinking, wild driving, petty theft, and vandalism, not one was officially arrested for any misdeed during the two-year study. The Roughnecks were constantly in trouble with police and community residents, even though their rate of delinquency was about equal to that of the Saints.

What was the cause of the disparity between these two groups? Community members, police, and teachers alike labeled the boys based on their perceptions of the boys' family backgrounds and social class. The Saints came from stable, white, upper-middle-class families; were active in school affairs; and were precollege students whom everyone expected to become professionals. The general community feeling was that the Roughnecks would amount to nothing. They carried around a label that was hard to change, and that label was realized.

Labels are powerful and can stigmatize—branding the target as disgraceful or reprehensible. This process can be extended to a number of issues, including fatness, as the "Sociology in Your Social World" on page 156 illustrates.

Thinking Sociologically

Reflect on the essay on the following page. In addition to body weight, what physical characteristics may affect one's self-perception or cause one to be labeled different or even deviant?

While labeling theory is mostly based on symbolic interactionism (a micro-level theory), labeling theorists also point to the role of macro-level social forces—social inequality or lack of access to power—in shaping whether people are labeled or whether they can avoid being labeled. Thus, labeling theory is also sometimes used by conflict theorists.

Meso- and Macro-Level Explanations of Deviance: Structural-Functionalist and Conflict Theories

While interpersonal interaction processes can result in deviance and being labeled deviant, many sociologists believe that meso- and macro-level analyses create greater understanding of the societal factors leading to deviance. Meso-level analysis focuses on ethnic subcultures, national organizations, and institutions inside a nation. Macro-level theories focus on societies and global social systems. We look first at structural-functionalist theories of deviance, those with the longest history in sociology. They include two themes: (a) anomie, the breakdown of the norms guiding behavior, which leads to social disorganization; and (b) strain created by the difference between definitions of success (goals) and the means available to achieve those goals.

Anomie and Social Disorganization

Villagers from industrializing countries in Africa, Asia, and Latin America were pushed off marginally productive rural lands and were pulled by the lure of the city to seek better lives and means for survival. They flocked to population centers with industrial opportunities, excitement, and a chance to change their lives, but when they arrived, they were often disappointed. Poor, unskilled, and homeless, they moved into crowded apartments or shantytowns of temporary shacks and tried to adjust to the new style of life that often included unemployment.

Many industrializing countries face structural changes as their economies move from agriculture to industry or service economies. Young men in particular leave behind strong bonds and a common value system that exist in the countryside. In cities, individuals melt into the crowd and live anonymously. Old village norms that have provided the guidelines for proper behavior crumble, sometimes without clear expectations emerging to take their place. The lack of clear norms in the rapidly changing urban environment leads to high levels of social disorganization and deviant behavior.

Sociologists use the term **anomie**, or normlessness, to describe the breakdown of norms caused by the lack of shared, achievable goals and lack of socially approved means to achieve goals (Merton 1968a). When norms are absent or conflicting, deviance increases as the previous example illustrates. Émile Durkheim (1858–1917) first described this normlessness as a condition of weak, conflicting, or absent norms and values that arise when societies are disorganized. This situation is typical in rapidly urbanizing, industrializing societies; at times of

Sociology in Your Social World

Stigmatizing Fatness

By Leslie Elrod

The United States is now the fattest country in the world. "American society has become 'obesogenic,' characterized by environments that promote increased food intake, nonhealthful food, and physical inactivity. . . . One in seven low-income preschool children are obese" (Centers for Disease Control and Prevention [CDC] 2009). The CDC has found that more than 60% of American adults are classified as overweight or obese, and 25% of children are classified as obese. Deviation from the idealized image of physical thinness allows others to judge and condemn nonconforming individuals, resulting in embarrassment, severe isolation, or alienation. According to Cooley's (1902, [1909] 1983) theory of the looking-glass self, because we tend to define ourselves by others' attitudes toward us and our interaction with others, individuals who are obese may suffer lower self-esteem and have negative self-images, thus creating heightened levels of psychological distress. When a physical attribute is assigned social significance, violators of this norm are likely to endure negative labeling because of perceived physical imperfections. The obese, labeled as self-indulgent, gluttonous, lazy, sloppy, and mean, experience social condemnation. Obese women tend to experience greater discrimination than obese men.

Women are taught that their physical appearance is a valuable commodity, in both the public and private spheres. Ascertaining the degree to which they fit the media models of female perfection, women attempt to adapt their appearance to reach this standard of beauty. Holding up an image of female perfection, such as slenderness, daintiness, or being demure, the media insinuate that women themselves somehow fall short of that perfection. This is exemplified in print ads, magazines, television programming, and movies as well as merchandising directed toward females of all ages. The "feminine failing" not only jeopardizes her happiness but also challenges her femininity. The appearance discrepancy is based on the fact that over the course of the past century, as real women grew heavier, models and "beautiful" women were portrayed as increasingly thinner. An example of the change in media imagery is that of the White Rock mineral girl, portrayed as 5 feet 4 inches tall and 140 pounds in 1950. More recently, she is 5 feet 10 inches tall, weighing 110 pounds (Phipher 1994).

While much of the research on women's obsession with weight assumes that all women and girls are affected by the culture of thinness, there is some evidence to suggest that not all women are affected equally. Obesity rates are highest among minorities, the poor, and the disenfranchised. Yet Powell and Kahn (1995) have noted that "few

black women seem to have eating disorders . . . and less emphasis [is placed] on eating and weight in general among black college students compared to white students" (p. 190). When compared to black females, white women are under significantly greater social pressure to be thin. Hesse-Biber (2007) find that eating disorders, an outcome of dissatisfaction with one's body, are no longer confined to upper-class white females, indicating that the effects of body mass on self-esteem may be changing.

This is not to suggest that people of color are not concerned about body image; rather, it suggests that the messages disseminated through the majority culture are mediated in various ways through the experiences and expectations that are associated with distinct social locations. Moreover, since media portrayals are still predominantly white, whether through the use of white models or those with white features—thin lips, thin hips, straight hair, and a light complexion—there is less pressure on minorities to relate and compare themselves to the given images because these images do not represent their reference groups.

Because the obese are victims of prejudice and discrimination resulting from social norm violation, they are less likely than their non-obese counterparts to be involved in various organizational and social activities, such as extracurricular participation, for fear of social rejection. Olson, Schumaker, and Yawn (1994) found that weight-based embarrassment was indicated as a reason that obese people shied away from social obligations. Several clinical studies have documented that obese women delay seeking medical care and participating in preventative medical techniques.

Some obese persons reject the stigmatizing "fat identity," using a variety of coping mechanisms such as avoidance of others who stigmatize them or coping through immersion in supportive subcultures. Those who suffer from low self-esteem may actively seek out activities and relationships that have the capacity to improve their self-esteem. While not all obese persons experience and internalize fat stigmatization, a statistically significant number of overweight and obese juveniles do indicate poor body image and diminished self-esteem. Many experts suggest that individuals should take a proactive approach by seeking medical or psychological help, joining support groups, and dealing with the problem if it is impairing their activities.

Leslie Elrod is an assistant professor of sociology at University of Cincinnati Raymond Walters College. Two of her specialties are studies of obesity and of teaching.

Chinese soldiers tried in vain to stop student protesters in China as the students demanded more freedom to make choices. The students were defined by the government as deviant, many were arrested, and some were shot in Tiananmen Square, the main square in Beijing.

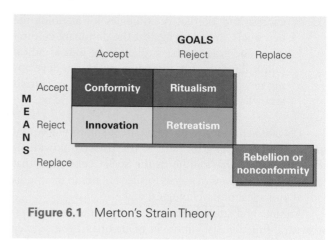

Figure 6.1   Merton's Strain Theory

sudden prosperity or depression; during rapid technological change; or when a government is overthrown. Anomie affects urban areas first but may eventually affect the whole society. Macro-level events, such as economic recessions or wars, show how important social solidarity is to an individual's core sense of values.

Strain Theory

Most people in a society share similar values and goals, but those with poor education and few resources have less opportunity to achieve those shared goals than others. When legitimate routes to success are cut off, frustration and anger result, and deviant methods may be used to achieve goals. Strain theory, primarily a macro-level theory, focuses on contradictions and tensions between the shared values and goals of a society on the one hand and the opportunity structures of the society on the other.

Strain theory (Merton 1968b) suggests that the difference between the society's definitions of goals and legitimate means, or ways, of attaining the goals can lead to strain in the society. Individuals may agree with society's definition of goals for success (say, financial affluence) but not be able to achieve it using the socially prescribed means of achieving that success. The strain that is created can lead to deviance. Merton uses U.S. society as an example because it places a heavy emphasis on success, measured by wealth and social standing. He outlines five ways individuals adapt to the strain. Figure 6.1 shows these five types and their relationship to goals and means.

To illustrate these, we trace a lower-class student who realizes the value of an education and knows it is necessary to get ahead, but who has problems financing education and competing in the middle class–dominated school setting.

1. *Conformity* means embracing the society's definition of success and adhering to the established and approved means of achieving success. The student works hard despite the academic and financial obstacles, trying to do well in school to achieve success and a good job placement. She uses legitimate, approved means—education and hard work—to reach goals that the society views as worthy.

2. *Innovation* refers to use of illicit means to reach approved goals. Our student uses illegitimate means to achieve her education goals. She may cheat on exams or get papers from Internet sources. Success in school is all that matters, not how she gets there.

3. *Ritualism* involves strict adherence to the culturally prescribed rules, even though individuals give up on the goals they hope to achieve. The student may give up the idea of getting good grades and graduating from college but, as a matter of pride and self-image, continue to try hard and to take classes. She conforms to expectations, for example, but with no sense of purpose. She just does what she is told.

4. *Retreatism* refers to giving up on both the goals and the means. The student either bides her time, not doing well, or drops out, giving up on future job goals. She abandons or retreats from the goals of a professional position in society and the means to get there. She may even turn to a different lifestyle—for example, becoming a user of drugs and alcohol—as part of the retreat.

5. *Rebellion* entails rejecting the socially approved ideas of success and the means of attaining that success. It replaces those with alternative definitions of

success and alternative strategies for attaining the new goals. Rebelling against the dominant cultural goals and means, the student may join a radical political group or a commune, intent on developing new ideas of how society should be organized and what a "truly educated" person should be.

Deviant behavior results from retreatism, rebellion, and innovation. According to Merton, the reasons individuals resort to these behaviors lie in the social conditions that lead to different levels of access to success, not in their individual biological or psychological makeup.

Anomie and strain theories fall under the macro-level structural-functionalist umbrella of theories because they help explain deviance from a social structural point of view, focusing on what happens if deviance disrupts the ongoing social order. They explore what causes deviance, how to prevent disruptions, how to keep change slow and non-disruptive, and how deviance can be useful to the ongoing society. However, anomie and strain theories fail to account for class conflicts, inequities, and poverty, which conflict theorists argue underlie deviance.

Conflict Theory

In 2011 in Libya, North Africa, thousands of citizens revolted against the regime of the leader, Moammar Gadhafi, because of what they felt was an oppressive and corrupt ruling elite. Many educated young people joined the rebel ranks, and many were killed in the uprising. Because of the Sea Shepherd Conservation Society ships, Japan suspended its whale hunt in 2011 due to disruptions. The Sea Shepherd harassment of Japanese whaling vessels in Antarctic waters prevented Japanese whalers from killing whales in the area ("Japanese Whaling Suspension From Sea Shepherd Harassment Not Good Enough for Conservationists" 2011). From 2002 to the present, protests against the Iraq (and later Afghanistan) war have been held in many major cities around the world. Are these protesters deviants and criminals, or are they brave heroes? The response depends on who answers the question.

Conflict theorists assume that conflict between groups is inevitable. Because many societies today are heterogeneous groupings of people, the differences in goals, resources, norms, and values between interest groups and groups in power often cause conflict. Conflict theory focuses on macro-level analysis of deviance, looking at deviance as a result of social inequality or of the struggle between groups for power.

Deviance is often related to social class status, interest groups, or cultural conflict between the dominant group and ethnic, religious, political, regional, or gender groups. Wealthy and powerful elites want to maintain their control

and their high positions (Domhoff 2009). They have the power to pass laws and define what is deviant, sometimes by effectively eliminating opposition groups. The greater the cultural difference between the dominant group and other groups in society, the greater the possibility of conflict. This is because minority groups and subcultures challenge the norms of the dominant groups and threaten the consensus in a society (Huizinga, Loeber, and Thornberry 1994).

Some conflict theorists blame capitalist systems for unjust administration of law and argue that the ruling class uses the legal system to further the capitalist enterprise (Quinney 2002). The dominant or ruling class defines deviance, applies laws to protect its interests, represses any conflict or protest, and, in effect, may force those in subordinate classes to carry out actions that it has defined as deviant. These actions are necessary to survive when legitimate avenues to resources are restricted by the affluent. This situation, in turn, supports the ideology that works against subordinate classes. Activities that threaten the interests and well-being of the wealthy capitalist class become defined as deviant. By subordinating certain groups and then defining them as deviant or criminal, the dominant group consolidates its powerful position. Because the dominant class is usually of one ethnic group and those of other races or ethnicities tend to be in the subordinate class, conflict often has racial and ethnic implications as well as social class dimensions. The fact that, for the same offenses, subordinate

A conservative group called the Tea Party Patriots protests against President Barack Obama's health care reform bill in September 2009. One sign with Obama's image calls him "Parasite in Chief," as thousands of people protest against his policies. The Tea Party has taken a strident oppositional position to stop governmental growth and a possible increase in taxes.

class or race members are arrested and prosecuted more often than dominant class or race members is provided as evidence to support this contention by conflict theorists (Quinney 2002).

To reduce deviance and crime, conflict theorists agree that we must change the structure of society. For instance, laws in many countries claim to support equal and fair treatment for all, but when one looks at the law in action, another picture emerges.

If the structure of society was changed and there were no dominant groups exploiting subordinate groups, would crime disappear? To answer this, we can look to patterns of crime in societies that have attempted to develop a communist economic classless structure. The rate of crime in China and Cuba, for instance, is lower than that in many Western democracies, in part because of less dramatic inequality and strict social controls on behavior (NationMaster 2010). On the other hand, deviance is still present in noncapitalist societies. So although capitalistic inequality contributes to deviance, it is only one of many variables at work.

Feminist Theory

The goal of most feminist theorists is to understand and improve women's status, including their treatment by men and those in power. Feminists argue that traditional theories do not give an adequate picture or understanding of women's situations. Although there are several branches of feminist theory, most see the macro level causes of abuses suffered by women as rooted in the capitalist patriarchal system. Feminist theorists look for explanations in gender relations and societal structures for violence against women and the secondary status of most women. They include the following ideas: (a) Women are faced with a division of labor resulting from their sex, (b) separation between public (work) and private (home) spheres of social activity create "we" versus "they" thinking between men and women, and (c) socialization of children into gender-specific adult roles has implications for how males and females perceive and relate to each other.

One result of women's status is that they are often victims of crime. The type of victimization varies around the globe, from sex trafficking to rape (Bales 2004, 2007). Women are less often in a position to commit crimes. In fact, deviant acts by women have traditionally fallen into the categories of shoplifting, credit card or welfare fraud, writing bad checks (in developed societies), prostitution, and in some countries adultery or inappropriate attire. Many Western feminist theorists contend that until women around the world are on an equal footing with men, crimes against women and definitions of various behaviors of women as deviant are likely to continue.

Consider the case of intimate partner violence, including rape. Until recent years, there have been few serious consequences for the offenders in many countries (especially during times of anomie or war), and women often were blamed or blamed themselves for "letting it happen" (Boy and Kulczycki 2008). In some cultures, women fear reporting violence and may blame themselves for the beating or rape. Yet, from limited studies, we know that 16% to 52% of married women in the Middle East were assaulted in one year, compared to 1.3% to 12% in Europe and North America (Krug et al. 2002).

Feminist theorists argue that we learn our gender roles, part of which is men learning to be aggressive. Women's status in society results in their being treated as sex objects, to be used for men's pleasure. Women's race and class identification become relevant in the exploitation of poor ethnically distinct women from developing countries as they become victims of human trafficking. Some branches of feminist theory argue that men exploit women's labor power and sexuality to continue their dominance. The system is reproduced through new generations that are socialized to maintain the patriarchy and to view inequality between the sexes as "normal" and "natural." Women who deviate from the cultural expectations of "normal" behavior are condemned.

No one theory of why deviance occurs can explain all deviance. Depending on the level of analysis of the questions

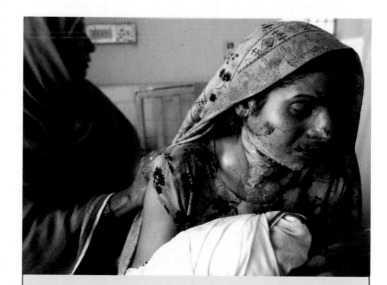

Ishrat Abdullah is comforted by her mother in a Pakistani hospital after her husband threw sulfuric acid over her during a domestic dispute. She has burns inflicted to more than 30% of her body. Her husband and mother-in-law repeatedly berate her but also plead that she not press charges. Such instances are not unusual in parts of South Asia, and they are a concern of feminist scholars.

sociologists wish to study (micro, meso, or macro), they select the theory that best fits the data they find. In the following sections, we explore in more detail the micro-meso-macro connections as they apply to one manifestation of deviance—crime. This illustrates how the sociological imagination can be applied to deviance at each level in the social system.

Thinking Sociologically

Meso- and macro-level social forces may be even more powerful than micro-level forces in explaining deviance. Pick a recent example of deviance discussed in your newspaper or on television. What might be the factors at higher levels in the social system that contribute to this deviance in your local community?

Crime and Individuals: Micro-Level Analysis

In this and the following two sections, we will be analyzing one type of deviance—crime. **Crime** is deviance that (if one is caught) involves formal sanctions from the government. A criminal justice system or court becomes involved in punishing wrongdoers and reinforcing conformity. Crimes that affect the individual or primary group seem most threatening to us and receive the most attention in the press and from politicians. Yet, these micro-level crimes are only a portion of the total crime picture and, except for hate crimes, are not the most dysfunctional or dangerous crimes. In the United States, more than 2,800 acts are listed as federal crimes. These acts fall into several types of crime, some of which are discussed below. First, we analyze how crime rates are measured.

How Much Crime Is There?

How do sociologists and law enforcement officials know how much crime there is, especially because not all crime is reported to the police? Each country has methods of keeping crime records. For instance, the official record of crime in the United States is found in the Federal Bureau of Investigation's *Uniform Crime Reports* (UCRs). The FBI relies on information submitted voluntarily by law enforcement agencies and divides crimes into two categories: Type I and Type II offenses. Type I offenses, also known as *FBI*

Index Crimes, include murder, forcible rape, robbery, aggravated assault, burglary, larceny theft, motor vehicle theft, and arson. Type II offenses include fraud, simple assault, vandalism, driving under the influence of alcohol or drugs, and running away from home. In fact, there are hundreds of Type II crimes. The Crime Clock in Figure 6.2 summarizes UCR records on Type I offenses.

To examine trends in crime, criminologists calculate a rate of crime, usually per 100,000 individuals. Recent data indicate that the rate of violent crime in the United States dropped after the mid-1990s but has begun to inch up again. In 1985 there were 558 violent crimes per 100,000 residents; in 1995 that number was 684.5; by 2000 the number had dropped to 506.5; and in 2007 and 2008 it had dropped well below 500. Moreover, black men were 6 times as likely to be homicide victims as white men (Stout 2009). Although the UCR data provide a picture of how much crime gets reported to the police and leads to arrest, it does not provide information on how much crime there is in the United States because not all crimes go through the criminal justice system. An alternative measurement is the *National Incident-Based Reporting System* (NIBRS) that reports incident-driven crime, meaning that the FBI gathers more detailed information and categories on victim and offender characteristics. Reports include the type of offense, whether a weapon was used, location, whether drugs were a factor, and any motivations related to race, religion, or gender. This system provides more detailed and accurate crime statistics.

Another technique to assess crime rates is *self-reporting surveys*—asking individuals what criminal acts they have committed. Criminal participation surveys typically focus on adolescents and their involvement in delinquency. Yet

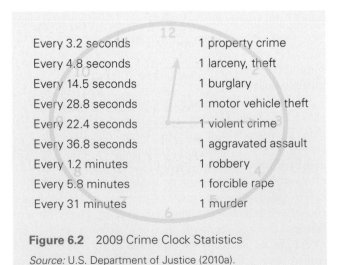

Every 3.2 seconds	1 property crime
Every 4.8 seconds	1 larceny, theft
Every 14.5 seconds	1 burglary
Every 28.8 seconds	1 motor vehicle theft
Every 22.4 seconds	1 violent crime
Every 36.8 seconds	1 aggravated assault
Every 1.2 minutes	1 robbery
Every 5.8 minutes	1 forcible rape
Every 31 minutes	1 murder

Figure 6.2   2009 Crime Clock Statistics
Source: U.S. Department of Justice (2010a).

another approach is *victimization surveys*—surveys that ask people how much crime they have experienced. The most extensively gathered victimization survey in the United States is the National Crime Victimization Survey, conducted by the Bureau of Justice Statistics. According to these records, the tendency to report crime to the police varies by the type of crime, with violent victimizations having the highest reporting rate.

How crime is measured affects what and how much crime is reported. Each measurement instrument provides a different portion of the total picture of crime. By using several data-gathering techniques (triangulation), a more accurate picture of crime begins to emerge. Most of the crimes that concern average citizens of countries around the world are violent crimes committed by individuals or small groups. The following are some examples of these micro-level crimes.

A woman smokes a marijuana pipe in Amsterdam, Netherlands, where smoking pot is legal. The Dutch think that, because there is no victim of the behavior, there should be no prohibition on it.

Thinking Sociologically

Which source of data discussed above do you think best tells us what the crime rates are? Why might there be a benefit to having more than one method? Why are FBI reports usually given attention by the press?

Predatory or Street Crimes

Predatory crimes committed against individuals or property are considered the most serious crimes by the public. In the United States, the Uniform Crime Reports list eight serious index crimes used to track crime rates: predatory acts against people (murder, robbery, assault, and rape) and property (burglary, arson, theft, and auto theft).

Citizens in the United States are increasingly afraid of violent predatory crime. Some people keep guns. Others, especially women, African Americans, older Americans, and low-income individuals, are afraid to go out near their homes at night (U.S. Bureau of Justice Statistics 2001). However, the percentage of U.S. households experiencing one or more crimes dropped from 25% in 1994 to 14% in 2005 (Klaus 2007). Most criminologists feel there are more serious crimes to be discussed under meso- and macro-level deviance.

Crimes Without Victims

Acts committed by or between individual consenting adults are known as **victimless** or **public order crimes**. Depending on the laws of countries, these can include

prostitution, homosexual acts, gambling, smoking marijuana and using drugs, drunkenness, and some forms of white-collar crime. Participants involved do not consider themselves to be victims, but the offense is mostly an affront to someone else's morals. These illegal acts may be tolerated as long as they do not become highly visible. Some prostitution is overlooked in major cities of the world, but if it becomes visible or is seen as a public nuisance, authorities crack down, and it is controlled. Even though these acts are called victimless, there is controversy over whether individuals are victims even when consenting to the act and whether others such as family members are victims dealing with the consequences of the illegal activities.

Thinking Sociologically

Can a person be victimized by drugs even if he willingly uses them? Many prostitutes only consent to sex acts because poverty leaves them with few other options and because, like many women without resources, they are vulnerable to domination by men. Are they victims?

Societies respond to victimless crimes such as using and selling drugs with a variety of policies, from execution in Iran and hanging in Malaysia to legalization in Holland. Long prison terms in the United States mean that 3 out of every 10 prison cells are now reserved for the user, the addict, and the drug seller—yet the problem has not diminished (Goode 1997, 2005). Proposals to legalize drugs, gambling, prostitution, and other victimless crimes meet with strong opinions both for and against. Although in many countries current policies and programs toward drugs are not working by almost any measure of success, the various consequences of alternative proposals are also uncertain.

Hate Crimes

Ethnic violence around the world results in reports of hate crimes in communities, at workplaces, and on college campuses. Hate crimes are criminal offenses committed against a person, property, or group that are motivated by the offender's bias against a religion, an ethnic or racial group, a national origin, a gender, or a sexual orientation. This is another micro-level crime since it affects primarily individuals. The Uniform Crime Reports in the United States indicate that hate crimes account for 11.5% of total criminal offenses. Most hate crimes are directed against property (84.4%) and involve destruction, damage, or vandalism; 31.3% include direct intimidation, totaling 9,035 hate crimes involving 9,528 victims (Federal Bureau of Investigation 2005; U.S. Bureau of Justice Statistics 2005). In 2008, the most recent data available, there were 7,783 hate crime incidents involving 9,168 offenses: 51% racial bias; 20% religious bias; 18% sexual-orientation bias; 12% nationality bias; and 1% disability bias (U.S. Department of Justice 2010b).

Research suggests that most hate crimes are spontaneous incidents, often a case of the victim being in the wrong place at the wrong time. Consider the case of Matthew Shepard, the gay college student who was robbed, tied to a fence post, beaten, and left to die in the cold Wyoming night. Nearly 16% of hate crimes are against those with different sexual orientations. Victims often form supportive in-groups to protect themselves from others who create a "culture of hate" (Jenness and Broad 1997; Levin and McDevitt 2003). Hate crimes are often vicious and brutal because the perpetrators feel rage against the victim as a representative of a group they despise. The crimes are committed by individuals or small vigilante groups (Blee 2008).

The examples above represent only three of the many micro-level crimes, characterized by individual or small-group actions. We now turn to crime in organizations and institutions. Not only is crime rooted in complex organizations at the meso level, but crimes themselves are committed within or by organizations.

Crime and Organizations: Meso-Level Analysis

As societies modernize, there is an almost universal tendency for crime rates to increase dramatically due to anomie affecting new migrants whose old norms are no longer relevant (Merton 1968a). Societies become more reliant on formal or bureaucratic mechanisms of control—in other words, development of a criminal justice system at the meso level of our social world.

Advocates of a hate-free America express their support for the gay and lesbian community. Others (right) express their sentiment that Matthew Shepard, a gay student who was beaten and left to die in a hate crime, was a deviant who had no moral standing and, seemingly, no right to live.

Crimes Involving Organizations and Institutions

Some crimes are committed by highly organized, hierarchically structured syndicates that are formed for the purpose of achieving their economic objectives in any way possible. For example, Medicare fraud is estimated at $60 billion a year. In South Florida, criminals open storefronts claiming to sell medical equipment; often fake offices are located in dying strip malls, and after the criminals have sent in false claims for Medicare payments, sometimes amounting to millions of dollars, they close the storefronts and cannot be traced. Although the federal government has a small number of investigators, they are hardly a match for the fast-moving criminals with their get-rich-quick schemes (CBS News 2010). Such fraudulent organizations intentionally flout the law. On the other hand, some crime is done by legitimate corporations that cross the line. Their crimes are very serious, but the purpose and the public image of the organization are not criminal. We will look first at organized criminal organizations and then at crime committed by people within their legitimate occupations.

Organized Crime

Organized crime, comprising ongoing criminal enterprises, has the ultimate purpose of personal economic gain through illegitimate means, using business enterprises for illegal profit (Siegel 2009). They engage in violence and corruption to gain and maintain power and profit (Adler et al. 2004). Our image of this type of crime is sometimes glamorized, coming from stereotypes in films such as *No Country for Old Men, The Godfather, Gangs of New York,* and many others. On television, *The Sopranos* is the ultimate media "mob" depiction. Despite the alluring view of these idealized stories, organized crime is a serious problem in many countries. It is essentially a counterculture with a hierarchical structure, from the boss down to underlings. The organization relies on power, control, fear, violence, and corruption. This type of crime is a particular problem when societies experience anomie and social controls break down.

Marginalized ethnic groups that face discrimination may become involved in a quest to get ahead through organized crime. Early in U.S. history, Italians were especially prominent in organized crime, but today, many groups are involved. Organized crime around the world has gained strong footholds in countries in transition (Siegel and Nelen 2008). For example, in Russia, the transition from a socialist economy to a market economy has provided many opportunities for criminal activity. The *Mafiya* is estimated to be 100,000 people strong, and some estimate members control 70% to 80% of all private business and 40% of the wealth in Russia (Lindberg and Markovic n.d.; Schmalleger 2009).

Organized crime usually takes one of three forms: (a) the sale of illegal goods and services, including gambling, loan sharking, trafficking in drugs and people, selling stolen goods, and prostitution; (b) infiltrating legitimate businesses and unions through threat and intimidation and using bankruptcy and fraud to exploit and devastate a legitimate company; or (c) racketeering, the extortion of funds in exchange for protection (i.e., not being hurt). Activities such as running a casino or trash collection service often appear to be legitimate endeavors on the surface but may be cover operations for highly organized illegal crime rings.

Although the exact cost of organized crime in the United States is impossible to determine exactly, the estimated annual gross income from organized crime activity is at least $50 billion, more than 1% to 2% of the gross national product. Some scholars estimate earnings as high as $90 billion per year ("Organized Crime" 2009; Siegel 2011).

Transnational organized crime takes place across national boundaries, using sophisticated electronic communications and transportation technologies. Experts identify several major crime clans in the world: (a) Hong Kong–based triads, (b) South American cocaine cartels, (c) Italian mafia, (d) Japanese *Yakuza,* (e) Russian *Mafiya,* and (f) West African crime groups. Organized crime is responsible for thousands of deaths a year through drug traffic and murders, and it contributes to a climate of violence in many cities (Siegel 2009). The value of the global illicit drug market, for example, is estimated at more than $13 billion at the production level, $94 billion at the wholesale level, and $322 billion at the retail level (based on retail prices and taking seizures and other losses into account). The largest market is cannabis, followed by cocaine, opiates, and other markets (methamphetamine, amphetamine, and ecstasy) (Common Sense for Drug Policy 2006; P. Smith 2009; United Nations Development Programme 2005). Many people survive off of the drug trade, from the poppy farmers in Afghanistan, who grow more than 75% of the world's poppies, to the street drug dealers. In 2007, the acreage devoted to growing poppies was the highest ever, especially in southern Afghanistan. This problem is not likely to end soon because farmers are in debt to the Taliban and are therefore forced to continue growing poppies. The Taliban profit from the sales by more than $100 million per year, which helps keep them in power. Thus, Westerners who use opium drugs help financially sustain this group (National Public Radio 2008).

Add other types of crimes (transporting migrants, trafficking in women and children for the sex industry, and sales of weapons and nuclear material) and the estimates of profits for international crime cartels are from $750 billion to more than $1.5 trillion a year (United Nations Office on Drugs and Crime 2005).

Occupational Crime

Occupational crime receives less attention than violent crimes because it is less visible and because it is frequently committed by people in positions of substantial authority and prestige, as opposed to reports of violent crimes that appear on the television news each night. Yet, it is far more costly in currency, health, and lives. Victims of financial scams who have lost their life savings are well aware of this. A violation of the law committed by an individual or a group in the course of a legitimate, respected occupation or financial activity is called a white-collar or *occupational crime* (Coleman 2006; Hagan 2007). Occupational crimes include embezzlement, pilfering, bribery, tax evasion, price fixing, obstruction of justice, and various forms of fraud.

Although the Madoff scandal became public knowledge in December 2008, the details unfolded for the next several months. Bernard (Bernie) Madoff, former chair of the NASDAQ stock exchange, had developed a Ponzi scheme that is probably the largest investment fraud Wall Street has ever seen. The scheme defrauded and wrecked thousands of investors, public pension funds, charitable foundations, and universities, with over $65 billion missing from investor accounts. Named after Charles Ponzi, the first to be caught (in 1919), Ponzi schemes involve promises of large returns on investments, paying old investors with money from new investors. Money is shifted between investors (*The Wall Street Journal* 2009).

White-collar crimes in the United States are "estimated to be ten times greater than all the annual losses from all the crimes reported to the police" (Coleman 2006:43). The collapse of the savings and loan industry cost the public billions of dollars. Antitrust violations cost $250 billion, tax fraud $150 billion, and health care industry fraud $100 billion. Employee theft adds an estimated 2% to the retail purchase price of products we buy (Coleman 2006).

Identity theft, embezzlement, international illegal transfers of money, illegal stock trades, sales of illegal or inferior products, creation of computer viruses, computer hackers—the list of cyber crimes is long and will become longer. "As technology advances, it facilitates new forms of behavior . . . new and as yet unimaginable opportunities for criminals positioned to take advantage of it and the power such technology will afford" (Schmalleger 2006:473).

Sociologists divide occupational crimes into four major categories: crimes against the company, crimes against employees, crimes against customers, and crimes against the general public (Hagan 2007). *Crimes against the company* include pilfering (using company resources such as the photocopy machine for personal business) and employee theft ("borrowing company property," taking from the till, and embezzlement).

Three types of corporate crime are often done on behalf of the company, and the victims are employees or members of the larger society. *Crimes against employees* refer to corporate neglect of worker safety. *Crimes against customers* involve acts such as selling dangerous foods or unsafe products, consumer fraud, deceptive advertising, and price fixing (setting prices in collusion with another producer). *Crimes against the general public* include acts by companies that negatively affect large groups of people. One example is hospitals or medical offices that overbill Medicare, which costs U.S. taxpayers an estimated $100 billion a year. The FBI estimates that 3% to 10% of total health care expenditures, both public and private, are fraudulent (Federal Bureau of Investigation 2006a).

The bottom line is that white-collar crime committed by company executives is by far our most serious crime problem. The economic cost of white-collar crime is vastly greater than the economic cost of street crime. White-collar criminals kill considerably more people than all violent street criminals put together (Coleman 2006).

Thinking Sociologically

Why are meso-level crimes considered by scholars to be more dangerous and more costly to the public than micro-level crimes? Why do they get so much less attention?

National and Global Crime: Macro-Level Analysis

Terrorism refers to "premeditated, politically motivated violence perpetrated against noncombatant targets by subnational groups or clandestine agents, usually intended to influence an audience" (Zalman 2009:1). Add to that international terrorism practiced in a foreign country, and terrorism can be seen as a worldwide problem. The number of international terrorist attacks in 2006 was down to 240 from 310 in 2005 and almost 400 in 2004 (Memorial Institute for the Prevention of Terrorism 2007). In 2004, 1,907 people were killed, and 9,300 were wounded (U.S. Department of State 2005). Terrorist groups can be religious, state-sponsored, left- or right-wing, or nationalist. Table 6.1 shows types of terrorist groups (Schmalleger 2006:347).

Crime is a national and global issue as illustrated by *state-organized crime,* acts defined by law as criminal but committed by state or government officials in the pursuit of their jobs. For example, a government might be complicit in smuggling, assassination, or torture, acting as an accessory to crime that is then justified in terms of "national defense." Government offices may also violate laws that restrict or limit government activities such as eavesdropping. In some countries, including the Guantánamo Bay detainees held by

Table 6.1 Types of Terrorist Groups

Nationalist	Irish Republican Army, Basque Fatherland and Liberty, Kurdistan Workers' Party
Religious	Al-Qaeda, HAMAS, Hezbollah, Aum Shinrikyo (Japan)
State-sponsored	Hezbollah (backed by Iran), Abu Nidal Organization (Syria, Libya), Japanese Red Army (Libya)
Left-wing	Red Brigades (Italy), Baader-Meinhof Gang (Germany), Japanese Red Army
Right-wing	Neo-Nazis, skinheads, white supremacists
Anarchist	Some contemporary antiglobalization groups

the United States, political prisoners have been held for long periods without charges, access to lawyers, and trials, or they are tortured, violating both national and international laws.

Bribery and corruption are the way of life in many governments and businesses. The percentage of persons who said they had paid a bribe to obtain services is as high as 87% in Liberia, 62% in Sierra Leone, and 55% in both Cameroon and Uganda (Transparency International 2009).

Thinking Sociologically

Sometimes, government officials and even heads of state are the perpetrators of crimes. Is there a difference in crime if it is done by an official and justified as necessary for national defense? Why or why not? Is it ever justified for a military or an intelligence agency to violate its own country's laws? Why are acts by someone acting on behalf of a nation or an institution acceptable when the same act by an individual at the micro level is considered criminal?

Cross-National Comparison of Crimes

The vending machine was on the corner near the Ballantines' house in Japan. The usual cola, candy, and sundries were displayed, along with cigarettes, beer, whiskey, sake, and pornographic magazines. Out of curiosity, Jeanne and her family watched to see who purchased what from the machines, and not once did they see teenagers sneaking the beer, cigarettes, or porn. It turns out the Ballantines were not the only ones watching! The neighbors also kept an eye on who did what, the neighborhood watch being an effective form of social control in Japan. Because of the **stigma**, the disapproval attached to disobeying the expected norms, teens understand the limits, and vigilant neighbors help keep the overall amount of deviance low.

Detainees sit in a holding area at the naval base in Guantánamo Bay, Cuba, during in-processing to the "temporary" detention facility. Five years later, 173 of these men were still being held without trial. In several U.S. Supreme Court rulings (2006 and 2008), the verdict was that the rights of these people have been violated by the U.S. administration. The Obama administration planned to shut down the facility by 2010, but legal complications have made this more difficult than President Obama had anticipated.

The neighborhood watch sends a signal that deviant behavior is unacceptable and provides social control of behaviors of those who might be tempted to commit crimes.

Japan and the United States are both modern, urban, industrial countries, but their crime rates and the way they deal with deviant behavior and crime differ dramatically. Japan had 637 murders for 127.1 million people in 2009 (5 for every 1 million people), whereas the United States had 12,658 murders for 307.2 million people or 41 per million (see Table 6.2). How can these differences in crime rates be explained? Researchers look at cultural differences: Japan's low violent crime rate is due in part to Japan's tight-knit, homogeneous society—inequality between citizens is not great. Further, success is not as focused on material possessions and consumption, Japanese people are loyal to a historic tradition of cooperation that provides a sense of moral order, and Japanese do not carry guns (Westermann and Burfeind 1991). The Japanese government spends far less of its gross national product on police, courts, and prisons, and police in Japan want to be thought of as kind and caring rather than strict. For many crimes in Japan, the offender may simply be asked to write a letter of apology. The humiliation of writing an apology and the fear of shame and embarrassing one's family are strong enough to curb deviant behavior (Lazare 2004).

Although comparing cross-national data on crime is difficult because there are variations in the definitions of crime and measurements used, comparisons do give us insight into what types of crimes are committed, under what circumstances, and how often. Two sources of international data are Interpol (the International Criminal Police Organization) and the United Nations. One problem is that they have no way to check accuracy of the data they receive from countries.

Table 6.2 in the next "Engaging Sociology" provides information on crimes in selected countries. Differences are due in part to the much higher disparity in income between rich and poor and the heterogeneity of populations. The size of the country is also provided since that is relevant to the comparison.

Global Crimes

Increasingly, crimes are global in nature. Some crimes are committed by transnational conglomerates and may involve organized crime and the smuggling of illegal goods and humans. Other crimes violate international laws, treaties, and agreements such as protection of the global environment, laws that are ignored when contrary to the self-interests of governments or corporations. The international community has the capacity to try people, organizations, and countries for violation of human rights or international laws, but the process is difficult and politically charged.

Some scholars use a world systems perspective, arguing that the cause of global crime lies in the global economy, inequalities between countries, and competition between countries for resources and wealth. As a result of the capitalist mode of production, an unequal relationship has arisen between core nations (developed, wealthy ones in the Global North) and peripheral nations (in the Global South) that results in inequality. Core nations often take unfair advantage of peripheral nations. Peripheral nations, in turn, must find ways to survive in this global system, and they sometimes turn to illegal methods to achieve their goals (Chase-Dunn and Anderson 2006). Semiperipheral nations benefit from extensive trade, making them less vulnerable than the

poorest nations. Map 6.1 on page 168 shows where core, peripheral, and semiperipheral nations are located.

As you look at this map, note that the developed or affluent countries are almost all located in the Northern Hemisphere. Although some poor countries are north of the equator, the pattern is obvious. Furthermore, the phrase "gone south" is often a colloquialism for an economy that is not strong. To avoid some misleading implications of the words *developed* and *developing,* some scholars prefer the term *Global South* to refer to less affluent nations. If you see or hear the phrase "Global South," this map should help you see why it refers to developing or poor countries.

One example of global crime and corruption is computer crimes. Internet deviance or cyberspace crime is growing faster than a cybergeek can move a mouse. This new world of crime ranges from online identity theft and gambling to cybersex and pornography to hate sites and stalking to hacking into files and terrorism (Thio 2007). As one team of researchers reports, "fraudsters can tap into an international audience from anyplace in the world" (Sager et al. 2006:261).

Of the fraud complaints received by the U.S. government, more than 70% involve the Internet. Financial fraud costs consumers $22 billion annually according to *Businessweek Online* (2002), and identity theft costs business and consumers $56.6 billion annually, according to the Identity Theft Resource Center (2011).

Even legitimate businesses such as Google, Yahoo!, or eBay can unwittingly support crime by connecting people to illegal operations. Of official complaints to the FBI (2007b), Internet auction fraud comprises 44.9%, undelivered merchandise or payment comprises 19%, and check fraud comprises 4.9%. Internet fraud in the United States increased by 33% between 2007 and 2008, with 275,284 fraud claims and the dollar loss at $265 million (Cratty 2008). Drug traffic is one of the most common forms of scam. For example, people purchase body-building drugs that are life threatening, find recipes for making illegal methamphetamines, or purchase the date rape drug GHB. Another venue of illegal Internet usage is music and video downloading and sales (Friedman 2005).

Other crimes have also become easier to commit: The FBI reported an increase of 1,789% in online child porn cases opened between 1996 and 2006. Between fiscal years 1996 and 2007, there was a 2,062% increase in the number of IINI cases opened (113 to 2,443) throughout the FBI. It is anticipated that the number of cases opened and the resources utilized to address the crime problem will continue to rise" (Federal Bureau of Investigation 2007a). One policy difficulty is that as many as five different federal U.S. agencies can be involved in preventing financial fraud on the Internet, not to mention the state and local agencies. Few local authorities feel that identity theft is within their jurisdiction, but government officials may not have the personnel or the interest to pursue these cases (Sager et al. 2006).

Engaging Sociology

Crime Comparison

The frequency of crimes varies across countries and is influenced by many variables. Examine this table and then answer the questions below.

Table 6.2 Crime Incidents in Selected Countries

Country	Murders	Rapes	Assaults	Burglaries	Auto Thefts	Population size (in millions)
Australia	302	15,630	141,124	436,865	139,094	21.3
Canada	489	24,049	233,517	23,065	160,268	33.5
Chile	235	1,250	53,133	13,375	n.a.	16.6
Denmark	58	497	9,796	1,297	32,203	5.5
Finland	148	579	27,820	370,993	16,391	5.3
France	1,051	8,458	106,484	1,885	301,539	64.1
Germany	960	7,499	116,912	n.a.	83,063	82.3
Italy	746	2,336	29,068	n.a.	243,890	58.1
Japan	637	2,260	43,229	3,027	309,638	127.1
New Zealand	45	861	30,177	65,675	21,992	4.2
United Kingdom	850	8,593	450,865	836,027	338,796	61.1
United States	12,658	89,110	2,238,480	2,099,700	1,147,300	307.2

n.a. = not available

Source: Interpol (2007). World Factbook (2010).

Engaging with Sociology:

1. Take the population (in millions) and divide it by the number of crimes in a given category (e.g., murder). If you do that for each country for a particular type of crime, you can see the ratio of crimes per person in the population.

2. Which countries have especially striking crime rates?

3. What can you tell about countries from studying their crime rates?

4. Do any particular rates stand out?

5. Why and how might you explain those rates?

Controlling Crime: Social Policy Considerations

When people are afraid to walk the streets because they might be assaulted and when a significant number of individuals are dropping out of society and taking up deviant lifestyles, deviance becomes a topic of great concern. Most governments pass laws and make policies to keep deviance from disrupting the smooth functioning of society. This section discusses mechanisms used by societies to control the amount of deviance.

Dealing With Crime: The Criminal Justice Process

Every society has a process for dealing with criminals. Sometimes the ground rules and processes of justice respect

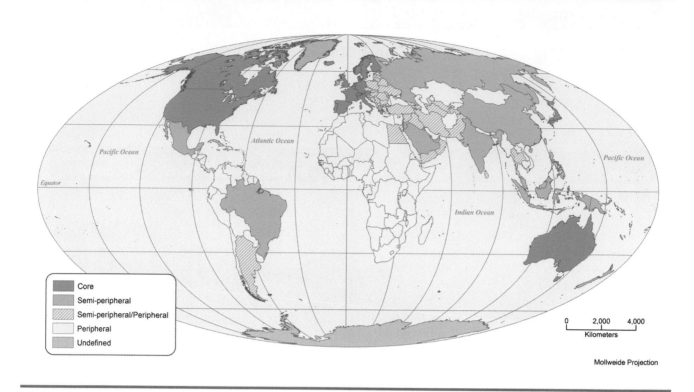

Map 6.1   Core, Semiperipheral, and Peripheral Countries of the World

Source: Map by Anna Versluis.

Note: Some countries are left undefined (in gray).

human rights and represent blind justice, meaning that all people are treated equally, but often they do not. Structural-functionalists see the justice system as important to maintaining order in society. Some conflict theorists argue that the criminal justice system depicts crime as a threat from poor people and minorities, which creates fear of victimization in members of society. It is in the interests of those in power to maintain the image that crime is primarily the work of outsiders and the poor. This deflects discontent and hostility from the powerful and helps them retain their positions (Reiman and Leighton 2010a, 2010b). Conflict theorists point out that there will always be a certain percentage of crime in society because the powerful will make sure that something is labeled deviant. Based on these theories, policy might focus on how to deter deviant acts or on how to correct the injustices of the system. In both cases, prisons and jails—penal institutions—are currently a primary means of controlling individual criminal behavior.

Prisons and Jails

Protecting the public from offenders often means locking criminals in prisons, a form of *total institution* that completely controls the prisoners' lives and regulates all of their activities. Inmates' lives are drastically changed through the processes of *degradation,* which marks the individual as deviant, and *mortification,* which breaks down the individual's original self as the inmate experiences *resocialization* (Goffman 1961; Irwin 1985). The inmate is allowed no personal property. There is little communication, and verbal abuse of inmates by guards is common. Heterosexual activity is prohibited, uniforms and standard buzz cuts are required, and the inmates' schedules are totally controlled.

Social systems that develop within the prison often involve rigid roles, norms, and privileges. In a famous study that simulated a prison situation, Zimbardo et al. (1973) illustrated the social organization that develops and the roles that individuals play within the prison system. Students were assigned to roles as prisoners or prison guards. Within a short time, the individuals in the study were acting out their roles. The students playing the role of "guard" became cruel and sadistic, causing Zimbardo to end the experiment prematurely because abuse was beginning to have alarming consequences (Zimbardo et al. 1973). The abuse of prisoners in Iraq and at Guantánamo prison in

Cuba parallel the findings from Zimbardo's (2010) study. In short, the authoritarian situation can lead to abuse; it is not a matter of maladjusted individuals in those roles.

Jails in local communities have been called "asylums for poor people." Most people in jails are there for their "rabble existence"—including petty hustlers, derelicts, junkies, "crazies," and outlaws, but mostly disorganized and economically marginal members of society (Irwin 1985). Jails in Europe and the United States house disproportionate numbers of immigrants (especially those with non-European features and skin tones), young men, and members of the poorest class of the citizenry. In the United States, only 34% of the incarcerated population is white, even though whites comprise 66.4% of the total population. African Americans make up 39% of all inmates and about 12.8% of the total population, and Hispanics make up 20% of all inmates and about 14.1% of the total population (Sabol, West, and Cooper 2009; U.S. Census Bureau 2009a). Among female prisoners, the number of African Americans declined as white women increased from 33 to 48 per 100,000 between 2000 and 2006. Table 6.3 breaks down the incarcerated population by race/ethnicity and sex. Conflict theorists believe that these figures are strong evidence that jails and prisons are mostly about controlling or "managing" the minorities and poor people, not about public safety. African American males, for example, are more than 6.6 times more likely to be incarcerated than white men, with over 10% serving or having served 25 to 39 years in prison or jail in June 2008 (Fathi 2009; The Sentencing Project 2006; West, Sabol, and Greenman 2009).

Jails or prisons are often the formal sanctions applied to enforce the rules passed by legitimate officials. This woman is being arrested for a serious legal infraction.

Table 6.3   U.S. State and Federal Prisoners by Gender, Race, and Hispanic Origin, 2008

	Number	% of total
Male	1,434,784	93.2%
White	562,800	36.5%
African American	477,500	31.0%
Hispanic/Latino	295,800	19.2%
Female	105,252	6.8%
White	29,100	1.9%
African American	50,700	3.3%
Hispanic/Latino	17,300	1.1%
Total	1,540,036	100%

Source: Sabol, West, and Cooper (2009).

Thinking Sociologically

From what you have learned so far in this chapter, why are the people who get sent to jail disproportionately young, poor, immigrants, or racial and ethnic minorities?

The Purposes of Prisons

From the functional perspective prisons serve several purposes for society: the desire for revenge or retribution; removing dangerous people from society; deterring would-be deviants; and rehabilitating through counseling, education, and work training programs inside prisons (Johnson 2002). However, in prison, inmates are exposed to more criminal and antisocial behavior, so rehabilitation and deterrence goals are often undermined by the nature of prisons. According to a recent Bureau of Justice Statistics report, roughly 5% of prison inmates are sexually assaulted by other prisoners or prison guards, and 12% of youth in juvenile detention facilities are victims ("New Federal Report" 2010), often in gang rapes (Banbury 2004). This ongoing problem of assault, rape, and threat of violence in prison so brutalizes inmates that it becomes difficult for them to reenter society as well-adjusted citizens ready to conform to the conventional society they feel has brutalized them (Hensley, Koscheski, and Tewksbury 2005). Many prisoners suffer from mental health problems and have difficulty reintegrating into society (Fathi 2009).

Although estimates indicate that only 3% of known criminals go to prison and the actual amount of crime has declined, incarceration rates in the United States have increased rather dramatically. In June 2008, the federal and state prison and local jail population reached more than 2.3 million prisoners (Fathi 2009). Incarceration rates were 762 of every 100,000 U.S. residents, a tremendous increase since the mid-1970s (Fathi 2009; Liptak 2008). At 762 per 100,000, the U.S. rate of locking up citizens is the highest in the world and five to eight times higher than that in other industrial nations (International

Table 6.4 World Rates of Incarceration (Rates per 100,000 People)

Top 10 Countries		Other Industrialized Countries	
United States	762[a]	United Kingdom	152[a]
Russian Federation	594	Netherlands	127
St. Kitts and Nevis	536	Australia	126
Bermuda	532	Canada	116[a]
Virgin Islands	521	Germany	97
Turkmenistan	489	Italy	97
Cuba	487	France	88
Palau	478	Switzerland	83
Belize	470	Sweden, Denmark, Finland	78–75
Bahamas	462	Japan	63[a]

Source: Fathi (2009) and International Centre for Prison Studies (2006).

a. Figures are for 2008.

Centre for Prison Studies 2011; The Sentencing Project 2006). For example, Canada has a rate of 116 per 100,000 residents and Japan 63. Table 6.4 shows the countries with the highest incarceration rates and compares rates of incarceration in industrial countries.

The disturbing reality is that, despite the high rates of incarceration in the United States, **recidivism rates**—the likelihood that someone who is arrested, convicted, and imprisoned will later be a repeat offender—are also very high in the United States. Three out of four men who do time in prison will be confined again for a crime. This means that as a specific deterrent or for rehabilitation, imprisonment does not work very well (Quinney 2002; Siegel 2009). What options does the government have? In the following sections, we discuss the death penalty and alternatives to prisons.

The Death Penalty

All modern societies provide some means of protecting individuals from criminals, especially those considered dangerous. Crimes of murder, assault, robbery, and rape usually receive severe penalties. The most controversial (and irreversible) method of control is for the state to put the person to death. The most common argument for using the death penalty, more formally known as capital punishment, is to deter people from crime. The idea is that not only the person who has killed but also others will be deterred because they know this is a possibility. Although most developed countries do not use the death penalty, the United States still does. In fact, as you can see from Map 6.2, the United States is one of the few countries outside of Africa and Asia with the death penalty. Capital punishment is most common in Asia, the Middle East, and parts of Africa.

By 2007, 133 of the 192 United Nations member states had abolished the death penalty in law or in practice, declaring it cruel and unusual punishment. In 2006, 86 countries had laws against the death penalty, 11 more forbid it except for exceptional crimes (e.g., war crimes), and 27 additional countries had not executed anyone in 10 or more years, although they had no laws forbidding capital punishment. Other countries retain the right to use the death penalty in extraordinary cases, although many of those countries actually do not follow through with the sentence (Amnesty International 2006). The United Nations is considering a resolution abolishing the death penalty in all UN nations (UN News Center 2010).

In 2010, there were 527 known executions in the world excluding China (no data available), a drop from 714 executions in 2009. About 90% of all executions occurred in six countries: China, Iran, Saudi Arabia, the United States, and Yemen. The regions responsible for the most deaths are Asia and the Middle East (Amnesty

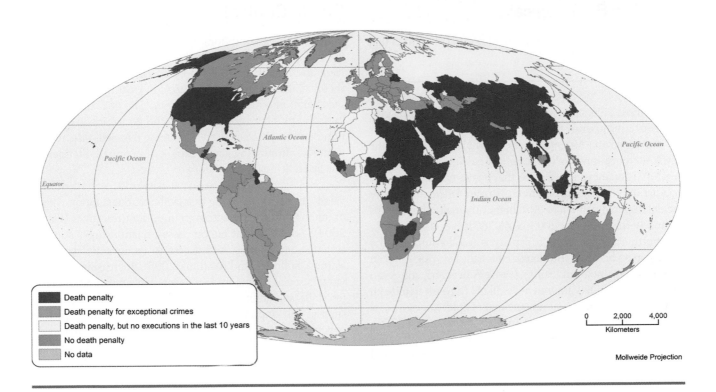

Map 6.2   Global Status of the Death Penalty in 2008

Source: Amnesty International. Map by Anna Versluis.

International 2011). Methods of killing included beheading, electrocution, hanging, lethal injection, shooting, and stoning to death.

U.S. states with the death penalty assume that those contemplating crimes will be deterred by severe penalties and those who have committed crimes will be justly punished. However, studies on the deterrent effects of capital punishment do not support the first assumption. Fewer than 25% of prison inmates believe the death penalty would deter violent crime (Steele and Wilcox 2003). Inmates who had committed three or more violent crimes indicated that their crimes were not planned; they "just happened," "things went wrong," and they were not thinking about the possible penalty when committing their crimes (Wilcox and Steele 2003). In 2009, the average murder rate per 100,000 people in states with the death penalty was 5.26, and the murder rate was much less—3.9—in states that do not have the death penalty (Death Penalty Information Center 2011; Hood 2002). So murder is more likely to happen in states with the death penalty, which is not very good evidence that this penalty has been a deterrent. However, it does serve as retribution (punishment for a crime) (Hood 2002).

Prisoners awaiting the death penalty are held in a separate isolation unit known as "death row." More than 70% of the members of the United Nations have abolished the death penalty, including all of the European nations. More than 90% of all executions in 2010 occurred in six countries: China, Iran, Saudi Arabia, the United States, and Yemen (Amnesty International 2011).

Thinking Sociologically

How can you explain higher murder rates in U.S. states that have the death penalty?

In most U.S. states with capital punishment, a disproportionate number of minority and lower-class individuals are put to death. African Americans make up 42% of the death sentence inmates (Bonczar and Snell 2005). This is a disproportionate number since they make up 12.5% of the population. In some jurisdictions, African Americans receive the death penalty at a rate 38% higher than all other groups (Amnesty International 2005; Sarat 2001; Siegel 2009).

Furthermore, the death penalty is usually imposed if a white person has been murdered. Homicides in which African Americans, Latinos and Latinas, Native Americans, and Asian Americans are killed are much less likely to result in a death penalty for the murderer, implying that people in society view their loss of life as less serious (Amnesty International 2006). In addition, mistakes are made. Since 1973, 123 prisoners have been released from death row because they were found innocent, some due to DNA tests. However, tests have also exonerated people who have already been executed.

A perhaps unexpected fact is that in the United States it costs more to put someone to death in most cases than to sentence him to life in prison. The national average cost for incarceration of an inmate is $25,000 a year, or about a million dollars for 40 years. In Florida, the average cost is $24 million for each execution. The death penalty in California costs taxpayers more than $250 million for each execution. In North Carolina an execution costs the state $2.16 million more than sentencing the person to life imprisonment (Death Penalty Information Center 2010). The majority of those costs occur at the trial level, with extensive preparations to try to be sure the right person is being convicted. The cost doesn't stop when the trial is over; it increases because death row only has one person per cell, and the laws require individual recreation and meal times, which means more supervision for death row inmates. The cost of death row incarceration jumps from $2 million to $3 million for the years while appeals are under way. New Jersey banned the death penalty in 2007 to help with its budget, and New Mexico followed in 2009. Maryland, Montana, and Colorado are among those considering similar legislation because of cost (Sunshine 2009).

The cost issue combined with the facts that the death penalty sometimes kills innocent people, that it is often racially discriminatory, and that it is not an effective deterrent has led to a search for alternative means to deter crime and has spurred policy analysts to rethink assumptions about what factors are effective in controlling human behavior.

Alternative Forms of Social Control

Based on the assumptions that most criminal behavior is learned through socialization, that criminals can be resocialized, and that tax dollars can be saved, several sociological theories of crime suggest methods of treatment other than incarceration without rehabilitation. The goal is to reduce both the number of individuals who go to prison and the number who are rearrested after being released—the recidivism rate. Improving social capital of potential offenders is one approach based in theory.

Social capital refers to social networks, shared norms, values, and understanding that facilitates cooperation within or among groups and access to important resources (Flavin 2004). Social capital encompasses relationships, support systems and services, and access to community resources (Jarrett, Sullivan, and Watkins 2005). Increasing an individual's social capital by increasing educational attainment, job skills, and ability to take advantage of available resources such as job opportunities can reduce the chances of that person going to prison in the first place and of recidivism, or repeat offenses and incarceration (Faulkner 2006). Consider the case study in "The Applied Sociologist at Work."

What works in one country cannot necessarily be imported directly into another country. Not many people think that the apology technique used in Japan would work as a deterrent to crimes in the United States. On the other hand, there is a wide range of options available for dealing with crime other than harsh (and expensive) punishments. Seeing how crime is controlled in other countries may challenge assumptions in one's own country, causing authorities to come up with creative new solutions that do work. For example, many criminologists argue that the United States should concentrate on the serious criminals and reduce the number of minor offenders in jails. They also suggest a number of alternatives to sending offenders to jail and prisons such as community service, work release, and educational training programs.

Thinking Sociologically

Do you think changing the "cost-benefit" ratio so the costs of crime to the criminal are higher will deter crime, or are other policies or methods more effective in enticing people to be responsible, contributing members of society?

Prison reforms, rehabilitation, training programs, shock probation, work release, halfway houses, and other alternative programs are intended to integrate the less serious offenders into the community in a productive way and help them regain social capital. If we can integrate the potential criminal into the community, reduce discrimination, and teach at-risk youth acceptable behavior patterns, we may reduce crime and gain productive citizens. Some state penal systems provide education from basic skills to college courses, provide counseling and therapy programs, and allow conjugal and family stays to help keep families together (U.S. Bureau of Justice Statistics 2003).

Another current trend is toward privatization of prisons in an attempt to run them in a more businesslike, cost-effective way. This trend is unlikely to improve the rehabilitation aspects in this age of overcrowded prisons and cost cutting, but several states are experimenting with private facilities, and some federal agencies such as the Immigration and Naturalization Service have contracted with private operators to run facilities. Because sociologists and criminologists study prison programs, make recommendations to improve the correction programs, and help find solutions to deviance, they are highly skeptical of privatization. There are many problems in having any public service administered by people whose primary and perhaps only goal is to make a profit. The current trend seems to be to "warehouse prisoners," with little effort to rehabilitate them or to help them reenter society (Irwin 2005).

Thinking Sociologically

Considering what you have read, how has this chapter affected some of your ideas about deviance, crime, and the criminal justice system?

The Applied Sociologist at Work— James Faulkner | **Reducing Recidivism by Increasing Social Capital**

James Faulkner is a police department detective. Part of his work is with adults and juveniles, trying to increase their chances of leading productive lives and staying out of prison by increasing their social capital. The following is a case study by Detective Faulkner, at the time a school resource officer, describing his work with one young man.

Ray was a 20-year-old male African American college student who started life with absolutely no social capital and continued to struggle to increase his chances for success. He was the type of offender who continually violated the law. Ray's father left when he was 7 years old, and his mother died when he was 8. The father's whereabouts were no longer known. With no information and no parents, this was a prime example of someone with limited resources.

As a school resource officer, I was approached by a school official who was concerned for Ray's long-term future, due to the complexities of his life. As a member of law enforcement and through my personal experiences and observations, I knew Ray was stealing frequently to eat and live. Ray was living with a disabled aunt in deplorable conditions with no food in the house. When I picked him up to help him move in with yet another relative, there was no electricity, water, or phone.

We used my phone as a light as we gathered his items, and I helped him pack what few clothes he had. I began taking Ray food with the assistance of others, a concept not known to Ray.

With his high school graduation approaching, Ray asked me to help him find a job. A simple endeavor proved to be very difficult. It began with a potential employer's request for his Social Security card. Not only did Ray not have one, but he had no idea how to get one. Initially thinking that this would be relatively simple to solve, I asked him to bring his birth certificate, and we would go apply for his Social Security card. It was then that he asked me how one goes about "getting one of those birth certificate things."

After graduation and the following summer and with the aid of the assistant school superintendent, who took Ray under her wing, we set out to increase Ray's level of social capital. I felt motivated to help Ray because he had been so motivated to try to help himself, yet he had met failure after failure. He continually fell, got back up, and tried something else. This encouraged me to invest in Ray. Small increases in his level of social capital and resources helped Ray establish a network of support for a more successful future. It is my belief that these small increases in Ray's resources will have significant impact on his future lifestyle and reduce his chances for a life of crime. Among other things, we taught him how to apply for loans to get financial help.

With the help of government loans and grants, Ray was admitted to and entered college that fall. We helped supply him with housing, toiletries, and clothing. We did our best not to do these things for him but rather to show him

(Continued)

how to do them himself. There was no family or celebration after his high school graduation, so some school officials and I took Ray to a local steakhouse to celebrate his graduation and entering college, a step he had once perceived as unobtainable. By investing a small amount of social capital in the form of relationships and resources, we were able to steer Ray in the direction of accumulating additional social capital in the form of education. This could be his ticket out of his previous way of life.

Note: James Faulkner has a master's degree in applied social science/criminal justice.

One of the dominant characteristics of modern society is social inequality, often an issue in criminal activity. Indeed, many of our social problems are rooted in issues of inequality. Extreme inequality may even be a threat to the deeper values and dreams of the society, especially ones that stress individualism and achievement. In the following three chapters, we look at three types of inequality: socioeconomic, ethnic or racial, and gender-based inequity.

What Have We Learned?

Perhaps the answers to some of the chapter's opening questions—what is deviance, why do people become deviant, and what should we do about deviance?—have now taken on new dimensions. Deviance has many possible explanations, and there are multiple interpretations about how it should be handled. Deviance and crime are issues for any society, for they can be real threats to stability, safety, and sense of fairness that undermine the social structure. The criminal justice system tends to be a conservative force in society and is often championed by the "haves" of society because of its focus on ensuring social conformity. Still, there may be positive aspects of deviance for any society, from uniting society against deviants to providing creative new ways to solve problems.

Key Points:

- Deviance—the violation of social norms, including those that are formal laws—is a complex behavior that has both positive and negative consequences for individuals and for society. (See pp. 146–148.)

- Deviance is often misunderstood because of simplistic and popular misconceptions. (See pp. 149–150.)

- Many theories try to explain deviance—rational choice, differential association, and labeling theories at the micro level, along with structural explanations including anomie and disorganization, strain theory, conflict theory, and feminist theory. (See pp. 151–160.)

- Many of the formal organizations concerned with crime (e.g., the FBI and the media) focus on crimes involving individuals—predatory crimes, crimes without victims, and hate crimes—but the focus on these crimes may blind us to crimes that actually are more harmful and more costly. (See pp. 160–162.)

- At the meso level, organized and occupational crimes may cost billions of dollars and create great risk to thousands of lives. Occupational crime may be against the company, employees, customers, or the public. (See pp. 162–164.)

- At the macro level, national governments sometimes commit state-organized crimes, sometimes in violation of their own laws or in violation of international laws. These crimes may be directed against their own citizens (usually minorities) or people from other countries. (See pp. 164–166.)

- Also at the macro level, some crimes are facilitated by global networks and by global inequities of power and wealth. (See pp. 166–167.)

- Controlling crime has generated many policy debates, from the use of prisons to the death penalty and even to alternative approaches to control of deviance. (See pp. 167–172.)

Contributing to Our Social World: What Can We Do?

At the Local Level:

- *LGBT Groups:* College campuses throughout the country have support groups for students who are lesbian, gay, bisexual, and/or transgender (often abbreviated as "LGBT"). The Consortium of Higher Education LGBT Resource Professionals, a national organization of such campus groups, maintains a website at http://www .lgbtcampus.org. Regardless of your identity/orientation, consider contacting your campus LGBT group, attending meetings, and participating in its support and public education activities.

- *Boys and Girls Clubs* and other organizations for youth need interns and volunteers to provide role models for youth. Consider volunteering to help children with homework or activities.

At the Organizational or Institutional Level:

The U.S. criminal justice system is an extensive and rapidly growing institution. Some aspects focus on crime prevention, some on law enforcement, some on corrections, and some on rehabilitation. Identify the aspect of the system that interests you most and, using faculty and community contacts, select an appropriate organization for volunteer work or an internship.

- *The criminal courts* are central actors in the administration of criminal justice, and trials are often open to the public. Attending a trial and/or contacting a judge or magistrate could provide a good introduction to a longer-term relationship.

- Volunteer in a prevention program such as *D.A.R.E.* (Drug Abuse Resistance Education) in high school. These are administered by local law enforcement agencies such as departments of police and might provide opportunities for volunteer work and for observing how the system works "from the inside."

- Organizations involved in corrections and rehabilitation include *halfway houses* for teens in trouble. These groups work one-on-one with juveniles in after-school or weekend programs to provide positive role models and increase their social capital.

- *Battered women's shelters* are among the many community programs that provide safe houses, counseling, and practical help for women and children in abusive situations.

- *Project Safe* provides information and research on date rape on college campuses and in other locations.

Working with any of these organizations will allow you to learn about how our social world operates while you contribute to making it a better world.

At the National and Global Levels:

One of the most serious forms of global organized crime is trafficking in human beings. This often involves kidnapping, slavery, sexual exploitation of children and adults, and even torture and murder.

- The international organization *Human Trafficking* is engaged in research and action programs intended to end such crimes. Its website, at http://www.human trafficking.org, discusses several programs in which you can participate, including a special section on activities at academic institutions.

- A similar organization, *Polaris Project* (www.polarisproject .org), works to reduce global trafficking of women and children. Volunteers participate in letter-writing campaigns, support antitrafficking legislation, and conduct research on the problem.

Visit **www.sagepub.com/oswcondensed2e** for online activities, sample tests, and other helpful information. Select "Chapter 6: Deviance and Social Control" for chapter-specific activities.

PART III

Inequality

The underlying question in the next three chapters is why some people rise to the top of society with wealth, power, and prestige at their fingertips and others languish near the bottom. Why are some individuals and countries rich and others poor? The focus of these chapters is inequality, the process of stratification through which some people "make it" and others do not. At the very bottom of the human hierarchy are those starving and diseased world citizens who have no hope of survival for themselves or their families. This compares with corporate executives, bankers, and some world politicians or royalty who have billions of dollars at their disposal.

Social inequality is one of the most important processes in modern societies, and the implications extend all the way to the global social network. Sometimes, the inequality is based on socioeconomic status, but the basis of differential treatment is often other characteristics: race, ethnicity, gender, sexual orientation, religion, or age. These differences often result in strong "we" versus "they" thinking. One factor runs throughout these patterns of inequality: They have implications for social interaction at the micro, meso, and macro levels of analysis. In this section, we do not try to cover all forms of inequality; rather, we illustrate the patterns by exploring issues of social class, race or ethnicity, and gender.

CHAPTER 7

Stratification

Rich and Famous—or Rags and Famine?

In rich countries, such as the United States, Canada, Japan, and Western European nations, we assume there are many economic opportunities, and we like to believe that anyone can become rich and famous. The reality, however, is that our social world is very brutal for many people, and what they experience is rags and famine.

Global Community

Society

National Organizations,
Institutions, and Ethnic Subcultures

Local Organizations
and Community

Me (and My
Rags or
Riches)

Micro: How I am regarded by my peers

Meso: Institutions support the
privilged. Ethnic subcultures often disadvantaged

Macro: The privileged control resources,
health care, economic markets, and tax rates

Macro: Rich and poor countries in global system

Think About It	
Me (and My Inner Circle)	Why do you buy what you buy, believe what you believe, and live where you live?
Local Community	Why are some people in your community rich and others poor?
National Institutions; Complex Organizations; Ethnic Groups	How do institutions—such as education, the family, religion, and the economy—help to keep people in the class they were born into?
National Society	Why are some nations affluent and others impoverished?
Global Community	How does the fact that we live in a global environment affect you and your social position?

In Newport, Rhode Island, spacious mansions are nestled along the coast, with tall-masted sailboats at the docks. These are the summer homes of the U.S. aristocracy. Members of this class have an elegant social life, engage in elite sports such as fencing and polo, patronize the arts, and are influential behind the scenes in business and politics.

Hidden from the public eye in each country are people with no known names and no swank addresses; some have no address at all. We catch glimpses of their plight through vivid media portrayals, such as those of refugees in Darfur, Sudan, and of impoverished victims of Hurricane Katrina in 2005 along the U.S. Gulf of Mexico coast. They are the poor; many of them live in squalor. Economic hard times have pushed some of them from their rural homes to cities in hopes of finding jobs. However, with few jobs for unskilled and semiskilled workers in today's postindustrial service economies, many of the poor are left behind and homeless. They live in abandoned buildings or sleep in unlocked autos, on park benches, under bridges, on beaches, or anywhere they can stretch out and hope not to be attacked or harassed. Beggars stake out spots on city sidewalks, hoping citizens and tourists will give them a handout. In the United States, cities such as Houston, Texas; Los Angeles; Washington, DC; and New York try to cope with the homeless by setting up sanitary facilities and temporary shelters, especially in bad weather. Cities rely on religious and civic organizations such as churches and the Salvation Army to run soup kitchens.

In some areas of the world, such as sub-Saharan Africa and India, the situation is much more desperate, and many families are starving. At daybreak, a cattle cart traverses the city of Kolkata (Calcutta), India, picking up bodies of diseased and starved homeless people who have died on the streets during the night. Mother Teresa, who won the Nobel Peace Prize for her work with those in dire poverty, established a home in India where these people could die with dignity. She also founded an orphanage for children who would otherwise wander the streets begging or die. These efforts are noble but only a drop in the world bucket of misery. Survival, just maintaining life, is a daily struggle for the 40% of the world's population that lives on 5% of the global income. Of the 2.2 billion children in the world, 1 billion live in poverty. Four hundred million children have no access to safe water (1 in 5 in the world), and 1.4 million children die each year due to lack of safe drinking water and adequate sanitation; 270 million children have no access to health services (1 in 7), and 2.2 million die because they are not immunized. Just today, as you are reading these facts, over 25,000 children will die (United Nations Development Programme 2007).

This chapter discusses (a) why stratification is important, (b) why people are rich or poor (stratification systems), (c) the importance and consequences of social rankings for individuals, (d) whether one can change social class positions (social mobility), (e) characteristics of major stratification systems, (f) poverty and social policies to address problems, and (g) the global digital divide (patterns of stability and change).

Newport, Rhode Island, has long been one of the most affluent cities in North America, a community where mansions and yachts line the seacoast. This vessel is the largest privately owned yacht in the world, docked in Newport in the summer of 2009.

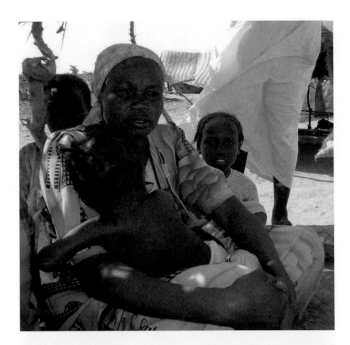

Poor people around the world find shelter wherever they can. A woman and her children displaced by war sit beneath a temporary shelter at a refugee camp in South Darfur. Even in affluent North America, some people are homeless and spend nights on sidewalks, in parks, or in homeless shelters.

The Importance of Social Stratification

Social stratification refers to how individuals and groups are layered or ranked in society according to how many valued resources they possess. Stratification is an ongoing process of sorting people into different levels of access to resources, with the sorting legitimated by cultural beliefs about why the inequality is justifiable. This chapter focuses on socioeconomic stratification, and subsequent chapters examine ethnic and gender stratification.

Three main assumptions underlie the concept of stratification: (a) People are divided into ranked categories; (b) there is an unequal distribution of desired resources, meaning that some members of society possess more of what is valued and others possess less; and (c) each society determines what it considers to be valued resources. In an agricultural society, members are ranked according to how much land or how many animals they own. In an industrial society, occupational position and income are two of the criteria for ranking. Most Japanese associate old age with high rank, whereas Americans admire and offer high status to some people for their youthful vigor and beauty.

What members of each society value and the criteria they use to rank other members depend on events in the society's history, its geographic location, its level of development in the world, its political philosophy, and the decisions of those in power. Powerful individuals are more likely to get the best positions, the most desirable mates, and the greatest opportunities. They may have power because of birth status, personality characteristics, age, physical attractiveness, education, intelligence, wealth, race, family background, occupation, religion, or ethnic group—whatever the basis for power is in that particular society. Those with power have advantages that perpetuate their power, and they try to hold onto those advantages through laws, custom, power, or ideology.

The social world model at the beginning of the chapter provides a visual image of socioeconomic stratification. The stratification process affects everything from individuals' social rankings at the micro level of analysis to positions of countries in the global system at the macro level.

Micro-Level Prestige and Influence

Remember how some of your peers on the playground were given more respect than others? Their high regard may have come from belonging to a prestigious family, having a dynamic or domineering personality, or owning symbols that distinguished them—"cool" clothing or shoes, a desirable bicycle, expensive toys, or a fancy car. This is stratification as children experience it at its beginning stage.

Property, power, and prestige are accorded to those individuals who have *cultural capital* (knowledge and access to important information in the society) and *social capital* (networks with others who have influence). Individual qualities such as leadership, personality, sense of humor, self-confidence, quick-wittedness, physical attractiveness, or ascribed characteristics—such as the gender or ethnicity or age—influence cultural and social capital.

These elite children attend a private school in England, and their privileges have been passed to them from their families. Note that even their typical school clothing acts as a symbol of social status and reflects a particular lifestyle.

Look at all those forks and knives. Some people know what to do with each of them! Knowing which fork or knife to use for each course of a meal could influence someone's chances of success on a job interview for certain kinds of positions.

Meso-Level Access to Resources

Often, our individual status in the society is shaped by our access to resources available through meso-level organizations and institutions. Our status is learned and reinforced in the family through the socialization process. Then we move on to learn grammar and manners that affect our success in school. Educational organizations treat children differently according to their social status, and our religious affiliation is likely to reflect our social status as well. Political systems, including laws, the courts, and police, reinforce the stratification system. Access to health care often depends on one's position in the stratification system. Our position and connections in organizations have a profound impact on how we experience life and how we interact with other individuals and groups.

Macro-Level Factors Influencing Stratification

The economic system, which includes the occupational structure, level of technology, and distribution of wealth in a society, is often the basis for stratification. Haiti, located on the island of Hispaniola, is the poorest country in the Western Hemisphere and one of the poorest countries in the world, with little technology, few resources, ineffective government, and an occupational structure based largely on subsistence farming. Even its forest resources are almost gone as desperately poor people cut down the last trees for firewood and shelters, leaving the land to erode (Diamond 2005). The economy is collapsing, leaving many already poor people still more destitute and on the lowest rungs of the world's stratification system. Add to the economic woes the 2010 earthquake and floods that drove people from what meager shelters they had. The economic position and geographic location of nations such as Haiti affect the opportunities available to individuals in those societies. There are simply no opportunities for Haitians to get ahead. Thus, macro-level factors can shape the opportunity structure and distribution of resources to individuals.

One problem for Haiti is that it has few of the resources that many other countries in the global system take for granted—a strong educational system, well-paying jobs in a vibrant economy, productive land, an ample supply of water, money to pay workers, and access to the most efficient and powerful technology. Almost all societies stratify members, and societies themselves are stratified in the world system, so each individual and nation experiences the world in unique ways related to its position. Stratification is one of the most powerful forces that we experience, but we are seldom conscious of how it works or how pervasive it is in our lives. This is the driving question sociologists ask when developing theories of stratification: How does it work?

Thinking Sociologically

Place yourself in the center of the social world model. Working outward from micro-level interactions toward the macro-level institutions, indicate what has influenced where you fall in the stratification system.

Theoretical Explanations of Stratification

Why do some people have more money, possessions, power, and prestige than others? We all have opinions about this question. Sociologists also have developed explanations—theories that help explain stratification. Recall that theories provide a framework for asking questions to be studied. Just as your interpretation of a question may differ from your friends' ideas, sociologists have developed different explanations for stratification and tested these with research data. These explanations of social rankings range from individual micro-level to national and global macro-level theories.

Micro-Level Theory

Symbolic Interaction

Most of us have been at a social gathering, perhaps at a swank country club or in a local bar, where we felt out of place. Each social group has norms that members learn through the socialization process. These norms are recognized within that group and can make clueless outsiders feel like space aliens. People learn what is expected in their groups—family, peer group, social class—through interaction with others. For instance, children are rewarded or punished for behaviors appropriate or inappropriate to their social position. This process transmits and perpetuates social rankings. Learning our social position means learning values, speech patterns, consumption habits, appropriate group memberships (including religious affiliation), and even our self-concept. In this way, children's home experiences and education help reproduce the social class systems (Ballantine and Hammack 2012).

Symbols often represent social positions. Clothing, for example, sets up some people as special and privileged. In the 1960s, wearing blue jeans was a radical act by college students to reject status differences. Today, the situation has changed, as young people wear expensive designer jeans that low-income people cannot afford. Drinking wine rather than beer, driving a Jaguar rather than a simpler mode of transportation, and living in a home that has six or eight bedrooms and 5,000-plus square feet is an expression of *conspicuous consumption*—displaying goods in a way that others will notice and that will presumably earn the owner respect (Veblen 1902). Thus, purchased products become symbols that are intended to define the person as someone of high status.

Interaction theories help us understand how individuals learn and live their positions in society. Next, we consider theories that examine the larger social structures, processes, and forces that affect stratification and inequality: structural-functionalist and various forms of conflict theory.

Meso- and Macro-Level Theories

Structural-Functionalist Theory

Structural-functionalists (sometimes simply called functionalists) view stratification within societies as an inevitable—and probably necessary—part of the social world. The stratification system provides each individual with a place or position in the social world and motivates individuals to carry out their roles. Societies survive by having an organized system into which each individual is born, where each is raised and where each contributes some part to the maintenance of the society.

The basic elements of the structural-functionalist theory of stratification were explained by Kingsley Davis and Wilbert Moore (1945), and their work still provides the main ideas of the theory today. Focusing on stratification by considering different occupations and how they are rewarded, Davis and Moore argue the following:

1. Positions in society are neither equally valued nor equally pleasant to perform. Some positions—such as physicians—are more highly valued because people feel they are very important to society and are dependent on this expertise. Therefore, societies must motivate talented individuals to prepare for and occupy the most important and difficult positions, such as being physicians.

2. Preparation requires talent, time, and money. To motivate talented individuals to make the sacrifices necessary to prepare for and assume difficult positions such as becoming a physician, differential rewards of income, prestige, power, or other valued goods must be offered. Thus, a doctor receives high income, prestige, and power as incentives.

3. The differences in rewards in turn lead to the unequal distribution of resources for occupations in society, and some people get richer. Therefore, stratification is inevitable. The unequal distribution of status and wealth in society motivates individuals to fill necessary positions—such as willingness to undertake the stress of being chief executive officer of a corporation in a highly competitive field.

In the mid-twentieth century, functional theory provided sociologists with a valuable framework for studying stratification (Tumin 1953), but things do change. In the twenty-first century, new criteria such as controlling information and access to information systems have become important for determining wealth and status, making scientists and technicians a new class of elites. The society also experiences conflict over distribution of resources that functionalism does not fully explain.

Conflict Theory

Conflict theorists see stratification as the outcome of struggles for dominance and scarce resources, with some individuals in society taking advantage of others. Individuals and groups act in their own self-interest by trying to exploit others, leading inevitably to a struggle between those who have advantages and want to keep them and those who want a larger share of the pie.

Conflict theory developed in a time of massive economic transformation. With the end of the feudal system, economic displacement of peasants, and the rise of urban factories as major employers, a tremendous gap between the rich and the poor evolved. This prompted theorists to ask several basic questions related to stratification: (a) How do societies produce necessities—food, clothing, shelter? (b) How are relationships between rich and poor people shaped by this process? and (c) How do many people become alienated in their routine, dull jobs in which they have little involvement and no investment in the end product?

Karl Marx (1818–1883), considered the father of conflict theory, lived during this time of industrial transformation.

In India, many people must bathe every day in public in whatever water supply they can find. Even many people with homes would not have their own water supply. Privacy for one's cleanliness and grooming is a symbol of privilege for the affluent.

Marx described four possible ways to distribute wealth, according to (a) what each person needs, (b) what each person wants, (c) what each person earns, or (d) what each person can take. It was this fourth way, Marx believed, that was dominant in competitive capitalist societies (Cuzzort and King 2002; Marx and Engels 1955).

Marx viewed the stratification structure as composed of two major economically based social classes: the haves and the have-nots. The haves consisted of the capitalist bourgeoisie, whereas the have-nots were made up of the working-class proletariat. Individuals in the same social class had similar lifestyles, shared ideologies, and held common outlooks on social life. The struggle over resources between haves and have-nots was the cause of conflict (Hurst 2006).

The haves control what Marx called the means of production—money, materials, and factories (Marx [1844] 1964). The haves dominate because the lower-class have-nots cannot earn enough money to change their positions. The norms and values of the haves dominate the society because of their power and make the distribution of resources seem "fair" and justified. Social control mechanisms including laws, religious beliefs, educational systems, political structures and policies, and police or military force ensure continued control by the haves.

The unorganized lower classes can be exploited as long as they do not develop a *class consciousness*—a shared awareness of their poor status in relation to those who own and control the production process. Marx contended that, with the help of intellectuals who believed in the injustice of the exploited poor, the working class would develop a class consciousness, rise up, and overthrow the haves, culminating in a classless society in which wealth would be shared (Marx and Engels 1955).

Thinking Sociologically

How can conflict theory help us understand the uprisings, led primarily by youth, in Egypt, Libya, and other North African countries?

Unlike the structural-functionalists, then, conflict theorists maintain that money and other rewards are not necessarily given to those in the most important positions in the society. Can we argue that a rock star or baseball player is more necessary for the survival of society than a teacher or police officer? Yet, the pay differential is tremendous. In addition, some high-prestige positions often go to sons and daughters of elites.

Not all of the predictions of Karl Marx have come true. No truly classless societies have developed, but some developments have moved in that direction. Labor unions arose to unite and represent the working class and put its members in a more powerful position vis-à-vis the capitalists, manager and technician positions emerged to create a large middle class, some companies moved to employee ownership, and workers gained legal protection from government legislative bodies in most industrial countries (Dahrendorf 1959).

Even societies that claim to be classless, such as China, have privileged classes and poor peasants. In recent years, the Chinese government has allowed more private ownership of shops, businesses, and other entrepreneurial efforts, motivating many Chinese citizens to work long hours at their private businesses to "get ahead." The only classless societies are a few small hunter-gatherer groups that lack extra resources that would allow some members to accumulate wealth and rise in economic status.

Some theorists criticize Marx for his focus on only the economic system, pointing out that noneconomic factors enter into the stratification struggle as well. Max Weber (1864–1920), an influential theorist, amended Marx's theory. He agreed with Marx that group conflict is inevitable, that economics is one of the key factors in stratification systems, and that those in power try to perpetuate their positions. However, he added two other influential factors that he argued determine stratification in modern industrial societies: power and prestige, discussed later in this chapter. Sometimes, these are identified as the "three Ps": property, power, and prestige.

Recent theorists suggest that using the three Ps, we can identify five classes—capitalists, managers, small business owners or the petty bourgeoisie, workers, and the underclass—rather than just haves and have-nots. *Capitalists* own the means of production, and they purchase and control the labor of others. *Managers* sell their labor to capitalists and manage the labor of others for the capitalists. The *petty bourgeoisie*, such as small shop or business owners, own some means of production but control little labor of others; nevertheless, they have modest prestige, power, and property (Sernau 2010). *Workers* sell their labor to capitalists and are low in all three Ps. The *underclass* has virtually no property, power, or prestige.

In the modern world, as businesses become international and managerial occupations continue to grow, conflict theorists argue that workers are still exploited but in different ways. Owners and chief executive officers (CEOs) get more income than many analysts feel is warranted by their responsibilities—for example, the CEO of Oracle received $84.5 million, of Boston Scientific $33.4 million, and of Occidental Petroleum $31.4 million in 2009 (Barr and Goldman 2010)—and better educated and skilled people get more income than is warranted by the differential in education (Wright 2000). The Obama administration's pay czar now publishes names and salaries of the 25 most highly paid employees in a number of companies, many of which received help from the bailout—Citigroup, Bank of

Taylor Swift (top) was the 2009 Entertainer of the Year, and although she is just 21 and has little education, she is paid millions of dollars each year to entertain the public. Meanwhile the public school teacher in the bottom photo would take 30 years of teaching hundreds of children to read before her total cumulative income for her entire career would add up to $1 million.

America, American International Group (AIG), GMAC, and others (Sorkin 2009). Moreover, the labor that produces our clothes, cell phones, digital cameras, televisions, and other products is increasingly provided by impoverished people around the world working for low wages at multinational corporations (Bonacich and Wilson 2005). In rich countries, service providers receive low wages at fast-food chains and box stores such as Wal-Mart (Ehrenreich 2001, 2005). One controversial question is whether multinational corporations are bringing opportunity to poor countries or exploiting them, as many conflict theorists contend (Wallerstein 2004).

The Evolutionary Theory of Stratification: A Synthesis

Evolutionary theory (Lenski 1966; Nolan and Lenski 2008) borrows assumptions from both structural-functionalist and conflict theories in an attempt to determine how scarce resources are distributed and how that distribution results in stratification. The basic ideas are as follows: (a) To survive, people must cooperate; (b) despite this, conflicts of interest occur over important decisions that benefit one individual or group over another; (c) valued items such as money and status are always in demand and in short supply; (d) there is likely to be a struggle over these scarce goods; and (e) customs or traditions in a society often prevail over rational criteria in determining distribution of scarce resources. After the minimum survival needs of both individuals and the society are met, power determines who gets the surplus: prestige, luxury living, the best health care, and so forth. Lenski believes that privileges (including wealth) flow from having power, and prestige usually results from having access to both power and privilege (Hurst 2006).

Lenski (1966; Nolan and Lenski 2008) tested his theory by studying societies at different levels of technological development, ranging from simple to complex. He found that the degree of inequality increases with technology until it reaches the advanced industrial stage. For instance, in subsistence-level hunting and gathering societies, little surplus is available, and everyone's needs are met to the extent possible. As surplus accumulates in agrarian societies, those who acquire power also control surpluses such as more land and servants, and they use this to benefit their friends and relations. However, even if laws are made by those in power, the powerful must share some of the wealth or fear being overthrown. Interestingly, when societies finally reach the advanced industrial stage, inequality is moderated. This happens because people in various social classes enjoy greater political participation and because more resources are available to be shared in the society.

Lenski's (1966; Nolan and Lenski 2008) theory explains many different types of societies by synthesizing elements of both structural-functionalist and conflict theory. For instance, evolutionary theory takes into consideration the structural-functionalist idea that talented individuals need to be motivated to make sacrifices by allowing private ownership to motivate them. Individuals will attempt to control as much wealth, power, and prestige as possible, according to this view, resulting in potential conflict as some accumulate more wealth than others (Nolan and Lenski 2008). The theory also recognizes exploitation leading to inequality, a factor that conflict theorists find in capitalist systems of stratification. The reality is that while some inequality may be useful in highly complex societies, extraordinary amounts of differential access to resources may even undermine productivity by making upward mobility so impossible that the most talented people are not always those in the most demanding and responsible jobs.

The amount of inequality differs in societies, according to evolutionary theorists, because of different levels of technological development. Because industrialization brings surplus wealth, a division of labor, advanced technology, and interdependence among members of a society, no one individual can control all the important knowledge, skills, or capital resources. Therefore, this eliminates the two extremes of haves and have-nots because resources are more evenly distributed.

The symbolic interaction, structural-functionalist, conflict, and evolutionary theories provide different explanations for understanding stratification in modern societies. These theories are the basis for micro- and macro-level discussions of stratification. Our next step is to look at some factors that influence an individual's position in a stratification system and the ability to change that position.

Thinking Sociologically

Using the theories discussed on pages 183–186, what are some reasons for *your* position in the stratification system?

Individuals' Social Status: The Micro Level

You are among the world's elite. Less than 7% of the world has a college degree ("6.7% of World Has College Degree" 2010). However, that number is expanding rapidly as countries such as China provide higher education opportunities to more students to support their growing economies. In 2006, 23 million Chinese students were enrolled, surpassing the U.S. enrollment, with an enrollment rate of 21% of the Chinese population (Jing 2007). Although China has more than 4,000 colleges and universities, the demand for a college education far exceeds the opportunity. Therefore, many Chinese college students are studying abroad, adding to their opportunities and enhancing China's knowledge of the world. In fact, Chinese students make up the largest number of international students studying in the United States. Worldwide, "the percentage of (college) age cohort enrolled in tertiary (college) education has grown from 19% in 2000 to 26% in 2007, with the most dramatic gains in upper middle and upper income countries. Over 150.6 million students are in tertiary education globally" (Altbach, Reisberg, and Rumbley 2009).

Yet, being able to afford the time and money for college is a luxury with little relevance to those struggling to survive each day. It is beyond the financial or personal resources of most people in the global south. Considered in this global perspective, college students learn professional skills and have advantages that billions of other world citizens will never know or even imagine. Does that give you a new respect for the opportunities you have?

In the United States, access to higher education is greater than in many other countries because there are more levels of entry—technical and community colleges, large state universities, and private four-year colleges, to name a few. However, with limited government help, most students must have financial resources to cover tuition and the cost of living. Many students do not realize that the prestige of the college they choose makes a difference in their future opportunities. Those students born into wealth can afford better preparation for entrance exams as well as tutors or courses to increase SAT or ACT scores. Some attend private prep schools and gain acceptance to prestigious colleges that open opportunities not available to those attending the typical state university or nonelite colleges (Persell 2005).

Ascribed characteristics, such as gender, can also affect one's chances for success in life. Imbedded gender stratification systems may make it difficult for women to rise in the occupational hierarchy. Many Japanese women, for example, earn college degrees but leave employment after getting married and having children (Globe Women 2011). The number of unmarried working women in Japan increased by 5% over 10 years, but married women's participation decreased by 3% during that time ("Japanese Women Face Difficulties Balancing Work and Family Life" 2007). Issues of gender stratification will be examined in more depth in Chapter 9, but they intersect with socioeconomic class and must be viewed as part of a larger pattern resulting in inequality in the social world.

Individual Life Chances

Life chances refer to one's opportunities based on achieved and ascribed status in society. That you are in college, that you probably have health insurance and access to health

For some people living in poverty, the standard of wealth is ownership of a single horse. The boys above travel to their village in Afghanistan, where they must work every day rather than going to school.

care, and that you are likely to live into your late 70s or beyond, compared to people in some poor countries whose average life span is in the 40s, are factors directly related to your life chances. Let us consider several examples of how placement in organizations and institutions at the meso level affects individual experiences and has global ramifications.

Education

Although education is valued by most individuals, the cost of books, clothing, shoes, transportation, child care, and time taken from income-producing work may be an insurmountable barrier to attendance from grade school through college. Economically disadvantaged students in most countries are more likely to attend less prestigious and less expensive institutions if they attend high school or university at all. Girls are particularly disadvantaged (Lewis and Lockheed 2006).

One's level of education affects many aspects of life, not just income. The higher levels of education affect political, religious, and marital attitudes and behavior: The more active individuals are in political life, the more mainstream or conventional their religious affiliation will be, the more likely they are to marry into a family with both economic and social capital, the more stable the marriage will be, and the more likely they are to have good health.

Health, Social Conditions, and Life Expectancy

Pictures on the news of children starving and dying dramatically illustrate global inequalities. The poorest countries in the world are in sub-Saharan Africa, where most individuals eat poorly, are susceptible to diseases, have great stress in their daily lives just trying to survive, and die at young ages compared to those in the developed world.

If you have a sore throat, you go to see your doctor. Yet, many people in the world will never see a doctor. Access to health care requires doctors and medical facilities, money for transportation and treatment, access to child care, and released time from other tasks to get to a medical facility. The poor sometimes do not have these luxuries. By contrast, the affluent eat better food, are less exposed to polluted water and unhygienic conditions, and are able to pay for medical care and drugs when they do have ailing health. Even causes of death illustrate the differences between people at different places in the stratification hierarchy. For example, in poor Global South countries, shorter life expectancies and deaths, especially among children, are due to controllable infectious diseases such as cholera, typhoid, AIDS, tuberculosis, and other respiratory ailments. By contrast, in affluent countries, coronary heart disease, stroke, and lung cancer are the most common causes of death, and most deaths are of people over the age of 65 (U.S. Census Bureau, Population Division, International Programs Center 2011; World Health Organization 2002a). So whether considered locally or globally, access to health care resources makes a difference in life chances. To an extent, the chance to have a long and healthy life is a privilege of the elite. Again, globally speaking, if you are reading this book as part of a college course, you are part of the elite.

By studying Table 7.1 in the next "Engaging Sociology" feature, you can compare life expectancy with two other measures of life quality for the poorest and richest countries: the gross national product (GNP) per capita income—the average amount of money each person has per year—and the infant mortality rates (death rates for babies). Note that average life expectancy in poor countries is as low as 32 years, income is as low as $200 a year (many of the people in these populations are subsistence farmers), and infant mortality is estimated to be as high as 180.2 deaths and as low as 2.8 deaths in the first year of life for every 1,000 births (Geocommons 2009; World Factbook 2010a, 2010b, 2010c). Numbers for the richest countries are dramatically different.

The United States has much larger gaps between rich and poor people than most other wealthy countries, resulting in higher poverty rates (more people at the bottom rungs of the stratification ladder). This is, in turn, reflected in health statistics, with infant mortality rates at 6.3 deaths in the first year of life per 1,000 births (World Factbook 2010c).

The world average infant mortality rate is 49.4 deaths per 1,000 live births (United Nations Department of Economic and Social Affairs: Population Division 2010). More than 25,000 children under five years old die each day on average; that is 1 child every 3 to 5 seconds, 17 to 18 children every minute, and approximately 9 million children

People living in poverty around the world often get health care at clinics or emergency rooms, if they have access to care, where they wait for hours to see a health care provider as shown in this mother-child health clinic in Kisumu, Kenya.

per year (Shah 2010). This evidence supports the assertion that health, illness, and death rates are closely tied to socioeconomic stratification. Race and gender interact with social class in ways that often have negative results as we shall see in the next two chapters.

Thinking Sociologically

What factors at the micro, meso, and macro levels affect your life expectancy and that of your family?

Individual Lifestyles

Your individual **lifestyle** includes how you think and act. Your attitudes, your values, your beliefs, your behavior patterns, and other aspects of your place in the world are shaped by socialization. As individuals grow up, the behaviors and attitudes consistent with their culture and family's status in society become internalized through the process of socialization. Lifestyle is not a simple matter of having money. Acquiring money—say, by winning a lottery—cannot buy a completely new lifestyle (Bourdieu and Passeron 1977). This is because values and behaviors are ingrained in our self-concept from childhood. You may gain material possessions, but that does not mean you have the lifestyle of the upper-class rich and famous.

Engaging Sociology

Life Expectancy, Per Capita Income, and Infant Mortality

Analyzing the meaning of data can provide an understanding of the health and well-being of citizens around the world. A country's basic statistics including life expectancy, per capita gross national product, and infant mortality tell researchers a great deal about its status.

1. What questions do the data in Table 7.1 raise regarding differences in mortality and life expectancy rates around the world?

2. Considering what you know from this and previous chapters and from Table 7.1, what do you think are some differences in the lives of citizens in the richest and poorest countries?

Table 7.1   Life Expectancy, Per Capita Income, and Infant Mortality for Selected Poor and Rich Countries

Poor Countries	Life Expectancy, 2009 (in Years)	Infant Mortality 2009	Per Capita GNP ($) 2008	Rich Countries	Life Expectancy, 2009 (in Years)	Infant Mortality 2009	Per Capita GNP ($) 2008
Swaziland	32.0	68.6	4,400	Japan	82.1	2.8	38,980
Angola	38.2	180.2	1,350	Hong Kong	81.9	2.9	43,800
Zambia	38.6	101.2	490	Australia	81.6	4.8	32,220
Mozambique	41.1	105.8	900	Canada	81.2*	5.0*	39,200
Sierra Leone	41.2	154.4	900	France	81.0	3.3	33,300
Liberia	41.8	138.2	500	Sweden	80.9	2.8	41,060
Afghanistan	44.6	152.0	800	Switzerland	81.9	4.2	42,000
Zimbabwe	45.8	32.3	200	Iceland	80.7	3.2	42,300
Chad	47.7	98.7	1,600	New Zealand	80.4	4.9	27,900
South Africa	49.0	44.2	10,100	Italy	80.7	5.5	31,400
Niger	52.6	116.7	700	United States	78.1**	6.3**	47,500

Source: World Factbook 2010b and 2010c for life expectancy and infant mortality; World Factbook 2010a for per capita income.

Note: Infant mortality is per 1,000 live births.

*Canada is 7th in life expectancy and 36th in infant mortality rates.

**United States is 50th in life expectancy and 45th in infant mortality rates.

Consider some examples of factors related to your individual lifestyle: attitudes toward achievement, political involvement, and religious membership.

Attitudes Toward Achievement

Attitudes differ by social status and are generally closely correlated with life chances. Motivation to get ahead and beliefs about what you can achieve are in part products of your upbringing and the opportunities you see as available to you. These attitudes differ greatly depending on the opportunity structure around you, including what your family and friends see as possible and desirable. Consider

the situation of children from poor countries and poor families. Their primary concern may be to help put food on the family's table. Even attitudes toward primary and secondary education reflect the luxury and inaccessibility that schooling is for some children. In Global North countries, opportunity is available for most children to attend school through high school and beyond. However, some students do not learn to value achievement in school due to poor self-concepts, difficulty in school, peer pressures, lack of support from family members and poor role models, poor schools and teaching, language differences, cultural differences, and many other factors (Ballantine and Hammack 2012).

Religious Membership

Religious affiliation also correlates with social status variables of education, occupation, and income. For instance, in the United States, upper-class citizens are found disproportionately in Episcopalian, Unitarian, Presbyterian, and Jewish religious groups, whereas lower-class citizens are attracted to Nazarene, Southern Baptist, Jehovah's Witness, and other holiness and fundamentalist sects. Each religious group attracts members predominantly from one social class, as will be illustrated in Chapter 11 in the section on religion. Those who attend the same religious group tend to share other values and attitudes about their lives and society, they tend to have different worship styles, and they have different ways of expressing religiosity. For example, people in higher socioeconomic statuses tend to know more about scripture and attend religious services more frequently, but those at the lower ends are more likely to pray daily (Roberts and Yamane 2012).

Political Behavior

What political preferences you hold and how you vote are also affected by social status. Around the world, upper middle classes are most supportive of elite or procapitalist agendas because these agendas support their way of life, whereas lower- or working-class members are least supportive (Wright 2000). Generally, the lower the social class, the more likely people are to vote for liberal parties (Kerbo 2008), and the higher the social status, the more likely people are to vote conservative on economic issues consistent with protecting wealth (Brooks and Manza 1997).

In the United States, members of the lower class tend to vote liberal on economic issues, favoring government intervention to improve economic conditions. However, those with lower levels of education and income vote conservatively on many social issues relating to minorities and civil liberties (e.g., rights for homosexuals, gay marriage, and abortion) (Gilbert and Kahl 2003; Jennings 1992; Kerbo 2008). In the 2008 election, many voters had to make choices about economic policies they liked and whether those policies were more important than their preferences on some of the social issues.

Status Inconsistency

The reality is that some people experience high status on one trait, especially a trait that is achieved, but may experience low status in another area. For example, a professor may have high prestige but low income. Max Weber called this unevenness in one's social standing *status inconsistency*. Individuals who experience such status inconsistency, especially if they are treated as if their lowest ascribed status is the most important one, are likely to be very liberal and to experience discontent with the current system (Weber 1946).

People tend to associate with others like themselves, perpetuating and reinforcing lifestyles. In fact, people often avoid contact with others whose lifestyles are outside their familiar and comfortable patterns. This desire for familiarity also means that most people remain in the same social class because they have learned the "subculture" and it is comfortable and familiar.

Life chances and lifestyles are deeply shaped by the type of stratification system that is prevalent in the nation. Such life experiences as hunger, the unnecessary early death of family members, or the pain of seeing one's child denied opportunities are all experienced at the micro level, but their causes are usually rooted in events and actions at other levels of the social world. This brings us to our next question: Can an individual change positions in a stratification system?

Thinking Sociologically

Describe your own lifestyle and life chances. How do these relate to your socialization experience and your family's position in the stratification system? What difference do they make in your life?

Social Mobility: The Micro-Meso Connection

For professional athletes, each hoop, goal, or touchdown throw is worth thousands of dollars. These riches give hope to those in rags that if they "play hard" on their local hoops, they too may be on the field or court making millions. The problem is that the chances of making it big are so small that such hopes are some of the cruelest hoaxes faced by young African Americans and others in the lower or working class. Therefore, it is a false promise to think of sports as the road to opportunity (Dufur and Feinberg 2007; Edwards 2000). The chances of success or even of moving up a little in the social stratification system through sports are very small.

Those few minority athletes who do "make it big" and become models for young people experience "stacking," holding certain limited positions in a sport. When retired from playing, few black athletes rise in the administrative hierarchy of the sports of football and baseball, although basketball has a better record of hiring black coaches and managers. Thus, when young people put their hopes and energies into developing their muscles and physical skills, they may lose the possibility of moving up in the social class system, which requires developing their minds and their technical skills.

The whole idea of changing one's social position is called social mobility. **Social mobility** refers to the "extent to which people move up or down in the class system, especially from one generation to the next" (Gilbert 2008:123).

What is the likelihood that your status will be different from that of your parents over your lifetime? Will you start a successful business? Marry into wealth? Win the lottery? Experience downward mobility due to loss of a job, illness, or inability to complete your education? What factors at different levels of analysis might influence your chances of mobility? These are some of the questions addressed in this section and the next.

Four issues dominate the analysis of mobility: (a) types of social mobility, (b) methods of measuring social mobility, (c) factors that affect social mobility, and (d) whether there is a "land of opportunity" for those wishing to improve their lot in life.

Types of Social Mobility

Mobility can be up, down, or sideways, as described below. *Intergenerational mobility* refers to change in class status compared to one's parents' status, usually resulting from education and occupational attainment. If you are the first to go to college in your family and you become a computer programmer, this would represent intergenerational mobility. The amount of intergenerational mobility in a society measures the degree to which a society has an **open class system** that allows movement between classes, meaning that you could move up or down in comparison with your parents in the

stratification system. This type of movement is not possible in many societies.

Intragenerational mobility (not to be confused with *intergenerational mobility*) refers to the change within a position in a single individual's life. For instance, if you begin your career as a teacher's aide and end it as a school superintendent, that is upward intragenerational mobility. However, mobility is not always up. *Vertical mobility* refers to movement up or down in the hierarchy and sometimes involves changing social classes. You may start your career as a waitress, go to college part-time, get a degree in engineering, and get a more prestigious and higher-paying job, resulting in upward mobility. Alternatively, you could lose a job and take one at a lower status, a reality for many when the economy is doing poorly. In the global economic downturn, people at all levels of the occupational structure are experiencing layoffs and often having problems finding new positions at comparable levels and pay.

How Much Mobility Is There? Measures of Social Mobility

Can one move up in the class system? One traditional method of measuring mobility is to compare fathers and sons. Surveys ask men about their occupations and those of their fathers or sons. Table 7.2 reflects questions asked of U.S. fathers about their sons' and daughters' mobility.

Table 7.2 Outflow From Father's Occupation to Son's or Daughter's Occupation

Father's Occupation	Upper-White Collar	Lower-White Collar	Upper Manual	Lower Manual	Farm	Total
Upper White-Collar	42	31	12	15	1	100
	54	33	9	3	*	
Lower White-Collar	34	33	13	19	1	100
	49	34	11	6	*	
Upper Manual	20	20	29	29	2	100
	35	37	18	8	1	
Lower Manual	20	22	20	36	12	100
	32	39	19	9	1	
Farm	16	18	19	35	3	100
	34	28	22	14	2	
Total (N = 3,398)	27	25	19	27	1	100
	27	25	19	27	3	

Son's Occupation (percentage) in Blue

Daughter's Occupation (percentage) in Red

Source: Gilbert 2008:124, 127.

Note: Rows but not columns add to 100%. For example, read across the row that begins with "lower manual" on the far left to trace the sons of fathers who held unskilled "lower manual" jobs. While 20% rose to upper white-collar (professional or managerial) positions, the largest group (36%) followed their fathers into unskilled manual jobs.

Several conclusions can be derived from Table 7.2:

1. There is a high level of occupational inheritance— sons following fathers into jobs at the very same occupational level.

2. The higher the father's occupation level, the better the son's chances for occupational achievement.

3. There is also considerable movement up and down the occupational ladder from one generation to the next.

4. Sons are more likely to move up than down.

5. Daughters are even more likely to move up than sons. (Gilbert 2008:124–25).

Thinking Sociologically

Indicate where your father falls in the occupational categories in Table 7.2. Compare this with your intended occupation and likely social class position in a few years. What can you conclude about your chances for mobility?

Determining the mobility of women is more difficult because they often have lower-level positions and their mothers may not have worked full-time, but a conclusion that can be drawn is that both women's and men's occupational attainment is powerfully influenced by class origins.

Factors Affecting an Individual's Mobility

Why are some people successful at moving up the ladder while others lag behind? Mobility is driven by many factors, from your family's *cultural capital* to global economic variables. One's chances to move up depend on micro-level factors—one's family background, socialization, personal characteristics, and education—and macro-level factors such as the occupational structure and economic status of countries, population changes, the numbers of people vying for similar positions, discrimination based on gender or ethnicity, and the global economic situation.

The study of mobility is complicated because these key variables are interrelated. The macro-level forces (e.g., the economy, opportunities available, and occupational structure in a country) are related to meso-level factors (e.g., access to education and job opportunities) and micro-level factors (e.g., socialization and family background and education) (Blau and Duncan 1967). An individual's background accounts for nearly half of the factors affecting occupational attainment (Jencks 1979).

Family Background, Socialization, and Education

Our family background socializes us into certain behavior patterns, language usage, and occupational expectations (Sernau 2010). Let us consider language. Parents in professional families use three times as many different words at home as parents in low-income families. By the time the children of professional families are three years old, they have a vocabulary of about 1,100 words and typically use 297 different words per hour. Children in working-class families have a 700-word vocabulary and use 217 words per hour, while children from low-income families have 500 words accessible to them and use 149 per hour. These numbers represent a gap in the range of words they hear at home (Hart and Risley 2003). How might that difference in one's vocabulary affect one's success in school? How might an expanded vocabulary affect one's opportunities in life?

College is one expectation key to upward mobility. Although not all those with a college degree are successful, few in Global North countries have a chance to be successful without a college degree (Lareau 2003). College education is the most important factor for high-income status, and the rewards of college degrees have increased. Those with degrees become richer than those without, largely because of changes in occupational structures creating new types of jobs for the computer information age. These "social-cultural specialists" work with ideas, knowledge, and technology rather than manufacturing (Florida 2002; Hurst 2006). We can see from Table 7.3 that most students in the United States with high ability from high-status families go to college, while high-ability students from low-status families go to college less often (83% vs. 51%).

When we look at actual college degrees awarded, the pattern is more extreme, with diplomas going disproportionately to students in the top status groups (50%), compared to only 10% of students from the bottom half of

Table 7.3 College Attendance by Social Class and Cognitive Ability (percent in college)

Cognitive Ability Quartile	Family Socioeconomic Status Quartile			
	Top	2nd	3rd	Lowest
Top	83	63	74	51
Second	69	42	51	33
Third	57	24	40	23
Lowest	35	13	20	13
Total (N = 3,398)	69	33	48	24

Source: Gilbert 2008:142.

These men hold very different positions in society; one can offer his family more cultural capital to spend in influential positions and can provide more opportunities because of his education, family background, and networks with others in positions of influence. One of the two has a law degree.

income levels. If American society was truly a meritocracy, as is so often claimed, one would expect cognitive ability to be the most important variable, and that is spread across social classes (Gilbert 2008).

Even when a young person is admitted to a college or university, she or he may be at a disadvantage in the classroom and alienated from past social ties. The culture of college is the culture of the well-educated upper middle class, and everything from values to knowledge base to sense of humor may be different and uncomfortable. This alienation is explored in more detail in the next "Engaging Sociology."

Many poor people lack education and skills such as interviewing for jobs and obtaining recommendations needed to get or change jobs in the postindustrial occupational structure (Ehrenreich 2001, 2005; McLeod 2008). Isolated from social networks in organizations, they lack contacts, or social capital, to help in the job search. The type of education system one attends also affects mobility. In Germany, Britain, France, and some other European countries, children are "streamed" (tracked) into either college preparatory courses or more general curricula, and the rest of their occupational experience usually reflects this early placement decision in school. In the United States, educational opportunities remain more open to those who can afford them—at least this is what is supposed to happen.

Occupational Structure and Economic Vitality

The economic vitality of a country affects the chances for individual mobility, since there will be fewer positions at the top if the economy is stagnant. As agricultural work is decreasing and technology jobs are increasing (Hurst 2006), these changes in the composition and structure of occupations affect individual opportunity. Thus, a country's economy and its place in the global system shape employment chances of individuals. The global economic downturn that started in 2008 illustrates the vulnerability of both macro-level nations whose banks and companies are failing and micro-level individuals living in those nations who are losing their jobs.

Population Trends

The fertility rates, or number of children born at a given time, are a macro-level trend that influences the number of people who will be looking for jobs. The U.S. nationwide baby boom that occurred following World War II resulted in a flood of job applicants and downward intergenerational mobility for the many who could not find work comparable to their social class at birth. By contrast, the smaller group following the baby boomer generation had

Engaging Sociology

First-Generation College Students: Issues of Cultural Capital and Social Capital

Socioeconomic classes develop subcultures that can be quite different from each other, and when one changes subcultures, it can be confusing and alienating. College campuses provide an example. Generally, they are dominated by middle-class cultures. Young people from blue-collar backgrounds and those who are first-generation college students often find themselves in a world as alien to them as visiting another country. Students whose parents went to college are more likely to have "cultural or social capital" that helps them adjust and helps them understand their professors who are generally part of the middle-class culture. Answer the following survey questions. How might your own cultural or social capital cause you to feel at home or alienated, privileged or disprivileged, hopeful or despairing?

1. Which of the following experiences were part of your childhood?

    ◊ Had a library of books (at least 50 adult books) at your childhood home
    ◊ Had a subscription to a newspaper that was delivered to your home
    ◊ Had news magazine subscriptions that came to your home *(Time, Newsweek, The Economist)*
    ◊ Listened to music as a family, including classical or instrumental music such as harp or flute
    ◊ Traveled to at least 20 other states or to at least 5 other countries
    ◊ Took regular trips to the library
    ◊ Took regular trips to museums
    ◊ Attended movies
    ◊ Attended plays (theater productions)
    ◊ Attended concerts
    ◊ Played a musical instrument
    ◊ Took dance lessons
    ◊ Listened to National Public Radio (NPR)
    ◊ Watched PBS (Public Broadcasting Station) on television

2. Which of the following *relationships* were part of your childhood?

    ◊ My parents knew at least two influential people in my community on a first name basis—such as the major, members of the city council, the superintendent of schools, members of the school board, the local county sheriff, the chief of police, the prosecuting attorney, the governor, and the district's representative to Congress.
    ◊ The regional leader of my religious group—church, temple, synagogue, or mosque—knew and respected my family.
    ◊ My parents knew on a first name basis at least three chief executive officers (CEOs) of corporations.
    ◊ When I entered new situations in high school, it was likely that my parents were known by the coaches, music directors, summer camp counselors/directors, or other authority figures who were "running the show."
    ◊ When I came to college, one or more professors and administrators at the college knew my parents, a sibling, or another family member.
    ◊ I have often interacted directly and effectively (in a nonadversarial way) with authority figures.

If you experienced many of the items in #1 at home, you had fairly high cultural capital. If you marked most of the items in #2, you had a lot of social capital. If you did not, you may find the culture of a college campus to be alien and even confusing.

Answer the three questions below and then read the sociological explanations.

1. Which of the following makes a first-generation college student feel most alienated at your college and even within this sociology course: *economic capital* (money), *social capital* (networks with those who have resources), or *cultural capital* (knowledge of important aspects of the culture)? Why?

2. What did "doing well" in school mean in your family? Did they stress education, and if so, how?

3. What did it mean within your family to be "independent" when you were in high school?

Note: The following ideas provide some sociological insight into the previous questions.

1. Students who do not have a middle-class cultural capital may find that they are in a strange culture at college as if they entered another country. Students who do not have a strong social capital may find that they do not know how to find advocates or support in difficult situations. If students lack financial capital then they may find themselves as social isolates, because they cannot afford to do the things that are part of the social life of the campus; in addition, they may have to work long hours to pay for their education.

2. Blue-collar definitions of *success in school* often stress

   - Obeying authority and memorizing material
   - Getting the reward (the diploma) is most important, with minimal necessary commitment of effort

   Middle-class definitions of *school success* often emphasize

   - Learn the material and get good grades
   - Become a committed and contributing member of the school

3. Blue-collar definitions of *independence* often emphasize

   - Supporting oneself financially
   - Not taking "crap" from anyone; that is, defending oneself physically

   Middle-class definitions of *independence* typically focus on

   - Being original and creative
   - Thinking things through on your own and challenging authority or "the common wisdom."

In short, social and cultural capital pervade the experiences of students pursuing a college education.

Source: Survey constructed in part using ideas from Morris and Grimes (1997).

fewer competitors for entry-level jobs. Baby boomers hold many of the executive and leadership positions today, so promotion has been hard for the next cohort. As baby boomers retire, opportunities will open up, and mobility should increase.

Gender and Ethnicity

Many women and ethnic minority groups, locked in a cycle of poverty, dependence, and debt, have little chance of changing their status. Women in the U.S. workforce, for instance, are more likely than men to be in dead-end clerical and service positions with no opportunity for advancement. In the past three decades, the wage gap between women and men has narrowed, and women now earn 77% of what men earn for similar jobs (Fitzpatrick 2010), compared to 60% in 1980 (U.S. Department of Labor 2005). African American women make 68% (Fitzpatrick 2010). Table 7.4 shows the earnings for white males and the percentage of those earnings for other gender and ethnic groups.

Special circumstances such as war have often allowed women and others who were denied access to good jobs to get a "foot in the door" and actually enhance their upward mobility. Another factor affecting career success of women is that today more females than males are earning college degrees.

Some people experience privilege, whereas some experience disprivilege due to socioeconomic status, ethnicity, gender, or a combination of these. Be aware that various forms of inequality intersect in many societies in highly complex ways.

Table 7.4   Median Annual Earnings by Race/Ethnicity and Sex

Race/Gender	Earnings	Wage Ratio
White men	$47,814	100.0%
White women	35,151	73.5
Black men	34,480	72.1
Black women	30,398	63.6
Hispanic men	27,490	57.5
Hispanic women	24,738	51.7
Total Gender Wage Gap		
All men	$42,210	
All women	$32,649	77.4%

Source: U.S. Census Bureau (2007b).

The Interdependent Global Market and International Events

If the Asian or Japanese stock markets hiccup, it sends ripples through world markets. If high-tech industries in Japan or Europe falter, North American companies in Silicon Valley, California, may go out of business, costing many professionals their lucrative positions. In ways such as these, the interdependent global economies affect national and local economies, and that affects individual families.

Whether individuals move from "rags to riches" is not determined solely by their personal ambition and work ethic. Mobility for the individual, a micro-level event, is linked to a variety of events at other levels of the social world, and one cannot assume that the unemployed individual is just lazy or incompetent.

Thinking Sociologically

Do you know individuals who have lost jobs because of economic slowdowns at the meso or macro level or who have gotten jobs because of economic booms and opening opportunities? What changes have occurred in their mobility and social class?

Is There a "Land of Opportunity"? Cross-Cultural Mobility

Countless immigrants have sought better opportunities in new locations. Perhaps, your parents or ancestors did just this. The question for this section concerns your chances for mobility: Do you have a better chance to improve your status in England, Japan, the United States, or some other country? The answer is not simple. If there is a land of opportunity where individuals can be assured of improving their economic and social position, it is not easy to identify, as many variables affect the opportunity structure. The reality of the "land of opportunity" depends on the historical period when immigrants first came to the country; current economic conditions, social events, and political attitudes toward foreigners when new immigrants arrive; immigrants' personal skills; and their ability to blend into the new society.

During economic growth periods, many immigrants have found great opportunities for mobility in the United States and Europe. Early industrial tycoons in railroads, automobiles, steel, and other industries are examples of success stories. Today, fortunes are being made in the high-tech industries of China, India, and other countries, and in energy resources in Russia. The number of millionaires and billionaires in the world is increasing dramatically, yet this is still a very small percentage of the world's population. In 2010, there were 1,011 billionaires globally. The richest man was Carlos Slim Helú with $53.5 billion from América Movíl company. Close behind was Bill Gates with $53 billion from Microsoft, and Warren Buffett with $47 billion from Berkshire Hathaway company (Miller and Kroll 2010). China is the dramatic success story, reaching an estimated 260 billionaires, including many women, in 2009, doubling the number in one year. It will soon top the list of countries with billionaires (France 24 International News 2009;

Spero News 2009). Such wealth eludes most people who immigrate and must work multiple low-paying jobs just to feed their families and stay out of poverty.

Opportunities for upward mobility have changed significantly with globalization. Many manufacturing jobs in the global economy have moved from the Global North to the Global South to take advantage of cheap labor; this has reduced the number of unskilled and low-skilled jobs available in Global North countries. Since the 1970s, the earnings differential between high school graduates and college graduates has widened, leaving high school graduates struggling to find work that will support a family. Multinational corporations look for the cheapest sources of labor, mostly in the Global South, which features low taxes, no labor unions, few regulations, and many workers needing jobs, thus draining away low-skilled jobs from the United States.

Why the changes in the job structure? The increase in international trade results in demand for high-tech products from the United States but fewer manufactured products, thus reducing manufacturing jobs. New technologies and automation leave low-skilled workers in Global North countries without jobs. The minimum-wage and low-wage jobs are all that is available to lower-skilled workers, and labor unions have less influence on the wages and working conditions of these workers today (Gilbert 2008). High-tech positions are good news for those with college degrees and technical skills, but the replacement of laboring positions with service jobs in fast-food and box store chains (e.g., Wal-Mart, Kmart, Home Depot, Lowe's) means a severe loss of genuine opportunity for living wages.

Although the new multinational industries springing up in Global South countries such as Malaysia, Thailand, and the Philippines provide opportunities for mobility to those of modest origins, much of the upward mobility in the world is taking place among those who come from small, highly educated families with individualistic achievement-oriented values and people who see education as a route to upward mobility (Blau and Duncan 1967; Featherman and Hauser 1978; Jencks 1979; Rothman 2005). They are positioned to take advantage of the changing occupational structure and high-tech jobs. As the gap between rich and poor individuals and countries widens, more individuals in the United States begin to move down rather than up in the stratification system. The next "Sociology in Your Social World" illustrates this problem.

Thinking Sociologically

What social factors in your society limit or enhance the likelihood of upward social mobility for you and your generation? Explain.

Sociology in Your Social World

Nickel and Dimed

Have you or has someone you know held a minimum-wage job? Did you have to support yourself and maybe other family members on that wage, or was it just pocket money? Millions in the United States live in this world of unskilled laborers, getting from $5.15 to $7 an hour for their work, and most rely on these wages to support themselves and their families.

Author Barbara Ehrenreich (2001) asked how individuals and families survive on these wages. To find out, she tried it herself. Living in several cities—Key West, Florida; Portland, Maine; and Minneapolis, Minnesota—she held jobs as a waitress, a retail clerk at Wal-Mart, a hotel housekeeper, and a nursing home aide. Each job was physically and mentally demanding.

After describing the details of each job and the lives of her coworkers, Ehrenreich (2001) concludes that people are earning far less than they need to live. Her coworkers struggled to meet minimum housing costs (sometimes living in hotels or cars), bought cheap food, and shopped at thrift stores. They were also consistently struggling with transportation, child care, and health care. Many were also managing large debts.

The number of "working poor" is growing. Many states added work requirements for mothers to remain eligible for welfare benefits. These reforms became federal law when President Bill Clinton replaced Aid to Families with Dependent Children (AFDC) with Temporary Assistance for Needy Families (TANF). Beyond work requirements, TANF also added time limits for welfare eligibility and eliminated increased benefits for families that grew while receiving assistance.

"The Economic Policy Institute recently reviewed dozens of studies of what constitutes a living wage and came up with an average figure of $30,000 a year for a family of one adult and two children, which amounts to a wage of $14 an hour" (Ehrenreich 2001:214). This budget includes health insurance, a telephone, and child care but no extras, and it is more than most minimum-wage workers have. Upward mobility for this large group in the population is difficult if not impossible. Ehrenreich's research makes it clear: To be a society of at least minimal opportunity for all, we must reconsider economic policies to better address underpaid workers as well as the unemployed. Although many states have raised the minimum wage since *Nickel and Dimed* was published—and federal minimum wage incrementally increased to $7.25 as of July 24, 2009—no state currently has a minimum wage that meets minimum living wage requirements by the Economic Policy Institute's calculation. Add to that the increases in costs of gas and food, and many U.S. citizens are struggling to make ends meet.

Major Stratification Systems: Macro-Level Analysis

Imagine being born into a society in which you have no choices or options in life because of your family background, age, sex, or ethnic group. You cannot select an occupation that interests you, you cannot choose your mate, and you cannot live in the part of town you choose. You see wealthy aristocrats parading their advantages and realize that this will never be possible for you. You can never own land or receive the education of your choice.

This situation is reality for millions of people in the world—they are born this way and will spend their lives in this plight. In **ascribed stratification systems**, social statuses that are assumed at birth or are assigned to one involuntarily later in life determine one's position in society. While race and sex may be mostly determined at birth, age or falsely being convicted of a crime may put one in a status that is not one's choice. In contrast, **achieved stratification systems**, such as class systems, allow individuals to earn positions through their ability and effort. Sometimes it is possible to achieve a higher ranking by working hard, obtaining education, gaining power, or doing other things that are highly valued in that culture.

The futures of these Aboriginal boys in Australia are determined by their ethnic group and family of birth, making it unlikely that they will ever experience much affluence in Australian society.

Ascribed Status: Caste and Estate Systems

Caste systems are the most rigid ascribed stratification systems. Individuals are born into a caste system, and they retain their birth status throughout life. Castes are maintained by cultural norms and social control mechanisms that are deeply imbedded in religious, political, and economic norms and institutions. Caste members have predetermined occupational positions, marriage partners, residences, social associations, and prestige levels. A person's caste is easily recognized through clothing, speech patterns, family name and identity, skin color, or other distinguishing characteristics. From their earliest years, individuals learn their place in society through the process of socialization. To behave counter to caste prescriptions would be to go against religion and social custom and to risk not fitting into society. That can be a death sentence in some societies.

Religious ideas dictate that one's status after death (in Christian denominations) or one's next *reincarnation* or rebirth (in the Hindu tradition) also might be in jeopardy. Stability in Hindu societies is maintained by the belief that people can be reborn into a higher status in the next life if they fulfill expectations in their ascribed position in this life. Thus, believers in both religions work hard in hopes of attaining a better life after death or in the next reincarnation. The institution of religion works together with the family, education, and economic and political institutions to shape (and sometimes reduce) both expectations and aspirations and to keep people in their prescribed places in caste systems.

The clearest example of a caste system is found in India. The Hindu religion holds that individuals are born into one of four *varna*, broad caste positions, or into a fifth group below the caste system, the *outcaste* group. The outcaste encompasses profoundly oppressed and broken people—"a people put aside"—referred to as untouchables, outcastes, *Chandalas* (a Hindu term), and *Dalits* (the name preferred by many "untouchables" themselves). Although the Indian Constitution of 1950 granted full social status to these citizens and a law passed in 1955 made discrimination against them punishable, deeply rooted traditions are difficult to change. Caste distinctions are still very prevalent, especially in rural areas, as seen in the discussion in the next "Sociology Around the World."

Estate systems are characterized by the concentration of economic and political power in the hands of a small minority of political-military elite, with the peasantry tied to the land (Rothman 2005). Based on ownership of land, the position one is born into, or military strength, estate systems are rigid as are caste systems in stratifying individuals. An individual's rank and legal rights are clearly spelled out, and arranged marriages and religion bolster the system. During the Middle Ages, knights defended the realms and the religion of the nobles. Behind every knight in shining armor were peasants, sweating in the fields and paying for the knights' food, armor, and campaigns. For farming the land owned by the nobility, peasants received protection against invading armies and enough of the produce to survive. Their life was often miserable. If the crops were poor, they ate little. In a good year, they might save enough to buy a small parcel of land. A very few were able to become independent in this fashion.

Estate systems existed in ancient Egypt, the Incan and Mayan civilizations, Europe, China, and Japan. Today, similar systems exist in some Central and South American, Asian, and African countries on large banana, coffee, cacao, and sugar plantations. Over time, development of a mercantile economy resulted in modifications in the early estate systems, and now peasants often work the land in exchange for the right to live there and receive a portion of the produce.

Achieved Status: Social Class Systems

Social class systems of stratification are based on achieved status—that means our life circumstances and social respect are based in large measure on what we achieve in

Sociology Around the World

The Outcastes of India

The Dalits—*sometimes called "untouchables"—are the most impoverished people in India and some of the most impoverished in the world. While they live in incredible hardship, they are beginning to mobilize and demand rights as citizens of India and the world.*

The village south of Madras (now known as Chennai) in the state of Tamil Nadu was on an isolated dirt road, one kilometer from the nearest town. It consisted of a group of mud and stick huts with banana leaf–thatched roofs. As our group of students arrived, the *Dalit* villagers lined the streets to greet us—and stare. Many had never seen Westerners. They played drums and danced for us and threw flower petals at our feet in traditional welcome.

Through our translator, we learned something of their way of life. The adults work in the fields long hours each day, plowing and planting with primitive implements, earning about eight cents from the landowner, often not enough to pay for their daily bowl of rice. Occasionally, they catch a frog or bird to supplement their meal. In the morning, they drink rice gruel, and in the evening, they eat a bowl of rice with some spices. Women and children walk more than a kilometer to the water well—but the water is polluted during the dry season. There are no privies but the fields. As a result of poor sanitation, inadequate diet, and lack of health care, many people become ill and die from health problems that are easily cured in Western societies. For instance, lack of vitamin A, found in many fruits and vegetables to which they have little access, causes blindness in many village residents. Although the children have the right to go to the school in the closest village, many cannot

do so because they have no transportation, shoes, or money for paper, pencils, and books. Also, the families need them to work in the fields alongside their parents or help care for younger siblings just to survive.

Many taboos rooted in tradition separate the *Dalits* from other Indians. For instance, they are forbidden to draw water from the village well, enter the village temple, or eat from dishes that might be used later by people of higher castes. The latter prohibition eliminates most dining at public establishments. About 95% are landless and earn a living below subsistence level.

Dalits who question these practices have been attacked and their houses burned. In one instance, 20 houses were burned on the birthday of B. R. Ambedkar, a leader in the *Dalit* rights movement. Official records distributed by the Human Rights Education Movement of India state that every hour, two *Dalits* are assaulted, three *Dalit* women are raped, two *Dalits* are murdered, and two *Dalit* houses are burned (Dalit Liberation Education Trust 1995; Thiagaraj 2006; Wilson 1993). Violence is used to control *Dalits* who try to uplift themselves and thus create threats to the social status and dominance of higher caste groups (Karthikeyan 2011). This group on the bottom rung of the stratification system has a long fight ahead to gain the rights that many of us take for granted.

A few *Dalits* have migrated to cities, where they blend in, and some of these have become educated and are now leading the fight for the rights and respect guaranteed by law. Recently, unions and interest groups have been representing the *Dalits*, and some members have turned to religious and political groups that are more sympathetic to their plight, such as Buddhists, Christians, or Communists.

One social activist, Henry Thiagaraj, has committed his life to improving conditions for the *Dalits*. Thiagaraj works on a micro, meso, and macro level. On the micro level, he suggests that those in power form *Dalit* youth and women's *sangams* (activist groups) and organize the people who live in *Dalit* slums. At the meso level, Thiagaraj and other *Dalit* activists work to initiate micro lending to the *Dalits* and improve their education and labor training. On the macro level, Thiagaraj works with nongovernmental organizations to increase support for *Dalit* interests. He also works to improve media coverage of the *Dalits* to raise awareness of their experiences. Thiagaraj's (2007) book, *Human Rights From the Dalits' Perspective,* provides an outstanding retrospective of how India has addressed caste discrimination since the 1980s.

our lives. Members of the same social class have similar income, wealth, and economic position and also share comparable styles of living, levels of education, culture, and patterns of social interaction. Most of us are members of class-based stratification systems, and we take advantage of opportunities available to our social class. Our families, rich or poor, educated or unskilled, provide us with an initial social ranking and socialization experience. We tend to feel a kinship and sense of belonging with those in the same social class—our neighborhood and work group, our peers and friends. We think alike, share interests, and probably look up to the same people as a reference group. Our social class position is based on the three main factors determining positions in the stratification system: property, power, and prestige.

This is this trio—the three "Ps"—that, according to Max Weber (1946, 1947), determines where individuals rank in relation to each other. By property (wealth), Weber refers to owning or controlling the means of production. Power, the ability to control others, includes not only the means of production but also the position one holds. Prestige involves the esteem and recognition one receives, based on wealth, position, or accomplishments. Households in the upper and lower social classes are affected by specific factors that determine a person's standing in each of the three areas listed above (Rossides 1997). Those are expressed in concrete areas:

Social Class (Economic) Variables

- Income
- Wealth
- Occupation
- Education
- Family stability
- Education of children

Prestige Variables

- Occupational prestige
- Respect in community
- Consumption
- Participation in group life
- Evaluations of race, religion, and ethnicity

Power (Political-Legal) Variables

- Political participation
- Political attitudes
- Legislation and governmental benefits
- Distribution of justice

Although these three dimensions of stratification are often found together, this is not always so. Recall the idea of *status inconsistency:* An individual can have a great deal of prestige yet not command much wealth (Weber 1946);

consider winners of the prestigious Nobel Peace Prize such as Wangari Maathai of Kenya; Rigoberta Menchú Tum of Guatemala (see "Sociology Around the World" on page 76 in Chapter 3); or Betty Williams and Mairead Corrigan of Northern Ireland. None of them is rich, but each has made contributions to the world that have gained her universal prestige. Likewise, some people gain enormous wealth through crime or gambling, but this wealth may not be accompanied by respect or prestige.

Compared with systems based on ascribed status, achieved status systems maintain that everyone is born with common legal status; everyone is equal before the law. In principle, all individuals can own property and choose their own occupations. However, in practice, most class systems pass privilege or poverty from one generation to the next. Individual upward or downward mobility is more difficult than the ideology invites people to believe.

The Property Factor

One's income, property, and total assets comprise one's **wealth**. These lie at the heart of class differences. The contrast between the splendor of aristocrats and royalty and the daily struggle for survival of those in poverty is an example of the differences extreme wealth creates. Another example is shown in the income distribution in the United States by quintiles (see Table 7.5). Note that there has been minimal movement between the groups over the years. In 2009, the median household income in the United States was $49,777 (U.S. Census Bureau 2011c). However, the poverty rate was 14.3% of the population, up from 12.5% of the population in 2007 (U.S. Census Bureau 2009d). Of children under 18, 20.7%, or 15,451, live in poverty (U.S. Census Bureau 2009d). Roughly 1 in 6 adults have no health insurance (Newport and Mendes 2009).

Table 7.5 Share of Household Income in Quintiles

	1995	2000	2007
Lowest quintile	3.7	3.6	3.4
Second quintile	9.1	8.9	8.7
Third quintile	15.2	14.8	14.8
Fourth quintile	23.3	23.0	23.4
Highest quintile	48.7	49.8	49.7
Ratio of highest fifth to lowest fifth	13.2	13.8	14.6

Source: U.S. Census Bureau 2009c.

Thinking Sociologically

Explain the trends found in Table 7.5. What evidence do you see of increased or decreased inequality?

Although income distribution in the United States has not changed dramatically over time, there has been an increase in overall inequality, as the middle class has decreased by more than 8% since 1969. A majority of the 8% have experienced a downward movement, although a few have moved up in the stratification system (Rose 2000).

The Power Factor

Power refers to the ability to control or influence others, to get them to do what you want them to do. Positions of power are gained through family inheritance, family connections, political appointments, education, hard work, or friendship networks.

We discussed previously the conflict theorists' view that those who hold power are those who control the economic capital and the means of production in society (Ashley and Orenstein 2009). Consistent with Marx, many recent conflict theorists have focused on a **power elite** in which power is held by top leaders in corporations, politics, and military. These interlocking elites make major decisions guiding nations (Domhoff 2001; Mills 1956). These people interact with each other and have an unspoken agreement to ensure that their power is not threatened. Each tends to protect the power of the other. The idea is that those who are not in this interlocking elite group do not hold real power and have little chance of breaking into the inner circles (Dye 2002a, 2002b).

Power pluralism theorists, on the other hand, argue that power is not held exclusively by an elite group but is shared among many power centers, each of which has its own self-interests to protect (Ritzer and Goodman 2004). Well-financed special interest groups (e.g., dairy farmers or truckers' trade unions) and professional associations (e.g., the American Medical Association) have considerable power through collective action. Officials who hold political power are vulnerable to pressure from influential interest groups, and each interest group competes for power with others. Creating and maintaining this power through networks and pressure on legislators is the job of lobbyists. For example, in the ongoing U.S. debate over health care legislation, interest groups from the medical community, insurance lobbies, and citizens' groups wield their power to influence the outcome, but because these major interests conflict and no one group has the most power, no resolution has been reached.

The Prestige Factor

Prestige refers to an individual's social recognition, esteem, and respect commanded from others. An individual's prestige ranking is closely correlated with the value system of society. Chances of being granted high prestige improve if one's patterns of behavior, occupation, and lifestyle match those that are valued in the society. Among high-ranked occupations across nations are scientists, physicians, military officers, lawyers, and college professors. Table 7.6 shows selected occupational prestige rankings in the United States. Note the correlation between recent events such as 9/11 and the increased rankings of occupations in which people have been portrayed as "heroes" in the United States (Schienberg 2006). Being a hero, obtaining material possessions, or increasing one's educational level can boost prestige but in itself cannot change class standing.

Thinking Sociologically

Describe your own wealth, power, and prestige in society. Does your family have one factor but not others? What difference does each of the factors make in your life?

Table 7.6 Prestige Rankings of 17 Professions and Occupations (Percents)

Occupations (Base: All Adults)	Very Great Prestige	Hardly Any Prestige at All
Scientist	51	2
Doctor	50	1
Military officer	47	3
Teacher	47	7
Police officer	40	7
Priest/minister/clergyman	36	11
Engineer	34	4
Architect	27	4
Member of Congress	27	11
Athlete	21	15
Entertainer	19	15
Journalist	19	12
Business executive	18	13
Lawyer	15	20
Banker	15	10
Union leader	14	23
Accountant	13	17

Source: Taylor (2002).

Social Classes in the United States

In the current U.S. economic structure, most people are middle class and identify themselves as such, but the middle class is shrinking. There is slight movement to the upper class and somewhat more movement to the lower class. As noted previously, the U.S. system allows for mobility within the middle class, but there is little movement at the very top and very bottom of the social ladder. People in the top rung often use their power and wealth to insulate themselves and protect their elite status, and the bottom group is isolated because of vicious cycles of poverty that are hard to break (Gilbert and Kahl 2003). Figure 7.1 illustrates the social class structure in the United States.

The middle class, as defined by sociologists, makes up about 30% of the population in the United States, depending on what economic criteria are used. Whereas most people in the United States identify themselves as middle

class, two thirds of the British population identify themselves as "working class" (Cashell 2007). How we identify ourselves expresses our feelings about our placement in the stratification system and also our class "culture." Classes have distinctive values, beliefs, and attitudes toward education, religion, politics, and what makes a good life in general. Often what we define as "normal" is actually what is affirmed by others in our socioeconomic status. A number of scholars have written about the cultural shock they experienced when moving from their blue-collar experiences as children to becoming professors. The shift in cultures was like entering a new country (Dews and Law 1995; Morris and Grimes, 1997). Even the differences in vocabulary usage, discussed earlier in this chapter, represent part of the cultural difference (Hart and Risley 2003).

Since the 1970s, wealth has become increasingly concentrated in the hands of the richest 1% of households. The income gap between the top 5% and bottom 40% of the U.S. population is increasing, and the number of full-time workers in poverty is rising. There has been no reduction in poverty since the 1970s, yet the proportion of families exceeding $100,000 in annual income has been growing (Gilbert 2008).

Wages and salaries in the middle classes have declined since the 1980s, but those of the upper classes have risen. Reasons for middle-class decline include downsizing and layoffs of workers, global shifts in production, technological innovations that displace laborers, competition, trade deficits between countries, and deregulation. All of these are macro-level economic forces that mean lower incomes for middle-class workers. The wealthiest 1% earned 21.5% of income gains in the last economic expansion, resulting in two thirds of income gains going to the top 1% (Feller and Stone 2009). Changes in class groups are due largely to changes in the occupational structure and transformations in the global economy. The upward movement among the few who have received huge gains in earning power is causing wages and earnings to become more unequal.

Upper-middle-class families typically have high income, high education, high occupational level (in terms of prestige and other satisfactions), and high participation in political life and voluntary associations. Families enjoy a stable life, stressing companionship, privacy, pleasant surroundings in safe neighborhoods, property ownership, and stimulating associations. They stress internalization of moral standards of right and wrong, taking responsibility for their own actions, learning to make their own decisions, and training for future leadership positions.

The *lower middle class* includes small-business owners and farmers; semiprofessionals (teachers, local elected officials, social workers, nurses, police officers, firefighters); middle-management personnel, both private and public; and sales and clerical workers in comfortable office settings. Families in this class are relatively stable. They participate in community life, and although they are less active in

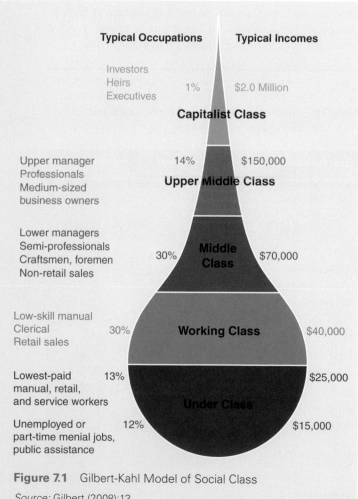

Figure 7.1   Gilbert-Kahl Model of Social Class

Source: Gilbert (2008):13.

political life than the upper classes, they are more politically involved than those in classes below them. Children are raised to work hard and obey authority. Therefore, child-rearing patterns more often involve swift physical punishment for misbehavior than talk and reasoning, which is typical of the upper middle class.

Thinking Sociologically

How does today's popular culture on TV and in films, magazines, and popular music reflect interests of different social classes? How are poor people (and those responsible for their poverty) depicted? Do any of these depictions question the U.S. class system? What kinds of music are class-based?

Poverty: Multilevel Determinants and Social Policy

Stories about hunger and famine in Global South countries fill the newspapers. Around the world, 925 million people, or 13.6%, were hungry in 2010 (World Hunger Education Service 2011), and that number has risen steadily since 1995. Figure 7.2 shows where most of the world hunger

occurs. One third of all deaths—18 million people each year or 50,000 per day—are due to poverty and hunger-related causes (United Nations Millennium Project 2007). By 2009, 1.02 billion people, or 15% of the world's population, went to bed hungry every night. Of the 10.9 million deaths of children each year, 5 million are directly related to undernutrition (World Hunger Education Service 2011).

One hardly expects to see hunger in rich countries, yet 4% of U.S. households did not have enough food and were skipping meals in 2008. Demand at food pantries was up 20% (Sells 2008), and the Supplemental Nutrition Assistance Program (SNAP) had a record 32.2 million recipients in January 2009, a 20% increase over the previous year (Food Research and Action Center 2009). Twenty-five percent of Americans are afraid they will not be able to afford food at some times in the next year (Casteel 2011). These poor families come from rural areas, urban slums, and disenfranchised groups such as the homeless, the unemployed, single parents, the disabled, the elderly, and migrant workers. With downturns in employment, low wages, and reduced aid to poor families, hunger and poor health in the United States are likely to continue.

The poor have no property-based income and no permanent or stable work, only casual or intermittent earnings in the labor market. They are often dependent on help from government agencies or private organizations to survive. In short, they have personal troubles in large part because they have been unable to establish linkages and networks in the meso- and macro-level organizations of our social world. They have no collective power and, thus, little representation of their interests and needs in the political system.

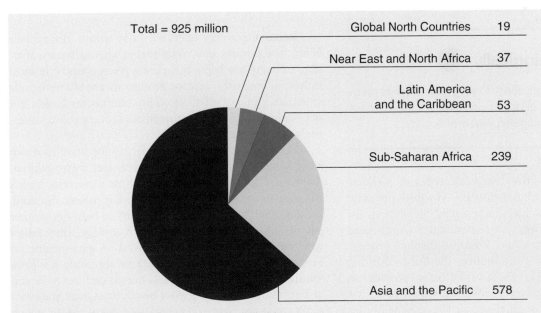

Figure 7.2 Distribution of Hunger in the World

Source: Map, "Global Hunger Declining, but Still Unacceptably High," from http://fao.org/hunger/en/. Food and Agriculture Organization of the United Nations.

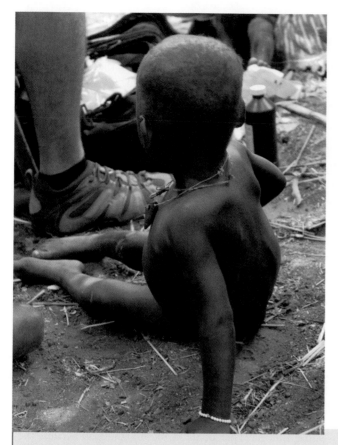

This little boy (left) sits at the feet of an aid worker. How much power do you imagine this little African boy's parents have to make sure their child's needs are met? This boy and his family have personal troubles because the meso and macro systems of his society have failed to work effectively for individuals and poor families. Compare his circumstances to those of the little girl (right) with her cell phone who is riding in a limousine.

Thinking Sociologically

Explain how your family's ability to provide food for its members at the micro level is largely dependent on its connectedness to the meso and macro levels of society.

Sociologists recognize two basic types of poverty: absolute poverty and relative poverty. **Absolute poverty**, not having resources to meet basic needs, means no prestige, no access to power, no accumulated wealth, and insufficient means to survive. Whereas absolute poverty in the United States is quite limited, the *Dalits* of India, described earlier in *The Outcastes of India,* provide an example of absolute poverty. Some die of diseases that might be easily cured in other people because the bodies of those in absolute poverty are weakened by chronic and persistent hunger and almost total lack of medical attention.

Relative poverty refers to those whose income falls below the poverty line, resulting in an inadequate standard of living relative to others in a given country. In most industrial countries, relative poverty means shortened life expectancy, higher infant mortality, and poorer health, but not many people die of starvation or easily curable diseases, such as influenza.

The *feminization of poverty* refers to the trend in which single females, increasingly younger and with children, make up a growing proportion of those in poverty. Vicki's situation provides one example. After her parents divorced, she quit high school to take odd jobs to help her mother pay the bills. At 18, she was pregnant, and the baby's father was out of the picture, so Vicki lived on government aid because without a high school degree she could not get a job that paid enough to support herself and her baby and certainly not enough to pay for health insurance. She eventually could not pay her rent and lived out of her car, which did not run because she lacked the money to repair it. Her life spiraled out of control, and her daughter has now been placed in foster care.

While these men may look impoverished by North American standards, they are relatively well off in comparison to many urban neighbors in India, for they own a method of transportation that can earn a cash income, providing food and shelter for their families. Their poverty is not absolute.

We do not like to admit that the feminization of poverty can happen in the affluent North American and European countries, and we try to blame the victim (why did Vicki not finish high school, not get pregnant, not have an abortion?). Still, people do what they must to survive. This problem is heightened as many middle-class women are pushed into poverty through divorce. Some of them have sacrificed their own careers for husbands and family, so their earning power is reduced, yet many divorced women are unable to collect child support from the fathers. The numbers in poverty are highest in and around large central cities such as Paris; Mexico City; São Paulo, Brazil; and Caracas, Venezuela. In the United States, numbers are highest in cities such as New York, Chicago, Detroit, Cleveland, and Los Angeles, where the percentage of poor African American families headed by females reaches close to 50% in some cities. Girls who grow up in female-headed households or foster homes, without a stable family model, are more likely to become single teen mothers and to live in poverty, causing disruption in their schooling and setting limits on their employment possibilities and marital opportunities. Conflict theorists argue that poor women, especially women of color, in capitalistic economic systems are used as a reserve labor force that can be called on when labor is needed and dismissed when not needed (Aguirre and Baker 2007; Ehrenreich

2001). They are an easily exploited group, living under constant stress that can cause mental or physical breakdowns and alienation from the social system. Some turn to alcohol or drugs to escape the pressure and failure or to crime to get money to pay the bills. Poor physical and mental health, inadequate nutrition, higher mortality rates, obesity, low self-esteem, feelings of hopelessness, daily struggle to survive, and dependence on others are a few of the individual consequences of poverty within our social world. Costs to the larger society are great:

- The loss of talents and abilities that these people could contribute.
- Expenditure of tax dollars to address their needs or to regulate their lives with social workers and police.
- The contradictions of their lives with cultural values: The United States claims that all citizens "are created equal" and are worthy of respect, yet not all can "make it" in U.S. society.

Welfare programs in affluent countries are only one of a number of kinds of government assistance programs. In the United States, some people think of such programs as unearned giveaways for the poor; however, there are massive programs of government support for people at all levels of the social system. Some of these include tax breaks for business owners, farmers, oil companies, and financial institutions. In other affluent countries, all citizens are provided with universal health insurance, and in some of these countries, university students pay little or nothing to attend college. Welfare in many countries is clearly not just for the poor. The question is whether affluence and prosperity are viewed as collectively created and shared, rather than as individual achievements.

Thinking Sociologically

Are tax breaks and bailouts for business owners and wealthy companies "unearned giveaways"? Are programs that help the poor "unearned giveaways"? What is the difference?

For most societies, poverty means loss of labor, a drain on other members in society to support the poor, and extensive health care and crime prevention systems. Those working with people in poverty generally argue that the elimination of poverty takes money and requires choices by policy makers to *do* something about poverty. Some argue that poverty will never be eliminated because poor people are needed in society. Consider the position put forth in the next "Sociology in Your Social World."

Sociology in Your Social World

The Functions of Poverty

Surely wealthy countries such as the United States have the means to eliminate poverty if they choose to do so. Its persistence invites debate. Some sociologists argue that poverty serves certain purposes or *functions* for society (Gans 1971, 1995), and these make it difficult to address the problem directly and systematically. Some people actually benefit from having poor people kept poor. Consider the following points:

1. The fact that some people are in poverty provides us with a convenient scapegoat—someone to blame for individual and societal problems. We have individuals to blame for poverty—the poor individuals themselves—and can ignore meso- and macro-level causes of poverty that would be expensive to resolve.

2. Having poor people creates many jobs for those who are not poor, including "helping" professions such as social workers, as well as law enforcement jobs such as police, judges, and prison workers.

3. The poor provide an easily available group of laborers to do work, and they serve as surplus workers to hire for undesirable jobs.

4. The poor serve to reinforce and legitimate our own lives and institutions. Their existence allows the rest of us to feel superior to someone, enhancing our self-esteem.

5. Their violation of mainstream values helps remind us of those values, thereby constantly reaffirming the values among the affluent.

This perspective can be extended to poverty on the global scale.

1. Just as poor U.S. laborers can be hired for undesirable jobs, the global poor work at very low wages to provide consumers in wealthy nations with low-cost goods.

2. The gifted and talented individuals in poor societies often migrate to wealthier nations, creating a brain drain that removes human capital from poor states and increases it in wealthy states.

3. The poor of the world are blamed for macro-level social problems such as overpopulation and terrorism. As with our local poor, their existence allows us to overlook the macro-level causes of poverty.

From this perspective, the poor serve a role in the structure of society. Therefore, some groups of people, be it individuals or nations, will always be at the bottom of the stratification ladder.

National and Global Digital Divide: Macro-Level Stratification

Mamadou from Niger and Eric from Ghana answer their cell phones to the sound of chimes from London's Big Ben clock tower and a Bob Marley song. One speaks in Kanuri and French and the other in Twi to friends thousands of miles away. They are the future generation of elites from the Global South, fluent in the languages of several countries, fluent in computer software, and leaders in their countries in the digital age. Many of their fellow citizens in Niger and Ghana in Africa are not so fortunate. They live subsistence lives and have little contact with the digital world swirling overhead through satellite connections. This represents the *digital divide*, the gap between those with access to information technology and those without it. The lines of the divide are drawn by socioeconomic status, minority group, and urban versus rural residence (Mehra, Merkel, and Bishop 2004).

The world economic and political institutions are increasingly based on producing and transmitting information through digital technology. Few tools are more important in this process than the computer, the Internet, cell phones, and iPods. In nearly every salaried and professional position, computer knowledge and ability to navigate the Internet are critical employment skills. Individuals with insufficient access to computers and lack of technical skills face barriers to many professions and opportunities. Because computer skills are important for personal success, this is an important micro-level issue. The digital divide is breaking down for some young and elite members of developing societies such as Mamadou and Eric, but many individuals and Global South countries have insufficient technology and educated population to participate in this new economy (Drori 2006; Nakamura 2004). Access to the Internet is not equal around the world (Gibson 1999).

The number of Internet users around the globe increased from 250 million in 2000 to 2 billion in 2011. Cell phone users around the world rose from 500 million in 2000 to 5 billion (Read 2011), mostly in affluent Global North countries, and especially among the young and well educated (Horrigan and Smith 2007). For most users, English is the language, although it is the second spoken language of more than 200 million users. Chinese is the next most popular language on the Internet, and its use is increasing at a high rate. In the United States, one of the most wired countries in the world, many poor people do not have access to computers or mentors to teach them how to use computers. This digital divide is beginning to close with active efforts by schools and libraries to provide accessibility, but it still creates barriers for many (Nakamura 2004; Shade 2004).

Comparing countries cross-nationally, in 2006 only 5% of the total world population was online, but over 50% of residents in North America and Scandinavia (Sweden, Norway, Finland, Denmark, and Iceland) were connected. About 79% of the online population lived in nations that belong to the Organisation for Economic Co-operation and Development—countries that hold 14% of the people on the planet. By contrast, in sub-Saharan Africa and in India, only 4 out of every 1,000 citizens used the Internet (Drori 2006). About 97% of Internet hosts were in the Global North (which has only 16% of the world's population), and 66% of income from royalties and licensing fees went to two countries: the United States (54%) and Japan (12%). In some countries, the monthly access fee for hooking up to the Internet was stunningly high compared to average monthly income: in Bangladesh, 191%; in Nepal, 278%; and in Madagascar, 614%. This may be why 35 of the poorest countries in the world in 2006 had less than 1% of the Internet users (Drori 2006).

An additional difficulty is that because most websites and e-mail services use English, computer keyboards are designed with a Western alphabet, and some of the digital systems in computers are established on the basis of English symbols and logic. It is difficult to use the Internet in other languages—a fact that many of us may not think about as

we use the system (Drori 2006). For people who are struggling for the very survival of their culture, the dominance of English may feel like a threat, one more example of Western dominance. So resistance to the use of computers and the Internet is more than a matter of finances or technology. There may be cultural objections as well.

Policy decisions at the international level affect the status of the Global South. The United Nations, the International Monetary Fund, and other international organizations have pressured countries to develop their Internet capacities. Indeed, this is sometimes used as a criterion for ranking countries in terms of their "level of modernization" (Drori 2006). Countries that have not been able to get "in the game" of Internet technology cannot keep pace with a rapidly evolving global economy.

Internet technology has created a digital divide, but it also has had some beneficial economic impact, stimulating jobs in poor countries. In 1999, computer component parts were a substantial percentage of production in some poor countries: 52% of Malaysia's exports, 44% of Costa Rica's, and 28% of Mexico's. The high-tech revenue for India increased from 150 million U.S. dollars in 1990 to 4 billion U.S. dollars in 1999. On the less positive side, e-waste from electronic equipment is extremely toxic, and it is almost always shipped to poor countries, where extremely poor people must deal with the consequences of toxic pollution (Drori 2006).

Digital technology is an example of one important force changing the micro- and macro-level global stratification system—a spectrum of people and countries from the rich and elite to those that are poor and desperate.

Thinking Sociologically

What evidence of the digital divide do you see in your family, community, nation, and world? For instance, can your grandparents program their DVD players? Do they know how to work a cell phone or navigate the Internet? Could they create their own Facebook page if asked to do so? If not, does this have any consequences for their lives?

The Global "Digital Divide" and Social Policy

Bangalore is home to India's booming digital industries, which are successfully competing in the global high-tech market. Yet, many villages and cities in India provide examples of the contrasts between the caste system and the emerging class system. In rural agricultural areas, change is extremely slow despite laws forbidding differential treatment

In the Global North, computers are seen as necessary equipment in homes. Yet in rural villages throughout much of the world, people may never see a computer and lack the reliable electricity and other support systems for Internet technology. This is part of the digital divide in the global system.

and others are vying for power. Within the world system, India is generally economically poor but developing certain economic sectors rapidly. Thus, India is in transition both internally and in the global world system.

As poor countries become part of the electronic age, some such as India are making policies that facilitate rapid modernization. They are passing over developmental stages that rich countries went through. As an illustration, consider the telephone. In 2002, 90% of all telephones in the world were cell phones, many using satellite connections. Some countries never did get completely wired for landlines, thus eliminating one phase of phone technology. With the satellite technology now in place, some computer and Internet options will be available without expensive intermediate steps (Drori 2006). In less than 30 years, cell phones that originated for the business elite have become a personal item. Use in poor countries has boomed, giving people access to health care and other services. By the beginning of 2007, 68% of the world's cell phone subscriptions were in Global South countries (Klein and Ember 2009; Rodgers 2009). Based on a United Nations report, 60% of the world uses mobile phones with 4.1 billion cell phone subscriptions worldwide (Barker 2009). In the United States, 20% of homes use cell phones exclusively, and 60% use both cell and landline phones (Corn 2009).

Positive efforts are under way to provide cheap technology to Global South countries. Currently 23% of the world's population uses the Internet; in Africa the figure is 1 in 20, and in rich nations the figure is much greater (Barker 2009). The "One Laptop per Child" foundation is helping fund efforts to distribute efficient small laptops to children around the world. Linux and Novatium, to name just two companies, are developing $100 computers, and some governments are buying large numbers for their schools (One Laptop per Child 2011; Rubenstein 2007). Engineers in India, working through an organization called Simputer Trust, have been designing a simple computer (a "simputer") that will be less expensive and will have more multilingual capacities than the PC. Such efforts will enhance access of poor countries to the computer and Internet and will provide the means for children in poor countries to become part of the competitive global stratification system. Technology is leveling the world playing field (Drori 2006; Friedman 2005).

of outcastes and mandating change. In urban industrial areas, new opportunities are changing the traditional caste structures, as intercaste and intracaste competition for wealth and power is increasing with the changes in economic, political, and other institutional structures. The higher castes were the first to receive the education and lifestyle that create industrial leaders. Now, shopkeepers, wealthy peasants, teachers,

We leave this discussion of stratification systems, including class systems, with a partial answer to the question posed at the beginning of this chapter: Why are some people rich and others poor? In the next two chapters, we expand the discussion to include other variables in stratification systems—race and ethnicity and gender. By the end of these chapters, the answer to the opening question should be even clearer.

What Have We Learned?

Perhaps you have a better understanding of why you are rich or poor—and what effect your socioeconomic status has on what you buy, what you believe, and where you live. Perhaps you have gained some insight into what factors affect your ability to move up in the social class system. The issue of social stratification calls into question the widely held belief in the fairness of our economic system. By studying this issue, we better understand why some individuals are able to experience prestige (respect) and to control power and wealth at the micro, meso, and macro levels of the social system while others have little access to those resources. Few social forces affect your personal life at the micro level as much as stratification. That includes the decisions you make about what you wish to do with your life or whom you might marry. Indeed, stratification influenced the fact that you are reading this book.

Key Points:

- Stratification—the layering or ranking of people within society—is one of the most important factors shaping the life chances of individuals. This ranking is influenced by micro, meso, and macro forces and resources. (See pp. 181–182.)

- Various theories of stratification disagree on whether inequality is functional or destructive to society and its members. The evolutionary perspective suggests ways stratification can be positive, but much of the inequality currently experienced creates problems for individuals and societies. (See pp. 183–186.)

- For individuals, personal respect (prestige) is experienced as highly personal, but it is influenced by the way the social system works at the meso and macro level—from

access to education and the problems created by gender and ethnic discrimination to population trends and the vitality of the global economy. (See pp. 186–190.)

- People without adequate capital and connections to the meso and macro levels are likely to experience less power, wealth, and prestige. (See pp. 190–196.)

- Some macro systems stress ascribed status (assigned to one, often at birth, without consideration of one's individual choices, talents, or intelligence). Other systems purport to be open and based on achieved status (dependent upon one's contributions to the society and one's personal abilities and decisions). Unlike the caste system, the class system tends to stress achieved status, although it does not always perform openly. (See pp. 197–202.)

- The elements of stratification are complex, with property, power, and prestige having somewhat independent influences on the system of inequality and on one's standing. (See pp. 200–201.)

- Poverty itself is a difficult problem, one that can be costly to a society as a whole. However, various solutions at the micro, meso, or macro levels have had mixed results, partially, perhaps, because it is in the interests of those with privilege to have an underclass to do the unpleasant jobs. (See pp. 203–206.)

- Technology is both a contributor and a possible remedy to inequality, as the digital divide creates problems for the poor, but electronic innovations may create new opportunities in the social structure for networking and connections to the meso and macro levels—even for those in the Global South, the poor regions of the world. (See pp. 206–208.)

Contributing to Our Social World: What Can We Do?

At the Local Level:

- *Community shelters for homeless people and soup kitchens* where homeless and very poor people can get shelter and a free meal often need volunteer help or interns.

- *Tip* service people in minimum-wage jobs generously— hotel housekeeping maids, meal servers at restaurants, and employees at fast-food restaurants.

At the Organizational or Institutional Level:

- *Habitat for Humanity* participants work with current and prospective homeowners in repairing or constructing housing for little or no cost, and they use volunteers extensively. See the organization's website for more details at www.habitat.org.

- *AmeriCorps* involves one year of service by people in a wide range of jobs, including teaching and community service. The pay is not good but is adequate to live on, and the experience and contribution to those in need provide other significant rewards. See the AmeriCorps website at www.americorps.gov. AmeriCorps is expanding significantly under the Obama administration and is a good way to get experience and begin your career.

At the National and Global Levels:

- *The Peace Corps* involves a serious, long-term commitment, but most who have done it agree that it was worth the effort. The Peace Corps is an independent agency of the U.S. government. Volunteers work in countries throughout the world, helping local people improve their economic conditions, health, and education. The Peace Corps website (www.peacecorps.gov) provides information on the organization, volunteer opportunities, and reports of volunteers.

- *Sociologists Without Borders* (or *Sociólogos sin Fronteras*) requires a shorter-term global commitment and has some paid internships (check www.sociologistswithoutborders.org). This organization promotes human rights globally.

- *Grameen Bank* (www.grameen-info.org or www.grameenfoundation.org) was started in Bangladesh by Professor Muhammad Yunus, winner of the 2006 Nobel Peace Prize. It makes small-business loans to people who live in impoverished regions of the world and who have no collateral for a loan. Consider doing a local fund-raiser with friends for the Grameen Bank or other microcredit organizations, such as FINCA or CARE International.

Visit **www.sagepub.com/oswcondensed2e** for online activities, sample tests, and other helpful information. Select "Chapter 7: Stratification" for chapter-specific activities.

CHAPTER 8

Race and Ethnic Group Stratification

Beyond "We" and "They"

As we travel around our social world, the people we encounter gradually change appearance. As human beings, we are all part of "we," but there is a tendency to define those who look different as "they."

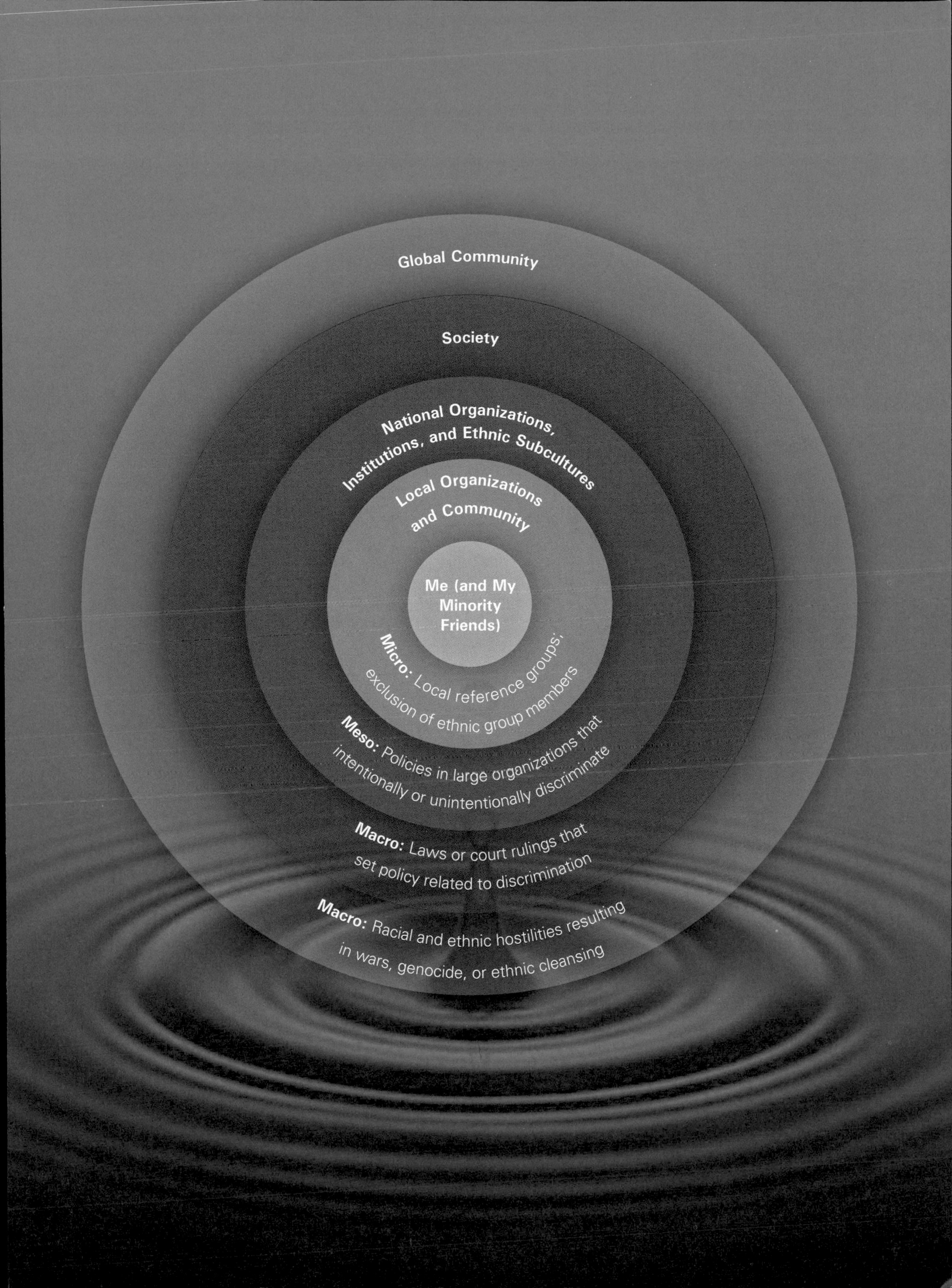

Global Community

Society

National Organizations, Institutions, and Ethnic Subcultures

Local Organizations and Community

Me (and My Minority Friends)

Micro: Local reference groups; exclusion of ethnic group members

Meso: Policies in large organizations that intentionally or unintentionally discriminate

Macro: Laws or court rulings that set policy related to discrimination

Macro: Racial and ethnic hostilities resulting in wars, genocide, or ethnic cleansing

Think About It	
Me (and My Minority Friends)	Why do you look different from those around you? What relevance do these differences have for your life?
Local Community	Why do people in the local community categorize "others" into racial or ethnic groups?
National Institutions; Complex Organizations; Ethnic Groups	How are advantages and disadvantages embedded in institutions—so that they operate independently of personal bias or prejudice?
National Society	Why are minority group members in most countries economically poorer than dominant group members?
Global Community	In what ways might ethnicity or race shape international negotiations and global problem solving? What can you do to make the world a better place for all people?

When Siri wakes, it is about noon. In the instant of waking, she knows exactly who and what she has become . . . the soreness in her genitals reminds her of the 15 men she had sex with the night before. Siri is 15 years old. Sold by her impoverished parents a year ago, she finds that her resistance and her desire to escape the brothel are breaking down and acceptance and resignation are taking their place. . . . Siri is very frightened that she will get AIDS . . . as many girls from her village return home to die from AIDS after being sold into the brothels. (Bales 2002:207–09)

In Calcutta's red-light district, more than 7,000 women and girls work as prostitutes. Only one group has a lower standing: their children. Zana Briski first began photographing prostitutes in Calcutta in 1998. Living in the brothels for months at a time, she quickly developed a relationship with many of the kids who, often terrorized and abused, were drawn to the rare human companionship she offered. Because the children were fascinated by her camera, Zana taught photography to the children of prostitutes. Learn more about her organization, Kids with Destiny, at http://www.kids-with-cameras.org/home/.

Siri is a sex slave, just like millions of other young women around the world. Slavery is not limited to poor countries: Dora was enslaved in a home in Washington, DC, and domestic slaves have been discovered in London, Chicago, New York, and Los Angeles. The Central Intelligence Agency (CIA) reports thousands of women and children are smuggled into the United States each year as sex and domestic slaves or are locked away in sweatshops (Bales 2004). International agencies estimate that more than one million children in Southeast Asia have been sold into bondage, mostly into the booming sex trade. The CIA estimates that 50,000 women and children are sold into slavery in the United States alone every year (Grey 2000).

It may surprise you to know that slavery is alive and flourishing around the world (Free the Slaves 2011). An estimated 27 million people, mostly women and children from poor families in poor countries, are slaves, auctioned off or lured into slavery each year by kidnap gangs, pimps, and cross-border syndicates (Bales 2004, 2007; Bales and Trodd 2008).

International events such as the Olympics and major soccer matches bring new markets for the sex trade. Young foreign girls are brought in from other countries—chosen for sex slaves because they are exotic, are free of AIDS, and cannot escape due to insufficient money and knowledge of the language or the country to which they are exported (Moritz 2001). Sometimes, poor families sell their daughters for the promise of high wages and perhaps money sent home. As a result, girls as young as six are held captive as prostitutes or as domestic workers. Child labor, a problem in many parts of the world, requires poor young children to do heavy labor for long hours in agriculture as well as brick-making, match-making, and carpet factories. Although they earn little, their income helps families pay debts. Much of the cacao (used to make chocolate) and coffee (except for Fair Trade Certified products) that we buy also supports slavery. Chocolate especially is grown and harvested at slave camps where young boys are given a choice of unpaid hard labor (with beatings for any disobedience) or death by starvation or shooting (Bales 2000, 2004). Very little chocolate is produced *without* slave labor.

Debt bondage is another form of modern-day slavery. Extremely poor families—often people with differences in appearance from those with power—work in exchange for housing and meager food. Severe debt, passing from generation to generation, may also result when farmers borrow money because they face drought or need cash to keep their families from starving. The only collateral they have on the loan is themselves—put up for bondage until they can pay off the loan. No one but the wealthy landowner keeps accounting records, which results in there being no accountability. The lack of credit available to marginal people contributes to slavery. Because those in slavery have little voice and no rights, the world community hears little about this tragedy (Bales 2007). A recent successful international movement in impoverished areas provides women with very small loans—called *microcredit*—to help them start small businesses and move out of desperate poverty and slavery. We will read more about this in future chapters.

In the slavery of the nineteenth century, slaves were expensive, and there was at least some economic incentive to care about their health and survival. In the new slavery, humans are cheap and replaceable. There is little concern about working them to death, especially if they are located in remote sugar, cacao, or coffee plantations (Bales 2000). By current dollars, a slave in the Southern United States prior to the Civil War would have cost as much as $40,000, but contemporary slaves are cheap. They can be procured from poor countries for an average of $90 (Bales 2004). The cost is $40 in Mali for a young male and $1,000 in Thailand for an HIV-free female (Free the Slaves 2011). Those slaves produce an estimated $1.4 billion in produce and profits for their owners each year.

What point does slavery make for our discussion of race and ethnic group stratification? What all of these human bondage situations have in common is that poor minority groups are victimized. Because many slaves are members of ethnic, racial, religious, tribal, gender, age, caste, or other minority groups and have obvious physical or cultural distinctions from the people who exploit them, they are at a distinct disadvantage in the stratification system. Although all humans have the same basic characteristics, few people have a choice about being born into a minority group, and it is difficult to change that minority status. Visible barriers include physical appearance, names, dress, language, or other distinguishing characteristics. Historical conditions and conflicts rooted in religious, social, political, and historical events set the stage for dominant or minority status, and people are socialized into their dominant or subservient group.

Minority or dominant group status affects most aspects of people's experiences in the social world. These include status in the community, socialization experience, residence, opportunities for success in education and occupation, the religious group to which they belong, and the health care they receive. In fact, it is impossible to separate minority status from position in the stratification system (Aguirre and Turner 2006; Farley 2010; Rothenberg 2007).

In this chapter, we explore characteristics of race and ethnic groups that lead to differential placement in stratification systems, including problems at the micro, meso, and macro levels—prejudice, racism, and discrimination. The next chapter considers ascribed status based on gender. The topics in this chapter and the next continue the discussion of stratification: who is singled out for differential treatment, why they are singled out, results for both the individuals and the society, and some actions or policies that deal with differential treatment.

In India, many slaves are used to extract slate from mines, and the slate then is used to make cheap pencils that can be exported around the world. The key to slavery is powerless humans exploited by those who need cheap labor. Young children who have no family are often sucked into slavery, as these children were.

The Fair Trade Certified symbol signifies that products such as coffee, tea, and chocolate (made from cacao) meet sustainable development goals, help support family farmers at fair prices, and are not produced by slave labor. Much of the work on cacao plantations is done by child slaves smuggled in from poor countries and therefore not fair trade.

What Characterizes Racial and Ethnic Groups?

Migration, war and conquest, trade, and intermarriage have left virtually every geographical area of the world populated by groups of people with varying ethnicities. In this section, we consider characteristics that set groups apart, especially groups that fall at the lower end of the stratification system.

Minority Groups

Several factors characterize **minority groups** and their relations with dominant groups in society (Dworkin and Dworkin 1999). Minority groups

1. are distinguishable from dominant groups due to factors that make them different from the group that holds power;

2. are excluded or denied full participation at the meso level of society in economic, political, educational, religious, health, and recreational institutions;

3. have less access to power and resources within the nation and are evaluated less favorably based on their characteristics as minority group members;

4. are stereotyped, ridiculed, condemned, or otherwise defamed, allowing dominant group members to justify and not feel guilty about unequal and poor treatment; and

5. develop collective identities among members to insulate themselves from the unaccepting world. This in turn perpetuates their group identity by creating ethnic or racial enclaves, intragroup marriages, and segregated group institutions such as religious congregations.

Thinking Sociologically

Look again at the preceding list of minority group characteristics. How might people be affected at the micro, meso, and macro levels of society, depending on their membership in dominant or minority groups?

Because minority status changes with time and ideology, the minority group may be the dominant group in a different time or society. Throughout England's history, wars and assassinations changed the ruling group from Catholic to Protestant and back several times. In Iraq, Shiite Muslims

are dominant in numbers and now also in power, but they were a minority under Saddam Hussein's Sunni rule. Map 8.1 indicates the location of minority groups in the United States today, although you should be aware that the density of groups in a particular location changes over time.

Dominant groups are not always a numerical majority. In the case of South Africa, advanced European weapons placed the native African Bantu population under the rule of a small percentage of white British and Dutch descendants in a system called *apartheid*. Until recently, each major group—white, Asian, colored, and black—had its own living area, and members carried identification cards showing the "race" to which they belonged. In this case, racial classification and privilege were defined by the laws of the dominant group.

The Concept of Race

Racial minority is one of the two types of minority groups most common in the social world. A **race** is a group identified by a society because of certain biologically inherited physical characteristics. However, in practice, it is impossible to accurately identify racial types. Most attempts at racial classifications have been based on combinations of appearance, such as skin color and shade, stature, facial features, hair color and texture, head form, nose shape, eye color and shape, height, and blood or gene type. Our discussion of race focuses on three issues: (a) origins of the concept of race, (b) the social construction of race, and (c) the significance of race versus class.

Origins of the Concept of Race

In the eighteenth and nineteenth centuries, scientists attempted to divide humans into four major groupings—Mongoloid, Caucasoid, Negroid, and Australoid—and then into more than 30 racial subcategories. In reality, few individuals fit clearly into any of these types.

From the earliest origins in East Africa more than 7 million years ago, humans slowly spread around the globe, south through Africa, north to Europe, and across Asia. Many scholars believe humans crossed the Bering Strait from Asia to North America around 20,000 BCE and continued to populate North and South America (Diamond 1999). Physical adaptations of isolated groups to their environments originally resulted in some differences in physical appearance—skin color, stature, hair type—but mixing of peoples over the centuries has left few if any genetically isolated people. Thus, the way societies choose to define race has come about largely through what is culturally convenient for the dominant group.

In the 1970s, the United Nations, concerned about racial conflicts and discrimination based on scientifically inaccurate beliefs, issued a "Statement on Race" prepared

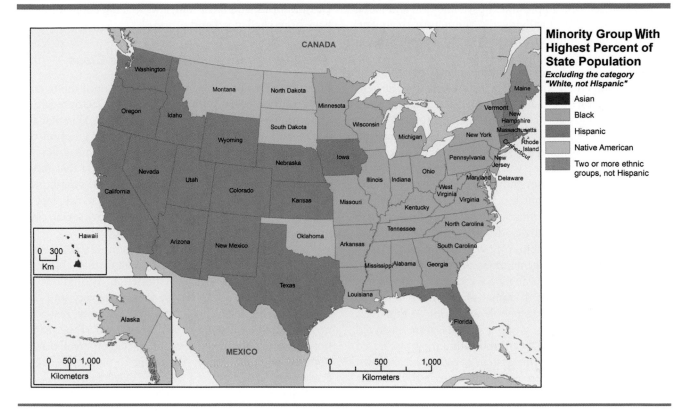

Map 8.1   Prevalence of Minority Ethnic Groups in the United States

Source: U.S. Census Bureau 2000. Map by Anna Versluis.

by a group of eminent scientists from around the world. This and similar statements by scientific groups point out the harmful effects of racist arguments, doctrines, and policies. The conclusion of this document upheld that (a) all people are born free and equal both in dignity and in rights, (b) racism stultifies personal development, (c) conflicts (based on race) cost nations money and resources, and (d) prejudice foments international conflict. Racist doctrines lack any scientific basis, as all people belong to the same species and have descended from the same origin. In summary, problems arising from race relations are social, not biological, in origin; differential treatments of groups based on "race" falsely claim a scientific basis for classifying humans. Biologically speaking, a "race" exists in any life form when the two groups cannot interbreed and where, if they do, the offspring are infertile or sterile. This is not true of any group of human beings. So what is the problem?

Social Construction of Race: Symbolic Interaction Analysis

Why are sociologists concerned about a concept that has little scientific accuracy and is ill defined? The answer is its social significance. The social reality is that people are

defined or define themselves as belonging to a group based in part on physical appearance. As individuals try to make meaning of the social world, they may learn from others that some traits—eye or nose shape, hair texture, or skin color—are distinguishing traits that make people different. Jean Piaget, the famous cognitive psychologist, described the human tendency to classify objects as one of our most basic cognitive tools (Piaget and Inhelder [1955] 1999). This inclination has often been linked to classifying "racial" groups. Once in place, racial categories provide individuals with an identity based on ancestry—"my kind of people have these traits."

Symbolic interaction theory contends that if people believe something is real, it may become real in its consequences. It does not matter whether scientists say that attempts to classify people into races are inaccurate and that the word is biologically meaningless. People on the streets of your hometown think *they* know what the word *race* means. However, race is not simple. People take on gradually different appearances as we traverse the globe, partly because migration throughout history has resulted in interbreeding. That people *think* there are differences based on appearance has consequences. As a social concept, race has not only referred to physical features and inherited genes but also carried over to presumed psychological and moral characteristics, thus justifying discriminatory treatment. The following examples illustrate the complex problems in trying to classify people into "races."

With the enactment of apartheid laws in 1948, the white government in South Africa institutionalized differential laws based on their definitions of racial groups and specified the privileges and restrictions allotted to each group (Marger 2012). Bantu populations (the native Africans) and coloreds (those of mixed blood) were restricted to separate living areas and types of work. Asians (mostly descendants of immigrants from India) received higher salaries than the Bantu groups but less than whites, while whites of European descent, primarily Dutch and English, had the highest living standard and best residential locations. Under the apartheid system, race was determined by tracing ancestry back for 14 generations. A single ancestor who was not Dutch or English might have caused an individual to be considered "colored" rather than white—the "one drop of blood" rule makes people a minority if they have any ancestry from another group. Physical features mattered little. Individuals carried a card indicating their race based on genealogy. Although this system began to break down in the 1990s due to international pressure and under the leadership of the first black president (Nelson Mandela, elected in 1994), vestiges of these notions of "reality" will take generations to change.

By contrast, in Brazil an individual's race is based on physical features—skin tone, hair texture, facial features, eye color, and so forth—rather than on the "one drop of blood" rule that existed in South Africa. Brothers and sisters who have the same parents and ancestors may be classified as belonging to different races. The idea of race is based on starkly contrasting criteria in Brazil and South Africa, illustrating the arbitrary nature of racial classification attempts (Kottak 2010).

Before civil rights laws were passed in the United States in the 1960s, a number of states had laws that spelled out differential treatment for racial groups. These were commonly referred to as Jim Crow laws. States in the South passed laws defining who was African American or Native American. In many cases, it was difficult to determine to which category an individual belonged. For instance, African Americans in Georgia were defined as people with any ascertainable trace of "Negro" blood in their veins. In Louisiana, one-thirty-second Negro blood defined one as black. Differential treatment was spelled out in other states as well. In Texas, for example, the father's race determined the race of the child. In West Virginia, a newborn was classified as "black" if either parent was considered black. Until recently, several U.S. states still attempted to classify the race of newborns by the percentage of black blood or parentage (Lopez 1996). Federal law now prohibits discrimination on the basis of "racial" classifications, and most state laws that are explicitly racial have been challenged and dropped.

The Significance of Race Versus Class

From the time of slavery in the Americas until the twenty-first century, race has been the determining factor regarding opportunities for people of African descent. Whether this is changing in the twenty-first century is a question that has occupied sociologists, politicians, educators, and other scientists in recent years. Some scholars argue that race is a primary cause of different placement in the stratification system, whereas others insist that race and social class are both at work, with socioeconomic factors (social class) more important than race.

Sociologist William Julius Wilson (1978, 1993a, 1993b) writes that the racial oppression that characterized the African American experience throughout the nineteenth century was caused first by slavery and then by a lingering caste structure that severely restricted upward mobility. However, the breakdown of the plantation economy and the rise of industrialism created more opportunities for African Americans to participate in the economy.

Wilson (1978) argues that after World War II, an African American class structure developed with characteristics similar to those of the white class structure. Occupation and income took on ever greater significance in social position, especially for the African American middle class. However, as black middle-class professionals moved up in the stratification structure, lower-class African American ghetto residents became more isolated and less mobile. Limited unskilled job opportunities for the lower class have resulted in poverty and stagnation so severe that some families are almost outside of the functioning economic system. Wilson (1978, 1984, 1993a) calls this group the *underclass*.

Some researchers assert that the United States cannot escape poverty because well-paid, unskilled jobs are disappearing from the economy and because the poor are concentrated in segregated urban areas (Massey 2007; Massey and Denton 1998). Poorly educated African American teenagers and young adults see their job prospects limited to the low-wage sector (e.g., fast-food work paying minimum wage), and they experience record levels of unemployment (W. J. Wilson 1987). Movement out of poverty becomes almost impossible.

Wilson's point is illustrated by the fact that more than two in five African Americans are now middle-class, compared to one in twenty in 1940. On the other hand, most adults in many inner-city ghetto neighborhoods are not employed in a typical week. Thus, children in these neighborhoods may grow up without ever seeing someone go to work (Wilson 1996). The new global economic system is a

This house in the Chicago slums was not atypical of the quality of homes in segregated America in the 1940s.

Table 8.1   Income (in Dollars) by Educational Level and Race/Ethnicity

Education	White	Black	Hispanic
Not a high school graduate	21,590	18,123	21,310
High school graduate	32,126	27,265	27,020
Some college, no degree	33,298	28,570	29,610
College graduate	59,866	46,527	48,081
Master's degree	72,125	58,311	74,122
Professional degree	127,968	104,656	81,968

Source: U.S. Census Bureau 2011c:Table 224. Figures are for 2008.

contributing factor as unskilled jobs go abroad to cheaper labor (Friedman 2005, 2008; Massey 2007). Without addressing these structural causes of poverty, we cannot expect to reduce the number of people in the underclass—regardless of race or ethnicity.

A big debate among scholars surrounds the following question: Has race declined in significance and class become more important in determining placement in the stratification system? Tests of Wilson's thesis present us with mixed results (Jencks 1992). For instance, African Americans' average education level (12.4 years in school) is almost the same as whites' (12.7 years), suggesting they have comparable qualifications for employment. However, this equity stops at the high school level; 29.9% of whites are college graduates, compared to 19.3% of blacks (U.S. Census Bureau 2011b). More important, African Americans earn less than whites in the same occupational categories. As Table 8.1 makes clear, income levels for African Americans and whites are not even close to being equal. So economics alone does not seem a complete answer to who is in the underclass.

Although racial bias has decreased at the micro (interpersonal) level, it is still a significant determinant in the lives of African Americans, especially those in the lower class. The data are complex, but we can conclude that for upwardly mobile African Americans, class may be more important than race. Still, physical traits such as skin color cannot be dismissed. They can be crippling for the underclass.

Thinking Sociologically

On what bases do you classify people into social groups? How do you describe someone to another person? Do you use racial terms? Why? Are people who look like you "just normal"? If so, what does that say?

Ethnic Groups

The second major type of minority group—the **ethnic group**—is based on cultural factors: language, religion, dress, foods, customs, beliefs, values, norms, a shared group identity or feeling, and sometimes loyalty to a homeland, monarch, or religious leader. Members are grouped together because they share a common cultural heritage, often connected with a national or geographical identity. Some social scientists prefer to call racial groups "ethnic groups" because the term *ethnic* encompasses most minorities and avoids problems with the term *race* (Aguirre and Turner 2006).

Visits to ethnic enclaves in large cities around the world give a picture of ethnicity. Little Italy, Chinatown, Greek Town, and Polish neighborhoods may have non-English street signs and newspapers, ethnic restaurants, culture-specific houses of worship, and clothing styles that reflect the ethnic subculture. Occasionally, an ethnic group shares power in pluralistic societies, but most often such groups hold a minority status with little power.

How is ethnicity constructed or defined? Many very different ethnic groups have been combined in government categories, such as censuses conducted by countries,

Ethnic enclaves have a strong sense of local community, holding festivals from the old country and developing networks in the new country. Such areas, called "ghettos," are not necessarily impoverished. This photo depicts a street on the Lower East Side of New York, which was once a transition station and ghetto for recent immigrants.

Biracial and Multiracial Populations: Immigration, Intermarriage, and Personal Identification

Our racial and ethnic identities are becoming more complex as migration around the world brings to distant shores new immigrants in search of safety and a new start. Keep in mind that our racial and ethnic identities come largely from external labels placed on us by governments and our associates but reinforced by our own self-identification.

Many European countries are now host to immigrants from their former colonies, making them multiracial. France hosts many North and West Africans, and Great Britain hosts large populations from Africa, India, and Pakistan. The resulting mix of peoples has blurred racial lines and created many multiracial individuals. Original migration patterns over thousands of years are shown in Map 8.2. These patterns illustrate that "push" factors drive people from some countries and "pull" them to other countries. The most common push-pull factors today are job opportunities, desire for security, individual liberties, and availability of medical and educational opportunities. The target countries of migrants are most often in North America, Australia, or Western Europe, and the highest emigration rates (leaving a country) are from Africa, Eastern Europe, Central Asia, and South and Central America.

The United States was once considered a biracial country, black and white (which, of course, disregarded the Native American population). However, the nation currently accepts more new immigrants than any other country (700,000 per year) and has the second highest rate of immigration (behind Canada) in terms of immigrants per 1,000 residents (Farley 2010). Immigration from every continent has led to a more diverse population, with up to 16% of the U.S. population being foreign born. With new immigration, increasing rates of intermarriage, and many more individuals claiming multiracial identification, the picture is much more complex today, and the color lines have been redrawn (DaCosta 2007; Lee and Bean 2004, 2007). One in forty individuals claims multiracial status today, and estimates are that one in five will do so by 2050 (Lee and Bean 2004). For the first time, the United States has a biracial president, although the application of the "one drop of blood" rule in the United States has caused many people to refer to President Obama as "black."

Census data are used in countries to determine many characteristics of populations. In the United States, questions about race and ethnic classification have changed with each ten-year study. The important point is that government-determined categories thereafter define the racial and ethnic composition of a country. In the 2000 census, citizens were for the first time given the option of picking more than one racial category. Seven million people or 2.3% of the U.S. population selected two or more racial categories, with

yet they speak different languages and often have very different religions. For example, in North America, ethnic group members often do not view themselves as "Indian" or "Native American." Instead, they use 600 independent tribal nation names to define themselves, including the Ojibwa (Chippewa), Dineh (Navajo), Lakota (Sioux), and many others. Likewise, in the U.S. census, Koreans, Filipinos, Chinese, Japanese, and Malaysians come from very different cultures but are identified as *Asian Americans*. People from Brazil, Mexico, and Cuba are grouped together in a category called *Hispanics* or *Latinos*. When federal funds for social services were made available to Asian Americans or American Indians, these diverse people began to think of themselves as part of a larger grouping for political purposes (Esperitu 1992). The federal government essentially created an ethnic group by naming and providing funding to that group. If people wanted services (health care, legal rights, etc.), they had to become a part of a particular group. This process of merging many ethnic groups into one broader category—called *panethnicity*—emphasizes that ethnic identity is itself socially shaped and created.

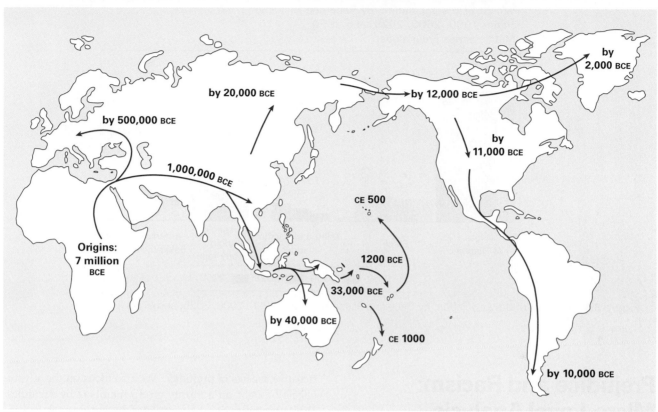

This map shows the historical spread of humans around the globe and the approximate time periods of the movements.

Map 8.2   The Spread of Humans Around the World

Source: Diamond 1999:37.

white/American Indian being the most common mixed category (Schaefer and Kunz 2008).

Latinos, sometimes called Hispanics, made up 16.3% (50.5 million) of the total U.S. population (308.7 million) in 2010 (Perez 2011). Hispanics accounted for 56% of the nation's growth since the 2000 census, up from 35.3% in 2000 (Pew Hispanic Center 2011). Among Latinos, Mexicans made up roughly 63%, Puerto Ricans 9%, Cubans 3.5%, Central Americans 7.6%, and South Americans 5.4% (U.S. Census Bureau 2009a). Many Latinos identify themselves as *panethnic*—that is, they identify with a broad ethnic category ("Hispanic" or "Latino") rather than with a specific ethnic group (e.g., Mexican American or Cuban American) (McConnell and Delgado-Romero 2004). Blacks follow Latinos with 12.6% (38.9 million) of the U.S. population. Non-Hispanic Whites make up 63.7% (196.8 million), Whites (including Hispanic Whites) 72.4%, Asians 4.8%, and Native Americans/Native Alaskans 0.9% of the total U.S. population (Day 2011; Humes, Jones, and Ramirez 2011). Figure 8.1 on the next page illustrates the ethnic group distribution and projections for the future for the United States.

Arbitrary socially constructed classifications of people into groups are frequently used as justification for treating individuals differently, despite the lack of scientific basis for such distinctions (Williams 1996). The legacy of "race" remains even in countries where discrimination based on race is illegal. The question remains: Why is a multiracial baby with any African, Native American, or other minority heritage classified by the minority status, not as a member of the majority? After all, the child has 50% genetic makeup from each parent.

Thinking Sociologically

Identify one dominant and one minority group in your community or on campus. How are each group's members regarded in the stratification or prestige system of the local community? How are the life chances of individuals in these groups influenced by factors beyond their control?

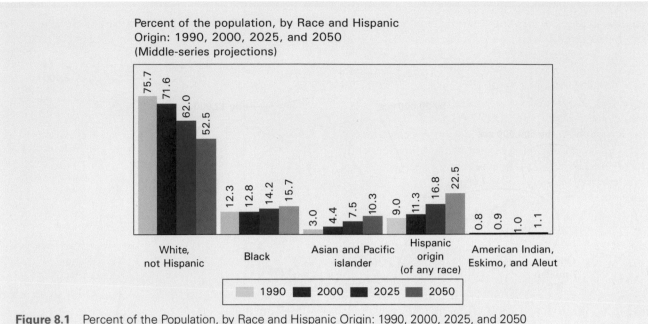

Percent of the population, by Race and Hispanic Origin: 1990, 2000, 2025, and 2050 (Middle-series projections)

Figure 8.1   Percent of the Population, by Race and Hispanic Origin: 1990, 2000, 2025, and 2050

Source: U.S. Census Bureau.

Prejudice and Racism: Micro-Level Analysis

Have you ever found yourself in a situation in which you were viewed as different, strange, undesirable, or "less than human"? Perhaps you have felt the sting of rejection, based not on judgment of you as a person but solely on the ethnic group into which you were born. Then again, you may have been insulated from this type of rejection if you grew up in a homogeneous community or in a privileged group; you may have even learned some negative attitudes about those different from yourself. It is sobering to think that where and when in history you were born determine how you are treated, your life chances, and many of your experiences and attitudes.

Prejudice

When minority groups are present within a society, prejudice influences dominant-minority group relations. **Prejudice** refers to attitudes that prejudge a group, usually negatively and not based on facts. Prejudiced individuals lump together people with certain characteristics as an undifferentiated group without considering individual differences. Although prejudice can refer to positive attitudes and exaggerations (as when patriots are prejudiced in thinking their society is superior), in this chapter we refer to the

negative aspects of prejudice. We also focus on the adverse effects brought on minority group members by prejudice. While prejudice can be stimulated by events such as conflicts at the institutional level and war at the societal level, attitudes are held by individuals and can be best understood as a micro-level phenomenon.

When prejudiced attitudes become actions, they are referred to as **discrimination**, differential treatment of and harmful actions against minorities. These actions at the micro level might include refusal to sell someone a house because of the religion, race, or ethnicity of the buyer or employment practices that treat groups differently based on minority status (Feagin and Feagin 2010). However, discrimination operates largely at the meso or macro level, discussed later in the chapter. For now, note that individual animosity toward those of another racial category is racial bigotry, and bigotry is not the same thing as racism.

The Nature of Prejudice

Prejudice is an understandable response of humans to their social environment. To survive, every social group or unit—a sorority, a sports team, a civic club, or a nation—needs to mobilize the loyalty of its members. Each organization needs to convince people to voluntarily commit energy, skills, time, and resources so the organization can meet its needs. Furthermore, as people commit themselves to a group, they invest a portion of themselves in the group.

Individual commitment to a group influences one's perception and loyalties, creating preference or even bias for the group. This commitment is often based on stressing distinctions from other groups and deep preference for one's own group. However, these loyalties may be dysfunctional for out-group members and the victims of prejudice.

One reason people hold prejudices is that it is easier to pigeonhole the vast amount of information and stimuli coming at us in today's complex societies and to sort information into neat, unquestioned categories than to evaluate each piece of information separately for its accuracy. Prejudiced individuals often categorize large numbers of people and attribute to them personal qualities based on their dress, language, skin color, or other identifying racial or ethnic features. This process is called **stereotyping**.

Stereotypes, or the pictures we form in our heads about members of a group, are distorted, oversimplified, or exaggerated ideas passed down over generations through cultures. They are applied to all members of a group, regardless of individual differences, and used to justify prejudice, discrimination, and unequal distribution of power, wealth, and opportunities. Often, the result is unfair and inaccurate judgments about individuals who are members of the stereotyped groups. The problem is that both those stereotyping and those being stereotyped come to believe the "pictures" and act accordingly.

Prejudice is difficult to change because it is rooted in traditions, cultural beliefs, and stereotypes. Individuals grow up learning these ingrained beliefs, which often go unchallenged. Yet when studied scientifically, stereotypes seldom correspond to facts.

Social scientists know that prejudice is related to the history and the political and economic climate in a region or country, part of the macro-level cultural and social environment. For instance, in some southern U.S. states where African Americans constitute a substantial percentage of the population, there is evidence that white racial attitudes are more antagonistic due to economic and political competition for jobs and power (Farley 2010; Glaser 1994).

In wartime, the adversary may be the victim of racial slurs, or members of the opponent society may be depicted in films or other media as villains. During World War II, American films often showed negative stereotypes of Japanese and German people, stereotypes that likely reinforced the decision to intern more than 110,000 Japanese Americans, the majority of whom were U.S. citizens, in detention camps following the bombing of Pearl Harbor. Similar issues and stereotypes have arisen for American citizens with Middle Eastern ancestry since the attacks on the New York World Trade Center on September 11, 2001.

Sometimes, minority group members incorporate prejudiced views about themselves into their behavior. This process, an example of a *self-fulfilling prophecy*, involves the adoption of stereotypical behaviors. (See Chapter 6, page 154.) No group is born dumb, lazy, dirty, or money hungry, but its members can be conditioned

to believe such depictions of themselves or be forced into acting out certain behaviors based on expectations of the dominant group.

Thinking Sociologically

Watch the Oscar-winning movie *Crash*. In what ways does this video raise issues of majority-minority stereotypes? How does it highlight labeling done by each group?

Explanations of Prejudice

We have all met people who express hostility toward others. They tell jokes about minorities, curse them, and even threaten action against them. Why do these individuals do this? The following theories have attempted to explain the prejudiced individual.

Frustration-aggression theory. In Greensboro, North Carolina, in 1978, a group of civil rights activists and African American adults and children listened as a guitarist sang freedom songs. A nine-car cavalcade of white Ku Klux Klan (KKK) and American Nazi Party members arrived. The intruders unloaded weapons from the backs of their cars, approached the rally, and opened fire for 88 seconds. Then

On December 7, 1941, Japan bombed Pearl Harbor in Hawaii, prompting President Franklin D. Roosevelt to sign an executive order designating the West Coast as a military zone from which "any or all persons may be excluded." Although not specified in the order, Japanese Americans were singled out for evacuation, and more than 110,000 were removed from many western states and sent to 10 relocation camps. Barber G. S. Hante points proudly to his bigoted sign against people of Japanese origin.

they left as calmly as they had arrived. Four white men and a black woman were dead (Greensboro Justice Fund 2005). According to frustration-aggression theory, many of the perpetrators of this and other heinous acts feel angry and frustrated because they cannot achieve their work or other goals. They blame any vulnerable minority group—religious, ethnic, sexual orientation—and members of that group become targets of their anger. Frustration-aggression theory focuses largely on poorly adjusted people who displace their frustration with aggressive attacks on others. Hate groups evolve from like-minded individuals, often because of prejudice and frustration (see Map 8.3).

Scapegoating. When it is impossible to vent frustration toward the real target—one's boss, one's teachers, the economic system—frustration can take the form of aggressive action against people who are vulnerable because of their low status. They become the scapegoats. The word **scapegoat** comes from the Bible, Leviticus 16:5–22. Once a year, a goat (which was obviously innocent) was laden with parchments on which people had written their sins. The goat was then sent out to the desert to die. This was part of a ritual of purification, and the creature took the blame for others.

Scapegoating occurs when a minority group is blamed for the failures of others. It is difficult to look at oneself to seek reasons for failure but easy to transfer the cause for one's failure to others. Individuals who feel they are failures in their jobs or other aspects of their lives may blame minority groups. From within such a prejudiced mind-set, even violence toward the out-group becomes acceptable. One example is the hostility represented in a notice distributed to a college campus's mailboxes: "Earth's Most Endangered Species: The White Male. Help Preserve It." The notice expressed frustration with the "plight" of white males and blamed policies favoring minorities for their perceived problems.

Today, jobs and promotions are harder for young adults to obtain than they were for the baby boom generation, but the reason is mostly demographic. The baby boom of the 1940s and 1950s resulted in a bulge in the population. There are so many people in the workforce at each successive step on the ladder that it will be another few years before those baby boomers retire in large numbers, and given the

Map 8.3   Active Hate Groups in 2010

Source: Reprinted by permission of the Southern Poverty Law Center 2011.

Klansmen in traditional white robes demonstrate in front of a courthouse in New York City in 1999. They carried a flag sewn together from parts of American and Confederate flags, a symbol of their blended loyalties.

economic downturn that may not happen at age 65. Until it does happen, there will be a good deal of frustration about apparent occupational stagnation. It is easier—and safer—to blame minorities or affirmative action programs than to vent frustration at the next-oldest segment of the population or at one's grandparents for having a large family. Blacks, Hispanics, and other minorities become easy scapegoats.

Although this theory helps explain some situations, it does not predict when frustration will lead to aggression, why only some people who experience frustration vent their feelings on the vulnerable, and why some groups become targets (Marger 2012).

Racial Bigotry and Its Forms

Racial bigotry at the micro level is real. We see it in practice when minority group members are viewed as different and treated negatively. One form of social bigotry is called *ideological bigotry*. It attempts to justify less respectful treatment of people who are "different" by using unscientific sets of ideas that are erroneously portrayed as science-based. It involves the belief that humans are divided into innately different groups, some of which are depicted as biologically inferior. People who hold these views see biological differences as the cause of most cultural and social differences. Remember Hitler's beliefs about the Jews and other groups? These illustrate bigotry based on ideological beliefs (Marger 2012). Hate groups in the United States, Europe, and many other countries justify themselves on the basis of ideological bigotry.

Another pattern of social bigotry, called *symbolic racism*, is more subtle. In this case, individuals insist that they are not prejudiced or racist—that they are color-blind and committed to equality. At the same time, they oppose any social policies

(such as scholarships specifically for minorities) that would reduce historically based disadvantage and make true equality of opportunity possible (Farley 2010). People who display symbolic racism claim to reject the idea that race and racism are present but also fail to correct any problems that are created because racism is still embedded in our social system.

Symbolic racism allows discrimination that is hidden within the society's institutions to remain in place. Symbolic racists reject ideological bigotry, the first type of social bigotry mentioned above, as blatant, crude, and ignorant, but fail to recognize that their actions may perpetuate inequalities at the institutional or meso level, discussed below. Note that many people without social science training see racism as a micro-level issue—one involving individual initiative and individual bigotry—whereas most social scientists see the problem as occurring at the meso and macro levels. We use the word *racism* in this book to refer to meso- and macro-level discrimination, but the word *symbolic racism* is used because it is a micro-level attitude that links to and helps perpetuate racial inequality at the meso and macro levels (Bonilla-Silva 2003).

Micro-level racial bigotry—in the form of either ideological bigotry or symbolic racism—has psychological and social costs, both to those on the receiving end and to the perpetrators. There is a waste of talent and energy of both minorities and those who justify and carry out discriminatory actions. In the 1990s, individual membership in white supremacy groups in Europe and North America grew, as did attacks on blacks, Jews, Muslims, immigrants, and those whose religious and cultural practices were different from those of the majority. For instance, in 2009 there were 1,211 anti-Semitic incidents involving vandalism, assaults, or threats directed at Jewish citizens or Jewish establishments (Anti-Defamation League 2010). Unfortunately, until there are better economic opportunities for more people, individual bigotry, symbolic racism, and institutional racism are likely to be results of economic competition for jobs (Feagin, Vera, and Batur 2001).

Although social-psychological theories shed light on the most extreme cases of individual or small-group prejudice and racism, there is much these theories do not explain. They say little about the everyday hostility and reinforcement of prejudice that most of us experience or engage in, and they fail to deal with institutional discrimination.

Discrimination: Meso-Level Analysis

Dear Teacher, I would like to introduce you to my son, Wind-Wolf. He is probably what you would consider a typical Indian kid. He was born and raised on the reservation. Like so many Indian children his age, he is shy and quiet in the classroom. He is 5 years old, in kindergarten, and I can't understand why you have already labeled him a "slow learner." He has already been through quite an education compared with his peers. He was bonded to his mother and to the Mother Earth in a traditional native childbirth ceremony. And he has been continuously cared for by his mother, father, sisters, cousins, aunts, uncles, grandparents, and extended tribal family since this ceremony. . . .

Wind-Wolf was strapped (in his baby basket like a turtle shell) snugly with a deliberate restriction on his arms and legs. Although Western society may argue this hinders motor-skill development and abstract reasoning, we believe it forces the child to first develop his intuitive faculties, rational intellect, symbolic thinking, and five senses. Wind-Wolf was with his mother constantly, closely bonded physically, as she carried him on her back or held him while breast-feeding. She carried him everywhere she went, and every night he slept with both parents. Because of this, Wind-Wolf's educational setting was not only a "secure" environment, but it was also very colorful, complicated, sensitive, and diverse.

As he grew older, Wind-Wolf began to crawl out of the baby basket, develop his motor skills, and explore the world around him. When frightened or sleepy, he could always return to the basket, as a turtle withdraws into its shell. Such an inward journey allows one to reflect in privacy on what he has learned and to carry the new knowledge deeply into the unconscious and the soul. It takes a long time to absorb and reflect on these kinds of experiences, so maybe that is why you think my Indian child is a slow learner. His aunts and grandmothers taught him to count and to know his numbers while they sorted materials for making abstract designs in native baskets. And he was taught to learn mathematics by counting the sticks we use in our traditional native hand game. So he may be slow in grasping the methods you use in your classroom, ones quite familiar to his white peers. It takes time to adjust to a new cultural system and learn new things. He is not culturally "disadvantaged," but he is culturally different. (Lake 1990:48–53)

This letter expresses the frustration of a father who sees his son being labeled and discriminated against by the school system without being given a chance. *Discrimination* refers to actions taken against members of a minority group. It can occur at individual and small-group levels but is particularly problematic at the organizational and institutional levels—the meso level of analysis.

Thinking Sociologically

How might the schools unintentionally misunderstand Wind-Wolf in such a way that this has negative consequences for his success?

Discrimination is based on race, ethnicity, age, sex, sexual orientation, nationality, social class, religion, or whatever other category members of a society choose to make significant (Feagin and Feagin 2010). Individual discrimination, actions taken against minority group members by individuals, can take many forms, from avoiding contact by excluding individuals from one's club, neighborhood, or even country to physical violence against minorities as seen in hate crime attacks on immigrant Americans perceived to be taking jobs from white Americans.

Racism is any meso-level institutional arrangement that favors one racial group over another; this favoritism may result in intentional or unintentional consequences for minority groups (Farley 2010). Racism is mostly embedded in institutions of society and often is supported by people who are not aware of the social consequence of their actions, as in the case of symbolic racism discussed above. So racism has nothing to do with being a nasty or mean-spirited person; it usually operates independently of racial bigotry (Bonilla-Silva 2003; Rothenberg 2011).

Racism, as sociologists use the concept, involves meso-level institutional discrimination—a normal or routine part of the way an organization operates that systemically disadvantages members of one group. It can include intentional actions, such as laws restricting minorities, as well as unintentional actions that have consequences restricting minorities. Institutional discrimination is built into organizations and cultural expectations in the social world. Even nonprejudiced people can participate in institutional racism quite unintentionally. For example, many schools place students in academic tracks based on standardized test results. Minority children end up disproportionately in lower tracks because the tests have biases that favor middle-class whites. Thus, a policy that is meant to give all children an equal chance ends up legitimizing the channeling of some minority group students into the lower-achieving classroom groupings.

Jim Crow laws, passed in the late 1800s in the United States, and laws that barred Jews in Germany from working in certain places are examples of purposeful discrimination embedded intentionally in organizations. Underlying these actions are individuals' ideological bigotry. By contrast, unintentional discrimination results from policies that have the unanticipated consequence of favoring one group and disadvantaging another. In this case there is discrimination "in fact" even if not in intent—entirely separated from personal ill will. This type of discrimination can be more damaging than that imposed by individuals because it is often done by people who are not the least bit prejudiced and may not recognize the effects of their actions (Merton 1949).

Unintentional discrimination usually occurs through one of two processes: side-effect discrimination or past-in-present discrimination (Feagin and Feagin 1986; Rydgren 2004). **Side-effect discrimination** refers to practices in one institutional area that have a negative impact because they are linked to practices in another institutional area. Figure 8.2 illustrates this idea. Each institution uses information from the other institutions to make decisions. Thus, discrimination in the criminal justice system, which has in fact been well documented, may influence discrimination in education or health care systems.

Consider the following examples of side-effect discrimination. The first is in the criminal justice and employment systems. In an interview conducted by one of the authors, a probation officer in a moderate-size city in Ohio said that he had never seen an African American in his county get a

not-guilty verdict and that he was not sure it was possible. He had known of cases in which minorities had pleaded guilty to a lesser charge even though they were innocent because they did not think they could receive a fair verdict in that city. When people apply for jobs, however, they are required to report the conviction on the application form. By using information about someone's criminal record, employers who clearly do not intend to discriminate end up doing so whether or not the individual was guilty. The side-effect discrimination is unintentional discrimination; the criminal justice system has reached an unjust verdict, and the potential employer is swayed unfairly.

The second example of side-effect discrimination shows that the Internet also plays a role in institutional discrimination and privilege. For example, in Alaska, 15.2% of the population is Native (U.S. Census Bureau 2011h), but Natives hold only 5% of state jobs and 27.3% of Native men and 16% of Native women are unemployed (AAANativeArts 2011; U.S. Census Bureau 2009b). Consider that the state of Alaska uses the Internet as its primary means of advertising and accepting applications for state jobs (State of Alaska 2006). No affordable Internet access is available in 164 predominantly Native villages in Alaska (Denali Commission 2005). Other options for application include requesting applications by mail, if a person knows about the opening. The usefulness of this process is limited, however, by the reliability and speed of mail service to remote villages and the often short application periods for state jobs. State officials may not intentionally use the mechanism to prevent Aleuts, Inupiats, Athabaskans, or other Alaska Natives from

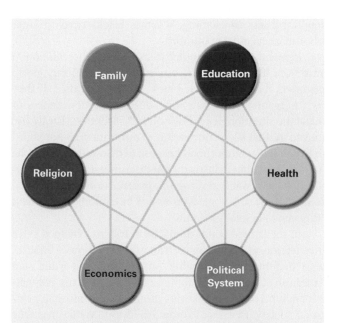

Figure 8.2   Side-Effect Discrimination

Each circle represents a different institution—family, education, religion, health, political-legal system, and economics. These meso-level systems are interdependent, using information or resources from the others. If discrimination occurs in one institution, the second institution may unintentionally borrow information that results in discrimination. In this way, discrimination occurs at the meso level without awareness by individuals at the micro level.

Children play on the porch of their rustic home with no plumbing in the rural Alaskan village of Akhiok, among the Aleutian Islands. Finding jobs through the Internet is not an option from this location.

gaining access to state jobs, but the effect can be institutionalized discrimination. Here, Internet access plays a role in participation of minorities in the social world (Nakamura 2004).

The point is that whites, especially affluent whites, benefit from privileges not available to low-income minorities. The privileged members may not purposely disadvantage others and may not be prejudiced, but the playing field is not level, even though discrimination may be completely unintentional (Rothenberg 2011). Consider the following privileges that most of us who are part of the dominant group take for granted (McIntosh 2002:97–101):

I can avoid spending time with people who mistrust people of my color.

I can protect my children most of the time from people who might not like them.

I can criticize our government and talk about how I fear its policies and behavior without being seen as a cultural "outsider."

I can easily buy posters, postcards, picture books, greeting cards, dolls, toys, and children's magazines featuring people of my race.

I can arrange my activities so that I will never have to experience feelings of rejection owing to my race.

Now imagine *not* being able to take these and many other privileges for granted. What would your reaction be, and what could you do about it?

Thinking Sociologically

Imagine someone in your hometown who runs a business, and the business is hiring people. In order to make a decision about whom to hire, she uses information that has been provided by another institution or organization. How might some of that information be a source of unintended side-effect discrimination for a minority group member? How might the employer discriminate in hiring against a minority group person without realizing it?

Past-in-present discrimination refers to practices from the past that may no longer be allowed but that continue to affect people today (Feagin and Feagin 1986; Verbeek and Penninx 2009). In Mississippi between 1951 and 1952, the average state expenditure to educate a white child was $147 per pupil, whereas the average was $34 per black pupil in segregated schools (Luhman and Gilman 1980). Such blatant segregation and inequality in use of tax dollars is no longer legal. This may seem like ancient history, yet some African Americans who were in school in the 1950s and 1960s are now trying to support a family and pay for their children's college expenses. To those who received a substandard education and did not have an opportunity for college, it is not ancient history because it affects their opportunities today.

Why do some minority groups do better than others? New immigrants to the United States from Southern, Central, and Eastern Europe did better than African Americans, but why? Some explanations have focused on skin color and discrimination, yet Japanese and Chinese have fared well. To address these questions, Stanley Lieberson (1980) did an extensive study and concluded that the new immigrants and blacks who were arriving in the U.S. North held similar aspirations for education and good jobs, but discrimination against blacks by employers, labor unions, realtors, and others was intense due to attitudes held over from the slave period, a case of *past-in-present discrimination*. Immigrants were given better jobs and chances for mobility. Also, Asians experienced less discrimination because their numbers were small and less threatening, whereas blacks tried to enter labor markets in large numbers.

Remember that prejudice is an attitude, discrimination an action. If neighbors do not wish to have minority group members move onto their block, that is prejudice. If they try to organize other neighbors against the newcomers or make the situation unpleasant once the minority family has moved in, that is discrimination. If minorities cannot afford to live in the neighborhood because of discrimination in the marketplace, that is institutional discrimination. An opportunity to clarify and to recognize interrelationships between some of these sociological terms can be explored in the next "Engaging Sociology" feature.

In the United States things seem to have changed since the election in 2008 because the president is now biracial, with half his heritage being African and half Caucasian American. Conservative commentators and many journalists are fond of saying that this means we are now in a postracial society, that race is now irrelevant. While it is true that President Obama is the nation's first African American president, it is also true that no senator (out of 100) is black in 2010. We have also seen in Table 8.1 on page 219 that college-educated African Americans earn $10,000 a year less than white college graduates, and whites with a professional degree earn about $118,000 per year while Hispanics with the same degree earn $89,000. Note also that on a typical Sunday morning, whites and blacks worship separately, with multiracial churches being rare (Emerson and Smith 2000; Emerson and Woo 2006; Marti 2005). As long as differences divide the United States, it is hard to support the notion that it is a "postracial" society.

Engaging Sociology

Using Concepts Correctly and Relating Them to One Another

A lot of terms in this chapter have related to issues of prejudice, discrimination, and racism and how they operate at micro and meso levels. This diagram indicates the levels at which each issue operates.

	Micro Level	Meso Level
Conscious and Intended	Prejudice: Ideological bigotry	Institutional discrimination (explicit)
Unconscious and Unintended	Prejudice: Symbolic racism	Indirect institutional discrimination Side-effect discrimination Past-in-present discrimination

See if you can

1. define each term,

2. provide an example of each,

3. identify which one or two cells represent "racism" as sociologists use the term, and

4. identify ways that each of these elements of intergroup conflict might foster the others.

Dominant and Minority Group Contact: Macro-Level Analysis

Economic hard times hit Germany in the 1930s, following that nation's loss in World War I. To distract citizens from the nation's problems, a scapegoat was found—the Jewish population. The German states began restricting Jewish activities and investments. Gradually, hate rhetoric intensified, but even then, most Jews had little idea about the fate that awaited them. Millions perished in gas chambers because the ruling Nazi party defined them as an undesirable race (although being Jewish is actually a religious or ethnic identification, not a biological category). The following examples illustrate group contact.

Japan has a relatively homogeneous population, but the Burakumin (a group also called the "invisible race") have been treated as outcasts. They make up 2% of the population. Because their ancestors were relegated to performing work considered ritually unclean—butchering animals, tanning skins, digging graves, and handling corpses—they lived in isolated hamlets. Today, discrimination is officially against the law, but customs persist. Ostracized and kept within certain occupations and neighborhoods, the Burakumin rarely intermarry or even socialize with other Japanese. However, today there is some intermarriage and blending into the larger society in cities (Alldritt 2000; International Humanist and Ethical Union 2009), but they remain a minority.

Mexico, Guatemala, and other Central American governments face protests by their Indian populations, descendants of Aztecs, Mayans, and Inca, who have distinguishing features and are today generally relegated to servant positions. These native groups have been protesting against government policies and their poor conditions—usurping of their land, inability to own land, absentee landownership, poor pay, and discrimination by the government (DePalma 1995). One result of discrimination against Central and South American native groups is that their numbers are diminishing and some groups, such as those in Tierra del Fuego, Chile, are dying out.

These examples illustrate contact between governments and minority groups. The Jews in Germany faced genocide; the Burakumin in Japan faced segregation; and Native Americans faced forced relocation to new geographical areas. The form these relations take depends on the following:

1. Which group has more power

2. The needs of the dominant group for labor or other resources (e.g., land) that could be provided by the minority group

3. The cultural norms of each group, including level of tolerance of out-groups

4. The social histories of the groups, including their religious, political, racial, and ethnic differences

5. The physical and cultural identifiers that distinguish the groups

6. The times and circumstances (wars, economic strains, recessions)

Where power between groups in society is unequal, the potential for differential treatment is always present. Yet, some groups live in harmony whether their power is equal or unequal (Kitano, Aqbayani, and de Anda 2005). Whether totally accepting or prone to conflict, dominant-minority relations depend on time, place, and circumstances. Figure 8.3 indicates the range of dominant-minority relationships and policies.

Genocide is the systematic effort of a dominant group to destroy a minority group. Christians were thrown to the lions in ancient Rome. Hitler sent Jews and other non-Aryan groups into concentration camps to be gassed. Iraqis used deadly chemical weapons against the Kurdish people within their own country. Members of the Serbian army massacred Bosnian civilians to rid towns of Bosnian Muslims, an action referred to as ethnic cleansing (Cushman and Mestrovic 1996). In Rwanda, people of the Tutsi and Hutu tribes carried out mass killings against each other in the late 1990s. In Darfur, a section of western Sudan in Africa, massive genocide occurred, and powerful nations of the world did little to stop it. An estimated 2 million Sudanese people have died, disappeared, or become refugees (Smith 2005). In 2011, some politicians argued for intervention in Libya because of the threat of genocide against civilians. Members of the international community intervened, but did so amid controversy. Genocide has existed at many points in history, and as illustrated, it still exists today. These examples show the lethal consequences of racism, one group systematically killing off another, often a minority, to gain control and power.

Subjugation refers to the subordination of one group to another that holds power and authority. Haiti and the Dominican Republic are two countries sharing the island of Hispaniola in the Caribbean. Because many Haitians are poor, they are lured by promises of jobs in the sugarcane fields of the Dominican Republic. However, they are forced to work long hours for little pay and are not allowed to leave until they have paid for housing and food, which may be impossible to do on their low wages.

Slavery is one form of subjugation that has existed throughout history. When the Roman Empire defeated other lands, captives became slaves. This included ancient Greeks, who also kept slaves at various times in their history. African tribes enslaved members of neighboring tribes, sometimes selling them to slave traders, and slavery has existed in Middle Eastern countries such as Saudi Arabia. As mentioned in the opening story for this chapter, slavery is flourishing in many parts of the world today (Bales 2000, 2004, 2007).

Segregation, a specific type of subjugation, keeps minorities powerless by formally separating them from the dominant group and depriving them of access to the dominant institutions. Jim Crow laws, instituted in the southern United States after the Civil War, legislated separation between groups—separate facilities, schools, and neighborhoods (Feagin and Feagin 2010; Massey and Denton 1998). Around the world, reservations, squatters' quarters, and favelas are maintained by the dominant group, usually unofficially but sometimes officially, and serve to contain and isolate minorities in poor or overcrowded areas. In the United States the most vivid example was the Jim Crow laws of the deep south from about the 1890s until the 1960s (Alexander 2010).

Domestic colonialism refers to exploitation of minority groups within a country (Blauner 1972; Kitano et al. 2005). African Brazilians and Native Americans in the United States and Canada have been "domestically colonized groups"— managed and manipulated by dominant group members.

Population transfer refers to the removal of a minority group from a region or country. Generally, the dominant group wants land or resources. In 1972, Ugandan leader General Idi Amin gave the 45,000 Asians in that country, mostly of Indian origin, 36 hours to pack their bags and leave under threat that they would be arrested or killed for not complying. Many found homes in England or India, but for the thousands who were born and raised in Uganda, this expulsion was a cruel act, barring them from their homeland. Because the Asian population had great economic resources, the primary motivation for their expulsion was

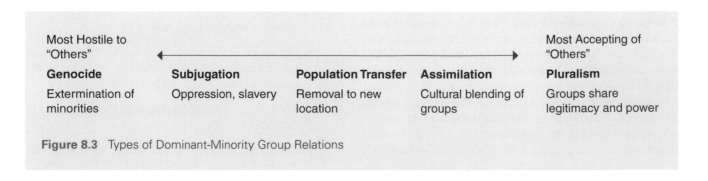

Most Hostile to "Others"				Most Accepting of "Others"
Genocide	**Subjugation**	**Population Transfer**	**Assimilation**	**Pluralism**
Extermination of minorities	Oppression, slavery	Removal to new location	Cultural blending of groups	Groups share legitimacy and power

Figure 8.3 Types of Dominant-Minority Group Relations

to regain power for Africans. Their departure, however, left the country with a void in the business class.

Examples of other population transfers are numerous: Native Americans in the United States were removed to reservations. The Cherokee people were forced to walk from Georgia and North Carolina to new lands west of the Mississippi—a "Trail of Tears" along which 40% of the people perished. During World War II, Japanese Americans were forcibly moved to "relocation centers" and had their land and property confiscated. Many Afghani people fled to Pakistan to escape oppression by the ruling Taliban and again in 2001 to escape U.S. bombing. Today, civilians along the Pakistan-Afghanistan border still suffer war and displacement.

The process by which immigrants create multi-national social relations that link together their original societies with their new locations is called **transnationalism** (Basch, Schiller, and Blanc 1994:7). This involves people who fully participate in and have loyalty to two nations and cultures and often hold dual citizenship (Levitt 2001). An increasing number of naturalized U.S. citizens are also tax-paying members of their countries of origin, and they return often or send money to help families and neighbors with financial needs or immigration plans (Levitt 2001, 2007; Levitt and Waters 2006). Yet, dual citizenship can create dilemmas of identity and sense of belonging.

Assimilation refers to the structural and cultural merging of minority and majority groups, a process by which minority members may lose their original identity (Kitano et al. 2005; Marger 2012). Forced assimilation occurs when a minority group is forced to suppress its identity. This happened in Spain around the time of World War II, when the Basque people were forbidden by the central government to speak or study the Basque language. For several centuries—ending only a few decades ago—the British government tried to stamp out the Welsh language from Wales. However, assimilation is often a voluntary process in which a minority chooses to adopt the values, norms, and institutions of the dominant group.

The notion that we should be a "color-blind" nation is really a call for assimilation, for the only way we can ignore real differences between people is to obliterate the differences. This means that other people would have to give up their cultures and become like the dominant Euro-American culture (Dalton 2012; Dyer 2012). It is for this reason that some minorities see assimilation policy as oppressive and an effort to destroy them.

Assimilation is more likely to occur when the minority group is culturally similar to the dominant group. For instance, in the United States, the closer a group is to being white, English speaking, and Protestant, or "WASP" (white Anglo-Saxon Protestant), the faster its members will be assimilated into the society, adopting the culture and blending in biologically through intermarriage.

Pluralism occurs when each ethnic or racial group in a country maintains its own culture and separate set

In February 2003, a massive genocide began in the Darfur region of Sudan (eastern Africa), where the international community has not been effective in stopping the slaughter. Some people have survived in refugee camps with almost no food and little water, in "homes" like these.

of institutions but has recognized equity in the society. For example, Switzerland has four dominant cultural language groups: French, German, Italian, and Rommansh (or Rumantsch). Four official languages are spoken in the government and taught in the schools. Laws are written in four languages. Each group respects the rights of the other groups to maintain a distinctive language and way of life. In Malaysia, three groups share power—Malays, Chinese, and Indians. Although the balance is not completely stable because Chinese and Indians hold more political and economic power than the native Malays, there is a desire to maintain a pluralistic society. While tensions do exist, both Switzerland and Malaysia represent examples of pluralist societies. Legal protection of smaller or less powerful groups is often necessary to have pluralism. In the United States, pluralism as a policy was first embraced by the nation's first President, George Washington, as explained in the next "Sociology in Your Social World."

Many individuals in the world face disruptions during their lifetimes that change their position in the social structure. The dominant-minority continuum illustrates the range of relations with dominant groups that can affect people's lives as transitions take place.

Thinking Sociologically

Think of examples from current news stories of positive and harmful intercultural contact. Where do your examples fit on the continuum from genocide to pluralism? What policies might address issues raised in your examples?

Sociology in Your Social World

Pluralism: A Long-Standing History in the United States

This Jewish synagogue, the oldest in the United States, proudly displays a letter from George Washington enshrining pluralism in the new nation's policies.

It is no mistake that the oldest Jewish synagogue in the United States is in Rhode Island, for separation of church and state and tolerance of other religious traditions was a founding principle of Rhode Island. After George Washington was elected president of the new nation, he received a letter from that early Jewish congregation in Newport, Rhode Island, asking about his policies of pluralism or multiculturalism (though those words had not been coined yet). In response in 1790, Touro Synagogue received a handwritten letter signed by President Washington (and now proudly on display by the synagogue) embracing an open and "liberal" policy to all American citizens, regardless of origins or religious affiliation. In this letter, George Washington affirmed a policy of pluralism from the very beginning of the country's existence as a nation. Passages from that letter follow.

The Citizens of the United States of America have a right to applaud themselves for having given to mankind examples of an enlarged and liberal policy: a policy worthy of imitation. . . . It is now no more that toleration is spoken of, as if it was by the indulgence of one class of people, that another enjoyed the exercise of their inherent natural rights. For happily the Government of the United States, which gives to bigotry no sanction, to persecution no assistance, requires only that they who live under its protection should demean themselves as good citizens.

. . . May the children of the Stock of Abraham, who dwell in this land, continue to merit and enjoy the good will of the other Inhabitants; while every one shall sit in safety under his own vine and figtree, and there shall be none to make him afraid. May the father of all mercies scatter light and not darkness in our paths, and make us all in our several vocations useful here, and in his own due time and way everlastingly happy.

G. Washington

Theoretical Explanations of Dominant-Minority Group Relations

Are human beings innately cruel, inhumane, greedy, aggressive, territorial, or warlike? Some people think so, but the evidence is not very substantial. To understand prejudice in individuals or small groups, psychological and social-psychological theories are most relevant. To understand institutional discrimination, studying meso-level organizations is helpful, and to understand the pervasive nature of prejudice and stereotypes over time in various societies, cultural explanations are useful. Although aspects of macro-level theories relate to micro- and meso-level analysis, their major emphasis is on understanding the national and global systems of group relations.

Conflict Theory

In the 1840s, as the United States set out to build a railroad, large numbers of laborers emigrated from China to do the hard manual work. When the railroad was completed and competition for jobs became tight, the once-welcomed Chinese became targets of bitter prejudice, discrimination, and sometimes violence. Between 1850 and 1890, whites in

California protested against Chinese, Japanese, and Chicano workers. Members of these minority groups banded together in towns or cities for protection, founding the Chinatowns we know today (Kitano 2005). Non-Chinese Asian groups suffered discrimination as well because the prejudiced generalizations were applied to all Asians (Son 1992; Winders 2004).

Why does discrimination occur? Conflict theorists argue that creating a "lesser" group protects the dominant group's advantages. Because privileges and resources are usually limited, those who have them want to keep them. One strategy used by privileged people, according to conflict theory, is to perpetrate prejudice and discrimination against minority group members. A case in point is the *Gastarbeiter* (guest workers) in Germany and other Western European countries, who immigrate from Eastern Europe, the Middle East, and Africa to fill positions in European economies. They are easily recognized because of cultural and physical differences and are therefore ready targets for prejudice and discrimination, especially in times of economic competition and slowing economies. This helps keep many of them in low-level positions. Today some European countries are considering laws to limit immigration, in part because of their weak economies.

Karl Marx argued that exploitation of the lower classes is built into capitalism because it benefits the ruling class. Unemployment creates a ready pool of labor to fill the marginal jobs, with the pool often made up of identifiable minority groups. This pool protects those in higher-level positions from others moving up in the stratification system and threatening their jobs.

Several theories stemming from the conflict tradition help explain minority relations. Two are discussed below. The first has to do with development of hostilities between groups, and the second involves "split labor markets."

Three critical factors contribute to animosity between groups, according to one conflict theorist (Noel 1968): First, if two groups of people are identifiably different in appearance, clothing, or language, then we-versus-they thinking and ethnocentrism may develop. However, this by itself does not establish long-term hostility between the groups. Second, if the two groups conflict over scarce resources that both want, hostilities are very likely to arise. The resources might be the best land, the highest-paying jobs, access to the best schools for one's children, energy resources such as oil, or positions of prestige and power. If the third element is added to the mix—one group having much more power than the other—then intense dislike between the two groups and misrepresentation of each group by the other are virtually inescapable. What happens is that the group with more power uses that power to ensure that its members (and their offspring) get the most valued resources. However, because they do not want to see themselves as unfair and brutish people, they develop

Chinese men were invited and encouraged to come to North America to help build railroads. However, prejudice was extremely prevalent, especially once the railroads were completed and the immigrants began to settle into other jobs in the U.S. economy.

stereotypes and derogatory characterizations of "those other people," so the lack of access provided to "them" seems reasonable and justified. Discrimination (often at the macro level) comes first, and ideological bigotry (at the micro level) comes later to justify the discrimination (Noel 1968). Thus, macro- and meso-level conflicts can lead to micro-level attitudes.

Split labor market theory, a branch of conflict theory, characterizes the labor market as having two main types of jobs: The primary market involves clean jobs, largely in supervisory roles, and provides high salaries and good advancement possibilities, whereas the secondary market involves undesirable, hard, and dirty work, compensated with low hourly wages and few benefits or career opportunities. Minorities, especially those from the urban underclass, are most likely to find dead-end jobs in the secondary market. For instance, when Mexicans work for little income picking crops as migrant laborers, they encounter negative stereotypes because they are poor and take jobs for low wages. Prejudice and discrimination build against the new, cheaper workers who threaten the next level of workers as the migrant workers seek to move up in the economic hierarchy. Thus, competition for lesser jobs pits minorities against each other and low-income whites against minorities. By encouraging division and focusing antagonism between worker groups, employers reduce threats to their dominance and get cheaper labor in the process (Bonacich 1972, 1976). This theory maintains that competition, prejudice, and ethnic animosity serve the interests of owner capitalists because that atmosphere keeps laboring classes from uniting.

Conflict theory has taught us a great deal about racial and ethnic stratification. However, conflict theorists often focus on people with power quite intentionally oppressing others to protect their own self-interests. They depict the dominant group as made up of nasty, power-hungry people. As we have seen in the meso-level discussion of side-effect and past-in-present discrimination, privilege and discrimination are often subtle and unconscious, which means they can continue without ill will among those in the dominant group. Their privilege has been institutionalized. Conflict theorists sometimes miss this important point.

Structural-Functionalist Theory

From the structural-functionalist perspective, maintaining a cheap pool of laborers who are in and out of work serves several purposes for society. Low-paying and undesirable jobs for which no special training is needed—busboys, janitors, nurse's aides, street sweepers, and fast-food service workers—are often filled by minority group members of societies, including immigrant populations.

Not only does this cheap pool of labor function to provide a ready labor force for dirty work or the menial unskilled jobs, these individuals also serve other functions for society. They make possible occupations that service the poor, such as social work, public health, criminology, and the justice and legal systems. They buy goods others do not want—day-old bread, old fruits and vegetables, secondhand clothes. They set examples for others of what not to be, and they allow others to feel good about giving to charity (Gans 1971, 1994).

Thomas Sowell (1994) contends that circumstances of the historical period and the situation into which one is born create the major differences in the social status of minority groups. He believes that minority individuals must work hard to make up for their disadvantages. His contentions are controversial in part because of the implication that institutional discrimination can be overcome by hard work. Conflict theorists counter his argument by saying that discrimination that reduces opportunities is built into institutions and organizations and must be dealt with through meso- and macro-level structural change. They argue that hard work is necessary, but not sufficient, for minorities to succeed.

Prejudice, discrimination, and institutional racism are dysfunctional for society, resulting in loss of human resources, costs to societies due to poverty and crime, hostilities between groups, and disrespect for those in power (Schaefer and Kunz 2008).

Thinking Sociologically

What are some micro-, meso-, and macro-level factors that enhance the chances that minority persons can move up the social ladder to better jobs?

The Effects of Prejudice and Discrimination

Pictures of starving orphans from Sudan and Ethiopia and broken families from war-torn Bosnia remind us of the human toll resulting from prejudice and discrimination. This section discusses the results of prejudice, racism, and discrimination for minority groups and for societies.

The Costs of Racism

Individual victims of racism suffer from the destruction of their lives, health, and property, especially in societies where racism leads to poverty, enslavement, conflict, or war. Poor self-concept and low self-esteem result from constant reminders of a devalued status in society.

Prejudice and discrimination result in costs to organizations and communities as well as to individuals. First, they lose the talents of individuals who could be productive and contributing members. Because of poor education, substandard housing, and inferior medical care, these citizens cannot use their full potential to contribute to society. In 2010, 50.7 million or 16.7% of U.S. citizens did not have health insurance (Wolf 2010). The number of uninsured children is approximately 8 million and growing, or 10.4% of children in the United States (Children's Defense Fund 2011). Yet the U.S. health expenditures consume 17.6% of the country's gross domestic product (GDP), the highest expenditure in the world (U.S. Department of Health and Human Services 2011). Still, the inequities in health care coverage are striking: 11.6% of whites (1 in 6) are without care, but the figure is 19.9% (1 in 5) for African Americans and 41.5% (more than 1 in 3) for Hispanics (Newport and Mendes 2009).

Second, government subsidies cost millions in the form of welfare, food stamps, and imprisonment, but they are made necessary in part by the lack of opportunities for minority individuals. Representation of ethnic groups in the U.S. political system can provide a voice for concerns of groups. Table 8.2 shows the representation of ethnic groups in Congress, but even this understates the lack of

Sudanese children wait in line to receive food in the Sudanese refugee camp of Narus. A worldwide study released May 2, 2006, by UNICEF reveals that some 5.6 million children die every year in part because they do not consume enough of the right nutrients, and 146 million children are at risk of dying early because they are underweight (Jane O'Brien 2006).

Table 8.2   Representation in the 112th United States Congress, 2011

	Native American	Asian	Black	Hispanic
Senate	0	1 (1%)	0	2 (2%)
House	1 (0.2%)	9 (2%)	44 (10%)	29 (6.7%)
% of Population	0.9%	4.8%	12.6%	16.3%

Source: Ethnic Majority 2010; Manning 2011.

representation. In the entire history of the United States, the number of senators from minority ethnic groups is extremely small: African American: 6; Asian American: 5; Hispanic American: 6; Native American: 3 (U.S. Senate 2011).

Thinking Sociologically

How might one's self-interests be underrepresented in policy decisions if there is low representation of one's ethnic group in the U.S. Congress or other government bodies?

Continued attempts to justify discrimination by stereotyping and labeling groups have cultural costs, too. There are many talented African American athletes who are stars on college sports teams, but very few of them have been able to break into the ranks of coaches and managers, although there has been more opportunity in basketball than in other sports (Eitzen and Sage 2003; Sage 2005). The number of African American and Mexican American actors and artists has increased, but the number of black playwrights and screenwriters who can get their works produced or who have become directors remains limited. African American musicians have found it much more difficult to earn royalties and therefore cannot compose full-time (Alexander 2003). Because these artists must create and perform their art "as a sideline," they are less able to contribute their talents to society. The rest of us are poorer for it.

Minority Reactions to Prejudice, Discrimination, and Racism

How have minority groups dealt with their status? Five different reactions are common: assimilation, acceptance, avoidance, aggression, and change-oriented actions directed at the social structure. The first four are micro-level responses. They do not address the meso- and macro-level issues.

Assimilation is an accommodation to prejudice and discrimination. Some minority group members attempt to pass or assimilate as members of the dominant group so as to avoid bigotry and discrimination. Although this option is not open to many because of their distinguishing physical characteristics, this strategy usually involves abandoning their own culture and turning their back on family roots and ties, a costly strategy in terms of self-esteem and sense of identity. People who select this coping strategy are forced to deny who they are as defined by their roots and to live their lives in constant anxiety, feeling as though they must hide something about themselves. In the 1960s, popular items advertised in African American magazines included "whitening creams" or "skin bleaches." Light-colored people with African ancestry would bleach their skin to pass as white. Skin-whitening creams can still be found today on pharmacy shelves in many countries. Dissatisfaction with one's body has an impact on one's self-concept.

Passing—pretending to be a member of the privileged group when one is not fully a part of that community—is one strategy used to enhance assimilation. It has also been a common response of gays and lesbians who are afraid to come out. Homosexuals experience the costly impact on self-esteem and the constant fear that they may be discovered. Likewise, assimilated Jews have changed their religion and their names to be accepted. Despite the wrenching from their personal history, passing has allowed some individuals to become absorbed into the mainstream and to lose the stigma of being defined as a minority. To these people, it is worth the high cost.

Acceptance is another common reaction to minority status. Some minority groups have learned to live with their minority status with little overt challenge to the system. They may or may not hold deep-seated hostility, but they ultimately conclude that change in the society is not very likely, and acceptance may be the rational means to survive within the existing system.

There are many possible explanations for this seeming indifference. For example, religious beliefs allow poor Hindus in India to believe that if they accept their lot in life, they will be reincarnated in a higher life-form. If they rebel, they can expect to be reincarnated into a lower life-form. Their religion is a form of social control.

Unfortunately, many children are socialized to believe that they are inferior or superior because minority group members are expected by the dominant group to behave in certain ways and often live up to that expectation because of the self-fulfilling prophecy (Farley 2010). Evidence to support stereotypes is easily found in individual cases—"inferior" kids live in shabby houses, dress less well, and speak a different dialect. At school and on the job, minority position is reaffirmed by these characteristics.

Avoidance means shunning all contact with the dominant group. This can involve an active and organized attempt to leave the culture or live separately as some political exiles have done. In the United States, Marcus Garvey organized a Back-to-Africa movement in the 1920s, encouraging blacks to give up on any hope of justice in American society and to return to Africa. Native Americans continually moved west in the nineteenth century—trying to or being forced to get away from white Anglo settlers who brought alcohol and deadly diseases. In some cases, withdrawal may mean dropping out of the society as an individual—escaping by obliterating consciousness in drugs or alcohol. The escape from oppression and low self-concept is one reason why drug use is higher in minority ghettos and alcohol abuse is rampant on Native American reservations.

Aggression resulting from anger and resentment over minority status and from subjugation may lead to retaliation or violence. Because the dominant group holds significant power, a direct route such as voting against the dominant group or defeating oppressors in war is not always possible. Indeed, direct confrontation can be very costly to those lacking political or economic power. Suicide bombers from Palestine represent the many Palestinians who are frustrated and angry over their situation in relation to Israel but have few options to express their anger.

Aggression usually takes one of two forms, indirect aggression or displaced aggression. Indirect aggression includes biting assertiveness in the arts—literature, art, racial and ethnic humor, and music—and in job-related actions such as inefficiency and slowdowns by workers. Displaced aggression, on the other hand, involves hostilities directed toward individuals or groups other than the dominant group, as happens when youth gangs attack other ethnic gangs in nearby neighborhoods. They substitute aggression against the dominant group by acting against the other minority group to protest their frustrating circumstances.

The four responses discussed thus far address the angst and humiliation that individual minorities feel. Each strategy allows an individual person to try to cope, but none addresses the structural causes of discrimination. The final strategy is change-oriented action: Minority groups pursue social change in the meso- and macro-level structures of society, as discussed below.

Nonviolent Resistance: A Strategy for Change

Another technique for bringing about change at the institutional and societal level is nonviolent resistance by minority groups. The model for this technique comes from India where, in the 1950s, Mahatma Gandhi led the struggle for

Mahatma Gandhi, leader of the Indian civil disobedience revolt, marched to the shore to collect salt, a clear violation of the law that he felt was inhumane and unjust. On the right is a woman lieutenant in his nonviolent resistance movement.

independence from Britain. Although Britain clearly had superior weapons and armies, boycotts, sit-ins, and other forms of resistance eventually led to British withdrawal as the ruling colonial power. Jesse Jackson, a U.S. presidential candidate in 1984 and 1988, led his Chicago-based organization, Operation PUSH (now the Rainbow PUSH Coalition), in economic boycotts against companies such as Coca-Cola to force them to hire and promote blacks. Cesar Chavez led boycotts against grape growers to improve the working conditions of migrant workers. This strategy has been used successfully by workers and students to bring about change in many parts of the world.

In the United States, Martin Luther King, Jr., followed in the nonviolent resistance tradition of India's Gandhi, who sought to change India's laws so minorities could have equal opportunities within the society. King's strategy involved nonviolent popular protests, economic boycotts, and challenges to the current norms of the society. The National Association for the Advancement of Colored People sought to bring about legal changes through lawsuits that create new legal precedents supporting racial equality. Often, these lawsuits address side-effect discrimination—a meso-level problem. Many other associations for minorities—including the Anti-Defamation League (founded by Jews) and La Raza Unida (a Chicano organization)—also seek to address problems both within organizations and institutions (meso level) and in the nation as a whole (macro level). Like Dr. King, who had an undergraduate degree in sociology, many sociologists have used their training to address the issues of discrimination and disprivilege through empowerment and change.

Some other minority individuals have used their sociology degrees in business, both to enhance their own competence in the business world and to help their ethnic communities. One example is the work of David Staddon who writes about his applications of sociology in consulting, administration, and business. See the next "Applied Sociologist at Work."

Sometimes, minority reactions result in assimilation, but often, the goal is to create a pluralistic society in which cultures can be different yet have economic opportunities open to all.

Thinking Sociologically

The preceding discussion presents five types of responses by minorities to the experience of discrimination and rejection. Four of these are at the micro level, and only one is at the meso and macro levels. Why do you suppose most of the coping strategies of minorities are at the micro level?

Policies Governing Minority and Dominant Group Relations

From our social world perspective, we know that no problem can be solved by working at only one level of analysis. A successful strategy must bring about change at every level of the social world—individual attitudes, organizational discrimination, cultural stereotypes, societal stratification systems, and national and international structures. However, most current strategies focus on only one level of analysis. The types of problems and their solutions at each level of the social system are discussed below and illustrated in Figure 8.4.

Individual or Small-Group Solutions

Programs to address prejudice and stereotypes through human relations workshops, group encounters, and therapy can achieve goals with small numbers of people. For instance, African American and white children who are placed in interracial classrooms in schools are more likely to develop close interracial friendships (Ellison and Powers 1994). Beyond that, education gives a broader, more universal outlook; reduces misconceptions and prejudices;

Types of Problems at Each Level	Types of Solutions or Programs
Individual level: stereotypes and prejudice	Therapy, tolerance-education programs
Group level: negative group interaction	Positive contact, awareness by majority of their many privileges
Societal level: institutional discrimination	Education, media, legal-system revisions
Global level: deprivation of human rights	Human rights movements, international political pressures

Figure 8.4   Problems and Solutions

The Applied Sociologist at Work— By David Staddon

Native American Cultures and Applied Sociology

I have had several positions in Indian Country, beginning with the YMCA of Michigan's Native American Outreach Program. The program worked with every tribe in Michigan and included urban youth leadership development, family enhancement, and the preservation of traditional cultural values and behaviors. I left that position to attend graduate school at Central Michigan University where I eventually became director of the Native American Programs office. Since then I have worked with a number of indigenous nations, including the Saginaw Chippewa in Michigan and the Northern Arapaho in Wyoming. A person with sociological/intercultural skills can have a distinct advantage in the marketplace, especially considering the changing demographics in the United States. This is especially true where there are cultural intersections involved, and I experienced many of those working with native nations.

One of the first challenges (of many) that I needed to overcome was the fact that 98% of all our casino customers in Wyoming were tribe members. That really did not help the Arapaho people since we were simply churning money through the local economy. We needed to diversify our customer base and bring in "outside" money. Many organizations currently talk about "reengineering their corporate culture" to be more friendly and accessible to minority groups. I was faced with the interesting challenge of creating an atmosphere where non-Indians felt safe, secure, and comfortable in our gaming environment, rather like "reverse engineering."

The situation was further complicated by the fact that I am from a different tribe than the Arapahos. Most people (including sociologists) have scant understanding of the intercultural differences between Indian tribes—an important factor in having a successful career in Indian Country. So I had to learn to deal with intersections between Arapaho-Ottawa-Mainstream-Male/Female-Corporate values, outlooks, and behaviors. My challenge was to build a corporate culture that took all these factors into consideration and led to financial success of the business. So some of my first priorities were in image development and customer service.

Having spoken with many white folks in Riverton, I came away with the distinct view that many (if not all) of them felt that the casino was an unsafe place. I was told, "If I win money, I'll just get knocked in the head in the parking lot." We had to do many things to change the image, including designing a new logo for the casino and providing snazzy uniforms for all the staff. The logo was on everything, so we could unify the corporate image. I instituted customer service training and standards for interaction with customers, installed more lights in the parking lot, and started an escort service (not *that* kind!) where, upon request, our security staff would escort customers to their cars in the parking lot. I also took pictures of our security staff and developed some advertising materials emphasizing friendliness, safety, and security. I got active with the local chamber of commerce, establishing relationships with local business and opinion leaders.

By the time I left, we had experienced a complete turnaround with the business—both from a financial standpoint and from the standpoint of our customer base. Our customer base is now over 90% nonnative, and we were bringing millions of dollars into the local native community. Prior to this, the casino had only had two years of profitability out of twelve years in business.

A lot of other peripheral efforts went into developing the organization and improving its image. In short, we worked with "image-management" ideas from Goffman ([1959] 2001) and notions of how people define a situation—a central idea in symbolic interactionism. I was doing applied sociology to help this business venture work—a business venture that also helped a minority community.

My background in social sciences was vital in melding the cultural considerations that contributed to an organizational culture conducive to employee creativity, success, and enjoyment. My training and education in social science has had direct relevance for my various jobs. One interesting aspect of applied social science in business is the examination of corporate culture and its relationship to behavior, public image, policies, planning, and other organizational behavior. A liberal arts education is becoming increasingly important in the U.S. workplace, especially one that emphasizes cross-cultural understanding. For me, coupling this knowledge with my business skills was the key to success.

David Staddon is a member of the Wikwemikong Band of Ottawa Indians, located on Manitoulin Island, Lake Huron, Georgian Bay, Southern Ontario. David has been working with "first nation" governments most of his working life. With a bachelor's degree in sociology/social science and a master's in administration, he is well prepared to deal with native issues. He now works with the St. Regis Mohawk Tribe near Massena, New York, as its director of public information.

An AmeriCorps volunteer supervises construction on a Habitat for Humanity building in St. Tammany Parish, Louisiana, for people displaced by Hurricane Katrina—many of them minorities. The solution proposed by some nongovernmental organizations such as Habitat is to address problems and suffering with volunteer labor and donations.

shows that many issues do not have clear answers; and encourages multicultural understanding. Two groups with strong multicultural education programs are the Anti-Defamation League and the Southern Poverty Law Center's Teaching Tolerance program. Both groups provide schools and community organizations with their literature, videos, and other materials aimed at combating intolerance and discrimination toward others.

However, these strategies do not address the social conditions underlying the problems because they reach only a few people achieving only limited results. They also do not begin to address dilemmas that are rooted in meso- and macro-level causes of problems.

Group Contact

Many social scientists advocate organized group contact between dominant and minority group members to improve relations and break down stereotypes and fears. Although not all contact reduces prejudice, many studies have shown the benefits of contact. Some essential conditions for success are equal status of participants, noncompetitive and nonthreatening contact, and projects or goals on which to cooperate (Farley 2010).

In a classic study of group contact, social psychologists Muzafer Sherif and Caroline Sherif (1953) learned that the most effective strategy to reduce group prejudice is to introduce a superordinate goal that can be achieved only if everyone cooperates. As groups work together, established stereotypes begin to fade away. The key, then, is to find common interests that can only be satisfied if all parties are seen as partners in solving some larger problem, and the

outcome is win-win for all. Positive contact experiences tend to improve relations in groups on a micro level by breaking down stereotypes, but to solidify gains we must also address institutionalized inequalities.

Institutional and Societal Strategies to Improve Group Relations

Sociologists contend that institutional and societal approaches to reduce discrimination get closer to the core of the problems and affect larger numbers of people than do micro-level strategies. For instance, voluntary advocacy organizations pursue political change through lobbying, watchdog monitoring, rallies, and boycotts (Minkoff 1995).

The Civil Rights Commission, Fair Employment Practices Commission, and Equal Employment Opportunity Commission are government organizations that protect rights and work toward equality for all citizens. These agencies oversee practices and hear complaints relating to racial, sexual, age, and other forms of discrimination. Legislation, too, can modify behaviors. Laws requiring equal treatment of minorities have resulted in increased tolerance of those who are "different" and have opened doors that previously were closed to minorities.

Affirmative action laws, first implemented during Lyndon Johnson's administration, have been used to fight pervasive institutional racism, but they are controversial (Crosby 2004; Farley 2010).

Affirmative Action

One of the most contentious policies in the United States has been affirmative action. The following discussion addresses the intentions and forms of the policy. A societal policy for change, affirmative action actually involves three different policies. Its simplest and original form, which we call *strict affirmative action,* involves affirmative or positive steps to make sure that unintentional discrimination does not occur. It requires, for example, that an employer who receives federal monies must advertise a position widely and not just through internal or friendship networks. If the job requires an employee with a college education, then by federal law, employers must recruit through minority and women's colleges as well as state and private colleges in the region. If employers are hiring in the suburbs, they are obliged to contact unemployment agencies in poor and minority communities as well as those in the affluent neighborhoods. After taking these required extra steps, employers are expected to hire the most qualified candidate who applies, regardless of race, ethnicity, sex, religion, or other external characteristics. The focus is on providing opportunities for the best qualified people. For many people, this is the meaning of affirmative action, and it is inconceivable that this could be characterized as reverse discrimination, for members of the dominant group will be hired if they are in fact the most qualified. These policies do not overcome the problem that qualified people who have been marginalized may be competent but do not have the traditional paper credentials that document their qualifications (Gallagher 2004).

The second policy is a *quota system,* a requirement that employers *must* hire a certain percentage of minorities. For the most part, quotas are now unconstitutional. They apply only in cases in which a court has found a company to have a substantial and sustained history of discrimination against minorities and in which the employment position does not have many requirements (if the job entails sweeping floors and cleaning toilets, there would not be an expectation of a specific academic degree or a particular grade point average).

The third policy and the one that has created the most controversy among opponents of affirmative action is *preference policies.* Preference policies are based on the concept of equity, the belief that sometimes people must be treated differently in order to be treated fairly. This policy was enacted to level the playing field, which was not rewarding highly competent people because of institutional racism.

The objectives of preference policies are to (1) eliminate qualifications that are not substantially related to the job but that unwittingly favor members of the dominant group and (2) foster achievement of objectives of the organization that are only possible through enhanced diversity. To overcome these inequalities and achieve certain objectives, employers and educational institutions take account of race or sex by making special efforts to hire and retain workers or accept students from groups that have been underrepresented. In many cases, these individuals bring qualifications others do not possess. Consider the following examples.

A goal of the medical community is to provide access to medical care for underserved populations. There is an extreme shortage of physicians on the Navajo reservation. Thus, a Navajo applicant for medical school might be accepted, even if her scores are slightly lower than those of another candidate, because she speaks Navajo and understands the culture. One could argue that she is more qualified to be a physician on the reservation than someone who knows nothing about Navajo society but has a slightly higher grade point average or test score. Some argue that tests should not be the only measure to determine successful applicants.

Likewise, an African American police officer may have more credibility in a minority neighborhood and may be able to defuse a delicate conflict more effectively than a white officer who scored slightly higher on a paper-and-pencil placement test. Sometimes, being a member of a particular ethnic group can actually make one more qualified for a position.

A 1996 proposition in California to eliminate affirmative action programs in the state passed in a popular referendum. The result was that colleges in California are not allowed to offer preference based on race, but can give preference on state residency, athletic competency, musical skill, having had a parent graduate from the school, and many other factors. Many colleges and universities admit students because they need an outstanding point guard on the basketball team, an extraordinary soprano for the college choir, or a student from a distant state for geographic diversity. These students were shown preference by being admitted with lower test scores than some other applicants.

A lawsuit filed in Detroit alleged that the University of Michigan gave unlawful preference to minorities in undergraduate admissions and in law school admissions. In this controversial case, the court ruled that undergraduate admissions were discriminatory because they used numbers rather than individualized judgments to make the admissions determination (Alger 2003). Consider the "Engaging Sociology" feature and decide whether you think the policy was fair and whether only race and ethnicity should have been deleted from the preferences allowed.

The question remains: Should preferences be given to accomplish diversity? Some people feel that programs involving any sort of preference are reverse discrimination. Others believe such programs have encouraged employers, educational institutions, and government to look carefully at hiring policies and minority candidates and that many more competent minority group members are working in the public sector as a result of these policies.

Thinking Sociologically

First, read the feature on the next page. Should colleges consider an applicant's state of origin, urban or rural background, ethnicity, musical or athletic ability, alumni parent, or other factors in admitting students if it helps the college achieve its goals? Should men with lower scores be admitted because the college wants to have balanced gender enrollment? Is ethnic diversity so central to the learning environment that it must be an admissions criterion? Why?

Global Movements for Human Rights

A unique coalition of world nations has emerged from a recent international event—the terrorist attack of September 11, 2001. In this attack on the World Trade Center in New York City, a center housing national and international businesses and workers, citizens from 90 countries were killed when two hijacked commercial jetliners flew into the towers. In addition to the world condemnation, many countries' governments pledged to fight against terrorism.

Engaging Sociology

Preference Policies at the University of Michigan

To enhance diversity on the campus—a practice that many argue makes a university a better learning environment and enhances the academic reputation of the school—many colleges have preference policies in admissions. However, the University of Michigan was sued by applicants who felt they were not admitted because others replaced them on the roster due to their racial or ethnic background.

The University of Michigan is a huge university where a numbering system is needed to handle the volume (tens of thousands) of applicants; admissions staff cannot make a decision based on personal knowledge of each candidate. Thus, they give points for each quality they deem desirable in the student body. A maximum of 150 points is possible, and a score of 100 would pretty much ensure admission. The university feels that any combination of points accumulated according to the following formula will result in a highly qualified and diverse student body.

For academics, up to 110 points are possible:

- 80 points for grades (a particular grade point average in high school results in a set number of points; a 4.0 results in 80 points; a 2.8 results in 56 points)
- 12 points for standardized test scores (ACT or SAT)
- 10 points for the academic rigor of high school (so all students who go to tougher high schools earn points)
- 8 points for the difficulty of the curriculum (e.g., points for honors curriculum vs. keyboarding courses)

For especially desired qualities, including diversity, up to 40 points are possible for any combination of the following (but no more than 40 in this "desired qualities" category):

- Geographical distribution (10 for Michigan resident; an additional 6 for underrepresented Michigan county)
- Legacy—a relative has attended Michigan (4 points for a parent; 1 for a grandparent or sibling)
- Quality of submitted essay (3 points)
- Personal achievement—a special accomplishment that is noteworthy (up to 5 points)
- Leadership and service (5 points each)
- Miscellaneous (only one of these can be used):

    __Socioeconomic disadvantage (20 points)
    __Racial or ethnic minority (20 points; disallowed by the court ruling)
    __Men in nursing (5 points)
    __Scholarship athlete (20 points)
    __Provost's discretion (20 points; usually the son or daughter of a large financial donor or of a politician)

In addition to ethnicity being given preference, athleticism, musical talent, having a relative who is an alum, or being the child of someone who is noteworthy to the university are also considered. Some schools also give points for being a military veteran. The legal challenge to this admissions system was based only on the racial and ethnic preference given to some candidates, not on the other items that are preferenced.

* * * * * * *

Engaging with Sociology:

1. Does this process seem reasonable as a way to get a diverse and highly talented incoming class of students? Why or why not?

2. Does it significantly advantage or disadvantage some students? Explain.

3. How would you design a fair system of admissions, and what other factors would you consider?

Yet, why did such a heinous act occur? Many social scientists attempting to identify a cause point to the disparities between rich and poor peoples of the world. The perpetrators likely felt that Muslims were treated as inconsequential players in the global world, and they struck out to make a dramatic impact on the world community and the United States. The point is that global issues and ethnic conflicts in the social world are interrelated.

Some civil rights or human rights movements have justice issues in many countries as their focal point. Amnesty International is one such movement, which has strong support at many college campuses.

often under the auspices of the United Nations, to deal with health issues, world poverty and debt, trade, security, and many other issues affecting world citizens—World Health Organization, World Bank, World Trade Organization, and numerous regional trade and security organizations.

The United Nations, several national governments (Britain, France, and Canada), and privately funded advocacy groups speak up for international human rights as a principle that transcends national boundaries. The most widely recognized private group is Amnesty International, a watchdog group that does lobbying on behalf of human rights and supports political prisoners and ethnic group spokespersons. When Amnesty International was awarded the Nobel Peace Prize in 1997, the group's visibility was dramatically increased. Even some activist sociologists have formed Sociologists Without Borders, or SSF (*Sociólogos sin Fronteras;* http://www.sociologistswithoutborders.org), a transnational organization committed to the idea that "all people have equal rights to political and legal protections, to socioeconomic security, to self-determination, and to their personality."

Everyone can make a positive difference in the world, and one place to start is in our community (see "Contributing to Our Social World"). We can counter prejudice, discrimination, and socially embedded racism in our own groups by teaching children to see beyond "we" and "they" and by speaking out for fairness and against stereotypes and discrimination.

The rights granted to citizens of any nation used to be considered the business of each sovereign nation, but after the Nazi Holocaust, German officers were tried at the Nuremberg trials, and the United Nations passed the Universal Declaration of Human Rights. Since that time, many international organizations have been established,

Inequality is not limited to social classes, race or ethnic groups, or religious communities. "We" and "they" thinking can invade some of the most intimate settings. It can infect relations between men and women in everything from the home to the boardroom and from the governance of nations to the decisions of global agencies. We will explore this in the next chapter: "Gender Stratification."

What Have We Learned?

Why are minority group members in most countries poorer than dominant group members? This and other chapter-opening questions can be answered in part by considering the fact that human beings have a tendency to create "we" and "they" categories and to treat those who are different as somehow less human. The categories can be based on physical appearance, cultural differences, religious differences, or anything the community or society defines as important. Once people notice differences, they are more inclined to hurt "them" or to harbor advantages for "us" if there is competition over resources that both groups want. Even within a nation, where people are supposedly all "us," there can be sharp differences and intense hostilities.

Key Points:

- Although the concept of race has no real meaning biologically, race is a social construction because people *believe* it is real. (See pp. 216–219.)

- Minority group status—having less power and less access to resources—may occur because of racial status or because of ethnic (cultural) factors. (See pp. 219–221.)

- Prejudice operates at the micro level of society and is closest to people's own lives, but it has much less impact on minorities than discrimination. Symbolic racism has become more of a problem—the denial of overt prejudice but the rejection of any policies that might correct inequities. (See pp. 223–225.)

- At the meso level, institutional discrimination operates through two processes: side-effect and past-in-present. These forms of discrimination are unintended and unconscious—operating quite separate from any prejudice of individuals in the society. (See pp. 225–229.)

- When very large ethnic groups or even nations collide, some people are typically displaced and find themselves in minority status. (See pp. 229–230.)

- The policies of the dominant group may include genocide, subjugation, population transfer, assimilation, or pluralism. (See pp.230–232.)

- The costs of racism to the society are high, including loss of human talent and resources, and the costs make life more difficult for the minority group members. (See pp. 234–235.)

- Coping devices used by minorities include five strategies, only one of which addresses the meso and macro causes. These strategies are assimilation, acceptance, avoidance, aggression, and organizing for societal change. (See pp. 235–237.)

- Policies to address problems of prejudice and discrimination range from individual and small-group efforts at the micro level to institutional, societal, and even global social movements. (See pp. 237–239.)

- Affirmative action policies are one approach, but the broad term *affirmative action* includes three different sets of policies that are quite distinct and have different outcomes. (See pp. 239–242.)

Contributing to Our Social World: What Can We Do?

At the Local Level:

- *African American, Arab American, and Native American student associations* on campuses fight bigotry and promote the rights of racial minorities. Identify one of these groups on your campus, attend a meeting, and, if appropriate, volunteer to help with its work.

At the Organizational or Institutional Level:

- *Leadership Conference on Civil and Human Rights* is a national coalition dedicated to combating racism and its effects. It maintains a website that includes a directory of more than 100 local chapters, listed by state (http://www.civilrights.org/about/internships/ and http://www.civilrights.org/career_center). Find a chapter in your area, and explore ways in which you can participate in its programs.

- *Teaching Tolerance* (www.splcenter.org/center/tt/teach.jsp), a program of the Southern Poverty Law Center, has curriculum materials for teaching about diversity and a program for enhancing cross-ethnic cooperation and dialogue in schools. Explore ways in which Teaching Tolerance can be incorporated into your campus curricula.

At the National and Global Levels:

- *The American Indian Movement* (AIM) has worked for many years to bring attention to the plight of Native Americans and to promote the civil/human rights of community members. Visit the AIM website at http://www.aimovement.org and explore ways in which you can learn more about Native Americans and assist in the community's educational and legislative initiatives.

- *Cultural Survival* and *the UN Permanent Forum on Indigenous Issues* (www.cs.org and www.un.org/esa/socdev/unpfii) are concerned with the plight of indigenous peoples. Organizations such as these also provide opportunities for combating racism globally. You should also consider purchasing only coffee and—especially—chocolate that are Fair Trade Certified (packages are clearly marked as such). Among other things, this will ensure that you are not supporting slavery someplace around the globe with your purchases.

- *Amnesty International* (www.amnestyusa.org or www.amnesty.org) is a worldwide movement of people who campaign for internationally recognized human rights. It relies heavily on volunteer workers.

Visit **www.sagepub.com/oswcondensed2e** for online activities, sample tests, and other helpful information. Select "Chapter 8: Race and Ethnic Group Stratification" for chapter-specific activities.

CHAPTER 9

Gender Stratification

She/He—Who Goes First?

Social inequality is especially evident in gender relations, and although in some societies women are treated with deference, they are rarely given first access to positions of significant power or financial reward. While they may hold many work roles, they often carry the load of child care by themselves, causing more role strains. The photos presented here focus on women's roles.

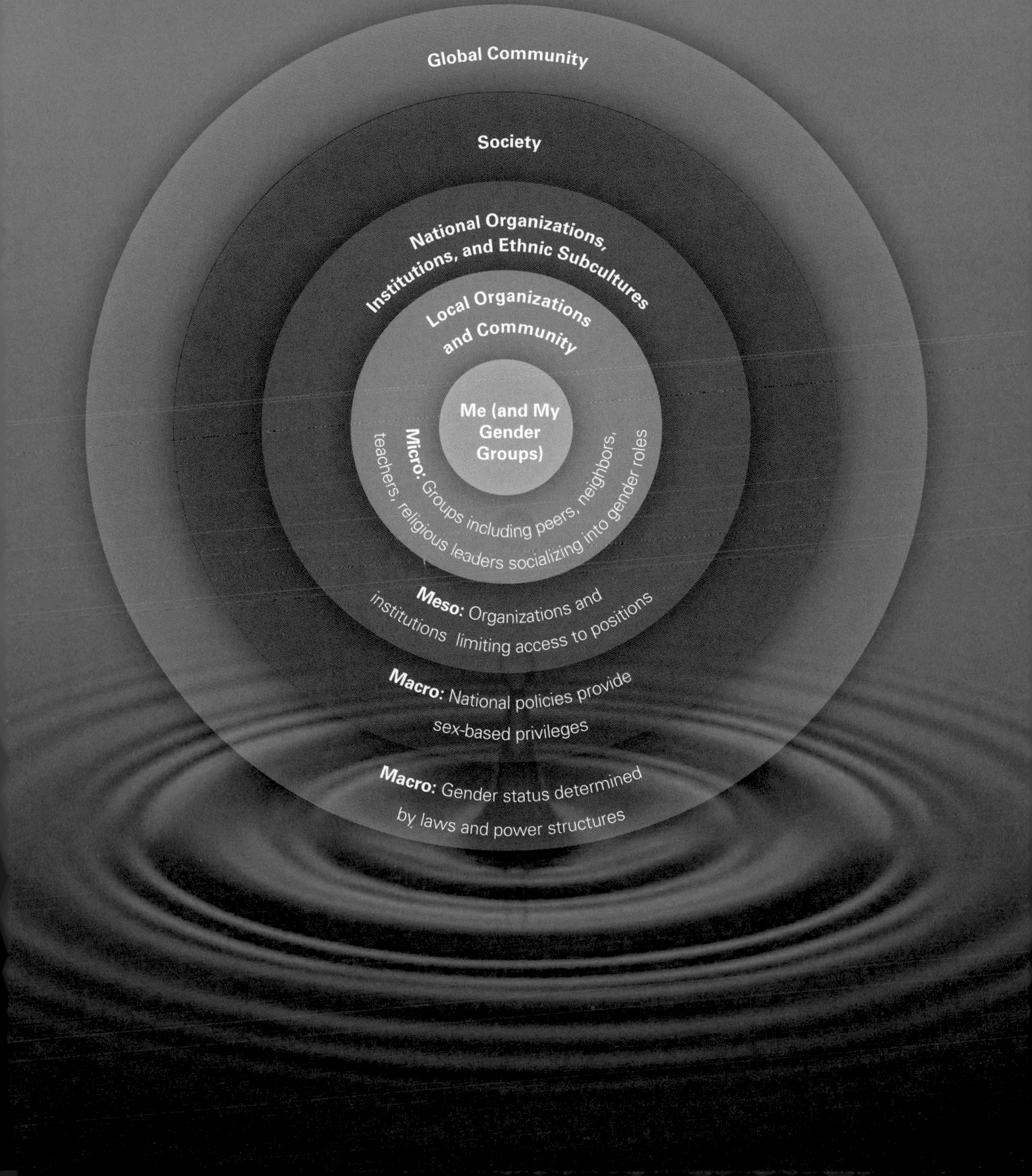

Global Community

Society

National Organizations,
Institutions, and Ethnic Subcultures

Local Organizations
and Community

Me (and My
Gender
Groups)

Micro: Groups including peers, neighbors,
teachers, religious leaders socializing into gender roles

Meso: Organizations and
institutions limiting access to positions

Macro: National policies provide
sex-based privileges

Macro: Gender status determined
by laws and power structures

Jocelyn is now retired, but she is having trouble making ends meet. After training in nursing, including a master's degree, she married and dropped her career to raise her family. The marriage did not work out, and 15 years after her college training she found herself with no credit, two children, little job experience, and mounting expenses. She is a conscientious and hard worker, but with two children and meager child support from their father, she could not put much away for retirement. Nursing does not pay well in her town in the Midwest, but there had been few other career options for females in the early 1960s when she was getting her education. Moreover, she had worked a full-time job and done all the housework for 22 years. Two decades does not build a very large retirement annuity, and she had never been able to buy a very adequate home on her income. If she had been a male with a master's degree, her lifetime earnings would have been more than $700,000 more (U.S. Census Bureau 2008a). Her life chances were clearly affected by the fact that she was female.

Jocelyn's granddaughter Emma will have a range of opportunities that were beyond consideration for her grandma. Ideas about sex, gender, and appropriate roles for men and women not only transform over time; they vary a great deal from one society to the next. Some practices of your own society may seem very strange to women and men in another society. Moreover, gender identities and roles are not stagnant; they change slowly over time, reflecting the economic, political, and social realities of the society. For instance, women in today's India seldom commit *sati* (suicide) on their husband's funeral pyre, but before the practice of *sati* was outlawed, it was a common way to deal with widows who no longer had a means of support (Ahmad 2009; Weitz 1995).

In this chapter, we will explore the concept of gender and the implications of one's sexuality and one's sex, combined with race and class in the stratification system. At the micro level, we will consider gender socialization, and at the meso and macro levels, gender stratification, or placement in the society's stratification system. The costs and consequences of gender stratification will end this chapter.

Sex, Gender, and the Stratification System

Variations around the world show that most roles and identities are not biological but rather socially constructed. In Chapter 7, we discussed factors that stratify individuals into social groups (castes and classes), and in Chapter 8, we discussed the roles race and ethnicity play in stratification. Add the concepts of sex and gender, and we have a more complex picture of how class, race and ethnicity, and gender together influence who we are and our positions in society.

Consider the following examples from societies that illustrate some human social constructions based on sex and gender that may seem unusual to most reading this text. These examples illustrate that gender roles are created by humans to meet needs of their societies. We will then move to more familiar societies.

Men of the Wodaabe society in Niger, Africa, are nomadic cattle herders and traders who would be defined as effeminate by most Western standards because of their behavior patterns. The men are like birds, showing their colorful feathers to attract females. They take great care in doing their hair, applying makeup, and dressing to attract women. They also gossip with each other while sipping their tea. Meanwhile, the women are cooking meals, caring for the children, cleaning, tending to the animals, planting small gardens, and preparing for the next move of this nomadic group (Beckwith 1993; Saharan Vibe 2007).

People in industrial societies might seem unacceptably aggressive and competitive to people of the Arapesh tribe in New Guinea, where gentleness and nonaggression are the rule for both women and men. Yet nearby, women of the Tchambuli people are assertive, businesslike, and the primary economic providers. Men of the Tchambuli exhibit expressive, nurturing, and gossipy behavior. The Mbuti and !Kung peoples of Africa value gender equality in their division of labor and treatment of women and men, and among the Agta of the Philippines, women do the hunting. In West African societies such as the Ashanti and Yoruba kingdoms,

Wodaabe men in Niger (Africa) go to great pains with makeup, hair, and jewelry to ensure that they are highly attractive, a pattern that is thought by many people in North America to be associated with females.

Muslim girls in some parts of the world cover their faces when in public. The display of skin, even in a college classroom, would be immoral to many Muslims. However, in other Muslim countries such coverings would be unusual.

women control much of the market system (Dahlberg 1981; Mead [1935] 1963; Turnbull 1962). Each tradition has evolved over time to meet certain needs of society.

Under the Taliban rule in Afghanistan at the turn of this century, women could not be seen in public without total body covering that meets strict requirements. Anyone not obeying could be stoned to death. If they became ill, women could not be examined by a physician because all doctors are male. Instead, they had to describe their symptoms to a doctor through a screen (Makhmalbaf 2003).

Why did groups in different corners of the globe develop such radically different ways of organizing their gender roles? Certain tasks must be carried out by individuals and organizations in each society for members to survive. Someone must be responsible for raising children, someone must provide people with the basic necessities (food, clothing, shelter), someone needs to lead, someone must defend the society, and someone must help resolve conflicts. One's sex and age are often used to determine who

holds what positions and who carries out what tasks. Each society develops its own way to meet these needs, and this results in variations from one society to the next.

Genetic, Cultural, and Structural Connections—and Divergences

At birth, when doctors say, "It's a . . . ," they are referring to the distinguishing primary characteristics that determine sex—the penis or vagina. **Sex** is generally seen as a term referring to ascribed genetic, anatomical, and hormonal differences between males and females. Right? Partly. Sex is also "a determination made through the application of socially agreed upon biological criteria for classifying persons as females or males" (West and Zimmerman 1987:127). In other words, occasionally this binary male-female categorization by biological criteria is not clear. Occasionally, babies are born with ambiguous genitalia, not fitting the typical definitions of male or female (the *intersexed*). About 1.7% of babies are born with unusual sex chromosomes, internal procreative organs, and external genitalia in a variety of combinations (Fausto-Sterling 2000). These babies often undergo surgeries to "clarify" their sex, with hormonal treatments later in life (Chase 2000). Whether male, female, or intersexed, anatomical differences or chromosomal typing results in cultural attempts to categorize sex. Still, the word *sex* when applied to a person refers largely to elements of one's anatomy.

Great lengths are taken to identify the sex of an infant. Why is this an issue? The reality is that sex constitutes a major organizing principle in most societies. People's roles and statuses and society's expectations guiding their behavior are largely determined by their sex. Our attraction to others is expressed by our sexuality and our sexual identity, with most people categorized as heterosexual (other sex), homosexual (same sex), bisexual (both sexes), or "varied." The term *heteronormative* defines the cultural expectations held in most societies that a "normal" girl or boy will be sexually attracted to and eventually have sex with someone of the other sex (Lorber and Moore 2007). However, the point is that sex is not always a straightforward distinction and is as social as it is biological.

In adolescence, secondary characteristics further distinguish the sexes, with females developing breasts and hips and males developing body hair, muscle mass, and deep voices. Individuals are then expected to adopt the behaviors appropriate to their anatomical features as defined by society. In addition, a few other physical conditions are commonly believed to be sex linked, such as a prevalence of color blindness, baldness, learning disabilities, autism, and hemophilia in males. Yet, some traits that members of society commonly link to sex are actually learned through socialization. There is little evidence, for instance, that emotions, personality traits, or ability to fulfill most social statuses is determined by inborn physical sex differences.

Although the terms *sex, gender,* and *sexuality* are often used interchangeably by the media, it is useful to understand the technical difference. A person's sex—male, female, or other—is a basis for stratification around the globe, used in every society to assign positions and roles to individuals. However, what is defined as normal behavior for a male, a female, or an intersexed person in one society could get one killed in another.

Gender, which is learned and created, refers to socially constructed notions of masculinity and femininity. *Gender identity* is how individuals form their identity using these categories and negotiating the constraints they entail. The examples at the beginning of this section illustrate some differences in how cultures are structured around gender.

These gender meanings profoundly influence the statuses we hold within the social structure and placement in the stratification system (Rothenberg 2007). Individuals are expected to fulfill positions appropriate for their sex category.

Statuses are positions within the structures of society, and roles are expected behaviors within those statuses. **Gender roles**, then, are those commonly assigned tasks or expected behaviors linked to an individual's sex-determined statuses (Lips 2007). Members of each society learn the structural guidelines and positions expected of males and females (West and Zimmerman 1987). Our positions—which affect access to power and resources—are imbedded in institutions at the meso level. However, socially appropriate statuses and roles vary greatly across cultures, with each different culture defining what is right and wrong. There is not some global absolute truth governing gender or gender roles. While both vary across cultures, gender is a learned cultural idea, while gender roles are part of the structural system of the society.

Sexuality refers to how cultures shape the meanings of sexual acts and how we experience our bodies in relation to others. Strange as it may seem, sexuality is also *socially constructed.* A sex act is a "social enterprise," with cultural norms defining what is normal and acceptable in each society, how we should feel, and hidden assumptions about what the act means (Steele 2005). Even what we find attractive is culturally defined. For a period in China, men found tiny feet a sexual turn-on—hence, bound feet in women. In some cultures, legs are the attraction, and in others, men are fascinated by breasts. In the United States and elsewhere, pornography is a moneymaker because of the way it stimulates people, yet as the next "Sociology in Your Social World" indicates, the stimulation itself is variable by gender.

Consider the ideal male body as depicted in popular magazines in contemporary Western cultures: "over 6 feet tall, 180 to 200 pounds, muscular, agile, with straight white teeth, a washboard stomach, six-pack abs, long legs, a full head of hair, a large penis (discreetly shown by a bulge), broad shoulders and chest, strong muscular back, clean shaven, healthy, and slightly tanned if White, or a lightish brown if Black or Hispanic" (Lorber and Moore 2007:114). We grow up learning what is appealing.

In the nineteenth century, medical science began to study sexuality and sexual behavior. Most often, sexuality was defined in binary terms (Colligan 2004), and behaviors falling outside heterosexual boundaries were defined as perverse. In the twentieth century, researchers such as Kinsey et al. (1953) and Masters and Johnson (1966, 1970) conducted studies that uncovered actual sexual behaviors, not just those defined as normal. These studies considered a range of sexual practices, from premarital sex and homosexuality to orgasm and masturbation (Lindsey 2008).

The struggles that individuals have with their sexual identity are reflected in the studies of *transgender:* when intersexed individuals do not fit clearly into female or male sex classifications (Leeder 2004). Transgender refers to "identification as someone who is challenging, questioning, or changing gender from that assigned at birth to a chosen gender—male-to-female, female-to-male, transitioning

Sociology in Your Social World

Sexuality and Pornography: What Turns You On?

By Michael Norris

Sexuality and arousal is complex and variable between men and women. Pornography is often used to titillate the viewers sexually, and it has become big business. Adult videos generate more revenue than Hollywood box office cinema; more people visit porn sites on the Internet than prominent news sites; 70% of those who visit porn sites admit doing so at work; and college students have more legal access to pornography than alcohol. So despite objections to the "thingification" of women (or of men) in pornography, it is widely used. However, there are some seldom understood variations in how people respond to pornographic videos.

Meredith Chivers and her colleagues (Chivers, Seto, and Blanchard 2007) were interested in gender differences in reaction to sexual videos. The researchers found that for women, sexual activity itself was arousing, whereas men tended to be influenced more by the gender of the actors in the videos. Women were aroused by sexual activity regardless of the gender of the participants and even when the actors were non-human primates. Women can apparently be genitally aroused just by cues of sexual activity. On the other hand, heterosexual men were aroused by videos of nude women engaged in nonsexual activities such as exercise. By contrast videos of nude males were no more arousing to heterosexual women than videos of the Himalayan mountain range (Newman 2008).

Chivers and her colleagues (2007) also discovered gender differences in biological arousal and subjective awareness of that arousal. Men were immediately aware of their physiological arousal while women were often not. This gender difference has been verified in previous research, which suggests that biological reactions precede subjective awareness of arousal in women, but not in men. Biological arousal may happen in women without any self-reported arousal at all.

This research adds to a growing body of literature suggesting greater flexibility of women's sexuality in terms of sexual identity, same-sex attraction, and same-sex behavior. Women's greater sexual flexibility may result from a tendency to identify with both male and female targets of sexual activity and greater same-sex emotional attachments than men.

This study also helps explain the fact that for heterosexuals viewing mainstream, commercially available pornography, watching two women having sex is more socially acceptable than watching two men having sex. A popular cultural belief, reinforced by comedians, is that men particularly enjoy "lesbian action" in adult videos. Research suggests that this is not the case. In fact, the adult video industry may include these vignettes because it stimulates women, and this may help to sell their product. Sexuality is complex and involves both biological reactions and socially constructed definitions of sexuality.

Michael Norris is a professor at Wright State University, where he teaches and does research in sociology and criminal justice.

---

between genders, or gender 'queer' (challenging gender norms)" (Lorber and Moore 2007:6). Transgendered individuals are of interest to sociologists because of their life on the boundaries. Due to the pressure to fit in, most transgendered people change themselves, sometimes through surgery, to fit into their chosen gender.

In summary, although the terms *sex, gender,* and *sexuality* are often used interchangeably, they do have distinct meanings. The connections between these social realities are not always as clear as the public thinks. One can be a masculine heterosexual female, a masculine homosexual male, or any of a number of possible combinations. As individuals

continually negotiate the meanings attached to gender and sexuality, they are *doing gender,* a process discussed later in this chapter (West and Zimmerman 1987).

Sex, Gender, and Sexuality: The Micro Level

"It's a boy!" brings varying cultural responses. In many Western countries, that exclamation results in blue blankets, toys associated with males, roughhousing, and gender socialization messages. In some societies, boys are sources for great rejoicing whereas girls may be seen as a burden. In China and India, *female infanticide* (killing of newborn girls) is sometimes practiced in rural areas, in part because of the cost to poor families of raising a girl and the diminished value of girls. In China, the male preference system is exacerbated by the government's edict that most couples may have only one child.

Beginning at birth, each individual passes through many stages. At each, there are messages that reinforce appropriate gender behavior in that society. These gender expectations are inculcated into children from birth by parents, siblings, grandparents, neighbors, peers, and even day care providers. If we fail to respond to the expectations of these significant people in our lives, we may experience negative sanctions: teasing, isolation and exclusion, harsh words, and stigma. To avoid these informal sanctions, children usually learn to conform, at least in their public behavior.

The lifelong process of gender socialization continues once we reach school age and we become more involved in activities separate from our parents. Other people—teachers, religious leaders, coaches—begin to influence us. We are grouped by sex in many of these social settings, and we come to think of ourselves as like *this* group and unlike *that* group: boys versus girls, we versus they. Even if our parents are not highly traditional, we still experience many influences at the micro level to conform to traditional gender notions.

With adulthood, differential treatment and stratification of the sexes take new forms. Men traditionally have more networks and statuses, as well as greater access to resources outside of the home. This has resulted in women having less power because they are more dependent on husbands or fathers for resources. Even spousal abuse is related to imbalance of power in relationships. Lack of connections to the larger social system makes it difficult for women to remove themselves from abusive relationships.

The subtitle of this chapter asks, "Who goes first?" When it comes to the question of who walks through a door first, the answer is that in many Western societies, *she* does—or at least, formal etiquette would suggest this is proper. The strong man steps back and defers to the weaker female, graciously holding the door for her (Walum 1974). Yet, when it comes to who walks through the metaphorical door to the professions, it is the man who goes first. Women are served first at restaurants and at other micro-level settings, but this seems little compensation for the fact that doors are often closed to them at the meso and macro levels of society.

Some scholars argue that language is powerful in shaping the behavior and perceptions of people, as discussed in the chapter on culture. Women often end sentences with tag questions, a pattern that involves ending a declarative statement with a short tag that turns it into a question: "That was a good idea, don't you think?" This pattern may cause male business colleagues to think women are insecure or uncertain about themselves. The women themselves may view it as an invitation to collaboration and dialogue. Yet, a perception of insecurity may prevent a woman from getting the job or the promotion. On the flip side, when women stop using these "softening" devices, they may be perceived by men as strident, harsh, or "bitchy" (Wood and Reich 2006).

Other aspects of language may also be important. The same adverb or adjective, when preceded by a male or female pronoun, can take on very different meanings.

Many women today are in major leadership positions. India's recently elected President Pratibha Patil (left) is the first woman to hold the post in her country, and she won the election with about twice the votes of the opposition candidate. Michelle Bachelet (center), the former president of Chile (now head of UN Women), won almost 54% of the popular vote in 2006. Songul Chapouk, trained engineer, teacher, and women's activist (right in the white coat), and Dr. Raja Habib al-Khuzaai, southern tribal leader (right in the yellow scarf), are members of the 25-member Iraqi National Assembly.

When one says, "He's easy" or "He's loose," it does not generally mean the same thing as when someone says, "She's easy" or "She's loose." Likewise, there are words such as *slut* for women for which there is no equivalent for men. There is no female equivalent for *cuckold,* the term describing a man whose wife is making a fool of him by having an affair. Why is that? To use another example, the word *spinster* is supposed to be the female synonym for *bachelor,* yet it has very different connotations. Even the more newly coined *bachelorette* is not usually used to describe an appealing, perhaps lifelong, role.

Those who invoke the biological argument that women are limited by pregnancy, childbirth, or breast-feeding from participating in public affairs and politics ignore the fact that in most societies these biological roles are time limited and women play a variety of social roles. They also ignore those societies in which males are deeply involved in nurturing activities such as child rearing.

Thinking Sociologically

Some people always write *he* first when writing "he and she." Others sometimes put *she* first. Does language influence how we view gender roles, or is this just fussiness about an insignificant matter? Explain.

Sex, Gender, and Sexuality: The Meso Level

By whatever age is defined as adulthood in our society, we are expected to assume leadership roles and responsibilities in the institutions of society. The roles we play in these institutions often differ depending on our gender (Brettell and Sargent 2005).

In most societies, sex and age stipulate when and how we experience *rites of passage,* rituals and formal processes that acknowledge a change of status. These include any ceremonies or recognitions that admit one to adult duties and privileges. Rites of passage are institutionalized in various ways: religious rituals such as the Jewish male bar mitzvah or female bat mitzvah ceremonies; educational celebrations such as graduation ceremonies, which often involve caps and gowns of gender-specific colors or place females on one side of the room and males on the other; and different ages at which men and women are permitted to marry.

Other institutions also segregate us by sex. Orthodox Jewish synagogues, for example, do not have families seated together. Men sit on one side of the sanctuary, and women on the other. Many institutions, including religious, political, and economic organizations, have historically allowed only males to have leadership roles. Only men are to teach the scriptures to the young among traditional Jews, but few men fill that role in contemporary Christian congregations.

Often, women's reduced access to power in micro-level settings has to do with a lack of power and status in meso-level organizations and institutions. This is why gender roles—structurally established positions defined by a person's female or male characteristics—are important. This is also a reason why policy makers concerned about gender equality have focused so much on inclusion of women in social institutions. For example, microcredit organizations around the world make small financial loans, primarily to women, to start small businesses to support their families. One of these organizations is described in the "Sociology Around the World" on the next page.

Thinking Sociologically

How might women's lack of positions and authority in organizations and institutions in some societies—the meso level—influence females at the micro level? How might it influence their involvements at the macro level?

Sex, Gender, and Sexuality: The Macro Level

Going to school, driving a car, voting, working—people around the world engage in these necessary activities. Yet, in some parts of the world, these activities are forbidden for women. When we turn to the national and global level, we again witness inequality between the sexes that is quite separate from any form of personal prejudice or animosity toward women. Patterns of social action that are imbedded in the entire social system may influence women and men, providing unrecognized privileges or disadvantages (McIntosh 1992).This is called *institutionalized privilege* or *disprivilege.*

In hunter-gatherer and agricultural societies, women increase their power relative to men as they age and as people gain respect for their wisdom. Especially within a clan or a household, women become masters over their domains. Men move from active roles outside the home to passive roles after retirement, whereas women often do the opposite.

The winds of change are influencing the roles of women in many parts of the world, as is seen in governing structures. Although women are still denied the right to vote in a few countries, voting is a right in most. In the United States, women have voted for a little more than 80 years. Yet, in its entire history, only 39 women

Sociology Around the World

Microcredits and Empowerment of Women

The 30 village women gather regularly to discuss issues of health, crops, their herds, the predicted rains, goals for their children, and how to make ends meet. They are from a subsistence farming village in southern Niger on the edge of the Sahara desert. Recently, a microcredit organization was established with a small grant of $1,500 from abroad. With training from CARE International, an international nongovernmental organization, the women selected a board of directors to oversee the loans. Groups of five or six women have joined together to explain their projects to the board and request small loans. Each woman is responsible for paying back a small amount on the loan each week, once the project is established and bringing in money.

The women are enthusiastic. In the past they had no funds, and banks charged enormous interest rates on loans. A loan of between $20 and $50 from the microcredit organization is a tremendous sum considering that for many of these women, it is equivalent to six months' earnings. Strong social norms are instituted to encourage repayment. Women who repay their loans promptly often decide who is eligible for future loans. Participation in the program often encourages women and grants them economic and social capital otherwise unavailable to them.

With the new possibilities for their lives, they have big plans: For instance, one group plans to buy a press to make peanut oil, a staple for cooking in the region. Currently, people pay a great deal for oil imported from Nigeria. Another group will buy baby lambs, fatten them, and sell them for future festivals at a great profit. Yet another group plans to set up a small bakery. Women are also discussing the possibility of making local craft products to sell to foreign fair trade organizations such as *Ten Thousand Villages* (a fair trade organization that markets products made by villagers and returns the profits to the villagers).

Some economists and social policy makers claim that grassroots organizations such as microcredits may be the way out of poverty for millions of poor families and that women are motivated to be small entrepreneurs to help support their families and buy education and health care for their children (KBYU-TV 2005). Indeed, in 2006, Muhammad Yunus, who founded Grameen Bank—a microcredit lender for the very poor—received the Nobel Prize for Peace.

Microcredit lenders build significant economic and social capital for their participants, but critics suggest that there is an underresearched downside to microlending. As they see it, despite its success, the solution is a micro-level attempt to address a macro-level problem. Macro economists such as Linda Mayoux (2002, 2008) question whether or not the program will address the gender inequalities in the developing nations they target. Moreover, such loans may shift relations between husbands and wives, leaving men in a less powerful position within marriages. Most economists agree that microlending works best alongside macro-level initiatives seeking to address similar problems.

have served in the U.S. Senate. As of 2011, the number is at an all-time high, with 17 out of 100 senators being women. Still, the United States is far behind many other countries in women's representation in governing bodies. In fact, the global average for women in national parliaments is 18.4%, so the United States is below the average of female representation in national governments. The nation with the highest percentage of women in national parliament or congress is Rwanda (sub-Saharan Africa) with 56% women—one of only two countries to exceed 50%. Canada is 40th among nations, and the United States ranks 90th—tied with Bosnia and Herzegovina and San Marino and well below Mozambique, Argentina, Afghanistan, Ethiopia, and Iraq (Inter-Parliamentary Union 2011). (See Table 9.1.)

Thinking Sociologically

What factors might affect the ranking of countries in Table 9.1 according to the number of women involved in government? What factors might explain why the United States and Canada rank so poorly in representation of women in their governments?

Several factors have been especially effective in increasing women's positions in national parliaments in Africa: the existence of a matriarchal culture (where women may

Table 9.1 Women in National Governments (Selected Countries), 2011

Rank	Country	Lower/Single House: % Women	Upper House/ Senate: % Women
1	Rwanda	56.3	34.6
2	Andorra	45.0	—
3	Sweden	45.0	—
4	South Africa	44.5	29.6
5	Cuba	43.2	—
6	Iceland	42.9	—
7	Finland	42.5	—
8	Norway	39.6	—
9	Belgium	39.3	36.6
9	Netherlands	39.3	36.0
11	Mozambique	39.2	—
12	Angola	38.6	—
12	Costa Rica	38.6	—
14	Argentina	38.5	35.2
15	Denmark	38.0	—
16	Spain	36.6	32.3
17	Tanzania	36.0	—
18	Uganda	34.9	—
19	New Zealand	33.6	—
20	Nepal	33.2	—
21	Germany	32.8	21.7
23	Ecuador	32.3	—
31	Ethiopia	27.8	16.3
32	Afghanistan	27.7	27.5
34	Mexico	26.2	22.7
36	Iraq	25.2	—
40	Australia	24.7	35.5
40	Canada	24.7	35.9
51	Pakistan	22.2	17.0
55	United Kingdom	22.0	20.1
60	China	21.3	—
60	Italy	21.3	18.4
67	Nicaragua	20.7	—
76	France	18.9	21.9
79	Bangladesh	18.6	—
81	Indonesia	18.0	—
87	Turkmenistan	16.8	—
90	United States of America	16.7	17.0
91	Albania	16.4	—

Source: Inter-Parliamentary Union 2011. Reprinted with permission of the Inter-Parliamentary Union.

Note: To examine the involvement of women in other countries or to see even more recent figures, go to http://www.ipu.org/wmn-e/classif.htm.

have increased authority and power in decision making), political systems that stress proportional representation, and the adoption of gender quotas for government positions (Yoon 2004, 2008). Yet, democratization of governmental systems is sometimes linked to a *decrease* in representation by women, a sad reality for those committed to establishing democracy around the world (Yoon 2001). Globally, women's access to power and prestige is highly variable, with African and Northern European countries having a position of leadership when it comes to gender equity in government.

Cross-cultural analyses confirm that gender roles either evolve over centuries or are transformed by sweeping reform laws such as voting rights. The fact that women in China generally work outside the home whereas women in some Muslim societies hardly venture from their homes is due to differences in cultural norms about gender roles that are dictated by governments or tradition and learned through the socialization process.

Gender Socialization: Micro- and Meso-Level Analyses

Socialization into gender is the process by which people learn the cultural norms, attitudes, and behaviors appropriate to their gender. That is, they learn how to think and act as boys or girls, women or men. Socialization reinforces the "proper" gender behaviors and punishes the improper behaviors. In many societies, traits of gentleness, passivity, and dependence are associated with femininity, whereas boldness, aggression, strength, and independence are identified with masculinity. For instance, in most Western societies, aggression in women is considered unfeminine, if not inappropriate or disturbed. Likewise, the gentle, unassertive male is often looked on with scorn or pity, stigmatized as a "wimp." Expectations related to these stereotypes are rigid in many societies (Pollack 1999). (See the next "Sociology in Your Social World.")

Thinking Sociologically

Read *The Boy Code* on page 256. Is there evidence of the boy code when you observe your friends and relatives? What is the impact of the boy code? Is there a similar code for girls?

Stages in Gender Socialization

Bounce that rough-and-tumble baby boy and cuddle that precious, delicate little girl. Thus begins gender socialization,

Sociology in Your Social World

The Boy Code

Boys and girls begin to conform to gender expectations once they are old enough to understand that their sex is rather permanent, that boys are not capable of becoming "mommies." They become even more conscious of adhering to norms of others in their gender category.

The Old Boy network in American society favors adult men over women through a system of networks. This system actually starts with "the boy code," the rules about boys' proper behavior. Young boys learn "the code" from parents, siblings, peers, teachers, and society in general. They are praised for adhering to the code and punished for violating its dictates. William Pollack (1999) writes that boys learn several stereotyped behavior models exemplifying the boy code:

1. "The sturdy oak": Men should be stoic, stable, and independent; a man never shows weakness.

2. "Give 'em hell": From athletic coaches and movie heroes, the consistent theme is extreme daring, bravado, and attraction to violence.

3. "The 'big wheel'": Men and boys should achieve status, dominance, and power; they should avoid shame, wear the mask of coolness, and act as though everything is under control.

4. "No sissy stuff": Boys are discouraged from expressing feelings or urges perceived as feminine—dependence, warmth, empathy.

The boy code is ingrained in society; by five or six years of age, boys are less likely than girls to express hurt or distress. They have learned to be ashamed of showing feelings and of being weak. This gender straitjacket, according to Pollack, causes boys to conceal feelings in order to fit in and be accepted and loved. As a result, some boys, especially in adolescence, become silent, covering any vulnerability and masking their true feelings. This affects boys' relationships, performance in school, and ability to connect with others. It also causes young males to put on what Jackson Katz (2006) calls the "tough guise," when young men and boys emphasize aggression and violence to display masculinity.

Pollack (1999) suggests that we can help boys reconnect to nongendered norms by

1. giving some undivided attention each day just listening to boys;

2. encouraging a range of emotions;

3. avoiding language that taunts, teases, or shames;

4. looking behind the veneer of "coolness" for signs of problems;

5. expressing love and empathy;

6. dispelling the "sturdy oak" image; and

7. advocating a broad, inclusive model of masculinity.

With the women's movement and shifts in gender expectations have come new patterns of male behavior. Some men are forming more supportive and less competitive relationships with other men, and there are likely to be continued changes in and broadening of "appropriate" behavior for men (Kimmel and Messner 2009; Nardi 1992; Stoltenberg 1993).

starting at birth and taking place through a series of life stages, discussed in Chapter 4 on socialization. Examples from infancy and childhood show how socialization into gender roles takes place.

Infancy

Learning how to carry out gender roles begins at birth. Parents in the United States describe their newborn daughters as soft, delicate, fine-featured, little, pretty, cute, and resembling their mothers. They depict their sons as strong, firm, alert, and well coordinated (Lindsey 2008; Rubin 1974). Clothing, room decor, and toys also reflect notions of gender. In Spain, parents and grandparents dress babies and their carriages in pink or blue depending on gender, proudly showing off the little ones to friends as they promenade in the evenings. Although gender stereotypes have declined in recent years, they continue to affect the way we handle and treat male and female infants (Karraker 1995).

Childhood

Once they are out of infancy, research shows that many boys are encouraged to be more independent and exploratory whereas girls are protected from situations that might prove harmful. More pressure is put on boys to behave in "gender appropriate" ways. Boys are socialized into "the boy code" that provides rigid guidelines (see "Sociology in Your Social World" on the previous page). Cross-cultural studies show boys often get more attention than girls because of their behavior, with an emphasis on achievement, autonomy, and aggression for boys (Kimmel and Messner 2009).

Stereotypes for girls in a majority of societies label feminine behaviors as soft, nonaggressive, and noncompetitive and favor diminutive women. Consider that boys act out their aggressive feelings, but girls are socialized to be nice, nurturing, and not aggressive. In a study about how school-girls express aggression, Simmons (2002) finds that girls express "relational aggression," aggression that affects girls' social contacts indirectly through rumors, name-calling, giggling, ignoring, backbiting, exclusion, and manipulation of victims. Friendship and needing to belong are the weapons, rather than sticks and stones. This form of bullying is subtle and hard to detect, but it can have long-lasting effects (Simmons 2002).

Names for children also reflect stereotypes about gender. Boys are more often given strong, hard names that end in consonants. The top 10 boys' names in 2009 included Jacob, Ethan, Michael, Alexander, William, Joshua, Daniel, Jayden, Noah, and Anthony. Girls are more likely to be given soft, pretty names with vowel endings such as most of the top 10: Isabella, Emma, Olivia, Sophia, Ava, Emily, Madison, Abigail, Chloe, and Mia (Social Security Online 2010).

Alternatively, girls may be given feminized versions of boys' names—Roberta, Jessica, Josephine, Nicole, Michelle, or Donna. Sometimes, boys' names are given to girls without first feminizing them. Names such as Lynn, Stacey, Tracey, Faye, Dana, Jody, Lindsay, Robin, and Carmen used to be names exclusively for men, but within a decade or two after they were applied to girls, parents stopped using them for boys. So a common name for males may for a time be given to either sex, but then it is given to girls only. The pattern rarely goes the other direction. Once feminized, the names seem to have become unacceptable for boys (Lieberson, Dumais, and Bauman 2000).

In the early childhood years, children become aware of their own gender identity. As they reach school age, they learn that their sex is permanent, and they begin to categorize behaviors that are appropriate for their sex. As children are rewarded for performing proper gender roles, these roles are reinforced. That reinforcement solidifies gender roles, setting the stage for gender-related interactions, behaviors, and choices in later life.

Meso-Level Agents of Gender Socialization

Clues to proper gender roles surround children in materials produced by corporations (books, toys, games), in mass media images, in educational settings, and in religious organizations and beliefs. In Chapter 4 on socialization, we learned about agents of socialization. Those agents play a major role in teaching children proper gender roles. The following examples demonstrate how organizations and institutions in our society teach and reinforce gender assumptions and roles.

Corporations

Corporations have produced materials that help socialize children into proper conduct. Publishers, for example, produce books that present images of expected gender behavior. Language and pictures in preschool picture books, elementary children's books, and school textbooks are steeped in gender role messages reflecting society's expectations and stereotypes. In a classic study of award-winning children's books from the United States that have sold more than 3 million copies, Weitzman et al. (1972) made several observations: (a) Males appear more often in stories as central characters; (b) activities of male and female characters in books differ, with boys playing active roles and girls being passive or simply helping brothers, fathers, or husbands; and (c) adult women are pictured as more passive and dependent, whereas males are depicted as carrying out a range of activities and jobs.

Although more recent books show some expansion in the roles book characters play, studies confirm a continuing pattern of gender role segregation in children's books (Anderson and Hamilton 2005; Diekman and Murmen

Girls and boys quickly pick up messages—from parents, other children, and the media—about what kinds of toys are appropriate for someone of their sex. Many toys, like Barbie dolls and construction toys, have very explicit gender messages.

2004). Studies show that males of any species are usually portrayed as adventurous, brave, competent, clever, and fun, whereas female counterparts may be depicted as incompetent, fearful, and dependent on others (Purcell and Stewart 1990). Moreover, even though authors and publishers have begun producing books presenting unbiased gender roles and strong girls and women, the tens of thousands of older, classic books in public and school libraries mean that a parent or child picking a book off of the shelf is still likely to select a book that has old stereotypical views of boys and girls.

Producers of toys and games also contribute to traditional messages about gender. Store-bought toys fill rooms in homes of children in the Western world. Each toy or game prepares children for future gender roles. Choices ranging from college major and occupational choice to activities that depend on visual-spatial and mathematical abilities appear to be affected by these early choices and childhood learning experiences (Tavris and Wade 1984).

Boys have more experience manipulating blocks, Tinkertoys, LEGO bricks, and Erector Sets—toys paralleling masculinized activities outside the home in the public domain, from constructing and building trades to military roles and sports. Girls prepare for domestic roles with toys relating to domestic activities. Barbie dolls stress physical appearance, consumerism, and glamour. Only a few Barbies are in occupational roles.

In 2009, Mattel produced Barbie's online dream house, a virtual house that can be decorated and furnished. Barbie's house was designed with girls as the target market. A complementary target boy program features Hot Wheels cars

with car races and use of virtual tools to customize cars. These models differed in more than simple appearance. The life lessons learned from these computer games reinforce gender stereotypes (Snider 2009).

Consider the example of Dungeons & Dragons, a popular game primarily among adolescent boys in Europe, Japan, and North America. The participants role-play their way through scenarios full of demons and dragons, using a vast array of dungeons, weapons, and magical spells. The boys develop characters that they impersonate throughout the game as they negotiate, bargain, create, imitate, and develop a variety of other social or cognitive skills. They must calculate complex mathematical formulae and use logic and imagination—all skills that will aid them in coping with the adult public world. A similar card game, Magic: The Gathering (MTG or Magic), and the online game, World of Warcraft (WoW), result in similar outcomes for boys. Although WoW is played with many online players, the individual players can sit in front of their computers alone, interacting only online. Although more girls are taking up gaming activities, the packaging of this product makes it clear it is an activity for boys.

Mass Media

Have you ever noticed the media coverage of female celebrities who have made poor choices? The men who get into trouble are slapped on the hand. "Boys will be boys," after all. However, actresses such as Paris Hilton, Lindsay Lohan, and Britney Spears experience intense curiosity, scrutiny, and even hostility. Is this because of their class, their gender, or their nontraditional roles and nonconforming behavior? Whatever variables affect the reactions to these women, they are subjects of mass media frenzy.

Mass media comes in many forms—magazines, ads, films, music videos, Internet sites—and are major agents of socialization into gender roles. Young men and women, desiring to fit in, are influenced by messages from the media. For instance, the epidemic of steroid use among boys and dieting among girls, driven by ads, is a health concern in the United States (Taub and McLorg 2010). Yet, manufacturers advertise such products with promises to remake teens into more attractive people.

Some action films produced within the past decade include adventurous and competent girls and women, helping counter media images. *Hanna* is about a young girl defending herself and her father. Hermione in *Harry Potter* is intelligent and creative. Eowyn, a noblewoman in *Lord of the Rings: The Return of the King,* dresses as a soldier and kills the Nazgul, fulfilling the prophecy. Queen of Naboo and Senator Amidala in *Star Wars* and Lyra Belacqua in *The Golden Compass* are strong, brave, and intelligent (and beautiful). Maggie Fitzgerald, boxer in *Million Dollar Baby,* is a gutsy and determined female fighter. Lara Croft in *Tomb Raider* holds her own against evil. However, videos depicting highly competent females are few, and most of those few women are unrealistically thin and highly attractive with large breasts.

Television is another powerful socializing agent. By the time they start school, typical English, Canadian, and U.S. children will have spent more time in front of a television than they will in classrooms in the coming 12 years of school, a behavior that contributes to obesity in the United States (Randerson 2008). Television presents a simple, stereotyped view of life, from advertisements to situation comedies to soap operas. Women in soap operas and ads, especially those working outside the home, are often depicted as having problems in carrying out their role responsibilities (Benokraitis and Feagin 1995; Boston Women's Health Book Collective 2006). Even the extraordinary powers of superheroes on Saturday morning television depict the female characters as having gender-stereotyped skills such as superintuition. Notice the next time you are in

Olsen twins Mary-Kate and Ashley (left) pose together in front of their new star on the Walk of Fame in Hollywood. Like most glamorous stars, they must conform to the image of very thin, shapely femininity.

a video game room that the fighting characters are typically male and often in armor. When fighting women do appear, they are usually clad in skin-revealing bikini-style attire—odd clothing in which to do battle!

Films and television series in the United States seldom feature average-size or older women (although more variety of ages and body types is seen in the British Broadcasting Corporation productions). Movie stars are attractive and thin, presenting an often unattainable model for young women (Taub and McLorg 2010). Women who are quite large are almost entirely depicted as comic figures in U.S. television.

How do social scientists know that television affects gender role socialization? Studies have shown that the more television children watch, the more gender stereotypes they hold. From cartoons to advertisements, television in many countries provides enticing images of a world in which youth is glorified, age is scorned, and female and male roles are stereotyped and/or unattainable (Kilbourne 2000; Tuchman 1996). However, research has also found that these media images have a much greater impact on white girls than on African American teens (Milkie 1999).

Thinking Sociologically

Think of recent mass media examples that you have seen. How do they depict women and men? How might these meso-level depictions affect young men and women at the micro level?

Educational Systems

Even centers of learning bolster sex-specific expectations and limitations. Educational systems are socialization agents of children through textbooks, classroom activities, playground games, and teachers' attitudes. For example, boys are encouraged to join competitive team sports and girls to support them (Gilligan 1982; Kramer 2010). These simulate hierarchical adult roles of boss/secretary and physician/nurse. Furthermore, the team sports that the boys learn teach them strategic thinking—a critical skill that involves anticipating the moves of the opponent and countering with one's own strategy. This is a very useful skill in the business world and is not explicitly taught in other places in the school curriculum. Girls' games rarely teach this skill.

Children's experiences in grade and middle school reinforce boundaries of "us" and "them" in classroom seating and activities, in the lunchroom, and in playground activities, as girls and boys are seated, lined up, and given assignments by sex (Sadker and Sadker 2005). Those who go outside the boundaries, especially boys, are ridiculed by peers and sometimes teachers, reinforcing stereotypes and separate gender role socialization (Sadker and Sadker 2005; Thorne 1993).

Part of the issue of male-female inequality in schools is tied to the issue of popularity, which seems to have less to do with being liked than with being known. If everyone knows a person's name, she or he is popular (Eder, Evans, and Parker 1995). At the middle school level, there are more ways that boys can become known. Even when there are both boys' and girls' basketball teams, many spectators come to the boys' contests and very few to the girls' games. Thus, few people know the female athletes' names. In fact, a far more visible position is cheerleader—standing on the sidelines cheering for the boys—because those girls are at least visible (Eder et al. 1995). The other major way in which girls are visible or known is physical appearance, a major standard of popularity and esteem (Eder et al. 1995). Generally, there are far more visible positions for boys than for girls in middle school.

Title IX of the U.S. Education Amendments of 1972 was a major legislative attempt to level the educational and sports playing field. Passed to bar gender discrimination in schools receiving federal funds, this legislation mandates equal opportunity for participation in school-sponsored programs (Lindsey 2011). The law has reduced or eliminated blatant discrimination in areas ranging from admissions to counseling and housing. However, the biggest impact of Title IX legislation has been in athletics.

Partly aided by Title IX, the number of women's athletic programs and scholarship opportunities grew in the late 1990s but has slowed since 2000 and still lags behind men's levels. From 1995 to 2005, the number of women in college sports at 738 National Collegiate Athletic Association schools surveyed grew from 26,000 to 205,492, while the number of men was 291,797 (Rosen 2007). Some men's sports have been cut back, but others have grown, resulting in a steady number of opportunities for men in sports. One third of high school women participate in sports compared to 45% of males (National Center for Education Statistics 2004). Yet participation in athletics, especially team sports, can foster skills in teamwork, thinking about strategy, and anticipating counteractions by a competitor, all of which are useful skills in business and government. Overall, 79% of the public supports Title IX and what it has accomplished (Women's Sports Foundation 2002).

Religious Beliefs

In some cases religious beliefs serve as agents of socialization by defining, reinforcing, and perpetuating gender role stereotypes and cultural beliefs. Religious teachings provide explanations of all aspects of life, including proper male/female roles. The three major monotheistic religions—Christianity, Islam, and Judaism—are traditionally patriarchal, stressing

High school and college athletics are much less reliable paths for women to become known on campus since so few people come to the games. Even this college game with a winning team has sparse attendance (top photo). By contrast, women in very sexy outfits are highly visible and can even become local celebrities when they perform as cheerleaders in front of 80,000 fans at men's sporting events (bottom photo).

separate female and male spheres (Kramer 2010). The following are examples of some of these traditional role expectations and the status of women in various religions.

Some interpretations of the Adam and Eve creation story in the Hebrew Bible (the Old Testament in the Christian Bible) state that because man was created first, men are superior. Because Eve, created from the rib of man, was a sinner, her sins keep women forever in an inferior, second-class position. Thus, in some branches of these religions, women are restricted in their roles within family and religious organizations. They cannot be priests in Catholic churches and cannot vote on business matters in some religious organizations. However, recent work by feminist scholars is challenging the notion of patriarchy in Judaic and Christian history, pointing out that women may have played a much broader role in religious development than is often recognized (Hunter College Women's Studies Collective 2005). Even *Yahweh*—the name for God in the Hebrew Bible—had both male and female connotations, and when God was referred to as a source of wisdom, feminine pronouns and references were used (Borg 1994). Increasingly, denominations are granting women greater roles in the religious hierarchies of major religious groups.

Women in Judaism lived for centuries in a patriarchal system where men read, taught, and legislated (Lindsey 2008). Today, three of the five main branches of Judaism allow women equal participation, illustrating that some religious practices change over time. However, Hasidic and Orthodox Jews have a division of labor between men and women following old laws, with designated gender roles.

Some Christian teachings have treated women as second-class citizens, even in the eyes of God. For this reason, some Christian denominations have excluded women from a variety of leadership roles and told them they must be subservient to their husbands. Other Christians point to the admonition by Saint Paul that, theologically speaking, "there is neither . . . male nor female, for you are all one in Christ Jesus" (Galatians 3:28). This suggests that women and men are not spiritually different.

Traditional Hindu religion painted women as seductresses, strongly erotic, and a threat to male spirituality and asceticism. To protect men from this threat, women were kept totally covered in thick garments and veils and seen only by men in their immediate families. Today, Hinduism comes in many forms, most of which honor the domestic sphere of life—mothers, wives, and homemakers—while accepting women in public roles (Lindsey 2011).

Traditional Islamic beliefs also portrayed female sexuality as dangerous to men, although many women in Islamic societies today are full participants in the public and private sphere. The Quran (also spelled Qur'an or Koran), the Muslim sacred scripture, includes a statement that men are superior to women because of the qualities God has given men. Hammurabi's Code, written in the Middle East between 2067 and 2025 BCE, is the earliest recorded legal system. The laws about women's status were written to distinguish between decent women, belonging to one man, and indecent or public women. Some aspects of these traditional beliefs have carried over to present times. Sharia law, or strict Islamic law, has been adopted in several Islamic

countries and has been used to punish women accused of violating its rules. Recent cases include a woman in Nigeria who had a child out of wedlock and another woman in Pakistan who was a rape victim. Both women were sentenced to death by stoning, but their sentences have not been carried out because of protests from within and outside the countries. These women suffer social humiliation, degradation, and potential death, but they also confirm the social expectations for others who might stray from the laws (Mydans 2002).

Women in fundamentalist Muslim societies such as Algeria, Iran, Syria, and Saudi Arabia are separated from men (except for fathers and brothers) in work and worship. They generally remain completely covered (Ward and Edelstein 2009). Today, some women point out that they wear the veil for modesty, for cosmetic purposes, or to protect themselves from stares of men. Others claim that the veil represents oppression and subservience, showing that women must keep themselves in submissive positions. Religious laws often provide the justifications to keep women servile, and public shaming and threat of severe punishments reinforce the laws.

Meso-level religious systems influence how different societies interpret proper gender roles and how sometimes these belief systems change with new interpretations of scriptures. (Further discussion of the complex relationship between religion and gender appears in Chapter 11.) All of these meso-level agents of socialization reinforce "appropriate" gender roles in each society.

Thinking Sociologically

What are some books, toys, games, television shows, school experiences, and religious teachings that influenced your gender role socialization? In what ways did they do so?

Gender Stratification: Meso- and Macro-Level Processes

"The **glass ceiling** keeps women from reaching the highest levels of corporate and public responsibility, and the 'sticky floor' keeps the vast majority of the world's women stuck in low-paid jobs" (Hunter College Women's Studies Collective 2005:393). Men, on the other hand, experience the "glass escalator," especially in traditionally female occupations. Even if they do not seek to climb in the organizational hierarchy, occupational social forces push them up the job ladder into higher echelons. However, minority men may not experience the same escalator effect as white men (Williams 1992; Wingfield 2009). Women around the world do two thirds of the work, receive 10% of the world's income, and own 1% of the world's means of production (Global Citizen Corps 2010). They make up more than 40% of the world's paid workforce but hold only about 20% of managerial jobs, and for those, they are often compensated at lower pay than their male counterparts. Only 5% of the top corporate jobs are held by women. However, companies with women in leadership positions do realize high profits (CNNMoney.com 2010; Hunter College Women's Studies Collective 2005). The next "Engaging Sociology" provides an exercise to think about how our ideas may subtly maintain the glass ceiling.

Women and Men at Work: Gendered Organizations

How can I do it all—marriage, children, career, social life? This is a question that women in our college classes ask. They already anticipate a delicate balancing act. Work has been central to the definition of masculinity in U.S. society, and for the past half-century, women have been joining the workforce in greater numbers (Kramer 2010). Today, women's proportion of the workforce is almost equal to men's in the United States and many other Western countries. Working is necessary for many women, especially single mothers, to support their families, and many women want to work and use their education and skills. Among countries of the Global North, Sweden has the highest percentage of women in the labor force (82%) and the lowest percentage of homemakers. Yet, even in Sweden, with its parental leave and other family-friendly policies, women feel pressures of work and family responsibilities (Eshleman and Bulcroft 2006). Dual-career marriages raise questions about child rearing, power relations, and other factors in juggling work and family.

Every workplace has a gendered configuration: ratios of female to male workers and matters of which workers are in the supervisory statuses and how positions are distributed. This, in turn, affects our experiences in the workplace. Consider the example of mothers who are breast-feeding their babies. Must they quit their jobs or alter their family schedules if the workplace does not provide a space for breastfeeding? Some workplaces accommodate family needs, but others do not. Men do not face the same issues because most corporations assume the average worker is male.

Research shows that workers are more satisfied when the sex composition of their work group and the distribution of men and women in power are balanced (Britton 2000). Feminists propose ideas to minimize gender differences in organizations—that is, to "degender" organizations so that all members have equal opportunities (Britton 2000).

Thinking Sociologically

If corporate structures were reversed so that women structured and organized the workplace, how might the workplace environment change?

Institutionalized Gender Discrimination

Gender stratification at the meso level—like race and ethnic stratification—can occur quite independently of any overt prejudice or ill will by others. It becomes part of the social system, and we are not even conscious of it, especially if we are one of the privileged members of society. Most of what has been discussed so far is de jure discrimination, done deliberately and justified with laws and ideological beliefs about women's inability to carry out certain tasks. The more subtle process of de facto discrimination (which is not intended) also needs attention. When inequality is woven into the web of the macro-level social structure and becomes taken for granted, it is called institutional discrimination. It can include intentional actions such as bank policies requiring single women, compared with single men, to have three times as much money up front for a house down payment to receive a homeowner's loan, but it can also include unintentional actions with consequences that disadvantage women (Feagin and Feagin 1986, 2010).

Recall from the previous chapter that side-effect discrimination involves practices in one institution that are linked to practices in another institution. The practices in the first institution have an effect in the second one that is not anticipated or even recognized by most people in the society. For example, if roles of women in family life are determined by rigid gender expectations, as research shows, then women find it more difficult to devote themselves to gaining job promotions. In addition, as long as little girls learn through socialization to use their voices and to hold their bodies and gesture in ways that communicate deference, employers assume a lack of the self-confidence necessary for major leadership roles. If women are paid less despite the same levels of education (see Table 9.2), they are less likely to have access to expensive health care or to be able to afford a $20,000 down payment for a house unless they are married. This makes women dependent on men in a way that most men are not dependent on women.

A factor affecting differences in income is the type of academic degree that men and women receive (engineering rather than education, for example). However, even when these differences are factored in, men still make considerably more on average than women with identical levels of experience and training.

Engaging Sociology

Masculinity and Femininity in Your Social World

1. Mark each characteristic with an "M" or an "F" depending on whether you think it is generally defined by society as a masculine or a feminine characteristic.

   achiever

   aggressive

   analytical

   caring

   confident

   deferential (defers to others; yields with courtesy)

   devious

   dynamic

   intuitive

   loving

   manipulative

   nurturing

   organized

   passive

   a planner

   powerful

   relationship-oriented (makes decisions based on how others will *feel*)

   rule oriented (makes decisions based on *abstract procedural rules*)

   sensitive

   strong

2. Next, mark an "X" just to the right of 10 characteristics that you think are the essential qualities for a leadership position in a complex organization (business, government, etc.). You might want to ask 20 of your acquaintances to do this and then add up the scores for "masculinity," "femininity," and "leadership trait."

3. Do you (and your acquaintances) tend to view leadership as having the same traits as those marked "masculine" or "feminine"? What are the implications of this for the "glass ceiling"?

4. How might correlations between the traits of leadership and gender notions help to explain the data on income in Table 9.2?

Table 9.2 U.S. Income by Educational Level and Sex (in Dollars)

Education	Men	Women
Not a high school graduate	24,831	14,521
High school graduate	36,753	24,329
College graduate (bachelor's)	72,868	44,078
Master's degree	88,450	54,517
Doctorate	116,574	70,898
Professional degree	147,518	87,723

Source: U.S. Census Bureau 2011f.

Little girls learn to use their voices and to hold their bodies and gesture in ways that communicate deference. This little girl does not look very powerful or confident. When women tilt their heads—either forward or to one side—they also look like they lack confidence, and this hurts their chances of promotion in the corporate world.

Men often get defensive and angry when people talk about sexism in society because they feel they are being attacked or asked to correct injustices of the past. However, the empirical reality is that the playing field is not level for men and women. Most men do not do anything to intentionally harm women, and they may not feel prejudiced toward women, but sexism operates so that men are given privileges they never asked for and may not even recognize.

Past-in-present discrimination refers to practices from the past that may no longer be allowed but that continue to affect people today. As we discussed in the previous section, men and women have different pay with similar levels of education. Note that a difference in the typical number of years women have spent in their professional fields is a small part of the difference in income levels. In other words, there are fewer women senior law partners in part because women who are now in their 50s or older were often not admitted to professional graduate programs in the 1960s and '70s. A policy of the past still influences women's positions and incomes today.

Let us consider another example of the past influencing the present. At an appliance industry in the Midwest investigated by one of the authors, there is a sequence of jobs one must hold to be promoted up the line to foreman. This requirement ensures that the foreman understands the many aspects of production at the plant. One of the jobs involves working in a room with heavy equipment that cuts through and bends metal sheets. The machine is extremely powerful and could easily cut off a leg or hand if the operator is not careful. Because of the danger, the engineers designed the equipment so it would not operate unless three levers were activated at the same time. One lever was triggered by stepping on a pedal on the floor. The other two required reaching out with one's hands so that one's body was extended. When one was spread-eagled to activate all three levers, there was no way one could possibly have a part of one's body near the blades. It was brilliant engineering. There was one unanticipated problem: The hand-activated levers were 5 feet, 10 inches off the ground and 5 feet apart. Few women had the height and arm span to run this machine, and therefore, no women had yet made it through the sequence of positions to the higher-paying position of foreman. The equipment cost millions of dollars, so it was not likely to be replaced. Neither the engineers who designed the machine nor the upper-level managers who established the sequence of jobs to become foreman had deliberately tried to exclude women. Indeed, they were perplexed when they looked at their employee figures and saw so few women moving up through the ranks. The cause of women's disadvantage was not mean-spirited men but features of the system that had unintended consequences. The barriers women face, then, are not just matters of socialization or other micro-level social processes. The nature of sexism is often subtle and pervasive in the society, operating at the meso and macro levels as institutional discrimination.

Gender Differences in Internet Use

One specific form of institutionalized difference in access to resources at the global level has to do with the Internet. Knowledge of events in the world, job skills, awareness of job openings, networks that extend beyond national borders, and other resources are available through the Internet. Women like the human connections created by the Internet, whereas men tend to like the experience and financial options it offers. Although women are catching up to men in usage in some Global North countries, there is a larger world digital divide in gender than in other categories (Drori 2006; Fallows 2005). The difference in Internet use by women as a percentage of all users varies significantly by country: Women are roughly 45% of Internet users in Sweden and Denmark, a bit more than 40% in Israel, and about 30% in China, 17% in Senegal, and 14% in Ethiopia. In the Arab world, women account for only 4% of the online population. In Kenya, a 1999 survey in two provinces found that 99% of the women had never heard of the Internet. Several other countries are close to equal in usage by women: Mexico at 46%, South

Africa and Thailand each at 49% (Drori 2006). Even in these countries, women's usage has climbed to that level only in the past few years.

However, the story is different in the United States, where women have caught up with or outpaced their male age and ethnic group peers. The 67% of the U.S. population that goes online includes 66% of all women and 68% of men. Between 18 and 29 years of age, more women use the Internet than men—86% compared to 80%. However, in the 65 and older cohort, more men (34%) than women (21%) are online. In addition, more African American women (60%) than men (50%) are web savvy. Men tend to use the Internet more often and for longer periods, and have more high-speed connections (Fallows 2005). The direction of U.S. usage parallels that of other Global North countries and may predict the future in other countries.

In Chapter 7, we explored global differences in Internet usage and found that the largest variance was based on wealth of the country and access to technology. The same principle holds for women's usage in the Global North and South. Many African countries may have only half of 1% of the population online, but in none of the African countries cited by Drori (2006) do women have more than 15% of that tiny share. So, of the women in most African countries, fewer than 7 out of 10,000 have used the Internet. Clearly, the majority of these women are "out of the digital loop." Yet competency in use of the Internet has become an extremely important resource in the contemporary global economy. Lack of access and savvy is a handicap.

Internet-related gender stratification exhibits itself not just in terms of *whether* a person uses the Internet but also in terms of *how* a person uses the Internet. Even within industrialized countries such as the United States, women tend to use mostly e-mail services, with the intent of keeping up with family and friends (Shade 2004). For men, the Internet is used to gather information and to exchange ideas and facts relevant to professional activities. Thus, "women are using the Internet to reinforce their private lives and men are using the Internet for engaging in the public sphere" (Shade 2004:63). The difference reflects differences in the professional positions men and women hold. For men, the Internet is enhancing their careers. This is far less true for women. However, things do change very rapidly in the world of the Internet.

Thinking Sociologically

Ask several people from different generations and different genders how they use computer technology and the Internet and how they learned these skills. What do you conclude?

In some countries, there is a huge gender digital divide. In others, nearly as many women as men use computers. However, research shows that women tend to use computers in different ways. These women in Iraq are learning computer skills.

Gender Stratification: Micro-to Macro-Level Theories

In recent years, some biologists and psychologists have considered whether there are innate differences in the makeup of women and men. For instance, males produce more testosterone, a hormone found to be correlated with aggression. Research shows that in many situations, males tend to be more aggressive and concerned with dominance, whether the behavior is biologically programmed or learned or both. Other traits, such as nurturance, empathy, or altruism, show no clear gender difference (Fausto-Sterling 1992; Pinker 2002).

Although biological and psychological factors are part of the difference between females and males, our focus here is on the major contribution that social factors make in social statuses of males and females in human society. This section explores social theories that explain gender differences.

Symbolic Interaction Theory: Micro-Level Analysis

Symbolic interactionists look at gender as socially constructed. Sex is the biological reality of different "plumbing" in our bodies, and interactionists are interested in how

Some cultures suggest that women are helpless and need the door held for them, and women's stylish attire—such as many forms of shoes and dresses—actually does make them more vulnerable and helpless in certain situations. In these decisions and behaviors, men and women are "doing gender."

those physical differences come to be symbols, resulting in different social rights and rewards. This chapter's discussions of micro-level social processes have pointed out that the meaning connected to one's sex produces notions of masculinity and femininity. The symbolic interaction perspective has been forceful in insisting that notions of proper gender behavior are not intrinsically related to a person's sex. The bottom line is that gender is a socially created or constructed idea (Mason-Schrock 1996).

Traditional notions of gender are hard to change. Admitting confusion over proper masculine and feminine roles creates anxiety and even anomie in a society. People want guidelines. Thus, it is easier to adhere to traditional notions of gender that are reinforced by religious dogmas, making those ideas appear sacred, absolute, and beyond human change. Absolute answers are comforting to those who find change disconcerting. Others believe the male prerogatives and privileges of the past were established by men to protect their rights. Any change in concepts of gender or of roles assigned to males and females will be hard to

bring about precisely because they are rooted in the meaning system and status and power structures in the social world.

Symbolic interaction—more than any other theory—stresses the idea of human agency, the notion that humans not only are influenced by the society in which they live but actively help create it (Charon 2007; Hewitt 2007). In a study of elementary children in classrooms and especially on playgrounds, Barrie Thorne (1993) found that while teachers influenced the children, the children themselves were active participants in creating the student culture that guided their play. As children played with one another, Thorne noticed the ways in which they created words, nicknames, distinctions between one another, and new forms of interaction. This is a very important point: Humans do not just passively adopt cultural notions about gender; they *do gender.* They create it as they behave and interact with others in ways that define "normal" male or female conduct (West and Zimmerman 1987). "Doing gender" is an everyday, recurring, and routine occurrence. It is a constant ongoing process that defines each situation, takes place in organizations and between individuals, and becomes part of institutional arrangements. Social movements such as civil rights and the women's movement challenge these arrangements and can bring about change. Understanding the process of doing gender helps us understand why we think and act as we do.

When children are in an ambiguous situation, they may spontaneously define their sex as the most relevant trait about themselves or others. Indeed, even when children act as if gender matters, they are helping make it a reality for those around them. This process helps make the notion of gender more concrete and real to the other children. As the next child adopts the "definition of reality" from the first, acting as if sex is more important than hair color or eye color or earlobe attachment to the cheek, this makes gender the most prominent characteristic in the mind of the next child. Yet each child, in a sense, could choose to ignore gender and decide that something else, like nationality, is more important. The same principle applies to adults. When a person "chooses" to recognize gender as a critical distinction between two individuals or two groups, that person is "doing gender" (Hewitt 2007; O'Brien 2006). Through interaction, people do gender, and although this process begins at a micro level, it has implications all the way to the global level.

Thinking Sociologically

How do you *do gender*? How did you learn these patterns of behavior? Are they automatic responses, or do you think about "who opens the door"?

Structural-Functionalist and Conflict Theories: Meso- and Macro-Level Analyses

Structural-Functionalist Theory

From the structural-functionalist perspective, each sex has a role to play in the interdependent groups and institutions of society. Some early theorists argued that men and women carry out different roles that have developed since early human history. Social relationships and practices that have proven successful in the survival of a group are likely to continue and be reinforced by society's norms and laws. Thus, relationships between women and men that are believed to support survival are maintained. In traditional hunter-gatherer, horticultural, and pastoral societies, for instance, the division of labor is based on sex and age. Social roles are clearly laid out, indicating who performs which everyday survival tasks. The females often take on the primary tasks of child care, gardening, food preparation, and other duties near the home. Men do tasks that require movements farther from home, such as hunting or fishing.

As societies industrialize, roles and relationships change due to structural changes in society. Durkheim ([1893] 1947) described a gradual move from traditional societies held together by *mechanical solidarity* (the glue that holds society together through shared beliefs, values, and traditions) to modern societies that hold together due to *organic solidarity*. The glue that holds modern society together is based less on common values and more on division of labor and the interdependence of statuses within the social world. According to early functionalists, gender division of labor exists in modern societies because it is efficient and useful to have different but complementary male and female roles. They believed this accomplishes essential tasks and maintains societal stability (Lindsey 2011).

More recent structural-functionalist theorists argue that two complementary types of roles are needed in the modern industrial and postindustrial family. The female plays the expressive role through childbearing, nursing, and caring for family members in the home. The male carries out the instrumental role by working outside the home to support the family (Anderson 1994; Parsons and Bales 1953). Although this pattern was relevant during the Industrial Revolution, it currently characterizes only a small percentage of families in Western industrial and postindustrial societies and less than 10% of families in the United States (Aulette 2002). In reality, gender segregation has seldom been total because in most cultures, women's work has combined their labor in the public sphere—that is, outside the home—with their work in the private sphere—inside the home (Lopez-Garza 2002). Poor minority women in countries around the world must often work in low-paying service roles in the public sphere and carry the major burden for roles in the private sphere. Gender analysis through a structural-functionalist perspective stresses efficiencies that are believed to be gained by specialization of tasks (Waite and Gallagher 2000).

Conflict Theory

Conflict theorists view males as the haves—controlling the majority of power positions and most wealth—and females as the have-nots. Women have less access to power and have historically depended on males for survival. This is the case even though they raise the next generation of workers and consumers, provide unpaid domestic labor, and ensure a pool of available, cheap labor during times of crisis, such as war. By keeping women in subordinate roles, males control the means of production and protect their privileged status.

A classical conflict explanation of gender stratification is found in the writings of Karl Marx's colleague, Friedrich Engels ([1884] 1942). In traditional societies, where size and strength were essential for survival, men were often dominant, but women's roles were respected as important and necessary to the survival of the group. Men hunted, engaged in warfare, and protected women. Over time, male physical control was transformed into control by ideology, by the dominant belief system itself. Capitalism strengthened

By contrast to the clothing and gestures that create vulnerability in women, men are encouraged to use gestures that communicate strength and self-assurance, and their clothing and shoes allow them to defend themselves or flee danger.

male dominance by making more wealth available to men and their sons. Women became dependent on men, and their roles were transformed into "taking care of the home" (Engels [1884] 1942).

Ideologies based on traditional beliefs and values have continued to be used to justify the social structure of male domination and subjugation of women. It is in the interest of the dominant group, in this case men, to maintain their position of privilege. Conflict theorists believe it is unlikely that those in power by virtue of sex, race, class, or political or religious ideology will voluntarily give up their positions as long as they are benefiting from them. By keeping women in traditional gender roles, men maintain control over institutions and resources (Collins 1971).

Feminist Theory

Feminist theorists agree with Marx and Engels that gender stratification is based on power struggles, not biology. Yet, some feminist theorists argue that Marx and Engels failed to include a key variable in women's oppression: patriarchy. Patriarchy involves a few men dominating and holding authority over all others, including women, children, and less powerful men (Arrighi 2000; Lindsey 2011). According to feminist theory, women will continue to be oppressed by men until patriarchy is eliminated.

A distinguishing characteristic of most feminist theory is that it actively advocates change in the social order, whereas many other theories we have discussed try only to explain the social world (Anderson 2006; Lorber 1998). There is a range of feminist theories. However, all feminist theories argue for bringing about a new and equal ordering of gender relationships to eliminate the patriarchy and sexism of current gender stratification systems (Kramer 2010).

Feminist theorists try to understand the causes of women's lower status and seek ways to change systems to provide educational and work opportunities, to improve the standard of living, and to give women control over their bodies and reproduction. Feminist theorists also feel that little change will occur until group consciousness is raised so that women understand the system that limits their options (Sapiro 2003). In addition to awareness, women need the networks that open occupational doors for many men.

As societies become technologically advanced and need an educated workforce, women of all social classes and ethnic groups around the world are likely to gain more equal roles. Women are entering institutions of higher education in record numbers, and evidence indicates they are needed in the world economic system and the changing labor force of most countries. Societies in which women are not integrated into the economic sphere generally lag behind other countries. Feminist theorists examine these global and national patterns, but they also note the role of patriarchy in interpersonal situations—such as domestic violence.

Men who play cards, games, or sports together or who join men's-only clubs develop networks that enhance their power and their ability to "close deals." When women are not part of the same networks, they are denied the same insider privileges. Even if it is not intentional, this works against women, people of color, and the laboring classes.

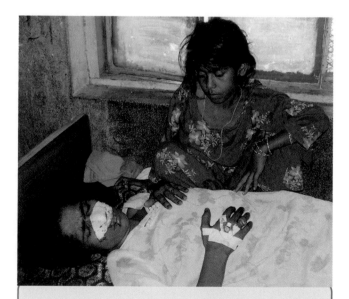

When women have less power at the meso level, they may also be more vulnerable at home. This girl attends to her injured mother at a hospital in Hyderabad, Pakistan, after the woman's husband chopped off her nose with an axe and broke all her teeth. According to the Pakistan Institute of Medical Sciences, more than 90% of married women report being severely abused when husbands are dissatisfied with their cooking or cleaning or when the women give birth to a girl instead of a boy or are unable to bear a child at all.

Violence against women perpetuates gender stratification, as is evident in the intimate environment of many homes. Because men have more power in the larger society, they often have more resources within the household as well (side-effect discrimination). Women are often dependent on the man of the house for his resources, meaning they are willing to yield on many decisions. Power differences in the meso- and macro-level social systems also contribute to power differentials and vulnerability of women in micro-level settings. In addition, women who lack stable employment have fewer options when considering whether to leave an abusive relationship. Although there are risks of staying with a violent partner, many factors enter into a woman's decision to stay or leave (Scott, London, and Myers 2002). The following "Sociology in Your Social World" discusses one form of violence around the world that is perpetuated predominantly against women: rape.

Sociology in Your Social World

Rape and the Victims of Rape

For many women around the world, rape is the most feared act of violence and the ultimate humiliation. Rape is a sexual act but closely tied to macho behavior. Rape is often a power play to intimidate, hurt, and dominate women ("Male Dominance Causes Rape" 2008; Sanday and Goodenough 1990) or the act of a hypersexual man who feels entitled to sex (Felson 2002). Some societies are largely free from rape whereas others are prone toward rape. What is the difference?

Rape provides an example of the impact of cultural practices, beliefs, and stereotypes. When a society is relatively tolerant of interpersonal violence, holds beliefs in male dominance, and strongly incorporates ways to separate women and men, rape is more common ("Male Dominance Causes Rape" 2008; Sanday 1981; Sanday and Goodenough 1990; Shaw and Lee 2005). Rape is also more common when gender roles and identities are changing and norms about interaction between women and men are unclear. What does this say, then, about the United States, where one in four college women will be raped or survive a rape attempt during their college years (Burn 2005; One in Four, Inc. n.d.) and where someone is sexually assaulted or raped every two minutes, averaging more than 2,000 rapes daily and 272,350 rapes or sexual assaults annually? These figures are probably much lower than reality because estimates are that only 1 in 10 rapes in the United States are reported (U.S. Department of Justice 2006).

An alarming problem is rape on college campuses in the United States. Researchers report that "nearly one of every four women on college campuses has experienced sexual violence" (Shaw and Lee 2005:424).

Women report being sexually assaulted whereas few males define their behavior as assault. This discrepancy points to the stereotypes and misunderstandings that can occur because of different beliefs and attitudes. Rape causes deep and lasting problems for women victims as well as for men accused of rape because of "misreading" women's signals. Some college men define gang rape as a form of male bonding. To them, it is no big deal. The woman is just the object and instrument. Her identity is immaterial (Martin and Hummer 1989; Sanday and Goodenough 1990).

Men who hold more traditional gender roles view rape very differently than women and less traditional men. They tend to attribute more responsibility to the female victim of a rape, believe sex rather than power is the motivation for rape, and look less favorably on women who have been raped. In a strange twist of logic, one study showed that some men actually believe that women want to be forced into having involuntary sex (Szymanski et al. 1993). Researchers show that societies with widespread gender stratification report more gendered violence (Sanday 1996, 2007).

Many citizens and politicians see rape as an individualized, personal act, whereas social scientists tend to see it as a structural problem that stems from negative stereotypes of women, subservient positions of women in society, and patriarchal systems of power. The rape culture in the war-torn parts of Sudan and the Congo are examples. A number of sociologists and anthropologists believe that rape will not be substantially reduced unless our macho definitions of masculinity are changed (McEvoy and Brookings 2008; Sanday and Goodenough 1990).

In summary, feminist analysis finds gender patterns imbedded in social institutions of family, education, religion, politics, economics, and health care. If the societal system is ruled by men, the interdependent institutions are likely to reflect and support this system. Feminist theory helps us understand how patriarchy at the meso and macro levels can influence patriarchy at the micro level and vice versa.

Thinking Sociologically

Why do women and men stay in abusive relationships? If these behaviors are hurtful or destructive, what might be done to change the situation, or what policies might be enacted to address the problems?

The Interaction of Class, Race, and Gender

Zouina is Algerian, but she was born in France to her immigrant parents. She lived with them in a poor immigrant suburb of Paris until she was forced to return to Algeria for an arranged marriage to a man who already had one wife. That marriage ended, and she returned to her "home" in France. Since she returned to France, Zouina has been working wherever she can find work. The high unemployment and social and ethnic discrimination, especially against foreign women, makes life difficult. An estimated 70,000 forced (and illegal) polygamous marriages occurred in 2006 in immigrant communities of France (Levy 2008).

The situation is complex. Muslim women from Tunisia, Morocco, and Algeria, former French colonies, living in crowded slum communities outside Paris face discrimination in the workplace and in French society more generally (da Silva 2004). Expected to be both good Muslim women and good family coproviders—which necessitates working in French society—they face ridicule when they wear their *hijabs* (coverings) to school or to work. However, they encounter derision in their community if they do not wear them. They are caught between two cultures.

Because of high unemployment in the immigrant communities, many youth roam the streets. Gang rapes by North African youth against young women have been on the rise in France and elsewhere abroad. These rapes mean that North African women are faced with rejection and disdain in both the immigrant community and their original North African communities. With the conflicting messages due to their ethnic differences (race), their poor status (class), and their gender, these women attempt to construct their identities under grueling circumstances (Killian 2006).

Feminist theory seeks explanations for the conditions women face (Walby 1990). One current trend in feminist

interpretation, illustrated in the previous example, is to view the social world as an intersection of class, race, and gender (Anderson and Collins 2006; Glenn 1999; Smith 1999). In this way, one can look at the variety of ways that many common citizens are controlled by those who have a monopoly on power and privilege.

The reality is that some women are quite privileged and wealthy. Not all women live in poverty. However, even privileged women often have less power than their husbands or other men in their lives. It is also true that some women in the world are privileged because of their race or ethnicity (Rothenberg 2007). In many respects, they have more in common with men of their own ethnicity or race than they do with women of less esteemed groups, and they may choose to identify with those statuses that enhance their privilege.

Consider another example of the intersection of race, class, and gender. Income for men and women varies significantly depending on ethnicity within the United States, and this means that minority women are even more disprivileged. However, even when ethnicity is held constant, women get paid less than men (see Table 9.3).

These examples illustrate that race, class, and gender have cross-cutting lines that may affect one's status in the society. Sexual orientation, age, nationality, and other factors may have the effect of either diminishing or increasing minority status of specific women, and theorists are paying increasing attention to these intersections (Rothenberg 2007). Chapter 8 discussed the fact that race and class lines may be either cross-cutting or parallel in a given society and the patterns affect ethnic relations. In the case of gender, there are always crosscutting lines with race and social class. Gender always affects one's prestige and privilege within that class or ethnic group. Thus, these three variables—race, class, and gender—should be considered simultaneously (Hossfeld 2006; Kirk and Okazawa-Rey 2007).

Table 9.3   Median Earnings (in Dollars) in Past 12 Months of Workers by Race, Sex, and Percent of Men's Earnings by Women

Race and Ethnicity	Male	Female	Women's % of Men's Earnings
Asian American	51,179	40,664	79.5
White	47,113	35,542	75.4
African American	35,652	31,035	87.1
Hispanic (any race)	29,239	25,454	87.1

Source: Bishaw and Semega 2008.

Gender, Homosexuality, and Minority Status

Sex and sexuality are not binary (male-female) concepts but embrace a broader range of combinations of masculine, feminine, heterosexual, homosexual, and other variations. This range is normal in human sexuality. It is societal expectations developed over time that impose categories on human sexuality.

One group of people who face particular discrimination in some countries is homosexuals. Homosexuality has always existed and has been accepted and even required at some times and places and rejected or outlawed in others. In some cases, homosexuals have been placed in a separate sexual category with special roles. Some societies ignore its existence; a few consider it a psychological illness or form of depraved immorality; and a very few even consider it a crime (as in some Muslim societies today and in most states in the United States during much of the twentieth century). In each case, the government or dominant religious group determines the status of homosexuals. The reality is that deviation from a society's gender norms, such as attraction to a member of the same sex, may cause one to experience minority status.

Lesbians are women attracted to other women. Because these women do not follow traditional gendered expectations of femininity in many societies, they often experience prejudice and discrimination. Women's status is typically based on their relationship with men, and in most societies, they are economically dependent on men. Therefore, lesbians—who support themselves and each other—may go against norms of societies and are in some instances perceived as dangerous, unnatural, or a threat to men's power (Burn 2005; Ward and Edelstein 2009). One author describes her daily struggles dealing with a society that does not fully accept who she is and that puts individuals in boxes with labels, even when those boxes do not fit the individual (Lucal 1999b).

Sexism affects the lives of gay men as well. *Homophobia* (intense fear and hatred of homosexuality and homosexuals, female or male) is highly correlated with and perhaps a cause of traditional notions of gender and gender roles (Lehne 1995; Pharr 1997; Shaw and Lee 2005). Some homosexuals deviate from traditional notions of masculinity and femininity and therefore from significant norms of many societies. This may result in hostile reactions and stigma from the dominant group. Indeed, homosexual epitaphs are often used to reinforce gender conformity and to intimidate anyone who would dare to be different from the norm (Lehne 1995). Still, the whole issue of homophobia focuses on prejudice held and transmitted by individuals.

Heterosexism is the notion that the social system itself reinforces heterosexuality and marginalizes anyone who does not conform to this norm. Heterosexism focuses on social processes that define homosexuality as deviant and legitimize heterosexuality as the only normal lifestyle (Oswald 2000, 2001). In short, heterosexism operates at the meso and macro levels of society, through privileges such as rights to health care and jobs granted to people who are heterosexual and denied to those who are not.

In the mass media inclusion of gay characters and themes in films such as *Brokeback Mountain, Innocent,* and *The Kids Are All Right,* plus television features such as *Ellen* and *Will and Grace,* shows popular culture's evolution in acceptance of portraying homosexuality. Gallup polls reveal that public acceptance of gays in the military grew to 80% by 2003 and approval of gays as elementary school teachers increased to 61% by 2003. Today about 59% of the U.S. public supports legalization of homosexual relations between consenting adults, up from 43% in 1977 (O'Brien and Westen 2007). Over 90% of the U.S. population now supports equal opportunity for homosexuals on the job (Johnson 2005). Despite these indications of greater openness in attitudes, in 2010, it was still legal in 29 states to fire someone based on his or her sexuality (Topix.com 2010).

Homosexuality has been an issue in recent political elections in the United States. As of 2007, almost all states had "Defense of Marriage Acts" or constitutional amendments prohibiting same-sex marriage. This has spawned debates about family life and whether homosexuals should be allowed to marry, to adopt or have children, or to have the same rights as heterosexuals. In short, many institutions in the social world have been influenced by this debate over sexuality and gender.

The notion of allowing gays and lesbians the right to a legal marriage has been extremely controversial in the United States, and more than 30 states have passed legislation defining marriage as between a man and a woman, and some have approved amendments to their constitutions to ensure that same-sex marriages will not happen in those states (Newman and Grauerholz 2002). As of June 2011, marriage licenses are available to same-sex couples in six states (Massachusetts, Connecticut, Iowa, Vermont, New Hampshire, and New York) and the District of Columbia. Vermont was the first to pass this policy through the legislature, followed later in 2009 by Maine and New Hampshire. (A referendum of the populace in Maine subsequently repealed that action by the legislature.) In the other states the law was implemented by court decisions. Similar court decisions have been made in Hawaii, Alaska, and California, but in each case a state referendum has overruled the courts. In California, the constitutionality of the referendum was then challenged. One reason the *lesbigay* community is so intent on having same-sex marriage is that same-sex partnerships are "insufficiently institutionalized" (Cherlin 1978; Stewart 2007), making them somewhat less stable and creating ambiguity about roles and rights. They have been denied the role of spouse and the rights that brings such as visits in the hospital. Note how this roots personal relationships in meso-level institutions.

Thinking Sociologically

What are ways in which the label one is given might influence one's behavior and sense of self, especially if the label is homosexuality?

The point here is not to argue for or against same-sex marriage but to note that homosexuals do not have many of the rights that heterosexuals have. The discrepancy is based on sexuality characteristics. Heterosexuals in the United States have a variety of rights—ranging from insurance coverage and inheritance rights for lifelong partners to jointly acquired property, hospital visitation rights as family, rights to claim the body of a deceased partner, and rights to have the deceased prepared for burial or cremation. The U.S. federal government confers 1,138 rights to heterosexuals that they normally take for granted, but often these rights do not extend to same-sex partners in lifelong relationships. Most states also bestow more than 200 specific rights to persons who "marry," but because most states do not allow same-sex marriages, homosexuals do not have these same rights (U.S. General Accounting Office 2004). In Canada and many European countries, citizens do have a right to same-sex marriage, and this has reduced the number of discrepancies in the rights of homosexuals and heterosexuals.

Liberals tend to see this issue as one of prejudice against persons for a characteristic that is unchangeable. They assume that homosexuality is an inborn trait—present from birth. Conservatives argue that homosexuality is a choice that has moral implications. They tend to see homosexuality as a behavior that is acquired through socialization; therefore, acceptance of homosexuality will likely increase the numbers of people who engage in this lifestyle. Social conservatives see political liberal notions about homosexuality and gender roles as a threat to society and family, a threat to the moral social order, whereas liberals often feel that sexuality within a committed relationship is not a moral issue. The real moral issue for them is bigotry—lack of tolerance of other lifestyles. Thus, the two groups have socially constructed the meaning of morality along different lines.

While some religions do not accept homosexuality, some mainstream Christian denominations have developed policies supportive of lesbian, gay, and bisexual persons in local churches: More Light (Presbyterian), Open and Affirming (United Church of Christ and Christian Church), Reconciling in Christ (Lutheran), and Reconciling Ministries (Methodist).

These designations apply to approximately 300 congregations that wish to be known as gay and lesbian friendly (www.mlp.org). A new denomination, the Universal Fellowship of Metropolitan Community Churches, affirms homosexuality as a legitimate lifestyle for Christians (Metropolitan Community Churches 2010; Rodriguez and Ouellette 2000). It has more than 300 local churches in the United States, with 17 in the conservative state of Texas. There are also congregations on six continents (Metropolitan Community Churches 2010). Since religious communities usually sanctify or sacralize the social norms and moral codes of society, these policies have been highly controversial.

Some ministers continue to perform same-sex marriage ceremonies as a protest against a policy they think is immoral. They may take a "don't ask, don't tell" position with their bishops. One retired minister estimated that even in his very conservative Midwestern state, roughly 50 homosexual holy unions are performed by Methodist pastors each year. He stated,

> Back in the 1950s, ministers could lose their ordination and ministerial privileges if they married someone in a church wedding who had been divorced. . . . If a divorced person wanted to marry a second time, a civil wedding was supposedly the only option. Many of us in the pastorate thought that the policy was inhumane. . . . By the 1960s, if the church purged all of us who had done this, the Methodist church would have lost more than half of its ministers. In another 20 years, the same thing will be true with holy unions. If you have two people in your congregation who really care for each other and want to ritually affirm their lifelong commitment, how can we not honor that commitment and caring? (Roberts and Yamane 2012:289)

If this minister is right, the change is very much in process. By contrast, for many in conservative churches, this issue is profoundly rooted in the moral and religious tradition of the church and in scripture, and it will never be acceptable. For them, this is a defining bedrock issue of the faith.

In short, notions about gender and sexuality—part of the culture—are influencing the structure of society at the meso level, including institutions such as family, politics, and religion. Gender is very personal and private but also a public issue with macro-level implications.

Thinking Sociologically

The concept of *homophobia* focuses on micro-level processes, whereas the notion of *heterosexism* is attentive to meso and macro forces. Which of these concepts reveals the most about issues faced by the lesbian and gay community? Why?

Costs and Consequences of Gender Stratification

In rapidly changing modern societies, role confusion abounds. Men hesitate to offer help to women, wondering if gallantry will be appreciated or scorned. Women are torn between traditional family roles on the one hand, and working to support the family and fulfill career goals on the other. As illustrated in the following examples, sex-based stratification limits individual development and causes problems in education, health, work, and other parts of the social world.

Psychological and Social Consequences: Micro-Level Implications

For both women and men, rigid gender stereotypes can be very constraining. Individuals who hold highly sex-typed attitudes feel compelled to behave in stereotypic ways, ways that are consistent with the pictures they have in their heads of proper gender behavior (Basow 1992, 2000). However, individuals who do not identify strongly with masculine or feminine gender types tend to be more flexible in thoughts and behavior, score higher on intelligence tests, have greater spatial ability, and have higher levels of creativity. Because they allow themselves a wider range of behaviors, they have more varied abilities and experiences and become more tolerant of others' behaviors. High masculinity in males sets up rigid standards for male behavior and has been correlated with anxiety, guilt, and neuroses, whereas less rigid masculine expectations are associated with emotional stability, sensitivity, warmth, and enthusiasm (Bellisari 1990). Rigid stereotypes and resulting sexism affect everyone and curtail our activities, behaviors, and perspectives.

Superwoman Image

The "superwoman syndrome," a pattern by which women assume multiple roles and try to do all well, takes its toll (Faludi 1993). The resulting strain can contribute to depression and certain health problems such as headaches, nervousness, and insomnia (Wood 2008). Women in many societies are expected to be beautiful, youthful, and sexually interesting and interested, while at the same time preparing the food, caring for the children, keeping a clean and orderly home, and sometimes bringing in money to help support the family. Some are also expected to be competent and successful in their careers. Multiple, sometimes contradictory, expectations for women cause stress and even serious psychological problems.

Women who work outside the home are often expected to keep up with the domestic tasks as well. Evidence indicates that when women enter the labor force, there are not parallel changes or redistribution of responsibilities in family life. On the contrary, working generally leads to an increase in what is expected of women.

Beauty Image

Beautiful images jump out at us from billboards, magazine covers, and TV and movie screens. Some of these images are unattainable by most women because they have been created through surgeries and eating disorders—and even the use of airbrushing on photographs. Disorders including anorexia and bulimia nervosa relate to societal expectations of the ideal woman's appearance. About 1 out of every 100 U.S. women (7 million women, compared to 1 million men) suffers from these severe eating disorders, and the numbers have been growing since the 1970s. This dangerous ailment occurs most frequently in females between 12 and 18 years of age, caused by distortions in body image

The beauty-image obsession in North America is far more of an issue with white Anglos than African Americans or Latinas, but especially with young white women it results in disorders relative to eating. Obsessed with thinness, some young women consider a small salad a complete dinner.

that are brought on by what the women, especially white middle- and upper-middle-class women, see as societal images of beauty, often images created in magazines that are unattainable in real life (Kilbourne 1999; Orobello 2008; Taub and McLorg 2010).

Thinking Sociologically

If females are encouraged to spend money, energy, and attention on how they look to ensure that they are physically appealing to others, how might this affect their view of themselves?

Men also suffer psychological costs of stereotyping. Some men also have eating disorders, and fitness magazines picture the "perfect" male body and advertise exercise equipment and steroids. In addition, men die earlier than women, in part due to environmental, psychological, and social factors. Problems in developed countries such as heart disease, stroke, cirrhosis, cancers, accidents, and suicides are linked in part to the male role that dictates that males should appear tough, objective, ambitious, unsentimental, and unemotional—traits that require men to assume great responsibility and suppress their feelings (Leit, Gray, and Pope 2002).

Societal Costs and Consequences: Meso- and Macro-Level Implications

Gender stratification creates costs for societies around the world in a number of ways. Poor educational achievement of female children leads to the loss of human talents and resources of half of the population. Lack of health care coverage affects not only the women but also their children. Discrimination and violence against women, whether physical or emotional, have consequences for all institutions in a society.

Consider how the ratio of women to men in an occupational field affects the prestige of the occupation. As more men enter predominantly female fields such as nursing and library science, the fields gain higher occupational prestige, and salaries tend to increase. It seems that men take their gender privilege with them when they enter female professions (Kramer 2010). However, the evidence indicates that as women enter male professions such as law, the status tends to become lower, making women's chances of improving their position in the stratification system limited.

Gender stratification has often meant loss of the talents and brainpower of women, and that is a serious loss to modern postindustrial societies because human capital—the resources of the human population—is central to social prosperity in this type of system. Yet, resistance to expanding women's professional roles is often strong. For example, Japanese women make up close to 50% of the workforce, but only 10.1% hold managerial positions (Fackler 2007). Although their education levels are generally high, Japanese women earn wages that are about 66% of their male counterparts' wages (Japan Institute for Labour Policy and Training 2009; Kumlin 2006). Breaking through the glass ceiling continues to be a barrier for individual women but also a challenge for entire societies that could benefit from abilities never fully maximized (French 2003).

Changing Gender Stratification and Social Policy

Women in factories around the globe face dangerous conditions and low pay. Sweatshops exist because poor women have few other job options to support their families and because people in rich countries want to buy the cheap products that perpetuate the multinational corporate system. For example, workers in maquiladoras (foreign-owned assembly plants) in Mexican border towns work for 2,000 multinational companies. The towns of Matamoros and Reynosa alone have 247 maquiladoras employing 174,000 people who are paid a minimum wage of $4.22 per *day*, compared to $7.25 per *hour* in the United States (CorpWatch 1999). Many workers earn more than the minimum, and they do have health care guaranteed by the government and a retirement benefit. Still, typical workers earn a fraction of what people on the northern side of the border receive, despite the fact that Mexico has a 48-hour work week. Two thirds of the workers at these maquiladoras are poor women (Burn 2005; Harlingen Economic Development Corporation 2011; Made in Mexico, Inc. 2005).

In El Salvador, women employees of the Taiwanese maquiladora, Mandarin, work shifts of 12 to 21 hours during which they are seldom allowed bathroom breaks. They are paid about 18 cents per shirt, which is later sold for $20. Mandarin makes clothes for Gap, J.Crew, and Eddie Bauer. In Haiti, women sewing clothing at Disney's contract plants are paid 6 cents for every $19.99 *101 Dalmatians* outfit they sew. They make 33 cents an hour. Meanwhile, Disney makes record profits and could easily pay workers a living wage for less than one half of 1% of the sales price of one outfit. In Vietnam, 90% of Nike's workers are females between the ages of 15 and 28. Nike's labor costs for a pair of basketball shoes (which retail for $149.50) are $1.50, 1% of the retail price (Burn 2005).

These women often earn less than half of what is needed to survive in their areas. Yet, corporations can pay low wages and, in many cases, continue poor and even dangerous working environments because these areas have high unemployment, and families have no other options.

What can be done about the abusive treatment of women around the world? This is a tough issue: Governments have passed legislation to protect workers, but governments also want the jobs that multinational corporations bring and therefore do little to enforce regulations. International labor standards are also difficult to enforce. Multinationals are so large that it is difficult to influence their practices. Trade unions have had little success attracting workers to join because companies squash their recruiting efforts immediately.

One way to begin considering fair wages and conditions for workers is to adopt practices that have worked for other groups facing discrimination in the past. Consider the following possibilities for bringing about change: holding nonviolent protests, sit-down strikes, and walkouts to protest unequal and unfair treatment; working together in support groups to help children and neighborhoods; using the Internet to carry a message to others; carrying out boycotts against companies that mistreat employees; following traditions that have succeeded in the past such as using the arts, preachers, storytellers, and teachers to express frustration and resistance, educate others, and provide ideas and strategies for resistance; and building on traditions of community and religious activism (Collins 2000). Most of these strategies require organized groups. However, such efforts face the danger of antagonizing the companies so that they move to other countries.

Thinking Sociologically

Take a look at the clothing in your closet and your drawers. Figure out where it was made, either by reading the label or looking up the company's factories on the Internet. Would you be willing to pay more for that clothing so that other people could have more humane living conditions? What about your friends? Should you be concerned about these workers' lives, or do you think you have no obligation to help them?

Women in Thailand produce shoes at extremely low rates. These jobs are better than no employment at all, but before pressures from Western societies changed their cultures, most people were able to feed their families on farms in small villages. Changes in the entire world system have made that form of life no longer feasible, but working for multinational corporations also keeps them impoverished.

Most United Nations member countries have at least fledgling women's movements fighting for improved status of women and their families. The movements attempt to change laws that result in discrimination, poverty, abuse, and low levels of education and occupational status. Although the goals of eliminating differential treatment of women, especially minority women, are jointly affirmed by most women's groups, the means to improve conditions for women are debated between women's groups. Whether any of these efforts will change women's individual lives and the lives of their children is unknown. We would be too optimistic to predict that grassroots efforts or boycotts against those who are enjoying the fruits of poor women's labor will change the system. The best hope may lie in increased opportunities for women as countries modernize and in efforts to enforce labor laws and improve conditions and wages for workers (Better Factories Cambodia 2006).

Inequality based on class, race, ethnicity, and gender is a process taking place at all levels of analysis, but is often entrenched at the meso level within institutions of family, education, religion, politics, and economics. In complex societies, there are other core institutions including health care, science and technology, sport, and the military. We turn now to a discussion of institutions in our social world.

What Have We Learned?

In the beginning of this chapter, we asked how being born female or male affects our lives. Because sex is a primary variable on which societies are structured and stratification takes place, being born female or male affects our public- and private-sphere activities, our health, our ability to practice religion or participate in political life, our educational opportunities, and just about everything we do. As sociologists with a focus on empirical ways of knowing and scientific methods, we usually focus on description and analysis of what is. Gender inequality clearly exists. What should be done to alleviate problems is a matter of debate.

Key Points:

- Whereas sex is biological, notions of gender identity and gender roles are socially constructed and therefore variable. (See pp. 248–251.)

- Notions of gender are first taught at the micro setting—the intimacy of the home—but they are reinforced and even sacralized at the meso and macro levels. (See pp. 251–253.)

- While "she" may go first in micro-level social encounters (served first in a restaurant or the first to enter a doorway), "he" goes first in meso and macro settings—with the doors open wider for men to enter leadership positions in organizations and institutions. (See pp. 253–255.)

- Greater access to resources at the meso level makes it easier to have entrée to macro-level positions, but it also influences respect in micro settings. (See pp. 255–261.)

- Much of the gender stratification today is unconscious and unintended—not caused by angry or bigoted men who purposefully oppress women. Inequality is rooted in institutionalized privilege and disprivilege. (See pp. 261–265.)

- Various social theories shed different light on the issues of sex roles and inequality. (See pp. 265–270.)

- For modern postindustrial societies, there is a high cost for treating women like a minority group—both individually for the people who experience it and for the society, which loses the intelligence, skills, and commitment of highly competent people. (See pp. 271–275.)

Contributing to Our Social World: What Can We Do?

At the Local Level:

- *Human resources* (employment) office: Schedule an interview with the director or another staff member to learn about your school's policies regarding gender discrimination. What procedures exist to ensure that women and men have an equal opportunity to be hired for faculty and staff positions? Do women and men receive the same salaries, wages, and benefits for equal work? Explore the possibility of your working as a volunteer or intern in the office specifically in the area of gender equity.

At the Organizational and National Levels:

- *National Organization for Women* (NOW): The website (www.now.org) contains information and opportunities for participation (by both women and men) on issues such as abortion and reproductive rights, legislative outreach, and economic justice, as well as activities aimed at ending sex discrimination and promoting diversity. Several internship programs are listed on the site, along with contact information and state and regional affiliates. Be sure to click on the "NOW on Campus" link to learn about programs oriented to undergraduate students.

At the Global Level:

- *MADRE* is an international women's rights organization working especially in the less developed countries. Visit its website, at http://www.madre.org, where you will find numerous opportunities for working on issues of justice, human rights, education, and health.

- *UN Women* (www.unwomen.org/) and the *United Nations Inter-Agency Network on Women and Gender Equality* (http://www.un.org/womenwatch/ianwge/) work on relevant global issues, including violence against women and women's working conditions. The WomenWatch website, at www.un.org/womenwatch, contains news, information, and ideas for contributing to the worldwide campaign for women's rights.

Visit **www.sagepub.com/oswcondensed2e** for online activities, sample tests, and other helpful information. Select "Chapter 9: Gender Stratification" for chapter-specific activities.

PART IV

Institutions

Picture a house, a structure in which you live. Within that house there are processes—the action and activities that bring the house alive. Flip a switch, and the lights go on because the house is well wired. Adjust the thermostat, and the room becomes more comfortable as the structural features of furnace or air-conditioning systems operate. If the structural components of the plumbing and water heating systems work, you can take a hot shower when you turn the knob. These actions taken within the structure make the house livable. If something breaks down, you need to get it fixed so that everything works smoothly.

Institutions, too, provide a framework or *structure* for society. *Processes* are the action dimension within institutions—the activities that take place. They include the interactions between people, decision making, conflict, and other actions in society. These processes are often dynamic and can lead to significant change within the structure—like a decision by a new homeowner to remodel the kitchen. Institutions are meso-level structures because they are larger in scope than the face-to-face social interactions of the micro level, yet they are smaller than the nation or the global system. Institutions—such as family, education, religion, politics, economics, and health care—are also interdependent and mutually supportive, just as the plumbing, heating system, and electricity in a house work together to make a home functional. However, breakdown in one institution affects the whole society, just as a malfunction in the electrical system may shut the furnace off and cool down the water heater.

The Importance of Institutions

Institutions are not anything concrete that you can see, hear, or smell. The concept of "institutions" is an abstraction—a way of understanding how society works. It refers to the behavior of thousands of people who—taken as a whole—form a social structure. Think of your own family. It has unique ways of interacting and raising children, but it is part of a community with many families. Those many families, in turn, are part of a national set of patterned behaviors we call "the family." This pattern meets basic needs of the society for producing and socializing new members and providing an emotionally supportive environment. Institutions do not dictate exactly how you will carry out the roles of family, but they do specify certain individual needs that families will meet and statuses (husband, wife, child) that will relate to each other in certain mutually caring ways to fill important family roles. An institution provides a blueprint (much like a local builder needs a blueprint to build a house), and in your local version of the institution you may make a few modifications to the plans. Still, through this society-encompassing structure and interlocking set of statuses, basic needs—for individuals at the micro level and for society at the macro level—are met.

Institutions, then, are organized, patterned, and enduring sets of social structures that provide guidelines for behavior and help each society meet its basic survival needs. While institutions operate mostly at the meso level, they also act to integrate micro and macro levels of society. Let us look more deeply at this definition.

1. *Organized, patterned, and enduring sets of social structures* do not represent the bricks and mortar of buildings but refer to a complex set of groups or organizations, statuses within those groups, and norms of conduct that guide people's behaviors. These structures ensure socialization of children, education of the young, sense of meaning in life to the distressed, or goods (food, clothing, automobiles, iPods, cell phones) to the members of the society. This is patterned behavior that is important; if it were missing, these needs might not be addressed. At the local level, we may go to a neighborhood school, or we may attend worship at a congregation we favor. These are local organizations—local franchises, if you will—of a much more encompassing structure (education and religion) that provides guidelines for education or addresses issues of meaning of life for an enormous number of people. The Catholic Church in your town, for example, is a local "franchise" of an organization that is transnational in scope and global in its concerns.

2. *Guidelines for behavior* help people know how to conduct themselves to meet basic needs. Individuals and local organizations, or "franchises," actually carry out the institutional guidelines in each culture; the exact ways the guidelines are carried out vary by locality. In local "franchises" of the political system, people know how to govern and how to solve problems at the local level because of larger norms and patterns—the political institution. Individual men and women operate a local hospital or clinic (a local "franchise" of the medical institution) because a national blueprint of how to provide health care informs local expectations and decisions. The specific activities of a local school, likewise, are influenced by the guidelines and purposes of the larger notions of "formal education" in a given nation.

3. *Meeting basic survival needs* is a core component of institutions because societies must meet needs of their members; otherwise the members die or the society collapses. Institutions, then, are the structures that support social life in a large bureaucratized society. Common to all industrialized societies are family, education, religion, economics, and politics. These institutions are discussed in the following three chapters.

4. *Acting to integrate micro and macro levels of society* is critical because one of the collective needs of society is coherence and stability—including some integration between the various

levels of society. Institutions help provide that integration for the entire social system. They do this by meeting needs at the local franchise level (food at the grocery, education at the local school, health care at the local clinic) at the same time they coordinate national and global organizations and patterns.

Again, if all of this sounds terribly abstract, that is because institutions are abstractions. You cannot touch institutions, yet they are as real as air or love or happiness. In fact, in the modern world institutions are as necessary to life as is air, and they help provide love and happiness that make life worth living.

The Development of Modern Institutions

If we go all the way back to early hunting and gathering societies, there were no meso or macro levels to their social experience. People lived their lives in one or two villages, and while a spouse might have come from another village or one might have moved to a spouse's clan, there was no national or state governance and certainly no awareness of a global social system. In those simple times, family provided whatever education was needed, produced and distributed goods, paid homage to a god or gods, and solved conflicts and disputes through a system of familial (usually patriarchal) power distribution. One social unit served multiple functions. As societies have become more complex and differentiated, not only have multiple levels of the social system emerged, but various new institutions have emerged. Sociology textbooks in the 1950s identified only "five basic institutions": family, economic systems, political systems, religion, and education (formal public education only having been created in the mid-nineteenth century). These five institutions were believed to be the core structures that met the essential needs of individuals and societies in an orderly way.

Soon thereafter, *medicine* moved from the family and small-town doctors to be recognized as an institution. Medicine had become bureaucratized in

hospitals, medical labs, professional organizations, and other complex structures that provided health care. *Science* is also now something more than flying kites in thunderstorms in one's backyard. It is a complex system that provides training, funding, research institutes, peer review, and professional associations to support empirical research in the sciences. New information is the lifeblood of an information-based or postindustrial society. Science, discussed in Chapter 14, is now an essential institution. Although it is arguable whether sports are an essential component for social viability, sports have clearly become highly structured in the past 50 years, and many sociologists consider them an institution. The mass media and military also fall into the category of institutions in more advanced countries. There are gray areas as to whether or not something is considered an institution, but the questions to ask are (a) whether the structure meets basic needs of the society for survival, (b) whether it has become a complex organization providing routinized structures and guidelines, and (c) whether it is national or even global in its scope, while also having pervasive local (micro) impact.

The Interconnections Between Institutions

Keep in mind that changes in one institution affect all other institutions, since they are interdependent. For example, the global economic crisis that began in 2008 illustrates the forces that bring about changes through interconnections between institutions. As described by "the sociological imagination," our individual problems such as loss of a job are tied to the macro-level changes in the economy. So the family is affected, and citizens expect the government to intervene and fix the problem. Religious congregations have increased demands at their food banks, soup kitchens, and thrift shops run for low-income people, yet religious contributions are more difficult in tough economic times. Schools also suffer from lack of income from the economic downturn. The table on the next page illustrates. As you study this table, see if

Table I.1   The Impact of Institutions at Each Level of Analysis

	Family	Education	Economic Systems	Political Systems	Religion	Medicine
Macro (national and global social systems and trends)	Kin and marriage structures, such as monogamy versus polygamy; global trends in family such as choice of partners rather than arranged marriages	National education system; United Nations Girls' Education Initiative	Spread of capitalism around the world; World Bank; International Monetary Fund; World Trade Organization	National government; United Nations; World Court; G8 (most powerful 8 nations in the world)	Global faith-based movements and structures: National Council of Churches; World Council of Churches; World Islamic Call Society; World Jewish Congress	National health care system; World Health Organization; transnational pandemics
Meso (institutions, complex organizations, ethnic subcultures, state/provincial systems)	The middle-class family; the Hispanic family; the Jewish family	State/provincial department of education; American Federation of Teachers*	State/provincial offices of economic development; United Auto Workers*	State/provincial governments; national political parties; each state or province's supreme court	National denominations/ movements: e.g., United Methodist Church or American Reform Judaism	HMOs; American Medical Association*
Micro (local "franchises" of institutions)	Your family; local parenting group; local Parents Without Partners; county family counseling clinic	Your teacher; local neighborhood school; local school board	Local businesses; local chamber of commerce; local labor union chapter	Neighborhood crime watch program; local city or county council	Your local religious study group or congregation	Your doctor and nurse; local clinic; local hospital

*These organizations are national in scope and membership, but they are considered meso level here because they are complex organizations *within* the nation.

you can place other institutions (mass media, military, science, sports) in the framework.

Interconnections between meso-level institutions are a common refrain in this book. The following two examples serve to illustrate the types of interconnections that hold our social world together.

Education, Family, and Other Linkages

When children enter kindergarten or primary school, they bring their prior experience, including socialization experiences from their families. Family background, according to many sociologists, is the single most important influence on children's school achievement (Jencks 1972). Children succeed in large part because of what their parents do to support them in education (MacLeod 2008; Schneider and Coleman 1993).

Most families stress the importance of education, but they do so in different ways. Middle-class parents in the Global North tend to manage their children's education, visiting schools and teachers, having educational materials in the home, and holding high expectations for their children's achievement. In these families, children learn the values of hard work, good grades, and deferred gratification for reaching future goals.

Involvement of parents from lower socioeconomic status and first-generation immigrant families can have significant impact on their children's educational outcomes (Bankston 2004; Domina 2005); however, many of these parents tend to look to schools as the authority. They are less involved in their children's schooling, leaving decisions to the school. Yet, children who must make educational decisions without familial guidance are more likely to do poorly or to drop out of school (Bridgeland, Dilulio, and Morison 2006; Kalmijn and Kraaykamp 1996).

Moreover, when economic times are rough, employment instability and uncertainty, economic strain, and deprivation can cause strained family relations. Low-income families, especially single-parent families headed by women, are particularly hard-hit and often have to struggle for survival (Staples 1999; Willie 2003). In some cases, families are so financially devastated that they become homeless—a difficult life, especially for children. Even if the family is not homeless, the trend toward the "feminization of poverty"—where single motherhood is widespread—can mean that children have less support for schoolwork at home (Williams, Sawyer, and Wahlstrom 2009). Single mothers experience dual roles as workers and mothers, lower earnings than men, and irregular paternal support payments, so they may have less time to be attentive to academic needs of their children. Thus, the economic health of the family affects its stability and the number of stressors it faces. These stressors, in turn, may affect the ability of the family to support a child's achievement in school. If the child does not get a good education, she or he is less likely to be a strong contributor to the economy in the future and is less likely to vote or be politically involved. Thus, family, education, economics, and politics become interrelated.

Religion and the Economy

Religion and the economy are intertwined and interdependent as you will note in Chapter 11. Here we illustrate how each can affect the other using a classic study of these two institutions. In fact, one sociologist felt that religious beliefs helped create the economic system that dominates most of the Global North. Why do most of us study hard, work hard, and strive to get ahead? Why are we sacrificing time and money now—taking this and other college courses—when we might spend that money on an impressive new car? Our answers probably have something to do with our moral attitudes about work, about those who lack ambition, and about convictions regarding the proper way to live. Max Weber ([1904–1905] 1958) believed that the economic system, particularly capitalism, and many of our attitudes about economic behavior are rooted in religious ideas about work and sacrifice.

Weber's study, *The Protestant Ethic and the Spirit of Capitalism*, became a classic in the field. Noting that the areas of Europe where the Calvinists had strong followings were the same areas where capitalism grew fastest, Weber ([1904–1905] 1958) argued that four elements in the Calvinist Protestant faith created the moral and value system necessary for the growth of capitalism: predestination, a calling, self-denial, and individualism.

1. *Predestination* meant that one's destiny was predetermined. Nothing anyone could do would change what was to happen. Because God was presumed to be perfect, he was not influenced by human deeds. Those people who were chosen by God were referred to as the *elect* and were assumed to be a small group. Therefore, people looked for signs of their status—salvation or damnation. High social status was sometimes viewed as a sign of being among the elect, so motivation was high to succeed in *this* life.

2. The *calling* referred to the concept of doing God's work. Each person was put on Earth to serve God, and each had a task to do in God's service. One could be called by God to any occupation, so the key was to work very hard and with the right attitude. Because work was a way to serve God, laziness or lack of ambition was viewed as a sin. These ideas helped create a society in which people's self-worth and their evaluation of others were tied to a work ethic. Protestants became workaholics.

3. *Self-denial* involved living a simple life. If one had a good deal of money, one did not spend it on a lavish home, expensive clothing, or various forms of entertainment. Such consumption would be offensive to God. Therefore, if people worked hard and began to accumulate resources, they simply saved them or invested them in a business. This self-denial was tied to an idea that we now call *delayed gratification,* postponing the satisfaction of one's present wants and desires in exchange for a future reward. The reward these people sought was in the afterlife.

4. *Individualism* meant that each individual faced his or her destiny alone before God. This stark individualism of Calvinistic theology stressed that each individual was on his or her own before God. Likewise, in the economic system that was emerging, individuals were on their own. The person who thrived was an individualist who planned wisely and charted his or her own course. Religious individualism and economic individualism reinforced one another.

For capitalism to develop, there was a need for individualistic entrepreneurs who had a strong work ethic, strong motivation based on delayed gratification and a hope for the future, and a pool of capital to invest (which was generated in part by the tendency to simple lifestyles). This combination of factors was fostered by the teachings of Luther, Calvin, and other early Protestant reformers. Religion contributed to major changes in cultural values, which, in turn, transformed the economic system. Weber ([1904–1905] 1958) saw the religious and economic systems as dynamic, interrelated, and ever-changing.

Gradually, the capitalistic system, stimulated by the Protestant Ethic, spread to other countries and to other religious groups. Many of the attitudes about work and delayed gratification no longer have supernatural focus, but they are part of our larger culture nonetheless. They influence our feelings about people who are not industrious and our ideas about why some people are poor.

Thus, religious beliefs may have had some influence in creating a particular type of economic system, but we know that the influence runs the other direction as well. People join religious groups that have values and attitudes compatible with their present socioeconomic status. Our social class standing influences our decisions about church, temple, or mosque membership (Bowles and Gintis 1976; Pew Forum on Religion and Public Life 2008; Weber 1946).

These examples give an idea of the interconnectedness between institutions at the meso level. We next consider theories that help us understand meso-level dynamics.

Theories Explaining Institutions

Various theoretical lenses look at institutions, and each has something worthwhile to teach us. Functional theorists speak of the essential *functions* carried out by institutions to meet societal needs. As society becomes more complex, there are more institutions needed, as in the case of science, an institution that has allowed humans to land on the moon and develop new forms of energy to power our technologies. If we do not have more energy sources, technology largely collapses, including many of our forms of health care and our economy. If any institution fails, the entire social system is in severe jeopardy. If all institutions fail, the basic needs of society would not be met, and the society would collapse—or there would be a massive overthrow of the existing system. While functionalists may overemphasize the role of stability, institutions do provide guidelines for each society. Functionalists help us identify the purposes of each institution.

Conflict theorists focus on the inequities and inequalities created by institutions that have been developed and run by powerful members of society. In this process, less powerful groups can be exploited. Powerful elites do try to maintain stability in societies as they have a vested interest in keeping things from changing too dramatically, because their own privileges are rooted in the existing structure.

We have pointed out that all institutions are interconnected and interdependent, with radical change in one potentially upsetting the balance in the others. This is a cause for concern for conflict theorists. For example, religious institutions help make family sacred through marriage ceremonies. In doing so, this sacred status makes it harder to change the system—even if the system treats women or homosexuals unfairly. Educational institutions build loyalty in political systems, for example, by requiring a pledge of allegiance at the outset of the day. The political system often makes the existing economic system seem indisputably right and fair. Note also how thoroughly sports have become involved in fostering patriotism through half-time shows and ongoing comments about national loyalty by the announcers. So institutions have a conservative bias because stability in the larger social system enhances stability of each institution making up the society. This serves the interests of those in leadership roles within each institution. Conflict theorists are concerned that institutions will maintain stability even if that stable system is oppressive to some members of the society. It is precisely that conserving role that is seen as dysfunctional to those who favor social change to enhance equality.

Interaction theorists argue that the micro-level interactions that shape our lives are based on patterns that we take for granted. These patterns are shaped by our perceptions of what is "normal," and these perceptions in turn come largely from the institutions that surround us. Generally we do what we have observed is right and acceptable in a particular status we occupy within an organization (officer in a corporation, for example) or within an institution (father, for example). So institutions affect the way we interact, how we treat people, and whether we expect deference from others. On the other hand, interaction theorists do not think our lives are totally determined by our surroundings. We do have *agency*—the possibility of being active and creative beings within our social settings. So a woman in a leadership role in an organization may redefine how leadership is understood by changing people's *definition of the situation*—their *social construction of reality*. Here we see, on the one hand, that people at the micro level are shaped by the meso-level norms; on the other hand, those meso-level norms can be changed by dynamic individuals who see things differently.

Rather than trying to be all-encompassing in this book, we attempt to illustrate how various aspects of society work. The next three chapters provide examples of institutions and how these structures make up societies. The first and most basic institution is family, discussed in Chapter 10. Next we consider two institutions that play a major role in our socialization, belonging, and meaning systems—education and religion. Finally, our example of the political-economic system illustrates institutions that impact the macro level but also affect us at the micro level. As you read these chapters, notice that change in one institution affects others. Sociologists studying the legal system, mass media, medicine, the military, science, and sports as institutions would raise similar questions and would want to know how it influences the micro, meso, and macro levels of a society. We begin with family—an institution that is such an intimate part of our lives and that is often called the "most basic" institution of society.

CHAPTER 10

Family

Partner Taking, People Making, and Contract Breaking

Appearing in rich variety, the family is often referred to as the "most basic" institution. In this social relationship, we take partners and "make people"—both biologically and socially speaking. In the modern world, these intimate "basic" unions often experience conflict, violence, and contract breaking as well.

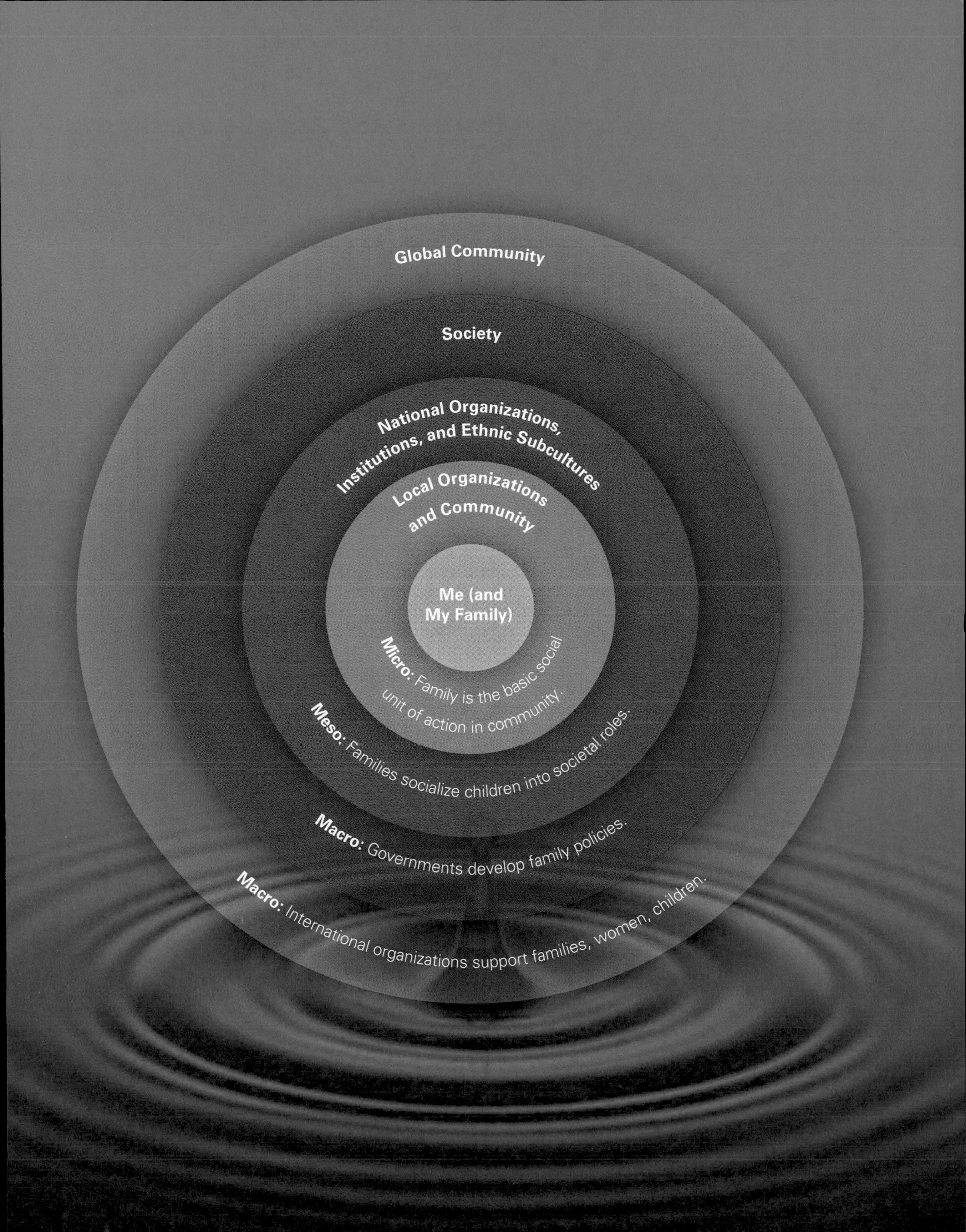

Global Community

Society

National Organizations, Institutions, and Ethnic Subcultures

Local Organizations and Community

Me (and My Family)

Micro: Family is the basic social unit of action in community.

Meso: Families socialize children into societal roles.

Macro: Governments develop family policies.

Macro: International organizations support families, women, children.

Think About It	
Self and Inner Circle	Why is family important to you?
Local Community	How do people find life partners?
National Institutions; Complex Organizations; Ethnic Groups	Why is family seen as the core or basic institution of society?
National Society	What, if anything, should be done by government to strengthen families?
Global Community	Why are families around the world so different?

A Guatemalan family in a rural village prepares for the day's chores. Maria prepares the breakfast as Miguel cares for the animals. The children fix their lunches of tortillas, beans, rice, and banana to take to school. When they return home, they will help with the farm chores. They live together with their extended family, several generations of blood relatives living side-by-side.

It is morning in Sweden. Anders and Karin Karlsson are rushing to get to their offices on time. The children, a 12-year-old son and an eight-year-old daughter, are being hurried out the door to school. All will return in the evening after a full day of activities and join together for the evening meal. In this dual-career family, common in many postindustrial societies, both parents are working professionals.

The gossip at the African village water well this day is about the rich local merchant, Abdul, who has just taken his fourth and last wife. She is a young, beautiful girl of 15 from a neighboring village. She is expected to help with household chores and bear children for his already extensive family unit. Several of the women at the well live in affluent households where the husband has more than one wife.

Tom and Jackson recently adopted Ty into their family. The couple share custody of the five-year-old boy; adoption by gay couples is not legal in all U.S. states, but it is legal for gay couples in some states and for single gay persons in many others (Johnson 2008). Ty's parents attend his school events and teach him what all parents are expected to teach their children. Dora, a single mom, lives next door with her two children. She bundles them off to school before heading to her job. After school she has an arrangement with Tom and Jackson to care for the children until she gets home.

What do the very different scenes described at the outset of this chapter have in common? Each describes a family, yet there is controversy about what constitutes a family. Those groupings that are officially recognized as families tend to receive a number of privileges and rights, such as health insurance and inheritance rights, but not every group that thinks it is a family is defined as "family" by the society (DeGenova, Stinnett, and Stinnett 2011). In this chapter, we will discuss characteristics of families, theoretical perspectives on family, family dynamics, family as an institution, family issues, and policies regarding marriage and family dissolution or divorce.

Families come in many shapes, sizes, and color combinations. We begin our exploration of this institution with a discussion of what *is* family.

What Is a Family?

Who defines what constitutes a family: each individual, the government, or religious groups? Is a family just Ma, Pa, and the kids? Let us consider several definitions. The U.S. government defines the family as "a group of two or more people (one of whom is the householder) related by birth, marriage, or adoption and residing together; all such people (including related subfamily members) are considered as members of the family" (U.S. Census Bureau 2010).

Thus, a family in the United States might be composed of siblings, cousins, a grandparent and grandchild, or other groupings. Some sociologists define **family** more broadly,

In recent decades, the definition of family has broadened. No longer is it necessarily limited to heterosexual couples. These gay men are parents to this baby.

saying a family is a unit comprising two or more people who share a residence for a substantial period of time and have legal or moral responsibilities for long-term care for each other; the unit typically shares one of the following: sex, pooling of incomes, care of children, and some forms of recreation (D'Antonio 1983; Newman 2009). This definition would include same-sex couples and many cohabiting heterosexuals as families. Some religious groups define family as a mother, a father, and their children, whereas others include several spouses and even parents and siblings living under the same roof.

How do you define the ideal family? Answering the questions in the next "Engaging Sociology" will indicate the complexity of this question.

The family is often referred to as the most basic *institution* of any society. First, the place where we learn many of the norms for functioning in the larger society is the family, so it serves as the institution that helps us function in all other institutions. This makes it pretty basic to us and to the society. Second, most of us spend our lives in the security of a family. People are born and raised in families, and many will die in a family setting. Through good and bad, sickness and health, most families provide for our needs, both physical and psychological. Therefore, families meet our primary, most basic needs. Third, major life events—marriages, births, graduations, promotions, anniversaries, religious ceremonies, holidays, funerals—take place within the family context and are celebrated with family members. In short, family is where we invest the most emotional energy and spend much of our leisure time. Fourth, in many nonindustrial societies and in some ethnic groups within modern cultures, the family is the key to social organization, for in such societies, one's status and identity are determined almost entirely by one's family. Finally, the family is capable of satisfying a range of social needs—belonging to a group, economic support, education or training, raising children, religious socialization, resolution of conflicts, and so forth. One cannot conceive of the economic system providing emotional support for each individual or the political system providing socialization and personalized care for each child. Family carries out these functions.

The family is the place where we confirm our partnerships as adults, and it is where we *make people*—not just biologically but also socially. In the family, we take an organism that has the potential to be fully human, and we mold this tiny bit of humanity into a caring, compassionate, productive person.

This Romanian family does not have much money, but the children learn many survival skills, and the most basic needs of the children, physical and psychological, are met.

In most Western societies, individuals are born and raised in the **family of orientation**, which consists of our parent(s) and possibly our sibling(s). In this family, we receive our early socialization and learn the language, norms, core values, attitudes, and behaviors of our community and society. When we find a life mate and/or have our own children, we establish our **family of procreation**. The transmission of values, beliefs, and attitudes from our family of orientation to our family of procreation preserves and stabilizes the family system. Because family involves emotional investment, we have strong feelings about what form it should take.

Whether we consider families at micro, meso, or macro levels, sociological theories can help us understand the role of families in the social world.

Theoretical Perspectives on Family

Consider the case of Felice, a young mother locked into a marriage that provides her with little satisfaction. For the first year of marriage, Felice tried to please her husband, Tad, but gradually he seemed to drift further away. He began to spend evenings out. Sometimes, he came home drunk and yelled at her or hit her. Felice became pregnant shortly after their marriage and had to quit her job. This increased the financial pressure on Tad, and they fell behind in paying the bills.

Thinking Sociologically

What purposes does your family serve for its members? What role does each member play in the family? Why might this be important for the larger society?

Engaging Sociology

The Ideal Family

What is "the ideal family"? Does it have one adult woman and one adult man? One child or many? Grandparents living with the family? First, complete the following survey. Then ask a friend or relative to answer the questions below.

1. How many adults should the ideal family contain? _____

2. How many children should the ideal family contain? _____

3. What should be the sex composition of the adults in an ideal family? (check all that apply)

   a. One female and one male
   b. Male-male or female-female
   c. Several males and several females
   d. Other (write in) _____

4. What should be the sexes of the child(ren) in the ideal family? _____

5. Who should select the marriage partner? (check all that apply)

   a. The partners should select each other.
   b. The parents or close relatives should select the partner.
   c. A matchmaker should arrange the marriage.
   d. Other _____

6. What is the ideal number of generations living in the same household?

   a. One generation: partners and no children
   b. Two generations: partners and children
   c. Three or more generations: partners, children, grandparents, and great-grandparents
   d. Other _____

7. Where should the couple live?

   a. By themselves
   b. With parents
   c. With brothers or sisters
   d. With as many relatives as possible

8. What should the sexual arrangements be? (Check all that apply.)

   a. Partners have sex only with each other.
   b. Partners can have sex outside of marriage if it is not "disruptive" to the relationship.
   c. Partners are allowed to have sex with all other consenting adults.
   d. Male partners can have sex outside marriage.
   e. Female partners can have sex outside marriage.
   f. Other _____

9. Which person(s) in the ideal family should work to help support the family?

   a. Both partners
   b. Male only
   c. Female only
   d. Both, but the mother only after children are in school
   e. Both, but the mother only after children graduate from high school
   f. All family members including children
   g. Other _____

10. Should the couple have sex before marriage if they wish? Yes ___ No ___ Other_____

11. Should physically disabled aging parents

    a. Be cared for in a child's home?
    b. Be placed in a nursing care facility?
    c. Other _____

Why do you hold these particular views of "the ideal family"? Are your answers different from those of your friend or relative? Why might others in your society have answered differently?

Note that all of these options can be found in some societies.

No other institution can fulfill the functions of the family, but the family can fulfill many functions of other institutions. On the left, a family works together as an economic team to produce food. On the right, a family in the United States prays together before lunch.

Then came the baby. They were both ecstatic at first, but Tad soon reverted to his old patterns. Felice felt trapped. She was afraid and embarrassed to go to her parents. They had warned her against marrying so young without finishing school, but she was in love and had gone against their wishes. She and Tad had moved away from their hometown, so she was out of touch with her old support network and had few friends in her new neighborhood. Her religious beliefs told her she should try to stick it out, suggesting that the trouble was partly her fault for not being a "good enough wife." Lacking a job or skills to get one, she could not live on her own with a baby. She thought of marriage counseling, but Tad refused to consider this and did not seem interested in trying to work out the problems. He had his reasons for behaving the way he did, including feeling overburdened with the pressure of caring for two dependents. The web of this relationship seems difficult to untangle. Sociological theories provide us with tools to analyze such family dynamics.

Micro-Level Theories of Family and the Meso-Level Connection

Symbolic Interaction Theory

Symbolic interaction theory can help us understand Felice's situation by explaining how individuals learn their particular behavior patterns and ways of thinking. Our role relationships are developed through socialization and interaction with others. Felice developed certain expectations and patterns of behavior by modeling her experiences on her *family of orientation* (the family into which she was born), observing others, and developing expectations from her initial interactions with Tad. He developed a different set of expectations for his role of husband, modeled after his father's behavior. His father had visited bars after work, had affairs with other women, and expected "his woman" at home to accept this without question.

Two related concepts in symbolic interaction theory are the *social construction of reality* and the *definition of a situation.* What we define as real or as normal is shaped by what significant others around us accept as ordinary or acceptable. Children who grow up in homes where adults hit one another or argue using sarcastic put-downs may come to think of this behavior as typical or a normal part of family life. They simply have known no other type of interaction. Thus, they may create a similar pattern of family interaction in their own *families of procreation* (the families they create).

Thinking Sociologically

Concepts such as *family, wife,* and *parenting* carry meaning to you and your siblings but may mean something very different to the person sitting beside you in class or to a potential mate. Ask several people you know what family means to them.

One of the great challenges of newlyweds is meshing their ideas about division of labor, family holidays, discipline of children, spousal relations, and economic necessities, along with their assumptions about being in a committed relationship. A new couple socially constructs a new relationship, blending the models of life partnership from their own childhood homes or creating an entirely new model as they jointly define their relationship. Furthermore, the meaning of one's identity and one's obligations to others

Introduction of a baby to a household changes the interpersonal dynamics, the topics of conversation, the amount of sleep people are able to get, the sense of responsibility for the future, relationships to the larger community (including schools), and many other aspects of social life.

changes dramatically when one becomes a parent. This brings us back to a central premise of symbolic interaction theory: Humans are active agents who create their social structure through interaction. We not only learn family patterns; we *do family* in the sense that we create roles and relationships and pass them on to others as "normal."

Our individual identities and family patterns are shaped by institutional arrangements at the meso level: Corporations, religious bodies, legal systems, and other government entities define the roles of "wife" and "husband." For example, each U.S. state actually spells out in its legal codes the duties of husbands and wives. Those who do not fulfill these duties may be in "neglect of duty." These family roles are embedded in the larger structure in ways that many people do not realize.

Rational Choice Theory

Rational choice theory can also shed light on Felice's situation, helping us understand why people seek close relationships and why they stay in abusive relationships. As discussed in Chapter 2, rational choice theory asserts that individuals evaluate the costs and rewards of engaging in

interaction. We look for satisfaction of our needs—emotional, sexual, and economic—through interaction. Patterns in the family are reinforced to the extent that exchanges are beneficial to members. When the costs outweigh the rewards, the relationship is unlikely to continue. Women in abusive relationships weigh the costs of suffering abuse against the rewards of having social legitimacy, income, religious approval, a home, and companionship. Many factors enter into the complex balance of the exchange. Indeed, costs and benefits of various choices are often established by meso-level organizations and institutions: insurance programs, health care options, and legal regulations that make partnering decisions easy or difficult.

According to rational choice theorists, even the mate selection process is shaped by a calculation of exchange. People estimate their own assets—physical, intellectual, social, and economic—and try to find the "best deal" they can make, with attention to finding someone with at least the level of resources they possess, even if those assets are in different areas. If someone marries a person with far more assets, the one with fewer assets is likely to have little power and to feel dependent on that relationship, often putting up with things that equal partners would not tolerate. Cost/benefit, according to this view, affects the forming of the relationship and the power and influence later in the relationship.

Thinking Sociologically

What situations can you identify within your family when cost/benefit calculations seemed to drive decisions? When is reciprocity the norm in family relationships?

Meso- and Macro-Level Theories of the Family

Structural-Functionalist Theory

Structural-functionalist theory points out the common purposes of family institutions in every society. Despite great variations in family forms, most human family systems satisfy similar needs, or functions, for their members and for society. Although families vary greatly, their members have a number of common needs and problems. For instance, they all must secure food and shelter, raise children, and care for dependents.

Why do all societies have families? One answer is that families fulfill certain purposes, or functions, for societies that enhance survival of individuals and societies. Traditionally, there have been at least six ways the family has helped stabilize the society, according to structural-functionalist theory:

Sexual regulation. Physically speaking, any adult human could engage in sex with any other human. However, in practice, no society allows total sexual freedom. Every society attempts to regulate the sexual behavior of its members in accordance with its own particular values; this is most often accomplished through marriage. Regulation ensures that the strong biological drive is satisfied in an orderly way that does not create ongoing disruption, conflict, or jealousy. Certain people are "taken" and "off-limits" (Ward and Edelstein 2009).

Reproduction and replacement. Societies need children to replace members who die, leave, or are incapacitated. Reproduction is controlled to keep family lineage and inheritance clear. Parent and caretaker roles are clearly defined and reinforced in many societies by ceremonies: baby showers, birth announcements, christenings, and naming ceremonies that welcome the child as a member of the family. In some places, such as New Guinea, procreation is so important that a young girl who has had children is more desirable because she has established her fertility.

Socialization. The family is the main training ground for children. In our families, we begin to learn values and norms, proper behavior, roles, and language. Later socialization in most societies is carried out by schools, religious organizations, and other institutions, but the family remains the most important initial socializing agent to prepare us for roles in society. Much of the socialization is done by parents, but siblings, grandparents, and other relatives are important as well.

Emotional support and protection. Families are the main source of love and belonging in many societies, giving us a sense of identity, security, protection, and safety from harm. The family is one place where people may experience unqualified acceptance and feelings of being cherished. Problems of children in youth shelters and incidents of family violence and neglect are reminders that this function is not always successfully provided in families. Still, the family is the environment most capable of meeting this need if all is well.

Status assignment. Our family of birth is the most important determinant of our social status, life chances, and lifestyles. It strongly affects our educational opportunities, access to health care, religious and political affiliations, and values. In fact, in societies with caste systems, the ascribed position at birth is generally the position at death. Although in class societies individuals may achieve new social statuses, our birth positions and the early years of socialization have a strong impact throughout life on who and what we are.

Economic support. Historically, the family was a unit of production—running a farm or a bakery or a cobbler shop. Although this function is still predominant in many societies, the economic function carried out in individual families has pretty much disappeared in most Global North families. However, the family remains an economic unit

This woman in New Guinea became more attractive to men as a potential wife after she had proven her fertility by having children.

Siblings often provide some of the emotional support and part of the socialization for younger siblings. This is especially true in the poor Global South, where mothers often must work or carry water for the family from distant water sources. This sister cares for her younger sister in Soweto, South Africa.

of consumption. Who paid for your clothing, food, and other needs as you were growing up? Who helps many of you pay your college tuition and expenses? Taxing agencies, advertising and commercial enterprises, workplaces, and other social organizations also treat the family as the primary economic unit.

Functional theorists recognize ways that the micro-level processes of the family (e.g., socialization of Japanese children to be cooperative and of members of groups and of U.S. children to be competitive and individualistic) are compatible with structural needs of society at the meso and macro levels (e.g., in most Global North countries, the need for motivated workers who thrive on competition). Each part of the system, according to functionalists, works with other parts to create a functioning society.

Changing family functions. In some societies, the family is the primary unit for bearing and educating children, practicing religion, structuring leisure time activities, caring for the sick and aged, and even conducting politics. However, as societies modernize, many of these functions are transferred to other institutions.

As societies change, so do family systems. The sociohistorical perspective of family tells us that changes in intimate relationships—sexuality, marriage, and family patterns—have occurred over the centuries. Major transitions from agricultural to industrial to postindustrial societal systems change all of the institutions within those societies. Families in agricultural societies are often large and self-sufficient, producing their own food and providing their own shelter, but this is not the case in contemporary urban societies.

Industrialization and urbanization typical in eighteenth- and nineteenth-century Europe and the United States created a distinct change in roles. The wife and child became dependent on the husband who "brought home the bread." The family members became consumers rather than independent and self-supporting coworkers on a farm. This male breadwinner notion of family has become uncommon in the Western world (Coontz 2005). In addition to evolving and changing roles, other changes in society have brought shifts to the family. Improved technology, for example, brought medical technology advances, new knowledge and skills to be passed on in schools, recreation organized according to age groups and outside of the family unit, and improved transportation.

As many families moved to urban areas, the economic function has moved to the factory, store, and office. Socialization is increasingly done in schools, and in some cases teachers have become substitute parents. The traditional protection and care function has been partially replaced by police, reform schools, unemployment compensation, Social Security, health care systems (e.g., Medicare and Medicaid), and other types of services provided by the state. Little League baseball, industrial

bowling teams, aerobic exercise groups, television, and computer games have replaced the family as the source of leisure activities and recreation. Although many would argue that the family still remains the center of the affection and is the only recognized place for producing children, one does not have to look far to discover that these two functions are also increasingly found outside the boundaries of the traditional family unit.

These changes have made the family's functions more specialized. Still, family remains a critical institution in society. Most families still function to provide stable structures to carry out early childhood socialization and to sustain love, trust, affection, acceptance, and an escape from the impersonal world.

Thinking Sociologically

Does reduction of traditional family functions mean a decline in the importance of family or merely an adaptation of the family to changes in society? Is the modern family—based largely on emotional bonds rather than structural interdependency—a healthier system, or is it more fragile?

Conflict Theory

Conflict theorists study both individual family situations and broad societal family patterns. They argue that conflict in families is natural and inevitable. It results from the struggle for power and control in the family unit and in the society at large. As long as there is an unequal allocation of resources, conflict will arise.

Family conflicts take many forms. For instance, conflicts occur over allocation of resources, a struggle that may be rooted in conflict between men and women in the society: Who makes decisions, who gets money for clothes or a car, who does the dishes? On the macro level, family systems are a source of inequality in the general society, sustaining class inequalities by passing on wealth, income, and educational opportunities to their own members or perpetuating disadvantages such as poverty and lack of cultural capital.

Yet, some conflict theorists argue that conflict within the family can be important because it forces constant negotiation among individual family members and may bring about change that can strengthen the unit as a whole. Believing that conflict is both natural and inevitable, these theorists focus on root causes of conflict and how to deal with the discord. For conflict theorists, there is no assumption of a harmonious family. The social world is characterized more by tension and power plays than by social accord.

Feminist Theory

The feminist approach is based on sociological studies done on, by, and for women. Because women often occupy very different places in society than men, feminist theorists argue the need for a feminist perspective to understand family dynamics. Feminist scholars begin by placing women at the center, not to suggest their superiority but to spotlight them as subjects of inquiry and as active agents in the working of society. The biases rooted in male assumptions are uncovered and examined (Eshleman and Bulcroft 2010).

One micro-level branch of feminist theory, the interpretive approach, considers women within their social contexts—the interpersonal relations and everyday reality facing women. It does not ignore economic, political, social, and historical factors but focuses on the ways women construct their reality, their opportunities, and their place in the community. According to feminist theorists, this results in a more realistic view of family and women's lives than many other theories provide. Applying this feminist approach to understand Felice's situation, the theorists would consider the way she views her social context and the way she assesses her support systems.

Many branches of feminist theory have roots in conflict theory. These theorists argue that patterns of patriarchy and dominance lead to inequalities for women. One of the earliest conflict theorists, Friedrich Engels (Karl Marx's close associate), argued that the family was the chief source of female oppression and that until basic resources were reallocated within the family, women would continue to be oppressed. However, he said that as women become aware of their collective interests and oppression, they will insist on a redistribution of power, money, and jobs (Engels [1884] 1942).

One vivid example of how women and men can be viewed as groups with competing interests is through a feminist analysis of domestic violence. In the United States, one in four women have experienced domestic violence in their lifetimes. With an estimated 1.3 million women assaulted by an intimate partner each year, there is an incident of domestic violence every 25 seconds in the United States. Eighty-five percent of the victims of those acts of domestic violence are women (Domestic Violence Resource Center 2009). Boys who witness domestic violence are more than twice as likely to abuse their own wives later in life. Each year, 1,204 women are murdered by boyfriends or husbands (AAUW Dialogue 2008).

In most Global North societies, men are under pressure to be successful. When they are not, home may be the place where they vent their frustrations. When socialization results in a hypermasculine sense of identity and this is combined with emotional dependency on a female, it is a lethal combination. Cultural messages tell men that they are to be in charge, but knowing they are emotionally and sometimes economically dependent on their wives may result in violence to regain control.

Women are most severely exploited in societies that treat them as property and in which the family is a key political unit for power and status. Where men are the heads of families and women are dependent, women may be treated as less than equal both within the family and in the labor market. In some societies such as the Masai of East Africa and in some communities in Brazil, a husband is expected to beat his wife if he is dissatisfied with her cooking, housekeeping, or child care. In some cultural situations, the woman's life is valuable only as it relates to her economic value and the needs of men. Changes in the patriarchal family structure, education and employment opportunities for women, and child care availability can lead to greater freedom of choice, equality, and autonomy for women, according to feminist theorists (Shaw and Lee 2005).

Family Dynamics: Micro-Level Processes

The Agabi family belongs to the Hausa tribe of West Africa. They share a family compound composed of huts or houses for each family unit of one wife and her young children, plus one building for greeting guests, one for cooking, one for the older children, and one for washing. The compound is surrounded by an enclosure. Each member of the family carries out certain tasks: food preparation, washing, child care, farming, herding—whatever is needed for the group. Wives live with their husbands' families. Should there be a divorce, the children generally belong to the husband's household because the family lineage is through the father's side.

The eldest male is the leader. He makes decisions for the group. When a child is born, the eldest male within the family presides over a ceremony to name and welcome the child into the group. When the child reaches marrying age, the eldest male plays a major role in choosing a suitable mate. Upon his death, the power he has held passes to his eldest son, and his property is inherited by his sons.

However, the Agabis' eldest son has moved away from the extended family to the city, where he works in a factory to support himself. He lives in a small room with several other migrants. He has met a girl from another tribe and may marry her, but he will do so without his family's blessing. He will probably have a small family because of money and space pressures in the city. His lifestyle and even his values have already altered considerably. In the social world, the global trend toward the Global North model of industrialization and urbanization is altering cultures around the world and changing family life.

Families can be studied at each level of analysis: as interdependent micro-level social units with mom, dad, and the kids; as meso-level institutions that can be seen as economic units; and as an influential force in macro-level social systems. Many individual family issues that seem very intimate and personal are actually affected by norms and forces at other levels (e.g., migration and urbanization), and decisions of individuals at the micro level affect meso- and macro-level social structures (e.g., size of families). Individual families are, in essence, local franchises of a larger social phenomenon.

Thinking Sociologically

What kinds of changes in families would you anticipate as societies change from agricultural to industrial to information technology economies and as individuals move from rural to urban areas? What changes has your family undergone over several generations?

Mate Selection: How Do New Families Start?

At the most micro level, two people get together to begin a new family unit. In 2011 the world population will reach 7 billion people, but it is highly unlikely that your mate was or will be randomly selected from the entire global population. Even in Global North societies, where we think individuals have free choice of marriage partners, mate selection is seldom an entirely free choice. Indeed, mate selection is highly limited by geographical proximity, ethnicity, age, social class, and a host of other variables. As we shall see, micro-level and macro-level forces influence each other in the mate selection process.

Norms Governing Choice of Marriage Partners: Societal Rules and Intimate Choices

A number of cultural rules—meso- and macro-level expectations—govern the choice of a mate in any society. Most are unwritten norms. **Exogamy** is a norm that requires individuals to marry outside of their own immediate group. The most universal form of exogamy is the incest taboo, including restrictions against father-daughter, mother-son, and brother-sister marriages. Some countries, as well as about half the U.S. states, forbid first cousins to marry, whereas others, such as some African groups and many Syrian villages, encourage first-cousin marriages to solidify family ties and property holdings. Some societies require

village exogamy (marriage outside the village) because it bonds together villages and reduces the likelihood of armed conflict between them.

The reasons for exogamy range from recognition of the negative biological results of inbreeding to necessity for families to make ties with outside groups for survival. One clear issue is that rights to sexual access can cause jealousy that rips a social unit apart. If father and son became jealous about who was sleeping with the wife/mother or sister, relationships would be destroyed and parental authority sabotaged. Likewise, if the father went to the daughter for sexual satisfaction, the mother and daughter bond would be severely threatened (Davis 1960; Williams et al. 2009). No society can allow this to happen to its family system. Any society that has failed to have an incest taboo self-destructed long ago.

On the other hand, norms of **endogamy** require individuals to marry inside certain boundaries, whatever the societal members see as protecting the homogeneity of the group. The purpose is to encourage group bonding and solidarity, and to help minority groups survive in societies with different cultures. Endogamous norms may require individuals to select mates of the same race, religion, social class, ethnic background, or clan (Williams et al. 2009). For example, strictly endogamous religious groups include the Armenian Iranians, Orthodox Jews, Old Order Amish, Jehovah's Witnesses, and the Parsi of India. The result is less diversified groups but protection of the minority identity (Belding 2004). Whether marriages are arranged or entered into freely, both endogamy and exogamy limit the number of possible mates. In addition to marrying within a group, most people choose a mate with similar social characteristics—age, place of residence, educational background, political philosophy, moral values, and psychological traits—a practice called *homogamy*.

Going outside the limits in mate selection can make things tough for newlyweds who need family and community support. Few take this risk. For instance, in the United States, close to 80% to 90% marry people with similar religious values (Newman 2009). About 80% to 90% of Protestants marry other Protestants, and 64% to 85% of Catholics marry within their religious faith. For Jews, the figure has been as high as 90% but has dropped in recent decades to as low as 50% for more liberal groups of Jews (Newman 2009). In Canada, only one person in five marries across religious boundaries (British Columbia Ministry of Labour & Citizens' Services 2006). However, with increased tolerance for differences, cross-denominational marriage is more likely today than a century ago.

Interracial marriages also challenge norms of endogamy, yet the practice is becoming more common with every passing year. Of marriages in the United States, 14.6% involve a spouse from a different race or ethnic group (Inniss 2010). Also, one in five has a close relative in a mixed-race marriage, and one half of the dating population has dated someone of a different race. Among Hispanics, Puerto Ricans are the most likely to marry exogamously. Black-white marriages usually involve black men and white women, with nearly 15% of black men and a bit less than 7% of black women in mixed-race marriages (Carroll 2010). By contrast, Asian women are more likely to marry outside the group than Asian American men, and there is an overall pattern of 40% marrying whites (Carroll 2010; Lee and Edmonston 2005).

Each group may have different definitions of where the exogamy boundary is. For Orthodox Hasidic Jews, marriage to a Reform Jew is exogamy—strictly forbidden. Marriage of a Hopi to a Navajo is also frowned on—even though many Anglos would think of this as an endogamous marriage of two Native Americans.

So cultural norms of societies limit individual decisions about micro-level matters such as choice of a spouse, and they do so in a way that most individuals do not even recognize. Exogamy and endogamy norms and expectations generally restrict the range of potential marriage partners, even though some of these norms are weakening. Still, the question remains: How do we settle on a life partner?

Finding a Mate

In most societies, mate selection is achieved through arranged marriages, free-choice unions, or some combination of the two. In either case, selection is shaped by cultural rules of the society.

Arranged marriages involve a pattern of mate selection in which someone other than the couple—elder males, parents, a matchmaker—selects the marital partners. This method of mate selection is most common in traditional, often patriarchal, societies. Some examples follow.

For many traditional girls in Muslim societies, marriage is a matter of necessity, for girls' support comes from the family system. Economic arrangements and political alliances between family groups are solidified through marriage. Daughters are valuable commodities in negotiations to secure these ties between families (Burn 2011). Beauty, youth, talent, and pleasant disposition bring a high price and a good match. Should the young people like each other, it is icing on the cake. Daughters must trust that the male elders in their families will make the best possible matches for them. Clearly, the men hold the power in this vital decision.

Seated front and center with the bride and groom at many Japanese weddings is the matchmaker, the person responsible for bringing the relationship into being. After both families agree to the arrangement, the couple meets over tea several times to decide whether the match suits them. Today, between 10% and 30% of marriages in Japan are still arranged this way, the rest being called "love marriages." However, some marriages are a mix, combining arrangements with "love" (About.com 2011).

Japanese weddings are very formal and colorful events, and in many cases they still involve the parental selection of spouse, often with the help of a matchmaker.

Where arranged marriages are the norm, love has a special meaning. The man and woman may never have set eyes on each other before the wedding day, but respect and affection generally grow over time as the husband and wife live together. People from these societies are assured a mate and have difficulty comprehending marriage systems based on love, romance, and courtship, factors that they believe to be insufficient grounds for a lifelong relationship. They wonder why anyone would want to place himself or herself in a marriage market, with all of the uncertainty and rejection. Such whimsical and unsystematic methods would not work in many societies, where the structure of life is built around family systems.

Free-choice marriage, in contrast to arranged marriage, involves the partners selecting each other based primarily on romance and love. Sonnets, symphonies, rock songs, poems, and plays have been written to honor love and the psychological and physiological pain and pleasure that the mating game brings. However impractical romance may seem, marriage choice based on *romantic love*, the idea that each person has the right to choose a partner with minimal interference from others (Eshleman and Bulcroft, 2006:254), is becoming more prevalent. As societies around the world become more Westernized, women gain more rights and freedoms, and families exert less control over their children's choice of mates (Eshleman and Bulcroft 2010). Industrial and postindustrial societies tend to value love and individualism and tend to have high marriage rates, low fertility rates, and high divorce rates (Levine et al. 1995).

Free-choice mate selection is found in most Western societies. Couples in the United States tend to put more emphasis on romantic love and the process of attracting a mate than most other societies. For example, 86% of U.S. college students say they would not marry without love, the figure being higher for men than for women. Romantic love is most common in countries where individualism is emphasized over community interests (Eshleman and Bulcroft 2010).

E-romance on the Internet is helping facilitate mate selection in modern societies. The Internet facilitates what those in arranged marriage systems find bewildering about free-choice systems: How do you meet possible mates? Many of the e-dating services claim that their profiles and processes are based on social science research. For example, eHarmony has participants fill out an extensive 436-question personality profile that it claims is a "scientifically proven" compatibility-matching system. More than 9 million users hope for one or more matches from the system. Pepper Schwartz, a sociologist who studies relationships, helped design PerfectMatch's system (www.perfectmatch.com). Many services have sprung up, including specialized services such as ConservativeDates.com.

Because online dating is relatively new, it is unclear whether it will become a dominant form of mate selection (Gottlieb 2006). Are these services successful in matching up potential mates? A recent study found that one in three respondents was unattached and 7% were actively looking for partners. About 37% of those had tried online dating sites, and about half of those had been out on a date as a result of online services. About one third had formed long-term relationships. Once again, the social world model helps us understand that even mate selection processes are shaped by macro-level influences that filter to the national, institutional, community, and individual family levels. A very private and personal process is becoming transformed by global forces and trends (Madden and Lenhart 2006).

Thinking Sociologically

Is e-dating a modern-day form of the matchmaker? Is it replacing other forms of finding a mate? Why or why not?

Starting with the assumption that eligible people are most likely to meet and be attracted to others who have similar values and backgrounds, sociologists have developed various mate selection theories, several of which view dating as a three-stage process. (See Figure 10.1.)

1. Stimulus: We meet someone to whom we are attracted by appearance, voice, dress, similar ethnic background, sense of humor, or other factors. Something serves as a stimulus that makes us take notice. Of course, sometimes the stimulus is simply knowing the other person is interested in us.

2. Value comparison: As we learn about the other's values, we are more likely to find that person compatible if she or he affirms our own beliefs and values toward life, politics, religion, and roles of men and women in society and marriage. If values are not compatible, the person does not pass through our filter. We look elsewhere.

3. Roles and needs stage: Another filter comes when the couple explores roles of companion, parent, housekeeper, and lover. This might involve looking for common needs, interests, and favored activities. If roles and needs are not complementary to one's own, desire for a permanent relationship wanes.

The mate selection process varies somewhat from person to person, but social scientists believe that a sequential series of decisions, in a pattern such as that described above, is often part of the process (Eshleman and Bulcroft 2010; Murstein 1987).

Who Holds the Power? Authority Relations in Marriage

Power relations, another micro-level issue shaped by cultural norms at the macro level, affect the interactions and decision making in individual families. Two areas that have received particular sociological attention are decision making in marriage and work roles.

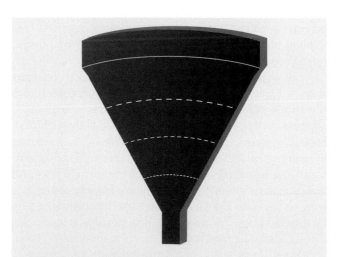

Figure 10.1 Mate Selection "Filtering"

The notion of mate selection described above is sometime referred to as a filter theory. It is as though you were filtering specs of gold, and the first filter holds out the large stones, the second filter holds back pebbles, the third filter stops sand, but the flakes of gold come through. Each stage in the mate selection process involves filtering some people out of the process. For you there may be other filter factors as well—such as religious similarities or common ethnicity.

Decision Making in Marriage

Cultural traditions establish the power base in society and family: patriarchy, matriarchy, or egalitarianism. The most typical authority pattern in the world is patriarchy, or male authority. Matriarchy, female authority, is rare. Even where the lineage is traced through the mother's line, males generally dominate decision making. Some analysts have suggested reasons for male dominance: Males are physically larger, they are free from childbearing, and they are not tied to one place by homemaking and agricultural responsibilities. However, social scientists find no evidence that there are any inherent intellectual or personality foundations for male authority as opposed to female authority (Kramer 2010; Ward and Edelstein 2009).

Egalitarian family patterns—in which power, authority, and decision making are shared between the spouses and perhaps with the children—are emerging, but they are not yet a reality in most households. For example, research indicates that in many U.S. families, decisions concerning vacation plans, car purchases, and housing are reached democratically. Still, most U.S. families are not fully egalitarian. Males generally have a disproportionate say in major decisions (Lindsey 2011).

Resource theory attempts to explain power relations by arguing that the spouse with the greater resources—education, occupational prestige, and income—has the greater power. In many societies, income is the most important factor because it represents identity and power. If only one spouse brings home a paycheck, the other is usually less powerful (Tichenor 1999). In families in which the wife is a professional, factors other than income, such as persuasion and egalitarian values, may enter into the power dynamic (Lindsey 2011). Regardless of who has greater resources, men in two-earner couples tend to have more say in financial matters and less responsibility for children and household tasks.

Who Does the Housework?

The *second shift,* a term coined by Hochschild (1989), refers to the housework and child care that employed women do after their first-shift jobs. Studies indicate that women work doing household activities 2.6 hours per day or about 18 hours per week, whereas men work 2.1 hours per day. Men spend more time doing leisure activities (U.S. Bureau of Labor Statistics 2011). On an average day, 83.7% of women and 66.8% of men spend some time doing housework (U.S. Bureau of Labor Statistics 2011). The next "Engaging Sociology" shows the breakdown in hours spent by men and women at various household activities.

Employment schedules also affect the amount of time each spouse contributes to household tasks. Husbands who are at home during hours when their wives are working tend to take on more tasks. Employment, education, and earnings give women more respect and independence and a power base for a more equitable division of labor across tasks (Cherlin 2010; Kramer 2010).

In many societies, couples exhibit highly sex-segregated family work patterns. When men do participate in household chores, they tend to do dishes, grocery shopping, repairs, and yard work, and they care for the car (Bianci et al. 2000). U.S. women spend on average 12 hours a week on child care, whereas men spend about half of that. Married mothers also spend a weekly average of just under 20 hours on housework other than child care, whereas married husbands devote just over 10 hours to such tasks (Newman 2009).

Interestingly, husbands who do an equitable share of the household chores actually report higher levels of satisfaction with the marriage (Stevens, Kiger, and Riley 2001). The success or failure of a marriage depends in large part on patterns that develop early in the marriage for dealing with the everyday situations including power relationships and division of labor.

While women generally are economically dependent on men, men are often dependent on women emotionally, for they are less likely to have same-sex friends with whom they share feelings and vulnerabilities. Men bond with one another, but they seldom develop truly intimate ties that provide support in hard times. So women are not entirely without power. It is just that their power frequently takes a different form (and results in fewer privileges) (Newman 2009). Thus, micro-level interaction between partners is shaped by a wide range of factors.

The Family as an Institution: Meso-Level Analysis

We experience family life at a very personal level, but the sum total of hundreds of thousands of families interacting in recognizable patterns results in "the family" as an institution.

Marriage and Family Structure: The Components of Family as an Institution

Although the family is an institution in the larger social system, it does vary in interesting ways from one society to another. One example is how many mates one should have. Some societies believe that several wives provide more hands to do the work and establish useful political and economic alliances between family groups. They bring more children into the family unit and provide multiple family members for emotional and physical support and satisfaction. On the other hand, having

Engaging Sociology

Household Maintenance by Gender

In many households, household tasks are highly gendered. As recently as the 1980s, wives and daughters spent two or three times as much time as fathers and sons in household tasks such as cleaning and laundry and yard work. However, the tides have been shifting, and while they are not entirely equal, they are more balanced.

Sociological Data Analysis:

1. What is the division of labor (by gender) for household maintenance in your family?
2. How did it evolve?
3. Is it considered fair by all participants?
4. How does it compare to the data in Figure 10.2?

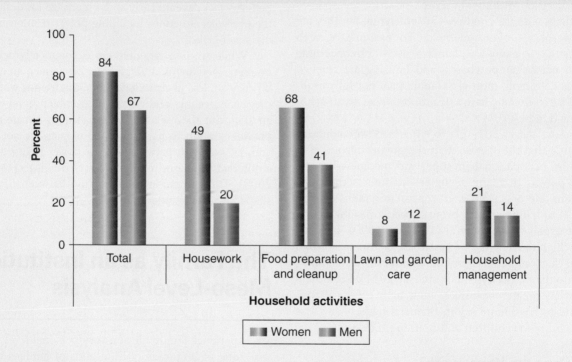

Figure 10.2   Percent of the Civilian Population Engaging in Household Activities, Averages per Day by Sex, 2010
Source: U.S. Bureau of Labor Statistics 2011.

one spouse per adult probably meets most individuals' social and emotional needs very effectively, is less costly, and eliminates the possibility of conflict or jealousy among spouses and shortage of wives for the remaining men. It is also easier to relocate a one-spouse family to urban areas, a necessity for many families in industrial and postindustrial societies. Let us examine the issue of adult partners in a family. Institutions lay out the general framework for families in any society and include types of marriages, extended and nuclear families, and other structural models of families. Individual families are local expressions of a larger corporate enterprise.

Types of Marriages

Monogamy and polygamy are the main forms of marriage found around the world. **Monogamy**, the most familiar form to those of us in Global North societies, refers to marriage of two individuals. **Polygamy**, in which a man or woman has more than one spouse, is most often found in agricultural societies where multiple spouses and children mean more help with the farm work. There are two main forms of polygamy—polygyny and polyandry. Anthropologist George Murdock (1967) found that *polygyny*, a husband having more than one wife, was allowed (although not always practiced) in 709 of the 849 societies he cataloged in his classic *Ethnographic Atlas*. Only 16% (136 societies) were exclusively monogamous (Barash 2002). Polygyny is limited because it is expensive to maintain a large family, brides must be acquired at high prices in some societies, and there is a global movement toward monogamy. Polygyny does increase at times of war when the number of men is reduced due to war casualties.

Polyandry, a wife having more than one husband, is practiced in less than 1% of the world's societies. Among the Todas of Southern India, for example, brothers can share a wife (O'Connel 1993). This usually happens when the men are poor and must share a single plot of land to eke out a meager livelihood, so they decide to remain a single household with one wife.

Members of Western societies often find the practice of polygamy hard to understand, just as those from polygamous societies find monogamy strange. Some societies insist on strict monogamy: Marriage to one other person is lifelong, and deviation from that standard is prohibited. Yet, most Western societies practice what could be called a variation of polygamy—serial monogamy. With high divorce and remarriage rates, Western societies have developed a system of marrying several spouses, but one at a time. One has spouses in a series rather than simultaneously.

Extended and Nuclear Families

The typical ma-pa-and-kids monogamous model that is familiar in many industrialized parts of the world is not as typical as it appears. From a worldwide perspective, it is only one of several structural models of family.

An **extended family** includes two or more adult generations that share tasks and living quarters. This may include brothers, sisters, aunts, uncles, cousins, and grandparents. In most extended family systems, the eldest male is the authority figure. This is a common pattern around the world, especially in agricultural societies. Some ethnic groups in the United States, such as Mexican Americans and some Asian Americans, live in extended monogamous families with several generations under one roof. This is financially practical and helps group members maintain their traditions and identity by remaining somewhat isolated from Anglo society.

As societies become more industrialized and fewer individuals and families engage in agriculture, the **nuclear family**, consisting of two parents and their children—or any two of the three—becomes more common. This worldwide

This polyandrous family poses for a photo in front of their tent in northwest China. Fraternal polyandry means that brothers share a common wife. When children are born, they call the oldest brother father and all other brothers uncle, regardless of who the biological father is. China's marriage law does not officially permit polyandry.

trend toward nuclear family occurs because more individuals live in urban areas where smaller families are more practical, mate selection is based on love, couples establish independent households after marriage, marriage is less of an economic arrangement between families, fewer marriages take place between relatives such as cousins, and equality between the sexes increases (Burn 2011; Goode 1970).

No matter in what form the family manifests itself structurally, a society's family institution is interdependent with each of the other major institutions. For example, if the health care institution is unaffordable or not functioning well, families may not get the care they need to prevent serious illness. If the economy goes into a recession and jobs are not available, families experience stress, abuse rates increase, and marriages are more likely to become unstable. When husbands lose jobs, it often makes their primary role in the family ambiguous, causing sense of failure by the husband and stress in the relationship. In single-parent families in which the mother is the custodial parent, the loss of her job can be financially devastating. In worst-case scenarios, families who lose their incomes may become homeless (Staples 1999; Willie 2003).

Thinking Sociologically

Under what social circumstances would an extended family be helpful? Under what circumstances would it be a burden? What are strengths and weaknesses of nuclear families?

Family is a diverse and complex social institution. It interacts with other institutions and in some ways reinforces them. Families prepare the next generation. Adult

In many societies, the family is still the primary unit of economic production. These family members in Myanmar are selling the goods they produced as a family.

family members teach national loyalty, tutor children in reading and math, and mentor their progeny on the use of money. Families pray together and provide care of disabled, infirm, or sick members. As a basic institution, the family plays a role in the vitality of the entire nation. So it should not be surprising that at the macro level, many national and global policy decisions concern how to strengthen the family.

National and Global Family Issues: Macro-Level Analysis

The most effective way to explore macro-level issues pertaining to families is to explore policy matters that affect the family or that are intended to strengthen families. After exploring issues of national concern—cohabitation, homosexual relationships, and divorce—we look at some global trends in marriage and family life.

Cohabitation

Cohabitation—living together in a sexual relationship without marriage—is a significant macro-level trend in many countries that has implications for national family laws,

tax laws, work benefits, and other macro-level issues. In the United States, the number of unmarried couples living together doubled in the 1990s, from 2.9 million in 1990 to more than 6.2 million in 2005 (U.S. Census Bureau 2006a). From 2007 to 2008, the increase in cohabitation was 5%, and from 2008 to 2009 it was down 2%. However, there was an unusually large 13% increase (868,000) in cohabiting couples between 2009 and 2010 (Kreider 2010a). The number of "unmarried households" has been rising dramatically for several decades. Two thirds of married couples lived together for an average of two years before marriage (Jayson 2005b).

Is cohabitation replacing dating? Some argue that a newly emerging pattern for some young adults between 25 and 34 is serial cohabitation (Jayson 2005a; Lichter, Turner, and Sassler 2010). As Figure 10.3 shows, the increases in cohabiting households (same-sex and different-sex couples) in the United States over 50 years are dramatic.

Why do heterosexual couples decide to cohabit? Cohabitants cite a number of reasons:

- Rejection of the superficial dating game
- A desire to enter more meaningful relationships with increased intimacy but with freedom to leave the union
- Emotional satisfaction and reduced loneliness
- A chance to clarify what individuals want in a relationship and try out a relationship before permanent commitment
- Financial benefits of sharing living quarters
- Sexual gratification (the latter cited more often by men than women) and some protection against disease by having one partner (Bumpass and Lu 2000; Lamanna and Riedmann 2008)

For some, cohabitation is part of the mate selection process. In 1987, 33% of the adult population between the ages of 19 and 44 had cohabited at some point, and by 2009, that percentage had increased to between 60% and 70% (McCarthy 2009; Stanley and Rhoades 2009). However, at any given point in time, only about 7% of the population in the United States is cohabiting (Benokraitis 2008). Countries with the highest percentages of women between 20 and 24 years of age in cohabiting relationships include Sweden (77%), New Zealand (67%), Austria (64%), France (63%), Netherlands and Norway (each 57%), and Canada (46%) (United Nations 2003). Latin American and Caribbean surveys have indicated that more than one in four women between the ages of 15 and 49 are in relationships they call consensual unions, living together without official sanction. However, such women typically have far less legal protection than European women during or after such unions. The rates of cohabitation seem to be declining in many African countries, where the rates are often below 15% of women (United Nations 2003).

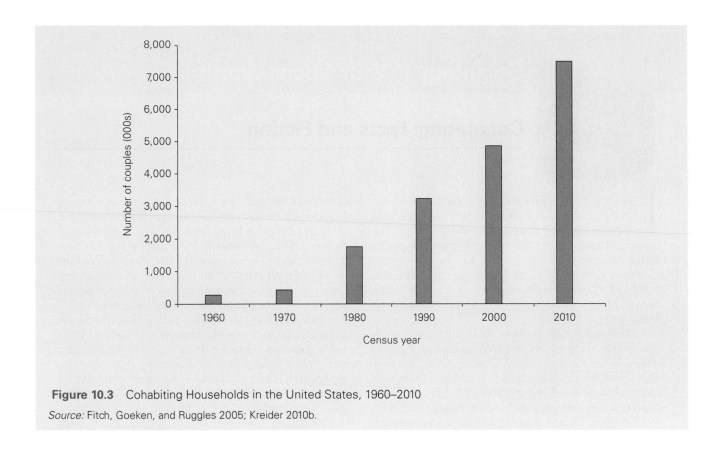

Figure 10.3   Cohabiting Households in the United States, 1960–2010

Source: Fitch, Goeken, and Ruggles 2005; Kreider 2010b.

Marriage versus cohabitation rates and reasons vary significantly by ethnicity. For whites in the United States, cohabitation is often a precursor to marriage. For African Americans, it may be an alternative to marriage. Financial problems encourage cohabitation in African American families because many African American men avoid marriage if they do not think they can support a family and fulfill the breadwinner role. Although childbearing increases the chance of marriage, it is a much stronger impetus for white than for black cohabitants.

We might assume that cohabiting would allow couples to make more realistic decisions about entering permanent relationships. However, studies show that this is not always the case. When couples have different objectives for cohabiting, or have not discussed the future of the relationship before moving in together, problems may arise and divorce may result as indicated in the next "Sociology in Your Social World."

Thinking Sociologically

First, read "Sociology in Your Social World." How do you evaluate the argument that cohabitation is a threat to the stability of marriages? Is reducing cohabitation an effective step to strengthening marriage?

Same-Sex Relationships and Civil Unions

A hotly debated macro-level policy matter is the official status of homosexual couples. Policy decisions affect rights and benefits for partners, but there is intense disagreement even over whether this issue is about human rights or about divinely determined rights and wrongs.

In the United States, the Defense of Marriage Act (U.S. General Accounting Office 2004) grants married couples 1,138 rights not available to unmarried partners. As same-sex relationships become more widely acknowledged in the Global North, many gay and lesbian couples are living together openly as families. Denmark was the first country to recognize same-sex unions in 1989, granting legal rights to couples. In 2001, the Netherlands was the first country to call those unions "marriages." Ten other countries now allow same-sex marriages, including Canada. However, some other countries have threatened punishment for openly gay individuals, and Uganda almost passed a bill authorizing the death penalty for homosexuals (Martin and Thompkins 2009).

The 2000 U.S. Census reports 601,209 declared gay or lesbian households in the United States. Most scholars acknowledge that this is probably a substantial underreporting because of the stigma in many communities of reporting that one is gay or lesbian (Smith and Gates 2001). The Human Rights Campaign estimates that 99.3% of all

Sociology in Your Social World

Cohabiting: Facts and Fiction

I t seems reasonable—and many people believe—that if a couple lives together before marriage, they are more likely to have a marriage that lasts. Yet, research shows that not all cohabiting unions are the same. While cohabitation can be a pathway to marriage for some (Smock 2004), other research suggests that cohabitation has primarily become a stage of dating instead of an inevitable path to marriage (Sassler and Miller 2011).

Many cohabiting couples never marry. It is more common for a cohabiting couple to have a child together than it is for them to marry (Cherlin 2010). However, some cohabiting couples do choose to marry. Those who had not talked about the possibility of marriage before moving in together (cohabitation as a stage of dating) are more likely to divorce than couples who saw cohabitation as a step to marriage or couples who did not cohabit prior to marriage (Stanley and Rhoades 2009).

The association between cohabitation before marriage and higher risk of divorce for some couples has intrigued researchers, and several attempts have been made to explain the pattern. Linda Waite and Maggie Gallagher (2000) offer an intriguing interpretation to explain high divorce rates among those who cohabited. They point out that marriage has many benefits. Married people have better physical and mental health, have more frequent and more satisfying sex, and are substantially better off financially by the time of retirement than single people (they have more than double the money invested per person). Interestingly, the same benefits do not accrue to people who cohabit.

Much of this is because marriage links two people together in a way that makes them responsible to each other. If one smokes or drinks excessively, the partner has the right to complain about it—to essentially nag that person into better health patterns—because one partner's health has a direct impact on the other's and vice versa. Likewise, if one partner likes to spend money freely on vacations, the more frugal person is likely to restrain these spending habits. This is acceptable because their financial futures are closely intertwined. As each person restrains the habits of the other, the couple is likely to end up with more savings. At the end of their careers, the couple is likely to have more money put away for retirement. Also, people tend to be more adjusted when they have unqualified and unambiguous emotional support from another person whose life is inextricably tied to their own.

However, there is little relationship between cohabitation and ensuing marital satisfaction, emotional closeness, sharing of roles, or amount of conflict. Some argue that this is because the cohabiting relationship is not grounded in a bargain created by a marriage contract. This is especially true for couples who live together without intending to marry or having discussed marriage. These cohabitors are more likely to keep their finances separate and their options open. Moreover, they do not feel as if they have the right to nag the other person about health habits or finances. Their bond is emotional, but their actions toward one another do not always show that they see their futures as being intertwined. Thus, many of the advantages that come with marriage are lost. Yet inertia may lead them into marriage. When cohabiting couples do marry, individual partners may be shocked to find the other person beginning to nag them or to restrain their spending habits. The relationship is quite different, and partners are irritated by the changed behavior. So cohabitation does not always tell us what life with this other person will be like, if the intent to marry was not present from the outset (Stanley and Rhoades 2009; Waite and Gallagher 2000).

Whether individuals cohabit as an alternative to marriage or as a path to marriage, relationships can be fulfilling and stable. For those who cohabit without considering marriage before moving in, relationships are less likely to result in marriage and are much more likely to end in divorce if they do marry. Then again, some view cohabiting as an alternative to marriage rather than a prelude (Benokraitis 2008).

counties in the United States have same-sex couples, and about 3.1 million people live in same-sex relationships. One in three lesbian couples and one in five gay male couples are raising children (Human Rights Campaign 2003). In some places, these unions are officially recognized by state or religious organizations, but recognition is controversial at the national level. Many religious groups believe homosexuality is unacceptable and condemn it in a variety of ways. Further, they deny the right to legal marriage unless the relationship is heterosexual (Laythe, Finkel, and Kirkpatrick 2001; Powell et al. 2010).

Civil unions and gay marriage are therefore highly controversial. In 2000, Vermont, spurred by a unanimous state Supreme Court ruling that the prohibitions against same-sex marriage were discriminatory, passed a law allowing civil unions for gays and lesbians. Other states followed with similar contract options: Connecticut in 2005, New Jersey in 2007, and New Hampshire and Oregon in 2008 (Human Rights Campaign 2008). Marriage was approved for gays and lesbians in Massachusetts in 2004. In 2008, the California Supreme Court overruled the state's ban on same-sex marriages, and later that year a state referendum reinstated the ban (Dolan 2008; Ontario Consultants on Religious Tolerance 2008).

In response to the controversies, many states have amended their state constitutions so that marriage is limited to legal unions between a man and a woman. As of March 2011, 41 states have passed laws prohibiting same-sex marriages. The U.S. public is evenly split on allowing gay and lesbian couples to legally form civil unions. By 2011, same-sex marriage was legal in Massachusetts (2004), Connecticut (2008), Iowa (2009), New Hampshire (2010), Vermont (2009), and Washington, DC (2010) (Gay Marriage Research Center 2011). Map 10.1 indicates states that allow or recognize gay marriage (Human Rights Campaign 2011). Thus, the picture is mixed.

Surveys conducted since 1977 have consistently shown a majority of Americans supportive of same-sex marriage, but they also reveal some remarkable shifts in attitudes toward homosexuals (O'Brien and Westen 2007). Surveys conducted since 1977 have consistently shown a majority of Americans against same-sex marriage, but they also reveal some remarkable shifts in attitudes toward homosexuals (O'Brien and Westen 2007). A recent study found that 59% of Americans define households with two men raising a child as a "family." Likewise, when a household involves two women raising children, 61% of Americans agree that this is a "family." Even when no children are involved, 32% of the public view homosexual couples to be a family. That figure is only a few percentage points below the number of people who view cohabiting heterosexual couples without children to be a "family." In

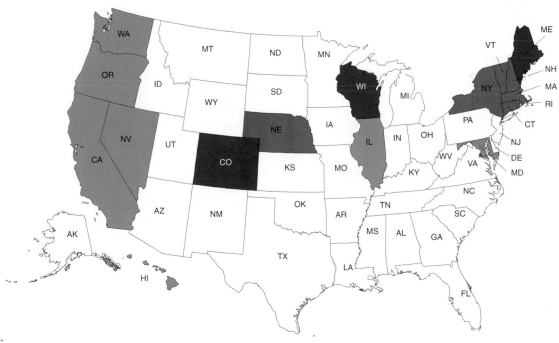

Legend:

❖ State issues marriage licenses to same-sex couples.

❖ State recognizes marriages by same-sex couples legally entered into in another jurisdiction.

❖ Statewide law providing the equivalent of state-level spousal rights to same-sex couples within the state (such as civil unions)

❖ Statewide law providing some statewide spousal rights to same-sex couples within the state

Map 10.1 Marriage Equality and Other Relationship Recognition Laws

Source: © 2011 the Human Rights Campaign. All rights reserved. Printed by permission.

In the backyard of their home in Beverly Hills, comedian Ellen DeGeneres, left, and Portia de Rossi got married during one of the short windows when same-sex marriage was legal in California.

2006, only 38% of the population was rigidly exclusionist regarding same-sex relationships never being recognized as a family (Powell et al. 2010). So the public, especially the younger generation, is increasingly willing to grant gay and lesbian households a status as legitimate families.

Those who favor allowing gay and lesbian marriages claim that supportive lifelong dyadic relationships are good for individuals and good for society. They see the fact that homosexuals want stable socially sanctioned relationships as an encouraging sign about how important the family is to society and how homosexuals want to fit in. Moreover, because many societies offer tax benefits, insurance coverage, and other privileges to married couples, the denial of marriage on the basis of one's gender attraction is seen as discriminatory and may be costly to partners who are denied rights and to societies that must care for the needs of uninsured or unemployed.

Those opposed argue that marriage has been a function of the church, temple, and mosque for centuries. None of the religious traditions have historically recognized gay relationships as legitimate (although a few are doing so now). Some opponents appeal to biology with the assertion that marriage is a legitimate way to propagate the species. Because homosexual unions do not serve this purpose, they do not serve the society, according to opponents of same-sex marriage. You may want to think about whether this seems to be an important criterion in what comprises a "legitimate" family.

Divorce—Contract Breaking

Is the family breaking down? Is it relevant in today's world? Although most cultures extol the virtues of family life, the reality is that not all partnerships work. Support is not forthcoming for the partner, trust is violated, abuse is present, and relationships deteriorate. So we cannot discuss family life without also recognizing the often painful side of family life that results in contract breaking.

Some commentators view divorce rates as evidence that the family is deteriorating. They see enormous problems created by divorce. There are costs to adults who suffer guilt and failure, to children from divided homes, and to the society that does not have the stabilizing force of intact lifelong partnerships. Many children around the world are raised without both natural parents present. For example, nearly half of all U.S. children today live at least part of their lives in single-parent families. About 9% of households in the United States are headed by single parents (U.S. Census Bureau 2007b). If we consider only households with children, just over 67% of children under 18 live with two married parents, 29.5% are in single-parent households, and roughly 4% are in other arrangements (U.S. Census Bureau 2011d).

Others argue that marriage is not so much breaking down as adapting to a different kind of social system. They claim that the family of the 1950s, depicted in television shows such as *Leave It to Beaver, The Adventures of Ozzie and Harriet,* and *Father Knows Best* as white, middle-class, and suburban, would be ill suited to the societies we have today. Indeed, more people today express satisfaction with marriage than at any previous time period, and there are more golden (50-year) wedding anniversaries now than ever before. Even late in the nineteenth century, the average length of a marriage was only 13 years—mostly because life expectancy was so short. "Till death do us part" was not such a long time then as it is today, when average life expectancy is in the late 70s (Coontz 2005). There are many misconceptions about divorce in the twenty-first century as well, and the next "Sociology in Your Social World" addresses some of those.

Thinking Sociologically

Micro-level issues of divorce may be easier to identify since they are usually rooted in the personalities and relationship factors of the individuals. Talk with friends and family who have divorced about micro-level factors that contributed. Now, based on the previous discussion, make a list of meso-level factors (e.g., religious, economic, legal, educational) that contribute to or reduce divorce rates.

Sociology in Your Social World

Debunking Misconceptions About Divorce

By David Popeneo and Barbara Dafoe Whitehead

1. Half of all marriages end in divorce.

That may have been the case several decades ago, but the divorce rate has been dropping since the early 1980s. If today's divorce rate continues unchanged into the future, the chances that a marriage contracted this year will end in divorce before one partner dies has been estimated to be between 40 and 45 percent.

2. Because people learn from their bad experiences, second marriages tend to be more successful than first marriages.

Although many people who divorce have successful subsequent marriages, the divorce rate of remarriages is in fact higher than that of first marriages.

3. Living together before marriage is a good way to reduce the chances of eventually divorcing.

Many studies have found that those who live together before marriage have a considerably higher chance of eventually divorcing. The reasons for this are not well understood. In part, the type of people who are willing to cohabit may also be those who are more willing to divorce. There is some evidence that the act of cohabitation itself generates attitudes in people that are more conducive to divorce, for example the attitude that relationships are temporary and easily can be ended.

4. Divorce may cause problems for many of the children who are affected by it, but by and large these problems are not long lasting and the children recover relatively quickly.

Divorce increases the risk of interpersonal problems in children. There is evidence, both from small qualitative studies and from large-scale, long-term empirical studies, that many of these problems are long lasting. In fact, they may even become worse in adulthood.

5. Having a child together will help a couple to improve their marital satisfaction and prevent a divorce.

Many studies have shown that the most stressful time in a marriage is after the first child is born. Couples who have a child together have a slightly decreased risk of divorce compared to couples without children, but the decreased risk is far less than it used to be when parents with marital problems were more likely to stay together "for the sake of the children."

6. Following divorce, the woman's standard of living plummets by seventy-three percent while that of the man's improves by forty-two percent.

This dramatic inequity, one of the most widely publicized statistics from the social sciences, was later found to be based on a faulty calculation. A reanalysis of the data determined that the woman's loss was twenty seven percent while the man's gain was ten percent. Irrespective of the magnitude of the differences, the gender gap is real and seems not to have narrowed much in recent decades.

7. When parents don't get along, children are better off if their parents divorce than if they stay together.

A recent large-scale, long-term study suggests otherwise. While it found that parents' marital unhappiness and discord have a broad negative impact on virtually every dimension of their children's well-being, so does the fact of going through a divorce. In examining the negative impacts on children more closely, the study discovered that it was only the children in very high conflict homes who benefited from the conflict removal that divorce may bring. In lower-conflict marriages that end in divorce—and the study found that perhaps as many as two thirds of the divorces were of this type—the situation of the children was made much worse following a divorce. Based on the findings of this study, therefore, except in the minority of high-conflict marriages it is better for the children if their parents stay together and work out their problems than if they divorce.

8. Because they are more cautious in entering marital relationships and also have a strong determination to avoid the possibility of divorce, children who grow up in a home broken by divorce tend to have as much success in their own marriages as those from intact homes.

Marriages of the children of divorce actually have a much higher rate of divorce than the marriages of children from intact families. A major reason for this, according to a recent study, is that children learn about marital commitment or permanence by observing their parents. In the children of divorce, the sense of commitment to a lifelong marriage has been undermined.

(Continued)

(Continued)

9. Following divorce, the children involved are better off in stepfamilies than in single-parent families.

The evidence suggests that stepfamilies are no improvement over single-parent families, even though typically income levels are higher and there is a father figure in the home. Stepfamilies tend to have their own set of problems, including interpersonal conflicts with new parent figures and a very high risk of family breakup.

10. Being very unhappy at certain points in a marriage is a good sign that the marriage will eventually end in divorce.

All marriages have their ups and downs. Recent research using a large national sample found that eighty-six percent of people who were unhappily married in the late 1980s, and stayed with the marriage, indicated when interviewed five years later that they were happier. Indeed, three fifths of the formerly unhappily married couples rated their marriages as either "very happy" or "quite happy."

ADDITIONAL MYTH

It is usually men who initiate divorce proceedings.

Two-thirds of all divorces are initiated by women. One recent study found that many of the reasons for this have to do with the nature of our divorce laws. For example, in most states women have a good chance of receiving custody of their children. Because women more strongly want to keep their children with them, in states where there is a presumption of shared custody with the husband the percentage of women who initiate divorces is much lower. Also, the higher rate of women initiators is probably due to the fact that men are more likely to be "badly behaved." Husbands, for example, are more likely than wives to have problems with drinking, drug abuse, and infidelity.

Source: Copyright 2002 by David Popenoe, the National Marriage Project, University of Virginia. Reprinted by permission of David Popenoe.

Macro-level social issues also contribute to divorce. These in turn result in micro-level individual family problems. The causes may be families separated by war or economic problems. One reason for divorce is family violence that can result from problems in the family's environment such as unemployment. Domestic abuse takes many forms: emotional abuse, denial and blame, intimidation, coercion or threats, use of power, isolation, using children as pawns, and economic abuse—including withholding of funds (Mayo Clinic 2011; Straus, Gelles, and Steinmetz 2006). Abuse is one of the major causes of divorce.

Children are sometimes the unwitting victims of abuse—beaten, bullied, abused, raped, targeted by predators, and neglected. The assault rate for children is over two times as high as the rate for the general population, 15.5 children per 1,000 (Finkelhor 2008). The stress on the family when violence is present means the family is not fulfilling its functions of security and belonging. Mothers may sometimes endure the pain themselves, especially if they feel they are trapped with no alternatives, but the same woman may take action when it is her child who suffers. No wonder families experiencing domestic abuse often end up with a divorce.

If we grant that not everyone is temperamentally suited to sustain a nurturing marriage for 50 years, and if we acknowledge that people do change over time, what would be an acceptable divorce rate for our society—5%

A security guard checks people for weapons before they board Miami-Dade County's Family Division Circuit Court bus, which was dubbed "The Divorce Bus." More than two dozen uncontested divorce cases were heard in fewer than 45 minutes. This does seem to give new meaning to the idea of divorce made convenient.

of all marriages? 20%? 40%? This is difficult for governments to decide, but many people feel that the current rate, someplace between 45% and 50% of all marriages (3.4% per 1,000 people in 2009), is too high (Centers for Disease Control and Prevention, National Vital Statistics System 2011; Divorce Magazine 2011; Religious Tolerance 2009a). When people say that more than half of today's marriages will end in divorce, they are pointing out that with upward divorce trends, by the time people now in their 20s reach their 70s, the rates could be 50% to 55%. However, it is also possible that such predictions will not come true. We know that social predictions can change as people and societies alter their behaviors and their policies. From the 1960s to the 1980s there was a very dramatic increase in divorce rates, but recently, the divorce trend has been moving in a downward direction (Coontz 2005; U.S. Census Bureau 2011e).

One reason for the dramatic increases in the U.S. divorce rate after the early 1970s was a policy change: no-fault divorce laws. For centuries in North America, one had to prove that the other party was in breach of contract. Marriage was a lifelong contract that could only be severed by one party having violated the terms of the contract. Each state spelled out those acts that were so odious that they justified ending such a sacred vow. Even then, if both parties had done something wrong (he was an adulterer, but she did not clean the house and was therefore in "neglect of duty"), the judge was obligated to rule that because they both violated the contract, the divorce would be denied. This resulted in ugly and contested divorces and in people being forced to live together in unloving, nonnurturing relationships.

In 1970, the state of California was the first to initiate no-fault divorce wherein a couple could end a marriage without proving that the other person was in breach of contract. A bilateral no-fault divorce (sometimes called *dissolution*) requires both parties to agree that they want out of the marriage. If they agree to the terms of settlement (child custody, child visitation rights, split in property, etc.), the marriage can be dissolved (Gilchrist 2003). A second form, unilateral no-fault divorce, allows one person to insist that the marriage features irreconcilable differences—the two do not have to agree. Many women feel this arrangement protects vulnerable women from staying in an unloving or abusive relationship. They do not have to give up everything to get their partners to sign the agreement.

All this makes divorce in the United States much easier to obtain than in previous years. However, some ask if divorces are being sought for the slightest offense. Some critics believe this ease has led to a *divorce culture*—a society in which people assume that marriages are fragile rather than assuming that marriages are for life (a *marriage culture*).

Thinking Sociologically

Is divorce really a problem, or is it a solution to a worse problem? Does divorce have lasting consequences? What evidence supports your position?

Divorce and Its Social Consequences

The highest rates of divorce are among young couples. In the United States, the highest rates are for women in their teens and men between 20 and 24. The rate of divorce has leveled off and even dropped since the U.S. high mark in 1981 and the Canadian high mark in 1987, as shown in Tables 10.1 and 10.2.

The emotional aspects of divorce are for many the most difficult. Divorce is often seen as a failure, rejection, and even punishment. Moreover, a divorce often involves a splitting with family and many close friends; with one's church, mosque, or synagogue; and from other social contexts in which one is known as part of a couple (Amato 2000). No

Table 10.1 U.S. Divorce Rate Trends

Year	Divorces per 1,000 Population
1950	2.6
1960	2.2
1970	3.5
1980	5.2
1981	5.3 (highest rate)
1990	4.7
2000	4.0
2009	3.5

Source: Centers for Disease Control and Prevention, National Vital Statistics System 2011.

Table 10.2 Canadian Divorce Rate Trends

Year	Divorces per 1,000 Population
1961	0.3
1969	1.2
1981	2.7
1987	3.6 (highest rate)
1990	2.9
1995	2.7

Source: DivorceRate.org n.d.

wonder divorce is so wrenching. Unlike simple societies, most modern societies have no ready mechanism for absorbing people back into stable social units such as clans.

Adjustment to divorced status varies by gender: Men typically have a harder time emotionally adjusting to single-hood or divorce than women. Divorced men must often leave not only their wives but also their children, and whereas many women have support networks, fewer men have developed or sustained friendships outside of marriage. Finances, on the other hand, are a bigger problem for divorced women than for men, especially if women have children to support. Many divorced and widowed women fall into poverty during the first five years of being single (McManus and DiPrete 2001). In fact, more than 25% of divorced women with children find themselves living below the poverty line (Newman 2009).

There are also costs for children, one million of whom experience their parents' divorce each year in the United States. These children's lives are often turned upside down: Many children move to new houses and locations, leave one parent and friends, and make adjustments to new schools and to reduced resources. Only 46.8% of parents who were owed child support in 2007 received the full child support payments due them. Nearly one third (29.5%) received less than the specified amount, and one quarter received none at all. Of the $34.1 billion owed in child support, 62.7% was paid (U.S. Census Bureau 2009d).

Adjustment depends on the age of the children and the manner in which the parents handle the divorce. Children in families with high levels of marital conflict may be better off in the long term if parents divorce (Sobolewski and Amato 2007). If the children are torn between two feuding parents or if they are the focus of a bitter custody battle, they may suffer substantial scars. Many studies indicate divorce lowers the well-being of children in the short term, affecting school achievement, peer relationships, and behavior. However, more important may be the long-term or lasting effects on their achievement and quality of life as these children become adults (Amato and Sobolewski 2001). The studies offer quite variable findings on this. One study that followed children of divorce for 15 years showed that through adolescence and into adulthood, many children continue to feel anxious and have fears, anger, and guilt (Sun 2001; Wallerstein and Blakeslee 2004). Adults may experience depression, lower levels of life satisfaction, lower marital quality and stability, more frequent divorce, poorer relations with parents, poorer physical health, and lower educational attainment, income, and occupational prestige (Eshleman and Bulcroft 2010). Other studies suggest that later in life, individuals whose parents divorced during their childhood have a higher probability of teen marriage, divorce, peer problems, delinquency, truancy, and depression (Chase-Lansdale, Cherlin, and Kiernan 1995; Newman 2009).

Men often find emotional adjustment to divorce especially difficult since they often leave the children and they have fewer intimate friendships outside of marriage.

On the other hand, some studies find that children who are well adjusted to begin with have an easier time with divorce, especially if they can remain in their home and in their familiar school, with both parents part of their lives, and if they maintain their friendship networks. Grandparents, too, can provide stability during these traumatic times. For instance, Ahrons (2004) found that adults whose parents were divorced are actually very well adjusted and happy. She found no general long-term negative consequences of parental divorce in her extensive longitudinal study that traced people into their middle years. About 79% of her respondents said their parents' decision to divorce was a good one, and 78% indicated that they were not affected by the divorce. The key, she found, is the nature of the postdivorce relationships. Large numbers of divorced partners have very civil relationships and continue to cooperate and collaborate on behalf of the children. In these cases, there are few, if any, long-term wounds.

Thinking Sociologically

Would making it harder to get a divorce create stronger and healthier families? Would it create more stable but less healthy and nurturing families? If you were making divorce policies, what would you do? What are the positive and negative aspects of your policy?

Global Family Patterns and Policies

Family systems around the world are changing in similar ways, pushed by industrialization and urbanization, by migration to new countries or refugee status, by changing kinship and occupational structures, and by influences from outside the family. The most striking changes include free choice of spouse, more equal status for women, equal rights in divorce, neolocal residency (when partners in a married couple live separate from either set of parents), bilateral kinship systems (tracing lineage through both parents), and pressures for individual equality (Sado and Bayer 2001). However, countermovements in some parts of the world call for strengthening of marriage through modesty of women, separation of the sexes (in both public and private spheres), and rejection of high divorce rates and other Western practices.

National policies that limit availability of birth control and knowledge about family planning affect the economic circumstances of families who have no choice but to raise large families. One result is an increase in out-of-wedlock births, which creates more single-parent families. Likewise, as individuals and couples around the world make decisions such as choosing a spouse based on love, rejecting multiple wives, or establishing a more egalitarian family, the collective impact may rock the foundations of the larger society. Because governments make policies that influence families, the interaction between policy makers and social scientists can lead to laws based on better information and more comprehensive analysis of possible consequences. This is part of the public contribution of applied sociology—providing accurate information and analysis for wise public policy decisions.

We see that family life, which seems so personal and intimate, is actually linked to global patterns. Global aid is activated when drought, famine, or another disaster affects communities and a country is not able to provide for families. In such cases, international organizations such as the United Nations, Doctors Without Borders, Oxfam, and the

Global forces, such as ethnic holocausts that create refugees, can strain and destroy families. This is a scene of refugees at a camp in the Central African Republic.

Red Cross mobilize to support families in crises. Support varies from feeding starving children to opposing the slavery that occurs when parents are reduced to selling their children to survive. International crises can lead to war, perhaps removing the main breadwinner from the family or taking the life of a son or daughter who was drafted to fight. Homes and cultivated fields may be destroyed and the families forced into refugee status.

Do marriage and divorce rates indicate the family is in crisis? To answer these questions, we need information on current patterns, historical trend lines, and patterns in other parts of the world. The next "Sociology Around the World" provides cross-cultural data on marriage and divorce ratios.

We have been talking about global and national trends (divorce rates) regarding an institution (the family) and the consequences they have for individuals. Processes at the macro and meso levels affect the micro level of society, and decisions at the micro level (i.e., to dissolve a marriage) affect the community and the nation. The various levels of the social world are indeed interrelated in complex ways.

Sociology Around the World

Cross-Cultural Differences in Family Dissolution

Is the institution of the family breaking down around the world? Perhaps this is the wrong question. We may, instead, need to consider how the family copes with changing national and global demands. Family conflict and disorganization occur when members of the family unit do not or cannot carry out roles expected of them by spouses, other family members, the community, or the society. This may be due to voluntary departure (divorce, separation, desertion), involuntary problems (illness or other catastrophe), a crisis caused by external events (death, war, depression), or failure to communicate role expectations and needs. Many of these role failures are a direct consequence of societal changes due to globalization. Once again, the social world model helps us understand macro-level trends and patterns that affect us in micro-level contexts.

Divorce is still very limited in some parts of the world, and it may be an option for only one gender. In some Arab countries, only the husband has had the right to declare "I divorce thee" in front of a witness on three separate occasions, after which the divorce is complete. The wife returns, sometimes in disgrace, to her family of procreation, while the husband generally keeps the children in the patriarchal family and is free to take another wife. Only recently is divorce initiated by the wife coming to be accepted in some countries, although the grounds for divorce by women may be restricted (Khazaleh 2009). Despite a seemingly easy process for men to divorce, the rate remains rather low in many Global South countries because family ties and allegiances are severely strained when divorces take place. Thus, informal pressures and cultural attitudes restrain tendencies to divorce.

Still, when family turmoil and conflict are too great to resolve or when the will to save the family disappears, the legal, civil, and religious ties of marriage may be broken. The methods for dissolving marriage ties vary, but most countries have some form of divorce. Table 10.3 compares marriage and divorce rates in selected industrial countries. Notice that while the divorce rate in countries such as the United States is quite high, the marriage rate is also high.

Table 10.3   Marriage and Divorce Rates in Selected Countries, 1980–2008

Country	Marriages per 1,000 Persons in Population				Divorces per 1,000 Persons in Population				
	1980	1990	2000	2008	1980	1990	2000	2008	2006
USA	15.9	14.9	12.5	10.6	7.9	7.2	6.2	5.2	
Canada	11.5	10.0	7.5	6.4	3.7	4.2	3.1	na	
Japan	9.8	8.4	9.2	na	1.8	1.8	3.1	na	
Denmark	8.0	9.1	10.8	10.3	4.1	4.0	4.0	4.1	
France	9.7	7.7	7.9	6.6	2.4	2.8	3.0	na	
Germany	X	8.2	7.6	6.9	X	2.5	3.5	3.5	
Ireland	10.9	8.3	7.6	6.3	na	na	1.0	na	
Italy	8.7	8.2	7.3	7.6	0.3	0.7	1.0	1.3	
Netherlands	9.6	9.4	8.2	7.6	2.7	2.8	3.2	2.9	
Spain	9.4	8.5	7.9	6.2	na	0.9	1.4	3.5	
Sweden	7.1	7.4	7.0	8.3	3.7	3.5	3.8	3.5	
United Kingdom	11.6	10.0	8.0	na	4.1	4.1	4.0	na	

Source: U.S. Census Bureau, 2011e, Table 1335.

Note: na = not available; X = country was two nations at that time.

The next two chapters continue to examine institutions and their interconnections. As children grow up and branch out from the embrace of the family, the social environments they experience first are usually the local school and a religious congregation. It is to education and religion that we turn next.

What Have We Learned?

Despite those who lament the weakening of the family, the institution of family is here to stay. Its form may alter as it responds and adapts to societal changes, and other institutions will continue to take on functions formerly reserved for the family. Still, the family is an institution crucial to societal survival, and whatever the future holds, the family will adapt in response to changes in other parts of the social world. It is an institution that is sometimes vulnerable and needs support, but it is also a resilient institution—the way we partner and "make people" in any society.

Our happiest and saddest experiences are integrally intertwined with family. Family provides the foundation, the group through which individuals' needs are met. Societies depend on families as the unit through which to funnel services. It is the political, economic, health, educational, religious, and sexual base for most people. These are some of the reasons family is important to us.

Key Points:

- Families are diverse entities at the micro level, having a wide range of configurations, but families also collectively serve as a core structure of society—institutions—at the meso and macro levels. (See pp. 287–289.)

- The family is sometimes called the most basic unit of society, for it is a core unit of social pairing into groups (partner taking), a primary unit of procreation and socialization (people making), and so important that when it comes unglued (contract breaking), the whole social system may be threatened. (See pp. 286–287.)

- Various theories—rational choice, symbolic interactionism, functionalism, conflict theory, and feminist theory—illuminate different aspects of family and help us understand conflicts, stressors, and functions of families. (See pp. 289–293.)

- At the micro level, people come together in partner-taking pairs, but the rules of partner taking (exogamy/endogamy, free choice/arranged marriage, polygamy/monogamy) are meso-level rules. (See pp. 293–297.)

- Power within a partnership—including distribution of tasks and authority—is assigned through intimate processes that are again largely controlled by rules imposed from another level in the social system. (See pp. 297–300.)

- At the macro level, nations and even global organizations try to establish policies that strengthen families. Issues that are of concern to some analysts include cohabitation patterns that seem a threat to family, same-sex households (including same-sex marriage), and contract breaking (divorce). (See pp. 300–309.)

Contributing to Our Social World: What Can We Do?

At the Local Level:

- *Support groups for married or partnered students:* An ever-increasing number of undergraduate students live on or near campus with spouses, partners, and children. If your campus has a support group, arrange to attend a meeting and work with members to help them meet the challenges associated with their family situation. If such a group does not exist, consider forming one.

At the Organizational or Institutional Level:

- *Support groups for immigrants:* Many communities of recent immigrants live in extended family households, in which grandparents, parents, children, and possibly other relatives occupy the same home or apartment. Identify a local social service agency that serves such communities. Arrange an interview with a professional to discuss the challenges that these extended households face. Explore the possibility of working with a support group serving these communities.

At the National and Global Levels:

- Select a family-related issue about which you feel strongly, such as no-fault divorce or same-sex marriage policies. Find out how the issue is treated in national or state/provincial laws. Next, identify a political representative whose record indicates an interest in the issue. Contact that person via letter or e-mail, stating your views.

- *The Association of Family and Conciliation Courts* is an advocacy organization specializing in international family law encompassing a wide range of global issues, including human rights, immigration policy, gender equity, adoption policies, and the rights of children. Its website, at www.afccnet.org, includes information about the field and opportunities for volunteer work. Hofstra University maintains a resource site on international family law at http://people.hofstra.edu/lisa_a_spar/intlfam/intlfam.htm where you can learn more about the field.

Visit **www.sagepub.com/oswcondensed2e** for online activities, sample tests, and other helpful information. Select "Chapter 10: Family" for chapter-specific activities.

CHAPTER 11

Education and Religion

Answering "What?" and "Why?"

Whether in schools or houses of worship, "What happened?" and "Why did that happen?" are commonly asked questions. In schools, the answers may have to do with causality (science) or with plot (literature). When the questions are asked in a religious setting, the answers are likely to focus on the meaning of the event for one's life or for human history.

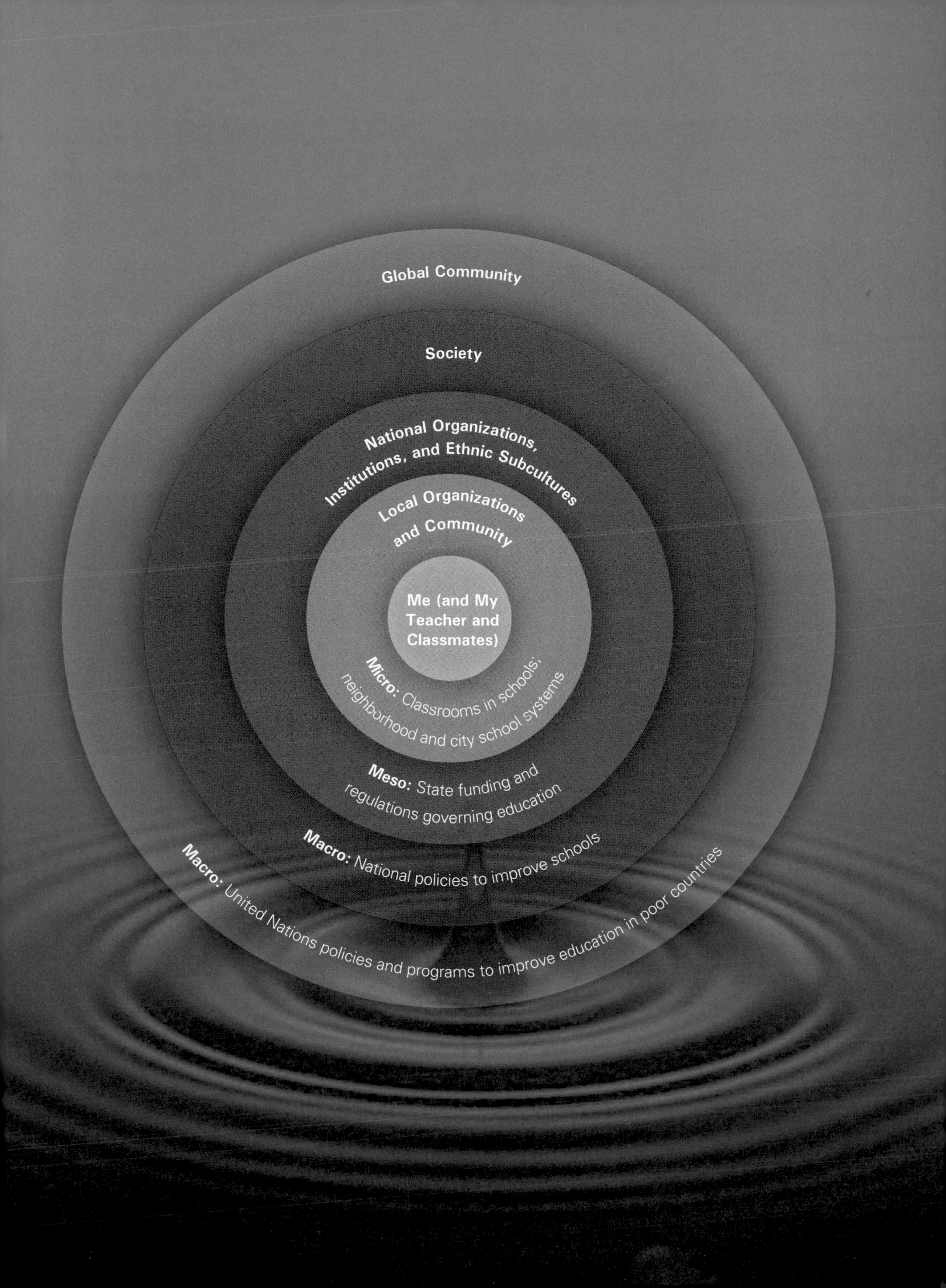

Global Community

Society

National Organizations, Institutions, and Ethnic Subcultures

Local Organizations and Community

Me (and My Teacher and Classmates)

Micro: Classrooms in schools; neighborhood and city school systems

Meso: State funding and regulations governing education

Macro: National policies to improve schools

Macro: United Nations policies and programs to improve education in poor countries

Think About It	
Self and Inner Circle	What did you personally learn—both formally and informally—in school?
Local Community	How do role expectations of people in a local school—student, teacher, principal—affect the learning that occurs in that school?
National Institutions; Complex Organizations; Ethnic Groups	Do families help or hurt children's school achievement?
National Society	How is education changing in your nation?
Global Community	Why is education a major concern around the world?

Aastik, in his home country of Nepal, goes to school most days of the week, where he learns not only skills like reading, writing, and mathematics, but the meaning of citizenship in his country and the history and geography of his nation. He also attends religious services with his family one day a week and on holiday occasions. With this faith community he learns about how those who share this religious perspective view the meaning of life, define the role of God in human existence, and characterize ethical behaviors and immoral prohibitions. In both of these institutions, Aastik is experiencing socialization.

Think about your own childhood. What were the most influential factors in shaping you into who you are today? You probably think of family members; your school, teachers, and best friends; and your religious group. This chapter discusses two institutions whose main purpose is to socialize us; after family, they are crucial to our early socialization. Both educational institutions and religious ones ask questions about the "what" and the "why" of our lives. Education tends to focus on answers that have to do with knowledge—causality and the relationship of facts—while religion tends to address ultimate meaning in life and values that shape our decision making. Both help shape who we are through socialization throughout our lives. We begin with an examination of education; the second half of the chapter explores religion.

Education: The Search for Knowledge

Tomás is a failure. At nine years old, he cannot read, write, or get along with his peers, and out of frustration, he sometimes misbehaves. He has been a failure since he was three, but his failure started earlier than that—his parents say so. They have told Tomás over and over that he will not amount to anything if he does not shape up. His teachers

have noticed that he is slow to learn and does not seem to have many friends. They, too, define him as a failure. Tomás believes his parents and teachers, for he has little evidence to contradict their judgment.

Two strikes against him are the judgments of his parents and his teachers. The third strike is Tomás's own acceptance of the label "failure." Tomás is an at-risk child, identified as having characteristics inclining him toward failure in school and society. Probably he will not amount to anything, and he may even get in trouble with the law unless caring people intervene, encouraging him to realize that he has abilities and is not worthless.

Tomás goes to school in Toronto (Ontario), Canada, but he could live in any country. In every community, there are Tomáses. Although successful children develop a positive self-concept that helps them deal with disappointments and failures, the Tomáses internalize failures. Successful children negotiate the rules and regulations of school, and school provides them with necessary skills for future occupations. Tomás carries a label with him that will largely determine his life chances because, next to home, schools play the biggest role in socializing children into their self-concepts and attitudes toward achievement. What factors could change educational outcomes for students like Tomás? The following discussion of the institution of education may shed some light on experiences of children in the educational system.

Schooling—learning skills such as reading and math in a building via systematic instruction by a trained professional—is a luxury some children will never know. On the other hand, in most urban areas around the world and in affluent countries, formal education is necessary for success—and even for survival. Education of the masses in a school setting is a modern concept that became necessary when literacy and math skills emerged as essential to many jobs, even if just to read instructions for operating machinery. Literacy is also necessary to democratic governments, where an informed citizenry elects officials and votes on public policies.

In this part of the chapter, we will explore the state of the world's education, theoretical perspectives on education, micro-level interactions in educational organizations, what happens in schools after the school bell rings, whether education is the road to opportunity, and global educational social policy issues.

State of the World's Education: An Overview

Every society educates its children. In most societies, national education systems carry out this task. Global organizations concerned with education also contribute. Over the past 50 years, UNESCO (the United Nations Educational, Scientific, and Cultural Organization) has become the "global center for discussion and implementation of educational ideas and organization models" (Boli 2002:307). It provides teacher training, curricular guidance, and textbook sources, and it gathers international statistics on educational achievement. Many countries in Africa, Asia, Europe, Latin America, and the Middle East have adopted UNESCO global standards, including the organizational model of six years of primary school and three years apiece for intermediate and secondary school, with an emphasis on comprehensive rather than specialized training (Institute for Statistics 2006a).

Education has become a global issue. What is considered essential knowledge to be taught in schools is based largely on a country's level of development, cultural values, political ideology, and guidelines from international standards. Country leaders believe that a literate population is necessary for economic development and expansion, a thriving political system, and the well-being of the citizenry. "Education has become a global social process that both reflects and helps create the global society that is under formation" (Boli 2002:312). Education plays a significant role in economic growth, social stratification, and political behaviors.

Formal education—schooling that takes place in a formal setting with the goal of teaching a predetermined curriculum—has expanded dramatically in the past several decades as higher percentages of students in many countries attend school. Although major educational gaps still exist between the elite and the poor and between females and males, these gaps are narrowing. Girls' enrollment worldwide is now close to half of students at the primary levels in most countries with 96 girls for every 100 boys worldwide in primary and secondary schools (World Bank 2011). Girls' secondary school enrollment rose from 32% in 1950 to 45% in the 1990s (Boli 2002). Some countries lag behind in the ratio of girls to boys in school: For every 100 boys, 270 girls are out of school in Yemen; 257 in Benin, Africa; 316 in Iraq; and 426 in India (UNESCO GMR 2007). In more developed countries, women are entering male-dominated fields of higher education and attending university at levels equal to or exceeding men.

Overcrowding in classrooms is not uncommon in poor countries, as seen in this very poor school in a Darfur refugee camp. Note that the students are primarily boys.

The quality of education is judged in part by student-teacher ratios and literacy rates. In secondary school, pupil-teacher ratios are lower than in primary school. The United Nations provides the pupil-teacher ratio in secondary schools for 189 countries and territories. Twenty-four countries have average class sizes of fewer than 10 students, but 13 countries average 30–39 and four countries average more than 40 students per classroom (Organisation for Economic Co-operation and Development 2007, 2008; UNESCO Institute for Statistics 2008). One concern is finding enough qualified teachers to handle the increasing global enrollments.

In poor Global South countries, 76% of the population is literate, including 64% of women (UNESCO 2006). However, there is variation among these countries: Only 38% of the women are literate in South and West Asia. Affluent countries have rates of 99% literacy for men and women (see Map 11.1 on page 318). In some math and science courses, males achieve at higher levels than females, especially in countries where women's status is lower, but females often have higher achievement in language and reading.

Economically developed countries influence levels and types of education worldwide. Beginning in the 1900s, mass education spread. The national education curriculum of many poor countries is similar to models of mass education used by colonial powers in Global North countries (Chabbott and Ramirez 2000; McEneaney and Meyer 2000). Mathematics, for instance, is taught universally, and science has been taught in most schools since World War II, although more science is taught in countries with a higher standard of living (Baker 2002).

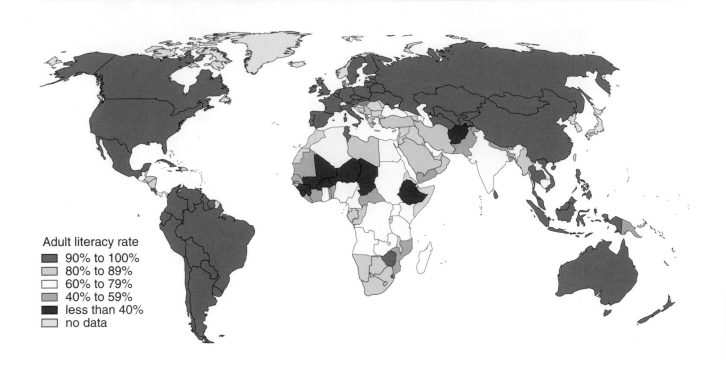

Map 11.1   Adult Literacy Rates by Country

Source: UNESCO Institute for Statistics, in Huebler 2008; United Nations Development Programme 2007/2008, p. 226.

Education can be studied at each level of analysis: micro, meso, and macro. The next section provides a summary of how various theories offer different perspectives for understanding education in society.

While Ethiopia shares the commitment to mass education of all young people, its resources are very meager, as we can see from this photo of a crowded impoverished school in an Ethiopian village.

Who Does What? Micro-Level Interactions in Educational Organizations

The process of education takes place at the micro level in the classrooms and corridors of local schools, with key players who enact the everyday drama of teaching and learning. You probably recall an important learning experience you had in which someone helped you master a skill—a scout leader, grandmother, choir director, coach, neighbor, or teacher. You know how important one-on-one mentoring can be to the learning process, how education often is an intimate exchange between two people. Education has the power to change the way people think about the world, influence their sense of competency, and affect their self-esteem and personal outlooks on self and society.

Schools are important organizations in local communities as well. The school is a source of pride and a unifying symbol of identity. Local communities rally around the success of their school. Moreover, in many communities, the school system is a large employer with real importance to the economic vitality of the area. At the micro level, much sociological analysis has focused on the school as a social setting within a community. At this level, sociologists look at roles and statuses in educational settings and the informal norms and interaction patterns that evolve in those settings. Let us begin by reviewing two micro-level theories, and then some common in-school interactions.

Micro-Level Theories: Individuals Within Schools

Symbolic Interaction Perspective and the Classroom. Symbolic interaction theory focuses on how people interact based on the meaning they have assigned to various traits, behaviors, or symbols (e.g., clothing). Children actively create distinctions among individuals and groups, becoming agents in determining the social reality in which they live. Popularity is a major issue for many children, especially in middle school years; it refers to being noticed and liked and having everyone know who you are. Students may increase their popularity by being attractive, representing the school in an athletic contest, or being seen in a leadership position. The difficulty is that there are few such positions, leading to a competition in which some individuals are losers. In the United States, the losers are more likely to be children from families who cannot afford to purchase popular clothing or other status symbols or to send their children to sports training or music camp. Winners have access to material and symbolic resources that give them high visibility. They are given special privileges in the school and are more likely to develop leadership skills and to feel good about themselves.

One's sense of self—an intensely personal experience—is shaped by the micro interactions of the school. Thus, for young people from 6 to 18 years old, the extensive time spent in school means the status of student has enormous impact on how individuals see themselves. The image that is reflected back to someone—as student or as teacher, for example—can begin to mold one's sense of competence, intelligence, and likability. The symbolic interaction perspective considers how symbols affect sense of self or shape social hierarchies (Eder, Evans, and Parker 1995).

The larger school organization creates a structure that influences how individuals make sense of their reality and interact with others. The Iowa School of symbolic interactionism emphasizes the link between the self and meso-level positions or statuses (Stryker 2000). Official school positions—such as president of the student council or senior class president or varsity team member—become important elements of one's *self*.

Thinking Sociologically

How do you think teachers affect the sense of self of students? How do students affect the sense of self, the confidence, and the effectiveness of each other and of teachers?

Males have many ways of becoming known and respected. Male athletic competitions draw bigger crowds than female athletics, and one can become a local celebrity based on one's skill on the field or court.

Rational Choice Theory and Educational Settings. Rational choice theory focuses on the cost-benefit analysis that individuals undertake in virtually everything they do. What are the costs—in terms of money, relationships, self-esteem, or other factors—and what are the benefits? If benefits outweigh costs, the individual is likely to decide to act in the most beneficial way, whereas if costs outweigh benefits, the individual is likely to seek other courses of action. How might weighing costs and benefits influence decisions about education? One example is teachers making rational choices about staying in the teaching profession. In recent years in the United States, approximately 85% of all teachers stayed in the same school, 7.6% moved to new schools, and 8% left teaching (5.3% of those because their contracts were not renewed) (Keigher and Cross 2010). Rational choice theorists explain teacher retention by looking at perceived benefits—rewarding professional practice, working with children or adolescents, time off in summers—and perceived costs, such as poor salary for a college graduate; lack of respect from parents, students, and administrators; 12- to 14-hour days for nine months of the year; and lack of professionalism in treatment of teachers. The costs today are seen by teachers in some countries as higher than they used to be for professionals in teaching, so the turnover rate in the profession is high.

Thinking Sociologically

Think of a teacher you know or have observed. What seem to be the costs and benefits of the various roles teachers fulfill? Do you think most teacher behavior is shaped by this kind of rational choice calculation?

Statuses and Roles in the Educational System

Students, teachers, staff, and administrators hold major statuses in educational systems. The roles associated with each status in educational organizations bring both obligations and inherent problems. When the status holders agree on expected behaviors (role expectations), schools function smoothly. When they do not agree, conflicts can arise. Let us look at several statuses and accompanying roles in schools.

Students and the Peer Culture of Schools. In a private Rwandan secondary school, 45 students are crowded onto benches. They are quiet, respectful, and very hardworking. They understand how lucky they are to be in secondary school and know the roles expected of them. Peer culture supports the rules and norms and reinforces their behavior. They know they are in a privileged position, and many students are lined up to take their place on the bench should they not carry out their roles, work hard, and succeed. Although they have no written texts, students write down

In Rwandan secondary schools as many as 45 students may be in a classroom with little more than wooden benches and tables.

the lectures in their notebooks and memorize the material. The lessons children learn in the classroom are supported by the *student peer culture*, "a stable set of activities or routines, artifacts, values, and concerns that children produce and share in interaction with peers" (Corsaro and Eder 1990:197).

In some countries, going to school is a privilege. In others, it is a necessary part of life that many students resist. In either case, children understand the school system well. They know how they must behave to be considered good students and that if they misbehave they will be viewed as bad students. However, they do not know how to change a negative label once it has been assigned to them.

Class, race, gender, and interests all affect children's experiences in school and in their peer culture. In schools around the world, students are assigned a variety of social identities, often given colorful labels. These identities are part of their peer cultures, which have a major effect on their school experience and future roles in society (Waller [1932] 1965). *Nerds* in the United States or *ear 'oles* in Britain are well-behaved, college-bound middle-class students who have an investment in the school system. *Burnouts* or *lads*, on the other hand, are working-class students who often feel hostile or alienated in the school environment. They may engage in behaviors that prevent them from succeeding in high school and can lead to dropping out (Jackson 1968; MacLeod 2008; Willis 1979). Micro-level interaction patterns in school also reflect society's racial and ethnic patterns. For example, studies in the United States have found that when choosing friends, students are one sixth as likely to choose a cross-race peer as a same-race peer (Grant 2004). Racial inequalities in early schooling affect everything from preparation for jobs to college attendance (Charles, Roscigno, and Torres 2007).

While many experts acknowledge that girls and boys have different experiences in school, a recent debate in Britain and the United States focuses on whether boys and girls should be taught in separate classrooms. Those who argue for separate classrooms, especially in the middle school years, point out the different interests and learning styles of girls and boys at these early adolescent ages. Others argue that equality requires mixed-gender classes. Some parochial schools have long been single-sex, and now public school districts are experimenting with single-sex classes. Research such as that presented in the next "Sociology in Your Social World" discusses concerns that have been raised about gender differences in U.S. schools (Sax 2005; Weil 2008).

Another gender issue in peer culture is sexual harassment. Four out of five students report some type of sexual harassment in school. Compared to boys, girls experience "hostile hallways" (American Association of University Women Educational Foundation 2001) in more physically

Sociology in Your Social World

Where the Boys Are: And Where Are the Boys?

"The Myth of the Fragile Girl"; "The War Against Boys"; "Failing at Fairness: How America's Schools Cheat Girls"; and "At Colleges, Women Are Leaving Men in the Dust" are just a few article titles in a debate about whether girls or boys have the biggest advantage or disadvantage in schools. For many years, concern focused on factors that inhibited minorities' educational attainment in school. Recently, some authors are turning the tables and focusing their concern on dominant groups. Gender has been one major focus of educational disparities. Statistics indicate that the state of educational achievement varies greatly by sex, age, race or ethnicity, and socioeconomic status. Why is this so? Among the many reasons for the differences, researchers point to the incredible gains made by women and the fact that women tend to study more. In the United States, African American, Hispanic, and low-income males lag behind all other groups, including females from their own ethnic group (King 2000). Asian or Pacific Islanders have a higher high school completion rate (96.6%) than whites (96.2%), blacks (90.1%), Hispanics (82.3%), and Native Americans (86.4%), but the gap has been narrowing in recent years (Stillwell 2010).

One in nine young black males in the United States ends up in prison or jail (National Association for the Advancement of Colored People [NAACP] 2008), further reducing their chances for education, good jobs,

and stable family life. The reasons are many, but the bottom line is that these young men feel disconnected from a society that helps women with children but ignores the vulnerabilities of men (Mincy 2006). They often feel alienated from their society. The following figures provide a partial picture:

- An estimated 2 million to 3 million youth ages 16 through 24 are without postsecondary education and are disconnected—neither in school nor employed (Mincy 2006).
- Among those between ages 16 and 24 who are not enrolled in school, only about half are working, and one third are involved with the criminal justice system (NAACP 2008).
- Thus, black teens have not witnessed education as a path to better jobs for their parents, siblings, and neighbors: In January 2011, unemployment for African Americans aged 20 or older was 14.6% compared to an 8.4% unemployment rate for whites of the same age (Allegretto, Amerikaner, and Pitts 2011).

A review of college attendance statistics shows that women are attending and graduating at a higher rate than men, and black men are at the bottom of the graduation rates. Table 11.1 compares these groups.

Women have surpassed men in college completion (Buchmann and DiPrete 2006), and men, regardless of

Table 11.1 College Attendance Rates and Graduation Rates

Race	College Attendance Rate % (2007)
White, non-Hispanic	72
Hispanic	64
Black	56
Males	66
Females	72

Source: U.S. Census Bureau 2011a.

(Continued)

(Continued)

race or class, get lower grades, take more time to graduate, and are less likely to get a bachelor's degree (Lewin 2006). However, men from the highest income groups attend college at a slightly higher rate than women in that group, and men from low-income families—disproportionately African American and Hispanic—are the most underrepresented in higher education. The gender gap in favor of females has been most pronounced among low-income whites and Hispanics.

The gender imbalance is of such concern to college admissions officers that some colleges are turning away more qualified females in favor of males (Britz 2006). Some colleges are even adding activities such as football to attract more male students (Pennington 2006).

Black women earned twice as many bachelor's degrees as black men (National Center for Education Statistics 2002); yet females who are racial or ethnic minorities lag behind white women (King 2000). Despite the sex differences, some researchers argue that the gender gap is not nearly as significant as the differences for race or ethnicity and social class. Furthermore, men still dominate math and science fields, where jobs pay more money and result in more power. Data indicate that although women do not score as high as men on achievement tests in math and science, women hold higher educational aspirations, are more likely to enroll in college, and in 2008 received 57.9% of undergraduate diplomas. In addition, in 2008, for the first time, women surpassed men in the number of doctoral degrees received in the United States (National Center for Education Statistics 2008a). Perhaps concern over boys' performance simply reflects nervousness about women's achievements (Lewin 2006).

What do we do? Girls and women face serious educational problems in many Global South countries, and women in Global North countries are still at a disadvantage in hiring for high-paid jobs and equal wages. Concern about boys is a relatively new twist in the equity issue. Ultimately, we hope to create an educational system that equally benefits all groups.

and psychologically harmful ways, and students who identified themselves as gay, lesbian, bisexual, or transgendered experienced high levels of bullying and assaults as well (Gay, Lesbian, and Straight Education Network 2006). While trying to deal with all the challenges of being a teenager, lesbian, gay, bisexual, and/or transgender (LGBT) teens additionally have to deal with harassment, threats, and violence directed at them on a daily basis. They hear antigay slurs such as "homo," "faggot," and "sissy" about 26 times a day or once every 14 minutes. Even more troubling, a study found that 31% of gay youth had been threatened or injured at school in the last year alone (Mental Health America 2011).

The environment outside the school also powerfully affects students' achievement and behavior within the school. Disorganization in the community and family is related to lack of school commitment and is reflected in delinquent behavior (Ogbu 1998). The students at highest risk for dropping out of school in Global North countries are also at higher risk for joining gangs and committing violent crimes. They often feel the system is stacked against them (Noguera 1996; Willis 1979). Educators are deeply concerned about disruptive students, not only because they disrupt learning and make school unsafe for others but because many are at risk of dropping out of school and becoming burdens to society.

Teachers: The Front Line. The degree of success students experience often depends on their role partners. Teachers are those role partners who serve as gatekeepers, controlling the flow of students, activities, resources, and privileges. One scholar estimated that teachers have more than a thousand interchanges a day in their roles as classroom managers (Jackson 1968). They also act as timekeepers and traffic managers and spend a great deal of time in nonteaching clerical work. Many teachers complain that they are so bogged down with paperwork and forms that they have little time to address the primary objectives of the school: student learning.

Teachers in the classroom occupy the front line in implementing the goals of the school, community, and society. As primary socializers and role models for students, they are expected to support and encourage students and at the same time to judge their performance—giving grades and recommendations as part of the selection and allocation functions of education. This creates role strain, which can interfere with the task of teaching and contribute to teacher burnout (Dworkin 2007; Dworkin and Tobe 2012).

In Japan, where education is considered extremely important for training future generations, teachers are treated with great respect and honor. They receive salaries competitive with those in industry and professions such as law and medicine (Ballantine and Hammack 2012). In Europe, many high schools are organizationally like universities. Teachers think of themselves as akin to professors. By contrast, studies in Australia and the United States show teachers feel they are unappreciated (Saha and Dworkin 2006). In the United States, secondary teachers think of themselves as more like middle school teachers than like university professors (Legters 2001). Many U.S. teachers complain that with each new presidential administration come new reforms requiring new lesson plans, curricula, classroom structure, discipline procedures, use of time, focus of lessons, goals for students—and many other changes. Add to that teachers' low salaries, the low status of teaching as a profession, unmotivated students, problems of discipline in the classroom, lack of support from students' families and the community, criticism of schools, and interruptions of classroom work for time-consuming special programs. U.S. teachers are held accountable for students' progress as measured on standardized tests, as well as for their own competency as measured in tests to determine their knowledge and skills (Dworkin and Tobe 2012; Grant and Murray 1999).

The organizational context of teachers' work is a key source of problems. Overcoming poor social standing and upgrading the status of teachers requires better recruiting and training (Ingersoll 2005; Ingersoll and Merrill 2012). Several sociologists suggest that teachers' dissatisfactions can be addressed through organizational and structural changes, allowing teachers more autonomy and control over their environments (Dworkin 2007; Ingersoll and Merrill 2012).

Most professionals have some sense of calling and commitment and are motivated not just by money but by a sense of contribution, prestige, and pride in belonging to the profession and serving others. Two key features of being a professional are autonomy on the job and self-regulation by the profession. Incentives from the macro-level federal government in the form of pay increases have been proposed by the new U.S. administration, along with other incentives to improve the quality of teaching. Yet control of teachers often comes from outside teacher organizations (Ingersoll 2004). When government regulates standards for teachers, quality might actually decrease as teachers—faced with lack of respect as professionals and lack of control over their work—opt instead for other occupations. This has the potential to leave schools with less capable and less committed instructors.

Thinking Sociologically

Who should enforce high teacher standards—the federal, state, or local government? Teacher unions? Community interest groups? Who should decide what these standards are?

Administrators: The Managers of the School System. Key administrators—superintendents, assistant superintendents, principals and assistant principals, or headmasters and headmistresses—hold the top positions in the educational hierarchy of local schools. They are responsible for a long

list of tasks: issuing budget reports; engaging in staff negotiations; hiring, firing, and training staff members; meeting with parents; carrying out routine approval of projects; managing public relations; preparing reports for boards of directors, local education councils, legislative bodies, and national agencies; keeping up with new regulations; making recommendations regarding the staff; and many other tasks. Some administrators specialize in overseeing discipline and acting as buffers between parents and teachers when conflicts arise. In some countries, a lay board of education oversees decisions, including hiring of administrators and expenditures.

Administrators operate one step removed from the actual educational functions of the classroom, and they mediate between the local community school and the larger bureaucracy at the state and national levels. They occupy high-profile leadership positions that may place them in conflict with the interests and goals of students, teachers, parents, or community groups.

The Informal System: What Really Happens Inside Schools?

The informal system of schooling includes the unspoken, unwritten, implicit norms of behavior that we learn in classrooms and from peers, whether in kindergarten or in college. We will discuss schools as complex formal organizations below, but every organization also has an informal system. The informal system does not appear in written goal statements or course syllabi but nevertheless

Schools have a formal structure and a culture that affect the classroom. Note the arrangement of desks and the norm of raising a hand before a student may speak. Yet this is far from a universal pattern in schools in North America or around the globe. The formal system is only part of the classroom environment, for every school and classroom also has an informal culture.

influences our experiences in school in important ways. Dimensions of the informal system include the educational climate, the value climate, and the hidden curriculum within a classroom.

Educational climate of schools refers to a general social environment that characterizes a group, an organization, or a community such as a school (Brookover, Erickson, and McEvoy 1996). Schools can feel comfortable, stimulating, and exciting with colorful pictures on the walls or cold and unfriendly with hall warnings posted and hall guards watching every interaction. The school's architecture, teachers' expectations, classroom layouts, student groupings by age and ability—all affect the climate of the school. Schools also have ceremonies and rituals that contribute to climate—logos, symbols, athletic events, pep rallies, and award ceremonies.

Classroom climates are influenced by the teachers' use of discipline and encouragement, the organization of tasks and opportunities for student interaction, the seating arrangement, and classroom furnishings. These nuances can create an atmosphere that celebrates or stifles student achievement. In addition, the friendships students make depend in part on how schools and classrooms are organized. Some schools track students on the basis of their tested ability in certain subjects, thus fostering the development of friendship groups within those tracks. Tracking systems also tend to create and maintain racial, ethnic, and social class segregation (Lucas and Berends 2002).

Teachers' responses to class, ethnicity or race, and gender differences create climates that have subtle but profound impacts on students' experiences and learning. Teachers call on boys more often than girls and give them instructions for accomplishing tasks independently. By contrast, teachers more often do the tasks *for* the girls in the class (Spade 2004). Moreover, teachers may unconsciously groom white girls for academic attainment based on dependence and loyalty, while African American girls are encouraged to emphasize social relationships over academic work. White boys are groomed for high attainment and high-status social roles, while African American boys are trained for conformity, being carefully monitored and controlled in the classroom (Grant 2004).

Teacher expectations influence how they react to different groups of students; some students are pushed harder than others to achieve based on their sex, socioeconomic status, race or ethnicity, appearance, language patterns, behavior, or tracking or grouping in the school and classroom. Even status of the school can affect expectations (Brookover et al. 1996). Low achievement is linked to low expectations for students. Do students have equal access to materials and technology? Are all students active and influential participants in the learning process? Goals of equitable teaching and learning are challenged in increasingly diverse classrooms (Lucas and Berends 2002).

The *value climate of schools* refers to students' motivations, aspirations, and achievements. Why is achievement significantly higher in some schools than in others? How much influence do the values and outlooks of peers, parents, and teachers have on students? Sociologists know that a student's home is influential in determining educational motivation. Recall the opening case of Tomás, who received only negative comments and little encouragement. Cultural components of the neighborhood racial, ethnic, and class composition also affect the value climate. Researchers have found, for example, that classrooms integrated along ethnic or class lines frequently raise the level of motivation and achievement for members of minorities (Lucas and Berends 2002). Also, students who are expected to do very well by teachers and parents generally rise to meet these expectations (Morris 2005).

Plans to improve school achievement need to take into consideration the educational and value climates of schools in order to be successful. Whether throughout the school or in a particular classroom, the atmosphere that pervades the learning environment and expectations has an impact on students' educational achievement.

Teachers try to elicit cooperation and participation from students by creating a cost-benefit ratio that favors compliance. Manipulating the classroom is one effective means of control: putting students at tables or in a circle, breaking up groups of chattering friends, or leading a discussion while standing beside the most disruptive child. All of this is part of the power-play and give-and-take interaction that shapes a particular classroom.

The *hidden curriculum* refers to the implicit "rules of the game" that students learn in school (Snyder 1970). It includes everything the student learns in school that is not explicitly taught, such as unstated social and academic norms. Students have to learn and respond to these to be socially acceptable and to succeed in the education system and with peers (Snyder 1970).

Children worldwide begin learning what is expected of them in preschool and kindergarten, providing the basis for schooling in the society (Neuman 2005). For example, Gracey (1967) describes early school socialization as "academic boot camp." Kindergarten teachers teach children to follow rules, to cooperate with each other, and to accept the teacher as the boss who gives orders and controls how time is spent. All of this is part of what young children learn; lessons are instilled in students even though it is not yet the formal curriculum of reading, writing, and arithmetic. These less formal messages form the hidden curriculum.

For conflict theorists, the hidden curriculum is a social and economic agenda that maintains class differences. More is expected of elites, and they are given greater responsibility and opportunities for problem solving that result in higher achievement (Brookover and Erickson 1975). Many working-class schools stress order and discipline, teaching

Four-year-old preschoolers recite the Pledge of Allegiance. Developing patriotism is part of the implicit and informal curriculum of schools.

students to obey rules and to accept their lot as responsible, punctual workers (Willis 1979). All of this brings us to a consideration of the formal organizational aspects of education—the rules, regulations, routines, and statuses.

Thinking Sociologically

What examples of the informal system can you see in the courses you are currently taking? How do these norms and strategies affect your learning experience?

After the School Bell Rings: Meso-Level Analysis

Schools can be like mazes, with passages to negotiate, hallways lined with pictures and lockers, and classrooms that set the scene for the educational process. Schools are mazes in a much larger sense as well. They involve complex interwoven social systems at the meso level where we encounter the formal organization of the school bureaucracy. The following sections focus on the formal systems, bureaucracy, and decision making in educational institutions.

Formal Education Systems

Formal education came into being in the Western world in sixteenth-century Europe when other social, political, economic, and religious institutions required new skills and knowledge that families did not necessarily possess. Schools were seen as a way for Catholics to indoctrinate people to

religious faith and for Lutherans to teach people to read so they could interpret the Bible for themselves. The first compulsory education system was in a Lutheran monastery in Germany in 1619. By the nineteenth century, schooling was seen as necessary to teach the European lower classes not only religion but also better agricultural methods, skills for the rapidly growing number of factory jobs, national loyalty, and obedience to authorities—all necessary for developing societies.

After 1900, national school systems were common in Europe and its colonial outposts. These systems shared many common organizational structures, curricula, and methods, as world leaders shared their ideas. The Prussian model, with strict discipline and ties to the military, became popular in Europe in the 1800s, for example, and the two-track system of education that developed—one for the rich and one for the poor—existed worldwide.

Industrializing societies required workers with reading and math skills; today this includes electronic and technological skills. Schooling that formerly served only the elite gradually became available to the masses, and some societies began to require schooling for basic literacy (usually third-grade level). Schools emerged as major formal organizations and eventually developed extensive bureaucracies. The postwar period from 1950 to 2005 brought about a rapid rise in education worldwide followed by a leveling off, as shown in Table 11.2.

Sometimes, differing goals for educational systems lead to conflict. What need, for example, does a subsistence farmer in Nigeria or Kenya have for Latin? Yet, Latin was often imposed as part of the standard curriculum in colonized nations. Some countries are now revising their curriculum based on goals and needs of agriculturally based economies.

Table 11.2 World Educational Enrollment Rates, 1950–1992 and 2000–2005 (in Percentages)

Year	Primary Schooling	Secondary Schooling	Tertiary Schooling (University)
1950	36	13	1.4
1960	77	27	6.0
1970	84	36	11
1980	96	45	11
1992	98	53	14
2000–2005*	90 (M), 86 (F)	61 (M), 60 (F)	

Source: Boli 2002:309 for figures to 1992; Institute for Statistics 2006b:Table 5.

*Number of children enrolled in primary/secondary school who are of official school age, expressed as a percentage of the total number of children of official school age.

M = male, F = female.

The Bureaucratic School Structure

The formal bureaucratic atmosphere that permeates many schools arose because it was cost-effective, efficient, and productive. Bureaucracy provided a way to document and process masses of students coming from different backgrounds. Recall Weber's (1947) bureaucratic model of groups and organizations, discussed in Chapter 5:

1. Schools have a *division of labor* among administrators, teachers, students, and support personnel. The roles associated with the statuses are part of the school structure. Individual teachers or students hold these roles for a limited time and are replaced by others coming into the system.

2. The *administrative hierarchy* incorporates a chain of command and channels of communication.

3. Specific *rules and procedures* in a school cover everything from course content to discipline in the classroom and use of the schoolyard.

4. Personal relationships are downplayed in favor of *formalized relations* among members of the system, such as placement on the basis of tests and grading.

5. *Rationality* governs the operations of the organization; people are hired and fired on the basis of their qualifications and how well they do their jobs.

These chairs represent the bureaucracy or hierarchy of the school, with every student facing the teacher. Look through other photos in this chapter, especially photos from nonindustrialized countries. Do you notice anything about the arrangement of students that suggests differences in relationships in the school?

One problem resulting from bureaucracy is that impersonal rules can lock people into rigid behavior patterns, leading to apathy and alienation (Kozol 2006; Sizer 1984). Children like Tomás in the opening example do not fit into neat cubbyholes that bureaucratic structures invariably create. These children view school not as a privilege but as a requirement imposed by an adult world. Caught between the demands of an impersonal bureaucracy and goals for their students, teachers cannot always give every child the personal help she or he needs. Thus, we see that organizational requirements of educational systems at a meso level can influence the personal student-teacher relationship at the micro level.

Educational Decision Making at the Meso Level

Who should have the power to make decisions about what children learn? In Africa, Asia, Japan, Latin America, and many European countries, centralized national ministries of education determine educational standards and funding for the whole country. By contrast, some heterogeneous societies, such as Canada, Israel, and the United States, with many different racial, ethnic, regional, and religious subcultures, have more local autonomy in decision making. Teachers, administrators, school boards, parents, and interest groups all claim the right to influence the curriculum. Consider influences from micro and macro levels on the meso-level educational organization.

Local-Level Influences. In U.S. communities, curriculum conflicts occur routinely over issues such as sex education courses and the selection of reading materials. Problems occur if content is thought to contain obscenity, sex, nudity, political or economic bias, profanity, slang or nonstandard English, racism or racial hatred, and antireligious or presumed anti-American sentiment (Delfattore 2004). For example, Family Friendly Libraries, an online grassroots interest group that started in Virginia, argues that the popular Harry Potter books should be banned from school libraries because they believe the series promotes the religion of witchcraft (DeMitchell and Carney 2005).

Banned books in the past have included the *Wizard of Oz* series, *Rumpelstiltskin, Anne Frank: The Diary of a Young Girl, Madame Bovary, Soul on Ice, The Grapes of Wrath, The Adventures of Huckleberry Finn,* Shakespeare's *Hamlet,* Chaucer's *The Miller's Tale,* and Aristophanes's *Lysistrata* (Ballantine and Hammack 2009). Table 11.3 shows the most frequently challenged books in 2010. Off the list that year but on for several years past were *The Catcher in the Rye* by J. D. Salinger, *Of Mice and Men* by John Steinbeck, and *Beloved* by Toni Morrison (American Library Association 2011).

National-Level Influences. Because the U.S. Constitution leaves education in the hands of each state,

Table 11.3 The 10 Most Challenged Books of 2010

The following are the 10 books that were most often challenged in public schools in 2010 and the reasons they were being contested.

1. *And Tango Makes Three* by Justin Richardson/ Peter Parnell. Reasons: Homosexuality, Religious Viewpoint, Unsuited to Age Group

2. *The Absolutely True Diary of a Part-Time Indian* by Sherman Alexie. Reasons: Offensive Language, Racism, Sex Education, Sexually Explicit, Unsuited to Age Group, Violence

3. *Brave New World* by Aldous Huxley. Reasons: Insensitivity, Offensive Language, Racism, Sexually Explicit

4. *Crank* by Ellen Hopkins. Reasons: Drugs, Offensive Language, Sexually Explicit

5. *The Hunger Games* by Suzanne Collins. Reasons: Sexually Explicit, Unsuited to Age Group, Violence

6. *Lush* by Natasha Friend. Reasons: Drugs, Offensive Language, Sexually Explicit, Unsuited to Age Group

7. *What My Mother Doesn't Know* by Sonya Sones. Reasons: Sexism, Sexually Explicit, Unsuited to Age Group

8. *Nickel and Dimed: On (Not) Getting By in America* by Barbara Ehrenreich. Reasons: Drugs, Inaccurate, Offensive Language, Political Viewpoint, Religious Viewpoint

9. *Revolutionary Voices* edited by Amy Sonnio. Reasons: Homosexuality, Sexually Explicit

10. *Twilight* by Stephenie Meyer. Reasons: Religious Viewpoint, Violence

Source: American Library Association 2011.

Note: For the top 100 banned books, go to www.ala.org/ala/ issuesadvocacy/banned/bannedbooksweek/index.cfm.

the involvement of the federal government has been more limited than in most countries. Yet, the federal government wields enormous influence through its power to make federal funds available for special programs, such as mathematics and science, reading, special education, or school lunch programs. The government may withhold funds from schools that are not in compliance with federal laws and the U.S. Constitution. For example, the federal government, courts, and public opinion forced all-male military academies to become coeducational, despite the schools' resistance to such change. School changes as a result of the Civil Rights Act and the Americans with Disabilities Act are other examples of federal government influence

on local schools through the enforcement of federal laws. Schools have had to accommodate people with various disabilities—people who in the past would have been left out of the system. With their classroom experiences and working with other teachers and children, the differently-abled can participate fully in society. However, as the "Sociology in Your Social World" on page 328 shows, this process is not always smooth.

Thinking Sociologically

Educational needs at the micro (individual) level, the meso (institutional or ethnic group) level, and the macro (national) level can be very different. This raises the question of who makes decisions and whether individual needs or societal needs take precedence. Where do you think the primary authority for decision making should be—at the local, state, or national level? Why?

In addition to the structural features of schools as organizations and part of the institutional structure of society, education interplays with national and global forces in interesting and complex ways. We turn next to a macro-level analysis.

Two young girls use the computers in a classroom during a Head Start program. The Head Start program introduces reading and writing to children who are about to enter elementary school. It is one federal effort in the United States to provide support to children from poor neighborhoods and families.

Sociology in Your Social World

Disability and Inequality

by Robert M. Pellerin

Living with a visual disability for over 40 years has provided me with lived experience. I also did research on experiences of others with disabilities as part of my PhD research. My results show that having a disability puts people at a disadvantage in education, employment, attitudes of others toward those with disabilities, and personal relationships. Technologies for those with disabilities have advanced, legislation has been introduced and sometimes passed, and advocacy for rights abound, but I still have to remind others, including professionals, that due to my visual impairment some methods of communication, such as print media, do not work. Additionally, most of the technology from which those with visual impairment could benefit is unaffordable, and most mainstream companies do not include features that would make products disability-friendly. The bottom line is that many of us are ready, willing, and able to be productive citizens, but we are often precluded from positions because of our disabilities and difficulty obtaining accommodation.

Disability in the United States has undergone a significant transition since the late 1800s. Most of those who have helped bring disability to the forefront have been American veterans who were wounded in war and notable figures like Helen Keller who influenced societal perceptions of disability. They have helped define disabilities as being a relative disadvantage instead of a tragedy. Historically, disability has been synonymous with an inability to engage in employment. During the 1800s and early 1900s, people deemed unfit were often warehoused in asylums or institutions and placed in residential schools where they were provided with less than adequate instruction. For several decades in the early twentieth century some countries—including the United States—sterilized people with disabilities. However, advocates campaigned during the late 1960s to change such laws and to close asylums for the disabled.

Media portrayals have contributed to false stereotypes that promote both false beliefs and lowered goal attainment for people with disabilities. Examples of images fostered by the media include miracle cures for people who become religious; foolish tales about superhuman hearing and "face feeling"; stories of blind or visually impaired males who are depicted as wise sages; and images of disabled women who are depicted as pure, vulnerable, and in need of rescue.

The most well-known legislation seeking to end exclusion and increase participation in areas of education and employment includes the Rehabilitation Act of 1973 and the Americans with Disabilities Act (ADA) of 1990. The 1973 regulations apply federally; the ADA applies to the states and public accessibility (i.e., access to jobs, public services, and telecommunications). Despite this legislation, approximately 70% of people with disabilities do not participate in the workforce. Unfortunately, people with disabilities are at higher risk for engaging in drug and alcohol use, sexual promiscuity, higher levels of abuse, and suicidal ideation than other groups.

Prior to the 1970s, people with disabilities were excluded from public education. Due to advocacy and legislation, today approximately 75% of school-aged students with disabilities attend public school. Legislation regarding full inclusion has not proven as useful as planned, although exposure to students with disabilities has increased comfort levels of teachers and peers who are not disabled. Recent deficiencies in providing equal education include lack of adequate technology and skills needed in college or in vocational or social settings.

Regarding employment, historically people with disabilities have gone from being evaluated on the same standards as individuals without disabilities to being assessed according to how well they know themselves, their accommodation needs, and their job qualifications. Despite legislation, courts have generally ruled in favor of employers in discrimination suits. Some social scientists advocate for improving the social capital or human capital of those with disabilities as a way to increase the level of workforce participation. Many employers believe the candidate is responsible for ensuring employment is attained, while some potential employers continue to believe people with disabilities are too difficult to hire and accommodate despite their academic achievement. Employees with disabilities, including myself, concur with recommendations such as improving social and human capital, educating employers about the benefits of hiring people with disabilities, and dispelling misconceptions and negative stereotypes.

The discrimination and inequality experienced by many with disabilities results in a decreased sense of health and well-being and increased feelings of isolation. The tragedy here is that people with disabilities, despite their efforts, continue to meet with virtual hoops posed by court decisions, changes in education, and negative labeling. I believe that until society views people as having abilities and strengths as opposed to disabilities, stigma will predominate, and society will be deprived of talents and qualities from which all can benefit.

Education, Society, and the Road to Opportunity: The Macro Level

Carlos Muñoz was born in San Diego, the only child of Rafael and Yolanda Muñoz. Both of Carlos's parents were born in Mexico, but they met and were married in the United States. They divorced several years ago, and Carlos lives alternately with his mother and father. His mother has long worked as a teacher's aide in the local school district. She lives in a tidy house in a working-class, mostly Mexican area near downtown San Diego. She periodically takes Carlos on trips to Mexico.

Mrs. Muñoz graduated from a high school in the same district where she is now employed. Carlos's high school is not overcrowded and understaffed as his junior high school was, but she is worried about his school performance. She says that he lacks *ganas* (desire) and is not spending enough time or effort on his studies. "He tries to do everything fast. . . . He's getting an F in biology and Cs in most other classes," she says. Of her career hopes for Carlos, she says, "It would be perfect for him to be a lawyer, but he needs to work harder. I don't think that he can do it." His father has now taken a more active role in Carlos's education, hoping to get him on the right track again (Portes and Rumbaut 2001:14–15).

Mrs. Muñoz and millions of other immigrant parents in countries around the world have high hopes for their children, hopes that can be realized through education. Immigrant families from around the globe view education as essential to success in their new cultures. Many groups push their children to achieve in order to improve their opportunities in the new country (Kao 2004). Note that immigration patterns, national rates of graduation, and patterns of divorce—all macro processes—shape the lives of individuals and their families. This section focuses on the role of education at the macro level, including its connection with the stratification system. Although it has implications regarding individual families, like the Muñoz family's chances to succeed, the reality is that education is deeply interwoven into the macro-level inequalities of the society.

Why Societies Have Education Systems: Macro-Level Theories

Wherever it takes place and whatever the content, education gives individuals the information and skills that their society regards as important and prepares them to live and work in their society. Education plays a more major role in the lives of children in some societies. For example, village children in Global South countries around the world go to the community school, tablets in hand, but when the family needs help in the fields or with child care, older children often stay at home. Even though attending several years of school is mandated by law in most countries, not all people become literate. Learning to survive—how to grow crops, care for the home, treat diseases, and make clothing—involves essential skills that come before formal education.

Functionalist Perspective on the Purposes of Education. Functional theorists argue that both formal and informal education serve certain crucial purposes in society, especially as societies modernize. The functions of education as a social institution are summarized in Figure 11.1. Note that some functions are planned and formalized (manifest functions) whereas others are not planned—informal results of the educational process (latent functions). First consider several manifest functions of education:

1. *Socialization: Teaching Children to Be Productive Members of Society.* Societies use education to pass on essential information of a culture—especially the values, skills, and knowledge necessary for survival. This process occurs sometimes in formal classrooms and other times in informal places. In postindustrial societies, family members cannot teach all the skills necessary for survival. Formal schooling emerged as a meso-level institution to meet the needs of macro-level industrial and postindustrial societies, furnishing the specialized training required by rapidly growing and changing technology.

2. *Selecting and Training Individuals for Positions in Society.* Students take standardized tests, receive grades at the end of the term or the year, and ask teachers to write recommendation letters. These activities are part of the selection process prevalent in competitive societies with formal educational systems. Individuals accumulate **credentials**—grade point

Manifest Functions (planned and formalized)

- Socialize children to be productive members of society
- Select and train individuals for positions in society
- Promote social participation, change, and innovation
- Enhance personal independence and social development

Latent Functions (unplanned and informal)

- Confine and supervise underage citizens
- Weaken parental controls over youths
- Provide opportunities for peer cultures to develop
- Provide contexts for the development of friendships and mate selection

Figure 11.1 Key Functions of Education

One role of formal education is training for positions in society.

averages, standardized test scores, and degrees—that determine the colleges or job opportunities available to them, the fields of study or occupations they can pursue, and ultimately their positions in society. In some societies, educational systems enact this social function through tracking, ability grouping, grade promotion and retention, high-stakes and minimum-competency testing, and pull-out programs that contribute to job training, such as vocational education and service learning.

Thus, education outfits people for making a living in their society and contributing to the economy. Individual, family, community, state, and national income and standard of living are linked to the level of education of the citizenry. For example, Map 11.2 in the next "Engaging Sociology" indicates the distribution of college degrees in the United States, a macro-level factor.

3. *Promoting Change and Innovation.* In multicultural societies such as Israel, France, and England, schools help assimilate immigrants by teaching them the language and customs, along with strategies for reducing intergroup tensions. In Israel, for example, many recent Jewish immigrants from Africa and Russia work hard to master Hebrew and to move successfully through the Israeli educational system. In most societies, providing educational opportunity to all groups is a challenge, but effective social participation requires education.

Institutions of higher education are expected to generate new knowledge, technology, and ideas and to produce students with up-to-date skills and knowledge to lead industry and other key institutions in society. In our high-tech age, critical thinking and analytical skills are more

essential for problem solving than rote memorization, and this fact is reflected in curriculum change.

India has top-ranked technical institutes, and the highly skilled graduates are employed by multinational companies around the world. Companies in Europe and the United States send information to India for processing and receive it back the next morning because of the time difference. Well-trained, efficient engineers and computer experts working in India for lower wages than workers in many highly developed countries have become an important part of the global economy (Friedman 2005).

4. *Enhancing Personal and Social Development.* Do you remember your first day of elementary school? Perhaps it marked a transition between the intimate world of the family and an impersonal school world that emphasized discipline, knowledge and skills, responsibility, and obedience. In school, children learn that they are no longer accepted unconditionally as they typically were in their families. Rather, they must meet certain expectations and compete for attention and rewards.

In school, children develop independence and are taught social skills and ethical conduct that will enable them to function in society. For example, they learn to get along with others, resolve disputes, stay in line, follow directions, obey the rules, take turns, be kind to others, be neat, tell the truth, listen, plan ahead, work hard, meet deadlines, and so on. Children worldwide begin learning what is expected of them in family, preschool, and kindergarten, providing the basis for schooling in each society (Neuman 2005).

Latent functions may be just as important to the society as manifest functions. Schools keep children off the streets

Education provides hope in this refugee camp in the Central African Republic. The adults created this school for children, even though families often lack food. They view education as functional.

Engaging Sociology

Consequences of High or Low Numbers of Bachelor's Degrees

Study the map below and answer the following questions.

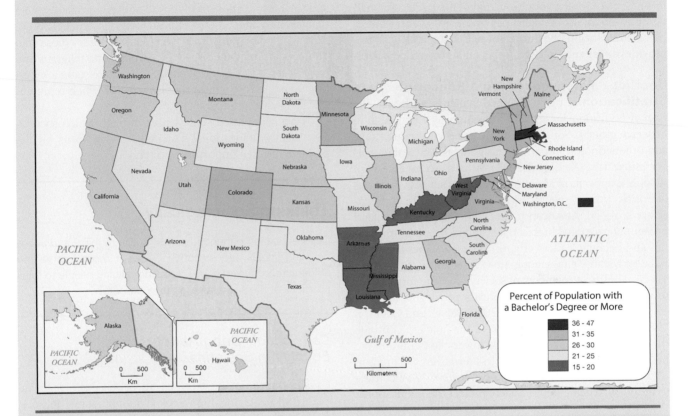

Map 11.2   Percentage of the U.S. Population Holding a Bachelor's Degree or Higher in 2004

U.S. Census Bureau. Map by Anna Versluis.

Engaging with Sociology:

1. How does your state rank?

2. How might the economy of a state be affected when an especially low percentage of the population has a college degree?

3. What kinds of businesses, industries, or professionals are more likely to locate in a state with a very high percentage of its population having a college education?

4. How might the politics, the health care system, or science be influenced by high or low levels of education within the state?

5. Look at the states that have especially high or especially low levels of the population with college degrees. What might be some causes of these high or low rates of college graduation? Some results? Check the statistics on your state or province.

6. What else can you learn or do you notice, or what questions does this map raise?

until they can be absorbed into productive roles in society. They provide young people with a place to congregate and interact among themselves, fostering a "youth culture" of music, fashion, slang, dances, dating, and cliques or gangs. At the ages when social relationships are being established, especially with the opposite sex, schools are the central meeting place for the young—a kind of mate selection market. Education also weakens parental control over youths, helps them begin the move toward independence, and provides experience in large, impersonal secondary groups.

Conflict Perspective on Education and Stratification. Critics of functionalism emphasize the role of education in social stratification and competition between groups. Some students receive elite educations, and others do not, in part because of class, race, and gender differences (Kerckhoff 2001). Attendance at elite high schools is a means of attaining high social status. Graduates of British "public" schools (similar to U.S. private or preparatory schools), American preparatory (prep) schools, and private international high schools attend the best universities and become leaders of government, business, and the military. Because elite schools are very expensive and highly selective, affluent members of society have the most access to them and thereby perpetuate class privilege and influence of powerful elites (Howard 2007; Persell and Cookson 1985). The sons and daughters of the "haves" continue their positions of privilege, while lower-class children are prepared for less prestigious and less rewarding positions in society (Bowles and Gintis 2002; Collins 2004). According to this view, when elites of society protect their educational

advantages, the result is **reproduction of class**—socioeconomic positions of one generation passing on to the next.

Proposed choice and voucher plans allowing students and parents to choose schools have been controversial in the United States, as well as other countries. Critics fear that public schools might be left with the least capable students and teachers, further stratifying an already troubled system (Chubb and Moe 1990; McKnight 2011; Torche 2005). Private schools could also become sanctuaries for those who do not want integrated schools, perpetuating religious and racial segregation. In the United States, more states are putting money into charter schools, some of which represent special interest groups.

Having explored some lenses through which sociologists analyze educational systems, we now focus on the problem of equal opportunity in schools.

Thinking Sociologically

Does education enhance upward social mobility and serve both individuals and society, or does it mostly serve the affluent, reproducing social class and training people to fulfill positions at the same level as their parents?

Can Schools Bring About Equality in Societies?

Equal opportunity exists when all people have an equal chance of achieving high socioeconomic status in society regardless of their class, ethnicity or race, or gender (Riordan 2004). James Coleman (1968, 1990) describes the goals of equal educational opportunity:

- To provide a common curriculum for all children regardless of background
- To ensure that children from diverse backgrounds attend the same school
- To provide equality within a given locality

Equal opportunity means that children are provided with equal facilities, financing, and access to school programs (Kozol 1991). Schools in poor neighborhoods or in rural villages in the United States and around the world, however, often lack the basics—safe buildings, school supplies and books, and funds to operate. Lower-class minority students who live in poor areas fall disproportionately at the bottom of the educational hierarchy. Many children face what seem to be insurmountable barriers to educational success: poverty, lack of health care and immunizations, and increased school absences or dropping out (Kozol 2005).

Zimbabwe once had one of the best education systems in Africa, but after severe economic crisis in the country, educational conditions have declined rapidly. Still, some children are privileged over other children, who do not get any education. This system that provides education only for the privileged protects the interests of the elite, according to conflict theorists.

Two widely cited classical studies on equal educational opportunity, the Coleman Report and Jencks's study of inequality (Coleman et al. 1966; Jencks 1972), have had a major impact on policy in education and stand out because of their comprehensive data collection, analysis, and contribution to the understanding of inequality. Coleman and his colleagues found that the differences in test scores between minority students and white students were due not only to in-school factors but also to parents' education levels and other environmental factors such as segregated schools, economic makeup of schools, and some differences in facilities. Based on these findings, the researchers recommended integration of schools to create a climate for achievement. That recommendation resulted in plans to integrate but also in controversy and court challenges (Orfield et al. 1997). Today, other methods to desegregate schools are in place, including charter schools, magnet schools (which draw students from a large area to special curricula or services), and choice plans.

Who Gets Ahead and Why? The Role of Education in Stratification

Education is supposed to be a **meritocracy**, a social group or organization in which people are allocated to positions according to their abilities and credentials. This, of course, is consistent with the principles of a rational or formal bureaucratic social system, where the most qualified person is promoted and decisions are impersonal and based on "credentials" (Charles et al. 2007). Still, in societies around the world, we see evidence that middle-class and elite children, especially boys in the Global South, receive more and better education than equally qualified children. Children do not attend school on an equal footing, and in many cases, meritocracy does not exist. Conflict theorists, in particular, maintain that education creates and perpetuates inequality. The haves hold the power to make sure that institutions, including schools, serve their own needs and protect their access to privileges (Sadovnik 2007). Elite parents have social and cultural capital—language skills, knowledge of how the social system works, and networks—to ensure that their children succeed (Kao 2004). In poorer countries education helps citizens participate and compete in global markets. International organizations such as the United Nations and transnational business communities have become more involved in educational development, especially in programs for training workers in technology.

Sources of Inequality. Three sources of inequality in schools—testing, tracking, and funding—illustrate how schools reproduce and perpetuate social stratification. They also give clues as to what might be done to minimize the repetitious pattern of poverty.

Assessing Student Achievement. Testing is one means of placing students in schools according to their achievement and merit and determining their progress. Critics of standardized testing argue that test questions, language differences, and testing situations are biased against lower-class, minority, and immigrant students, resulting in lower scores and relegating these students to lower tracks in the educational system. In addition, in the case of IQ tests, scientists know that intelligence is complex and that paper-and-pencil tests measure only selected types of intelligence (Gardner 1987, 1999; M. Smith 2008). Other scholars question whether the tests are unbiased for children from different class and ethnic backgrounds; higher-class students with better schooling and enriched backgrounds generally do better on the tests. Nonetheless, testing is the means used to evaluate student achievement locally, nationally, and internationally. The U.S. Obama administration has broadened assessment to include factors in addition to test scores for placement of students. Table 11.4 in the "Engaging Sociology" on page 334 shows differences in ACT (American College Test) and SAT (Scholastic Assessment Test) scores depending on sex, race, and ethnic group. Answer the questions that are posed as you engage with sociological data.

Governments compare their students' educational test scores on the National Assessment of Educational Progress (NAEP) and the International Association for the Evaluation of Educational Achievement (IEA) with other countries' scores to determine how their students and educational systems are performing (NAEP 2006). These are tests of children around the world in literacy, mathematics, science, civic education, and foreign language. For example, recently IEA tested students in the fourth Trends in International Mathematics and Science Study, in which more than 60 countries participated, testing children at the 4th-, 8th-, and 12th-grade levels (Ballantine and Spade 2008; IEA 2007). Another international comparison, the Third International Mathematics and Science Study, found that in mathematics, Canadian 8th graders scored 18th among nations and the United States was 28th. In science, the 12th graders in the two nations were 5th and 16th, respectively (Forgione 2011). However, simply copying educational systems of the more successful countries is not necessarily the answer to improvement, for education occurs in different cultural contexts (Kagan and Stewart 2005; Zhao 2005).

1. *Student Tracking. Tracking* (sometimes called *streaming*) places students in ability groups, allowing educators to address students' individual learning needs. It also contributes to the stratification process that perpetuates inequality. Research finds that tracking levels correlate directly with factors such as the child's class background and ethnic group, language skills, appearance, and other socioeconomic variables (Rosenbaum 1999; Wells and Oakes 1996). In other words, track placement is not always a measure of

Engaging Sociology

Test Score Variations by Gender and Ethnicity

Evaluate your testing experiences and compare them to those of other groups:

Table 11.4   ACT and SAT Scores by Sex and Race/Ethnicity

ACT Scores: 2007	Average	SAT Scores: 2008	Average
Composite, total scores	21.2	SAT Writing, all students	497
Male	21.2	Male	491
Female	21.0	Female	502
White	22.1	White	519
African American	17.0	Black/African American	424
American Indian	18.9	Mexican American	447
Hispanic	18.7	Puerto Rican	445
Asian American	22.6	Other Hispanic	449
		Asian American	516
SAT Scores: 2008	**Average**	American Indian	470
SAT Critical Reading, all students	502	Other	494
Male	504	SAT Math, all students	509
Female	500	Male	527
White	528	Female	499
Black/African American	430	White	537
Mexican American	454	Black/African American	426
Puerto Rican	456	Mexican American	463
Other Hispanic	455	Puerto Rican	453
Asian American	513	Other Hispanic	461
American Indian	495	Asian American	591
Other	496	American Indian	491
		Other	512

Source: Inside Higher Ed 2007 for ACT scores; Inside Higher Ed 2008 for SAT scores.

Sociological Data Analysis:

1. Were your scores an accurate measure of your ability or achievement? Why or why not?

2. What other factors such as your gender or ethnicity enter in?

3. Have your scores affected your life chances? Are there ways in which you have been privileged or disprivileged in the testing process?

4. What might be some causes of the variation in test scores between groups or categories of students?

a student's ability but can be based on teachers' impressions or questionable test results. Over time, differences in children's achievement become reinforced. Students from lower social classes and minority groups are clustered in the lower tracks and complete fewer years and lower levels of school (Lucas and Berends 2002; Oakes et al. 1997; Zehr 2009), resulting in school failure in early adolescence (Chen and Kaplan 2003).

Thinking Sociologically

Were you tracked in any subjects? What effect, if any, did this have on you? What effect did tracking have on friends of yours? How might tracking shape friendship networks?

2. *School Funding.* The amount of money available to fund schools and the sources providing it affect the types of programs offered, an important issue for nations that must compete in the global social and economic system. Money for education comes in some societies from central governments and in others from a combination of federal, state, and local government and private sources, such as religious denominations and philanthropies. In countries such as Uganda, the government runs the schools, but most funding comes from tuition paid by each student or by the student's parents. Whatever the source, schools sometimes face budget crises and must trim programs.

In the United States, unequal public school spending results from reliance on unequal local property taxes as well as state and federal funds. About 49% of the money for U.S. schools comes from state funds, and 43% comes from property taxes. Federal funds provide only about 8.5% of total funding (National Center for Education Statistics 2006). Spending is closely related to the racial and class composition of schools and to student achievement levels. Schools in low-income communities are particularly disadvantaged by smaller tax bases and fewer local resources (Condron and Roscigno 2003). Wealthier districts, on the other hand, can afford better education for their children because more money is collected from property taxes. Higher-class students have advantages not available in poor districts (Kozol 2005). Controversies over school funding in the United States have reached the courtroom in a number of states, yet overall, the United States spends less money per student on education than most other Global North nations.

Public and Private Schools

In the United States, Catholic parochial schools are the most prevalent type of private schools. Other religiously affiliated schools are the next most common. Private preparatory schools are third. About 10% of U.S. students attend private schools.

Studies of low-income and minority students have found that those attending Catholic schools, especially in inner cities, perform more like white middle-class students than those attending public schools (Morgan 2001). In general, private schools are more academically demanding, more stringent, more disciplined, and more orderly. They assign more homework, and they demand more parental involvement than do public schools (Cookson and Persell 1985; Smarick 2009). They can also expel children who do not perform or are disruptive, an option the public schools generally do not have.

However, much of the reason that private schools typically have higher levels of achievement is the select population they admit. A massive study released in the summer of 2006 by the U.S. Department of Education compared 7,000 public schools and 530 private schools (National Center for Education Statistics 2006). The study controlled for student socioeconomic backgrounds and found that children performed roughly on par in the two types of schools, private and public. The only group that lagged significantly was children attending conservative Protestant Christian schools (Schemo 2006).

Proposed choice and voucher plans allowing students and parents to choose their schools have been controversial in part because public schools might well be left with the least capable students and teachers, further stratifying an already troubled system (Chubb and Moe 1990). Private schools could also become sanctuaries for those who do not want integrated schools and could perpetuate religious and racial segregation.

This is a somewhat affluent school in South Africa. Children like this, whose parents can afford to send them to school, will likely grow up with enough education to have better economic prospects.

Education and Social Policy Issues

In the first part of the twenty-first century, school systems around the world face dramatic changes: Pressures for equal educational opportunity for all groups—classes and castes, race and ethnic groups, females and males; the need for technological training of citizenry; increased demand for access to higher education; changes in the student composition of the classroom; and accountability and testing of students and teachers are but a few of the many issues. Policy decisions by lawmakers influence what and how children learn. However, because education reflects societal politics and problems, policies swing from conservative to liberal and back again, depending on who is in power and what their agenda is for change. Reform programs have variously stressed more rigorous high school graduation requirements, raising college entrance requirements, emphasizing basic skills, requiring a longer school day and school year, improving the training and status of teachers, holding educators responsible for students' performance, providing the funds necessary to improve the educational system, and holding families accountable for children's achievement.

Since 1965 and the establishment of the Department of Education in the United States, each president has put his mark on the U.S. education system. President George W. Bush's controversial No Child Left Behind (NCLB) initiative tied school performance to federal funds and required annual competency testing of students (Center for Education Policy 2006). However, critics argue that the NCLB policy made no provision for differences in family backgrounds, socioeconomic status, preschool education, or the community context (Bracey 2005). NCLB also ignored much of what educators and social scientists know from research about teaching and learning; overemphasizing testing penalizes schools that have proportionally more low-income students and students with disabilities, demoralizes schools through unrealistic timetables, and does not provide the funding necessary to carry out the mandates, further disadvantaging small, rural, and poor school districts (Bracey 2005; Karen 2005).

President Obama and his education secretary, Arne Duncan, are changing the No Child Left Behind plan by funding parts of the law that were left unfunded, adding new assessment measures of student learning, taking some emphasis off of high-stakes testing, and nationalizing some of the standards that were previously in the hands of states. They plan to place new emphasis on "zero to five" education and expand Head Start funding for preschoolers. Changes are also in store for higher education where some propose that community college should become free. Also, tuition may be waived for those who become long-term teachers, and tax credits may be given for community service and college tuition. What both Presidents Bush and Obama have in common is increased centralization of power in the federal government, establishing standards of learning.

The greatest barrier to equal education in the twenty-first century, both within and between countries, may prove to be socioeconomic, with many minorities represented in lower classes. The value of early childhood education for low-income children has been established, but access is far from universal. Still, applied sociologist Geoffrey Canada has had success with implementing programs for poor children in Harlem, from preschool to high school, as discussed in the next "Applied Sociologist at Work."

Global Policy Issues in Education

Most societies view the education and training of young people as an economic investment in the future. Countries with capitalist economic systems are more likely to have an educational system that stresses individualism and competition, pitting students against one another for the best grades and the best opportunities. The elites often ensure that their own children get a very different education than the children of the laboring class. Socialist and traditional economic systems often encourage cooperation and collaboration among students, with the collective needs of society viewed as more important than those of individuals. The social economic values of the society are reflected in approaches to learning and in motivation of students (Rankin and Aytaç 2006).

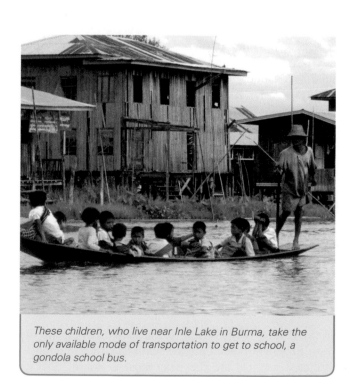

These children, who live near Inle Lake in Burma, take the only available mode of transportation to get to school, a gondola school bus.

The Applied Sociologist at Work— Geoffrey Canada and the Harlem Children's Zone

The United States is "the land of opportunity," but not for residents of Harlem, according to Geoffrey Canada. In Harlem the goal is to avoid being beaten, shot, or raped. Canada has an ambitious agenda—to break the cycle of poverty and have all young people in the Harlem Children's Zone graduate from college.

Geoffrey Canada's life experience prepared him to work as a social activist and educator. He grew up in the South Bronx, was raised by a single mom, and then had a lucky break. He moved to the suburbs with his grandparents, and from there he received a college degree from Bowdoin College and a master's from the Harvard Graduate School of Education.

Having been given an opportunity for education, he is now giving back to the community. As president and CEO of the Harlem Children's Zone in New York, he works with students to increase high school and college graduation rates. The Harlem Children's Zone started out as a 24-block area of Harlem, but has grown to 97 blocks due to its success. In the 97-block neighborhood, his center follows the academic careers of youth, providing social, medical, and educational services that are free to the 10,000 children who live in the Zone. He has built his own charter school, the Promise Academy, with 1,200 students in grades K–10, soon to be K–12. Tuition to the school is free, and admission is done by lottery. For those who do not win the school's admissions lottery, Harlem Children's Zone still provides services to everyone in the Zone, including parenting classes, preschool language classes, school preparation classes, and SAT tutoring.

The Promise Academy has long days and a short summer vacation, a dress code, and strict discipline. The student-teacher ratio is 6 to 1. For students who work hard and achieve, there are rewards. Canada is not apologetic about "buying" the students' cooperation. Some get free trips to Disneyland for good grades, and others get paid for good high school grades.

The Zone is not cheap to run, but Canada points out that the costs of a child ending up in the criminal justice system are much greater, for if a child fails, the community and society have also failed. Therefore, he believes the investment is sound. The program costs $76 million a year, or $5,000 per child. Much of the funding comes from the business community and Wall Street. That may sound like a lot of money, but the national average in 2006 was over $9,000 per student, and the average expenditure per student for the state of New York was nearly $15,000 (U.S. Census Bureau 2008b). To test the effectiveness of the school and program, a Harvard economics professor studied the data from tests of achievement and other academic indicators. He found that the Promise Academy elementary school had closed the gap in math and reading between its students and students in white or mixed schools, and outperformed many of the comparison schools. Those middle school students who started the Promise Academy were behind, but they caught up to students in comparison schools. According to the evaluator, the results were "stunning."

The project has been called "one of the biggest social experiments of our time" (Tough 2008). Because of the proven success of the Harlem Children's Zone, President Obama has taken notice and plans to replicate the model in 20 other cities across the nation.

Geoffrey Canada received his undergraduate degree from Bowdoin College and his master's in education from Harvard. He has written several books and articles, including *Reaching Up for Manhood: Transforming the Lives of Boys in America* (1998).

Findings generally support a convergence of curricular themes across nations, reflecting the interdependence of nations. Still, many researchers question whether that convergence is good for all people in all societies, especially students from peripheral Global South countries.

Political and economic trends outside of a country can also have an impact on the educational system within the country. Examples of external influences include world educational and technological trends, new inventions, and new knowledge.

Education of Girls Around the Globe. "One of the silent killers attacking the developing world is the lack of quality basic education for large numbers of the poorest children in the world's poorest countries—particularly girls" (Sperling 2005:213). In 2011, 55% of out-of-school children were girls, and two thirds of illiterate adults were women (Global Campaign for Education 2011). Another 150 million dropped out of primary school. About 50% of girls in sub-Saharan Africa do not complete primary school, and only 17% are in secondary school (UNESCO 2005). Of the one sixth of the world's population that is illiterate, two

Sociology in Your Social World

Distance Education: Breaking Access Barriers?

Amy J. Orr

Technology has had a tremendous impact on education. Computers are used in a variety of ways in classrooms around the world for both the consumption and the production of knowledge. The Internet, as well as programs such as WebCT and Blackboard, has made the distribution of education around the world much more feasible. Today, thousands of individuals worldwide can obtain a college degree while sitting in front of their computers.

According to the U.S. Department of Education (2004–2005), distance learning is defined as "an option for earning course credit at off-campus locations via cable, television, Internet, satellite classes, videotapes, correspondence courses, or other means." In 2008, about two thirds (66%) of postsecondary institutions in the United States offered distance education courses (National Center for Education Statistics 2008b), and the number continues to increase. Around the world, distance education is available in both Global North and Global South countries. One of the primary purposes of distance learning is to increase access by learners who might not ordinarily be able to obtain a traditional postsecondary education. For example, distance education courses are often heavily used by nontraditional students who wish to continue their education but face barriers such as family, employment, or distance. Questions arise, however, regarding the level of accessibility that is truly provided to underrepresented groups throughout the United States and the world. Does distance education provide new opportunities for these groups to attain postsecondary education? Or are new barriers created due to the increasing use of technology to deliver a college curriculum?

In an attempt to address these questions, a study of California's higher education system finds that distance education can increase students' access to college and efficiency of gaining an education (Chapman 2010). The National Postsecondary Education Cooperative (2004) concluded that (a) distance education increases overall access to postsecondary education; (b) a "digital divide" still remains for some groups, such as blacks and Hispanics; (c) a "dearth of germane information, literacy barriers, and limited diversity of content are significant barriers to getting lower income users online"; and (d) low-income students are less likely than those with higher incomes to have opportunities to gain the skills required to use the technology.

Similar issues arise on a global level. Access to technology differs significantly by nation. Like disadvantaged students in the United States, individuals in the least developed nations may face literacy barriers or find the curriculum irrelevant. An additional concern arises with regard to diversity of cultural contexts and perspectives. Numerous studies report cross-cultural differences in student learning styles as well as diverse preferences for teaching approaches. For example, on reviewing a number of cross-cultural studies, researchers noted that Asian students are accustomed to individualized learning techniques and tend to prefer "teacher-centered" teaching styles (Weisenberg and Stacey 2005). Online pedagogies tend to emphasize collaborative learning and "student-centered" techniques. Because distance learning rests on the technological foundations of Western culture, minority cultures around the world may be further marginalized, as the dominant pedagogies may not be responsive to their cultural uniqueness (Smith and Ayers 2006).

Overall, distance education presents a dilemma with regard to the accessibility of postsecondary education. Although it provides unprecedented opportunities to expand access to education, distance education also runs the risk of creating new divides.

Professor Amy Orr teaches at Linfield College in Oregon. She specializes in race and ethnicity, discrimination, diversity, the impact of wealth and poverty on academic achievement, and gender and the "boy crisis" in schools. Her PhD is from University of Notre Dame.

Religion: The Search for Meaning

Abu Salmaan, a Muslim father and shopkeeper in Syria, prays frequently in keeping with the commands of the holy book of this faith. Like his neighbors, when the call to prayer is heard, he comes to the village square, faces Mecca, and prostrates himself, with his head to the ground, to honor God and to pray for peace. Doing this five times a day is a constant reminder of his ultimate loyalty to God—whom he calls Allah. As part of the larger Abrahamic religious tradition (which includes Judaism, Christianity, and Islam), he believes in one God and accepts the Hebrew Bible and the authority of Jeremiah, Isaiah, Amos, and Jesus as prophets. He believes that God also revealed Truth in another voice— that of Muhammad. He is devoted, worshipping daily, giving generously to charities, and making business decisions based on moral standards of a God-loving Muslim.

Trevor Weaver is a Presbyterian living in Louisville, Kentucky. He attends worship and prays to God in church and in emergencies when he feels helpless. Trevor had theological studies at a church-related college, and he has a strong knowledge of the scriptures. When he makes daily decisions, he thinks about the ethical implications of his behavior as "a member of the larger family of God." He opposes prayer in schools because this would make some children feel ostracized; he values diversity and acceptance of other traditions, and he believes each person needs to "work out his own theology." He thinks of himself as a person of faith, but his evangelical neighbor thinks he is a fallen soul.

Tuneq, knowledgeable Netsilik Eskimo hunter that he is, apologizes to the soul of the seal he has just killed. He shares the meat and blubber with his fellow hunters, and he makes sure that every part of the seal is used or consumed—skin, bones, eyes, tendons, brain, and muscles. If he fails to honor the seal by using every morsel, or if he violates a rule of hunting etiquette, an invisible vapor will come from his body and sink through the ice, snow, and water. This vapor will collect in the hair of Nuliajuk, goddess of the sea. In revenge, she will call the sea mammals to her so the people living on the ice above will starve. Inuit religion provides rules that help enforce an essential ecological ethic among these arctic hunters to preserve the delicate natural balance.

These are but three examples from the world's many and varied religious systems. What they have in common is that each system provides directions for appropriate and expected behaviors and serves as a form of social control for individuals within that society. Indeed, these directions are made *sacred*, a realm of existence different from mundane everyday life. Religion, according to Andrew M. Greeley (1989), the well-known Catholic priest and sociologist, pervades the lives of people of faith. It cannot be separated from the rest of the social world. Sociologists are interested in these relationships—in the way social relationships and

Religion takes many forms and is expressed in many ways, but it is always about a sense of meaning. That sense of meaning generates sacredness and makes other aspects of life meaningful. Here a Jewish girl reads the Torah during her bat mitzvah, the initiation ceremony into the faith.

structures affect religion and in the consequences of religion for individuals and for society as a whole.

In this part of the chapter, we explore religion as a complex social phenomenon, one that is interrelated with other processes and institutions of society. We investigate what religion does for individuals, how individuals become religious, and how religion and modern societies interact.

What Does Religion Do for Us?

We began this chapter by looking at examples of daily experiences in which religion and society have enormous power over people. Why do people engage in religious practices, beliefs, and organizations? In short, they do so because religion meets certain very basic needs.

Human questions about the meaning of life, the finality of death, or whether injustice and cruelty will ever be ended cannot normally be answered by science or by everyday experience. Religion helps explain the meaning of life, death, suffering, injustice, and events beyond our control. As sociologist Émile Durkheim (1947) pointed out, humans generally view such questions as belonging to a realm of existence different from the mundane or profane world of experience. He called this separate dimension the *sacred realm*. This sacred realm elicits feelings of awe, reverence, and even fear. It is viewed as being above normal inquiry and doubt. Religious guidelines, beliefs, and values dictate "rights and wrongs," provide answers to the big questions of life, and instill moral codes and ideas about the world in members of each society or subculture. For that reason, religions are extremely important in controlling everyday behavior of individuals (Durkheim [1915] 2002).

Thus, religion is more than a set of beliefs about the supernatural. It often *sacralizes* (makes sacred and unquestionable) the culture in which we live, the class or caste position to which we belong, the attitudes we hold toward other people, and the morals to which we adhere. Religion is a part of our lifestyle, our gender roles, and our place in society. We are often willing to defend it with our lives. Many different people around the world believe quite as strongly as we do that they have found the Truth—and are willing to die for their faith.

Although the root causes of various wars such as those between Hindus and Muslims in India, Catholics and Protestants in Northern Ireland, and Sunnis and Shiites in the Muslim country of Iraq are often political and economic, religious differences help polarize we-versus-they sentiments. Such conflicts are never exclusively about religion, but religion can convince each antagonist that God is on its side. Religion is an integral part of most societies and is important in helping individuals define reality and answer difficult questions.

Thinking Sociologically

Consider your own religious tradition. Which of the purposes or functions of religion mentioned above does your religious faith address? If you are not part of a faith community or do not hold religious beliefs, are there other beliefs or groups that fulfill these functions for you?

Components of Religion

Religion normally involves at least three components: a faith or worldview that provides a sense of meaning and purpose in life (which we will call the *meaning system*), a set of interpersonal relationships and friendship networks (which we will call the *belonging system*), and a stable pattern of roles, statuses, and organizational practices (which we will call the *structural system*) (Roberts and Yamane 2012).

Meaning System

The meaning system of a religion includes the ideas and symbols it uses to provide a sense of purpose in life and to help explain why suffering, injustice, and evil exist. It provides a big picture to explain events that would otherwise seem chaotic and irrational. For example, although the loss of a family member through death may be painful, many people find comfort and hope and a larger perspective on life in the idea of life after death. Most religious people find that love of God gives deep satisfaction to life.

Because each culture has different problems to solve, the precise needs reflected in the meaning system vary. Hence, different societies have developed different ways of answering questions and meeting needs. In agricultural societies, the meaning systems revolve around growing crops and securing the elements necessary for crops—water control, sunlight, and good soil. Among the Zuni of New Mexico and the Hopi of Arizona, water for crops is a critical concern. These Native American people typically grow corn in a climate that averages roughly 10 inches of rain per year, so it is not surprising that the central focus of the dances and the supernatural beings—*kachinas*—is to bring rain. In many societies, the death rate is so high that high fertility has been necessary to perpetuate the group. Thus, fertility goddesses take on great significance. In other societies, strong armies and brave soldiers have been essential to preserve the group from invading forces; hence, gods or rituals of war have been popular. Over time, meaning systems of religion have reflected needs of the societies in which the religion is practiced.

Belonging System

Belonging systems are profoundly important in most religious groups. Many people remain members of religious groups not so much because they accept the meaning system of the group but because that is where their belonging system—their friendship and kinship network—is found. Their religious group is a type of extended family. A prayer group may be the one area in their lives in which people can be truly open about their personal pain and feel safe to expose their vulnerabilities (Wuthnow 1994). Irrespective of the meaning system, a person's sense of identity may be very much tied up with being a Buddhist, a Christian, or a Muslim.

The religious groups that grow the fastest are those that have devised ways to foster friendship networks within the group, including emphasis on *endogamy*, marrying within one's group. If a person is a member of a small group in which interpersonal ties grow strong—a church bridge club or a Quran study group—he or she is likely to feel a stronger commitment to the entire organization. The success of megachurches is due in part to their attention to the belonging system, making people feel they belong to smaller groups within the larger organization. In short, the belonging system refers to the interpersonal networks and the emotional ties that develop between adherents to a particular faith community.

Structural System

A religion involves a group of people who share a common meaning system. However, if each person interprets the beliefs in his or her own way and if each attaches his or her own meanings to the symbols, the meaning system becomes so individualized that *sacralization* of common values can no longer occur. Therefore, some system of control and screening of new revelations must be developed. Religious leaders in designated statuses must have the authority to interpret the theology and define the essentials of the faith. The group also needs methods of designating leaders, of raising funds to support their programs, and of ensuring continuation of the group. To teach the next generation the meaning system, members need to develop a formal structure to determine the content and form of their educational materials, and then they must produce and distribute them. In short, if the religion is to survive past the death of a charismatic leader, it must undergo institutionalization.

Religious institutions embrace several interrelated components: the meaning system, mostly operating at the micro level; the belonging system, critical at the micro level but also part of various meso-level organizations; and the structural system, which tends to have its major impact at the meso and macro levels. These may reinforce one another and work in harmony toward common goals, or there may be conflict between them. When change occurs, it usually occurs because of disruption in one of the systems. A group cannot survive in the modern world unless it undergoes *routinization of charisma*. That is, the religious organization must develop established roles, statuses, groups, and routine procedures for making decisions and obtaining resources (Weber 1947).

At the local level, too, a formal religious structure develops, with committees doing specific tasks, such as overseeing worship, maintaining the building, recruiting religious educators, and raising funds. These committees report to an administrative board that works closely with the clergy (the ordained ministers) and has much of the final responsibility for the life and continued existence of the congregation. The roles, statuses, and committees make up the structural system. Religious organizations are, among other things, bureaucracies, with the same sort of dysfunctions of any formal organization (as discussed in Chapter 5).

Thinking Sociologically

Think about the meaning, belonging, and structural systems of a religion with which you are familiar. Illustrate how these elements influence and are influenced by the social world, from individual to national and global systems.

How Do Individuals Become Religious? Micro-Level Analysis

We are not born religious, although we may be born into a religious group. We learn our religious beliefs through socialization, just as we learn our language, customs, norms, and values. Our family usually determines the religious environment in which we grow up, whether it is an all-pervasive message or a one-day-a-week lesson in religious socialization. We start imitating religious practices such as prayer before we understand these practices intellectually. Then, as we encounter the unexplainable events of life, religion is there to provide meaning. Gradually, religion becomes an ingrained part of many people's lives.

It is unlikely that we will adopt a religious belief that falls outside the religions of our society. For instance, if we are born in India, we will be raised in and around the

Hindu, Muslim, or Sikh faiths. In most Arab countries, we will become Muslim; in South American countries, Catholic; and in many Southeast Asian countries, Buddhist. Even if we are born into the family next door, our religion and politics might be different from our neighbors'. Indeed, although our religious affiliation may seem normal and typical to us, none of us is part of a religion that is held by a majority of the world's people, and we may in fact be part of a rather small minority religious group when we think in terms of the global population. Table 11.5 shows this vividly.

Table 11.5   Religious Membership Around the Globe

Religion	Membership (in Millions)	Percentage of World Population
Christian (Total)	2,199.8	33.3
Roman Catholic	1,121.5	17.0
Independents	433.1	6.5
Protestant	381.8	5.8
Orthodox	233.1	3.5
Anglican	82.6	1.2
Unaffiliated	19.5	1.8
Muslim	1,387.5	21.0
Hindu	875.7	13.2
Nonreligious	776.8	11.7
Ethnic/Tribal Religions	652	9.8
Universist (Chinese Folk Religion)	385.6	5.8
Buddhist	385.6	5.8
Atheist	153.4	2.3
New Religious Movements	106.5	1.6
Sikh	22.9	0.3
Jewish	15.0	0.2
Spiritist	13.5	0.2
Baha'i	7.7	0.1
Confucianism	6.4	0.1
Jain	5.3	0.1
Taoism	3.4	0.1
Shintoism	2.8	a
Zoroastrian	.2	a

Source: Barrett, Johnson, and Crossing 2008. Reprinted with permission from *Encyclopedia Britannica Almanac 2008.* Copyright © 2008 by Encyclopedia Britannica, Inc.

Note: Numbers add up to more than the six billion people in the world because many people identify themselves with more than one religious tradition. Thus, the percentages will also add up to more than 100%. Percentages are rounded.

a = Less than 0.05%.

these stories, myths transmit values and a particular outlook on life. If a story such as the exodus from Egypt by ancient Hebrew people elicits some sense of sacredness, communicates certain values, and helps life make sense, then it is a myth. The Netsilik Eskimo myth of the sea goddess Nuliajuk (explained earlier in the chapter) reinforces and makes sacred the value of conservation in an environment of scarce resources. It provides messages for appropriate behavior in that group. Thus, whether a myth is factual or not is irrelevant. Myths are always "true" in some deeper metaphorical sense.

Rituals are group activities in which myths are reinforced with music, dancing, kneeling, praying, chanting, storytelling, and other symbolic acts. A number of religions, such as Islam, emphasize devotion to orthopraxy (conformity of behavior in rituals and in morality) more than orthodoxy (conformity to beliefs or doctrine) (Preston 1988; Tipton 1990). Praying five times a day while facing Mecca, mandated for the Islamic faithful, is an example of orthopraxy.

Often, rituals involve an enactment of myths. In some Christian churches, the symbolic cleansing of the soul is enacted by actually immersing people in water during baptism. Likewise, Christians frequently reenact the last supper of Jesus (Communion or Eucharist) as they accept their role as modern disciples. Among the Navajo, rituals enacted by a medicine man may last as long as five days. An appropriate myth is told, and sand paintings, music, and dramatics lend power and unique reality to the myths.

The group environment of the ritual is important. Ethereal music, communal chants, and group actions such as kneeling or taking off one's shoes when entering the shrine or mosque create an aura of separation from the everyday world and a mood of awe so that the beliefs seem eternal and beyond question. They become sacralized. Rituals also make ample use of symbols, discussed in Chapter 4.

A **symbol** is anything that can stand for something else. Because religion deals with a transcendent realm, a realm that cannot be experienced or proven with the five senses, sacred symbols are a central part of religion. They have a powerful emotional impact on the faithful and reinforce the sacredness of myths.

Sacred symbols have been compared to computer chips, which can store an enormous amount of information and deliver it with force and immediacy (Leach 1979). Seeing a cross can flood a Christian's consciousness with a whole series of images, events, and powerful emotions concerning Jesus and his disciples. Tasting the bitter herb during a Jewish Seder service may likewise elicit memories of the story of slavery in Egypt, recall the escape under the leadership of Moses, and send a moral message to the celebrant to work for freedom and justice in the world today. The mezuzah, a plaque consecrating a house on the doorpost of a Jewish home, is a symbol reminding the occupants of their commitment to obey God's commandments and reaffirming God's commitment to them as a people. Because symbols are often heavily laden with emotion and can elicit strong feelings, they are used extensively in rituals to represent myths.

Myths, rituals, and symbols are usually interrelated and interdependent (see Figure 11.3). Together, they form

Muslims pray to God (whom they call Allah) five times a day, removing their shoes and prostrating themselves as they face Mecca. This is an important ritual, and it illustrates that orthopraxy is central for Muslims. Personal devotion, which is expressed in actions, is emphasized in Islam.

The yarmulke (skullcap or kippah) has been worn by Jewish men since roughly the second century CE. It symbolizes respect for and fear of God and serves to remind the wearer of the need for humility: There is always some distance between himself and God. These caps are also a sign of belonging and commitment to the Jewish community.

These shoes outside a mosque are a symbol of awe for the holy. Removal of shoes is an expression of sacred respect for God among Muslims.

the meaning system—a set of ideas about life or about the cosmos that seem uniquely realistic and compelling. They reinforce rules of appropriate behavior and even political and economic systems by making them sacred. They can also control social relationships between different groups. The Gypsy revulsion at the filthy *marime* practices of non-Gypsies, like the Kosher rules for food preparation among the Jews, create boundaries between "us" and "them" that greatly reduce prospects of marriage outside of the religious

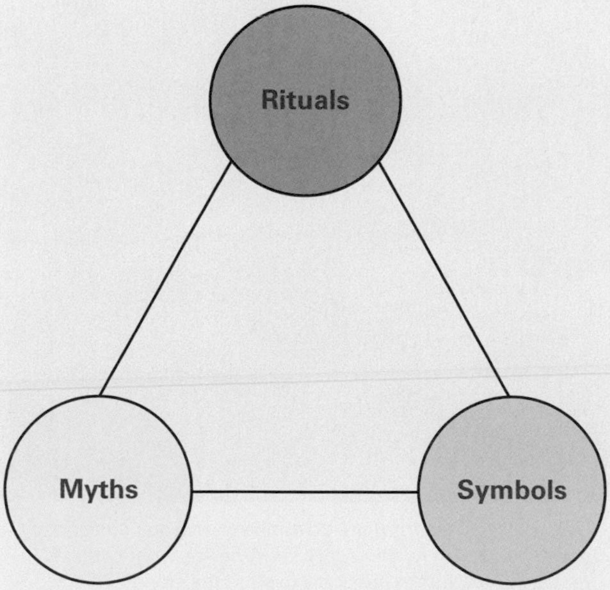

Figure 11.3 The Meaning System of Religion Is Composed of Three Interrelated Elements

[Diagram showing three circles connected in a triangle: "Rituals" at top, "Myths" at bottom left, "Symbols" at bottom right]

community. Some scholars think these rituals and symbolic meanings are the key reason why Gypsies and Jews have survived for millennia as distinct groups without being assimilated or absorbed into dominant cultures. The symbols and meanings created barriers that prevented the obliteration of their cultures.

When symbolic interactionists study religion, they tend to focus on how symbols influence people's perception of reality and on the role rituals and myths play in defining what is "really real" for people. Symbolic interactionists stress that humans are always trying to create, determine, and interpret the meaning of events. Clearly, no other institution focuses as explicitly on determining the meaning of life and its events as religion.

Thinking Sociologically

In a tradition with which you are familiar, how do symbols and sacred stories reinforce a particular view of the world or a particular set of values and social norms?

Seeking Eternal Benefits: A Rational Choice Perspective

People with religious freedom make rational choices to belong to a religious group or change to another religion after weighing the costs (financial contributions, time involved) and benefits of belonging (eternal salvation, a belonging system, and a meaning system) (Finke and Stark 2005; Warner 1993). Rational choice theory is based on an economic model of human behavior. The basic idea is that the process people use to make decisions is at work in religious choices: What are the benefits, and what are the costs? Do the benefits outweigh the costs? The benefits, of course, are nonmaterial when it comes to religious choices—feeling that life has meaning, confidence in an afterlife, sense of communion with God, and so forth. This approach views churchgoers as consumers who are out to meet their needs or obtain a "product." It depicts churches as entrepreneurial establishments, or "franchises," in a competitive market, with "entrepreneurs" (ministers) as leaders. Competition for members leads churches to "market" their religion to consumers. Converts and religious people generally are thus regarded as active and rational agents pursuing self-interests, and growing churches are those that meet "consumer demand" (Finke 1997; Finke and Stark 2005; Iannaccone 1994). Religious groups produce religious "commodities" (rituals, meaning systems, sense of belonging, symbols, etc.) to meet the "demands" of consumers (Sherkat and Ellison 1999).

Rational choice theorists believe that aggressive religious entrepreneurs who seek to produce religious products that appeal to a target audience will reap the benefits of a large congregation. Churches, temples, and mosques are competitive enterprises, and each must make investments of effort, time, and resources to attract potential buyers. There are many religious entrepreneurs seeking to meet individual needs. The challenge is for the various groups to beat the competition by better meeting the demand (Finke and Stark 2005).

Rational choice theorists believe that when more religious groups compete for the hearts and minds of members, concern about spiritual matters is invigorated, and commitment is heightened. Religious pluralism and spiritual diversity increase rates of religious activity as each group seeks its market share and as more individual needs are met in the society (Finke and Stark 2005; Iannaccone 1995).

Thinking Sociologically

Does the rational choice approach seem to you to make sense of religious behavior? Is religious behavior similar to self-interested economic behavior?

Religion and Modern Life: Meso-Level Analysis

As we think of religion at the meso level, two categories come to mind. First, religion is an institution, as described in the introduction to institutions. Second, the institution of religion is composed of many real organizations at the meso level—our denominational affiliation (Lutheran, Presbyterian), regional Catholic diocese, or Jewish movement (Hasidic, Orthodox, Conservative, Reform). The next section gives an overview of types of religious organizations.

Types of Religious Associations

From denominations to sects to new religious movements (NRMs), religious organizations take many forms. The following discussion explores *ideal types*—that is, models that summarize the main characteristics of types of religious organizations. Any specific religious group may not fit all of the characteristics exactly. Furthermore, these organizational patterns are particularly relevant to understanding the evolution of religious traditions, especially Western Christianity. One major world religion that displays a distinctive pattern is Islam, as is discussed in the next "Sociology in Your Social World."

The Ecclesia. Official state religions are called **ecclesia**, referring to religious groups that claim as members everybody within the boundaries of a particular society (Roberts and Yamane 2012). Ecclesia include all members of a society. They try to monopolize religious life in that society, have a close relationship with the power structure, have a formal structure with officially designated full-time clergy, and have membership based on birth into the society.

Many countries have official state religions. Norway and Sweden are Lutheran. Spain, France, Italy, and many Latin American countries are Roman Catholic. Greece is Greek Orthodox. Iran, Egypt, and other Middle Eastern countries are Islamic. England is Anglican. India is predominantly Hindu, although Muslims share power. Because ecclesia represent the interests of the state and those in power, disfranchised groups often seek other religious outlets. In the early twentieth century in Britain, 80% of the Welsh people (Wales is the westernmost of the British Isles) affiliated with nonconformist

Clan totems are part of the cultural and religious tradition of native peoples in northwest North America, from Oregon to Alaska, and totem poles are used to tell stories. This photo from St. John United Methodist Church in Anchorage, Alaska, shows the blending of local cultural and religious symbols with a Christian symbol (the cross) in a Christian sanctuary. This one tells the Easter story.

Sociology in Your Social World

Islam, Mosques, and Organizational Structure

Most mosques in the United States are relatively new, with 87% having been founded since 1970 and 62% since 1980. At the turn of the century, there were 1,209 mosques in the United States (Bagby 2003). Christian congregations tend to be somewhat autonomous entities, supported by members but linked organizationally to a denomination. According to Islamic scholar Ihsan Bagby (2003:115), "Most of the world's mosques are simply a place to pray. . . . A Muslim cannot be a member of a particular mosque" because mosques belong to God, not to the people. The role of the *imam*—the minister—is simply to lead prayers five times a day and to run the services on the Sabbath. The imam does not operate an organization and does not need formal training at a seminary. Mosques were historically government supported in other countries, so they had to make major changes in how they operate when they opened in North America. Because they could not depend on government funding, Islamic mosques needed to adapt to the congregational model: members who would support a particular mosque. They also began to put more emphasis on religious education (which had been done largely by extended families in the "old country"), religious holidays at the mosque rather than with families and dinners at a fellowship hall, and more life-cycle celebrations (births and marriages solemnized). This is a major change in the role of the mosque and the imam for many Muslims.

Bagby (2003) reports that there are two main categories of mosques: those attended primarily by African Americans and those attended primarily by immigrants. African American mosques represent only 27% of the total, and they are in some important ways different in organizational structure from those attended by immigrants (28% of the remaining American Muslims being from South Asia, 15% being Arab, and 30% having a mixture of backgrounds). Even the makeup of mosques is unlike anything familiar to most Americans. Only 28% of American mosques depend on an imam as the final authority, whereas the majority are led by an executive committee or board of directors (called the *majlis*), which has authority. More important, 93% of African American mosques are imam-led compared with only 38% of immigrant-attended mosques.

About 33% of all mosques in the United States have a paid, full-time imam, and 16% of those imams have to work a second job. By comparison, 89% of other congregations (Christian and Jewish) have paid ministers. Only 13% of imams have a master's degree in theology, which is the standard expected for most mainstream Christian and Jewish clergy. So despite having exceptionally high levels of adherents in management and the professions, having unusually high incomes, and having 58% of adherents with college degrees, Islamic mosques are less bureaucratized, with less emphasis on professional credentials, membership roles, or denominational connections for imams. It is likely that mosques will begin to assimilate to the religious organizational pattern of the larger society, but at this point, it is important to recognize that the discussion of "denominations" and "sects" does not apply to Islam in America.

Christianity, not with the Church of England, in which the queen is officially the head of the church. They explicitly rejected the meshing of religious and national loyalty, feeling that religious bodies should have a free and independent voice that is not compromised by politics or patriotism (Davies 2007; Morgan 1987). In societies that do not have official religions such as Canada and the United States, religion takes several forms: denominations, sects, and NRMs (or cults).

Denominations. In the United States, Congregationalist, Episcopalian, Presbyterian, Lutheran, Baptist, Methodist, and other mainline Christian religious groups are denominations.

Denominations are voluntary religious associations (not part of a state church) that accept members without strict prerequisites such as a particular national or ethnic membership or literal interpretation of scriptures. Still, denominations have certain characteristics in common. The denomination is seen as a legitimate form of religious expression but does not have religious dominance or monopoly in the society. Each religious group appeals to a particular segment of the population, often related to class, race, ethnicity, and sometimes regional area. Denominations coexist with and usually are accepting of others (Christiano, Swatos, and Kivisto 2008; Roberts and Yamane 2012).

Most denominations have formal bureaucracies with hierarchical positions, specialization of tasks, and official creeds. This hierarchy extends even to the international level in some denominations, such as the Anglicans and Catholics. The leadership is trained, and there is often a professional church staff. Worship typically involves formal, ritualized, and prescribed ceremonies. Note that the Roman Catholic Church is an ecclesia in Spain because it is the official state religion, but it is a denomination in the United States and Canada, where it is one of many religious groups and is not affiliated with or supported by the state.

Although denominations vary somewhat in the values and beliefs they advocate, they usually support the basic values and social arrangements of the larger society. They may work for moderate changes, but they seldom call for radical or revolutionary restructuring of the society. Thus, the membership tends to be made up of those people who have benefited from the existing social arrangements and who hold positions of power and influence. Most of these denominations are large and financially secure, in part because they attract a middle- and upper-class clientele. This comfortable accommodation between denominations and the existing social order is precisely what made Marx feel that religion was used to keep the social power structure from changing.

The key characteristic of a denomination is its accommodation to the state. Disfranchised members of society are more likely to be attracted to sects or cults.

Sects. John Wesley (1703–1791), founder of a movement now known as the Methodist Church, began a renewal movement in the Church of England. He wanted to call people back to the basics of the faith. His evangelical revival called followers to lead Christian lives through a method of strict discipline—prayer, worship, study, and mutual support groups. Hence, the name **Methodists** arose—a term that was originally derogatory. It referred to people whose worship was too "enthusiastic" to be decent and orderly—the standard of worship among Anglicans and Presbyterians of the day. Because church authorities disapproved of Wesley's new methods, he was marginalized and refused an Anglican church to lead. He took his movement to the streets. As a skilled organizer, Wesley formed associations in Great Britain and North America. Today, Methodists, once a sect, have transformed into one of the largest Christian bodies and a mainstream denomination in the United States.

A sect is a voluntary and exclusive religious group with demanding membership standards requiring high levels of commitment. **Sects** are religious groups that break away to protest against their parent religion, rejecting their parent churches as "compromised" and the social order as ungodly (Roberts and Yamane 2012; Stark and Bainbridge 1985). They believe that the true religious doctrines are being abandoned, people are becoming contaminated by worldly ways, and the group must save itself by returning to the true religion. Often, the break has other underlying social dimensions. Sect members may feel the denomination has come to embrace the existing power structure too closely. In other instances, the splinter group is simply alienated by the bureaucracy and hierarchical structure of the religious group itself.

Sects are often characterized by their separation from other religious and even social groups. An example of an extreme isolationist sect—a type of counterculture—is the Old Believers of the Russian Orthodox Church. In the mid-1600s, the Old Believers were so adamant about not adhering to the reforms of the tsar and the head of the Russian Orthodox Church that when troops approached, they barricaded themselves inside their churches and set themselves on fire rather than accommodate to change (Crummey 1970; Pentikänen 1999; Robson 1995).

Members of Christian sects in North America are often socially separated from those of higher social classes. Sectarians substitute religious status for social status (Pope [1942] 1965). For those who are less well-off, a sect gives the feeling that members are among God's chosen, even if they do not have much social prestige in this world. Because most people who join sects feel deprived, there is a high degree of tension between the values of members and those of the larger society (Stark 1985).

There are now a number of Old Believer communities in the Kenai Peninsula of Alaska, like this Church of St. Nicholas. Some of these Old Believer sects are so isolationist that they are accessible almost entirely by horse or by foot, and they have large "Keep Out" signs as one approaches the community. Commitment to the beliefs of the in-group is intense in this Russian Orthodox sect.

As sects grow in membership, coordination becomes necessary, the group must have a process for deciding who succeeds the leader when she or he dies, a hierarchy develops, and members become less alienated from the society as social class positions improve. Any of these factors can be the catalyst to move the sect toward becoming a denomination. This process—which we call institutionalization—occurs and helps survival, but it is fraught with difficulties. For example, the intimacy of the smaller group is replaced by the impersonality of the emerging large organization.

NRMs (new religious movements or cults), like sects, are protest or splinter groups. However, unlike sects, if NRMs survive for several generations, become established, and gain some legitimacy, they become new religions in the society rather than new denominations of the existing faith. *Cult* was once the common term for this kind of movement, but the media and the public have so completely misused the word, making it evaluative, that its meaning has become unclear and often negative. Most sociologists of religion now prefer *NRMs* to describe these religious forms (Christiano et al. 2008).

NRMs are founded on a new revelation (or insight) or on a radical reinterpretation of an old teaching. They are usually out of the mainstream religious system, at least in their early days. Christianity, Buddhism, and Islam all began as NRMs or cults. For example, either Jesus started a new cult, or else we must conclude that Christianity is a denomination of Judaism—an idea that neither Christians nor Jews would likely accept. The estimated number of NRMs in North America at the turn of the century was between 1,500 and 2,000 (Mather and Schmidt 2006; Melton 1992). There could be 10,000 more NRMs in Africa and an undetermined but large number in Asia (Hadden 2006; Religious Worlds 2007).

An NRM is often started by a charismatic leader, someone who claims to have received a new insight, often directly from God. For example, Reverend Sun Myung Moon founded an NRM called the Unification Church, the members of which are often referred to as *Moonies.* Some sensational NRMs have ended in tragedy. A group suicide occurred in 1997 in California by a band called Heaven's Gate, whose members thought supernatural beings were coming to take them away in a flying saucer (Wessinger 2000).

Most religious groups that are now accepted and established were stigmatized as weird or evil when they started. Early Christians were characterized by Romans as dangerous cannibals, and in the early decades of the twentieth century in the United States, Roman Catholics were depicted in the media as dangerous, immoral, and anti-American (Bromley and Shupe 1981). When we encounter media reports about NRMs, we should listen to and read these with a good dose of skepticism and recognize that not all cults are like the sensational ones.

Many NRMs are short lived, lasting only as long as the charismatic leader does. Only those groups that institutionalize and prepare for their futures, as did the early Christians, are likely to survive. So both sects and NRMs find that developing a complex organization enhances survival, but in the process, they are usually transformed as religious groups. Figure 11.4 illustrates the parallel evolution of these two types of religious groups.

Figure 11.4   The Evolution of Sects and Cults (New Religious Movements: NRMs)

Source: Roberts and Yamane 2012.

Religion in Society: Macro-Level Analysis

As an integral part of society, religion meets the needs of individuals and of the society. In this section, we explore functionalist and conflict theories as we consider some functions of religion in society and the role of religion in stratification systems of society.

The Contribution of Religion to Society: A Functionalist Perspective

Regardless of their personal belief or disbelief in the supernatural, sociologists of religion acknowledge that religion has important social consequences. Functionalists contend that religion has positive consequences—helping people answer questions about the meaning of life and providing part of the glue that helps hold a society together. Let us look at some of the social functions of religion, keeping in mind that the role of religion varies depending on the structure of the society and the time period.

Social Cohesion. Religion helps individuals feel a sense of belonging and unity with others, a common sense of purpose with those who share the same beliefs. Durkheim's ([1897] 1964) widely cited study of suicide stresses the importance of belonging to a group. Research shows that a high rate of congregational membership and religious homogeneity in a community is associated with lower rates of suicide (Ellison, Burr, and McCall 1997). Thus, religion serves society well as long as religious views are consistent with other values of society (Bainbridge and Stark 1981).

In complex and heterogeneous societies, however, no single religion can provide the core values of the culture. In such circumstances, a **civil religion**—a shared public faith in the nation and what the nation stands for based on a country's history, social institutions, set of beliefs, symbols, and rituals that pervade secular life and institutions—provides a theology of the nation that serves to bless the nation and to enhance conformity and loyalty. Civil religion is based on a set of beliefs, symbols, and rituals that pervade many aspects of secular life and institutions: making the Pledge of Allegiance, saluting the flag, and singing the national anthem. This civil religion often serves as an alternative religion to Christianity or to whatever religion is most dominant. It involves a shared public faith in the nation and what that nation stands for. Civil religion is supported by various types of patriotic groups that legitimate the governmental system (Bellah 1992; Roberts and Yamane 2012).

Legitimating Social Values and Norms. The values and norms in a culture must be seen as compelling to members of the society. Religion often sacralizes social norms—grounds them in a supernatural reality or a divine command that makes them beyond question. Whether those norms have to do with care for the vulnerable, the demand to work for peace and justice, the sacredness of a monogamous heterosexual marriage, or proper roles of men and women, foundations of morality from scripture create guidelines and feelings of absoluteness. This, in turn, lends stability to the society. Of course, the absoluteness of the norms also makes it more difficult to change them as the society evolves. This inflexibility is precisely what pleases religious conservatives and distresses theological liberals, the latter often seeking new ways to interpret the old norms. Note also that what people say they believe and how they behave are not always compatible, as the next "Sociology in Your Social World" makes clear.

Civil religion blends reverence for the nation with more traditional symbols of faith. Pictured is the chapel at Punchbowl, the Pacific cemetery for U.S. military personnel, located in Hawaii. Note that two U.S. flags are inside the altar area and are more prominent than two of the three religious symbols that also adorn the chancel: the Christian cross, the Jewish Star of David, and the Buddhist Wheel of Dharma.

Thinking Sociologically

In order to determine the role of religious organizations in your community, answer the following: Which religious groups in your community have a stabilizing influence, and which ones seem to have a disruptive influence? Which ones make the existing system seem sacred and beyond question? Which groups push for more social equality and less ethnocentrism toward others? Are there mosques, temples, or churches that oppose the government's policies, or do they foster unquestioning loyalty?

Social Change. Religion can work for or against social change, depending on the time and place. Some religions fight to maintain the status quo or return to simpler times. This is true of many fundamentalist religions, be they branches of Christian, Jewish, Hindu, or Islamic faiths, that seek to simplify life in the increasingly complex industrial world. Other religious traditions support or encourage change. Japan was able to make tremendous strides in industrialization in a short time following World War II, in part because the Shinto, Confucian, and Buddhist religions provided no obstacles and, in fact, supported the

Sociology in Your Social World

Red Sex, Blue Sex

Even sex has political-religious guidelines and implications. Reactions at the 2008 Republican convention to the pregnancy of Sarah Palin's unmarried evangelical 17-year-old daughter were far from what some people might have expected. One delegate said, "I think it's great that she instilled in her daughter the values to have the child and not to sneak off someplace and have an abortion." Another added that "even though young children are making that decision to become pregnant, they've also decided to take responsibility for their actions and . . . get married and raise this child" (Talbot 2008).

For social liberals in the U.S. "blue states," sex education is the key. They are not particularly bothered by teens having sex before marriage but would regard a teenage daughter's pregnancy as devastating news. On the other hand, social conservatives in "red states" generally advocate abstinence-only education and denounce sex before marriage. However, they are relatively unruffled when a teenager does become pregnant, as long as she does not choose to have an abortion.

From a national survey of 3,400 teens between 13 and 17 years old and from a government study of adolescent sexual behavior, the authors of the National Longitudinal Study of Adolescent Health conclude that "religion is a good indicator of attitudes toward sex, but a poor one of sexual behavior, and that this gap is especially wide among teenagers who identify themselves as evangelical" (Regnerus 2007). The vast majority of white evangelical adolescents—74%—say that they believe in abstaining from sex before marriage. (Only half of mainline Protestants and a quarter of Jews say that they believe in abstinence.) Moreover, among the major religious groups, evangelical virgins are the least likely to anticipate that sex will be pleasurable and the most likely to believe that having sex will cause their partners to lose respect for them. Yet the adolescent health research indicated that evangelical teenagers are more sexually active than Mormons, mainline Protestants, and Jews. On average, white evangelical Protestants make their "sexual debut" shortly after turning 16 (Regnerus 2007; Shriver 2007).

Another key difference between evangelical and other teens "is that evangelical protestant teenagers are significantly less likely than other groups to use contraception. This could be because evangelicals are also among the most likely to believe that using contraception will send the message that they are looking for sex" (Talbot 2008) or that condoms will not really protect them from pregnancy or venereal disease.

The disconnect between belief ideals and actual behavior is obvious when we examine the outcomes of abstinence-pledge movements. Roughly 2.5 million people have taken a pledge to remain celibate until marriage, usually under the auspices of religiously based movements such as True Love Waits or the Silver Ring Thing. However, more than half of those who take such pledges end up having sex before marriage—and not usually with their future spouse. While those who take the pledge tend to delay their first sexual intercourse for 18 months longer than nonpledgers and have fewer partners, communities with high rates of pledging also have very high rates of sexually transmitted diseases. This could be because fewer people in these communities use condoms when they break the pledge (Regnerus 2007; Talbot 2008).

The main point is that sexual attitudes and behaviors can be linked to our religious beliefs and affiliations but perhaps not in the ways we expect.

changes. In the United States, the central figures in the civil rights movement were nearly all African American religious leaders. As the first nationwide organizations controlled by black people, African American religious organizations established networks and communication channels that were used by those interested in change (Lincoln and Mamiya 1990; McAdam 1999).

The Link Between Religion and Stratification: A Conflict Perspective

Our religious ideas and values and the way we worship are shaped not only by the society into which we are born but also by our family's position in the stratification system. Religion serves different primary purposes for individuals,

Contributing to Our Social World: What Can We Do?

At the Local Level:

- *Programs to identify and assist at-risk students:* Early intervention programs can curb high dropout rates.

- *FIG (first-year college interest group) programs:* Entering students identified as at-risk take special courses, receive tutoring and peer mentoring, meet together to discuss common problems, and in some cases live in the same residence hall. If there is a FIG or similar program on your campus, contact its office to learn how it works and if you can volunteer to work as a tutor or in another capacity. Consider setting up a program on your campus. (Resources are available from the University of Wisconsin [www.lssaa.wisc.edu/figs], the University of Washington [http://depts.washington.edu/figs/], and the University of Missouri [http://admissions.missouri.edu/housing/academic-interest-based-communities.php].)

- *Campus religious foundations or ministries:* The Newman Foundation (Roman Catholic) and Hillel Foundation (Jewish) are branches of national organizations. Attend a meeting to learn about the group's activities and participate in outreach work in the community or to other denominations: soup kitchens, food pantries, or thrift stores for the poor.

At the Organizational or Institutional Level:

- *Reading volunteers:* Volunteer to read to young students and tutor them in reading. Contact a faculty member on your campus who specializes in early childhood education and investigate the opportunities for such volunteer work. Community Learning Centers use volunteers and may even have paid, part-time job openings.

- *Habitat for Humanity (www.habitat.org):* This organization makes extensive use of volunteers and would welcome your participation.

- *American Atheists (www.atheists.org):* "Now in its fourth decade," according to its website, "American Atheists is dedicated to working for the civil rights of atheists, promoting separation of state and church, and providing information about atheism." Volunteers assist in this work through contributions, research, and legal support.

At the National and Global Levels:

- *Teach for America (www.teachforamerica.org):* Modeled along the lines of AmeriCorps and Peace Corps, Teach for America places recent college graduates in short-term (approximately two-year) assignments teaching in economically disadvantaged neighborhood schools. Consider working for the organization following graduation.

- *UNICEF and CARE International (www.globaltesol.com, www.soyouwanna.com, www.teachabroad.com, and related websites):* At the international level, you can support these and other educational organizations. Consider teaching English abroad through one of many organizations that sponsor teachers.

- *Tikkun Community (www.tikkun.org):* This ecumenical organization was started in the Jewish community to "mend, repair, and transform the world." According to its website, Tikkun is an "international community of people of many faiths calling for social justice and political freedom in the context of new structures of work, caring, communities and democratic social and economic arrangements."

- *The American Friends Service Committee (www.afsc.org):* According to its website, AFSC "is a practical expression of the faith of the Religious Society of Friends (Quakers). Committed to the principles of nonviolence and justice, it seeks in its work and witness to draw on the transforming power of love, human and divine."

Visit **www.sagepub.com/oswcondensed2e** for online activities, sample tests, and other helpful information. Select "Chapter 11: Education and Religion" for chapter-specific activities.

CHAPTER 12

Politics and Economics

Penetrating Power and Privilege

The phrase in the subtitle of this chapter refers to the fact that power and privilege penetrate every aspect of our lives. The phrase has a double meaning, however, for sociology helps us to penetrate the sources and the consequences of power and privilege—both political and economic.

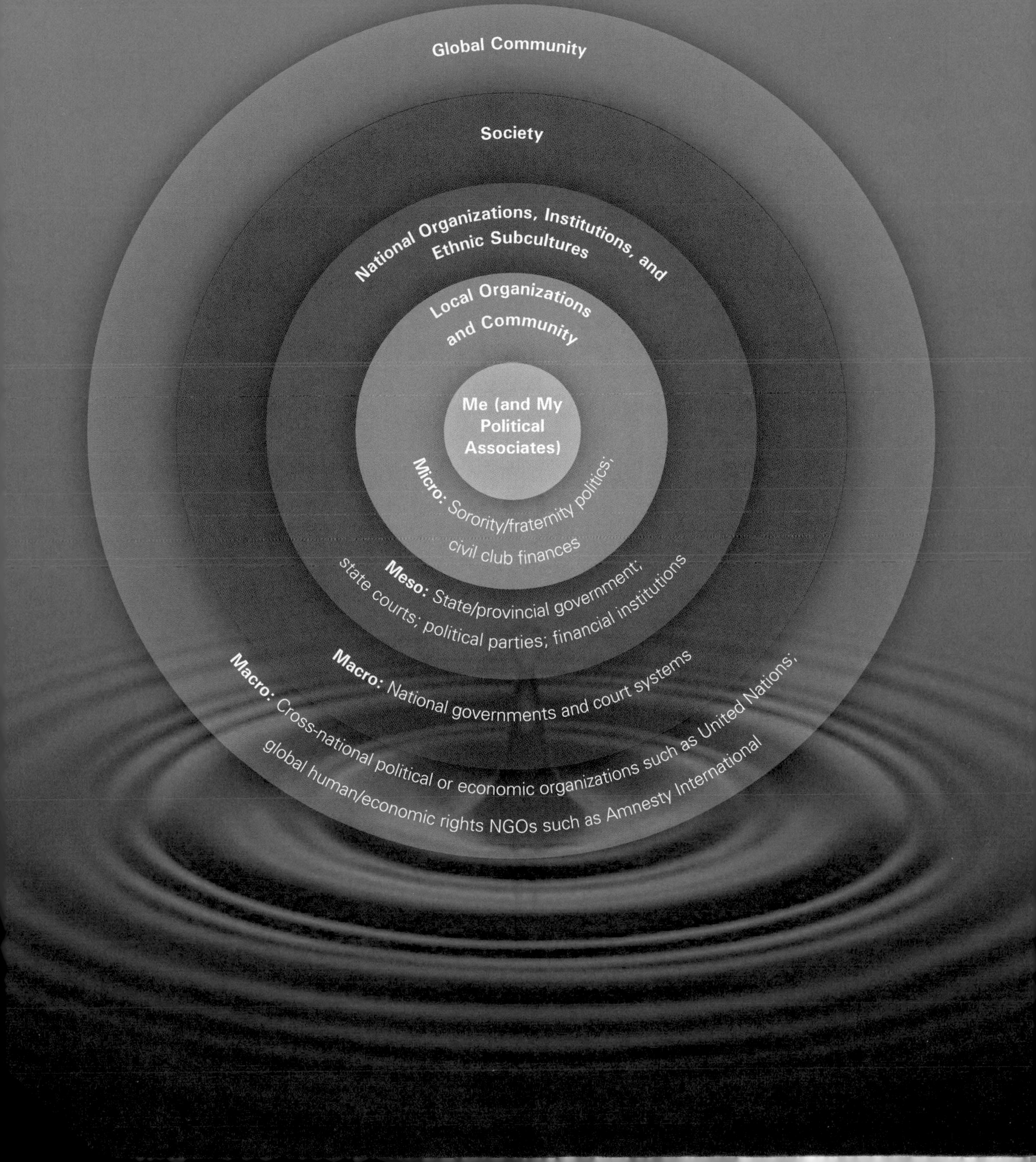

Global Community

Society

National Organizations, Institutions, and Ethnic Subcultures

Local Organizations and Community

Me (and My Political Associates)

Micro: Sorority/fraternity politics; civil club finances

Meso: State/provincial government; state courts; political parties; financial institutions

Macro: National governments and court systems; Cross-national political or economic organizations such as United Nations; global human/economic rights NGOs such as Amnesty International

Think About It	
Me (and My Political and Economic Life)	How do political and economic power penetrate my own life, even in the privacy of my home?
Local Community	How do people in my local community exercise power in constructive or destructive ways?
National Institutions; Complex Organizations; Ethnic Groups	How does the political institution interact with the economic institution?
National Society	Why does economic instability threaten a government?
Global Community	Why do struggles over power and privilege often evolve into war and terrorism?

Imagine that a nuclear disaster has struck. The mortality rate is stunning. The few survivors gather together for human support and collectively attempt to meet their basic survival needs. They come from varying backgrounds and have diverse skills. Before the disaster some—the stockbroker and the business executive, for instance—earned more money and held higher social status than the others, but that is in the past. Faced with the new and unfamiliar situation, different skills seem more immediately important for survival.

Where should this group begin? Think about the options. Some sort of organization seems essential, a structure that will help the group meet its needs. Food and shelter are paramount. Those with experience in agriculture and building trades are likely to take leadership roles to provide these initial necessities. As time goes on, the need for clear norms and rules emerges. The survivors decide that all members must work—must contribute their share of effort to the collective survival. At first, these norms are unwritten, but gradually some norms and rules are declared more important and are recorded, with sanctions (penalties) attached for noncompliance. Committees are formed to deal with group concerns, and a semblance of a judicial system emerges. Someone is appointed to coordinate work shifts, and others are chosen to oversee emerging aspects of this small society's life. This scenario could play out in many ways.

What is happening? A social structure is evolving. Not everyone in the group will agree with the structure, and some people will propose alternatives. Whose ideas are adopted? Leadership roles may fall to the physically strongest, perhaps the most persuasive, or those with the most skills and knowledge for survival. Those who are most competent at organizing may become the leaders, but that outcome is by no means ensured. In our world of power and privilege, a war, an invading power, or revolutionary overthrow of an unstable government can change the form of a political system overnight, necessitating rapid reorganization.

The opening scenario and the political activity in our modern society share a common element—power. The concept of power is critical to understanding many aspects

Politics is about power and about mobilizing support to lead. The 2008 presidential primaries generated a lot of interest as these two U.S. senators—John McCain (left) and Barack Obama (right)—battled for the most powerful political office on the planet—the U.S. presidency.

of our social world. Our primary focus in this chapter is the political and economic dimensions of society, since both enforce the distribution of power in societies. Political systems involve the relationships between individuals and between the individual and larger social institutions. Economic systems produce and distribute goods and services, and not everyone gets an equal share; thus, some citizens are given privileges that others do not have. While politics and economics are intertwined, we give more attention to political systems in this chapter because we have discussed economics in several other chapters such as those in Part III (on inequality). We consider the nature of power, politics, and economics at each level in our social world; theoretical perspectives on power and privilege; individuals and power; the distribution of power and privilege through economic and political systems; and national and global systems of governance, including international conflicts, war, and terrorism.

What Is Power?

Power is an age-old theme in many great scholarly discussions. Social philosophers since Plato, Aristotle, and Socrates have addressed the issue of political systems and power. Machiavelli, an early sixteenth-century Italian political philosopher, is perhaps best known for his observation that "the ends justify the means." His understanding of how power was exercised in the fifteenth, sixteenth, and seventeenth centuries significantly influenced how monarchs used the powers of the state (the means) to obtain wealth, new territories, and trade dominance (the ends).

The most common definition of power used in social sciences today comes from Max Weber (1947), who saw **power** as the ability of people or groups to realize their own will in group action, even against resistance of others who disagree.

Building on Weber's idea of power, one perspective is that there are various *power arenas*. First, the nation-state (national governments) attempts to control the behavior of individuals through (a) *physical control* (police force) or *outright coercion* (threats and actual violence), (b) *symbolic control* such as intimidation or manipulation of people, and (c) *rules of conduct* that channel behavior toward desired patterns, such as workplace rules.

Second, Weber's definition explains power as the ability to influence social life. Wherever people interact or participate in activities or organizations, power is a consequence (Olsen 1970). Therefore, individuals who have an understanding of interorganizational dynamics and can manipulate organization members are likely to have more power than others in organizational settings.

A third perspective focuses on a traditional Marxist approach to class structures, arguing that the control of economic resources and production allows the ruling class to keep ruling (Therborn 1976). People who control economic resources also protect their self-interests by controlling political processes through ideology, economic constraints, and physical coercion or political resources. Among many recent examples, Zimbabwe's leader, Robert Mugabe, used all of these methods of controlling political processes to hold onto power. Thus, power is found in all parts of the social world and is an element of every social situation (Domhoff 1998). So how does power work at each level?

Power and Privilege at Various Levels in the Social World

Power operates at the most micro levels of interaction, from individuals to family groups. In family life, husband-wife relations often involve negotiation and sometimes involve conflict over how to run a household and spend money. Interactions between parents and children also involve power issues as parents socialize their children. Indeed, the controversy over whether spanking is effective discipline or abusive imposition of pain is a question of how parents use their power to teach their children and control their behavior.

At the meso level, power operates in cities, counties, and states and provinces. Governments make decisions about which corporations receive tax breaks to locate their plants within the region. They pass laws that regulate everything from how long one's grass can be before a fine is imposed to how public schools will be funded. State and provincial governments in Western democracies also can control the way people live and make their living. Therefore, people have an interest in influencing governments by selecting their leaders, contributing to political campaigns, and helping elect the people who support their views. Interest groups such as ethnic or minority groups and national organizations and bureaucracies also wield power and try to influence the political process at the meso level.

At the macro level, international organizations such as the United Nations and World Bank; nongovernmental organizations such as Doctors Without Borders; and military, political, and economic alliances such as the North Atlantic Treaty Organization are parts of the global system of power. Locally organized groups can force change that influences politics at the local, state/provincial, national, or global level. Provincial or state laws shape what can and cannot be done at the local level. Laws at the national level influence state, province, or county politics and policies. Global treaties affect national autonomy.

Individuals can work to elect the party—and the candidates—that they think will make their lives better and improve conditions in the world. They work at the micro level to influence the meso and macro systems.

Power can be studied in political structures such as governments and political parties and in organizations such as the auto and banking industries. It can also be understood in terms of the allocation of economic resources in a society and what factors influence patterns of resource distribution. Both economic and political systems are important in the sociologist's consideration of power distribution in any society. Let us first consider the theoretical lenses that help us understand power and politics.

Theoretical Perspectives on Power and Privilege

Do you and I have any real decision-making power? Can our voices or votes make a difference, or do leaders hold all the power? Many sociologists and political scientists have studied these questions and found several answers to who holds power and the relationship between the rulers and the ruled.

Micro- and Meso-Level Perspectives: Legitimacy of Power

Interaction theorists focus on symbols and constructions of reality that allow some people to assume power. For symbolic interactionists, a central question is how loyalty to the power of the state is created—a loyalty that is so strong that citizens are willing to die for the state in a war. In the founding years

of the United States, loyalty tended to be mostly to individual states. Even as late as the Civil War, Northern battalions fought under the flag of their own state rather than that of the United States. The Federalist Party, which stressed centralized government in early U.S. history, faded from the scene. The Democratic-Republican Party, which evolved into the current Democratic Party, had downplayed the power of the federal government. This has changed. Today, the Democratic Party generally supports a larger role for the federal government than its rival, although Republican President George W. Bush expanded federal powers substantially.

Most people in the United States think of themselves as U.S. citizens more than Virginians or Pennsylvanians or Oregonians, and they are willing to defend the whole country. National symbols such as anthems and flags help create loyalty to nations. The treatment of flags is an interesting issue that illustrates the social construction of meaning around national symbols. The next "Sociology in Your Social World" explores this issue.

Thinking Sociologically

First, read "The Flag, Symbolism, and Patriotism." Is wearing a shirt or sweater with the U.S. stars and stripes in some sort of artistic design an act of desecration of the flag or a statement of patriotism? Does flying a Confederate flag symbolize disrespect for the national U.S. flag? Explain.

Socialization of individuals at the micro level generally includes instilling a strong sense of loyalty to the government in power and loyalty to a flag or a monarch that represents the nation.

Legitimacy, Authority, and Power: Social Constructions

Max Weber (1946) distinguished between legitimate and illegitimate power. Those people subject to legitimate power, which he referred to as **authority**, view it as lawful and just. Governments are given legitimate power when citizens acknowledge that the government has the right to exercise power over them. This is measured by two factors: whether the state can govern without the use or threat of forceful coercion and the degree to which challenges to state authority are processed through channels such as the legal system rather than overthrow of the government (Jackman 1993). Citizens of Western societies recognize elected officials and laws made by elected bodies as legitimate authority. They adhere to a judge's rulings because they recognize that court decrees are legitimate. In contrast, illegitimate power, or coercion, includes living under force of a military regime or being imprisoned without charge. (See Figure 12.1.) These distinctions between legitimate and illegitimate power are

Sociology in Your Social World

The Flag, Symbolism, and Patriotism

Flags have become pervasive symbols of nations, creating a national identity (Billig 1995). In some countries, loyalty to the nation is taught with daily pledges to the flag at work or school. National loyalty becomes sacred—as does the flag itself. Indeed, the nation is reified—it becomes a concrete material reality through this symbol. Durkheim (1947) maintained that one's larger group elicits sacredness and thereby becomes sacralized. He believed that sacredness actually is a form of respect for that which transcends the individual, including the state. In the United States, the flag and its construction illustrate key ideas in symbolic interaction theory. For many decades after the nation was founded, Americans had more loyalty to their state than to a federal government, but in the aftermath of the Civil War, a sense of nation began to gel (Answers .com 2011a).

Symbolic interactionists sometimes speak of the externalization, objectification, and internalization of important symbols (Berger and Luckmann 1966). In this case, the symbol was created (externalized by Betsy Ross in the summer of 1776), it came to have a life of its own separate from its creator (objectified by presidential order in 1912 when the official arrangement of stars was established), and it came to be incorporated by people as a symbol that was meaningful (internalized as a symbol of "us" for Americans) (Independence Hall Association 2010). In places like Britain and India, a national flag does not have the same symbolic power and internal resonance as a national symbol. This is not because those countries are less loyal to the nation or less proud of their heritage. It is that other symbols, such as royal status, work just as well.

Care of the U.S. flag is an interesting example of symbolism and respect for that symbol. Flag etiquette instructions make it clear that flying a flag that is faded, soiled, or dirty is considered an offense to the flag. We are told to either burn or bury a damaged flag as a way to honor and respect it.

Some propose a constitutional amendment prohibiting burning of the U.S. flag to prevent protesters from using the flag as a protest statement against certain American policies. Because protests show disrespect, some patriots have a visceral reaction of outrage. As recently as June 2006, the Senate came within one vote of sending the flag burning amendment to the individual states for ratification (CNN.com 2006). Supporters want flag burners punished and disrespect for the flag out-lawed. For many, the flag is dear and symbolizes all that is good about the United States (Billig 1995).

Those who oppose this amendment feel that only tyrannical countries limit freedom of speech. They feel that the principle of free speech, central to democracy, must be allowed even if a sacred symbol is at stake. Indeed, opponents of the amendment think passing such a law would be a desecration of what that flag stands for. The two sides have each attached different meanings to what is considered desecration of the national symbol. In the meantime, if you have a tattered or fading flag, burning it is the way you honor that flag—as long as you do so in private!

Other aspects of the U.S. Flag Code (Sons of Union Veterans of the Civil War 2010), which specifies what is considered official respect for or desecration of the flag, are interesting precisely because many people violate this code while they believe themselves to be displaying their patriotism.

1. The flag should *never* be used for advertising in any manner whatsoever. It should not be embroidered on cushions, handkerchiefs, or scarves, nor reproduced on paper napkins, carry-out bags, wrappers, or anything else that will soon be thrown away.

2. No *part* of the flag—depictions of stars and stripes that are in any form other than that approved for the flag design itself—should ever be used as a costume, a clothing item, or an athletic uniform.

3. Displaying a flag after dark should not be done unless it is illuminated, and it should not be left out when it is raining.

4. The flag should never be represented flat or horizontally (as many marching bands do). It should *always* be aloft and free.

5. The flag should under no circumstances be used as a ceiling covering.

According to the standards established by U.S. military representatives and congressional action, any of these forms of display may be considered a desecration of the flag, yet the meaning that common people give to these acts is quite different. Symbolic interactionists are interested in the meaning people give to actions and how symbols themselves inform behavior.

This man no doubt feels he is expressing his patriotism, yet technically he is violating the U.S. Flag Code and "desecrating" the American flag. During the Vietnam War, protesters risked attack for dressing this way, which was viewed as disrespect for the flag and the country.

Force + Consent = Power

Force < Consent = Legitimate Power (authority)

Force > Consent = Illegitimate Power (e.g., dictatorship)

Figure 12.1 Weber's Formula Regarding Power

important to our understanding of how leaders or political institutions establish the right to lead. To Weber (1946), illegitimate power is sustained by brute force or coercion. Authority is granted by the people who are subject to the power, which means that no coercion is needed.

How Do Leaders Gain Legitimate Power?

In constitutional democracies, those with power do not have the right to hold people against their will, to take their property, to demand they make unauthorized payments, or to kill them to protect others. Yet, even in democracies, certain people in power have the right to carry out such duties against people who are determined to be threats to society.

How do leaders get these rights? To establish legitimate power or authority, leaders generally gain their positions in one of three ways:

1. *Traditional authority* is passed on through the generations, usually within a family line, so that positions are inherited. Tribal leaders in African societies pass their titles and power to their sons. Japanese and many European royal lines pass from generation to generation. Authority is seen as "normal" for a family or a person to hold because of tradition. It has always been done that way, so no one challenges it. When authority is granted based on tradition, authority rests with the position rather than the person. The authority is easily transferred to another heir of that status such as a king, queen, or tribal chief.

2. *Charismatic authority* is power held by an individual resulting from a claim of extraordinary, even divine, personal characteristics. Charismatic leaders often emerge at times of change when strong, new leadership is needed. Some vivid historical examples of charismatic religious leaders are Jesus, Muhammad, and the founder of the Mormon Church, Joseph Smith. Charismatic political leaders include Mao Zedong in China, Nelson Mandela in South Africa, and Mahatma Gandhi in India. These men led their countries to independence and had respect from citizens that bordered on "awe." Some women have also been recognized as charismatic leaders, such as Burma's (Myanmar's) Aung San Suu Kyi, prodemocracy

His Royal Highness Charles, Prince of Wales, at his investiture in Caernarvon Castle. Charles is next in line to inherit the throne of England, and his position has legitimacy because the citizenry of England consent to the system. The authority of the throne is traditional.

activist, widely recognized prisoner of conscience, leader of the National League for Democracy of Burma, and winner of the Nobel Peace Prize.

The key point is that for charismatic leaders, unlike traditional authority leaders, the right to lead rests with the person, not the position. This is a change-oriented and unstable form of leadership because authority resides in a single person and at death the leadership also dies. Ultimately, as stability reemerges, power will become institutionalized—rooted in stable routine patterns of the organization. Charismatic leaders are effective during transitional periods but are often replaced by rational-legal leaders once affairs of state become stable.

3. *Rational-legal authority* is most typical in modern nation-states. Leaders have the expertise to carry out the duties of their positions, and the leadership structure is usually bureaucratic and rule bound. Individuals are granted authority because they have proper training or have proven their merit. This is the form of authority most familiar to individuals living in democracies. The rational-legal form of authority often seems entirely irrational and an invitation to chaos to people in tradition-oriented societies. It is important that authority in this system is divided between the position (which establishes criteria and credentials for the position) and the person (who has achieved those credentials for the position).

Each of these three types of authority, according to Weber (1947, 1958), is a "legitimate" exercise of power because the people being governed give their consent to those in power. However, on occasion leaders overstep their legitimate bounds and rule by force.

Macro Perspectives: Who Rules?

Pluralist Theory

Pluralist theory holds that power is distributed among various groups so that no one group rules. It is primarily through interest groups that you and I, average people, can influence decision-making processes. Our interests are represented by groups such as unions or environmental organizations that act to keep power from being concentrated in the hands of an elite few (Dahl 1961; Dye and Zeigler 1983).

Politics involves negotiation and compromise between competing groups. Interest groups can veto policies that conflict with their own interests by mobilizing large numbers against certain legislative or executive actions. Witness the efforts to influence health care reform in the United States and to reform government and industry practices. Greenpeace,

Pluralist theories see value in having many sources of power in a society so that no one group can dominate. For workers, this usually means uniting to have a strong voice through unions. Here we see several hundred ironworkers and their supporters stage a protest against poor pay and working conditions in Hong Kong in 2007.

Common Cause, Earth First!, the Christian Coalition, Focus on the Family, various labor unions, and other consumer, environmental, religious, and political action groups have had impacts on policy decisions. According to pluralists, shared power is found in each person's ability to join groups and influence policy decisions and outcomes.

National or international nongovernmental organizations (NGOs) can have a major impact on global issues and policy making, as exemplified by the Grameen Bank and other microcredit organizations (Yunus and Jolis 1999). NGOs exert influence on power holders because of the numbers they represent, the money they control, the issues they address, and the effectiveness of their spokespeople or lobbyists. Sometimes they form coalitions around issues of concern such as the environment, human rights, and women's and children's issues. According to pluralists, multiple power centers offer the best chance to maintain democratic forms of government because no one group dominates and many citizens are involved. Although an interest group may dominate decision making on a specific issue, no one group dictates policy on all issues.

Another major theory counters the pluralists, arguing that the real power centers at the national level are controlled by an elite few and that most individuals like you and me have little power.

Elite Theory

Power elite theorists believe it is inevitable that a small group of elite individuals will rule societies. Individuals have limited power through interest groups (Domhoff 1998, 2008; Dye 2002b; Mills 1956), but real power is held by the power elite. They wield power through their institutional roles and make decisions about war, peace, the economy, wages, taxes, justice, education, welfare, and health issues—all of which have serious impact on citizens. These powerful elite attempt to maintain, perpetuate, and even strengthen their rule. This influence eventually leads to abuse of their power (Michels [1911] 1967).

The social philosopher Pareto ([1911] 1955) expanded on this idea of abuse of power, pointing out that abuse would cause a countergroup to challenge the elite for power. Eventually, as the latter group gains power, its members become corrupt as well, and the cycle—a circulation of the elite—continues. Corruption in many countries illustrates this pattern. Consider the long-lasting regimes in the Middle East and North Africa that have amassed power and money over many years and are now being challenged by members of their public that no longer tolerate corruption, abuses, and the poverty into which many are relegated.

C. Wright Mills (1956) points out another angle, that there is an invisible but interlocking power elite in U.S. society, consisting of leaders in military, business, and political spheres; they make the key political, economic, and social decisions for the nation. This group manipulates what the public hears (Mills 1956). At the turn of the century, the top U.S. elite included 7,314 people from these three spheres (Dye 2002b). For example, in the business sphere, the top corporations control more than half of the nation's industrial assets, transportation, communication, and utilities. They also manage two thirds of the insurance assets. These corporations are controlled by 4,500 presidents and directors. According to the power elite theory, the U.S. upper class provides a cohesive economic/political power structure that represents upper-class interests (Domhoff 1998, 2008).

Private preparatory schools are one example of how elite status is transmitted to the next generation (Howard 2007; Persell and Cookson 1985). Many of those who hold top positions on national committees and boards or in the foreign policy-making agencies of national government attended the same private preparatory schools and Ivy League colleges—Brown, Columbia, Cornell, Harvard, University of Pennsylvania, Princeton, and Yale.

Key government officials come from industry, finance, law, and universities. They are linked with an international elite that helps shape the world economy. According to this theory, Congress ultimately has minimal power. Elite theorists believe that government seldom regulates business. Instead, business co-opts politicians to support its interests by providing financial support needed to run political election campaigns.

Pluralist theorists, however, disagree, believing that one reason we have big government is that a very powerful government serves as a balance to the enormous power of the corporate world. Big business and big government are safety checks against tyranny—and each is convinced that the other is too big.

Thinking Sociologically

Is your national society controlled by pluralist interest groups or a power elite? Can individuals influence the power elite? What evidence supports your view?

Micro-Level Analysis: Individuals, Power, and Participation

Karin signed up for every credit card available in the United States. She would max out one card and move on to the next, always paying just enough to keep the debt collectors from the door—until the day it all came crashing down on her. By then she was $53,000 in debt. She had transferred money from one no-interest card to another to avoid paying the interest. Then the monetary crisis of late 2008 hit, and soon banks were in such serious trouble that they had to tighten policies. Loans were hard to get, credit card companies became more selective, and when she could not pay the minimum, Karin's interest rate jumped to 26%. She had little choice but to declare bankruptcy, even though it would devastate her credit rating for at least seven years. The bankruptcy provision is governed by laws passed by her government and administered by the courts. Little does Karin know that credit cards did not even exist until the 1950s, but the trajectory of her life for the next few years will be shaped by the innovations of an entrepreneur, by global economic forces, and by legislative and judicial political systems that define her options.

Whether you experience credit or home mortgage problems, have health insurance, or are subject to a military draft depends in part on the political and economic decisions made by the government in power. Political systems influence our personal lives in myriad ways, some of which are readily apparent: health and safety regulations, taxation, military draft, regulations on food and drugs that people buy, and even whether the gallon of gas pumped into one's car is really a full gallon. In this section, we explore the impact individuals have on the government and the variables that influence participation in political and economic policy-making processes. A key issue at the micro level is decisions by individuals to vote or otherwise participate

A Pakistani woman casts her ballot in the 2008 Pakistani national and provincial elections. The elections were considered a crucial step in moving Pakistan from military to civilian rule. Voting is viewed as a wonderful privilege and a source of hope in this setting.

in the political system. This private decision is, in turn, affected by where those individuals fall in the stratification system of society, not just by personal choices.

Participation in Democratic Processes

Citizens in democratic countries have the power to vote. Most countries, even dictatorships, have some form of citizen participation. In only a handful of countries are there no elections. Sociologists ask many questions about voting patterns, such as what influences voting and why some individuals do not participate in the political process at all. Social scientists want to know how participation affects (and is affected by) the individual's perception of his or her power in relationship to the state.

Ideology and Attitudes About Politics and Economics

Political ideology affects how people think about power. Let us consider several ways that our beliefs and attitudes affect our political ideas. First, what do we believe about the power of the individual versus the power of the state? If we believe that individuals are motivated by selfish considerations and desire for power, we may feel as the seventeenth-century English philosopher Thomas Hobbes did: Humans need to be controlled, and order must be imposed by an all-powerful

sovereign. This is more important than individual freedom and liberty. On the other hand, we might believe, as did John Locke, another seventeenth-century political philosopher, that human nature is perfectible and rational, that we are not born selfish but we learn selfishness through experience with others. Humans, Locke argued, should have their needs and interests met, and among these needs are liberty, ability to sustain life, and ownership of property. He felt the people should decide who governs them. Thus, we can see that support for democracy is influenced by one's core assumptions about what it means to be human.

Second, do we believe in equal distribution of resources, or do we think that those who are most able or have inherited high status should receive more of the wealth? Some social scientists, politicians, and voters think that individuals have different abilities and are therefore entitled to different rewards. Some are successful, and some are not. Others think society should facilitate more equal distribution of resources simply because all persons are equally deserving of dignity. Conflict theorists tend to support this view.

In the United States, for example, Republicans (and others on "the right") tend to believe individuals and local communities should take more responsibility for education, health care, welfare, child care, and other areas of common public concern, feeling that this protects rights to local control and prevents creation of a powerful bureaucracy. Democrats (and those on "the left") are more likely to argue for the federal government's social responsibility to the people. For instance, Democrats have been concerned that leaving policies such as school integration to local communities would perpetuate inequality and discriminatory patterns in some communities. National government involvement, they feel, protects the rights of all citizens. The ongoing debates about the national welfare system and health care policies in the United States reflect these different philosophies. Both parties support government spending and national laws; the question is for what: Military? Social programs? Education? Health care? Abortion? Prisons?

Third, do we believe that change is desirable? Generally, these views fall into two camps: change as a potential threat to stability versus policy change to benefit the general population or segments of the population. Views on change affect how people vote.

Voters in many countries are influenced by issues such as the environment or immigration rather than traditional party ideology. Party affiliation based on ideology is becoming relatively weak in the United States, and an increasing number of people are identifying themselves as Independents rather than Democrats or Republicans, either because they do not want to commit themselves to one ideology or because they are more interested in specific issues than in an overriding philosophy of government.

Thinking Sociologically

How might your decision about how to vote (a micro-level decision) make a difference in your life at the state/provincial, national, or global level?

Levels of Participation in Politics

The majority of people in the world are uninvolved in the political process because there are few opportunities for them to be meaningfully involved (especially in non-democratic countries). They feel that involvement can have little relevance for them (apathy) or that they cannot affect the process (alienation from a system that does not value them). However, political decisions may affect people directly, and they may be drawn unwittingly into the political arena. Peasants making a subsistence living may be forced off the land and into refugee camps by wars over issues that have little relevance to them. Their children may be drafted and taken away to fight and be killed in these battles. Religious or ideological factions may force them to help pay for conflicts in which they see no purpose or have no stake. In recent years, such situations have drawn the uninvolved into politics in Guatemala, Uganda, Cambodia (Kampuchea), Haiti, Rwanda, Somalia, India, Iraq, Lebanon, Gaza, Afghanistan, the Middle East, North Africa, and Darfur (Sudan).

In representative systems, citizens are encouraged to have a voice, although their levels of involvement vary greatly. While some remain uninvolved, some vote in most elections. Others have contact with government representatives only when they have issues of personal concern. Still others are involved in local politics, actively working on issues or local elections. The most involved engage in both local and national campaigns. In some countries voting is mandatory.

Political participation is affected by election laws, including those that enfranchise people, that stress voting as a requirement of citizenship, and that structure elections to facilitate representation by historically underrepresented groups. In some countries, voting is an obligation of citizenship, and voter turnout is above 90%. For example, in Argentina, Australia, Brazil, Chile, and Congo, to name a few, it is a violation of the law not to vote (World Factbook 2011). Fines, community service, and even jail time are penalties for not voting. Elections are held over many days to ensure that people can get to the polls. Some other countries also make sure that ethnic minorities and women have a voice by structuring elections to ensure broad representation. The next "Sociology Around the World" examines the reasons that an African country—the war-torn nation of

Rwanda—emerged early in the twenty-first century with the highest percentage of women in government of any nation in the world.

In the United States, voter turnout went down from 63% of eligible voters in 1960 to 50% in 1996 to 47.3% in 2000. However, in the 2008 presidential election, turnout was about 62% (Bergman 2005; U.S. Election Project 2010). More than 70% of citizens age 45 or older voted in 2004, and 67% of whites voted in 2004 compared to 60% of African Americans, 47% of Hispanics, and 44% of Asian Americans (Bergman 2005). All of this means the 2004 and 2008 elections had the highest turnout since 1968 (Information Please Database 2008; McDonald 2009). African American participation increased from 11.1% of the total electorate to 13% (Short 2009).

Participation in elections in the United States is the second lowest of the Western democracies, as indicated in Table 12.1 on page 378. This means that citizens are not exercising their right to vote. The unusually high number of "inactives" is not an encouraging sign for the vitality of a democracy (Orum 2001). Still, there was a resurgence of interest in politics among those under 30 years old during the 2008 presidential campaign with the primary elections for nomination of candidates and the November election for president yielding record-breaking turnouts. This is discussed in the next "Sociology in Your Social World" feature on page 379.

Thinking Sociologically

How important do you think voting blocs were in the 2008 U.S. presidential election—the youth vote, women, African Americans, Latinos, the religious right, blue-collar families, white males?

Meso-Level Analysis: Distributions of Power and Privilege Within a Nation

A village within the Bantu society of southern Africa has lost its chief. Bantu societies provide for heirs to take on leadership when a leader dies. However, there is no male heir to the position, so a female from the same lineage is appointed. This woman must assume the legal and social roles of a male husband, father, and chief by acting as a male and taking a "wife." The wife is assigned male sexual partners, who become the biological fathers of children. This provides heirs for the lineage, but the female "husband" is

Sociology Around the World

Women and Political Change in Postgenocide Rwanda

By Melanie Hughes

In 2003, Rwanda became the new global leader in women's political representation. In 2011, women were elected to 56% of the seats in Rwanda's Chamber of Deputies, plus the speaker's chair and cabinet positions, making it the most gender balanced of any national legislature in the history of the world (McCrummen 2008; Paxton and Hughes 2007; see also Table 9.1). For the first time since 1988, a country outside of Scandinavia garnered the top spot in women's political representation, and for the first time in history, the position was held by an African country. From just 10 years earlier, the number of women serving in Rwanda's parliament almost tripled.

Many were particularly surprised about Rwanda's women's involvement, given the country's recent history of economic upheaval and civil war. The instability culminated in 1994, when during a span of 100 days, an estimated 800,000 Rwandans died at the hands of their countrymen and -women in a horrific ethnic genocide. So, how did Rwanda bounce back within a decade to lead the world in women's political representation?

At the micro level, research suggests that the behavior of individual women during and after the Rwandan genocide generated support for their empowerment (Hughes 2004). During the civil war, women served on the front lines with men, led military actions, and worked as mediators to help end the insurgency. After conflict subsided, women played key roles in the reconstruction effort (UNIFEM 2002). Interviews with Rwandans suggest that the burdens taken on by women during this period generated both the political will and the public support necessary to advance women in politics (Mutamba 2005).

Important changes also occurred at the meso level. Immediately after the killing subsided, women's associations, both new and old, began to step into the void (Longman 2005). Women's organizations took action early on to shape the new state. For example, in 1994, an organization of women's associations drafted a document addressing Rwanda's postconflict problems and suggesting how women could foster reconciliation (Powley 2003). Building up to the adoption of the new constitution in 2003, women's organizations served as a bridge, taking suggestions from women at the grassroots level into meetings with the transitional government.

Women at all levels were supported by international organizations and foreign aid. Rwanda's economic troubles meant that dependence on international funding was unavoidable, and women were well situated to take advantage of foreign monies. The empowerment of women, especially in the Global South, was on the agenda of the United Nations and other global bodies. Therefore, many international organizations helped advance the idea that women's incorporation into political decision-making positions was essential for sustainable peace (Hughes 2004).

Actions by individual women, native women's associations, and international organizations all helped encourage the transitional government to adopt female-friendly political institutions. Women's councils and women-only elections were established to guarantee female representation down to the grassroots level. The new constitution mandated that women fill 30% of all policy-making posts in Rwanda.

Rwanda today has a democratically elected government, but with a fairly authoritarian leader. However, since the election in 2003, women have still been able to revise inheritance laws, pass a law banning discrimination against women, and strengthen rape laws (Longman 2005). Rwanda has come a long way toward giving women a political voice.

their social father because she has socially become a male. This pattern has been common practice in many southern Bantu societies and among many other populations in four separate geographic areas of Africa (O'Brien 1977). Ruling groups in society, in this case a meso-level tribal society under the jurisdiction of a nation-state, have mechanisms

Table 12.1   Voter Turnout Percentages for Elections Since 1945*

Country	Voter Participation %
Italy	92.5
Iceland	89.5
New Zealand	86.2
South Africa	85.5
Austria	85.1
Netherlands	84.8
Australia	84.4
Denmark	83.6
Sweden	83.3
Germany	80.6
Greece	80.3
Guyana	80.3
Israel	80.0
Suriname	77.7
United Kingdom	74.9
Turkey	73.5
Argentina	70.6
Uruguay	70.3
Philippines	69.6
Japan	69.0
Dominican Republic	68.7
Canada	68.4
France	67.3
Bolivia	61.4
India	60.7
USA	48.3
Mexico	48.1
Brazil	47.9
Thailand	47.4
Chile	45.9
Ecuador	44.7

Source: International Institute for Democracy and Electoral Assistance 2008.

Note: To see the voting participation figures for 172 countries in the world, go to http://www.idea.int/vt/.

*The figures are averages of voter participation for all elections over a 60-year period. Note that enfranchisement of women and various ethnic minorities has changed in some countries during that time, so these should be viewed as very crude overall indicators of voting patterns.

Among the Bantu of southern Africa, if a chief dies it is possible for a woman within the family lineage to succeed him, but she must take a "wife" and fulfill the leadership role normally established for males.

for ensuring a smooth transition of power and in this case maintaining male dominance to keep the controlling structure functioning.

Meso-level political institutions include state or provincial governments, national political parties, and large formal organizations within the nation. Equally important is that meso-level political institutions influence and are influenced by other institutions: family, education, religion, health care, and economics.

What Purposes Do Political and Economic Institutions Serve?

Most of us have had an argument over ownership of property, have been in an accident, have met people who needed help to survive, or have been concerned about wars raging around the world. The political institution addresses these issues and serves a variety of other purposes, or functions, in societies. The economic institution determines the statuses, groups, and processes that ensure the production of goods and services that people in the society need to survive and to thrive.

For most people, interaction with the government begins with a record of their birth and ends with a record of their death. In between, the government is the institutional structure that collects taxes, keeps records of the work activities and wages of citizens, and keeps fingerprints and other personal information on file.

We have learned in earlier chapters that each institution has purposes or functions it serves. Just as family, education, medicine, and religion meet certain societal needs,

Sociology in Your Social World

The Youth Vote in the U.S. Presidential Election of 2008

Despite rising levels of education, "young Americans," those under 30 years old, have had a lower rate of turnout than their older counterparts for several decades (Abramson, Aldrich, and Rhode 2007). Political scientists argue that the low rates of youth voters are the primary reason for the generally low turnout rates in American elections (Rosenstone and Hansen 1993). For this reason, the nonpartisan organization Rock the Vote was launched in 1990 in an attempt to stimulate political involvement of young people. Rock the Vote's strategies involved depicting voting as "fashionably subversive" and encouraging young adults to register, including facilitating voter registration online (Rampell 2008). Another nonpartisan movement—Declare Yourself—has used similar strategies. Rock the Vote and Declare Yourself each registered roughly a million young people, resulting in 4.2 million more people in this age group voting in 2004. Rock the Vote estimated that it registered at least 2 million in time for the 2008 presidential election (H. Smith 2008).

Yet Rock the Vote alone cannot claim all of the credit for the major increases in new voters in 2008. According to the Youth Vote Coalition, 64% of young adults were registered to vote in the 2008 presidential election (Romano 2008). In the 2008 presidential primaries, the rate of participation among 18- to 29-year-old voters nearly quadrupled in Tennessee and Missouri (Rampell 2008). What was the reason for this increase in participation?

The 2008 election cycle involved candidates who excited many younger voters. Although he was the oldest person ever to be nominated for a first term as president, John McCain was deeply admired not only as a war hero but as an independent thinker and a "maverick" who stood up against the powerful, regardless of party, when he thought they were wrong (Powell 2008b). To add to the excitement, McCain selected Governor Sarah Palin, a relatively young woman—only the second in the history of the country to be nominated for the vice presidency—as his running mate. She brought to the ticket a reputation for being a courageous reformer and for standing up to the "old boy" system in her state.

On the Democratic side, many young women were drawn to Hillary Clinton as the first woman who seemed to have a realistic chance at the presidency. However, Barack Obama was the primary beneficiary of youth participation. Like John Kennedy in 1960 and his brother, Robert Kennedy, in 1968, Obama entered the race as a young candidate and an agent of change— visionary, charismatic, transcending the rhetoric of we-they politics, and exuding a message of hope (Bristow 2008). Equally important, an exit poll by NBC showed that voters between the ages of 18 and 29 were more likely than their elders to favor government intervention to solve problems, to oppose the Iraq War, and to select energy policy as the most important national issue (Keeter, Horowitz, and Tyson 2008). It is notable that all of these preferences correspond with Obama's political strengths, ensuring that Obama's policies communicated the same visionary ideas of change that his personality did.

A third factor is the Obama campaign's incredible success at mobilizing younger voters. Obama's "Yes, we can" mantra was a YouTube video created by Will.i.am of the Black Eyed Peas, and many of Obama's speeches were put on YouTube (Geist 2008). The Obama campaign also used Myspace and Facebook heavily, including use of a Facebook group to encourage users to watch videos, donate money, and volunteer for the campaign (Geist 2008; Sarno 2008). Exit polls by NBC show that these efforts at mobilization paid off: 25% of younger voters were contacted by the Obama campaign, whereas just 13% were contacted by the McCain campaign (Keeter et al. 2008). Once they were contacted, many younger voters became a part of the campaign. David Plouffe (2009), Obama's campaign manager, wrote, "At least 95 percent of our six thousand employees were under the age of thirty, most under the age of twenty-five" (p. 370).

Together, these various appeals added up to a major win among youth for the Obama campaign. Obama secured 66% of the vote among those under 30, compared to just 50% of the vote among those older than 30 (Keeter et al. 2008). Younger voters also rose as a proportion of total voters, particularly in key battleground states such as Indiana, North Carolina, and Virginia (Keeter et al. 2008). It is notable that just 62% of 18- to 29-year-olds identify their race as "white," compared to 79% among those 45–64 (Keeter et al.

(Continued)

(Continued)

2008). This suggests that Obama's strength among youth may have been compounded by the higher percentages of African American and Hispanic voters among the youngest age cohort. Younger voters were also the least likely to identify themselves as part of a religious tradition or attend church regularly, further suggesting that sociological differences played a role in the election's outcome. The 2008 election results show the largest gap between younger and older presidential voters since exit polling began in 1972, leading researchers to declare that "a significant general shift in political allegiance is occurring" (Keeter et al. 2008).

A study from the Harvard University Institute of Politics reports that this is not a one-time phenomenon, but rather the "civic reawakening of a new generation" (Cillizza and Murray 2008). The idealism, the desire for national unity, the avoidance of red state–blue state polemics, the desire for hope and optimism, and the rejection of explicit racial or gender categories as defining characteristics of leadership have all been traits of this group of younger voters, and those attitudes have matched well with the emphasis of Barack Obama's campaign.

Written with the assistance of Jeremiah Castle.

so does the political system. The following six activities are typical purposes (functions) of meso-level political and economic institutions. They set the stage for power and privilege carried out at the macro level in national and international arenas.

1. *To maintain social control.* We expect to live in safety, to live according to certain "rules," to be employed in meaningful work, and to participate in other activities prescribed or protected by law. Ideally, governments help clarify expectations and customs and implement laws that express societal values. However, in some cases, governments rule with an iron hand, and people live in fear because of the social control imposed by government. This has been the case in Afghanistan under the Taliban, when leaders used armed militia to terrorize the country by imprisoning, torturing, and killing suspected dissenters.

2. *To serve as an arbiter in disputes.* When disputes arise over property or the actions of another individual or group, a judicial branch of government can intervene. In some systems, such as tribal groups mentioned above, a council of elders or powerful individuals performs judicial functions. In other cases, elected or appointed judges have the right to hear disputes, make judgments, and carry out punishment for infractions.

3. *To protect citizens of the group.* Governments are responsible for protecting citizens from takeover by external powers or disruption from internal sources. However, they are not always successful. Cities are often violent, gangs roam the streets, terrorists threaten lives, minority groups receive unfair treatment, and governments lose territory or even control of their countries to external forces. While

Venezuelan President Hugo Chavez has nationalized some of the nation's oil and other industries to prevent foreign interests from controlling the country's resources, Mexico's President Felipe Calderón is fighting a difficult war against drug cartels (Herman and Peterson 2006; Hispanic-Americans.com 2011).

4. *To represent the group in relations with other groups or societies.* Individuals cannot negotiate agreements with foreign neighbors. Official representatives deal with other officials to negotiate arms and trade agreements, protect the world's airways, determine fishing rights, and establish military bases in foreign lands, among other agreements.

The four functions listed thus far are rather clearly political in nature, but the last two are areas of contention between political and economic realms.

5. *To make plans for the future of the group.* As individuals, we have little direct impact on the direction our society takes, but the official governmental body shares responsibility with economic institutions for planning in the society. In some socialist societies, this planning dictates what each individual will contribute to the nation: how many engineers, teachers, or nurses these societies need. They then train people according to these projections. In capitalist systems, for instance, supply and demand is assumed to regulate the system, and there is less governmental planning—especially in economic matters. The question of who plans for the future is often a source of stress between the political and economic institutions: Are the planners elected politicians or private entrepreneurs?

6. *To provide for the needs of their citizens.* Governments differ greatly in the degree to which they attempt to meet the material needs of citizens. Some provide

for many of the health and welfare needs of citizens, whereas others tend to leave this largely to individuals, families, and local community agencies. Not everyone agrees that providing needs is an inherent responsibility of the state. The debates over a health care system and welfare system in the United States point to the conflicts over who should be responsible—the state or private entrepreneurs. Should such services be coordinated by the government or left to "the invisible hand" of market forces?

The ways in which governments carry out these six functions are largely determined by their philosophies of power and political structures. Political and economic institutions, like family and religious institutions, come in many forms. In essence, these variations in political institutions reflect variations in human ideas of power. The point is that functionalists focus on the positive role of each institution in the lives of people and for the stability of society.

Thinking Sociologically

In an era of terrorist threats, how do you think the "protecting the safety of the citizens" function has affected the ability of governments to meet their other functions?

Meso- and Macro-Level Systems of Power and Privilege

While *politics* refers to the social institution that determines and exercises power relations in society, *economics* is the social institution that deals with production and distribution of goods and services. Both politics and economics focus on questions related directly to power relationships among individuals, organizations, nation-states, and societies. How goods are distributed to the members of society is often determined by who has power.

Government officials have a vested interest in the well-being of the economy, for should the economy fail, the state is likely to fail as well. Recessions, depressions, and high rates of inflation put severe strains on governments that need stable economies to run properly. When problems occur, government officials are inclined to increase their roles in the economic sector. Witness the volatile money markets in 2008 and 2009 and measures such as bailouts taken by the governments to stabilize the economy.

In September 2008, the Dow Jones index dropped an unprecedented 777 points in one day and fell another 782 points one week later. The New York Stock Exchange, usually bustling, was nearly empty, and President Bush and Congress immediately began to look for "stimulus packages"

to keep the economy from going into a recession. Turning the economy around has become the key challenge for President Obama. The problem of a dramatic drop in financial markets is a loss of confidence so that no one invests, lending institutions are not able to loan money easily, business stagnates, unemployment rates skyrocket, and the entire government may be held responsible for lack of economic vitality. In February 2009, Elkhart, Indiana, had an unemployment rate of over 15%—a figure that approached the unemployment rates of the Great Depression in the 1930s (S. Smith 2009). When a country goes into a recession, the party in power is often held responsible and will not likely be reelected if the recovery takes too long. Economic recessions can destroy the careers of politicians but can also create such dissatisfaction that the entire government may be at risk.

Indeed, Dennis Blair, the U.S. director of national intelligence, said in February 2009 that the global economic crisis is the most serious national security issue facing the nation (Miller 2009). He pointed out that in the past, countries have been able to export more goods to work their way out of a recession, but that is not possible when the economic slump is global. The result could be a "backlash against American efforts to create free markets" (Miller 2009), unstable governments, and waves of refugees crossing national borders (Haniffa 2009). The economic crisis could mean that governments cannot meet their defense

A stable economic system is essential to political stability, and extreme fluctuations in the economy are frightening to those in power. On the day the Dow Jones index dropped dramatically, the New York Stock Exchange, usually bustling with activity, was nearly empty.

obligations or provide the services necessary to the citizens. Therefore Blair called the extreme economic downturn a "bigger threat than Al Qaeda terrorists" (Haniffa 2009).

In many countries, the government is the largest employer, purchaser of goods, controller of exports and imports, and regulator of industry and of interest rates. In the United States, many government regulatory agencies, such as the Food and Drug Administration, Department of Agriculture, and Justice Department, watch over the economic sector to protect consumers.

Types of Political Systems

The major political systems in the world range from fascist totalitarianism to democracy. However, each culture puts its own imprint on the system it uses, making for tremendous variation in actual practice. Two broad approaches are discussed below to illustrate the point.

Authoritarian Systems. The government of Saudi Arabia is a hereditary monarchy based on traditional leadership. **Authoritarian political systems** such as this are controlled by absolute monarchs or dictators who allow limited or no participation of the population in government and control much of what happens in the lives of individuals. They often have the backing of the military to keep them in power. Authoritarian regimes have been common forms of government in the world. Some are helpful to citizens—benevolent dictators—but most control and discourage dissent. The Castro brothers—Fidel and more recently Raúl—have maintained absolute control of Cuba since 1959, and while they are despised in the United States, they are admired by many Cubans as ruling in the best interests of the people, even if autocratically.

In one type of authoritarian system, a **totalitarian political system** such as that found in Libya under Col. Gadhafi, the state regulates all aspects of people's lives and can give or deny rights. The leaders use propaganda and force to monitor and control people's actions. Such systems are often based on a specific political ideology and run by a single ruling group or party of elites, referred to as an *oligarchy*. These elites hold total control over the population, while preserving inequalities between classes in the society. Russia under Joseph Stalin and Germany under Adolf Hitler were totalitarian dictatorships that followed and demanded adherence to strong ideologies. The state typically controls the workplace, education, the media, and other aspects of life. Acts of dissent or opposition are eliminated. Interrogation by secret police, imprisonment, and torture are used to quiet dissenters. Terror is used as a tactic to deal with both internal and external dissents, but when it is used by the state to control the citizenry or to terrorize those of another nation, it is called *state terrorism*.

Throughout history, most people have lived under authoritarian or totalitarian systems. Under certain conditions, totalitarian regimes can turn into democratic ones, and of course, democratically elected leaders can become self-proclaimed dictators, as was the case of Ferdinand Marcos in the Philippines. The next "Sociology Around the World" provides an example of one totalitarian regime.

Democratic Systems. In contrast to totalitarian regimes, democratic systems are characterized by accountability of the government to the citizens and a large degree of control by individuals over their own lives. Democracies always have at least two political parties that compete in elections for power and that generally accept the outcome of elections. Mechanisms for the smooth transfer of power are laid out in a constitution or another legal document. "Ideal-type" democracies share the following characteristics, although few democracies fit this description exactly:

1. *Citizens participate in selecting the government.* There are free elections with anonymous ballots cast, widespread suffrage (voting rights), and competition between members of different parties running for offices. Those who govern do so by the consent of the majority, but the political minorities have rights and representation.

2. *Civil liberties are guaranteed.* These usually include freedom of association, freedom of the press, freedom of speech, and freedom of religion. Such individual rights ensure dissent, and dissent creates more ideas about how to solve problems. These freedoms are therefore essential for a democracy to thrive.

3. *Constitutional limits are placed on governmental powers.* The government can intrude only into certain areas of individuals' lives. Criminal procedures and police power are clearly defined, thus prohibiting harassment or terrorism by the police. The judicial system helps maintain a balance of power. These limits have caused dissent in the United States in recent years, with debates over whether the Patriot Act inappropriately limits freedoms.

4. *Governmental structure and process are spelled out.* Generally, some officials are elected whereas others are appointed, but all are accountable to the citizens. Representatives are given authority to pass laws, approve budgets, and hold the executive officer accountable for activities.

5. *Written documents such as constitutions are the basis for the development of legal systems.* The constitutions describe activities in which the government

Sociology Around the World

The Khmer Rouge Revolution: A Totalitarian Regime

These skulls are the remains of people massacred by the government in the Killing Fields of the Khmer Rouge in Cambodia.

When the Khmer Rouge faction took over the government of Cambodia in 1978, it abolished private property; relocated urban dwellers to rural areas; seized personal property; classified some people as peasants, workers, or soldiers—and killed the rest. The group's most amazing feat was the total evacuation of the capital city, Phnom Penh. This was done to remove urban civilization and isolate Cambodia from other political influences, such as democracy.

This complete social and economic revolution under the leadership of Pol Pot was planned in Paris by a small group of intellectual revolutionaries. They believed it would allow Kampuchea, their former name for the country, to rebuild from scratch, eliminating all capitalism, private property, and Western culture and influence.

After urban dwellers were resettled in rural camps, the totalitarian regime tried to break down the family system by prohibiting contact between members, including sexual relations between husbands and wives. Many people, including defeated soldiers, bureaucrats, royalty, businesspeople, intellectuals with opposing views, Muslims, and Buddhist monks, were slaughtered for minor offenses—hence the term *the Killing Fields* to describe the execution sites. The death toll is estimated at more than 1.25 million. However, the Cambodian Genocide Program has uncovered meticulous records kept by Khmer Rouge leaders, and combined with evidence from mass graves, these data may double that number (Crossette 1996; Mydans 2009).

Famine followed the killings, causing many Cambodians to flee their land, traveling by night and hiding by day to reach refugee camps across the border in Thailand. Today, the economy in Cambodia is growing, especially in the areas of garment work and tourism. Thousands of tourists visit Angkor Wat and the Killing Fields each year.

The economy in Cambodia mostly provides sweatshop employment. More than 200,000 young women work in the 200 export garment factories, making clothes for companies such as the Gap. Their wages were $35 a month in 2002. To keep their jobs, they must often work overtime, up to 80 hours a week, with no extra pay (Sine 2002). While these may seem like cruel conditions, some who are knowledgeable about Cambodia argue that these jobs are coveted by many because they are far better than other options (Kristof 2009).

Cambodia has never been a country at peace, and that seems to be true today. Violence still exists. The Human Rights Center reports chaos, corruption, poverty, and a reign of terror in Cambodia, as the military kills and extorts money from citizens. Killings, violence, and intimidation surrounded elections in 2002, with more than 14 killings of political activists and candidates (U.S. Department of State 2002). The Khmer Rouge is still a threat to any hopes of democracy, but since the U.S. government officially recognizes the current government in Cambodia, it will not grant political refugee status to any new refugees from Cambodia (Tomsen 1994).

must—or must not—engage. Constitutions can provide some protection against tyrants and arbitrary actions by government. The two main forms of democratic constitutional government are the parliamentary and presidential systems. In typical parliamentary governments, the head of state is often a monarch, and the head of government is a prime minister, chancellor, or premier. These are two different people. Belgium, Canada, Denmark, Great Britain, Japan, the Netherlands, Norway, and Sweden have this model. Examples of presidential governments include France, Italy, the United States, and Germany. The presidents in these countries tend to have more autonomy than the heads of parliamentary governments.

Proportional representation means that each party is given a number of seats corresponding to the percentage of votes it received in the election. In winner-takes-all systems, the individual with more than 50% of the votes gets the seat. In the United States, the "winner-takes-all" system has come under attack because the winner of the popular vote can lose the electoral vote to an opponent who wins several of the most populous states by narrow margins. This actually happened in a presidential election in the United States: Al Gore received the most votes for president in 2000, but because all but two states (Maine and Nebraska) had a winner-takes-all system for electing the Electoral College, George W. Bush became the next president. Defenders of

The British system of government is a democracy and parliamentary in form. This is the opening ceremony in 2007 for the Parliament. Note the pomp and circumstance that are used to create a sense of awe for power and authority.

this system of choosing the president argue that this protects the voice of each state, even if each individual voice is not given the same weight.

Constitutional governments may have from two to a dozen or more parties, as has been the case in Switzerland. Most have four or five viable ones. In European countries, typical parties include Social Democrats, Christian Democrats, Communists, Liberals, and other parties specific to local or state issues, such as green parties.

In the modern world, there are new challenges and issues that face democracies. Electronic technology—the Internet and other telecommunications technologies—can be a boon to democracy, an opportunity for people around the world to gain information necessary to be an informed electorate, or a burden that hinders thoughtful debate and civic engagement in ideas. The latter are essential ingredients of a functioning democracy (Barber 2006).

The Internet, fax machines, camcorders, and other telecommunications devices have been major instruments for rural poor and indigenous people, linking them to the outside world and combating oppressive governments. On the other hand, blogs, talk shows, webpages, and Internet discussions are often known more for sound bites and polemical attacks on opponents than for reasoned debates in which opposing sides express views.

The key contribution that these technologies bring is speed. However, speed is not necessarily good for democracy:

> Democracy takes thought, patience, and reconsideration. That is why parliamentary procedure often requires several readings of a legislative bill prior to passage. The aim is to require time before precipitous action is taken. . . . Both representative and direct democracy are speed-averse, requiring time and patience to implement civic judgment. (Barber 2006:64)

Representative democracy involves citizens electing officials periodically and then letting them make the decisions. In a direct democracy, the voters make major policy decisions, and citizens work in communities to govern their social life, develop civic trust, and create social capital. This, of course, requires a well-informed electorate, which does not exist in all countries.

Rapid communication can help individuals stay in touch with their representatives in a representative democracy—witness the mass text messaging at the 2008 Democratic convention. However, several features of technology can have less than positive consequences for democracy: (a) a confounding of information (with which we are sometimes overloaded) with wisdom; (b) the tendency of digital media to reduce everything to simplistic opposites, as though only two choices are possible; (c) the tendency to isolate individuals behind their own keyboards and monitors so that collaborative skills may wane; (d) the inclination to use pictorial images that sensationalize issues, making reasoned deliberation less central to decision making; (e) the rise of immoderate rhetoric and divisive attacks by people who know little or nothing about the history of the problem; and (f) the tendency for the Internet (and many other media) to be primarily about commerce—creating a consumer mentality rather than a place for debate and problem solving. Social policy consideration—careful deliberative reflection—is necessary if we are to have technology benefit, and not undermine, democratic systems (Barber 2006).

Thinking Sociologically

How can the issues of technology enumerated above be problems for representative democracy? What positive effect might electronic technology have on direct participatory democracy?

Types of Economic Systems

As societies become industrialized, one of two basic economic systems evolves: a planned or a market system. **Planned or centralized systems** involve state-based planning and control of property, whereas **market systems/capitalism** stress individual planning and private ownership of property, with much less governmental coordination or oversight. These basic types vary depending on the peculiarities of the country and its economy. For instance, China has a highly centralized planned economy with strict government control, yet some private property and incentive plans exist, and these are expanding. The United States is a market system, yet the government puts many limitations on business enterprises and regulates the flow and value of money. For example, the Federal Reserve is the overseer of the U.S. banking system, which entails a Board of Governors—five economists—who control the flow of money so as to regulate inflation and recession. This is a form of regulation of the economy by "planners." Distinctions between planned and market systems rest on the degree of centralized planning and the ownership of property. In each type of system, decisions must be made concerning which goods to produce (and in what quantity), what to do in the event of shortages or surpluses, and how to distribute goods. Who has power to make these decisions helps determine what type of system it is.

Market Systems/Capitalism. The goal of capitalism is profit, made through free competition between competitors for the available markets. It assumes that the laws of supply and demand will allow some to profit while others fail.

Needed goods will be made, and the best product for the price will win out over the others. No planning is needed by any oversight group because the invisible hand of the market will ensure sufficient quality control, production, and distribution of goods. This system also rewards innovative entrepreneurs who take risks and solve problems in new ways, resulting in potential growth and prosperity.

The goal of capitalist manufacturers is to bring in more money than they pay out to produce goods and services. Because workers are a production cost, getting the maximum labor output for the minimum wage is beneficial to capitalists. Thus, for example, multinational corporations look for the cheapest world sources of labor with the fewest restrictions on employment and operations. Marx predicted that there would be victims in such a system—those whom the system exploited. This potential for exploitation leads most governments to exercise some control over manufacturing and the market, although the degree of control varies widely.

Capitalism was closest to its pure form during the Industrial Revolution, when some entrepreneurs gained control of the capital and resources to manipulate those who needed work and became laborers. Using available labor and mechanical innovations, these entrepreneurs built industries. Craftspeople such as cobblers could not compete with the efficiency of the new machine-run shops, and many were forced to become laborers in new industries to survive.

Marx predicted that capitalism would cause citizens to split into two main classes: the *bourgeoisie*, capitalists who own the means of production (the "haves"), and the *proletariat*, those who sell their labor to capitalists (the "have-nots"). He argued that institutions such as education, politics, laws, and religion would evolve to preserve the privileges of the elite. Religious ideology often stresses hard work and deference to authority, allowing entrepreneurs to increase profits that benefit the owners. Furthermore, members of the economic and political elite usually encourage patriotism to distract the less privileged from their conflicts with the elite. According to Marx, the elite want the masses to draw the line between "us" and "them" based on national loyalty, not based on lines of economic self-interests (Gellner and Breuilly 2009). So in Marxist thought, even patriotism is a tool of the elite to control the workers. However, Marx believed that ultimately the workers would realize their plight, develop political awareness, and rebel against their conditions. They would overthrow the "haves" and bring about a new and more egalitarian order.

The revolutions that Marx predicted have not occurred in most countries. Labor unions have protected workers from the severe exploitation that Marx witnessed in the early stages of industrialization in England, and capitalist governments have created and expanded a wide array of measures to protect citizens, including social security systems, unemployment compensation, disability programs, welfare systems,

PART V

V

Social Dynamics

Social structures such as institutions—family, education, religion, health, politics, and economics—tend to resist change. Yet, this entire book shows that societies are dynamic and changing. Institutions and organizations come alive with *processes* that are fluid and vibrant. Globalization itself, a major theme in this book, is a process bringing transformation to our social world. We do not live in the same sort of world our grandparents inhabited. The macro- and meso-level dimensions of the world have become increasingly powerful, which is exactly why we need a sociological perspective (imagination) to understand how the events in our own micro worlds are influenced by the larger society.

This section looks at some of those processes that are dynamic, fluid, and vibrant—population changes, expansion of technology, social movements, and more. When we are in periods of rapid change, understanding how that change occurs and what processes are involved is key to influencing change. For example, from global climate change to terrorism and from new digital technologies to immigrants in our communities, we need to understand what the actual facts are regarding causes and consequences before we can respond constructively. We live in exciting and challenging times, and we will thrive best if we understand the micro-, meso-, and macro-level dimensions of change in our lives and the linkages between parts of our social world.

CHAPTER 13

Population and Health

Living on Spaceship Earth

The human population is limited to this one moderate-sized sphere on which it depends for survival. The question is whether human groups can cooperate with each other and use the resources of the planet responsibly. The challenges of controlling population size, migration patterns, and disease epidemics add to the need for global cooperation. The night shot in the background of this page illustrates bright lights that beam from urban centers in Europe but from fewer places in Africa.

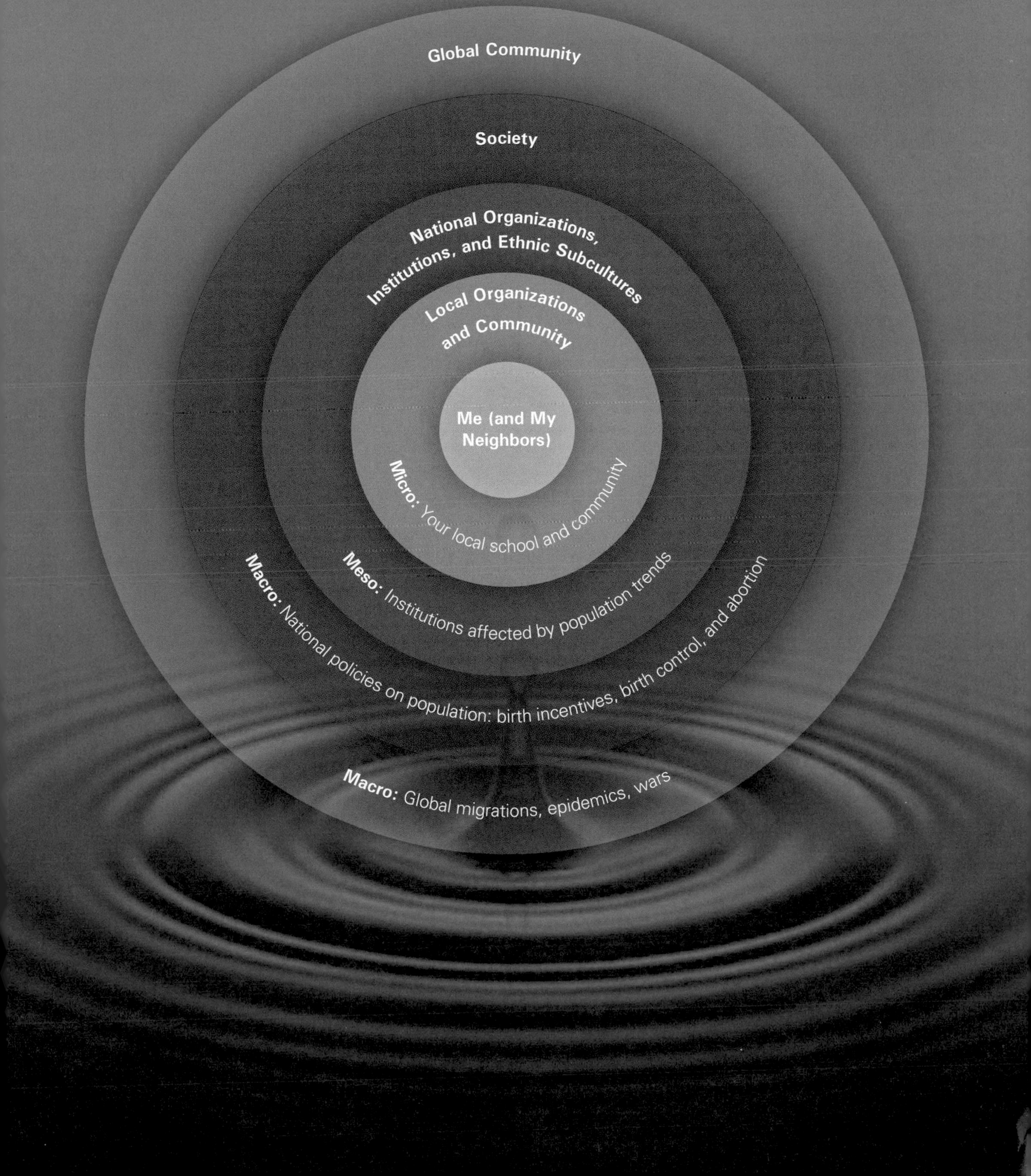

Global Community

Society

National Organizations,
Institutions, and Ethnic Subcultures

Local Organizations
and Community

Me (and My
Neighbors)

Micro: Your local school and community

Meso: Institutions affected by population trends

Macro: National policies on population: birth incentives, birth control, and abortion

Macro: Global migrations, epidemics, wars

Think About It	
Self and Inner Circle	Why is your family the size it is?
Local Community	What characterizes the population composition of your hometown? Where do you fit into that composition?
National Institutions; Complex Organizations; Ethnic Groups	Why do people move from rural areas to urban areas? What problems does this movement pattern create?
National Society	How might immigration affect the makeup of a nation, and what effect could this have on a nation's policies?
Global Community	How do global issues relating to urbanization, the environment, and technology affect your family and your local community?

Imagine you are from outer space looking down at a spherical object drifting through space. It appears to be a beautiful mix of greens and blues, and at night, parts of the sphere glow with lights while other parts are dark. That relatively small planet is home to earthlings. The controlling inhabitants of the planet are humans, almost 7 billion of them (Population Reference Bureau 2009b), increasing by 211,090 people daily (Helium 2011). The topic of this chapter is the life, death, spread, and distribution of those humans living on spaceship Earth (Diamond 2005).

Since the emergence of *Homo sapiens* in East Africa, human populations have grown in uneven surges and declines due to births, deaths, and migrations. The World Population Clock (see Table 13.1) illustrates the current state of the human population.

The world's human population has grown sporadically over the millennia, so the explosion of human beings on the planet in the past two and a half centuries is stunning. If we collapsed all of human history into one 24-hour day, the time period since 1750 would consume one minute. Yet, 25% of all humans have lived during this one-minute time period. In the 200 years between 1750 and 1950, the world's population mushroomed from 800 million to

2.5 billion. On October 12, 1999, the global population reached 6 billion. It has now expanded to almost 7 billion, growing by over 82 million in 2009. By mid-2011, the world population was projected to reach 7 billion with most growth in the poorest countries (Population Reference Bureau 2010). This means the world is growing each year by the number of people in Germany, the Philippines, or Vietnam. Every minute, 267 children are born and 108 people die around the world, resulting in a net increase of 159 people per minute (Population Reference Bureau 2010). Between 2000 and 2050, virtually all of the world's growth will occur in Africa, Asia, and Latin America, where 81% of the world's population lives and 90% of each year's births occur (Population Reference Bureau 2008).

Let us start by focusing on one area of our world: Kenya. We focus here partly because East Africa, where Kenya and Tanzania are located, was home to spaceship Earth's earliest human inhabitants. Scientists believe bones found in the dry Olduvai Gorge area are the oldest remains of *Homo sapiens* ever found. We also focus here because today Kenya is making human history for another reason. With a population of nearly 40 million people and a natural increase of 2.58% annually (Central Intelligence Agency 2011), Kenya has one

Table 13.1 Population Clock, 2010 (in Thousands)

	World	Global North (More Developed)	Global South (Less Developed)
Population	6,892,319	1,236,646	655,673
Births per day	384	39	345
Births per year	140,213	14,246	125,968
Deaths per day	156	33	123
Deaths per year	56,898	12,115	44,783
Infant deaths per day	17	220	17
Natural annual increase	83,315	2,130	81,185

Source: Population Reference Bureau 2010a.

of the most rapidly growing populations on Earth. Kenya is made up of many tribal groups of people. With different religions and value systems, the people have clashed in power struggles in recent years. Still, there are several themes that pervade most Kenyan subcultures.

Wengari, like Kenyan girls of most tribal affiliations, married in her teens. She has been socialized to believe that her main purpose in life is to bear children, to help with the farming, and to care for her parents in their old age. Children are seen as an asset in Kenya. Religious beliefs and cultural value systems encourage large families. However, the population of Kenya is 45% *dependent:* people under 15 or over 64 in societies where the population is living longer, who rely on working-age citizens to support them (Central Intelligence Agency 2011). The working-age population is becoming scarce and cannot continue to feed the growing dependent population. Further, severe droughts ravage parts of the country—droughts that are killing animal herds and preventing growth of crops. These facts, however, have little meaning to young women like Wengari, who have been socialized to conform to the female role within their society.

By contrast, far to the north of Kenya—in the industrialized, urbanized countries of Europe—birthrates are below population replacement levels, meaning population size eventually will begin to drop. Germany, Hungary, and Latvia, for instance, are losing population. The U.S. Population Reference Bureau (2011b) concludes:

> While Asia's share of world population may continue to hover around 60 percent through 2050, Europe's portion has declined sharply and is likely to drop even more during the 21st century. Africa would gain part of Europe's portion. . . . The more developed countries in Europe and North America, as well as Japan, Australia, and New Zealand, are growing by less than 1 percent annually.

In industrialized societies, children in the middle class and above are dependent until they leave home. Typical European young people wait until their late 20s or even 30s to start a family, postponing children until their education is complete and a job is in hand. Many limit their family size because societal values support small families. It is difficult to house large families in small urban apartments where the majority of the population lives. Workers must support and feed their families on earned wages rather than through farming. Both mother and father often work, and unlike many children in the Global South, most children born in Europe will survive to old age. Life expectancy, the average age at death, is 58.8 years in Kenya (Central Intelligence Agency 2011). In some European countries, it is older than 81 years.

On yet another continent, China, the country with the largest population in the world (1.33 billion people), had the greatest drop in population growth in the late twentieth

Changes in the environment and in the global economy make it difficult for this Kenyan family to provide for itself. Some of these children and their cousins may find it necessary to move to urban centers, a worldwide migration trend. Much of the socialization they receive in villages will not be relevant to their adult urban lives.

century due to strict governmental family planning practices (Central Intelligence Agency 2011). India, the second-largest country, has a population growth rate (increase in a country's population during a specified period of time) of 1.38% a year, just over replacement level (Central Intelligence Agency 2011).

Although some countries have birthrates below population replacement levels, the world's population continues to grow because of the skyrocketing growth rate in other countries and because of *population momentum* caused by the large number of individuals of childbearing age having children. Even though birthrates per couple drop, the number of women of childbearing age is still very high, resulting in continued growth in population size. Unfortunately, most of the countries with the highest growth rates are in the Global South, where there are fewer resources to support the additional population.

What we have been discussing is the study of human populations, called **demography**. All permanent societies, states, communities, adherents of a common religious faith, racial or ethnic groups, kinship or clan groups, professions, and other identifiable categories of people can be referred to as **populations**. The size, geographical location, and spatial movement of the population; its concentration in certain geographical areas including urban areas; and changing characteristics of the population are important elements in the study of demography. With growing populations and limited farmland to support the population, hungry people move to cities in hopes of finding jobs. This pattern of movement is called **urbanization**.

Hanoi, Vietnam, is a crowded Asian city. The overcrowding in some cities means that governments have a difficult time providing the infrastructure and services needed for the growing urban population.

The previous chapters have been organized by moving from micro- to macro-level analysis. Because demographic work has focused on societies and global trends, we will reverse the order and discuss macro-level patterns in world population growth first, followed by meso-level institutional influences on population, and finally micro-level population patterns.

Thinking Sociologically

Do you have a choice in how many children you have? What factors go into your decision? How might your decision differ if you were in a different country? Should macro-level global patterns—which include food shortages and climate change—be a consideration in the size of your family and your neighbors' families? Why or why not?

Macro-Level Patterns in World Population Growth

Early humans roamed the plains of Africa for thousands of years, their survival and growth depending on the environment in which they lived. They mastered fire and tools,

then domesticated animals and invented agriculture, and with these skills slowly increased control over the environment, allowing their numbers to expand. This evolution in the growth patterns of human populations is worth closer examination.

Patterns of Population Growth Over Time

Members of the small band of early *Homo sapiens* who inhabited the Olduvai Gorge moved gradually, haltingly, from this habitat into what are now other parts of Africa, Asia, and Europe. The process took thousands of years. At times, births outnumbered deaths and populations grew, but at other times, plagues, famines, droughts, and wars decimated populations. From the beginning of human existence, estimated from perhaps one million years ago until modern times, the number of births and deaths balanced each other over the centuries (Diamond 2005). The large population we see today results from population evolution that consisted of three phases:

1. Humans, because of their thinking ability, competed satisfactorily in the animal kingdom to obtain the basic necessities for survival of the species.

2. With the agricultural revolution that occurred about 10,000 years ago and the resulting food surplus, mortality rates declined, and the population grew as more infants survived and people lived longer.

3. The biggest increase came with the Industrial Revolution, beginning about 300 years ago. Improved medical knowledge and sanitation helped bring the death rate down.

When industrialization made its debut, it brought the social and economic changes discussed in Chapter 3 (e.g., machines replaced human labor, and mass production used resources in new ways), but it also augmented urbanization of societies. The population explosion began with industrialization in Europe and spread to widely scattered areas of the globe. With trade and migrations came the diffusion of ideas and better medical care, influencing population growth rates in all parts of the world by keeping people alive longer. Figure 13.1 shows population growth throughout history. The worldwide *rate* of population growth, or how fast the population increases, reached its peak in the 1960s. Although it has dropped to a current rate of about 1.14% per year, the population will continue to increase until large baby booms level off, resulting in a leveling off of increases in world population (Central Intelligence Agency 2011).

In the urban Global South, overcrowding is so severe that many people are homeless and must bathe every day in public in whatever water supply they can find. Here in Kolkata (Calcutta), India, even some people with homes would not have their own water supply and would have to use fire hydrants on the streets, as this family is doing.

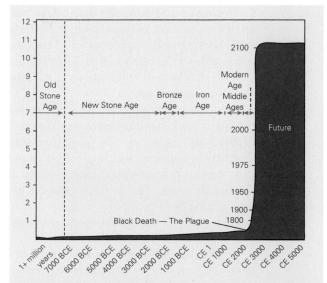

Figure 13.1   The Exponential World Population Growth From About 8000 BCE to Twenty-First Century

Source: Abu-Lughod 2001:50.

Predictors of Population Growth

In some villages in sub-Saharan and East African countries, which have the highest rates of HIV/AIDS in the world, children are forced to fend for themselves. With large percentages of the working-age population dead (or dying) from AIDS, orphaned children take care of their younger siblings. In some villages in Uganda, for instance, social workers visit periodically to bring limited food for survival and see that children are planting crops. These children must learn survival skills and gender roles at a very young age. They have little chance to experience a childhood typical in other places or receive schooling.

Think for a moment about the impact that your age and sex have on your position in society and your activities. Are you of childbearing age? Are you dependent on others for most of your needs, or are you supporting others? Your status is largely due to your age and sex and what they mean in your society. In analyzing the impact of age and sex on human behavior, three sets of concepts can be very useful: youth and aged dependency ratios, sex ratios, and age-sex population pyramids.

The *youth dependency ratio* is the number of children under age 15 compared to the number between 15 and 64. The number of those older than 64 compared to those between 15 and 64 is called the *aged dependency ratio.* Although many of the world's young people under 15 help support themselves and their families and many over 64 are likewise economically independent, these figures have been taken as the general ages when individuals are not contributing to the labor force. They represent the economic burden (especially in wealthy countries) of people in the population who must be supported by the working-age population. The **dependency ratio**, then, is the ratio of those in both the young and aged groups compared to the number of people in the productive age groups between 15 and 64 years old.

In several resource-poor countries nearly half of the population is under 15 years of age. These include Niger (50.1% under age 15), Uganda (48.7%), Congo (46.4%), and Afghanistan and Malawi (45.9%) (Population Reference Bureau 2010b). Working adults in less privileged countries have a tremendous burden to support the dependent population, especially if a high percentage of the population is urban and not able to be self-supporting through farming.

Similarly, high percentages of dependent people older than 64 are found in most Global North countries. In the European countries of Norway, Sweden, Denmark, the United Kingdom, and Germany, between 15% and 20% of the population is in the age group over 64. These countries have low death rates, resulting in the average life expectancy at birth being as high as 82 years (Central Intelligence Agency 2011; Human Development Report 2009).

Japan is an aging society, with life expectancy exceeding 81 years and more than 20% of the population over age 65. Ninety-two-year-old Toshi Uechi practices a traditional Japanese dance in Okinawa. An active lifestyle and a Spartan diet have helped make Okinawa the home of an exceptionally high percentage of centenarians (those over 100 years)—39.5 per 100,000 residents.

The percentage of dependent elderly people is growing, especially in affluent countries. Consider the case of Japan, which faces the problem of its "graying" or aging population. In 2009, 22.7% of its population was 65 or older, and the average life expectancy was over 82 years (Statistical Handbook of Japan 2010). The Japanese population is graying nearly twice as fast as the population in many other nations, in large part because the birthrate is very low, only 1.2 children per woman. There simply are not enough replacement workers to support the aging population (Pearce 2010). Japan provides a glimpse into the future for other rapidly aging societies, including Germany, the United States, and China. (See Figure 13.2 for a vivid depiction of the expected transformation in the age composition of Japan in one century.) Japan's percentage of population over 65 is growing faster than that of any other nation, and in one century will have become transformed.

The **sex ratio** refers to the ratio of males to females in the population. For instance, the more females there are, especially in their fertile years, the more potential there is for population growth. The sex ratio also affects population growth patterns by determining the supply of eligible spouses. Economic cycles, wars in which the proportion of males to females may decrease, and migrations that generally take males from one area and add them to another affect marriage patterns. **Population pyramids** illustrate sex ratios and dependency ratios (see Figure 13.3).

The graphic presentation of the age and sex distribution of a population tells us a great deal about that population. The structures are called pyramids because that is the shape they took until several decades ago. By looking up and down the pyramid, we can see the proportion of population at each age level. Looking to the right and left of the centerline tells us the balance of males to females at each age. The bottom line shows us the total population at each age.

The first pyramid shows populations that have fairly low birth- and death rates, typical of Global North countries. The second pyramid illustrates populations with high birthrates and large dependent youth populations, typical of the Global South. The world population has been getting both younger (the Global South) *and* older (the Global North), resulting in large numbers of dependent people.

As Global South nations have more and more children, they are creating more potential parents in later years, adding momentum to the world's population growth. Today young children survive their early years, whereas in the past they might have died of disease and malnutrition. Fewer deaths of infants and

Japan's percentage of population over 65 is growing faster than that of any other nation, and in one century will have become transformed.

Figure 13.2 Japan Grows Old

Source: Statistical Handbook of Japan 2010.

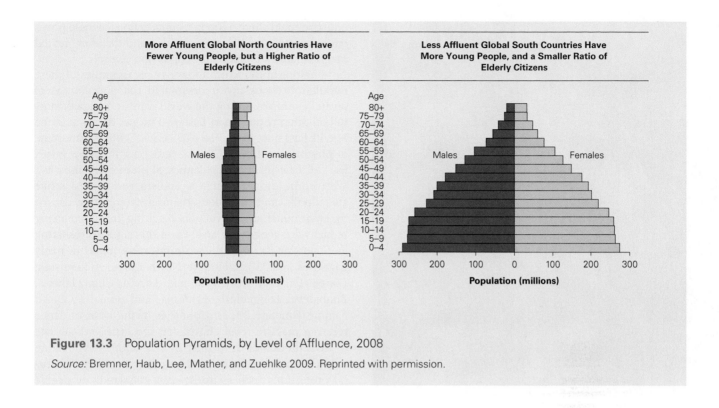

Figure 13.3 Population Pyramids, by Level of Affluence, 2008

Source: Bremner, Haub, Lee, Mather, and Zuehlke 2009. Reprinted with permission.

children, which can be credited to immunizations and disease control, result in lower mortality rates, younger populations, and higher potential numbers of births in the future. Ask yourself what might change rapid population growth in some countries.

Thinking Sociologically

Consider your own country's population pyramid. You can find it at http://www.census.gov/ipc/www/idb/pyramids .html. What can you tell about your country's level of development by studying the population pyramid? How might societies differ if they have a young population versus an old one?

Population Patterns: Theoretical Explanations

We find evidence of interest in population size and growth from the earliest historical writings. Scriptures such as the Quran and the Bible have supported population growth to increase the ranks of the faithful. Of course, population expansion made sense at the time these holy tracts were written. Government leaders throughout the ages have

adopted various philosophies about the best size of populations. The ancient Greek philosopher Plato ([circa 350 BCE] 1960) argued that the city-state should have 5,040 citizens and that measures should be taken to increase or decrease the population to bring it in line with this figure. However, the first significant scholarly analysis that addressed global population issues came from Thomas Malthus (1766–1834), an English clergyman and social philosopher.

Malthus's Theory of Population

In *An Essay on the Principle of Population,* Malthus ([1798] 1926) argued that humans are driven to reproduce and will multiply without limits if no checks are imposed. An unchecked population increases geometrically: 2 parents could have 4 children, 16 grandchildren, 64 great-grandchildren, and so forth—and that is simply with a continuous average family size of 4 children. Because the means of subsistence (food) increases at best only arithmetically or lineally (5, 10, 15, 20, 25), the end result is a food shortage and possible famine.

Malthus recognized several positive checks on populations, factors that would keep populations from excessive growth, including wars, disease epidemics, and famine (a drastic, wide-reaching shortage of food). He suggested preventive checks on rapid population growth, primarily in the form of delayed marriage and practice of sexual abstinence

Malthus predicted that if left unchecked, population increases would result in massive famine and disease. Actress/activist Mia Farrow took this photo of food distribution in famine-plagued Sudan, Africa.

Near Alem Ketema in Ethiopia, scarcity of freshwater and other resources threatens survival. In the nineteenth century, Thomas Malthus predicted such shortages due to population increases.

until one could afford a family. Contraception technology was crude and often unrealistic in his day, and therefore, he did not present it as an option for population control.

Looking at the world today, we see examples of these population checks. War decimated the populations of several countries during the world wars and has taken its toll on other countries in Eastern Europe, Africa, and the Middle East since then. The AIDS virus, SARS (severe acute respiratory syndrome), Ebola, and bird flu have raised fears of new plagues (epidemics of often fatal diseases). Waterborne diseases such as cholera and typhus strike after floods, and the floods themselves are often caused by environmental destruction resulting from too many humans in a geographic area. Food shortages necessitating food aid due to impending famines were found in many countries in 2010, including North Korea, Afghanistan, Congo, Burundi, Eritrea, Sudan, Angola, Chad, Liberia, Zimbabwe, Bangladesh, Ethiopia, and Somalia (World Famine Timeline 2011). Countries in the horn of Africa received massive relief efforts, but too little and too late to save thousands who died of starvation. Famines are caused in part by erosion and stripping the earth of natural protections such as forests and grasslands by people in need of firewood to cook or more land to cultivate crops or to graze animals. Today, economic factors are also affecting populations as imported cheap food is driving local farmers out of business in some areas.

Four main criticisms have been raised about Malthus's theory. First, Malthus did not anticipate the role capitalism would play in exploiting raw materials and encouraging excessive consumption patterns in wealthy industrial nations, escalating the environmental impact (Robbins 2011). Second, Malthus's idea that food production would grow arithmetically and could not keep up with population growth must be modified in light of current agricultural techniques, at least in some parts of the world. Third, Malthus saw abstinence from sex, even among the married, as the main method of preventing births and did not recognize the potential for contraception. Fourth, poverty has not always proven to be an inevitable result of population growth.

Two neo-Malthusians, scientists who accept much of his theory but make modifications based on current realities, are Garrett Hardin and Paul Ehrlich. Hardin (1968), a biologist, argues that individuals' personal goals are not always consistent with societal goals for population growth. If people act solely on their own and have many children, social tragedy may well ensue.

Ehrlich and Ehrlich (1990) add to the formula of "too many people and too little food" the problem of a "dying planet" caused by environmental damage. To hold on to economic gains, population must be checked, and to check population, family planning is necessary, they assert. Their ideas can be summed up as follows:

America and other rich nations have a clear choice today. They can continue to ignore the population problem and their own massive contributions to it. Then they will be trapped in a downward spiral that may well lead to the end of civilization in a few decades. More frequent droughts, more damaged crops and famines, more dying forests, more smog, more international conflicts, more epidemics . . . will mark our course. (Ehrlich and Ehrlich 1990:23)

The Ehrlichs also acknowledge that much of the environmental damage is caused by corporate pollution and excessive consumption habits in affluent areas such as the United States, Canada, and Europe (Weeks 2011).

Thinking Sociologically

What are contemporary examples of Malthus's population checks of war, disease, and famine? Are family planning and contraception sufficient to solve the problem of global overpopulation by humans? Can you think of other alternatives?

Demographic Transition: Explaining Population Growth and Expansion

Why should a change in the economic structure such as industrialization and movement from rural agricultural areas to urban cities have an impact on population size? One explanation is found in the **demographic transition theory**.

The idea of demographic transition involves comparing countries' stages of economic development with trends in birth- and death rates. Three stages of development are identified in this theory:

Stage 1:   These populations have high birth- and death rates that tend to balance each other over time. Births may outpace deaths until some disaster diminishes the increase. This has been the pattern for most of human history.

Stage 2:   Populations still have high birthrates, but death rates decline (i.e., more people live longer) because of improvements in health care and sanitation, establishment of public health programs, disease control and immunizations, and food availability and distribution. This imbalance between continuing high births and declining deaths means that the population growth rate is very high.

Stage 3:   Populations level off at the bottom of the chart with low birthrates and low death rates. Most industrial and postindustrial societies are in this stage. Population growth rates in these countries are very low because Global North urban nuclear families are small.

These stages are illustrated in Figure 13.4.

Demographic transition theory helps explain the developmental stage and population trends in countries around the world, but it does not consider some other important factors that affect the size of populations: (a) People's age at marriage determines how many childbearing years they have (late marriage means fewer years until menopause); (b) contraceptive availability determines whether families can control their number of children; (c) a country's resources and land may determine how much population a country can support; (d) the economic structure of a country, religious beliefs, and political philosophies affect attitudes toward birth control and family size; and (e) economic expansion rates influence a country's need for labor.

Critics argue that there is a built-in assumption that modernization in the second and third stages will result in rational choices about family size. Yet, unless women gain status by having smaller families, they are likely to continue to have large families (Robbins 2011). However, economic development and education of girls generally result in a decline in the birthrate. The process of modernization that

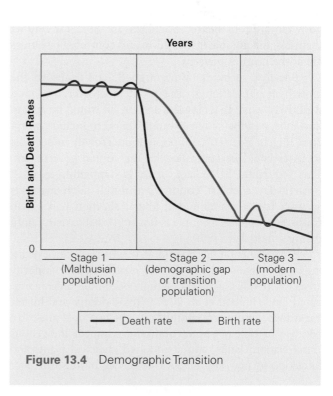

Figure 13.4   Demographic Transition

parallels economic development puts pressure on extended families to break apart into smaller nuclear family units, especially as families move to crowded urban areas. Urban families tend to have fewer children because children are a liability and cannot help support the family. Economic development, modernization, and urbanization did not occur together in all parts of the world, so the outcome of the three-stage transition has not always occurred as predicted in the theory.

The *wealth flow theory* suggests that two strategies are operating in couples' personal decisions about their family size. When wealth flows from children to parents—that is, when children are an asset working on the family farm or laboring—parents have larger families. When wealth flows from parents to children, families are likely to have fewer children (Caldwell 1982). To raise a child to 18 years in the United States, for instance, costs on average more than $222,360 (Belkin 2011).

Conflict Theorists' Explanations of Population Growth

Karl Marx and Friedrich Engels did not agree with Malthus's idea that population growth outstrips food and resources because of people's fertility rates, resulting in poverty. They felt that social and structural factors built into the economic system were the cause of poverty. Capitalist structures resulted in wealth for the capitalists and created overpopulation and poverty for those not absorbed into the system. Workers were expendable, kept in competition for low wages, used when needed, and let go when unprofitable to capitalists. In short, for conflict theorists, inequitable distribution and control of resources are at the heart of poverty.

Socialist societies, Marx argued, could absorb the growth in population so the problem of overpopulation would not exist. In a classless society, all would be able to find jobs, and the system would expand to include everyone. Engels asserted that population growth in socialist societies could be controlled by the central government. This regulation is, in fact, what is happening in most present-day socialist countries through such methods as strict family planning and liberal abortion policies. In China, for instance, a couple is not supposed to marry until their combined age is 50.

Environmental racism and justice has become a pressing issue in many neighborhoods, especially poor minority areas where housing has knowingly been built on contaminated land (Bullard et al. 2007). Toxic dumps and burns, hazardous waste sites, landfills on which people must live because of lack of space, dumping waste in indigenous "First Nation" lands, and abandoned chemical plants and mines occur not only in poor countries but also in poor areas of Global North countries. Regulations in the United States to prevent housing being built on contaminated land are controversial because they require testing and delay development. Conflict between haves and have-nots is clear in such an issue.

One award-winning study in Chicago focused on efforts to have a more eco-friendly or "green" city by doing more recycling. However, there were substantial problems of pollution, disease risk, and other costs to the neighborhoods where this recycling was done. It is sobering to realize that environmentally friendly policies have often been implemented at a cost to those who have fewest resources—people living in poverty and minorities (Pellow 2002).

Overpopulation presents a challenge to food and water resources, and large populations damage the environment and provide little ecological recovery time. The top photo shows pollution of a stream in Yunnan, China. The bottom photo pictures a billboard in Shanghai advertising China's "One Child Only" policy.

Policy Implications: Population Patterns and Economic Development

Does rapid population growth retard the economic development of a country? This question has been a subject of debate among demographers and policy makers in recent years. It is an important issue because the beliefs of decision makers affect the policies and solutions they advocate. For instance, if policy makers feel population growth retards economic development, family planning efforts are more likely because economic prosperity is of more immediate concern to political office holders (Solow 2000).

This issue of population growth has caused heated debate at several World Population Conferences. Some socialist and Catholic countries argue that capitalistic economic exploitation and political control, not population growth, cause poverty in Global South countries. Multinational companies and foreign countries exploit poor countries' resources, sometimes with payoffs to government officials, leaving the citizens with no gains. The majority of demographers agree that in most nations, high population growth contributes to poverty because countries cannot adjust quickly enough to provide the infrastructure (housing, health care, sanitation, education) for so many additional people (United Nations Population Division 2007). Better sex education, access to contraceptives, and birth control advice help reduce population growth and spread of disease. Also important to limiting population growth is providing opportunities for citizens, especially women, to obtain education and jobs.

Meso-Level Institutional Influences on Population Change

Populations change in size (overall number of people), composition (the makeup of the population, including sex ratio, age distribution, and religious or ethnic representation in the population), and distribution (density or concentration in various places, especially urban areas). The key demographic variables that cause changes are **fertility** (the birthrate), **mortality** (the death rate), and **migration** (movement of people from one place to another). Populations change when births and deaths are not evenly balanced or when significant numbers of people move from one area to another. Migration does not change the size or composition of the world as a whole but can affect size or makeup in a local micro-level community or national

macro-level population. The most unpredictable yet potentially controllable population factor is fertility.

Factors Affecting Fertility Rates

Jeanne, one of the coauthors, was riding in the back of a "mammy wagon," a common means of transport in Africa. Crowded in with the chickens and pigs and people, she did not expect the conversation that ensued. The man in his late 20s asked if she was married and for how long. Jeanne responded, "Yes, for three years." The man continued, "How many children do you have?" Jeanne answered, "None." The man commented, "Oh, I'm sorry!" Jeanne replied, "No, don't be sorry. We planned it that way!" This man had been married for 10 years to a woman three years younger than he and had eight children. The ninth was on the way. In answer to his pointed questions, Jeanne explained that she was not being cruel to her husband and that birth control was what prevented children, and no, it did not make sex less enjoyable. He expressed surprise that limiting the number of children was possible and rather liked the idea. He jumped at the suggestion that he visit the family planning clinic in the city. With his meager income, he and his wife were finding it hard to feed all the little mouths. The point is that knowledge of available family planning options is not always available.

Demographers consider micro-, meso-, and macro-level factors in attempting to understand fertility rates around the world. We know that individuals' personal decisions are key. People deciding to marry, couples' decisions to use contraception, their ideas about the acceptability of abortion, and whether they choose to remain childless can have an impact on national and global rates of population change. So choices at the micro level do make a difference at the macro level.

Economic Factors

Fertility also fluctuates with what is happening in meso-level institutions such as economic and political systems. During depressions, for example, the rate of fertility tends to drop. However, macro-level structural factors also affect fertility: (a) level of economic prosperity within the nation, (b) the government's commitment to providing (or restricting) contraception, (c) changes in norms and values about sexuality within a society, and (d) health care factors, including the availability of food and water.

We know that one of the most significant distinguishing characteristics between rich Global North and poor Global South countries is their fertility rates. The worldwide fertility rate has fallen in every major world region, but the rate in some places still remains very high. In sub-Saharan Africa, the average number of children per woman is 5.4, which is high, but that number has actually declined since 1950, when the average was 6.7. Worldwide, the number of children per woman fell from 5.0 in 1950 to 2.6 by 2008 (Population Reference Bureau 2008). Figure 13.5 compares Global North regions of the world with the poorest Global South regions.

Thinking Sociologically

What does Figure 13.5 (fertility rates by development levels) tell you about the lives of individuals in these different regions of the world?

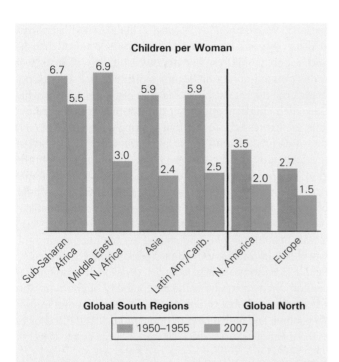

Figure 13.5 Fertility Rates: Comparing Various Regions of the World in the Early 1950s With 2007

Source: Population Reference Bureau 2007a. From Population Reference Bureau 2007b:1. Reprinted with permission.

Note: Fertility rates have fallen in every major world region but are still highest in sub-Saharan Africa.

Women's fertile years are roughly from 15 to 44. One woman dies every minute in pregnancy or birth—that is 10 million women over a generation. Most of these live in Global South countries (United Nations Population Fund n.d.). With care, these women would not be at such risk. Women in Global South countries have 80% of their children between the ages of 20 and 35, with a few below age 20 (United Nations Statistics Division 2010). Still, close to 20% of births in developing regions occur at age 35 or older, ages when there is greater risk in pregnancy, labor,

Extremely dense populations are often found in poverty-stricken areas. This slum in Kolkata (Calcutta), India, houses many rural migrants who are the lucky ones, having found a spot in the overpopulated sliver of land by a highway where they put up a shelter of whatever materials are available. Others sleep on sidewalks and highway medians.

and delivery. Government programs can influence fertility rates and birthrates, a factor discussed next.

Government Influence

Government *pronatalist policies* (those that encourage fertility) or *antinatalist policies* (those that discourage fertility) take several forms: manipulating contraceptive availability; promoting change in factors that affect fertility such as the status of women, education, and degree of economic development; using propaganda for or against having children; creating incentives (maternity leaves, benefits, and tax breaks) or penalties (such as fines); and passing laws governing age of marriage, size of family, contraception, and abortion.

Antinatalist policies arise out of concern over available resources and differences in birthrates among population subgroups. Singapore, a country in Asia located off the Malay Peninsula, consists of one main island and many smaller islands. It is one of the most crowded places on Earth, with 18,645 people per square mile. This is compared to 84 in the United States, 9 in Canada, and 836 in Japan (Information Please Database 2007). The entire population of Singapore is urban, and 90% live in the capital city. The country has little unemployment and one of the highest per capita incomes in Asia. However, it is dependent on imports from other countries for most of its raw materials and food.

Some years ago, the central government in Singapore started an aggressive antinatalist plan. Birth control was made available, and residents of Singapore who had more than one or two children were penalized with less health care, smaller housing, and higher costs for services such as education. Singapore now claims one of the lowest natural increase rates (the birthrate minus the death rate) in Southeast Asia, at 0.8% a year, just behind China. Singapore's governmental policies have controlled the natural increase rate.

China's antinatalist policy has been in effect since 1962. The government discourages traditional preferences for early marriage, large families, and many sons by using group pressure, privileges for small families, and easy availability of birth control and abortion. The government has reduced the natural increase rate in China, the most populous nation on Earth, to only 0.5% annually (Population Reference Bureau 2008). Unfortunately, there are side effects to such a stringent policy among a people who value male children. There has been an increase in selective abortions by couples hoping to have sons, and there are instances of female infanticide—killing of female infants when they are born—so that families can try for a male child.

An example of pronatalist government policies can be seen in Romania. Many Romanian men were killed in World War II, creating a sex imbalance. Marriage rates and birthrates plummeted. Concerned about the low birthrate, in 1966, the government banned abortions and the importation of most contraceptives. Within 8 months, the birthrate doubled, and within 11 months it tripled.

In the United States, citizens like to think that decisions about fertility are entirely a private matter left to the couple. Indeed, it is sometimes hard to pin a simple label of antinatalist or pronatalist on the administration in power. President George W. Bush reinstated the "gag rule" that Presidents Ronald Reagan and George H. W. Bush had implemented and Presidents Bill Clinton and Barack Obama each eliminated. The rule limits the availability of birth control for teens in the United States unless parents are informed that the teen has applied for contraception. Many experts argue that this policy has contributed to the increased teen pregnancy rates in the United States, but others argue that it is a parent's responsibility to deal with such matters. The Bush administration also blocked U.S. funding to international family planning groups that, among their many services, "promote abortions and abortion counseling" and proposed a 19% cut in international family planning contributions for 2006 ("Overseas Population Spending Threatened" 2006). The Obama administration reversed this abortion-funding policy, allowing international family planning clinics to receive U.S. funding even if they provide counseling on abortion (CNN Politics 2009). Congress proposed federal funding for Planned Parenthood, but this did not pass. However, some U.S. states are banning state funding for Planned Parenthood.

A condom mascot offers leaflets to teenagers in Bangkok, Thailand, during a promotional campaign to educate Thai youths on how to use condoms. The issue is related partially to birth control and partially to HIV/AIDS prevention. The effect is an antinatalist effort to lower the fertility rate.

Both limits to contraceptive availability and prohibitions on abortion are pronatalist because they increase fertility. While promoting births may not be the intention of those who oppose birth control and abortion, the policy has the latent consequence of encouraging population increases. Other governmental policies that might encourage larger families, such as family tax breaks or access to day care centers, are much less available in the United States. Each new administration brings its own policy initiatives.

Thinking Sociologically

Do you think it is appropriate for governments to use enticements or penalties to encourage or discourage fertility decisions? Why or why not? What are positive factors and problems with different policies?

Religious and Cultural Norms

Religion is a primary shaper of morality and values in most societies. Norms and customs of a society or subculture also influence fertility. In some cultures, pronatalist norms support a woman having a child before she is married so she can prove her fertility. In other societies, a woman can be stoned to death for having a child or even sex out of wedlock.

In Indonesia, Nur Azizah binti Hanafiah, 22, receives a caning, having been found by a citizen having illegal sex with her boyfriend at her house. This Aceh region of Indonesia has practiced Islamic Sharia law since 2001. In some societies, she would have been stoned to death for having premarital sex.

Some religious groups oppose any intervention (birth control or abortion) in the natural processes of conception and birth. Roman Catholicism, for example, teaches that large families are a blessing from God and that artificial birth control is a sin. The Roman Catholic Church officially advocates the rhythm method to regulate conception, a less reliable method in lessening birthrates than contraception technology. However, many Catholics in the Western world are not following these teachings. Islam also encourages large family size to increase the faithful. These policies were created when population growth was kept in check by high death rates.

Nonreligious cultural customs affect fertility as well. Couples may be pressured to delay marriage until they are in their late 20s or even 30s when they are economically secure. In Ireland, the mean age at first marriage for men was 30 and for women 28.2 years (Office of National Statistics [UK] 2007). Although Ireland is a predominantly Catholic country, delayed marriage helps keep the birthrate down. Another antinatalist custom in some polygamous groups is sexual abstinence after birth of a child, usually during the lactating (breast-feeding) period, which lasts anywhere from one to five years.

Education

Women with a "secondary school education have substantially smaller families than women with less education," illustrating that the higher women's status in society as measured by education level and job opportunities, the lower their fertility (Population Reference Bureau 2011a). If a country wants to control population growth, raising the status and education level of women is key. Figure 13.6 shows the relationship between education and family size in five Global South countries (Population Reference Bureau 2007a). Note that the higher the educational level, the lower the fertility rate and population growth. So, again, education and reduction of poverty are major variables contributing to moderation.

Studies repeatedly show that investing in education of girls and women raises every index of a country's progress toward economic growth and development. Yet, 300 million children are without access to education, two thirds of them girls. Of the estimated 1 billion illiterate adults in the world, two thirds are women. Although this gender gap has narrowed, it persists in sub-Saharan Africa, Arab states, and South and West Asia where only 6 in 10 adults are literate (Institute for Statistics 2005). Two thirds of the world's illiterate adults live in nine countries, and 45% of these adults live in India and China.

Family planning programs and contraceptive availability have resulted in a significant decline in global fertility rates, over and above such factors as education (Rama Rao and Mohanam 2003). For instance, worldwide use of contraception increased from 10% in the 1960s to 62% in 2007, with sharp contrasts between regions (Population Reference Bureau 2007a; United Nations Population Division 2008). An estimated 200 million women do not have access to contraception, resulting in 76 million

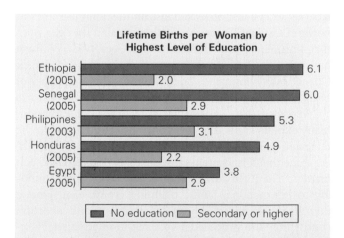

Figure 13.6 Women's Education and Family Size in Selected Countries, 2007

Source: Demographic and Health Surveys n.d. From Population Reference Bureau 2007a.

Note: Among women in developing countries, more education often leads to lower fertility.

unplanned births a year (Medical News Today 2009). Usage doubled in Global South countries, resulting in a decline of fertility in all regions and a total world decline from 3.2% to 2.7%. However, availability of contraception is spotty. In Africa, only 27% of married women are using any method of contraception, and only 20% are relying on modern methods. In some parts of Africa, fewer than 5% use modern methods. As a result, the region has the highest levels of fertility (and AIDS) in the world. Globally, 201 million women lack access to effective contraceptives, but many would practice family planning if given the option (United Nations Population Fund 2005).

As seen above, many factors affect fertility rates. Lower population growth means less pressure on governments to provide emergency services for booming populations and more attention to services such as schools, health care, and jobs. Most population experts encourage governments and other meso-level institutions in fast-growing countries to act aggressively to control population size. Critics argue that wealthy countries also need to help by curbing excesses.

There are consequences of population fluctuations for affluent parts of the world as well as for poor parts. The impact of the baby boom in the United States illustrates this, as discussed in the next "Sociology in Your Social World."

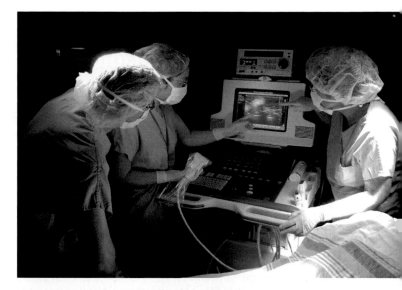

Thinking Sociologically

First, read the essay on page 420. What impact have the baby boom and the baby boomlet had on your opportunities for education and a career? How will retirement of baby boomers affect your opportunities?

Mortality Rates: Social Patterns of Health, Illness, and Death

Imagine living in Sierra Leone, West Africa, with a population of 5,245,695, where the life expectancy at birth is 55.7 years and women have an average of 5 children. Moreover, of every 1,000 babies born alive, 89 will die within the first year (Central Intelligence Agency 2011). A majority of the population lives on less than $2 a day (United Nations Development Programme 2007/2008). At 25, life is almost half over. Close to 70% of men and women are subsistence farmers, working small plots that may not provide enough food to keep their families from starving. When one plot is overfarmed and the soil is depleted so that plants will no longer grow, the family moves to another plot and clears the land, depleting more arable land. Shortages of food result in malnutrition, making the population susceptible to illnesses and

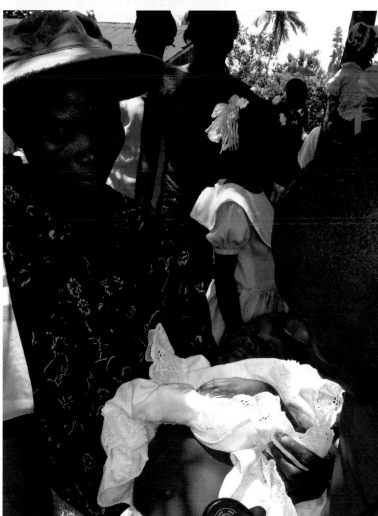

High-tech medical equipment is expensive (top), but in places like Haiti (bottom), physicians must use what little technology they have available.

Engaging Sociology

Population Pyramids and Predicting Community Needs and Services

Study these three Population Pyramid graphs. Based on what you see, answer the following questions:

1. Which community would be likely to have the lowest crime rate? Explain.

2. Which would be likely to have the most cultural amenities (theaters, art galleries, concert halls, etc.)? Explain.

3. Imagine you were an entrepreneur planning on starting a business in one of these communities.

   a. Name three businesses that you think would be likely to succeed in each community. Explain.
   b. Name one business that you think would be unlikely to succeed in each community. Explain.

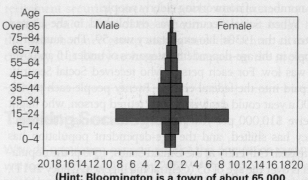

Bloomington, Indiana

(Hint: Bloomington is a town of about 65,000 and is the home of Indiana University, a large Big Ten university.)

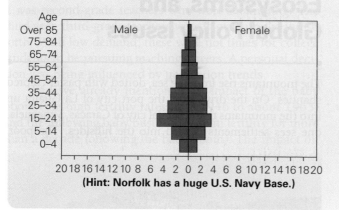

Norfolk, Virginia

(Hint: Norfolk has a huge U.S. Navy Base.)

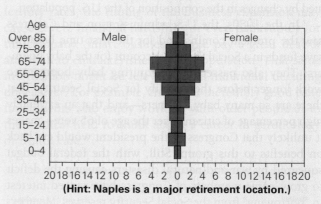

Naples, Florida

(Hint: Naples is a major retirement location.)

who have come from throughout the country to find opportunities in the capital, make shelters in the hills surrounding Caracas, often living with no running water, electricity, or sewage disposal. The laundry list of urban problems is overwhelming: excessive size and overcrowding; shortages of services, education, and health care; slums and squatters; traffic congestion; unemployment; and effects of global restructuring, including loss of agricultural land, environmental degradation, and resettlement of immigrants and refugees (Brunn et al. 2003). This section considers several of the many problems facing urban areas such as Caracas.

In the past three decades, the urban populations in Africa, Asia, and Latin America have grown rapidly, and many of the largest cities in the world are now in the Global

South. Rural-to-urban migration and development of megacities dominate the economic and political considerations in many countries. The newcomers spill out into the countryside, engulfing towns along the way.

A study involving 1,360 scientists from 95 nations asserts: "Sixty percent of the ecological systems that sustain life on Earth are being degraded or used unsustainably" ("Ecosystems Report Links Human Well-Being With Health of Planet" 2006:1). Extinction of species, lack of water and water pollution, resource exploitation, collapse of some global fisheries, and new diseases are the likely results of this breakdown. Urbanization by 2050 will stretch resources to their limits. Cities will have no way to dispose of wastes, resulting in epidemics. Humans are contributing to the problems that are killing people now and will kill many in the future. The problems are exacerbated by an increase in natural disasters (e.g., hurricanes), which scientists believe to be a result of global climate change. An increase in quality of life in some regions of the world is actually magnifying the problems and hastening the demise of the environments that support cities ("Ecosystems Report Links Human Well-Being With Health of Planet" 2006).

Additional infrastructure problems threaten to immobilize cities in the Global South. Traffic congestion and pollution are so intense in some cities that the slow movement of people and goods reduces productivity, jobs, health, and vital services. Pollution of the streets and air is a chronic problem, especially with expansion of automobile use around the world. Older cities face deteriorating infrastructures, with water, gas, and sewage lines in need of replacement. Even in affluent cities like Tokyo and London, many people are forced to wear face masks in July and August due to pollution. In Global North countries, concern for these problems has brought some action and relief, but in impoverished countries where survival issues are pressing, environmental contamination is a low priority. Thus, the worst air pollution is now found in major cities in the Global South. This air pollution, in turn, contributes to global climate change that has the potential to have devastating effects on life on this planet.

The reality is that the demographic migration pattern of urbanization is connected to the health and illness of a population—and ultimately to mortality. Millions of individual decisions to move to cities have resulted in meso and macro problems—which then impact the individuals who moved. In demography as in all other areas of our social world, the micro, meso, and macro levels are inextricably linked.

Many factors create change in a society. Some of them contribute to change in a particular direction, and others retard change. The next chapter examines the larger picture of social transformation in our complex and multi-leveled social world.

What Have We Learned?

Population trends, including migration resulting in urbanization, provide a dynamic force for change in societies. Whether one is interested in understanding social problems such as environmental degradation, social policy, or factors that may affect one's own career, demographic processes are critical forces. We ignore them at our peril—as individuals and as a society. Family businesses can be destroyed, retirement plans obliterated, and the health of communities sabotaged by population factors if they are overlooked. If they are considered, however, they can enhance planning that leads to prosperity and enjoyment of our communities.

Key Points:

- Population analysis (called demography) looks at the makeup of a population and how the trends and the composition of a population affect the society at each level of analysis. (See pp. 406–408.)

- The planet-wide increase in the human population's fertility is stunning. Implications for adequate resources to support life are illustrated in the population pyramids. (See pp. 408–411.)

- Various theories explain causes of the rapid growth, ranging from medical technology to cultural factors, but the demographic transition is key. (See pp. 411–415.)

- Many institutions affect and are affected by fertility and mortality rates at the meso level—especially health care issues such as infant mortality, spread of various diseases, and life expectancy. (See pp. 415–425.)

- Migration is also an important issue for the society—whether the migration is international or internal—for it can change the size, distribution, and composition of a nation's citizenry. (See pp. 425–429.)

- Population patterns can also affect individual decisions at the micro level, from career choices to business decisions to programs that will affect retirement possibilities. (See pp. 429–435.)

- The migration pattern that causes urbanization has created a series of problems that are related to health, mortality, environmental destruction, and global climate change, yet cities remain a draw because of prospects for jobs. (See pp. 435–436.)

Contributing to Our Social World: What Can We Do?

At the Local Level:

- *U.S. Census Bureau:* Invite an official to your campus to discuss the bureau's activities. Consider working for a community project.

- *Local department of urban planning, urban and regional development, or community development:* Invite a representative to campus. Discuss how population information is used in planning and service delivery contexts.

- Volunteer to work with a *neighborhood recreation department, crime watch, or community-organizing agency* to provide services or to help neighbors solve problems and work with local government.

At the Organizational or Institutional Level:

- *Urban planning, urban and regional development, community development, and similar municipal, regional, and state departments* often use volunteers. Contact one of these organizations and explore the possibility of making a contribution and/or exploring interest in this field.

At the National and Global Levels:

- *Planned Parenthood* (www.plannedparenthood.org) is a national organization that promotes family planning education and outreach programs throughout the United States. The organization uses volunteers and interns, as well as providing long-term employment opportunities.

- *The Population Council* (www.popcouncil.org) conducts research worldwide to improve policies, programs, and products in three areas: HIV/AIDS; poverty, gender, and youth; and reproductive health. Its website discusses several ways to get involved locally as well as globally.

Visit **www.sagepub.com/oswcondensed2e** for online activities, sample tests, and other helpful information. Select "Chapter 13: Population and Health" for chapter-specific activities.

CHAPTER

14

The Process of Change

Can We Make a Difference?

Humans are profoundly influenced by the macro structures around them, but people are also capable of creating change, especially if they band together with others and approach change in an organized way. Social movements such as those depicted in the pictures are one powerful way to bring about change.

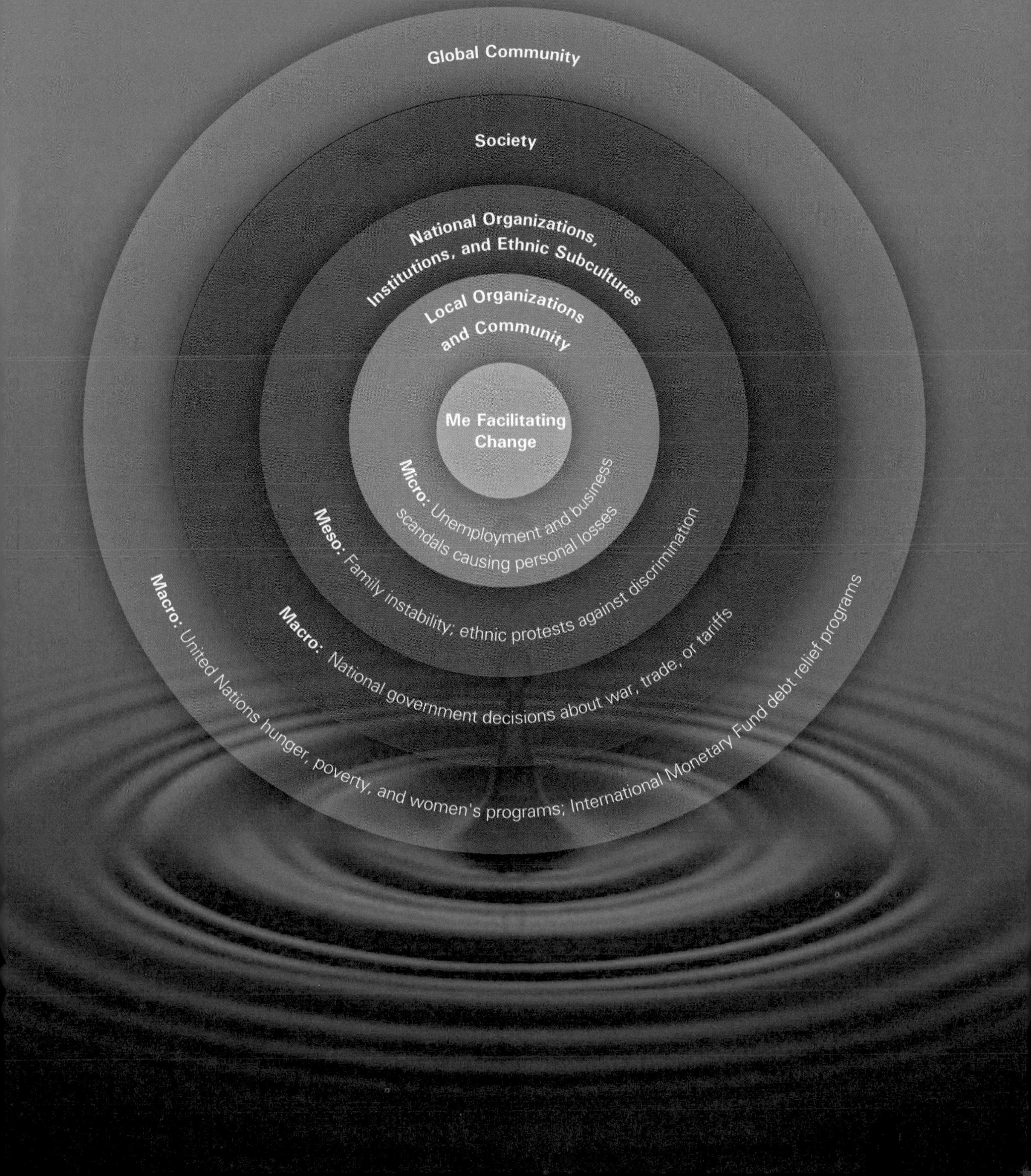

Global Community

Society

National Organizations,
Institutions, and Ethnic Subcultures

Local Organizations
and Community

Me Facilitating
Change

Micro: Unemployment and business
scandals causing personal losses

Meso: Family instability; ethnic protests against discrimination

Macro: National government decisions about war, trade, or tariffs

Macro: United Nations hunger, poverty, and women's programs; International Monetary Fund debt relief programs

Think About It	
Self and Inner Circle	Can you as an individual bring about change in the world?
Local Community	What do you think needs to change in your community?
National Institutions; Complex Organizations; Ethnic Groups	What organizational or ethnic community factors enhance or retard social change?
National Society	How does the training and support for technological innovation affect the process of change in your country?
Global Community	How do global changes—such as climate change—impact people and societies at each level?

The planet is in peril, according to evidence from the Asian subcontinent to the Arctic and from Africa to the Americas (Intergovernmental Panel on Climate Change 2007; United Nations Climate Change Conference 2009). A major part of this problem is the waste humans create. Wet, dry, smelly, and sometimes recyclable, garbage is a problem, and we are running out of space to dispose of our refuse. We dump it in the ocean and see garbage surfacing on beaches and killing fish. We bury it in landfills, and the surrounding land and water resources become toxic. We sort and recycle it, creating other problems such as where to dispose of recycled materials (Pellow 2002). Perhaps your community or campus has separate bins for glass, cans, paper, and garbage. Recycling is a relatively new movement in response to the urgent pleas from environmentalists about our garbage and trash that pollute water sources, cause areas of the oceans to die from trash dumping, and deplete renewable resources.

Recycling, salvaging items that can be reused, is part of a social reform movement—the environmental movement. However, few issues have simple solutions. The dumping and recycling have to take place somewhere. Many of the recycling plants and trash dumps are located in areas where poor people and minorities live. Some have referred to this as *environmental racism,* in which ethnic minorities are put at risk by the diseases and pollutants that recycling entails (Black Politics on the Web 2009; Pellow 2002). This illustrates the complexity of solving global problems: While individuals and small groups may help solve problems at the micro level of analysis, others may be created. At the micro level, each of us can do our part in the environmental movement to save the planet through responsible personal actions. At the meso level, the environmental movement can help local and regional governments enact policies and plans to reduce the garbage problem. At the macro level, world leaders need to find responsible ways to dispose of environmental wastes. Yet, the solution to this macro-level issue of protecting the Earth's resources may have micro-level implications for local minority families. Our social world is, indeed, complex and interdependent.

Turn on the morning or evening news, and there are lessons about other aspects of our changing social world. We see headlines of medical advances and cures for disease; biological breakthroughs in cloning and the DNA code; terrorist bombings in Israel, Iraq, Pakistan, Chechnya, and other parts of the world; famine in drought-afflicted sub-Saharan Africa; disasters such as earthquakes, hurricanes, tsunamis, and floods; and social activists calling for boycotts of Wal-Mart, chocolate, coffee, oil companies, or other multinational corporations. Some events seem far away and hard to imagine: thousands killed by a tidal wave in India or hundreds swept away by mud from an erupting volcano in Colombia or a rise in terrorism reflecting divisions in world economic, political, and religious ideologies. Some of these are natural events; others are due to human actions.

Social change is defined as variations or alterations over time in the structure, culture (including norms and values), and behavior patterns of a society. Some change is controllable, and some is out of our hands, but change is inevitable and ubiquitous. Change can be rapid, caused by some disruption to the existing system, or it can be gradual and evolutionary. Very often, change at one level in the social world occurs because of change at another level. Micro, meso, and macro levels of society often work together in the change process but are sometimes out of sync.

In this chapter, we explore the process of change, causes of change, and some strategies for bringing about desired change. We consider the complexity of change in our social world; explanations and theories of social change; the role of collective behavior in bringing about change; planned change in organizations; and macro-level social movements, technology, and environment as they affect and are affected by change.

Our social world model is based on the assumption that change, whether evolutionary or revolutionary, is inevitable and ever present in the social world. The impetus for change may begin at the micro, meso, or macro level of analysis. Studies of the change process are not complete, however, until the level under study is understood in relation to other levels in the model, for each level affects the others in multiple ways.

Thinking Sociologically

In what ways do you take actions to lessen your impact on the environment? Might your activities be linked to improving conditions for the planet or worsening conditions for ethnic minorities?

The Complexity of Change in Our Social World

The Yir Yoront, a group of Australian Aborigines, have long believed that if their own ancestors did not do something, then they must not do it. It would be wrong and might cause evil to befall the group (Sharp 1990). Obviously, this is not a people who favor change or innovation. By contrast, *progress* is a positive word in much of Australia and in other countries where change is seen as normal, even desirable. The traditions, cultural beliefs of a society, and internal and external pressures all affect the degree and rate of change in society.

Change at the Individual Level: Micro-Level Analysis

One of the nation's top entrepreneurs, Microsoft's Bill Gates, combines intelligence, business acumen, and philanthropy, qualities that appeal to American individualism. Gates has power to influence others because of his fame, wealth, and personal charisma. He is able to bring about change in organizations through his ability to motivate people and set wheels in motion. Many people have persuasive power to influence decision making, but it is not always based on charisma. For some, it is due to expertise, wealth, privileged positions, access to information, or the ability to use coercive force. On the other hand, common people, if they feel strongly about an issue, can rally others and bring about

Mother Teresa, an Italian nun, who devoted her life to helping the dying and destitute in India, established the orphanage shown here in Kolkata (Calcutta). In the photo at the right, Bill Gates holds a child who is receiving a trial malaria vaccine at a medical research center in Mozambique. Gates announced a grant of $168 million to fight malaria, a disease that kills more than one million people a year, 90% of them children in Africa. Sometimes social change occurs because of individual initiatives.

People at the local (micro) level will often try to influence policies at the meso level. At this local community event, people lobby their neighbors on a state referendum in an upcoming election.

change in a society. Each individual in society has potential to be an agent of change.

Most organizations—schools, businesses, volunteer associations—use one or more of the following strategies to persuade individuals to accept change: They appeal to individuals' values, they use persuasion by presenting hard data and logic, they convince individuals that the existing benefits of change outweigh the costs, they remove uncooperative individuals from the organization ("addition by subtraction"), they provide rewards or sanctions for acceptance of change to alter the cost-benefit ratio, or they compel individuals to change by an order from authority figures. Thus, individuals are active agents who initiate and bring about change.

Change at the Institutional Level: Meso-Level Analysis

The terrorist attacks of September 11, 2001, which killed 3,025 innocent people from 68 nations, resulted in repercussions in many U.S. institutions. Local, state, and national governments responded by putting in place measures to deter any further attacks and to seek out and punish the guilty parties. Religious services had high attendance in the weeks following 9/11 as religious leaders and U.S. citizens tried to make sense of the brutal attacks. Some families lost

loved ones. Other families were soon separated as the military and the National Guard were called into action. Security measures were stepped up at airports and other transportation centers. These institutions at the meso level—government, religion, family, the military—guided the response to the attacks.

Terrorism refers to "the use of indiscriminate violence to cause mass fear and panic to intimidate a population and advance one's political goals, whatever they may be" (Nolan 2002a:1648). This usually refers to acts of violence by private nonstate groups. In 2009, there were 10,999 terrorist attacks in 83 countries around the world, which killed 14,971 people. These figures were down from 2008 (11,725 attacks and 15,727 deaths) and from 2007 (14,435 attacks and 22,736 deaths), showing a decline for the past three years (Kellerhaus 2010). Contemporary terrorism is a meso-level phenomenon, even though it has both personal and global ramifications. Think of the additional hassles we must endure when flying—all in reaction to terrorist threats. Most modern terrorist organizations are not nations. They are ethnic, political, or religious subgroups that have elicited passionate loyalty from followers—even to the point of suicide on behalf of the group and its ideology.

Terrorism also has consequences for economies and other institutions. Not only do terrorist acts destabilize economies, but they change the kinds of jobs that are available. Consider the new jobs created in airport and seaport

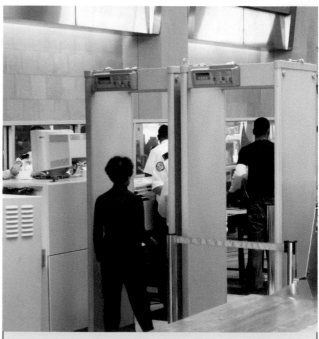

Only a decade ago family members could walk to the airplane departure gate to meet or send off a family member or friend. The terrorist attacks of 9/11 changed that, with extensive security checks that can take an hour to get through.

Revolutionary groups with terrorist tactics operate at the meso level, trying to change the society by intimidating the citizens and government so they will change policies. In this photo, we see the memorial site for victims of the bombing of a federal building in Oklahoma City on April 19, 1995—a terrorist act by members of a "patriot group" that was opposed to governmental policies. The small chairs are for deceased children who were in a day care center in the destroyed building.

security. In addition to new jobs in security, investors hesitate to invest their money if they think the economy will be negatively affected by security threats, and this lack of investment can spawn a recession. Economic and political disruptions or ripple effects are a core motivating factor for terrorists who commit horrific acts. Further, after 9/11, health care professionals began to make plans for biological and chemical terrorist attacks, changing the way monies are allocated in the health care industry (O'Toole and Henderson 2006). Thus, terrorism is a concern of the medical institution as well.

Change at the National and Global Levels: Macro-Level Analysis

To understand why terrorism occurs around the world, we must look at some of the driving forces. The unequal distribution of world resources has inspired deep anger toward affluent countries among those poor who believe the rich are using their power to maintain the inequities. High-income countries consume over $22,000 billion in resources (World Resources Institute 2007). That amounts to 76.6% of the world's resources for 20% of the world's

people. The world's middle-income countries with 60% of the world's people consume 21.9% of the world's resources, and the world's poorest people consume 1.5% in resources (Shah 2008). U.S. citizens make up only 4.6% of the total world population, yet the United States accounted for 33% of the consumed resources in the world in 2004 (World Resources Institute 2007), and 25% of the world's energy resources (Shah 2008). The average U.S. citizen consumes six times more energy than the world average (Energy Information Administration 2006). This seems grossly unfair to people who can barely feed their families and who have limited electricity, water, and sanitation and minimal shelter for their children. Furthermore, wealthy countries have considerable economic influence over nations of the Global South because poor countries are dependent on the income, employment, and loans from the core of affluent countries. Citizens of poor countries may work for multinational corporations, often for very low wages. However, the profits are returned to wealthy countries, helping perpetuate the gross inequity in distribution of resources. Whether or not we think it is justified, this inequity leads to hostilities and sometimes terrorism as a means of striking out against more powerful countries.

The 9/11 attack was partially motivated by young men who were intensely pro-Palestinian and anti-Israeli. Of the hijackers, 15 of the 19 were originally from Saudi Arabia. None were from Iraq or Palestine, but they sympathized with those who had been displaced by the establishment of Israel. From the terrorist perspective, the attack on the New York City World Trade Center was an effort to strike out at the United States. September 11 was a symbolic date for Palestinians, the day when Britain declared control of Palestine. That declaration set off a chain of events leading to the United Nations granting the land to Jews to establish Israel. Many Middle Easterners felt this was unjust because the Palestinians lost the country in which they had been living. In addition, the Camp David Accords, which established Israel's right to exist in the Middle East, were signed on September 11, 1979. So on two counts, this date had powerful symbolic meaning to the people who were displaced from Palestine.

The events of September 11 changed the United States as a nation and the core issues, priorities, and spending of the Bush presidency. George W. Bush's administration proposed, and Congress approved, a Patriot Act that channeled resources into heightened security and military preparedness. The provisions of the bill also greatly restricted civil liberties and allowed the government to snoop into the private lives of citizens in ways that had never before been tolerated, from monitoring home telephone connections of Americans to scrutinizing the books they checked out of the library. Fear for their security meant that many U.S. citizens welcomed efforts to prevent further attacks, although the root causes of terrorism were not addressed.

Following September 11, many state governments mandated more intense patriotism training and rituals in the schools, with additions to the curriculum and daily loyalty ceremonies. The No Child Left Behind education bill, which was passed by Congress in 2002, also mandated that school personnel turn over personal information about students to military recruiters, including private information that had previously not been available to anyone but the student and school personnel (Ayers 2006; Westheimer 2006).

Thinking Sociologically

How has terrorism affected institutions and countries with which you are familiar? How is your life being changed by threats of terrorism?

Societal Level Change

Take a look at the impact of humans on the global environment and the constant change we are bringing to our planet. To illustrate the increasingly complex and biologically interdependent social world, consider that pollution of the environment by any one country now threatens other countries. Carcinogens, acid rain, and other airborne chemicals carry across national boundaries (Brecher, Costello, and Smith 2012). People in the United States comprise about 5% of the world's population but emit 25% of the heat-trapping gases (Lindsay 2006). Pollutants affect the air that surrounds the entire planet, destroying the ozone that protects us from intense sun rays and warming the planet in ways that could threaten all of us. In the past century, scientists claim, the Earth's surface has warmed by one degree. That does not sound like much until one considers that during the last ice age, the Earth's surface was only seven degrees cooler than today. Small variations can make a huge difference, and we do not know what all of those consequences might be if the Earth's surface temperature increases by another two or three degrees. Currently, massive blocks of sea ice are melting each year at a rate that equals the size of Maryland and Delaware combined (Cousteau 2008; Lindsay 2006). It is alarming to visit the glaciers on the South Island of New Zealand (closest to Antarctica) and realize that glaciers there are melting so fast that they have receded by as much as 10 or 12 miles in just a couple of decades. In the northern hemisphere, Greenland is home to many glaciers. Warming caused one of the largest to lose a 100-mile square section that is now a free-floating iceberg (BBC News 2010).

In a warmer world, there is less snowfall, smaller mountain icecaps, and a resultant smaller spring runoff of crucial freshwater (Barnett, Adam, and Lettenmaier 2005; Cyranoski 2005; Struck 2007). This causes major water shortages. While some of the environmental change may be rooted in natural causes, the preponderance of the evidence suggests that human activity—the way we consume and the way we live our lives—is the primary cause (Gore 2006). Even if humans were not a significant cause, the global climate change has consequences that mean change for many parts of our lives. What happens when people do not have enough water in their current location to survive? What happens when they try to move into someone else's territory to gain access to needed resources?

Obviously, this is a global issue with implications for nations that must work together for change. Yet, some nations do not want to change because they feel controls will impede progress. The Kyoto Protocol on global warming requires commitment by nations to curb carbon dioxide and other emissions, but U.S. President Bush rejected it because it "does not make economic sense" (Lindsay 2006:310–11). As of 2009, 187 countries and the European Union have approved the treaty. The exceptions are the United States, China, India, and Brazil—all major polluters. The Obama administration has indicated support for dramatic efforts to curb climate change and global warming, but passage of a bill is still pending.

Environmental destruction and global climate change are problems that necessitate cooperation of nations. Air pollution around China's Three Gorges Dam reduces visibility and creates health hazards.

Fixing the environmental issues will be expensive, it may hinder the economy and slow the rate of growth, and it may even contribute to continuation of the recession. Because recessions are terrifying for any elected politician who aims to keep the public happy, change is not easy. Still, most of the rest of the world's nations have signed the Kyoto Protocol, and there is continuing pressure on the United States to get on board. Because nations are still the most powerful units for allocating resources and for setting policy, changes in national policies that address the issues of a shared environment are of critical importance. The two big quandaries are (a) the costs and benefits to various nations of participating in a solution and (b) the matter of time—will nations respond before it is too late to make a difference? Currently, most of the cost of pollution is accruing to impoverished countries, while rich nations benefit from the status quo.

Global Systems and Change

As the world becomes increasingly interconnected and interdependent, impetus for change comes from global organizations, national and international organizations and governments, and multinational corporations. New and shifting alliances between international organizations and countries link together nations, form international liaisons, and create changing economic and political systems. The following international alliances between countries, for example, are based primarily on economic ties:

- SADC: Southern African Development Community
- NAFTA: North American Free Trade Agreement
- CEFTA: Central European Free Trade Agreement
- CAFTA: Central America Free Trade Agreement
- ASEAN: Association of Southeast Asian Nations
- WIPO: World Intellectual Property Organization
- G8: Group of Eight
- OPEC: Organization of the Petroleum Exporting Countries
- ADP: Asian Development Bank
- EFTA: European Free Trade Association
- EEA: European Economic Area
- APEC: Asia-Pacific Economic Cooperation
- EU: European Union
- SEATO: Southeast Asia Treaty Organization

Consider NAFTA, which was initiated in 1993 to establish a free trade area between Canada, the United States, and Mexico to facilitate trade in the region. Promoters, including many global corporations, promised the agreement would create thousands of new high-wage jobs, raise living standards in each of the countries, improve environmental conditions, and transform Mexico from a poor developing country into a booming new market. Opponents (including labor

unions, environmental organizations, consumer groups, and religious communities) argued the opposite—that NAFTA would reduce wages; destroy jobs, especially in the United States; undermine democratic policy making in North America by giving corporations free rein; and threaten health, environment, and food safety (American Cultural Center Resource Service 2004; Public Citizen's Global Trade Watch 2003; U.S. Trade Representative 2011).

Analyses of the agreement show mixed results. There is some indication that tariffs are down and U.S. exports have increased. The treaty countries produced $15.3 billion in goods and services, and trade tripled between 1993 and 2007 to $903 billion. NAFTA has been more effective in increasing trade in agricultural commodities than in nonagricultural products. Some analysts argue that there is improvement in areas of environmental protection and labor rights, but others have challenged these optimistic assessments (Amadeo 2009).

In the above discussion of changes at different levels of analysis, the principle is that change at one level leads to change in other levels as it has done in the global cases of terrorism, climate change, and NAFTA. Changes at the macro level affect individuals, just as change at the micro level has repercussions at the meso and macro levels.

Social Change: Process and Theories

The Process of Social Change

Something always triggers a social change. The impetus may come from within the organization, a source of change known as *strain*. Sometimes, it comes from outside the organization, in what sociologists call *stress*. Strain may be caused by conflicting goals or by contrasting belief systems within the organization. For example, conflicting goals are seen in the case of the steel industry and its workers. Individual workers' goals are to meet their basic needs for food and shelter for their families by holding jobs. Company goals focus entirely on being profitable in a competitive environment. In Pittsburgh and Cleveland, many steel companies have closed down or moved to less costly sites due to lack of profits from changing economic demands and competition. This created massive unemployment, and people who had created hopeful futures for themselves and their families were left without jobs.

Contrasting belief systems (political, religious, economic, and social) within a society can also have a major effect on the type and rate of change. For example, some religious groups oppose stem cell research, which uses cells of fetuses, most of which were created in test tubes. Others within the same congregation believe this research will alleviate suffering of loved ones and save lives. Although both sides believe they are pro-life, the internal strain in the religious group emanates from events and forces in science, medicine, and other institutions.

Stresses, those pressures for change that come from the organization's external environment, can be traced to several sources: the natural environment and natural disasters, population dynamics, actions of leaders, new changing technologies, changes in other institutions, and major historical events.

The natural environment can bring about either slow or dramatic change in a society. Natural disasters such as floods, hurricanes, tsunamis, heavy snows, earthquakes, volcanic eruptions, mudslides, tornadoes, and other sudden events are not planned occurrences, but they can have dramatic consequences. Disease epidemics, such as the cholera outbreak in Haiti that is killing many people each day and the cholera epidemic in Zimbabwe, Africa, that killed more than 4,000 people in 2008–2009, are often unpredictable (Rusere 2009).Epidemics such as the SARS (severe acute respiratory syndrome) threat have brought about change in the World Health Organization, global medical reporting systems, and response networks. For instance, the Global Public Health Intelligence Network scans Internet communications for rumors and reports of suspicious diseases. This way, health organizations from the local to global levels can act quickly to contain the spread of deadly epidemics.

Less rapid natural changes also can have incremental but dramatic effects. For example, most scientists predict that the climatic changes resulting from the greenhouse effect are warming our atmosphere due to a buildup of carbon dioxide or other gases (Environmental Protection Agency 2009). These scientists are convinced that global warming is a human-generated problem that, at worst, is a threat to our survival and, at best, will increase cost of living in ways that drastically change the lifestyle of ordinary people (Lindsay 2006).

Population dynamics—birth- and death rates, size of populations, age distribution, and migration patterns—can be important contributors to external stress on organizations. Where populations are growing at extremely rapid rates, strains on government systems result in inability to meet basic needs of the people. Values and beliefs regarding childbearing, knowledge of birth control, and the position of women in society are some of the crucial social variables in addressing the ability to meet needs. Immigration due to political upheavals or motivated by anticipated economic opportunities creates stress on the societies that receive the newcomers as they attempt to meet the immigrants' needs. For example, many refugees from the conflicts in the Darfur region in Sudan are fleeing to camps in the nearby country of Chad. Map 14.1 shows the global hot spots for both internal refugees (those displaced within their countries) and external ones (those being displaced to other lands).

Leaders influence change through their policy decisions or the social movements they help generate. Mohandas K. (Mahatma) Gandhi in India taught the modern world nonviolent methods of bringing about change in political systems. Policies of Charles Taylor, former military dictator of Liberia, created long-term war and resulted in thousands of deaths. President Robert Mugabe of Zimbabwe locked

Natural disasters—floods, hurricanes, tornadoes, earthquakes, volcanic eruptions—can be the cause of major social changes in a community. As shown in this photo from a community on Lake Pontchartrain near New Orleans, Hurricane Katrina took its toll on Mississippi and Louisiana.

New arrivals at Al Salaam camp, in Sudan's Darfur region, make temporary shelters out of household goods they were able to carry with them. A USAID-supported program at this one site is helping register more than 10,000 people who were displaced by violence in their home regions. Migration is a major factor in social change, and in cases like this, it is associated with great suffering and hardship.

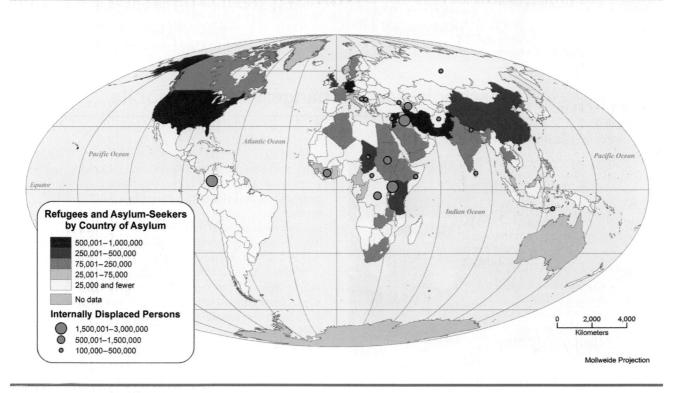

Map 14.1   Countries Receiving Refugees, Asylum Seekers, and Internally Displaced People, 2006

Source: UNHCR 2006a, 2006b. Map by Anna Versluis.

his country in a downward spiral of economic turmoil and disease, killing thousands. These leaders' actions created internal strains in their own countries and external stressors resulting in change in the international community.

Technology also influences societal change. William F. Ogburn (1933) compiled a list of 150 social changes in the United States that resulted from the invention of the radio, such as instant access to information. Other lists could be compiled for the telephone, television, automobile, computer, and new technologies such as iPhones. Some of these changes give rise to secondary changes. For example, automobile use led to the development of paved highways, complex systems of traffic patterns and rules, and need for gasoline stations. The "Sociology in Your Social World" on page 448 explores several issues involving the automobile and change.

Thinking Sociologically

What might be some long-term social consequences for our individual lives and societies of the expanded use of the computer, cell phones and iPhones, or the microwave oven?

The diffusion or spread of the technology throughout the world is likely to be uneven, especially in the early stages of the new technology. For example, computer technology is advancing rapidly, but those advances began in corporate boardrooms, on military bases, and in university laboratories. Policies of governing bodies—such as funding for school computers—determine the rate of public access. Thus, only gradually are computers reaching the world's citizenry through schools, libraries, and eventually private homes.

Major historical events—wars, economic crises, assassinations, political scandals, and catastrophes—can change the course of world events. For instance, the triggering event that actually started World War I was the assassination of Archduke Franz Ferdinand of the Austro-Hungarian empire. This assassination resulted in the German invasion of several other countries and the beginning of the war. So a micro-level act, the murder of an individual, had global ramifications. Clearly, internal strains and external stressors give impetus to the processes of change. The question is: How do these processes take place?

Theories of Social Change

Social scientists seek to explain the causes and consequences of social change, sometimes with the hope that change can be controlled or guided. Theories of change often reflect

Sociology in Your Social World

Technology and Change: The Automobile

O nly a century ago, a newfangled novelty was spreading quickly from urban areas to the countryside: the automobile. At the turn of the twentieth century, this strange horseless carriage was often referred to in rural areas as the "devil wagon." The introduction of this self-propelled vehicle was controversial, and in the 1890s and early 1900s, some cities and counties had rules forbidding motorized vehicles. In Vermont, a walking escort had to precede the car by an eighth of a mile with a red warning flag, and in Iowa, motorists were required to telephone ahead to a town they planned to drive through to warn the community lest their horses be alarmed (Berger 1979; Clymer 1953; Glasscock 1937; Morris 1949). In most rural areas, motorists were expected to pull their cars to a stop or even to shut down the motor when a horse-drawn buggy came near. "Pig and chicken legal clauses" meant the automobile driver was liable for any injury that occurred when passing an animal near the road, even if the injury was due to the animal running away (Scott-Montagu 1904).

Automobiles were restricted to cities for nearly a decade after their invention because roads were inadequate outside of the urban areas and they often slid off muddy roads into ditches. These conditions had not deterred horses. In 1915, Owen G. Roberts invented a seat belt because on a daylong 80-mile drive westward from Columbus, Ohio, Mary Roberts was knocked unconscious when the rough roads caused her head to smack the roof of the car. Paving of roads became

a necessity for automobile travel and, of course, made automobile travel much faster and more common. The expansion was stunning. Roughly 85,000 motored vehicles were in use in America in 1911. By 1930, the number was nearly 10 million (Berger 1979), and in 2007 the estimated number of registered passenger vehicles was 254.4 million (Tilley 2009).

Forms of entertainment began to change when people were able to be more mobile. As the Model T made cars affordable, families no longer had only each other for socializing, and entertainment became available virtually any night of the week (Berger 1979; McKelvie 1926). Thus, dependence on family was lessened, possibly weakening familial bonds and oversight (Berger 1979). Even courting was substantially changed, as individuals could go farther afield to find a possible life partner, couples could go more places on dates, and two people could find more privacy.

Transportation that made traversing distances more possible changed how people related to a number of other institutions as well. Motorized buses made transportation to schools possible, and attendance rates of rural children increased substantially (U.S. Department of Interior Office of Education 1930). Because people could drive farther to churches, they often chose to go to city churches, where the preachers were more skilled public speakers and the music was of higher quality. Some people found that a country drive was a more interesting way to spend Sunday mornings, and preachers often condemned cars for leading people away from church (Berger 1979). Many country churches consolidated or closed. Still, once pastors could afford cars, rural people received new services such as pastoral calls (Wilson 1924). The automobile was also a boon to the mental health of isolated farm women, allowing them to visit with neighbors (Berger 1979; McNall and McNall 1983).

As people could live in less congested areas but still get to work in a reasonable amount of time via an automobile or public transport, the suburbs began to develop around major cities. No longer did people locate homes close to shopping, schools, and places of worship. Still, a dispersed population needs to use more gasoline, thereby creating pollution. As the wealthy moved to expensive suburbs and paid higher taxes to support outstanding schools, socioeconomic and ethnic stratification between communities increased.

> When Owen G. Roberts built one of the first automobiles in Ohio and established a large automobile dealership, it was not his intent to heighten segregation, to create funding problems for poor inner-city areas, or to pollute the environment. Yet, these are some of the *unintended consequences* of the spread of the automobile. It sometimes takes decades before we can identify the consequences of the technologies we develop and adopt.

the events and belief systems of particular historical time periods. For example, conflict theory developed during periods of change in Europe; it began to gain adherents in the United States during the 1960s when intense conflict over issues of race and ethnic relations, the morality of the Vietnam War, and changes in social values peaked. Theories that focused on social harmony were of little help.

Major social change theories tend to focus on the micro level (symbolic interaction and rational choice theories) or the meso and macro levels (evolutionary, functional, conflict, and world systems theories). As we review these theories, many will be familiar from previous chapters. However, here they are related to the process of change.

Micro-Level Theories of Change

Symbolic Interactionism. According to symbolic interaction theory, a micro-level theory, human beings are always trying to make sense of the things they experience, figure out what an event or interaction means, and determine what action is required of them. Humans construct meanings that agree with or diverge from what others around them think. This capacity to define one's situation, such as concluding that one is oppressed even though others have accepted the circumstances as normal, can be a powerful impetus to change. It can be the starting point of social movements, cultural changes, and revolutions.

Some sociologists believe that individuals are always at the core of any social trends or movements, even if those movements are national or global (Blumer 1986; Giddens 1986; Simmel [1902–17] 1950). After all, it is individuals who act, make decisions, and take action. There are a number of leaders, for example, who have changed the world for better or worse, including Adolf Hitler and Mahatma Gandhi. Neither corporations nor nations nor bureaucracies make decisions—people do. The way in which an individual defines the reality he or she is experiencing makes a huge difference in how that person will respond.

Social institutions and structures are always subject to maverick individuals "thinking outside the box" and changing how others see things. Individual actions can cause riots, social movements, planned change in organizations, and a host of other actions that have the potential to transform the society. That people may construct reality in new ways can be a serious threat to the status quo, and those who want to protect the status quo try to ensure that people will see the world the same way they do. If change feels threatening to some members who have a vested interest in

the current arrangements, those individuals who advocate change may face resistance.

Leaders often provide opportunities for group members to participate in suggesting, planning, and implementing change to help create acceptance and positive attitudes toward change. This collaborative process is often used when a firm or a public agency is planning a major project, such as the development of a shopping mall or a waste disposal site, and cooperation and support by other parts of the community become essential. Symbolic interactionists would see this as an effort to build a consensus about what the social changes mean and to implement change in a way that is not perceived as threatening to the members.

Rational Choice. To rational choice theorists, behaviors are largely driven by individuals seeking rewards and limiting costs. Because of this, most individuals engage in those activities that bring positive rewards and try to avoid the negative. A group seeking change can attempt to set up a situation in which desired behavior is rewarded. The typology presented in Figure 14.1 on page 450 shows the relationship between behaviors and sanctions.

Bringing about change may not require a change in costs or rewards. It may be sufficient simply to change the people's perception of the advantages and disadvantages of certain actions. Sometimes, people do not know all the rewards, or they have failed to accurately assess the costs of an action. For example, few citizens in the United States realize all the benefits of marriage. To change marriage rates, we may not need more benefits to encourage marriage. We may do just as well to change the population's appraisal of the benefits already available.

Meso- and Macro-Level Theories of Change

Social Evolutionary Theories. Social evolutionary theories at the macro level assume that societies change slowly from simple to more complex forms. Early unilinear theories maintained that all societies moved through the same steps and that advancement or progress was desirable and would lead to a better society. These theories came to prominence during the Industrial Revolution when European social scientists sought to interpret the differences between their own societies and the "primitive societies" of other continents. Europe was being stimulated by travel, exposure to new cultures, and a spawning of new philosophies, a period called the Enlightenment. Europeans witnessed the developments

		Sanction	
		Formal	**Informal**
Behavior	**Positive**	Bonuses, advances, fringe benefits, recognition	Praise, smile, pat on the back
	Negative	Demotion, loss of salary	Ridicule, exclusion, talk behind back

Figure 14.1 Relationship Between Behaviors and Sanctions

of mines, railroads, cities, educational systems, and rising industries, which they defined as "progress" or "civilization." World travelers reported that other peoples and societies did not seem to have these developments.

In recent versions of evolutionary theory, Nolan and Lenski (2008) discuss five stages through which societies progress: hunter-gatherer, horticultural, agrarian, industrial, and postindustrial (see Chapter 3). This does not mean that some stages are "better" than others but means that this is the typical pattern of change due to new technologies and more efficient harnessing of energy.

Many modern cases do not fit this pattern because they skip steps or are selective about what aspects of technology they wish to adopt. Countries such as India and China are largely agricultural but are importing and developing the latest technology. Furthermore, advocates of some religious, social, and political ideologies question the assumption that "material progress" (which is what technology fosters) is desirable.

Even the phrase *developing countries* has been controversial with some scholars because it might imply that all societies are moving toward the type of social system characterized by the affluent or "developed" societies. Many now use the term *Global South* because poor countries are disproportionately south of the 20th latitude north whereas affluent nations are typically north of that line. Note that the term is a metaphor for all poor countries, north or south. The term is meant to avoid an assumption of inevitable evolution toward Western cultures.

Contemporary evolutionary theories are multilinear, acknowledging variations in the way change takes place. The rapid spread of ideas and technologies means that societies today may move quickly from simple to complex, creating modern states. Consider the mass of contradictions of the Middle East today. Due to the world demand for their oil, several countries in this region now have the highest per capita incomes in the world. In 2010, Qatar had an income of $179,000 for each individual; the United Arab Emirates $49,600; Kuwait $48,900; and Bahrain $40,300 (Central Intelligence Agency 2011). The urban elite in these countries have access to modern conveniences such as the latest technology, jets, and cell phones. Yet other Middle Eastern people still live traditional lives as nomads or herders in small villages or earn their living from the desert. Not all segments of society change at the same rate, making categorization of some societies difficult. A single nation may be both postindustrial and pastoral.

The efficiency and speed of modes of transportation for goods and people vary around the world, often reflecting the level of development of the region or country. "Premodern" modes of transportation leave less pollution and sometimes move more easily through congested streets than trucks. Sometimes, technological progress has a high cost, and resistance to that "progress" may make sense.

Functionalist Theories. Functional theorists assume that societies are basically stable systems held together by the shared norms and values of their members. The interdependent parts work together to make the society function smoothly. A change in one part of the society affects all the other parts, each changing in turn until the system resumes a state of equilibrium. Change can come from external or internal sources, from stresses in contact with other societies, or from strains within.

Slow, nondisruptive change occurs as societies become more complex, but any change may be seen as threatening to the equilibrium of a system. Rapid change is seen as especially dysfunctional or disruptive. Because sudden, disruptive change is difficult to explain using functional theory and because any major change is viewed with some suspicion, some sociologists have turned to conflict theories to help explain change, especially rapid or violent changes.

Conflict Theories. Conflict theorists assume that societies are dynamic and that change and conflict are inevitable. According to Karl Marx, socioeconomic class conflict is the major source of tension leading to change in any society. Karl Marx and Friedrich Engels ([1848] 1969) argued that the antagonistic relationship they saw developing between the workers (proletariat) and the owners of the production systems (bourgeoisie) in nineteenth-century England would lead to social revolution. From this, they thought a new world order would emerge in which the workers themselves would own the means of production. Thus, conflict between the owners and the workers would be the central factor driving social change.

Other conflict theorists study variables such as gender, religion, politics, and ethnic and interest group problems in their analyses, feeling that these factors can also be the grounds for oppression and "we" versus "they" differences (Dahrendorf 1959). Some see conflict as useful for society because it forces societies to adapt to new conditions and leads to healthy change (Coser 1956). Conflict over slavery or over gender inequality is an example of a problem that causes stresses and strains, eventually resulting in improved society. The current conflict over health care in the United States may also eventually lead to a better system.

World Systems Theory of Global Change. World systems theorists focus on the historical development of the whole world and how that development has influenced individual countries today. Capitalist economies first appeared about 1500. Since then, except for a few isolated tribal groupings, almost all societies have been at least indirectly influenced by dominant capitalist world economic and political systems (Wallerstein 1974).

This theory divided the world system into three main parts: the core, semiperipheral, and peripheral areas (see Figure 14.2). The core areas are economically and politically powerful. Core countries include most European states,

Australia and New Zealand, Japan, Canada, the United States, and a few others (Wallerstein 1974). Historically, they have controlled global decision making, received the largest share of the profits from the world economic system, and dominated peripheral areas politically, economically, and culturally by controlling the flow of technology and capital into those countries. Peripheral countries, many of which are in Africa and Asia, provide cheap labor and raw materials for the core countries' needs. The semiperipheral countries are in an intermediate position, trading with both the core and the peripheral countries. The Baltic regions of Eastern Europe, Brazil, Argentina, South Korea, South Africa, India, the Philippines, Iran, and Mexico are among the semiperipheral areas. Because most semiperipheral countries are industrializing, they serve as areas to which core-country businesses and multinational corporations can move for continued growth, often in partnerships, as semiperipheral states aspire to join the core countries. The core and semiperipheral countries process raw materials, often from peripheral countries, and may sell the final products back to the peripheral countries. The semiperipheral countries and the peripheral countries need the trade and the resources of the core countries, but they are also at a severe disadvantage in competition and are exploited by those at the core, resulting in an uneasy relationship.

These basic relationships between countries have endured since the 1700s. However, South Korea, Thailand, Taiwan, India, and China may challenge the existing relationships with their rapidly expanding economies (Friedman 2005; Kristoff and WuDunn 2000). Further, since the 1960s, production processes have modified relationships between regions of the world. Changes in technology and in international global institutions have allowed corporations to break their production processes into smaller segments. These segments are then scattered over the world to take advantage of the lower manufacturing costs in the periphery. This process creates commodity chains—worldwide networks of labor resources and production processes that create a product. Each piece of the chain can be located in a core,

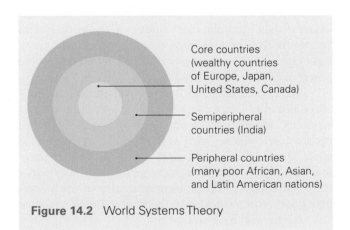

Figure 14.2 World Systems Theory

Core countries (wealthy countries of Europe, Japan, United States, Canada)

Semiperipheral countries (India)

Peripheral countries (many poor African, Asian, and Latin American nations)

In this typical situation, these South African miners are all black and work for low wages, whereas the supervisors and managers are white. These gold miners from poor peripheral countries are part of a multinational corporation and the larger world economic system, with stockholders from the core capitalist world system.

semiperipheral, or peripheral country. Because manufacturing processes are often performed in semiperipheral countries, their share of the world's manufacturing and trade production has risen sharply. By contrast, the distribution of profits from multinational corporations still benefits the core countries.

In one sense, world systems theory is a conflict theory that is global in nature, with core countries exploiting the poor countries. As we might expect from conflict theory, some groups of noncore countries have increased their collective power by forming alliances such as OPEC (Organization of the Petroleum Exporting Countries), OAS (Organization of African States), and SEATO (Southeast Asia Treaty Organization). These alliances present challenges to the historically core countries of the world system because of their combined economic and political power. For example, the price we pay at the gas pump reflects the power of OPEC to set prices.

When we understand international treaties and alliances as part of larger issues of conflict over resources and economic self-interests, the animosity by noncore countries toward core countries such as the United States begins to make sense. Likewise, the mistrust of the United States toward countries that seem to be getting U.S. jobs is not entirely unfounded. The problem is an extraordinarily complex system that always leaves the most vulnerable more at risk and the wealthiest even richer.

Sometimes behavior that results in change is unplanned and even spontaneous as described in the following section.

Thinking Sociologically

Where is your clothing made? Did a multinational corporation have it assembled in the Global South? Who benefits from companies buying cheap labor from the Global South: You? The workers? Governments? The companies that manufacture the products? Who, if anyone, is hurt?

Collective Behavior: Micro-Level Behavior and Change

The following incident occurred in Chicago during riots in the 1960s and provides an example of collective behavior, a form of unplanned action that can be understood at a micro level.

On Saturday, July 15, [Director of Police Dominick] Spina received a report of snipers in a housing project. When he arrived he saw approximately 100 National Guardsmen and police officers crouching behind vehicles, hiding in corners and lying on the ground around the edge of the courtyard. Since everything appeared quiet and it was broad daylight, Spina walked directly down the middle of the street. Nothing happened. As he came to the last building of the complex, he heard a shot. All around him the troopers jumped, believing themselves to be under sniper fire. A moment later a young guardsman ran from behind a building.

The Director of Police went over and asked him if he had fired the shot. The soldier said yes, he had fired to scare a man away from a window; his orders were to keep everyone away from windows. Spina said: "Do you know what you just did? You have now created a state of hysteria. Every guardsman up and down this street and every state policeman and every city policeman that is present thinks that somebody just fired a shot and that it is probably a sniper."

A short time later more "gunshots" were heard. Investigating, Spina came upon a Puerto Rican sitting on a wall. In reply to a question as to whether he knew "where the firing is coming from?" the man said: "That's no firing. That's fireworks. If you look up to the fourth floor, you will see the people who are throwing down these cherry bombs."

By this time four truckloads of National Guardsmen had arrived and troopers and policemen were again crouched everywhere looking for a sniper. The Director of Police remained at the scene for three hours, and the only shot fired was the one by the guardsman. However, at six o'clock that evening two columns of National Guardsmen and state troopers were directing mass fire at the Hayes Housing Project in response to what they believed were snipers. (Report of the National Advisory Commission on Civil Disorders 1968:3–4)

Collective behavior refers to unplanned, spontaneous, unstructured, disorganized actions that often violate norms. It arises when people are trying to cope with stressful situations and unclear or uncertain conditions (Goode 1992; Smelser 1963, 1988). Collective behavior falls into two main types: crowd behavior and mass behavior. It often starts as a response to an event or a stimulus as in the perceived threat from snipers in the above example. It could begin with a shooting or beating, a speech, a sports event, or a rumor. The key is that as individuals try to make sense of the situations they are in and respond based on their perceptions, collective social actions emerge.

Crowd behaviors—mobs, panics, riots, and demonstrations—are all forms of collective behavior in which a crowd acts, at least temporarily, as a unified group (LeBon [1895] 1960). Crowds are often made up of individuals who see themselves as supporting a just cause. Because the protesters are in such a large group, they may not feel bound by the normal social controls—either internal (normal moral standards) or external (fear of police sanctions).

Mass behavior occurs when individual people communicate or respond in a similar manner to ambiguous or uncertain situations, often based on common information from the news or on the Internet. Examples include public opinion, rumors, fads, and fashions. Unlike social movements, these forms of collective behavior generally lack a hierarchy of authority and clear leadership, a division of labor, and a sense of group action.

Theories of Collective Behavior

Social scientists studying group and crowd dynamics find that most members of crowds are respectable, law-abiding citizens, but faced with specific situations, they act out (Berk 1974; Turner and Killian 1993). Several explanations of individual involvement dominate the modern collective behavior literature.

Based on principles of rational choice theory, the *minimax strategy* (Berk 1974) suggests that individuals try to minimize their losses or costs and maximize their benefits. People are more likely to engage in behavior if they feel the rewards outweigh the costs. Individuals may become involved in a riot if they feel the outcome—drawing attention to their

Suspicion about fraudulent vote counts that reelected an unpopular president of Iran—Mahmoud Ahmadinejad—resulted in extensive rioting, as well as more peaceful protests.

plight, the possibility of improving conditions, solidarity with neighbors and friends, looting goods—will be more rewarding than the status quo or the possible negative sanctions.

Emergent norm theory (Turner and Killian 1993) points out that individuals in crowds have different emotions and attitudes that guide their decisions and behaviors than if they act alone. The theory addresses the unusual situations and breakdown of norms in which most collective behavior takes place. Unusual situations may call for the development of new norms and even new definitions of what is acceptable behavior. The implication of this theory is that in ambiguous situations, people look to others for clues about what is happening or what is acceptable, and norms emerge in ambiguous contexts that may be considered inappropriate in other contexts.

Value-added theory describes conditions for crowd behavior and social movements. Key elements are necessary for collective behavior, with each new variable adding to the total situation until conditions are sufficient for individuals to begin to act in common. At this point collective behavior emerges (Smelser 1963). These are the six factors Smelser (1963) identified that can result in collective behavior:

1. *Structural conduciveness:* Existing problems create a climate that is ripe for change (e.g., tensions between religious and ethnic groups in Iraq).

2. *Structural strain:* The social structure is not meeting the needs and expectations of the citizens, which creates widespread dissatisfaction with the status quo—the

On the southern outskirts of Basra in Iraq, British soldiers monitor a checkpoint leading into the city, checking people for weapons. A young Iraqi girl experiences the tense and hostile realities of war. This kind of military presence is often scary for residents and is very dangerous work for soldiers. This is an example of the precariousness of maintaining social control in volatile situations.

the mobilization, a social movement or other crowd behavior (e.g., a riot or mob) is likely to develop. (Some analysts have argued that civil war has erupted in Iraq because of a lack of a trained police force.)

When all six factors are present, some form of collective behavior will emerge. Those trying to control crowds that are volatile must intervene in one or more of these six conditions (Kendall 2004; Smelser 1963).

Thinking Sociologically

Think of an example of crowd behavior or a social movement, preferably one in which you have been involved. Try to identify each of the six factors from Smelser's (1963) theory as they operate in your example.

Types of Collective Behavior

Collective behavior ranges from spontaneous violent mobs to temporary fads and fashions. Figure 14.3 shows the range of actions.

Mobs are emotional crowds that engage in violence against a specific target. Examples include lynchings, killings, and hate crimes. Near the end of the U.S. Civil War, self-appointed vigilante groups roamed the countryside in the South looking for army deserters, torturing and killing both those who harbored deserters and the deserters themselves. There were no courts and no laws, just "justice" in the eyes of the vigilantes. Members of these groups constituted mobs. The film *Cold Mountain* (Frazier 1997) depicts these scenes vividly. Unless curbed, mobs often damage or destroy their target.

Riots—an outbreak of illegal violence committed by individuals expressing frustration or anger against people, property, or both—begin when certain conditions occur. Often, a sense of frustration or deprivation sets the stage for a riot—hunger, poverty, poor housing, lack of jobs, discrimination, poor education, or an unresponsive or unfair judicial system. If the conditions for collective behavior are present, many types

current arrangements (the Iraqi government is unable to control violence and provide basic services).

3. *Spread of a generalized belief:* Common beliefs about the cause, effect, and solution of the problem evolve, develop, and spread (U.S. troops begin to leave; militias are killing members of other groups).

4. *Precipitating factor:* A dramatic event or incident occurs to incite people to action (groups of men from different religious groups are kidnapped, bound, and shot).

5. *Mobilization for action:* Leaders emerge and set out a path of action, or an emergent norm develops that stimulates common action (citizens gather to protest the killings and angry spokespersons enrage the crowd).

6. *Social controls are weak:* If police, military, or strong political or religious leaders are unable to counter

Spontaneous and often violent					Less spontaneous and seldom violent	
Crowd behavior				**Mass behavior**		
Mob	Riot	Panic		Rumor	Fad	Fashion

Figure 14.3   Types of Collective Behavior

of incidents can be the precipitating factor setting off a riot. For example, in late 2010 as a response to the rumor that Nepalese United Nations soldiers brought the cholera epidemic from Nepal, frustrated Haitian citizens rioted, wounding a number of UN peacekeeping troops. The distinction between riots and mobs is illustrated in Figure 14.4.

Panic occurs when a large number of individuals become fearful or try to flee threatening situations that are beyond their control, sometimes putting their lives in danger. Panic can occur in a crowd situation, such as a restaurant or theater in which someone yells "fire," or it can occur following rumors or information spread by the media. Panic started by rumors set off the run on the stock market in October 1929. A large number of actions by individuals caused the stock market crash in the United States and repercussions around the world. In 2008, the collapse of global investment banking and securities trader Bear, Stearns, and Co. resulted in turmoil in the financial markets. Only with radical intervention by the federal government was the immediate panic abated. Panics can result in collapse of an organization, destruction, or even death as a result of the group action.

Rumors are a form of mass behavior in which unsupported or unproven reports about a problem, an issue, or a concern circulate widely throughout the public. Rumors may spread only in a local area, but with electronic means available, rumors are spreading more widely and rapidly. Without authoritative information, ambiguous situations can produce faulty information on which decisions are made and actions are based. *Urban legends*, one example of widely spread but unverified messages, are unsubstantiated stories that sound plausible and become widely circulated. People telling them usually believe them (Mikkelson and Mikkelson 2011). Go to www.snopes.com/college/college.asp for some entertaining urban legends about professors, pranks, exam scams, embarrassments, and other college folly.

Fads are temporary items or activities that spread rapidly and are copied enthusiastically by large numbers of people. Body modification, especially tattooing, appeals mostly to young people of all social classes. Tattoo artists emblazon IDs, secret society and organization emblems, fraternity symbols, and decorations to order on all parts of the customers' bodies. Body modification has taken place for centuries, but it goes through fads (University of Pennsylvania 2010). Sometimes, fads become institutionalized—that is, they gain a permanent place in the culture. Other fads die out, replaced by the next hot item.

Fashions refer to social patterns favored by a large number of people for a limited period of time. Here today, gone tomorrow. Examples include clothing styles, music genres, color schemes in home décor, types of automobiles, and architectural designs. Fashions typically last longer than

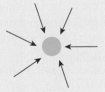

Riots involve dispersed actions expressing frustration (e.g., urban riots over poor conditions).

Mobs involve a group collectively focusing their action on a single individual or location (e.g., a lynch mob).

Figure 14.4 The Difference Between Riots and Mobs

Fashions are established largely at fashion shows, where designers introduce new clothing styles. Fashions cannot occur unless there is a very high level of affluence, where people can afford to throw away perfectly good clothing for something more stylish. For many people in the Global South, it is a gift just to have clean, warm clothing, and the very existence of such displays of consumerism is amazing, appealing, and sometimes appalling.

fads but sometimes survive only a season, as can be seen in the clothing industry. Music styles such as "hardcore techno," "acid," "alternative hip-hop," and "UK 2-step garage" that were popular among some groups as this book was being written may be passé by the time you read this, replaced by new styles resulting from change in mass behavior.

Each of these forms of collective behavior involves micro-level individual actions that cumulatively become collective responses to certain circumstances. However, ripples are felt in other levels of the social world. Insofar as these various types of collective activity upset the standard routines of the society and the accepted norms, they can unsettle the entire social system and cause lasting change.

The separation of each of the forms of change into levels is somewhat artificial, of course, for individuals are also acting in organizations and in national social movements. However, when we move to meso- and macro-level analyses, the established structures and processes of the society become increasingly important. Much of the change at these levels is planned change.

Planned Change in Organizations: Meso-Level Change

The board of trustees of a small liberal arts college has witnessed recent drops in student enrollments that could cause the college to go out of business, but the college has a long tradition of fine education and devoted alumni. How does the college continue to serve future students and current alumni? The problem is how to plan change to keep the college solvent.

A company manufactures silicon chips for computers. Recently, the market has been flooded with inexpensive chips, primarily from Asia, where they are made more cheaply than this North American firm can possibly make them. Does the company succumb to the competition, figure out ways to meet it, or diversify its products? What steps should be taken to facilitate the change? Many companies in Silicon Valley, California, face exactly this challenge.

A Native American nation within the United States faces unemployment among its people due in large measure to discrimination by Anglos in the local community. Should the elders focus their energies and resources on electing sympathetic politicians, boycotting racist businesses, filing lawsuits, becoming entrepreneurs as a nation so they can hire their own people, or beginning a local radio station so they will have a communication network for a social movement? What is the best strategy to help this proud nation recover from centuries of disadvantage?

All these are real problems faced by real organizations. Anywhere we turn, organizations face questions involving change, questions that arise because of internal strains and external stresses. How organizational leaders and applied sociologists deal with change will determine the survival and well-being of the organizations.

How Organizations Plan for Change

When working for an organization, you will engage in the process of planning for change. Some organizations spend time and money writing long-range strategic plans and doing self-studies to determine areas for ongoing change. Sometimes change is desired, and sometimes it is forced on the organization by stresses from society, more powerful organizations, or individuals (Kanter 1983, 2001a, 2001b; Olsen 1968). Moreover, a problem solved in one area can create unanticipated problems someplace else.

Planned change such as strategic planning is the dream of every organizational leader. It involves deliberate, structured attempts, guided by stated goals, to alter the status quo of the social unit (Bennis, Benne, and Chin 1985; Ferhansyed 2008). There are several important considerations as we think about planned change: How can we identify what needs to be changed? How can we plan or manage the change process successfully? What kind of systems adapt well to change? Here, we briefly touch on the topic and outline three approaches advocated by experts to plan change. Keep in mind the levels of analysis as you read about change models.

Models for Planning Organizational Change

Change models fall into two main categories: closed system models, which deal with the internal dynamics of the organization, and open system models (e.g., our social world model), which consider the organization and its environment. Let us sample a couple of these models.

Closed system models, often called classical or mechanistic models, focus on the internal dynamics of the organization. The goal of change using closed models is to move the organization closer to the ideal of bureaucratic efficiency and effectiveness. An example is time and motion studies, which analyze how much time it takes a worker to do a certain task and how it can be accomplished more efficiently. Each step in the McDonald's process of getting a hamburger to you, the customer, has been planned and timed for greatest efficiency (Ritzer 2010). In some closed system models, change is legislated from the top executives and filters down to workers.

Both the human relations approach and the organizational development movement state that participants in the organization should be involved in decision making leading to change. The leadership is more democratic and supportive of workers, and the atmosphere is transparent—open, honest, and accountable to workers and investors. This model emphasizes that change comes about through adjusting workers' values, beliefs, and attitudes regarding new demands on the organization. Many variations on this theme have evolved, with current efforts including team building and change of the organizational culture to improve worker morale. Closed system models tend to focus on group change that occurs from within the organization.

Open system models combine both internal processes and the external environment. The latter provides the organization with inputs (workers and raw materials) and feedback (acceptability of the product or result). In turn, the organization has outputs (products) that affect the larger society. There are several implications of this model: (a) Change is an ever-present and ongoing process, (b) all parts of the organization and its immediate environment are linked, and (c) change in one part has an effect on other parts. The model in Figure 14.5 illustrates the open system.

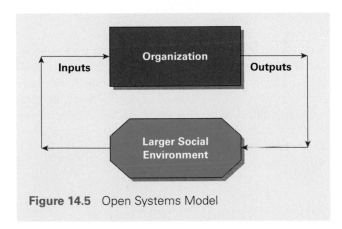

Figure 14.5   Open Systems Model

Thinking Sociologically

Using the model in Figure 14.5, fill in the parts as they relate to your educational institution. For example, inputs might include students and federal student aid grants.

The Process of Planned Change

A huge issue causing conflict in the Global South is the availability of clean drinking water. Nongovernmental organizations (NGOs) plan ways to improve the lives of individuals in many parts of the world. For example, in parts of Africa, women must spend as much as six hours a day carrying water to the home. Because daughters are needed to care for younger siblings while the mother is away, many girls are unable to attend school. This, in turn, has implications for the continuation of poverty. Further, if estimates are correct that 443 million school days are lost each year due to water-related diseases, the crisis of continuing poverty is aggravated (WaterAid 2008).

Clean water is essential for life and for health, but one out of six people—more than a billion humans on this planet—do not have access to it. This affects the quality of their lives and results in more than two million deaths per year. With global climate change, the glaciers on top of mountains like Mount Kenya are melting. Although that mountain peak has been snow-covered for more than 10,000 years, the glaciers are expected to be completely gone in perhaps 20 years (Cousteau 2008). When the mountaintop snow disappears, the water supply for hundreds of thousands of people and animals will disappear (Barnett et al. 2005; Cyranoski 2005; Struck 2007). An example of one British nongovernmental organization that is bringing about change in this area is WaterAid. It has grown to become an international NGO that focuses entirely on water and sanitation issues, including hygiene. Communities in poor countries throughout the world are assisted in developing the most appropriate technologies for clean water, given the geographical features and resources of the area.

This example shows one of the many types of organizations that both change within themselves to meet new conditions and bring about change in the world, adding to our understanding of change. The process of planned change is like a puzzle with a number of pieces that differ for each organization but must fit together for the smooth operation of the organization. The goal of most organizations is to maintain balance and avoid threats or conflict. Slow, planned change is generally perceived as the desirable way to bring about change. Unplanned change can be disruptive to the system.

At the societal and global macro levels, change is often stimulated by individuals and events outside the chambers of power, and there is much less control over how the change evolves. We turn next to an exploration of change at the macro level.

Social Movements: Macro-Level Change

Beijing, China, hosted the prestigious summer 2008 Olympics, athletic games that promote peace and goodwill around the globe. The Olympic torch traversed the globe to prepare for and celebrate the event. However, its travels and the opening of the games were far from peaceful and

Protesters hold signs outside the White House in Washington, DC, in 2008 during a demonstration organized by Amnesty International. The group was calling on China to respect human rights, using the Olympic Games in Beijing as an occasion to pressure China by affecting its international reputation.

full of goodwill. Protests about Chinese human rights and autonomy for Tibet (now a part of China) dogged the torch, and major political leaders refused to attend the opening ceremonies. A social movement was spurred by an uprising in Tibet, a region that was an independent country until the 1950s. The issues—human rights violations by China and autonomy of Tibet—created protests and demonstrations around the world. Will this social movement bring about change in the status of Tibetans or in human rights in China? Worldwide attention may aid the cause of those in Tibet seeking improved life conditions, or it may not. However, those people who are making an issue of the games in China are hoping to change more serious life and death policies for people. The point is that a social movement tries to bring about change. Let us explore several questions: What is a social movement, what brings it about, and what might be the results?

What Is a Social Movement?

From human rights and women's rights to animal rights and environmental protection, individuals seek ways to express their concerns and frustrations. **Social movements** are consciously organized attempts outside of established institutions to enhance or resist change through group action. Movements focus on a common interest of members, such as abortion policy. They have an organization, a leader, and one or more goals that aim to correct some perceived wrongs existing in the society or even around the globe.

Social movements are most often found in industrial or postindustrial societies, although they can occur anyplace groups of people have a concern or frustration. Generally such movements are made up of diverse groups, each advocating for its own goals and interests.

Movements are usually begun by individuals outside the power structure who might not otherwise have an opportunity to express their opinions (Greenberg 1999). The problems leading to social movements often result from the way resources—human rights, jobs, income, housing, money for education and health care, and power—are distributed. In turn, countermovements—social movements against the goals of the original movement—may develop, representing other opinions (McCarthy and Zald 1977).

Many individuals join social movements to change the world or their part of the world and affect the direction of history. In fact, social movements have been successful in doing just that. Consider movements around the world that have protected lands, forests, rivers, and oceans, seeking environmental protection for the people whose survival depends on those natural resources. For example, the Chipko Movement in a number of areas of India has been fighting the logging of forests by commercial industries. Villagers who depend on the forests, mostly women, use Gandhi-style nonviolent methods to oppose the deforestation. These women have set an example for environmentalists in many parts of the world who wish to save trees (Scribd 2011). Another example is a movement by local peasants in Bihar, India, that fought efforts by government officials to control their fishing rights in the Ganges River. Those controls would have limited their livelihoods, but the citizens stood up against the power of the state.

Types of Social Movements

Stonewall is a gay, lesbian, and bisexual rights movement that began when patrons of a gay bar fought back against a police raid in New York in 1969. After that incident, the concern about gay rights erupted from a small number of activists into a widespread movement for rights and acceptance. Stonewall now has gone global, with chapters in other countries and continents. Proactive social movements, such as the gay rights movement, promote change. The 2010 lifting of the ban on openly gay and lesbian soldiers in the military is one example of a recent victory for the gay rights movement. Reactive social movements resist change. James Dobson's Focus on the Family has organized lobbying efforts and rallies against the acceptance of homosexuality or giving gays equal status to heterosexual married relationships, although this is not the organization's only concern. The reactive movement had success in the 1990s with the Defense of Marriage Act that allowed states to honor same-sex marriages that were contracted in another state. Whether proactive or reactive, there are five main types of social movements:

Expressive movements take place in groups, but they focus on changing individuals and saving people from corrupt lifestyles. Many expressive movements are religious, such as the born-again Christian movements, Zen Buddhism, Scientology, the Christian Science Church, and Transcendental Meditation. Expressive movements also include secular psychotherapy movements and self-help or self-actualization groups.

Social reform movements seek to change some aspect of society. Movement members generally are concerned about specific social issues but support the society as a whole. These movements focus on a major issue such as environmental protection, women's rights, same-sex marriage, "just and fair" globalization, reducing the national debt, or national abortion policy. Typically, these movements use nonviolent, legislative means or appeals to the courts to accomplish their goals, although splinter groups may resort to more aggressive methods such as marches or threats.

The Sojourners Community was started in the early 1970s by highly committed theology students at Trinity Evangelical Divinity School in Deerfield, Illinois. Their group was theologically evangelical and conservative Christian, but wanted their faith to engage the social crisis that surrounded them, including civil rights, poverty, and the Vietnam War. They began a newspaper that eventually became known as *Sojourners* magazine. They advocate for a consistent pro-life policy (antiabortion, antiwar, and anti–death penalty). Their widely known and charismatic leader, Jim Wallis, writes and speaks around the country, arguing that on matters of faith and social policy, "the right gets it wrong and the left doesn't get it" (Wallis 2005). They have been a voice from the evangelical community, but they are also highly critical of many on the political "left." They advocate for policies they believe stem from the compassion and love of the Christian message, and they are also unapologetic about their traditional stance on the core of the faith.

Revolutionary movements attempt to transform society, to bring about total change in a society by overthrowing existing power structures and replacing them with new ones. These movements often resort to violent means to achieve their goals, as has been the case with many revolutions throughout history. When we read in the paper that there has been a coup, we are learning about a revolutionary movement that has ousted the government in power. Recently there was an attempted coup d'état in Ecuador that killed four people, but did not succeed in overthrowing the government in power (Carroll 2010). Although it was not violent, Nelson Mandela's African National Congress succeeded in taking power in South Africa in 1994, rewriting the constitution, and purging a racially segregated apartheid social system that had oppressed four fifths of the South African population.

Many environmental and peace groups use nonviolent protest, legislative means, and appeals to the public or to the courts to accomplish the desired goals. This is a silent protest on the beach in Santa Monica, California, where protesters set up crosses to represent all of the deaths of U.S. soldiers in Iraq.

Resistance or regressive movements try to protect an existing system, protect a part of that system, or return to what a system had been by overthrowing current laws and practices. Members of such movements see societal change as a threat to values or practices and wish to maintain the status quo or return to a former status by reversing the change process (Inglehart and Baker 2001). The movement to pass a referendum in California to ban gay marriage is one of the association's recent efforts. The Taliban religious movement in Afghanistan was a regressive movement against modernization, especially against the Western pattern of giving freedom and autonomy to women. The movement was successful in gaining power in the 1990s, imposing a harsh brand of Islamic law in the sections of Afghanistan under its control. The Taliban insists that its version of Islam is pure in that it follows a literal understanding of the Muslim holy book. This means that someone believed to have committed adultery should be stoned to death; the hands or arms of thieves should be amputated; and women who deviate from their interpretation of Muslim law should be mutilated, publically beaten, and sometimes executed (Antonowicz 2002).

Global transnational movements focus on large-scale, often global issues and take place across societies as international organizations seek change in the status of women, child labor, rights of indigenous peoples, environmental degradation, global warming, disease pandemics, and other issues that affect the global community. Movements to improve the working conditions and environmental impact of multinational corporations are examples (The Durban Accord 2003; Flavin 2001; Mackie 2003). Another example is Free the Slaves, a global antislavery organization started by sociologist Kevin Bales. This organization has researched and written about the plight of millions of indentured people in the world (Bales 2007; Bales, Trodd, and Williamson 2009; Bales and Soodalter 2010). Figure 14.6 summarizes the types of movements and focus of each, from the micro to the macro level.

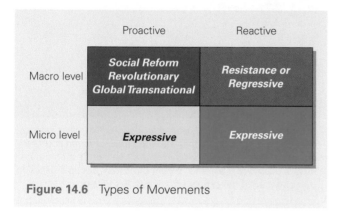

Figure 14.6   Types of Movements

Thinking Sociologically

Consider a social movement with which you are familiar. What type of movement is it, and what was or is it trying to accomplish?

Globalization and Social Movements

Social movements are about people trying to improve their situations within their societies. They are intriguing because they provide compelling evidence that humans make choices and are capable of countering macro- and meso-level forces. As Eitzen and Zinn (2012:269) put it:

> Powerful social structures constrain what we do, but they can never control entirely. Human beings are not passive actors. Individuals acting alone, or with others, can shape, resist, challenge, and sometimes change the social structures that impinge on them. These actions constitute human agency.

Since World War II, power in the global system has been dominated by a group of industrial giants recently calling themselves the Group of 8 (or the "G8"). These nations control world markets and regulate economic and trading

policies (International Encyclopedia of the Social Sciences 2008). Included among these elites are the dominant three (Japan representing the East, Germany representing Central Europe, and the United States representing the American continents) and five other important but less dominating powers (Canada, France, Great Britain, Italy, and now Russia). The G8 are the core countries that have the most power in the World Trade Organization, the World Bank, the International Monetary Fund (IMF), and other regulatory agencies that preside over the global economy. These agencies have often required that poor countries adhere to their demands or lose the right to loans and other support. These policies imposed from above have sometimes been disastrous for poor countries, causing situations in which a debt burden is created that can never be paid off due to World Bank and IMF policies. The IMF (2011) points out that the loans are the only way poor countries can meet the UN Millennium Development Goals, but critics argue that the loans make poor countries dependent on wealthy countries and that the loans should be forgiven. They argue that if these were individuals rather than countries, we would call them indentured servants or slaves (Brecher et al. 2012; Weller and Hersh 2006). This creates nations where hopelessness would seem to reign supreme, yet social movements are arising in precisely these places and are often joining forces across national boundaries (Stein 2006; Weidenbaum 2006). Even some groups within the G8 nations—labor unions, college student groups, and religious bodies concerned about social justice—are joining the movements.

Some of the poorest nations have now formed their own organization to counter what was then called the G7 (now the G8). Calling themselves the G77, these countries are now uniting, rather like a labor union seeking collective unity among workers, in attempts to gain some power and determine their own destinies by challenging what they experience as the tyranny of the G8 (Brecher et al. 2012; Hayden 2006). Map 12.1 on page 394 displays the location of G8 and G77 countries.

In the competition to find cheaper labor for higher profits, there is a "race to the bottom," as communities must lower standards or else lose jobs to some other part of the world that is even more impoverished. Globalization as it now exists—with corporate profits as the ruling principle of most decision making—has lessened environmental standards, consumer protection, national sovereignty and local control of decisions, and safety protections for workers. Yet when jobs move elsewhere, the people who had come to depend on those jobs are devastated and often thrust into poverty and homelessness, and the country loses tax revenue.

Although workers risk losing jobs by participating in protests for higher wages and better working conditions, one result of globalization has been a rise in countermovements. "Globalization from below" refers to the efforts by common people in small groups and protest movements to fight back. Rather than globalization being controlled solely by the pursuit of profits, these countermovements seek to protect workers, to defend the environment, and to combat the bone-crunching poverty that plagues so much of the Global South. The argument goes like this:

> It is the activity of people—going to work, paying taxes, buying products, obeying government officials, staying off private property—that continually re-creates the power of the powerful. . . . [The system, for all its power and resources, is dependent on common people to do the basic jobs that keep the society running.] This dependency gives people a potential power over the society—but one that can be realized only if they are prepared to reverse their acquiescence. . . . Social movements can be understood as the collective withdrawal of consent to established institutions. (Brecher et al. 2012:279)

The movement against globalization can be understood as taking away consent for such globalization. There are thousands of small resistance actions to the oppressive policies of G8 transnational corporations. Consider the following examples:

> Under heavy pressure from the World Bank, the Bolivian government sold off the public water system of its third-largest city, Cochabamba, to a subsidiary of the San Francisco–based Bechtel Corporation, which promptly doubled the price of water for people's homes. Early in 2000, the people of Cochabamba rebelled, shutting down the city with general strikes and blockades. The government declared a state of siege, and a young protester was shot and killed. Word spread all over the world from the remote Bolivian highlands via the Internet. Hundreds of e-mail messages poured into Bechtel from all over the world, demanding that it leave Cochabamba. In the midst of local and global protests, the Bolivian government, which had said that Bechtel must not leave, suddenly reversed itself and signed an accord accepting every demand of the protestors. (Brecher et al. 2012:284)

Concerned groups in Europe, Japan, and the United States found that global corporations were increasingly fostering growth of sweatshops in the Global South. They pressured companies like the Gap and Nike to establish acceptable labor and human rights conditions in their factories around the world. Their efforts gradually grew into an antisweatshop movement with strong labor and religious support and tens of thousands of active participants. In the United States, college students took up the antisweatshop cause on hundreds of campuses, ultimately holding sit-ins on many campuses to force their colleges to ban the use of college logos on products not produced under acceptable labor conditions (Brecher et al. 2012:275). Many concerned citizens in the Global North now buy Fair Trade Certified goods such as coffee, cacao, and fruit.

Individuals in G8 countries have been extremely generous in donating to international organizations helping victims of disasters and in joining in solidarity with those in dire straits in the poor Global South. However, governments have often been unsympathetic. In 2005, the United States gave .22% of its gross national product to the United Nations; of the 22 major donor states, only Portugal was lower in its donation (Millennium Project 2006). The UN Millennium Goals that address poverty, famine, illiteracy, and disease around the world call for all major donor countries to give .7% of their gross national product to UN programs to meet the need. The United States gave $6,347,415 in 2009, but this was far from the .7% needed according to the United Nations (WhiteHouse .gov 2010). The Obama administration made a campaign promise to double foreign assistance to $50 billion by 2012 and has requested increases in the U.S. budget. However, with the current economic situation these increases may not come to pass (Schaefer and Kim 2010).

In all, there has been a decrease of nearly 90% from the Kennedy administration era, roughly 50 years ago. However, many activists believe that actions by individuals and small groups, globalization from below, can have a real impact on global problems. Consider how effective you think actions taken by individuals and local groups can be by thinking about the issues in the "Engaging Sociology" on the next page.

In summary, some social change is planned by organizations, some is initiated by groups that are outside the organizational structure (social movements), and some is unplanned and spontaneous (collective behavior). The most important point, however, is that actions by individuals can affect the larger social world, sometimes even having global ramifications. Likewise, national and international changes and social movements influence the lives of individuals.

Technology, Environment, and Change

At the edge of the town of Bhopal, India, looms a subsidiary plant of the America-based Union Carbide Corporation. The plant provides work for many of the town's inhabitants. However, on December 3, 1984, things did not go as normal. A storage tank from the plant, filled with the toxic chemical liquid methyl isocyanate, overheated and turned to gas, which began to escape through a pressure-relief

Engaging Sociology

Micro to Macro: Change From the Bottom Up

The idea of globalization "from the bottom up" suggests that actions of lots of people at the micro or local level can have a significant impact on how things develop at the most macro level of the social world we inhabit. Think about that process and what forces can enhance or retard that kind of change.

1. Are you familiar with cases in which "globalization from below" has made a difference in local, national, or international events? If so, what are those?

2. Identify three structural challenges that might make it hard for people at the micro level to change the national and global forces that interfere with the quality of their lives.

3. Identify three reasons to be optimistic about why change from the bottom up can be successful.

4. Do a Google search of the Zapatistas or of their leader, Subcomandante Marcos. What are the pros and the cons of this movement? Do the Zapatistas have any chance of bringing change to the poor disenfranchised people of southern Mexico?

valve. The gas formed a cloud and drifted away from the plant. By the time the sirens were sounded, it was too late for many people. The deadly gas had done its devastating work: More than 3,000 were dead and thousands ill, many with permanent injuries from the effects of the gas. Put in perspective, the number of casualties was about equal to the number in the terrorist attack on the Twin Towers of New York City on September 11, 2001.

This example illustrates change at multiple levels of analysis. We see a global multinational company (Union Carbide) in a society (India) that welcomed the jobs for its citizens. A community within that larger society benefited from the jobs until many residents were killed or disabled, leaving shattered families and devastated individual lives. The accident also spawned a number of forms of collective behavior. The immediate aftermath of the accident at Bhopal included panic, as people tried to flee the deadly gas. Later it resulted in several social movements as activists demanded accountability and safety measures. The courts, after 25 years, are rendering decisions about the legal ramifications of the leak. The disaster also brought about planned change in the way Union Carbide does business and protects workers and citizens.

Technology refers to the practical application of tools, skills, and knowledge to meet human needs and extend human abilities. Technology and environment cannot be separated as we see at Bhopal. The raw products that fuel technology come from the environment, the wastes return to the environment, and technological mistakes affect the environment. This section discusses briefly the development and process of technology, the relationship between technology and environment, and the implications for change at each level of analysis.

Throughout human history, there have been major transition periods when changes in the material culture brought about revolutions in human social structures and cultures (Toffler and Toffler 1980). For example, the agricultural revolution resulted in the plow to till the soil, establishing new social arrangements and eventually resulting in food surpluses that allowed cities to flourish. The Industrial Revolution brought machines powered by steam and petrol, resulting in mass society, divisions of labor in manufacturing, and socialist and capitalist political-economic systems. Today, postindustrial technology, based on the microchip, is fueling the spread of information, communication, and transportation on a global level to explore space and analyze, store, and retrieve masses of information in seconds. However, each wave affects only a portion of the world, leaving other people and countries behind and creating divisions between the Global North and the Global South.

Sociologist William Ogburn ([1922] 1938, 1961, 1964) has argued that change is brought about through three processes: discovery, invention, and diffusion. Discovery is a new way of seeing reality. The material objects or ideas have been present, but they are seen in a new light when the need arises or conditions are conducive to the discovery. It is usually accomplished by an individual or a small group, a micro-level activity.

Invention refers to combining existing parts, materials, or ideas to form new ones. There was no light bulb or combustion engine lying in the forest waiting to be discovered. Human ingenuity was required to put together something that had not previously existed. Technological innovations often result from research institutes and the expansion of science, increasingly generated at the meso level of the social system.

Global protests often focus on actions of Western multinational corporations. Women demonstrators, including Bhopal gas victims in India, hold a "Wanted" poster of former Union Carbide Chairman Warren Anderson, arguing that he should be tried for crimes for the deaths of more than 3,000 people and the injuries of tens of thousands in Bhopal. The lawsuit was finally settled in 2010.

Diffusion is the spread of an invention or a discovery from one place to another. The spread of ideas such as capitalism, democracy, and religious beliefs has brought changes in human relationships around the world. Likewise, the spread of various types of music, film technology, telephone systems, and computer hardware and software across the globe has had important ramifications for global interconnectedness. Diffusion often involves expansion of ideas across the globe, but it also requires individuals to adopt ideas at the micro level.

Technology and Science

The question "How do we know what we know?" is often answered, "It's science." **Science** is the systematic process of producing human knowledge. Whether social, biological, or physical, science provides a systematic way to approach the world and its mysteries. It uses empirical research methods to discover facts and test theories. Technology applies scientific knowledge to solve problems. Early human technology was largely a result of trial and error, not scientific knowledge or principles. Humans did not understand why boats floated or fires burned. Since the Industrial Revolution, many inventors and capitalists have seen science and technology as routes to human betterment and happiness. Science has become a major social institution in industrial and postindustrial societies, providing the bases of information and knowledge for sophisticated technology.

Indeed, one of the major transformations in modern society is due to science becoming an institution. Prior to the eighteenth century, science was an avocation. People like Benjamin Franklin experimented in their backyards with whatever spare cash they had to satisfy their own curiosity.

Institutionalization means creation of the organized, patterned, and enduring sets of social structures that provide guidelines for behavior and help the society meet its needs. Like all other institutions, science in the contemporary world is both a structure and a social process. Modern science involves mobilizing financial resources and employing the most highly trained people (which in turn has required the development of educational institutions). Innovation resulting in change will be very slow until a society has institutionalized science—providing extensive training and paying some people simply to do

Scientists have developed technologies to ease the looming energy crisis and climate change. Solar panels at this restaurant in Portugal provide independence from other sources of energy.

Thomas Edison had more than a thousand patented inventions, including the light bulb, recorded sound, and movies, but perhaps his most influential invention was the research lab—in which people were paid to invent and to conduct research at Menlo Park in New Jersey. This was the seminal step in the institutionalization of science.

research. Specialization in science speeds up the rates of discovery. A researcher focuses in one area and gets much more in-depth understanding. Effective methods of communication across the globe mean that we do not need to wait six years for a research manuscript to cross the ocean and to be translated into another language. Competition in science means that researchers move quickly on their findings. Delaying findings may mean that the slowpoke does not get a permanent position, called tenure, at his or her university; promotion in the research laboratory; or awards for innovation.

We would not have automobiles, planes, missiles, space stations, computers, the Internet, and many of our modern conveniences without the institutionalization of science and without scientific application (technology). Science is big business, funded by industry and political leaders. University researchers and some government-funded science institutes engage in *basic research* designed to discover new knowledge, often on topics that receive funding. Industry and some governmental agencies such as the military and the Department of Agriculture employ scientists to do *applied research* and to discover practical uses for existing knowledge.

Scientific knowledge is usually cumulative, with each study adding to the existing body of research. However, radical new ideas can result in scientific revolutions (Kuhn 1970). Galileo's finding that the Earth revolves around the Sun and Darwin's theory of evolution are two examples of radical new ideas that changed history. More recently, cumulative scientific knowledge has resulted in energy-efficient engines that power cars and computer technology that has revolutionized communication.

Thinking Sociologically

Imagine what your life would have been like before computers, e-mail, and the Internet. What would be different? (Note that you are imagining the world from only 10 to 20 years ago.) Ask your parents or grandparents what this past world was like.

Technology and Change

The G8 countries have yearly meetings to regulate global economic policy and markets. The group's power enables those eight countries to dominate technology by controlling raw products such as oil. In the process, they profoundly influence which countries will be rich or poor. Although some politicians like to tell us they believe in a free market economy, uninhibited by governmental interference, they actually intervene regularly in the global market.

New technological developments can be a force for world integration but also for economic and political disintegration (Schaeffer 2003). For example, research and development (R&D) is a measure of countries' investment in the future. For many years the biggest investors have been the European Union, the United States, and Japan. However, between 2002 and 2007, China increased its R&D while the top research countries reduced their share. In fact, in 2002 Global North countries provided 83% of the research and development, but today the European Union, the United States, and China each provide 20% of the world's researchers, with Japan at 10% and Russia at 7% (UNESCO 2010).

The technological revolution in communications has resulted in fiber-optic cable and wireless microwave cell phones and satellite technologies that make it easier to communicate with people around the world. We now live in a global village, a great boon for those fortunate enough to have the education and means to take advantage of it (Drori 2006; Howard and Jones 2004; Salzman and Matathia 2000).

However, the changes in technology do not always have a positive effect on less affluent countries. For example, by substituting fiber-optic cable for old technologies, the demand for copper, used for more than 100 years to carry electrical impulses for telephones and telegraphs, has bottomed out. Countries such as Zambia and Chile, which depended on the copper trade, have seen major negative impacts on their economies. New developments resulting in artificial sweeteners reduced demand for sugar, the

Technology has had some unexpected consequences, including a staggering increase in the amount of trash generated in the society. Depicted here is trash from the fast-food industry in Egypt. The volume of trash has created problems of disposal and pollution control.

major source of income for 50 million people who work in the beet and cane sugar industries around the globe. As new technologies bring substantial benefits to many in the world, those benefits actually harm people in other parts of the interconnected world.

Some societies are attempting to replace the old economies with new service industries such as tourism. However, with violence and terrorist threats, many potential tourists prefer not to travel to peripheral countries. Money spent on travel often goes to airlines, first-class hotels, and cruise ship lines based in core countries rather than to local businesses in the peripheral countries (Schaeffer 2003). Changes in technology and the economy have forced many individuals to leave their native villages in search of paid labor positions in the urban factories and the tourism industry, disrupting family lives.

Rapid technological change has affected generational relationships as well. It is not uncommon for younger generations to have more technological competence than their elders, and this sometimes creates a generational digital divide. In some settings younger people are looked to for their expertise, whereas in the past the elders were the source of knowledge and wisdom. Often the older members of a business or group feel devalued and demoralized late in their careers because those with technical competence view the older members of the group as "outdated has-beens." Even networks and forms of entertainment such as the Wii further the social distance between generations. With the changes brought about by technology come changes in the nonmaterial culture—the values, political ideologies, and human relationships. Clearly technology can have a variety of social impacts—both positive and negative.

In the opening questions, we asked whether you as an individual can make a difference in your community, country, or world. In closing, we present in the final "Engaging Sociology" a plan you can follow to make a difference.

Thinking Sociologically

First, read "Engaging Sociology," *Making a Difference* on page 466. Then use the steps to plan how you would bring about a change that would make a difference in your workplace, college, community, region, nation, or world. Go to *Contributing to Our Social World: What Can We Do?* at the end of this and other chapters to find ways that you could be actively involved in bettering the society.

Engaging Sociology

Making a Difference

Do you want to develop sociological interventions that will bring about needed change? Applied sociologists and individuals like you can take certain steps to facilitate change (Glass 2004). Because bringing about change requires cooperation, working in a group context is often essential. Flexibility, openness to new ideas, and willingness to entertain alternative suggestions are also key factors in successful change. The following steps provide a useful strategy for planning change:

1. *Identify the issue:* Be specific and focus on what is to be changed. Without clear focus, your target for change can get muddied or lost in the attempt.
2. *Research the issue and use those findings:* Learn as much as you can about the situation or problem to be changed. Use informants, interviews, written materials, observation, existing data (such as Census Bureau statistics), or anything that helps you understand the issues. That will enable you to find the most effective strategies to bring about change.
3. *Find out what has already been done and by whom:* Other individuals or groups may be working on the same issue. Be sure you know what intervention has already taken place. This can also help determine whether attempts at change have been tried, what has been successful, and whether further change is needed.
4. *Change must take into account each level of analysis:* When planning a strategy, you may focus on one level of analysis, but be sure to consider what interventions are needed at other levels to make the change effective or to anticipate the effects of change on other levels.
5. *Determine the intervention strategy:* Map out the intervention and the steps to carry it out. Identify resources needed, and plan each step in detail.
6. *Evaluate the plan:* Get feedback on the plan from those involved in the issue and from unbiased colleagues. If possible, involve those who will be affected by the intervention in the planning and evaluation of the change. When feasible, test the intervention plan before implementing it.
7. *Implement the intervention:* Put the plan into effect, watching for any unintended consequences. Ask for regular feedback from those affected by the change.
8. *Evaluate the results:* Assess what is working, what is not, and how the constituents that experience the change are reacting. Sociological knowledge and skills should help guide this process.

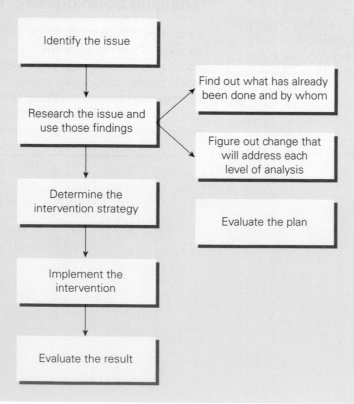

What Have We Learned?

Can you change the world? The underlying message of this chapter and the text is that choices we make facilitate change at each level in our social world. Understanding sociology provides knowledge and tools to make informed decisions and allows us to work with groups to make a difference. One of the founders of sociology, Émile Durkheim ([1893] 1947), argued that sociology is not worth a single hour's work if it does not help change the world for the better.

As you face individual challenges to bring about change in your social world, keep in mind this message: Change at one level affects all others. Sociology as a discipline is focused on gathering accurate information about the society in which we live. Still, sociologists as citizens often use their knowledge to advocate for changes that they think will make a better society, and applied sociologists help organizations bring about change. Although you are not yet a sociologist, we hope you have gained important insights that will help you contribute to the dialogue about how to make our social world a better, more humane place.

Key Points:

- Social change—variation or alteration over time in behavior patterns, culture, or structure in a society—typically involves change at one level of the social system that ripples through the other levels, micro, meso, and macro. (See pp. 440–447.)

- Strains within an organization or a group can induce change—as can stress that is imposed from the outside environment. (See pp. 447–449.)

- Sociological theories—whether micro or macro—offer explanations for the causes of change. (See pp. 449–454.)

- At the micro level, change is often initiated through collective behavior, which can take several forms: crowds, mobs, riots, rumors, fads, and fashions. (See pp. 454–457.)

- At the meso level, change of organizations is often managed through a planned process. (See pp. 457–459.)

- Social movements often provide impetus for change at the macro level. Social changes can be induced at the micro level, but they have implications even at the global level. (See pp. 459–463.)

- Science and technology can also stimulate change, but science has its greatest impact for change when it is institutionalized. (See pp. 463–468.)

- Social structures constrain what we do, but individuals, especially when acting in concert with others, can challenge, resist, and change the social systems that constrain them. We can change our society because of *human agency*. (See p. 462.)

Contributing to Our Social World: What Can We Do?

At the Local Level: Campus-wide Movements

- *A wide range of social movements,* including international peace, environmental issues, human rights, and specific student concerns such as campus safety and rising costs of higher education, may have representatives on your campus. Contact the participants, arrange to attend a meeting, and consider participating in its activities. If you feel strongly about an issue for which no movement exists, consider organizing one with a few like-minded students.

- *Many communities have movements to bring about change.* Check with your professors, service learning opportunities, and chambers of commerce to find these.

At the Organizational or Institutional Level:

- *Invite a movement leader to campus.* Consider inviting movement leaders at the state or regional level to your campus for a lecture, or organize a conference that features several experts in a particular field. Environment, civil rights, and alternative health care are topics that have wide appeal. Consider getting involved with one of these groups so you can be proactive in shaping the society in which you live.

- *VolunteerMatch* is an organization that seeks to connect individuals with movements of interest to them. Its website (www.volunteermatch.org) has suggestions for getting involved in your local area.

At the National and Global Levels:

- *International debt burden.* A major obstacle to improvement in living conditions in the less developed countries is the enormous amount of money they owe to banks and governments in the industrialized nations for past-due development-oriented loans. The leading organization in the promotion of debt forgiveness—a solution that the U.S. government has embraced in principle—is Jubilee. Named for the commandment in the Old Testament that people are to cancel all debt owed to them every Jubilee Year (once every 49 years), the organization has legislative programs, research work, and educational outreach activities throughout the world. The website of the U.S. branch of Jubilee (www.jubileeusa.org) provides information on ways in which you can get involved.

Visit **www.sagepub.com/oswcondensed2e** for online activities, sample tests, and other helpful information. Select "Chapter 14: The Process of Change" for chapter-specific activities.

References

"6.7% of World Has College Degree." 2010. *The Huffington Post,* July 19. Retrieved March 11, 2011 (*www.huffingtonpost.com/2010/05/19/percent-of-world-with-col_n_581807.html*).

AAANativeArts. 2011. "Facts About Alaskan Natives." Retrieved April 19, 2011 (www.aaanativearts.com/alaskan-natives/index.html).

AAUW Dialogue. 2008. "Violence Against Women." October 28. Retrieved February 21, 2009 (http://blog-aauw.org/2008/10/28/violence-against-women/).

Abbott, Carl. 1993. *The Metropolitan Frontier: Cities in the Modern American West.* Tucson: University of Arizona Press.

Abma, J. C., G. M. Martinez, W. D. Mosher, and B. S. Dawson. 2004. "Teenagers in the United States: Sexual Activity, Contraceptive Use, and Childbearing, 2002." *Vital Health Statistics* 23(24):1–48. Atlanta, GA: National Center for Health Statistics.

About.com. 2011. "Getting Married in Japan." Retrieved August 6, 2011 (http://japanese.about.com/library/weekly/aa080999.htm).

Abramson, Paul R., John H. Aldrich, and David W. Rhode. 2007. *Change and Continuity in the 2004 and 2006 Elections.* Washington, DC: CQ Press.

Abu-Lughod, Janet L. 1991. *Changing Cities: Urban Sociology.* New York: HarperCollins.

———. 2001. *New York, Chicago, Los Angeles: America's Global Cities.* Minneapolis: University of Minnesota Press.

Acker, C. 1993. "Stigma or Legitimation? A Historical Examination of the Social Potentials of Addition Disease Models." *Journal of Psychoactive Drugs* 25:13–23.

Acker, Joan. 1990. "Hierarchies, Jobs, Bodies: A Theory of Gendered Organizations." *Gender and Society* 4:139–58.

———. 1992. "Gendered Institutions: From Sex Roles to Gendered Institutions." *Contemporary Sociology* 21:565–69.

Adams, Mike S. 1996. "Labeling and Differential Association: Towards a General Social Learning Theory of Crime and Deviance." *American Journal of Criminal Justice* 20(2):147–64.

Ad Hoc Committee on Environmental Stewardship. 2000. "Green Facts for the U.S. and the World at Large." Atlanta, GA: Emory University.

Adler, Emily Stier and Roger Clark. 2003. *How It's Done: An Invitation to Social Research,* 2nd ed. Belmont, CA: Wadsworth.

Adler, Freda, Gerhard O. W. Mueller, William S. Laufer, and E. Mavis Hetherington. 2004. *Criminology,* 5th ed. New York: McGraw-Hill.

Adler, Patricia A. and Peter Adler. 1991. *Backboards and Blackboards: College Athletes and Role Engulfment.* New York: Columbia University Press.

———. 2004. "The Gloried Self." Pp. 117–26 in *Inside Social Life,* 4th ed., edited by Spencer E. Cahill. Los Angeles: Roxbury.

Agrell, Siri. 2008. "Youth Vote Goes Solidly to Obama." *The Globe and Mail,* January 5, A15.

Aguirre, Adalberto and David V. Baker. 2007. *Structured Inequality in the United States,* 2nd ed. Englewood Cliffs, NJ: Prentice Hall.

Aguirre, Adalberto, Jr., and Jonathan H. Turner. 2006. *American Ethnicity: The Dynamics and Consequences of Discrimination,* 5th ed. Boston: McGraw-Hill.

Ahmad, Nehaluddin. 2009. "Sati Tradition—Widow Burning in India: A Socio-legal Examination." *Web Journal of Current Legal Issues.* Retrieved April 7, 2011 (http://webjcli.ncl.ac.uk/2009/issue2/ahmad2.html).

Ahrons, Constance. 2004. *We're Still Family.* New York: HarperCollins.

Akbar, Imran. 2006. "Internet Users Currently Spend More on Online Dating Than Music." Retrieved March 13, 2008 (www.newswiretoday.com/news/5114/).

Akers, Ronald. 1992. *Deviant Behavior: A Social Learning Approach,* 3rd ed. Belmont, CA: Wadsworth.

———. 1998. *Social Learning and Social Structure: A General Theory of Crime and Deviance.* Boston: Northeastern University Press.

Akers, Ronald L., Marvin D. Krohn, Lonn Lanza-Kaduce, and Marcia Radosevich. 1979. "Social Learning and Deviant Behavior." *American Sociological Review* 44(August):635–54.

Alaska State Library. 2005. "Library Services & Technology Act, Alaska State Plan, 2003–2007." Retrieved August 7, 2011 (http://www.library.state.ak.us/pub/lsta2003_2007.html#goa11).

Alatas, Syed Farid. 2006. "Ibn Khaldun and Contemporary Sociology." *International Sociology* 21(6):782–95.

Albrecht, Gary. 2002. *The Disability Business: Rehabilitation in America.* Thousand Oaks, CA: Sage.

Albrecht, Stan L. and Tim B. Heaton. 1984. "Secularization, Higher Education, and Religiosity." *Review of Religious Research* 26(September):43–58.

Alexander, Michelle. 2010. *The New Jim Crow: Mass Incarceration in the Age of Color Blindness.* New York: The New Press.

Alexander, Victoria D. 2003. *Sociology of the Arts: Exploring Fine and Popular Forms.* Malden, MA: Blackwell.

Alger, Jonathan R. 2003. "*Gratz/Grutter* and Beyond: the Diversity Leadership Challenge." University of Michigan. Retrieved August 7, 2011 (http://www.vpcomm.umich.edu/admissions/overview/challenge.html).

Alldritt, Leslie D. 2000. "The Burakumin: The Complicity of Japanese Buddhism in Oppression and an Opportunity for Liberation." *Journal of Buddhist Ethics* 7(July). Retrieved July 23, 2009 (http://www.buddhistethics.org/7/alldritt001.html).

Allegretto, Sylvia, Ary Amerikaner, and Steven Pitts. 2011. "Black Employment and Unemployment: Teen Employment Population Ratios by Race" (Chart 14, p. 14). *Work in the Black Community,* February 4. UC Berkeley Labor Center. Retrieved April 17, 2011 (http://laborcenter.berkeley.edu/blackworkers/monthly/bwreport_2011–02–04_27.pdf).

Altbach, Philip G. and Todd M. Davis. 1999. "Global Challenge and National Response: Note for an International Dialogue on International Higher Education." *International Higher Education* 14:2–5.

Altbach, Philip G., Liz Reisberg, and Laura E. Rumbley. 2009. *Trends in Global Higher Education: Tracking an Academic Revolution.* New York: UNESCO World Conference on Higher Education.

Altheide, David, Patricia A. Adler, Peter Adler, and Duane Altheide. 1978. "The Social Meanings of Employee Theft." P. 90 in *Crime at the Top,* edited by John M. Johnson and Jack D. Douglas. Philadelphia: Lippincott.

Amadeo, Kimberley. 2009. "NAFTA Pros and Cons." Retrieved December 20, 2009 (http://useconomy.about.com/b/2008/04/24/nafta-pros-and-cons.htm).

Amato, Paul R. 2000. "The Consequences of Divorce for Adults and Children." *Journal of Marriage and the Family* 62(November):1269–87.

Amato, Paul R. and Bruce Keith. 1991. "Parental Divorce and the Well-Being of Children: A Meta-Analysis." *Psychological Bulletin* 110(1):26–46.

Amato, Paul R. and Juliana M. Sobolewski. 2001. "The Effects of Divorce and Marital Discord on Adult Children's Psychological Well-Being." *American Sociological Review* 63(December):697–713.

American Academy of Pediatrics. 2011. "Fact Sheet: Medicaid and Children." Retrieved July 26, 2011 (www.aap.org/research/factsheet.pdf).

American Association of University Women Educational Foundation. 2001. "Hostile Hallways: Bullying, Teasing, and Sexual Harassment in School." Retrieved August 26, 2006 (http://www.aauw.org/research/hostile.cfm).

American Cancer Society. 2006. *Cancer Facts and Figures for African Americans 2005–2006.* Atlanta, GA: Author.

American Cultural Center Resource Service. 2004. "North American Free Trade Agreement." Retrieved September 29, 2006 (http://usinfo.org/law/nafta/chap-01.stm.html).

American Library Association. 2011. "'And Tango Makes Three' Waddles Its Way Back to the Number One Slot as America's Most Frequently Challenged Book." Retrieved July 12, 2011 (http://ala.org/ala/news-presscenter/news/pr.cfm?id=6874).

American Medical Association. 1988. "It's Over, Debbie." *Journal of the American Medical Association* 259(14):2094–98.

American Psychological Association. 2007. "APA Resolution on Religious, Religion-Based, and/or Religion-Derived Prejudice." August. Retrieved December 24, 2009 (http://wthrockmorton.com/apa-resolution-on-religious-religion-based-andor-religion-derived-prejudice).

———. 2008. "Report From: Working Group on Assisted Suicide and End-of-Life Decisions." Retrieved August 16, 2008 (http://www.apa.org/pi/aseol/introduction.html).

American Sociological Association. 2002. *Careers in Sociology,* 6th ed. Washington, DC: American Sociological Association.

———. 2006. *What Can I Do with a Bachelor's Degree in Sociology? A National Survey of Seniors Majoring in Sociology.* Washington DC: American Sociological Association Research and Development Department.

———. 2009. *21st Century Careers with an Undergraduate Degree in Sociology.* Washington, DC: Author.

American Sociological Association Task Force on Institutionalizing Public Sociologies. 2005. *Public Sociology and the Roots of American Sociology: Re-establishing Our Connection to the Public.* Washington, DC: American Sociological Association.

Ammerman, Nancy Tatom. 1988. *Bible Believers: Fundamentalists in the Modern World.* New Brunswick, NJ: Rutgers University Press.

———. 1990. *Baptist Battles: Social Change and Religious Conflict in the Southern Baptist Convention.* New Brunswick, NJ: Rutgers University Press.

Amnesty International. 2005. "Abolish the Death Penalty: The Death Penalty Is Racially Biased." Retrieved August 4, 2006 (http://www.amnestyusa.org/abolish/factsheets/racialprejudices.html).

———. 2006. "Facts and Figures on the Death Penalty." Retrieved July 21, 2009 (http://www.amnesty.org/en/library/info/ACT50/006/2006/en).

———. 2007. "Stop the Death Penalty: The World Decides." Retrieved July 21, 2009 (http://www.amnesty.org/en/news-and-updates/news/stop-the-death-penalty-the-world-decides-20071010).

———. 2011. "Death Penalty in 2010." Retrieved July 7, 2011 (http://www.amnesty.org/en/news-and-updates/report/death-penalty-2010-executing-countries-left-isolated-after-decade-progress).

Amodeo, Karrin. 1999. "Japan Nuke Update." Federation of American Scientists. Retrieved September 29, 2006 (www.fas.org/news/japan/990930-jpn.htm).

Ananova. 2001. "Death Workers Sell Body Parts for Extra Christmas Cash." Retrieved December 4, 2005 (www.ananova.com/news/sm_465113.html).

———. 2002. "Students Willing to Sell Body Parts to Fund Education." April 8. Retrieved August 26, 2006 (www.ananova.com).

Andersen, Ronald M. 1995. "Revisiting the Behavioral Model and Access to Medical Care: Does It Matter?" *Journal of Health and Social Behavior* 36(1):1–10.

Andersen, Ronald M., Thomas H. Rice, and Gerald F. Kominski. 2001. *Changing the U.S. Health Care System: Issues in Health Services, Policy and Management,* 2nd ed. San Francisco: Jossey-Bass.

Anderson, Benedict. 2006. *Imagined Communities: Reflections on the Origin and Spread of Nationalism,* Rev. ed. London: Verso.

Anderson, David A. and Mykol Hamilton. 2005. "Gender Role Stereotyping of Parents in Children's Picture Books: The Invisible Father." *Sex Roles: A Journal of Research* 52(3/4):145.

Anderson, Elijah. 2000. *Code of the Street: Decency, Violence, and the Moral Life of the Inner City.* New York: W. W. Norton.

Anderson, Margaret L. 2006. *Thinking About Women: Sociological Perspectives on Sex and Gender,* 7th ed. Boston: Allyn & Bacon.

Anderson, Margaret and Pat Hill Collins. 2006. *Race, Class, and Gender: An Anthology,* 6th ed. Belmont, CA: Wadsworth.

Anderson, Michael. 1994. "What Is New About the Modern Family?" Pp. 67–90 in *Time, Family, and Community: Perspectives on Family and Community History,* edited by Michael Drake. Oxford, UK: The Open University.

Andrew, Caroline and Beth Moore Milroy, eds. 1991. *Life Spaces: Gender, Household, Employment.* Vancouver: University of British Columbia Press.

Anson, Ofra. 1998. "Gender Difference(s) in Health Perceptions and Their Predictors." *Social Science and Medicine* 36(4):419–27.

Answers.com. 2011a. "Flag." Retrieved July 22, 2011 (http://www.answers.com/flag&rr=67).

———. 2011b. "Refugee." Retrieved May 23, 2011 (http://www.answers.com/topic/refugee).

Anti-Defamation League. 2010. "ADL Audit: 1,211 Anti-Semitic Incidents Across the Country in 2009." Retrieved April 7, 2011 (www.adl.org/PresRele/ASUS_12/5814_12.htm).

Antonowicz, Anton. 2002. "Zarmina's Story." *The Mirror,* June 19. Retrieved January 24, 2011 (www.freerepublic.com/focus/news/702415/posts).

Antoun, Richard T. 2001. *Understanding Fundamentalism: Christian, Islamic, and Jewish Movements.* Walnut Creek, CA: AltaMira.

Anyon, Jean. 1980. "Social Class and the Hidden Curriculum of Work." *Journal of Education* 162(1):67–92.

Armstrong, Karen. 2000. *The Battle for God.* New York: Knopf.

Arnold, David O., ed. 1970. *The Sociology of Subcultures.* Berkeley, CA: Glendessary.

Aronson, Ronald. 1978. "Is Busing the Real Issue?" *Dissent* 25(Fall):409.

Arrighi, Barbara A. 2000. *Understanding Inequality: The Intersection of Race, Ethnicity, Class, and Gender.* Lanham, MD: Rowman & Littlefield.

Arulampalam, Wiji, Alison L. Booth, and Mark L. Bryan. 2007. "Is There a Glass Ceiling Over Europe—Exploring the Gender Pay Gap Across the Wages Distribution." *Industrial and Labor Relations Review* 60(2):163–86.

Aseltine, Robert H., Jr. 1995. "A Reconsideration of Parental and Peer Influences on Adolescent Deviance." *Journal of Health and Social Behavior* 36(2):103–21.

Ashley, David and David Michael Orenstein. 2009. *Sociological Theory,* 7th ed. Boston: Allyn & Bacon.

Asian Development Bank. 2002. "Cambodia's Economy Continues to Gain Ground" (News Release No. 042/02, April 9).

Association for Applied and Clinical Sociology. 2006. "Code of Ethics." Retrieved August 8, 2011 (*www.uvu.edu/besc/student/ethicalcodes/aacscodeofethics.doc*).

Astone, Nan Marie and Sara S. McLanahan. 1991. "Family Structure, Parental Practices and High School Completion." *American Sociological Review* 56(3):318–19.

Attewell, Paul. 2001. "The First and Second Digital Divides." *Sociology of Education* 74(3):252–59.

Aulette, Judy Root. 2002. *Changing American Families.* Boston: Allyn & Bacon.

Avert. 2009. "HIV and AIDS in Africa" (updated February 20, 2009). Retrieved February 21, 2009 (http://www.avert.org/aafrica.htm).

———. 2010. "Global HIV and AIDS Estimates, End of 2009." Retrieved May 14, 2011 (www.avert.org/worldstats.htm).

Ayalon, Hanna and Yossi Shavit. 2004. "Educational Reforms and Inequalities in Israel: The MMI Hypothesis Revisited." *Sociology of Education* 77(2):103–20.

Ayers, William. 2006. "Hearts and Minds: Military Recruitment and the High School Battlefield." *Phi Delta Kappan* 87(8):594–99.

Ayers, William and Michael Klonsky. 2006. "Chicago's Renaissance 2010: The Small Schools Movement Meets the Ownership Society." *Phi Delta Kappan* 87(6):453–56.

Azumi, Koya. 1977. "Japan's Changing World of Work." *The Wilson Quarterly* (Summer):72–80.

Bagby, Ihsan A. 2003. "Imams and Mosque Organization in the United States: A Study of Mosque Leadership and Organizational Structure in American Mosques." Pp. 113–34 in *Muslims in the United States,* edited by Philippa Strum and Danielle Tarantolo. Washington, DC: Woodrow Wilson International Center for Scholars.

Bahney, Anna. 2009. "Don't Talk to Invisible Strangers." *The New York Times,* March 9. Retrieved August 13, 2009 (http://www.nytimes .com/2006/03/09/fashion/thursdaystyles/09parents.html).

Bainbridge, William S. and Rodney Stark. 1981. "Suicide, Homicide, and Religion: Durkheim Reassessed." *Annual Review of the Social Sciences of Religion* 5:33–56.

Baker, David P. 2002. "International Competition and Education Crises: Cross-National Studies of School Outcomes." Pp. 393–98 in *Education and Sociology: An Encyclopedia,* edited by David L. Levinson, Peter W. Cookson, Jr., and Alan R. Sadovnik. New York: RoutledgeFalmer.

Balch, Robert W. 1982. "Bo and Peep: A Case Study of the Origins of Messianic Leadership." Pp. 13–72 in *Millennialism and Charisma,* edited by Roy Wallis. Belfast, Northern Ireland: The Queen's University.

———. 1995. "Waiting for the Ships: Disillusionment and the Revitalization of Faith in Bo and Peep's UFO Cult." Pp. 137–66 in *The Gods Have Landed: New Religions from Other Worlds,* edited by James R. Lewis. Albany: State University of New York Press.

Bales, Kevin. 2000. *New Slavery: A Reference Handbook,* 2nd ed. Santa Barbara, CA: ABC-CLIO.

———. 2002. "Because She Looks Like a Child." Pp. 207–29 in *Global Woman: Nannies, Maids, and Sex Workers in the New Economy,* edited by Barbara Ehrenreich and Arlie Russell Hochschild. New York: Henry Holt.

———. 2004. *Disposable People: New Slavery in the Global Economy,* Rev. ed. Berkeley: University of California Press.

———. 2007. *Ending Slavery: How We Free Today's Slaves.* Berkeley: University of California Press.

Bales, Kevin and Ron Soodalter. 2010. *The Slave Next Door: Human Trafficking and Slavery in America Today.* Berkeley: University of California Press.

Bales, Kevin and Zoe Trodd. 2008. *To Plead Our Own Cause: Personal Stories by Today's Slaves.* Ithaca, NY: Cornell University Press.

Bales, Kevin, Zoe Trodd, and Alex Kent Williamson. 2009. *Modern Slavery: The Secret World of 27 Million People.* Oxford, UK: Oneworld Press.

Ballantine, Jeanne H. 1991. "Market Needs and Program Products: The Articulation Between Undergraduate Applied Programs and the Market Place." *Journal of Applied Sociology* 8(3/4):1–19.

———. 2001. "Education." Pp. 277–93 in *Social Problems: A Case Study Approach,* edited by Norman A. Dolch and Linda Deutschmann. Dix Hills, NY: General Hall.

Ballantine, Jeanne H. and Floyd M. Hammack. 2009. *The Sociology of Education: A Systematic Analysis,* 6th ed. Upper Saddle River, NJ: Prentice Hall.

———. 2012. *The Sociology of Education: A Systematic Analysis,* 7th ed. Upper Saddle River, NJ: Prentice Hall.

Ballantine, Jeanne H. and Joan Z. Spade. 2008. *Schools and Society,* 3rd ed. Thousand Oaks, CA: Pine Forge Press.

Banbury, Samantha. 2004. "Coercive Sexual Behavior in British Prisons as Reported by Adult Ex-Prisoners." *The Howard Journal* 43(2):113–30.

Bankston, Carl L., III. 2004. "Social Capital, Cultural Values, Immigration, and Academic Achievement: The Host Country Context and Contradictory Consequences." *Sociology of Education* 77(2):176–80.

Barash, David. 2002. "Evolution, Males, and Violence." *The Chronicle Review,* May 2, B7.

Barber, Benjamin R. 2006. "The Uncertainty of Digital Politics: Democracy's Relationship With Information Technology." Pp. 61–69 in *Globalization: The Transformation of Social Worlds,* edited by D. Stanley Eitzen and Maxine Baca Zinn. Belmont, CA: Wadsworth.

Barber, Bonnie L., Jacquelynne S. Eccles, and Margaret R. Stone. 2001. "Whatever Happened to the Jock, the Brain, and the Princes? Young Adult Pathways Linked to Adolescent Activity Involvement and Social Identity." *Journal of Adolescent Research* 16(5):429–55.

Barefoot, Coy. 2003. "Our Childhood Memories." *U.S. Airways,* November.

Barker, Colin. 2009. "60% of the World Uses Mobile Phones." *ZD Net UK,* March 3. Retrieved November 12, 2009 (http://news.zdnet.co.uk/communications/0,1000000085,39621541,00.htm).

Barnes, Patricia M., Eve Powell-Griner, Kim McFann, and Richard L. Nahin. 2004. "Complementary and Alternative Medicine Use Among Adults: United States, 2002." *CDC Advance Data Report,* No. 343.

Barnett, T. P., J. C. Adam, and D. P. Lettenmaier. 2005. "Potential Impacts of a Warming Climate on Water Availability in Snow-Dominated Regions." *Nature* 438(November 17):303–9.

Barr, Colin and David Goldman. 2010. "20 Highest Paid CEOs." CNNMoney, April 6. Retrieved July 6, 2011 (http://money.cnn.com/galleries/2010/news/1004/gallery.top_ceo_pay/).

Barraclough, Geoffrey, ed. 1986. *The Times Concise Atlas of World History,* Rev. ed. London: Times Books Limited.

Barrett, David B., Todd M. Johnson, and Peter F. Crossing. 2008. "The 2007 Annual Megacensus of Religions." Pp. 600–605 in *Time Almanac 2008.* Chicago: Encyclopaedia Britannica.

Barry, Ellen. 2009. "Protests in Moldova Explode, With Help of Twitter." *The New York Times,* April 8, A1. Retrieved December 22, 2009 (www .nytimes.com/2009/04/08/world/europe/08moldova.html).

Bartkowski, John P. and Christopher G. Ellison. 1995. "Divergent Models of Childrearing in Popular Manuals: Conservative Protestants vs. the Mainstream Experts." *Sociology of Religion* 56(1):21–34.

Bartlett, Donald L. and James B. Steele. "How the Little Guy Gets Crunched." *Time,* February 7, 38–41.

Basaglia, Franca Ongaro. 1992. "Politics and Mental Health." *International Journal of Social Psychiatry* 38(1):36–39.

Basch, Linda, Nina Glick Schiller, and Cristina Szanton Blanc. 1994. *Nations Unbounded: Transnational Projects, Postcolonial Predicaments, and Deterritorialized Nation-State.* Amsterdam, Netherlands: Gordon and Breach.

Basow, Susan A. 1992. *Gender: Stereotypes and Roles,* 3rd ed. Belmont, CA: Wadsworth.

———. 2000. "Gender Stereotypes and Roles." Pp. 101–15 in *The Meaning of Difference: American Constructions of Race, Sex and Gender, Social Class, and Sexual Orientation,* 2nd ed., edited by Karen E. Rosenblum and Toni-Michelle C. Travis. New York: McGraw-Hill.

Basso, Keith H. 1979. *Portraits of the Whiteman: Linguistic Play and Cultural Symbols Among the Western Apache.* Cambridge, UK: Cambridge University Press.

Batavia, A. I. 2000. *So Far So Good: Observations on the First Year of Oregon's Death With Dignity Act.* Washington, DC: American Psychological Association.

Bayley, David. 1991. *Forces of Order: Policing Modern Japan.* Berkeley: University of California Press.

BBC News. 2000. "Profile: Vicente Fox." *World News,* July 3.
———. 2004. "Q and A: Muslim Headscarves." Retrieved July 21, 2009 (http://news.bbc.co.uk/2/hi/europe/3328277.stm).
———. 2005. "Mukhtar Mai—History of a Rape Case." Retrieved July 13, 2008 (http://news.bbc.co.uk/2/hi/south_asia/4620065.stm).
———. 2008. "US Elections Map: State-by-State Guide." March 5. Retrieved March 21, 2008 (http://news.bbc.co.uk/2/hi/in_depth/629/629/7223461.stm).
BBC News: Science and Environment. 2010. "Huge Ice Island Breaks From Greenland Glacier." Aug. 7. Retrieved December 2, 2010 (www.bbc.co.uk/news/science-environment-10900235).
BBC's Science and Nature. 2009. "The Ghost in Your Genes." Retrieved November 12, 2009 (http://www.bbc.co.uk/sn/tvradio/programmes/horizon/ghostgenes.shtml).
Beckwith, Carol. 1983. "Niger's Wodaabe: People of the Taboo." *National Geographic* 164(4):483–509.
———. 1993. *Nomads of Niger.* New York: Harry N. Abrams.
Belding, Theodore C. 2004. "Nobility and Stupidity: Modeling the Evolution of Class Endogamy." Retrieved August 7, 2008 (http://arxiv.org/abs/nlin.AO/0405048).
Belkin, Lisa. 2011. "The Cost of Raising a Child." *The New York Times,* July 25. Retrieved July 27, 2011 (http://parenting.blogs.nytimes.com/2010/06/25/the-cost-of-raising-a-child/).
Bell, Daniel. 1973. *The Coming of Post-Industrial Society: A Venture in Social Forecasting.* New York: Basic Books.
———. [1976] 1999. *The Coming of Post-Industrial Society: A Venture in Social Forecasting,* Special anniversary edition. New York: Basic Books.
———. 1989. "The Third Technological Revolution." *Dissent* 36(2):169.
Bellah, Robert N. 1970. "Civil Religion in America." Pp. 168–215 in *Beyond Belief: Essays on Religion in a Post-Traditionalist World.* New York: Harper & Row.
———. 1992. *The Broken Covenant: American Civil Religion in Time of Trial.* Chicago: University of Chicago Press.
Bellah, Robert N. and Frederick E. Greenspahn, eds. 1987. *Uncivil Religion: Interreligious Hostility in America.* New York: Crossroad.
Bellah, Robert N., Richard Madsen, William M. Sullivan, Ann Swindler, and Steven M. Tipton. 1996. *Habits of the Heart: Individualism and Commitment in American Life,* Updated edition. Berkeley: University of California Press.
Bellisari, Anna. 1990. *Biological Bases of Sex Differences in Cognitive Ability and Brain Function.* Unpublished paper.
Bem, Sandra L. 1974. "The Measurement of Psychological Androgyny." *Journal of Consulting and Clinical Psychology* 42(2):155–62.
———. 1983. "Gender Schema Theory and Its Implications for Child Development: Raising Gender-Aschematic Children in a Gender-Schematic Society." *Signs: Journal of Women in Culture and Society* 8(Summer):598–616.
Benavot, Aaron. 1992. "Curricular Content, Educational Expansion, and Economic Growth." *Comparative Education Review* 36(2):150–74.
Benavot, Aaron, John Meyer, and David Kamens. 1991. "Knowledge for the Masses: World Models and National Curricula: 1920–1986." *American Sociological Review* 56(1):85–100.
Benderly, Beryl Lieff. 1982. "Rape Free or Rape Prone." *Science* 82(3):40–43.
Benevolo, Leonardo. 1995. *The European City.* Cambridge, MA: Blackwell.
Benguigui, Yamina. 2000. *The Perfumed Garden.* Brooklyn, NY: First Run Icarus Films.
Bennett, Neil, Ann Blanc, and David Bloom. 1988. "Commitment and the Modern Union: Assessing the Link Between Premarital Cohabitation and Subsequent Marital Stability." *American Sociological Review* 53(1):127–38.
Bennis, Warren G., Kenneth D. Benne, and Robert Chin. 1985. *The Planning of Change,* 4th ed. New York: Holt, Rinehart, and Winston.
Benokraitis, Nijole V. 2008. *Marriages and Families: Changes, Choices, and Constraints,* 6th ed. Englewood Cliffs, NJ: Prentice Hall.

Benokraitis, Nijole V. and Joe R. Feagin. 1995. *Modern Sexism: Blatant, Subtle, and Covert Discrimination,* 2nd ed. Englewood Cliffs, NJ: Prentice Hall.
Berger, Helen A. 1999. *A Community of Witches: Contemporary Neo-Paganism and Witchcraft in the United States.* Columbia: University of South Carolina Press.
Berger, Helen A. and Douglas Ezzy. 2007. *Teenage Witches: Magical Youth and the Search for the Self.* New Brunswick, NJ: Rutgers University Press.
Berger, Michael L. 1979. *The Devil Wagon in God's Country: The Automobile and Social Change in Rural America, 1893–1929.* Hamden, CT: Archon.
Berger, Peter L. 1961. *The Noise of Solemn Assemblies.* Garden City, NY: Doubleday.
———. 1990. *The Sacred Canopy.* Garden City, NY: Anchor Books.
Berger, Peter L. and Thomas Luckmann. 1966. *The Social Construction of Reality.* Garden City, NY: Doubleday.
Bergman, Mike. 2005. "U.S. Voter Turnout Up in 2004." *U.S. Census Bureau News.* Retrieved August 8, 2011 (http://indymedia.us/en/2005/06/7963.shtml).
Berk, Richard A. 1974. *Collective Behavior.* Dubuque, IA: Brown.
Berry, Brian J. L. and John Kasarda. 1977. *Contemporary Urban Ecology.* New York: Macmillan.
Bertman, Stephen. 1998. *Hyperculture: The Human Cost of Speed.* Westport, CT: Praeger.
Better Factories Cambodia. 2006. Retrieved September 29, 2006 (www.betterfactories.org).
Bettie, Julie. 2003. *Women Without Class: Girls, Race, and Identity.* Berkeley: University of California Press.
Beyer, Peter. 2000. "Secularization From the Perspective of Globalization." Pp. 81–93 in *The Secularization Debate,* edited by William H. Swatos, Jr., and Daniel V. A. Olson. Lanham, MD: Rowman and Littlefield.
Bianci, Suzanne M., Melissa A. Milkie, Liana C. Sayer, and John P. Robinson. 2000. "Is Anyone Doing the Housework? Trends in the Gender Division of Household Labor." *Social Forces* 79(1):191–228.
Bibby, Reginald W. 1987a. *Fragmented Gods: The Poverty and Potential of Religion in Canada.* Toronto, Canada: Irwin.
———. 1987b. "Religion in Canada: A Late Twentieth Century Reading." Pp. 263–67 in *Yearbook of American and Canadian Churches,* edited by Constant H. Jacquet, Jr. Nashville, TN: Abington.
———. 2002. *Restless Gods: The Renaissance of Religion in Canada.* Toronto, Canada: Stoddard.
Billig, Michael. 1995. *Banal Nationalism.* Thousand Oaks, CA: Sage.
Billings, Dwight B. and Shaunna L. Scott. 1994. "Religion and Political Legitimation." *Annual Review of Sociology* 20:173–202.
Birdwhistell, Raymond L. 1970. *Kinesics and Context: Essays on Body Motion Communication.* Philadelphia: University of Pennsylvania Press.
Bishaw, Alemayehu and Jessica Semega. 2008. "Income, Earnings, and Poverty Data From the 2007 American Community Survey." U.S. Census Bureau. Retrieved July 19, 2011 (www.census.gov/prod/2008pubs/acs-09.pdf).
Black Politics on the Web. 2009. "Environmental Racism: An Unfortunate Reality." Retrieved December 22, 2009 (http://blackpoliticsontheweb.com/2009/03/13/environmental-racism-an-unfortunate-reality/).
Blank, Rebecca M. 2002. "Evaluating Welfare Reform in the United States." *Journal of Economic Literature* 40(4):1105–1166.
Blasi, Joseph Raphael and Douglas Lynn Kruse. 1991. *The New Owners: The Mass Emergence of Employee Ownership in Public Companies and What It Means to American Business.* New York: HarperCollins.
Blasi, Joseph, Douglas Kruse, and Aaron Bernstein. 2003. *In the Company of Owners.* New York: Basic Books.
Blau, Peter M. 1956. *Bureaucracy in Modern Society.* New York: Random House.
———. 1964. *Exchange and Power in Social Life.* New York: John Wiley.
Blau, Peter and Otis Dudley Duncan. 1967. *The American Occupational Structure.* New York: John Wiley.
Blauner, Robert. 1964. *Alienation and Freedom.* Chicago: University of Chicago Press.

————. 1972. *Racial Oppression in America.* New York: Harper & Row.

Blee, Kathleen M. 2008. "White Supremacy as Extreme Deviance." Pp. 108–17 in *Extreme Deviance,* edited by Erich Goode and D. Angus Vail. Thousand Oaks, CA: Pine Forge Press.

Blendon, Robert J., Cathy Schoen, Catherine M. DesRoches, Robin Osborn, Kimberly L. Scoles, and Kinga Zapert. 2002. "Inequities in Health Care: A Five-Country Survey." *Health Affairs* 21(3):182–91.

Blumer, Herbert. 1969. *Symbolic Interactionism: Perspective and Method.* Englewood Cliffs, NJ: Prentice Hall.

————. 1986. *Symbolic Interactionism: Perspective and Method.* Berkeley: University of California Press.

Blumstein, Philip and Pepper Schwartz. 1985. *American Couples: Money, Work, Sex.* New York: Pocket Books.

Bobo, Lawrence and Ryan A. Smith. 1998. "From Jim Crow Racism to Laissez-Faire Racism: An Essay on the Transformation of Racial Attitudes in America." Pp. 182–220 in *Beyond Pluralism: Essays on the Conceptions of Groups and Identities in America,* edited by Wendy Freedman Katkin, Ned C. Landsman, and Andrea Tyree. Urbana: University of Illinois Press.

Bohannan, Paul. 1959. "The Impact of Money on an African Subsistence Economy." *Journal of Economic History* 19:491–503.

Bolduan, Kate. 2009. "The Plight of Young Uninsured Americans." CNN Politics, March 7. Retrieved January 10, 2010 (www.cnn.com/2009/POLITICS/03/07/young.uninsured/index.html).

Boli, John. 2002. "Globalization." Pp. 307–13 in *Education and Sociology: An Encyclopedia,* edited by David L. Levinson, Peter W. Cookson, Jr., and Alan R. Sadovnik. New York: RoutledgeFalmer.

Bonacich, Edna. 1972. "A Theory of Ethnic Antagonism: The Split Labor Market." *American Sociological Review* 37(October):547–59.

————. 1976. "Advanced Capitalism and Black-White Race Relations in the United States: A Split Labor Market Interpretation." *American Sociological Review* 41(February):34–51.

Bonacich, Edna and Jake B. Wilson. 2005. "Hoisted by Its Own Petard: Organizing Wal-Mart's Logistics Workers." *New Labor Forum* 14:67–75.

Bonczar, Thomas P. and Tracy L. Snell. 2005. "Capital Punishment, 2004." *Bureau of Justice Statistics Bulletin* (NCJ 211349). Washington, DC: U.S. Department of Justice.

Bonilla-Silva, Eduardo. 2003. *Racism Without Racists: Color-Blind Racism and the Persistence of Racial Inequality in the United States.* Berkeley: University of California Press.

Booker, Salih and William Minter. 2006. "Global Apartheid: AIDS and Murder by Patient." Pp. 517–22 in *Beyond Borders: Thinking Critically About Global Issues,* edited by Paula S. Rothenberg. New York: Worth.

Booth, Alan and Paul R. Amato. 2001. "Parental Predivorce Relations and Offspring Postdivorce Well-Being." *Journal of Marriage and the Family* 63(February):197–212.

Borg, Marcus J. 1994. *Meeting Jesus Again for the First Time.* San Francisco: Harper San Francisco.

Boston Women's Health Book Collective. 2006. *Our Bodies, Ourselves: Menopause.* New York: Touchstone.

Bottomore, Tom. 1979. *Political Sociology.* New York: Harper & Row.

Boulding, Elise with Jennifer Dye. 2002. "Women and Development." In *Introducing Global Issues,* 2nd ed., edited by Michael T. Snarr and D. Neil Snarr. Boulder, CO: Lynne Rienner.

Bourdieu, P. and J. C. Passeron. 1977. *Reproduction in Education, Society and Culture.* London: Sage.

Bowen, Debra. 2011. "History Behind California's Primary Election System." Retrieved August 8, 2011 (http://www.sos.ca.gov/elections/npp.htm).

Bowler, I. R., C. R. Bryant, and M. D. Nellis, eds. 1992. *Contemporary Rural Systems in Transition, Volume 1: Agriculture and Environment, Volume 2: Economy and Society.* Wallingford, Oxfordshire, UK: C.A.B. International.

Bowles, Samuel. 1977. "Unequal Education and the Reproduction of the Social Division of Labor. P. 137 in *Power and Ideology in Education,* edited by Jerome Karabel and A. H. Halsey. New York: Oxford University Press.

Bowles, Samuel and Herbert Gintis. 1976. *Schooling in Capitalist America.* New York: Basic Books.

————. 2002. "Schooling in Capitalist America Revisited." *Sociology of Education* 75(1):1–18.

Boy, Angie and Andrzej Kulczycki. 2008. "What We Know About Intimate Partner Violence in the Middle East and North Africa." *Violence Against Women* 14(1):53–70.

Bracey, Gerald W. 2003. "April Foolishness: The 20th Anniversary of A Nation at Risk." *Phi Delta Kappan* 84(8):616–21.

————. 2005. "The 15th Bracey Report on the Condition of Public Education." *Phi Delta Kappan* 87(2):138–53.

Brady, Henry E., Sidney Verba, and Kay Lehman Schlozman. 1995. "Beyond SES: A Resource Model of Political Participation." *American Political Science Review* 89(2):271–85.

Brandon, Mark E. 2005. "War and American Constitutional Order." In *The Constitution in Wartime: Beyond Alarmism and Complacency,* edited by Mark Tushner. Durham, NC: Duke University Press.

Brasher, Brenda E. 2004. *Give Me That On-Line Religion.* New Brunswick, NJ: Rutgers University Press.

Brecher, Jeremy, Tim Costello, and Brendan Smith. 2012. "Globalization and Social Movements." Pp. 272–90 in *Globalization: The Transformation of Social Worlds,* 3rd ed., edited by D. Stanley Eitzen and Maxine Baca Zinn. Belmont, CA: Wadsworth.

Bremner, Jason, Carl Haub, Marlene Lee, Mark Mather, and Eric Zuehlke. 2009. "World Population Highlights." *Population Bulletin* 64(3):3. Population Reference Bureau. Retrieved July 23, 2011 (http://www.prb.org/pdf09/64.3highlights.pdf).

Brettell, Caroline B. and Carolyn F. Sargent. 2001. *Gender in Cross-Cultural Perspective,* 3rd ed. Englewood Cliffs, NJ: Prentice Hall.

————. 2005. *Gender in Cross-Cultural Perspective,* 4th ed. Englewood Cliffs, NJ: Prentice Hall.

Bridgeland, John M., John J. Dilulio, and Karen Burke Morison. 2006. *The Silent Epidemic: Perspectives of High School Dropouts.* March. Retrieved July 11, 2011 (http://www.civicenterprises.net/pdfs/thesilentepidemic3–06.pdf).

Brier, Noah Rubin. 2004. "Coming of Age." *American Demographics* 26(9):16.

Brint, Steven, Mary F. Contreras, and Michael T. Matthews. 2001. "Socialization Messages in Primary Schools: An Organizational Analysis." *Sociology of Education* 74(July):157–80.

Bristow, Jason. 2008. "Can Canadian Youth Be Inspired to Vote?" *The Toronto Sun,* March 6, 17.

British Columbia Ministry of Labour & Citizens' Services. 2006. *BC Stats Infoline* (6)40. October 6. Retrieved January 3, 2010 (http://www.bcstats.gov.bc.ca/releases/info2006/in0640.pdf).

Britton, Dana M. 2000. "The Epistemology of the Gendered Organization." *Gender and Society* 14(3):418–34.

Britz, Jennifer Delahunty. 2006. "Are Today's Girls Too Successful?" *Dayton Daily News,* March 31, A7.

Brody, Gene H., Zolinda Stoneman, Douglas Flor, and Chris McCrary. 1994. "Religion's Role in Organizing Family Relationships: Family Process in Rural, Two-Parent African American Families." *Journal of Marriage and the Family* 56(November):878–88.

Bromley, David G. and Anson D. Shupe, Jr. 1981. *Strange Gods: The Great American Cult Scare.* Boston: Beacon.

Brookover, Wilbur B., C. Beady, P. Flood, J. Schweitzer, and J. Wisenbaker. 1979. *School Social Systems and Student Achievement: Schools Can Make a Difference.* New York: Praeger.

Brookover, Wilbur B. and Edsel L. Erickson. 1975. *Sociology of Education.* Homewood, IL: Dorsey.

Brookover, Wilbur B., Edsel L. Erickson, and Alan McEvoy. 1996. *Creating Effective Schools: An In-Service Program.* Holmes Beach, FL: Learning Publications.

Brooks, Clem and Jeff Manza. 1997. "Social Cleavages and Political Alignments: U.S. Presidential Elections, 1960–1992." *American Sociological Review* 62(December):937–46.

Broom, Leonard and Philip Selznick. 1963. *Sociology: A Text With Adapted Readings,* 3rd ed. New York: Harper & Row.

Brophy, Jere E. 1983. "Research on the Self-Fulfilling Prophesy and Teacher Expectations." *Journal of Educational Psychology* 75:631–61.

Brown, Dee. 2001. *Bury My Heart at Wounded Knee: An Indian History of the American West,* 30th anniversary ed. New York: Holt.

Brown, Donald E. 1991. *Human Universals.* Philadelphia: Temple University Press.

Brown, Louise. 2001. *Sex Slaves: The Trafficking of Women in Asia.* London: Virago/Little, Brown.

Bruner, Jerome. 1996. *The Culture of Education.* Cambridge, MA: Harvard University Press.

Brunn, Stanley D., Jack F. Williams, and Donald J. Zeigler. 2003. *Cities of the World: World Regional Urban Development,* 3rd ed. Lanham, MD: Rowman and Littlefield.

Bryant, A. L. 1993. "Hostile Hallways: The AAUW Survey on Sexual Harassment in America's Schools." Washington, DC: AAUW Foundation.

Brym, Robert J. and John Lie. 2007. *Sociology: Your Compass for a New World,* 3rd ed. Belmont, CA: Wadsworth.

Buchmann, Claudia and Thomas A. DiPrete. 2006. "Gender Specific Trends in the Value of Education and the Emerging Gender Gap in College Completion." *Demography* 43:1–24.

Buechler, Steven. 2008. "What Is Critical About Sociology?" *Teaching Sociology* 36(4):318–30.

Bukhari, Zahid H. 2003. "Demography, Identity, Space: Defining American Muslims." Pp. 7–18 in *Muslims in the United States,* edited by Philippa Strum and Danielle Tarantolo. Washington, DC: Woodrow Wilson International Center for Scholars.

Bullard, Robert D., Paul Mohai, Robin Saha, and Beverly Wright. 2007. "Toxic Wastes and Race at Twenty: 1987–2007." Environmental Justice/Environmental Racism. Retrieved April 22, 2009 (http://www.ejnet.org/ej/twart.pdf).

Bumiller, Elisabeth. 1990. *May You Be the Mother of a Hundred Sons: A Journey Among the Women of India.* New York: Fawcett Columbine.

Bumpass, Larry L. and Hsien-Hen Lu. 2000. "Trends in Cohabitation and Implications for Children's Family Contexts in the United States." *Population Studies* 51(1):29–41.

Bumpass, Larry L., James A. Sweet, and Andrew Cherlin. 1991. "The Role of Cohabitation in Declining Rates of Marriage." *Journal of Marriage and the Family* 53 (November):913–27.

Burawoy, Michael. 2005. "For Public Sociology." *American Sociological Review* 56(2):4–28.

Burda, D. 2001. "Hospital Operating Margins Hit 5.2%." *Modern Healthcare* 31(45):10.

Burgess, Robert and Ronald L. Akers. 1966. "A Differential Association-Reinforcement Theory of Criminal Behavior." *Social Problems* 14:363–83.

Burgos-Debray, Elisabeth, ed. 1984. *I, Rigoberta Menchu: An Indian Woman in Guatemala,* translated by Ann Wright. London: Verso.

Burn, Shawn Meghan. 2000. *Women Across Cultures: A Global Perspective.* Mountain View, CA: Mayfield.

———. 2005. *Women Across Cultures: A Global Perspective,* 2nd ed. New York: McGraw-Hill.

———. 2011. *Women Across Cultures: A Global Perspective,* 3rd ed. New York: McGraw-Hill.

Burt, Ronald S. 1992. *Structural Holes: The Structure of Competition.* Cambridge, MA: Harvard University Press.

Businessweek Online. 2002. "The Underground Web." September 2. Retrieved July 7, 2011 (http://www.businessweek.com/magazine/content/02_35/b3797001.htm).

Caldwell, John C. 1982. *Theory of Fertility Decline.* New York: Academic Press.

Calhoun, 2002. *Dictionary of the Social Sciences.* New York: Oxford University Press.

Campbell, A., P. Converse, W. Miller, and D. Stokes. 1960. *The American Voter.* New York: Wiley.

Campbell, Ernest Q. and Thomas F. Pettigrew. 1959. *Christians in Racial Crisis.* Washington, DC: Public Affairs Press.

Canada, Geoffrey. 1998. *Reaching Up for Manhood: Transforming the Lives of Boys in America.* Boston: Beacon Press.

Canadian Health Care. 2007. "Introduction." Retrieved May 14, 2011 (www.canadian-healthcare.org/).

Cancian, Francesca M. 1992. "Feminist Science: Methodologies That Challenge Inequality." *Gender and Society* 6(4):623–42.

Carroll, Lizz. 2010. "Interracial Marriage: Which Groups Are More Likely to Wed?" *DiversityInc,* May 27. Retrieved July 11, 2011 (http://www.diversityinc.com/article/7719/Interracial-Marriage-Which-Groups-Are-More-Likely-to-Wed/).

Carroll, Rory. 2010. "Ecuador Declares State of Emergency as Country Thrown Into Chaos." September 30. Retrieved December 2, 2010 (http://www.guardian.co.uk/world/2010/sep/30/ecuador-chaos-police-rafael-correa).

Carrothers, Robert M. and Denzel E. Benson. 2003. "Symbolic Interactionism in Introductory Textbooks: Coverage and Pedagogical Implications." *Teaching Sociology* 31(2):162–81.

Cashell, Brian W. 2007. "CRS Report for Congress: Who Are the 'Middle Class'?" Retrieved February 28, 2008 (http://opencrs.cdt.org/rpts/RS22627_20070320.pdf).

Cashmore, Ellis and Barry Troyna. 1990. *Introduction to Race Relations.* London: Routledge.

Casper, Lynne M., Sara S. McLanahan, and Irwin Garfinkel. 1994. "The Gender-Poverty Gap: What We Can Learn From Other Countries." *American Sociological Review* 59(August):594–605.

Casasanto, Daniel. 2008. "Who's Afraid of the Big Bad Whorf? Crosslinguistic Differences in Temporal Language and Thought." *Language Learning,* 58(1):63–79.

Casteel, Chris. 2011. "New Survey on Hunger in America Measures Problems, Perceptions." March 11. Retrieved March 18, 2011 (http://newsok.com/new-survey-on-hunger-in-america-measures-problems-perceptions.article/3547666).

Castells, Manuel. 1977. *The Urban Question: A Marxist Approach.* Cambridge, MA: MIT Press.

———. 1989. *The Informational City: Information Technology, Economic Restructuring and the Urban-Regional Process.* Cambridge, MA: Basil Blackwell.

Castiello, Umberto, Cristina Becchio, Stefania Zoia, Cristian Nelini, Luisa Sartori, Laura Blason, . . . Vittorio Gallese. 2010. "Wired to Be Social: The Ontogeny of Human Interaction." *PLoS ONE* 5(10). Retrieved October 19, 2010 (www.plosone.org/article/info%3Adoi%2F10.1371%2Fjournal.pone.0013199).

Caulfield, Jon. 1994. *City Form and Everyday Life: Toronto's Gentrification and Critical Social Practice.* Toronto: University of Toronto Press.

CBS News. 2010. "Medicare Fraud: A $60 Billion Crime." September 5. Retrieved June 30, 2011 (http://www.cbsnews.com/stories/2009/10/23/60minutes/main5414390.shtml).

Center for Defense Information. 2002. "Military Almanac 2001–2002." Retrieved July 21, 2009 (http://www.scribd.com/doc/12928711/Military-Almanac-20012002).

Center for Education Policy. 2006. *Survey on Hours Spent on Subjects.* Menlo Park, CA: SRI International.

Center for Voting and Democracy. 2008. "Understanding Super Tuesday: State Rules on February 5 and Lessons for Reform." Retrieved March 21, 2008 (http://www.fairvote.org/?page=27&pressmode=showspecific&showarticle=185).

Center on Congress at Indiana University. 2009. "Learn About Congress." Retrieved January 12, 2009 (www.centeroncongress.org).

Centers for Disease Control and Prevention. 2009. "Overweight and Obesity." Retrieved November 4, 2009 (http://www.cdc.gov/obesity/index.html).

———. 2011. "U.S. Divorce Rate 0.68% for 2009." *Divorce Statistics and Studies Blog,* February 7. Retrieved April 12, 2011 (http://familylaw.typepad.com/stats/2011/02/us-divorce-rate-068-for-2009-table-of-state-totals.html).

Centers for Disease Control and Prevention, National Vital Statistics System. 2011. "National Marriage and Divorce Rate Trends." Retrieved April 12, 2011 (http://www.cdc.gov/nchs/nvss/marriage_divorce_tables.htm).

Centers for Medical and Medicaid Services. 2004. *Health Care Indicators.* Baltimore, MD: Office of the Actuary, Office of National Health Statistics.

Central Intelligence Agency. 2011. *The World Factbook.* Retrieved August 1, 2011 (https://www.cia.gov/library/publications/the-world-factbook/index.html).

Centre for Research on the Epidemiology of Disasters. 2011. "EM-DAT: The International Disaster Database." Retrieved August 8, 2011 (http://www.emdat.be/).

Chabbott, Colette and Francisco O. Ramirez. 2000. "Development and Education." Pp. 163–87 in *Handbook of Sociology of Education,* edited by Maureen T. Hallinan. New York: Kluwer Academic Publishers/Plenum.

Chaffins, Stephanie, Mary Forbes, and Harold E. Fuqua, Jr. 1995. "The Glass Ceiling: Are Women Where They Should Be?" *Education* 115(3):380–87.

Chalfant, H. Paul, Robert E. Beckley, and C. Eddie Palmer. 1994. *Religion in Contemporary Society,* 3rd ed. Palo Alto, CA: Mayfield.

Chalfant, H. Paul and Charles W. Peck. 1983. "Religious Affiliation, Religiosity, and Racial Prejudice: A New Look at Old Relationships." *Review of Religious Research* 25(December):155–61.

Chambliss, William J. 1973. "The Saints and the Roughnecks." *Society* 11(December):24–31.

Chapman, Paige. 2010. "Report Calls for Distance Learning to Improve Higher Education Access and Efficiency in California." *The Chronicle of Higher Education,* October 29. Retrieved April 17, 2011 (http://chronicle.com/blogs/wiredcampus/distance-learning-can-improve-higher-education-access-and-efficiency-in-california/27978).

Charles, Camille Z., Vincent J. Roscigno, and Kimberly C. Torres. 2007. "Racial Inequality and College Attendance: The Mediating Role of Parental Investments." *Social Science Research* 36(1):329–52.

Charon, Joel M. 2007. *Symbolic Interactionism: An Introduction, an Interpretation,* 9th ed. Upper Saddle River, NJ: Prentice Hall.

Chase, Cheryl. 2000. "Genital Surgery on Children Below the Age of Consent: Intersex Genital Mutilation." In *Psychological Perspectives on Human Sexuality,* edited by L. Szuchman and F. Muscarella. New York: John Wiley.

Chase-Dunn, Christopher and E. N. Anderson. 2006. *The Historical Evolution of World-Systems.* New York: Palgrave Macmillan.

Chase-Lansdale, P. Lindsay, Andrew J. Cherlin, and Kathleen E. Kierman. 1995. "The Long-Term Effects of Parental Divorce on the Mental Health of Young Adults: A Developmental Perspective." *Child Development* 66(6):1614–34.

Chaves, Mark. 1993. "Denominations as Dual Structures: An Organizational Analysis." *Sociology of Religion* 54(20):147–69.

———. 1999. *Ordaining Women: Culture and Conflict in Religious Organizations.* Cambridge, MA: Harvard University Press.

———. 2004. *Congregations in America.* Cambridge, MA: Harvard University Press.

Chaves, Mark and Philip S. Gorski. 2001. "Religious Pluralism and Religious Participation." *Annual Review of Sociology* 27:261–81.

Chen, Zeng-Yin and Howard B. Kaplan. 2003. "School Failure in Early Adolescence and Status Attainment in Middle Adulthood: A Longitudinal Study." *Sociology of Education* 76(2):110–27.

Cherlin, Andrew. 1978. "Remarriage as an Incomplete Institution." *American Journal of Sociology* 84(3):634–650.

———. 2005. *Public and Private Families: An Introduction,* 4th ed. Boston: McGraw-Hill.

———. 2007. *Public and Private Families: An Introduction,* 5th ed. Boston: McGraw-Hill.

———. 2010. *Public and Private Families: An Introduction,* 6th ed. Boston: McGraw-Hill.

Child Trends Data Bank. 2005. "Percentage of Births to Unmarried Women." Retrieved August 22, 2006 (http://www.childtrendsdatabank.org/pdf/75_PDF.pdf).

Children's Defense Fund. 2011. "Number and Percentage of Uninsured Children in Each State." Retrieved April 7, 2011 (www.childrensdefense.org/policy-priorities/childrens-health/uninsured-children/uninsured-children-state.html).

Chivers, M. L., M. C. Seto, and R. Blanchard, R. (2007). "Gender and Sexual Orientation Differences in Sexual Response to Sexual Activities Versus Gender of Actors in Sexual Films." *Journal of Personality and Social Psychology* 93(6):1108–21.

Christiano, Kevin J., William H. Swatos, Jr., and Peter Kivisto. 2008. *Sociology of Religion: Contemporary Developments,* Rev. ed. Walnut Creek, CA: AltaMira.

Chubb, John E. and Terry M. Moe. 1990. *Politics, Markets, and America's Schools.* Washington, DC: Brookings Institution.

Cillizza, Chris and Shailagh Murray. 2008. "This Time We Mean It: The Youth Vote Matters." *The Washington Post,* April 27, A8.

Clark, Burton R. 1985. "The High School and the University: What Went Wrong in America?" *Phi Delta Kappan* 66:391–97.

Clarke, Adele E. and Virginia L. Olesen. 1999. *Revisioning Women, Health, and Healing.* London: Routledge.

Clausen, John A. 1986. *The Life Course: A Sociological Perspective.* Englewood Cliffs, NJ: Prentice Hall.

Clawson, Dan, Alan Newstadt, and Denise Scott. 1992. *Money Talks: Corporate PACs and Political Influence.* New York: Basic Books.

Clinard, Marshall B. and Robert F. Meier. 2004. *Sociology of Deviant Behavior,* 12th ed. Belmont, CA: Wadsworth/Thomson Learning.

Clymer, Floyd. 1953. *Those Wonderful Old Automobiles.* New York: Bonanza.

CNN.com. 2006. "Flag-Burning Amendment Fails by a Vote." June 28. Retrieved August 16, 2008 (http://www.cnn.com/2006/POLITICS/06/27/flag.burning/).

———. 2007. "Report: Global Terrorism Up More Than 25 Percent." April 30. Retrieved April 14, 2008 (www.cnn.com/2007/US/04/30/terror.report/index.html).

CNNMoney.com. 2010. "Fortune 500: Women CEOs." Retrieved July 9, 2011 (http://money.cnn.com/magazines/fortune/fortune500/2010/womenceos/).

CNN Politics. 2009. "Obama Reverses Abortion Funding Policy." Retrieved July 27, 2011 (http://articles.cnn.com/2009–01–23/politics/obama.abortion_1_abortion-counseling-family-planning-family-planning?_s=PM:POLITICS).

Cockburn, Andrew. 2006. "21st Century Slaves." Pp. 299–307 in *Globalization: The Transformation of Social Worlds,* edited by D. Stanley Eitzen and Maxine Baca Zinn. Belmont, CA: Wadsworth.

Cockerham, William C. 2007. *Medical Sociology,* 10th ed. Englewood Cliffs, NJ: Prentice Hall.

Cohen, Elizabeth G. 1997. "Equitable Classrooms in a Changing Society." Chapter 1 in *Working for Equity in Heterogeneous Classroom: Sociological Theory in Practice,* edited by E. Cohen and R. Lotan. New York: Columbia University Press.

Coile, Russell C., Jr. 2002. "Top 10 Trends in Health Care for 2002." *Health Trends* 14(3):2–12.

Cole, Stephen and Robert Lejeune. 1972. "Illness and the Legitimation of Failure." *American Sociological Review* 37(3):347–56.

Coleman, James S. 1968. "The Concept of Equality of Educational Opportunity." *Harvard Education Review* 38(Winter):7–22.

———. 1990. *Equality and Achievement in Education.* Boulder, CO: Westview.

Coleman, James S., Ernest Q. Campbell, Carol J. Hobson, James McPartland, Alexander M. Mood, Frederic D. Weinfeld, and Robert L. York. 1966. *Equality of Educational Opportunity.* Washington, DC: Government Printing Office.

Coleman, James William. 2006. *The Criminal Elite: Understanding White Collar Crime,* 6th ed. New York: Worth.

Colligan, S. 2004. "Why the Intersexed Shouldn't Be Fixed: Insights From Queer Theory and Disability Studies." Pp. 45–60 in *Gendering Disability,* edited by B. G. Smith and B. Hutchinson. New Brunswick, NJ: Rutgers University Press.

Collins, Patricia Hill. [1990] 2000. *Black Feminist Thought: Knowledge, Consciousness, and Politics of Empowerment.* Boston: Unwin Hyman.

———. 2000. *Black Feminist Thought: Knowledge, Consciousness, and the Politics of Empowerment,* 2nd ed. New York: Routledge.

———. 2005. *Black Sexual Politics: African Americans, Gender, and the New Racism.* New York: Routledge.

Collins, Randall. 1971. "A Conflict Theory of Sexual Stratification." *Social Problems* 19(Summer):2–21.

———. 2004. "Conflict Theory of Educational Stratification." *American Sociological Review* 36:47–54.

Collins, Randall and Scott L. Coltrane. 2001. *Sociology of Marriage and the Family: Gender, Love, and Property,* 5th ed. Belmont, CA: Wadsworth.

Coltrane, Scott. 2000. "Research on Household Labor: Modeling and Measuring the Social Embeddedness of Routine Family Work." *Journal of Marriage and the Family* 62(November):1208–233.

Comarow, Murray. 1993. "Point of View: Are Sociologists Above the Law?" *The Chronicle of Higher Education,* December 15, A44.

Committee for the Study of the American Electorate. 2000. *Participation in Elections for President and U.S. Representatives, 1968–2000.* Washington, DC: Author.

Common Sense for Drug Policy. 2006. "Estimating the Illicit Drug Market." Retrieved February 10, 2008 (www.drugwardistortions.org/distortion19.htm).

Condron, Dennis J. and Vincent J. Roscigno. 2003. "Disparities Within: Unequal Spending and Achievement in an Urban School District." *Sociology of Education* 76(January):18–36.

Conrad, Peter and Joseph Schneider. 1992. *Deviance and Medicalization: From Badness to Sickness.* Philadelphia: Temple University Press.

Cook, Karen S., Jodi O'Brien, and Peter Kollock. 1990. "Exchange Theory: A Blueprint for Structure and Process." Pp. 158–81 in *Frontiers of Social Theory: The New Syntheses,* edited by George Ritzer. New York: Columbia University Press.

Cookson, Peter W., Jr., and Caroline Hodges Persell. 1985. *Preparing for Power: America's Elite Boarding Schools.* New York: Basic Books.

Cooley, Charles Horton. 1902. *Human Nature and the Social Order.* New York: Scribner.

———. [1909] 1983. *Social Organization: A Study of the Larger Mind.* New York: Schocken Books.

Coontz, Stephanie. 1997. *The Way We Really Are: Coming to Terms With America's Changing Families.* New York: Basic Books.

———. 2005. *Marriage, a History: From Obedience to Intimacy, or How Love Conquered Marriage.* New York: Viking.

Corbett, Thomas. 1994–95. "Changing the Culture of Welfare." *Focus* 16(2):12–22.

Corn, Andrew. 2009. "Wireless vs. Landlines: Past the Point of No Return." Seeking Alpha, May 8. Retrieved July 18, 2011 (http://seekingalpha.com/article/136416-wireless-vs-landlines-past-the-point-of-no-return).

CorpWatch. 1999. "Maquiladoras at a Glance." Retrieved March 7, 2008 (www.corpwatch.org/article.php?id=1528#wages).

Corsaro, William A. and Donna Eder. 1990. "Children's Peer Cultures." *Annual Review of Sociology* 16:197–220.

Coser, Lewis A. 1956. *The Functions of Social Conflict.* New York: The Free Press.

Council for a Livable World. 1998. *Nuclear Weapons: Still High Costs and Huge Stockpiles.* Washington, DC: Author.

Council of Europe. 2001, 2004. *Recent Demographic Developments in Europe 2001, 2004.* Strasbourg, France: Council of Europe Publishing.

"Cousin Marriages in the United States." 2008. Retrieved January 20, 2008 (www.marriage.about.com/cs/marriagelicenses/a/cousin.htm).

Cousteau, Jacques-Yves. 2008. "The Great Ocean Adventure." Lecture at Hanover College, January 15.

Crafts, William A. 1876. *Pioneers in the Settlement of America,* Vol. 1. Boston: Samuel Walker.

Cratty, Carol. 2008. "Internet Crime Jumps by a Third Last Year." Retrieved July 1, 2009 (www.CNN.com/2009/TECH/03/30/internet.crime/index.html).

Creswell, John W. 2009. *Research Design: Qualitative, Quantitative, and Mixed Methods Approaches,* 3rd ed. Thousand Oaks, CA: Sage.

Crosby, Faye J. 2004. *Affirmative Action Is Dead; Long Live Affirmative Action.* New Haven, CT: Yale University Press.

Crossette, Barbara. 1996. "Angkor Emerges From the Jungle." *The New York Times,* January 28. Retrieved September 9, 2008 (http://query.nytimes.com/gst/fullpage.html?res=9500e0d91139f93ba15752c0a960958260&sec=travel&spon=&page wanted=1).

Crummey, Robert O. 1970. *The Old Believers and the World of Antichrist: The Vyg Community and the Russian State, 1694–1855.* Madison: University of Wisconsin Press.

Curtiss, S. 1977. *Genie: A Psycholinguistic Study of a Modern-Day "Wild Child."* New York: Academic Press.

Cushman, Thomas and Stjepan G. Mestrovic. 1996. *This Time We Knew: Western Responses to Genocide in Bosnia.* New York: New York University Press.

Cuzzort, R. P. and Edith W. King. 2002. *Social Thought Into the Twenty-First Century,* 6th ed. Belmont, CA: Wadsworth.

Cyranoski, David. 2005. "The Long-Range Forecast." *Nature* 438(November 17):275–76.

DaCosta, Kimberly McClain. 2007. *Making Multiracials: State, Family, and Market in the Redrawing of the Color Line.* Stanford, CA: Stanford University Press.

Dahl, Robert A. 1961. *Who Governs?* New Haven, CT: Yale University Press.

Dahlberg, Frances. 1981. *Woman the Gatherer.* New Haven, CT: Yale University Press.

Dahrendorf, Ralf. 1959. *Class and Class Conflict in Industrial Societies.* Palo Alto, CA: Stanford University Press.

Dalit Liberation Education Trust. 1995. *10th Anniversary Newsletter,* May. Madras: Human Rights Education Movement of India.

Dalton, Harlon. 2012. "Failing to See." Pp. 15–18 in *White Privilege,* edited by Paula S. Rothenberg. New York: Worth.

Danesi, Marcel. 2008. *Popular Culture.* Lanham, MD: Rowman and Littlefield.

D'Antonio, William V. 1983. "Family Life, Religion, and Societal Values and Structures." Pp. 81–108 in *Families and Religion: Conflict and Change in Modern Society,* edited by William V. D'Antonio and Joan Aldous. Beverly Hills, CA: Sage.

Darr, Kurt. 2007. "Assistance in Dying: Part II. Assisted Suicide in the U.S." *Hospital Topics* 85(2):31–36.

Darwin, Charles. [1858] 1909. *The Origin of Species.* New York: P. F. Collier.

da Silva, Marina. 2004. "France: Outsider Women." *Le Monde Diplomatique,* October. Retrieved July 4, 2006 (http://mondediplo.com/2004/10/12women).

Davidhizar, R., R. Shearer, and J. N. Giger. 1997. "Pain and the Culturally Diverse Patient." *Today's Surgical Nurse* 19(6):36–39.

Davidman, Lynn. 1990. "Accommodation and Resistance to Modernity: A Comparison of Two Contemporary Orthodox Jewish Groups." *Sociological Analysis* (Spring):35–51.

Davies, James C. 1962. "Toward a Theory of Revolution." *American Sociological Review* 27(1):5–19.

———. 1974. "The J-Curve and Power Struggle Theories of Collective Violence." *American Sociological Review* 39(August):607–19.

Davies, John. 2007. *A History of Wales,* Revised and updated ed. London: Penguin.

Davis, Kingsley. 1940. "Extreme Social Isolation of a Child." *American Journal of Sociology* 45:554–65.

————. 1947. "A Final Note on a Case of Extreme Isolation." *American Journal of Sociology* 52:432–37.

————. 1960. "Legitimacy and the Incest Taboo." Pp. 398–402 in *A Modern Introduction to the Family,* edited by Norman W. Bell and Ezra F. Vogel. Glencoe, IL: The Free Press.

Davis, Kingsley and Wilbert Moore. 1945. "Some Principles of Stratification." *American Sociological Review* 10(April):242–45.

Davis, Nancy J. and Robert V. Robinson. 1999. "Their Brothers' Keepers? Orthodox Religionists, Modernists, and Economic Justice in Europe." *American Journal of Sociology* 104(6):1631–65.

Day, Jennifer Cheeseman. 2011. "Population Profile of the U.S.: Percentage of the Population, by Race and Hispanic Origin." Census Bureau, Population Division. Retrieved April 19, 2011 (http://www.census.gov/population/www/pop-profile/natproj.html).

Death Penalty Information Center. 2010. "Facts About the Death Penalty." Retrieved August 5, 2011 (www.deathpenaltyinfo.org/FactSheet.pdf).

————. 2011. "Nationwide Murder Rates." Retrieved August 8, 2011 (http://www.deathpenaltyinfo.org/murder-rates-nationally-and-state).

DeGenova, Mary Kay and Philip F. Rice. 2002. *Intimate Relationships, Marriages, and Families,* 5th ed. New York: McGraw-Hill.

DeGenova, Mary Kay, Nick Stinnett, and Nancy Stinnett. 2011. *Intimate Relationships, Marriages, and Families,* 8th ed. New York: McGraw-Hill.

Delfattore, Joan. 1992. *What Johnny Shouldn't Read: Textbook Censorship in America.* New Haven, CT: Yale University Press.

————. 2004. "Romeo and Juliet Were Just Good Friends." Pp. 177–83 in *Schools and Society,* 2nd ed., edited by Jeanne H. Ballantine and Joan Z. Spade. Belmont, CA: Wadsworth.

DeMaris, Alfred and K. Vaninadha Rao. 1992. "Premarital Cohabitation and Subsequent Marital Stability in the United States: A Reassessment." *Journal of Marriage and the Family* 54(February):178–90.

DeMartini, Joseph R. 1982. "Basic and Applied Sociological Work: Divergence, Convergence, or Peaceful Coexistence?" *The Journal of Applied Behavioral Science* 18(2):205–6.

Deming, W. E. 2000. *Out of the Crisis.* Cambridge, MA: MIT Press.

DeMitchell, Todd A. and John J. Carney. 2005. "Harry Potter and the Public School Library." *Phi Delta Kappan* 87(2):159–65.

Demographic and Health Surveys. n.d. "Measure DHS STATcompiler." Retrieved June 15, 2007 (www.measuredhs.com).

Denzin, Norman K. 1992. *Symbolic Interactionism and Cultural Studies: The Politics of Interpretation.* Cambridge, MA: Blackwell.

DePalma, Anthony. 1995. "Racism? Mexico's in Denial." *The New York Times,* June 11, E4.

Deutsch, M. and R. M. Krauss. 1960. "The Effect of Threat on Interpersonal Bargaining." *Journal of Abnormal and Social Psychology* 61:181–89.

Deutsch, Morton and Roy J. Lewicki. 1970. "'Locking-In' Effects During a Game of Chicken." *Journal of Conflict Resolution* 14(3):367–78.

Dews, C. L. Barney and Carolyn Leste Law, eds. 1995. *This Fine Place So Far From Home: Voices of Academics From the Working Class.* Philadelphia: Temple University Press.

Diamond, Jared. 1999. *Guns, Germs, and Steel: The Fates of Human Societies.* New York: W. W. Norton.

————. 2005. *Collapse: How Societies Choose to Fail or Succeed.* New York: Viking.

Diamond, Larry. 1987. "Ethnicity and Ethnic Conflict." *Journal of Modern African Studies* 25(3):117–28.

————. 1992. "Introduction: Civil Society and the Struggle for Democracy." Pp. 1–28 in *The Democratic Revolution: Struggles for Freedom and Pluralism in the Developing World,* edited by Larry Diamond. New York: Freedom House.

————. 1994. "The Global Imperative: Building a Democratic World Order." *Current History* 93(January):1–7.

————. 2003. "Universal Democracy?" *Policy Review,* June/July. Retrieved August 28, 2008 (www.hoover.org/publications/policyreview/3448571.html).

————. 2009. *The Spirit of Democracy: The Struggle to Build Free Societies Throughout the World.* New York: Times Books/Henry Holt & Co.

Dickens, Charles. 1948. *Bleak House.* London: Oxford University Press.

Diekman, Amanda B. and Sarah K. Murmen. 2004. "Learning to Be Little Women and Little Men: The Inequitable Gender Equality of Nonsexist Children's Literature." *Sex Roles: A Journal of Research* 50(5/6):373.

Dieter, Richard C. 2004. *Innocence and the Crisis in the American Death Penalty.* Retrieved July 6, 2008 (www.deathpenaltyinfo.org/article.php?scid=45&did=1149).

Directorate for Employment, Labour and Social Affairs. 2011. "OECD Health Data 2011." Retrieved August 5, 2011 (www.oecd.org/document/16/0,3343,en_2649_34631_2085200_1_1_1_1,00.html).

DiversityInc. 2008. "Belonging Nowhere: The Biracial Children of Vietnam Veterans." Retrieved October 16, 2009 (http://www.diversityinc.com/content/1757/article/4574/).

Divorce Magazine. 2011. "U.S. Divorce Statistics." Retrieved July 11, 2011 (www.divorcemag.com/statistics/statsUS.shtml).

DivorceRate.org. "Divorce Rate in Canada." Retrieved July 11, 2011 (http://www.divorcerate.org/divorce-rates-in-canada.html).

Dobbelaere, Karel. 1981. *Secularization: A Multidimensional Concept.* Beverly Hills, CA: Sage.

————. 2000. "Toward an Integrated Perspective of the Processes Related to the Descriptive Concept of Secularization." Pp. 21–39 in *The Secularization Debate,* edited by William H. Swatos, Jr., and Daniel V. A. Olson. Lanham, MD: Rowman and Littlefield.

Dobriansky, Paula. 2006. *The Education of Girls in the Developing World.* Washington, DC: U.S. Department of State.

Dodds, Peter Sheridan, Roby Muhamad, and Duncan J. Watts. 2003. "An Experimental Study of Search in Global Social Networks." *Science* 301(5634):827–29.

Dolan, Maura. 2008. "Gay Marriage Ban Overturned." *Los Angeles Times,* May 17. Retrieved August 5, 2011 (http://www.latimes.com/news/la-me-gay-marriage17-2008may17,0,7229587.story).

The Dollars and Sense Collective. 2006. "The ABCs of the Global Economy." Pp. 82–92 in *Globalization: The Transformation of Social Worlds,* edited by D. Stanley Eitzen and Maxine Baca Zinn. Belmont, CA: Wadsworth.

Domestic Violence Resource Center. 2009. "Domestic Violence Statistics." Retrieved November 28, 2009 (http://www.dvrc-or.org/domestic/violence/resources/C61/).

Domhoff, G. William. 1967. *Who Rules America.* New Jersey: Prentice Hall.

————. 1971. *Higher Circles: The Governing Class in America.* New York: Random House.

————. 1983. *Who Rules America Now? A View for the Eighties.* Englewood Cliffs, NJ: Prentice-Hall.

————. 1998. *Who Rules America? Power and Politics in the Year 2000,* 3rd ed. Mountain View, CA: Mayfield.

————. 2001. *Who Rules America? Power, Politics, and Social Change,* 4th ed. New York: McGraw-Hill.

————. 2005. *Who Rules America? Power, Politics, and Social Change,* 5th ed. New York: McGraw-Hill.

————. 2008. "Who Rules America.net: Power, Politics, and Social Change." Retrieved March 24, 2008 (http://sociology.ucsc.edu/who-rulesamerica/).

————. 2009. *Who Rules America? Challenges to Corporate and Class Dominance,* 6th ed. Upper Saddle River, NJ: Prentice Hall.

Domina, Thurston. 2005. "Leveling the Home Advantage: Assessing the Effectiveness of Parental Involvement in Elementary School." *Sociology of Education* 78(3):233–49.

Dotzler, Robert J. and Ross Koppel. 1999. "What Sociologists Do and Where They Do It—The NSF Survey on Sociologists' Work Activities and Work Places." *Sociological Practice: A Journal of Clinical and Applied Sociology* 1(1):71–83.

Douglas, Mary. 1966. *Purity and Danger.* London: Routledge and Kegan Paul.

Drafke, Michael and Stan Kossen. 2002. *The Human Side of Organizations,* 8th ed. Englewood Cliffs, NJ: Prentice Hall.

Drori, Gili S. 2006. *Global E-Litism: Digital Technology, Social Inequality, and Transnationality.* New York: Worth.

Du Bois, W. E. B. [1899] 1967. *The Philadelphia Negro: A Social Study.* New York: Schocken.

Dudley, R. L. 1996. "How Seventh-Day Adventist Lay Members View Women Pastors." *Review of Religious Research* 38(2):133–41.

Dufur, Mikaela J. and Seth L. Feinberg. 2007. "Artificially Restricted Labor Markets and Worker Dignity in Professional Football." *Journal of Contemporary Ethnography* 36(5):505–36.

Duncan, Arne. 2006. "Chicago's Renaissance 2010: Building on School Reform in the Age of Accountability." *Phi Delta Kappan* 87(6):457–58.

The Durban Accord. 2003. "Our Global Commitment for People and Earth's Protected Areas." World Commission on Protected Areas. Retrieved July 22, 2009 (http://www.danadeclaration.org/pdf/durbanaccordeng.pdf).

Durkheim, Émile. [1893] 1947. *The Division of Labor in Society,* translated by George Simpson. New York: The Free Press.

———. [1897] 1964. *Suicide.* Glencoe, IL: The Free Press.

———. [1915] 2002. *Classical Sociological Theory,* edited by Craig Calhoun. Malden, MA: Blackwell.

———. 1947. *Elementary Forms of Religious Life.* Glencoe, IL: The Free Press.

———. 1956. *Education and Society,* translated by Sherwood D. Fox. Glencoe, IL: The Free Press.

Dworkin, Anthony Gary. 2001. "Perspectives on Teacher Burnout and School Reform." *International Education Journal* 2(2):69–78.

———. 2007. "School Reform and Teacher Burnout: Issues of Gender and Gender Tokenism." Pp. 69–78 in *Gender and Education: An Encyclopedia,* edited by Barbara Banks, Sara Delamont, and Catherine Marshall. New York: Greenwood.

Dworkin, Anthony Gary and Rosalind J. Dworkin. 1999. *The Minority Report: An Introduction to Racial, Ethnic, and Gender Relations,* 3rd ed. Fort Worth, TX: Harcourt Brace.

Dworkin, Anthony Gary and Pamela F. Tobe. 2012. "Teacher Burnout in Light of School Safety, Student Misbehavior, and Changing Accountability Standards." Pp. 199–211 in *Schools and Society: A Sociological Approach to Education,* edited by Jeanne H. Ballantine and Joan Z. Spade. Thousand Oaks, CA: Pine Forge Press.

Dye, Thomas R. 2002a. *Who's Running America? The Bush Restoration.* Englewood Cliffs, NJ: Prentice Hall.

———. 2002b. *Who's Running America? The Clinton Years.* Upper Saddle River, NJ: Prentice Hall.

Dye, Thomas and Harmon Zeigler. 1983. *The Irony of Democracy.* Massachusetts: Duxbury Press.

Dyer, Richard. 2012. "The Matter of Whiteness." Pp. 9–14 in *White Privilege,* edited by Paula S. Rothenberg. New York: Worth.

Earls, Felton M. and Albert J. Reiss. 1994. *Breaking the Cycle: Predicting and Preventing Crime,* p. 49. Washington, DC: National Institute of Justice.

Ebaugh, Helen Rose Fuchs. 1988. *Becoming an Ex: The Process of Role Exit.* Chicago: University of Chicago Press.

———. 2005. *Handbook of Religion and Social Institutions.* New York: Springer.

Eckert, Penelope. 1989. *Jocks and Burnouts: Social Categories and Identity in High School.* New York: Teacher's College Press.

Eckholm, Erik. 2006. "City by City, an Antipoverty Group Plants Seeds of Change." *The New York Times,* June 26, A12.

The Economist. 2006a. "Immigration: Don't Fence Us Out." April 1–7, 41.

———. 2006b. "The United States and Mexico: Sense, Not Sensenbrenner." April 1–7, 10.

"Ecosystems Report Links Human Well-Being With Health of Planet." 2006. *Popline* 28(January/February):1.

Eder, Donna, Catherine Colleen Evans, and Stephen Parker. 1995. *School Talk: Gender and Adolescent Culture.* New Brunswick, NJ: Rutgers University Press.

Edwards, Harry. 1994. "Black Youth's Commitment to Sports Achievement: A Virtue-Turned-Tragic-Turned-Virtue." *Sport* 85(7):86.

———. 2000. "Crisis of the Black Athlete on the Eve of the 21st Century." *Society* 37(3):9–13.

Ehrenreich, Barbara. 2001. *Nickel and Dimed: On (Not) Getting By in America.* New York: Henry Holt.

———. 2005. *Bait and Switch: The (Futile) Pursuit of the American Dream.* New York: Henry Holt.

Ehrlich, Paul and Ann Ehrlich. 1990. *The Population Explosion.* New York: Simon and Schuster.

Eitzen, D. Stanley and George H. Sage. 2003. *Sociology of North American Sport.* Boston: McGraw-Hill.

Eitzen, D. Stanley and Maxine Baca Zinn, eds. 2006. *Globalization: The Transformation of Social Worlds.* Belmont, CA: Wadsworth.

———. 2012. *Globalization: The Transformation of Social Worlds,* 3rd ed. Belmont, CA: Wadsworth.

Eitzen, D. Stanley, Maxine Baca Zinn, and Kelly Eitzen Smith. 2009. *In Conflict and Order.* Upper Saddle River, NJ: Pearson.

Ellens, G. F. S. 1971. "The Ranting Ranters: Reflections on a Ranting Counter-Culture." *Church History* 40(March):91–107.

Ellison, Christopher G., John P. Bartkowski, and Michelle L. Segal. 1996. "Do Conservative Protestant Parents Spank More Often?" *Social Science Quarterly* 77(3):663–73.

Ellison, Christopher G., J. A. Burr, and P. L. McCall. 1997. "Religious Homogeneity and Metropolitan Suicide Rates." *Social Forces* 76(1):273–99.

Ellison, Christopher G. and Daniel A. Powers. 1994. "The Contact Hypothesis and Racial Attitudes Among Black Americans." *Social Science Quarterly* 75(2):385–400.

Emerling, Gary and Christina Bellantoni. 2008. "Youth Enthusiastic for Voice in Primary." *The Washington Times,* February 12, A1.

Emerson, Michael O. and Christian Smith. 2000. *Divided by Faith: Evangelical Religion and the Problem of Race in America.* New York: Oxford University Press.

Emerson, Michael O. with Rodney Woo. 2006. *People of the Dream: Multiracial Congregations in the United States.* Princeton, NJ: Princeton University Press.

Energy Information Administration. 2006. "Energy Kid's Page." Retrieved July 22, 2009 (http://www.eia.doe.gov/kids/).

Engels, Friedrich. [1884] 1942. *The Origin of the Family, Private Property, and the State.* New York: International Publishing.

Enloe, Cynthia. 2006. "Daughters and Generals in the Politics of the Globalized Sneaker." In *Beyond Borders: Thinking Critically About Global Issues,* edited by Paula S. Rothenberg. New York: Worth.

Environmental Protection Agency. 2009. "Terms of Environment." Retrieved July 27, 2011 (http://www.epa.gov/OCEPAterms/).

Erikson, Erik H. 1950. *Childhood and Society.* New York: W. W. Norton.

Erikson, Kai T. [1966] 2005. *Wayward Puritans: A Study in the Sociology of Deviance.* Boston: Pearson Education.

———. 1976. *Everything in Its Path: Destruction of Community in the Buffalo Creek Flood.* New York: Simon & Schuster.

———. 1987. "Notes on the Sociology of Deviance." Pp. 9–21 in *Deviance: The Interactionist Perspective,* 5th ed., edited by Earl Rubington and Martin S. Weinberg. New York: Macmillan.

Eshleman, J. Ross and Richard A. Bulcroft. 2006. *The Family,* 11th ed. Boston: Allyn & Bacon.

———. 2010. *The Family,* 12th ed. Boston: Allyn & Bacon.

Esperitu, Yen Le. 1992. *Asian American Panethnicity: Bridging Institutions and Identities.* Philadelphia: Temple University Press.

Esposito, L. and J. W. Murphy. 2000. "Another Step in the Study of Race Relations." *The Sociological Quarterly* 41(2):171–87.

Ethnic Majority. 2010. "African, Hispanic (Latino), and Asian American Members of Congress." Retrieved April 22, 2011 (www.ethnicmajority.com/congress.htm).

Etounga-Manguelle, Daniel. 2000. "Does Africa Need a Cultural Adjustment Program?" Pp. 65–77 in *Culture Matters: How Values*

Shape Human Progress, edited by Lawrence E. Harrison and Samuel P. Huntington. New York: Basic Books.

Etzioni, Amitai. 1975. *A Comparative Analysis of Complex Organizations.* New York: The Free Press.

Europa World Year Book. 2005. London: Europa Publications Limited.

Evans, Peter and John Stephens. 1988. "Development and the World Economy." Pp. 739–773 in *Handbook of Sociology,* edited by Neil Smelser. Newbury Park, CA: Sage.

Facebook.com. 2011. "Statistics." Retrieved August 5, 2011 (http://www .facebook.com/press/info/php?statistics).

Fackler, Martin. 2007. "Career Women in Japan Find a Blocked Path." *The New York Times,* August 6. Retrieved November 8, 2009 (www .nytimes.com/2007/08/06/world/asia/06equal.html).

Fadiman, Anne. 1997. *The Spirit Catches You and You Fall Down.* New York: Noonday.

Fainstein, Susan S., Ian Gordon, and Michael Harloe, eds. 1992. *Divided Cities: New York and London in the Contemporary World.* Oxford, UK: Blackwell.

Fallows, Deborah. 2005. "How Women and Men Use the Internet." Washington, DC: Pew Internet and American Life Project.

Faludi, Susan. 1993. *Backlash: The Undeclared War Against American Women.* New York: Anchor.

Farley, John E. 2010. *Majority-Minority Relations,* 6th ed. Englewood Cliffs, NJ: Prentice Hall.

Farley, Reynolds. 2000. *Strangers to These Shores,* 6th ed. Boston: Allyn & Bacon.

Farmer, Paul. 1999. *Infections and Inequalities: The Modern Plagues.* Berkeley: University of California Press.

Farrer, Claire R. 1996. *Thunder Rides a Black Horse: Mescalero Apaches and the Mythic Present,* 2nd ed. Prospect Heights, IL: Waveland.

Fathi, David C. 2009. "Prison Nation." Human Rights Watch, April 9. Retrieved April 15, 2009 (www.hrw.org/en/news/2009/04/09/prison-nation).

Faulkner, James B. 2006. *Social Capital, Social Services, and Recidivism.* Thesis for Applied Behavioral Science Program, Wright State University.

Fausto-Sterling, Anne. 1992. *Myths of Gender. Biological Theories About Women and Men,* 2nd ed. New York: Basic Books.

———. 1993. "The Five Sexes: Why Male and Female Are Not Enough." *The Sciences* (March/April):20–24.

———. 2000. *Sexing the Body: Gender Politics and the Construction of Sexuality.* New York: Basic Books.

Feagin, Joe R. 1983. *The Urban Real Estate Game: Playing Monopoly With Real Money.* Englewood Cliffs, NJ: Prentice Hall.

Feagin, Joe R. and Clairece Booher Feagin. 1986. *Discrimination American Style: Institutional Racism and Sexism.* Malabar, FL: Krieger.

———. 2010. *Racial and Ethnic Relations,* 9th ed. Englewood Cliffs, NJ: Prentice Hall.

Feagin, Joe R. and Hernan Vera. 2001. *Liberation Sociology.* Boulder, CO: Westview.

Feagin, Joe R., Hernan Vera, and Pinar Batur. 2001. *White Racism: The Basics.* New York: Routledge.

Featherman, David L. and Robert Hauser. 1978. *Opportunity and Change.* New York: Academic Press.

Federal Bureau of Investigation. 2005. *Fact Sheet for Hate Crime Statistics* (2004 Uniform Crime Reports). Washington, DC: Author.

———. 2006a. "Financial Crimes Report to the Public: Fiscal Year 2006." Retrieved February 8, 2008 (www.fbi.gov/publications/financial/fcs_report2006/financial_crime_2006.htm).

———. 2007a. "Innocent Images: Online Child Pornography/Child Sexual Exploitation Investigations." Retrieved August 8, 2011 (http://www.fbi.gov/stats-services/publications/innocent-images-1).

———. 2007b. "Internet Fraud." Retrieved April 15, 2009 (www.fbi.gov/majcases/fraud/internetschemes.htm).

Federal Register. 2008. "2008 HHS Poverty Guidelines." Retrieved March 12, 2009 (http://aspe.hhs.gov/poverty/08Poverty.shtml).

Feller, Avi and Chad Stone. 2009. "Top 1% of Americans Reaped Two-Thirds of Income Gains in Last Economic Expansion." Center on Budget and Policy Priorities, September 9. Retrieved January 7, 2010 (www.cbpp.org/cms/index.cfm?fa=view&id=2908).

Felson, Richard B. 2002. *Violence and Gender Reexamined.* Washington, DC: American Psychological Association.

Ferhansyed. 2008. "Fundamental Terminology of Planned Change." Retrieved December 21, 2009 (http://organizationdevelopment.wordpress.com/2008/08/10/fundamental-terminology-of-organization-development/).

Fernando, Suman. 2002. *Mental Health, Race and Culture,* 2nd ed. New York: Palgrave.

Ferree, Myra Marx. 1991. "The Gender Division of Labor in Two-Earner Marriages: Dimensions of Variability and Change." *Journal of Family Issues* 12(2):158–80.

Ferree, Myra Marx and David A. Merrill. 2000. "Hot Movements, Cold Cognition: Thinking About Social Movements in Gendered Frames." *Contemporary Sociology: A Journal of Reviews* 12:626–48.

Field, Mark. 1953. "Structured Strain in the Role of the Soviet Physician." *American Journal of Sociology* 58(5):493–502.

———. 1957. *Doctor and Patient in Soviet Russia.* Cambridge, MA: Harvard University Press.

Fields, Jason and Lynne Casper. 2001. "America's Family and Living Arrangements: March 2000." *Current Population Reports.* Washington, DC: U.S. Census Bureau.

Fine, Gary Alan. 1990. "Symbolic Interactionism in the Post-Blumerian Age." Pp. 117–57 in *Frontiers of Social Theory: The New Synthesis,* edited by George Ritzer. New York: Columbia University Press.

Finfacts. 2005. "From World Bank Development Indicators 2005." Retrieved August 21, 2006 (http://www.finfacts.com/biz10/global-worldincomepercapita.htm).

Fingarette, Herbert. 1988. *Heavy Drinking: The Myth of Alcoholism as a Disease.* Berkeley: University of California Press.

Finke, Roger. 1997. "The Consequences of Religious Competition: Supply-Side Explanations for Religious Change." Pp. 45–64 in *Rational Choice Theory and Religion: Summary and Assessment,* edited by L. A. Young. New York: Routledge.

Finke, Roger and Rodney Stark. 2005. *The Churching of America, 1776–2005: Winners and Losers in Our Religious Economy,* 2nd ed. New Brunswick, NJ: Rutgers University Press.

Finkelhor, David. 2008. *Childhood Victimization: Violence, Crime, and Abuse in the Lives of Young People.* New York: Oxford University Press.

"First Born Children." 2008. Retrieved June 29, 2008 (http://social.jrank.org/pages/261/Firstborn-children.html).

Fischer, Claude S. 1984. *The Urban Experience,* 2nd ed. San Diego: Harcourt Brace Jovanovich.

Fish, Virginia Kemp. 1986. "The Hull House Circle: Women's Friendships and Achievements." Pp. 185–227 in *Gender, Ideology, and Action: Historical Perspectives on Women's Public Lives,* edited by Janet Sharistanian. Westport, CT: Greenwood.

Fishbein, H. D. 1996. "Experimenting on Social Issues: The Case of School Desegregation." *American Psychologist* 40:452–60.

Fiske, Alan Page. 1991. *Structures of Social Life: The Four Elementary Forms of Human Relations.* New York: The Free Press.

Fitch, Catherine, Ron Goeken, and Steven Ruggles. 2005. "The Rise of Cohabitation in the United States: New Historical Estimates." Retrieved July 14, 2011 (http://www.hist.umn.edu/~ruggles/cohab-revised2.pdf).

Fitzpatrick, Laura. 2010. "Why Do Women Still Earn Less Than Men?" *Time Magazine,* April 20. Retrieved March 18, 2011 (www.time.com/time/nation/article/0,8599,1983185,00.html).

"Five Rules for Online Networking." 2005, April 1. CNN. Retrieved July 25, 2006 (www.cnn.com/2005/US/Careers/03/31/online.networking).

Flanagan, William G. 1993. *Contemporary Urban Sociology.* New York: Cambridge University Press.

————. 2001. *Urban Sociology: Images and Structure,* 4th ed. Boston: Allyn & Bacon.

Flavin, Christopher. 2001. "Rich Planet, Poor Planet." Pp. 1–410 in *State of the World 2001,* edited by Lester R. Brown, Christopher Flavin, and Hilary French. New York: W. W. Norton.

Flavin, Jeanne. 2004. "Employment, Counseling, Housing Assistance and Aunt Yolanda? How Strengthening Families' Social Capital Can Reduce Recidivism." *Fordham Urban Law Journal* 3(2):209–16.

Florida, Richard. 2002. *The Rise of the Creative Class.* New York: Basic Books.
————. 2004. *Cities and the Creative Class.* New York: Routledge.

Flynn, Stephen. 2006. "Why America Is Still an Easy Target." Pp. 246–52 in *Globalization: The Transformation of Social Worlds,* edited by D. Stanley Eitzen and Maxine Baca Zinn. Belmont, CA: Wadsworth.

Foner, Nancy. 2005. *In a New Land: A Comparative View of Immigration.* New York: New York University Press.

Food and Agriculture Organization of the United Nations. 2008. "Climate Change, Biofuels, and Land." Retrieved March 7, 2011 (ftp.fao.org/nr/HLCinfo/Land-Infosheet-En.pdf).

Food Research and Action Center. 2009. "New SNAP/Food Stamp Record: 32.2 Million." April 1. Retrieved April 20, 2009 (www.frac.org/).

Ford, Clennan S. 1970. *Human Relations Area Files: 1949–1969—A Twenty Year Report.* New Haven, CT: Human Relations Area Files.

Forgione, Pascal D., Jr. 2011. "International Test Scores." Retrieved May 13, 2011 (http://4brevard.com/choice/international-test-scores.htm).

Fox, Mary F. 1995. "Women and Higher Education: Gender Differences in the Status of Students and Scholars." Pp. 220–37 in *Women: A Feminist Perspective,* 5th ed., edited by Jo Freeman. Mountain View, CA: Mayfield.

France 24 International News. 2009. "Increase in Number of Billionaires Despite Credit Crisis." October 13. Retrieved July 18, 2011 (http://www.france24.com/en/20091013-global-crisis-china-billionaires-economy-asia-wealthy-rich-list).

Frank, Kenneth A., Yong Zhao, and Kathryn Borman. 2004. "Social Capital and the Diffusion of Innovation Within Organizations: The Case of Computer Technology in School." *Sociology of Education* 77(2):148–71.

Frank, Mark G. and Thomas Gilovich. 1988. "The Dark Side of Self- and Social Perception: Black Uniforms and Aggression in Professional Sports." *Journal of Personality and Social Psychology* 54(1):74–85.

Frankl, Razelle. 1987. *Televangelism: The Marketing of Popular Religion.* Carbondale: Southern Illinois University Press.

Frazier, Charles. 1997. *Cold Mountain.* New York: Atlantic Monthly Press.

Free the Slaves. 2011. "Modern Slavery." Retrieved April 19, 2011 (www.freetheslaves.net).

Freedom House. 2002. *Freedom in the World 2001–2002.* New York: Author.

Freeman, Jo. 1975. *The Politics of Women's Liberation: A Case Study of an Emerging Social Movement and Its Relation to the Policy Process.* New York: McKay.

Freese, J., B. Powell, and L. C. Steelman. 1999. *Rebel Without a Cause or Effect: Birth Order and Social Attitudes.* Washington, DC: American Sociological Association.

French, Howard W. 2003. "Challenges for Japan: The Glass Ceiling Starts at the Floor." *International Herald Tribune,* June 25, 1, 4.

Freud, Sigmund. [1923] 1960. *The Ego and the Id.* New York: W. W. Norton.

Freund, Peter E. S. and Meredith B. McGuire. 1999. *Health, Illness, and the Social Body,* 3rd ed. Englewood Cliffs, NJ: Prentice Hall.

Frey, Bruno S. 2004. *Dealing With Terrorism—Stick or Carrot?* Cheltenham, UK: Edward Elgar.

Friedman, Thomas L. 2005. *The World Is Flat: A Brief History of the Twenty-First Century.* New York: Farrar, Straus, and Giroux.
————. 2006. "Opening Scene: The World Is Ten Years Old." Pp. 21–29 in *Globalization: The Transformation of Social Worlds,* edited by D. Stanley Eitzen and Maxine Baca Zinn. Belmont, CA: Wadsworth.
————. 2008. *Hot, Flat and Crowded: Why We Need a Green Revolution—and How It Can Renew America.* New York: Farrar, Straus and Giroux.

Fuentes, Annette and Barbara Ehrenreich. 1983. *Women in the Global Factory.* Institute for New Communications; South End Press, INC Pamphlet No. 2.

Furnham, Adrian. 1994. "Explaining Health and Illness: Lay Perceptions on Current and Future Health, The Causes of Illness, and the Nature of Recovery." *Social Science and Medicine* 39(5):715–25.

Furstenberg, F. F. 2003. "Growing Up in American Society: Income, Opportunities, and Outcomes." Pp. 211–33 in *Social Dynamics of the Life Course: Transitions, Institutions, and Interrelations,* edited by W. R. Heinz and V. W. Marshall. New York: Aldine de Gruyter.

Future for All. 2011. "Future Technology and Society." Retrieved August 5, 2011 (www.futureforall.org/).

Gallagher, Charles. 2004. "Transforming Racial Identity Through Affirmative Action." Pp. 153–70 in *Race and Ethnicity: Across Time, Space and Discipline,* edited by Rodney D. Coates. Leiden, Holland: Brill.

Gallagher, Eugene B. and Janardan Subedi, eds. 1995. *Global Perspectives on Health Care.* Englewood Cliffs, NJ: Prentice Hall.

Gallagher, Eugene B., Thomas J. Stewart, and Terry D. Stratton. 2000. "The Sociology of Health in Developing Counties." Pp. 389–97 in *Handbook of Medical Sociology,* 5th ed., edited by C. Bird, P. Conrad, and A. Fremont. Englewood Cliffs, NJ: Prentice Hall.

Gallagher, Timothy J. and Robert J. Johnson. 2001. "A Model for Unmet Medical Care Needs." Paper presented at the American Sociological Association meetings, August, Anaheim, CA.

Gallaway, Brad. 2003. "Dead or Alive: Xtreme Beach Volleyball" (Review). Retrieved August 5, 2008 (http://www.gamecritics.com/review/doaxvolleyball/main.php).

Gallup, George, Jr., and D. Michael Lindsay. 1999. *Surveying the Religious Landscape: Trends in U.S. Beliefs.* Harrisburg, PA: Morehouse.

GameSpy. 2002. "Top Ten Shameful Games." Retrieved July 9, 2007 (http://archive.gamespy.com/top10/december02/shame/index4.shtml).

Gamoran, Adam. 1987. "The Stratification of High School Learning Opportunities." *Sociology of Education* 37:19–35.
————. 2001. "American Schooling and Educational Inequality: A Forecast for the 21st Century." *Sociology of Education* (Extra Issue: Currents of Thought: Sociology of Education at the Dawn of the 21st Century) 60(3):135–55.

Gamson, Zelda F. 1998. "The Stratification of the Academy." Pp. 67–73 in *Chalk Lines: The Politics of Work in the Managed University,* edited by Randy Martin. Raleigh, NC: Duke University Press.

Gans, Herbert J. 1962. *The Urban Villagers: Group and Class in the Life of Italian-Americans.* New York: The Free Press.
————. 1971. "The Uses of Poverty: The Poor Pay All." *Social Policy* 2(2):20–24.
————. 1991. *People, Plans, and Policies: Essays on Poverty, Racism, and Other National Urban Problems.* New York: Columbia University Press.
————. 1994. "Positive Functions of the Undeserving Poor: Uses of the Underclass in America." *Politics and Society* 22(3):269–83.
————. 1995. *The War Against the Poor.* New York: Basic Books.
————. 2007. "No, Poverty Has Not Disappeared." Reprinted in *Sociological Footprints,* edited by Leonard Cargan and Jeanne Ballantine. Belmont, CA: Wadsworth.

Gardner, Howard. 1987. "The Theory of Multiple Intelligences." *Annual Dyslexia* 37:19–35.
————. 1999. *Intelligence Reframed: Multiple Intelligences for the 21st Century.* New York: Basic Books.

Garfield, Richard and Glen Williams. 1992. *Health Care in Nicaragua: Primary Care Under Changing Regimes.* New York: Oxford University Press.

Garner, Catherine L. and Stephen W. Raudenbush. 1991. "Neighborhood Effects on Educational Attainment: A Multilevel Analysis." *Sociology of Education* 64(4):251–60.

Garreau, Joel. 1991. *Edge City: Life on the New Frontier.* New York: Doubleday.

Garrett, Laurie. 2000. *Betrayal of Trust: The Collapse of Global Public Health.* New York: Hyperion.

Gaustad, E. S. and Phillip L. Barlow. 2001. *New Historical Atlas of Religion in America.* Oxford, UK: Oxford University Press.

Gay, Lesbian, and Straight Education Network (GLSEN). 2006. "GLSEN's 2005 National School Climate Survey Sheds New Light on Experiences of Lesbian, Gay, Bisexual and Transgendered (LGBT) Students." Retrieved August 8, 2008 (http://www.glsen.org/cgi-bin/iowa/all/library/record/1927.html).

Gay Marriage Research Center. 2011. "Gay Facts and Statistics in 2011." January 3. Retrieved July 11, 2011 (http://www.gaymarriageresearch.com/gay-facts-statistics-2011/).

Gehrig, Gail. 1981. "The American Civil Religion Debate: A Source of Theory Construction." *Journal for the Scientific Study of Religion* 20(1):51–63.

Geist, Michael. 2008. "How Obama's Using Tech to Triumph." *The Toronto Star,* February 4, B1.

Gelles, Richard J. 1995. *Contemporary Families: A Sociological View.* Thousand Oaks, CA: Sage.

Gellner, Ernest. 1987. *Culture, Identity, and Politics.* Cambridge: Cambridge University Press.

———. 1993. "Nationalism." Pp 409–11 in *Blackwell Dictionary of Twentieth Century Thought,* edited by William Outhwaite and Tom Bottomore. Oxford, UK: Basil Blackwell.

Gellner, Ernest and John Breuilly. 2009. *Nations and Nationalism,* 2nd ed. Ithaca, NY: Cornell University Press.

Geocommons. 2009. "Infant Mortality Rates, World by Country, 2009." Retrieved July 7, 2011 (http://finder.geocommons.com/overlays/11932).

Gettleman, Jeffery. 2008. "Mob Sets Kenya Church on Fire, Killing Dozens." *The New York Times,* January 2. Retrieved June 3, 2008 (www.nytimes.com/2008/01/02/world/africa/02kenya.html?pagewanted=1&ref=africa).

Gibbs, Jack P. 1989. *Control: Sociology's Central Notion.* Urbana: University of Illinois Press.

Gibson, William. 1999. "Science and Science Fiction." *Talk of the Nation,* National Public Radio, November 30.

Giddens, Anthony. 1986. *The Constitution of Society.* Berkeley: University of California Press.

———. 1987. *Social Theory and Modern Sociology.* Cambridge, UK: Polity Press.

Gilbert, Dennis. 2008. *The American Class Structure in an Age of Growing Inequality,* 7th ed. Thousand Oaks, CA: Pine Forge Press.

Gilbert, Dennis and Joseph A. Kahl. 2003. *The American Class Structure in an Age of Inequality: A New Synthesis,* 6th ed. Belmont, CA: Wadsworth.

Gilchrist, John. 2003. *Anderson's Ohio Family Law.* Cincinnati, OH: Anderson.

Gilligan, Carol. 1982. *In a Different Voice: Psychological Theory and Women's Development.* Cambridge, MA: Harvard University Press.

Ginsberg, E. 1999. "U.S. Health Care: A Look Ahead to 2025." *Annual Review of Public Health* 20(1):55–67.

Gitelson, Alan, Margaret Conway, and Frank Feigert. 1984. *American Political Parties: Stability and Change.* Boston: Houghton Mifflin.

Glaser, James M. 1994. "Back to the Black Belt: Racial Environment and White Racial Attitudes in the South." *The Journal of Politics* 56(1):21–41.

Glass, John. 2003. "Hunting for Bambi. Hoax? Reality? Does It Matter?" *Common Dreams* (online journal). Retrieved August 5, 2008 (http://www.commondreams.org/views03/0801–05.htm).

———. 2004. "Developing Sociological Interventions." Retrieved September 29, 2006 (http://iws.collin.edu/jglass/How%20to%20Frame%20Sociological%20Interventions.pdf).

Glasscock, C. B. 1937. *The Gasoline Age: The Story of the Men Who Made It.* Indianapolis, IN: Bobbs-Merrill.

Glasser, Susan B. 2005. "U.S. Figures Show Sharp Global Rise in Terrorism." *The Washington Post,* April 25, A01.

Glenn, Evelyn Nakano. 1999. "The Social Construction and Institutionalization of Gender and Race: An Integrative Framework." Pp. 3–43 in *Revisioning Gender,* edited by Myra Marx Ferree, Judith Lorber, and Beth B. Hess. Thousand Oaks, CA: Sage.

Global Campaign for Education. 2011. "100th Anniversary of International Women's Day." Retrieved April 17, 2011 (www.campaignforeducation.org/en/news/gces-news).

Global Citizen Corps. 2010. "Women's Rights." December 6. Retrieved April 8, 2011 (www.globalcitizencorps.org/groups/issue-human-rights/14178).

Globe Women. 2011. "WEXPO: Women's Online Marketplace." Retrieved August 8, 2011 (http://www.wexpo.biz//).

Goffman, Erving. [1959] 2001. *Presentation of Self in Everyday Life.* New York: Harmondsworth: Penguin.

———. 1961. *Asylums: Essays on the Social Situation of Mental Patients and Other Inmates.* New York: Anchor.

———. 1967. *Interaction Ritual.* New York: Anchor.

Gold, Marsha R. 1999. "The Changing U.S. Health Care System: Challenges for Responsible Public Policy." *Milbank Quarterly* 77(1):3–37.

Gold, Marsha R., L. Nelson, T. Lake, R. Hurley, and R. Berenson. 1998. "Behind the Curve: A Critical Assessment of How Little Is Known About Arrangements Between Managed Care Plans and Physicians." Pp. 67–100 in *Contemporary Managed Care: Readings in Structure, Operations, and Public Policy,* edited by M. R. Gold. Chicago: Health Administration Press.

Goldberg, David Theo, ed. 1990. *Anatomy of Racism.* Minneapolis: University of Minnesota Press.

Goldberg, Gertrude Schaffner and Eleanor Kremen. 1990. *The Feminization of Poverty.* New York: Praeger.

Golden, Tim. 1994. "Health Care, Cuba's Pride, Falls on Hard Times." *The New York Times,* October 30, 1.

Goode, Erich. 1992. *Collective Behavior.* New York: Harcourt Brace Jovanovich.

———. 1997. *Between Politics and Reason: The Drug Legalization Debate.* New York: St. Martin's.

———. 2005. *Drugs in American Society,* 6th ed. Boston: McGraw-Hill.

Goode, William J. 1970. *World Revolution and Family Patterns.* New York: The Free Press.

———. 1990. "Encroachment, Charlatanism, and the Emerging Profession: Psychiatry, Sociology, and Medicine." *American Sociological Review* 25(6):902–14.

Goodman, Marc D. and Susan W. Brenner. 2002. "The Emerging Consensus on Criminal Conduct in Cyberspace." *International Journal of Law and Information Technology* 10(2):139–223.

Gordon, Milton. 1970. "The Subsociety and the Subculture." Pp. 150–63 in *The Sociology of Subcultures,* edited by David O. Arnold. Berkeley, CA: Glendessary.

Gore, Al. 2006. *An Inconvenient Truth.* Emmaus, PA: Rodale.

Gottdiener, Mark. 1987. *The Decline of Urban Politics: Political Theory and the Crisis of the Local State.* Newbury Park, CA: Sage.

Gottdiener, Mark and Ray Hutchison. 2006. *The New Urban Sociology,* 3rd ed. Boston: McGraw-Hill.

Gottfredson, Michael R. and Travis Hirschi. 1990. *A General Theory of Crime.* Palo Alto, CA: Stanford University Press.

Gottlieb, Lori. 2006. "How Do I Love Thee?" *The Atlantic* 297(2):58–70.

Gould, Stephen J. 1997. *The Mismeasure of Man.* New York: W. W. Norton.

Gouldner, Alvin W. 1960. "The Norm of Reciprocity: A Preliminary Statement." *American Sociological Review* 25(2):161–178.

Gracey, Harry L. 1967. "Learning the Student Role: Kindergarten as Academic Boot Camp." Pp. 215–226 in *Readings in Introductory Sociology,* 3rd ed., edited by Dennis Wrong and Harry L. Gracey. New York: Macmillan.

Grandpa Junior. 2006. "If You Were Born before 1945." Retrieved July 20, 2006 (www.grandpajunior.com/1945.shtml).

Granovetter, Mark. 2007. "Introduction for the French Reader." *Sociologica* 1(S):1–10.

Grant, Gerald and Christine E. Murray. 1999. *Teaching in America: The Slow Revolution.* Cambridge, MA: Harvard University Press.

Grant, Linda. 2004. "Everyday Schooling and the Elaboration of Race-Gender Stratification." Pp. 296–308 in *Schools and Society: A Sociological Approach to Education,* 2nd ed., edited by Jeanne H. Ballantine and Joan Z. Spade. Belmont, CA: Wadsworth.

Greeley, Andrew M. 1972. *The Denominational Society.* Glenview, IL: Scott, Foresman.

———. 1989. *Religious Change in America.* Cambridge, MA: Harvard University Press.

Green, John C. and Mark Silk. 2005. "Why Moral Values Did Count." *Religion in the News* 8(1):5–8.

Greenberg, Edward S. 1999. *The Struggle for Democracy,* 3rd ed. New York: Addison-Wesley.

The Green Papers. 2008a. "Presidential Primaries, Caucuses, and Conventions." Retrieved March 21, 2008 (http://www.thegreen papers.com/P08/CO-R.phtml).

———. 2008b. "Presidential Primaries 2008: Republican Delegate Selection and Voter Eligibility." Retrieved March 21, 2008 (http://www.thegreenpapers.com/P08/R-DSVE.phtml?sort=a).

Greensboro Justice Fund. 2005. "Courage from the Past." *GJF Newsletter* 17(Summer):1.

Greimel, Hans. 2007. "Outbreak of Violent Crime Unnerves Japan." *The Washington Post,* May 18. Retrieved July 21, 2009 (http://www.washingtonpost.com/wp-dyn/content/article/2007/05/18/AR2007051801059.html).

Grey, Barry. 2000. "Leaked CIA Report Says 50,000 Sold Into Slavery in US Every Year." World Socialist Web Site, April 3. Retrieved April 7, 2011 (http://www.wsws.org/articles/2000/apr2000/slav-a03.shtml).

Guest, Avery M. and Keith R. Stamm. 1993. "Paths of Community Integration." *Sociological Quarterly* 34(4):581–95.

Gumperz, John J. and Stephen C. Levinson. 1991. "Rethinking Linguistic Relativity." *Current Anthropology* 32(5):613–23.

———, eds. 1996. *Rethinking Linguistic Relativity.* Cambridge, UK: Cambridge University Press.

Habermas, Jurgen and Thomas McCarthy. 1984. *The Theory of Communicative Action.* Boston: Beacon Press.

Hadden, Jeffrey K. 1987. "Toward Desacralizing Secularization Theory." *Social Forces* 65:587–611.

———. 2006. "New Religious Movements." Retrieved August 14, 2008 (http://www.hirr.hartsem.edu/denom/new_religious_movements.html).

Hadden, Jeffrey K. and Anson Shupe. 1988. *Televangelism: Power and Politics on God's Frontier.* New York: Henry Holt.

Hafferty, Frederic. 2000. "Reconfiguring the Sociology of Medical Education: Emerging Topics and Pressing Issues." Pp. 238–57 in *Handbook of Medical Sociology,* 5th ed., edited by C. Bird, P. Conrad, and A. Fremond. Englewood Cliffs, NJ: Prentice Hall.

Hagan, Frank E. 2007. *Introduction to Criminology,* 6th ed. Belmont, CA: Wadsworth.

Hagan, John L. 1993. "The Social Embeddedness of Crime and Unemployment." *Criminology* 31:465–91.

———. 1994. *Crime and Disrepute.* Thousand Oaks, CA: Pine Forge.

Hagan, John, Edward T. Silva, and John H. Simpson. 1977. "Conflict and Consensus in the Designation of Deviance." *Social Forces* 56(2):320–40.

Hagopian, Mark N. 1984. *Regimes, Movements, and Ideologies,* 2nd ed. New York: Longman.

Hall, Edward T. 1959. *The Silent Language.* New York: Doubleday.

———. 1983. *The Dance of Life.* Garden City, NY: Anchor Books/Doubleday.

Hall, Edward T. and Mildred Reed Hall. 1992. *An Anthropology of Everyday Life.* New York: Doubleday.

Hall, Richard H. 2002. *Organizations: Structures, Processes, and Outcomes,* 7th ed. Englewood Cliffs, NJ: Prentice Hall.

Hallinan, Maureen T. 1994. "Tracking: From Theory to Practice." *Sociology of Education* 67(2):79–91.

Hallinan, Maureen T. and Richard A. Williams. 1989. "Interracial Friendship Choices in Secondary Schools." *American Sociological Review* 54(February):67–78.

———. 1990. "Students' Characteristics and the Peer-Influence Process." *Sociology of Education* 63(2):122–32.

Hamilton, Brady E., Stephanie J. Ventura, Joyce A. Martin, and Paul D. Sutton. 2006. "Final Births for 2004." National Center for Health Statistics. Retrieved September 15, 2006 (www.cdc.gov/nchs/products/pubs/pubd/hestats/finalbirths04/finalbirths04.htm).

Hammersley, Martyn and Glenn Turner. 1980. "Conformist Pupils." In *Pupil Strategies: Explorations in the Sociology of the School,* edited by Peter Woods. London: Croom Helm.

Handel, Gerald, Spencer Cahill, and Frederick Elkin. 2007. *Children and Society: The Sociology of Children and Childhood Socialization.* New York: Oxford University Press USA.

Handwerk, Brian. 2004. "Female Suicide Bombers: Dying to Kill." *National Geographic News,* December 13. Retrieved July 21, 2009 (http://news.nationalgeographic.com/news/2004/12/1213_041213_tv_suicide_bombers.html).

Haniffa, Aziz. 2009. "Financial Crisis Bigger Threat Than Al Qaeda, Says U.S. Intelligence Czar." *Rediff India Abroad,* February 15. Retrieved March 17, 2009 (www.rediff.com/money/2009/feb/15bcrisis-financial-crisis-bigger-threat-than-al-qaeda-says-us-intel-chief.htm).

Hannigan, John. 1998. *Fantasy City: Pleasure and Profit in the Postmodern Metropolis.* London: Routledge.

Hansen, Randall and Katharine Hansen. 2003. "What Do Employers Really Want? Top Skills and Values Employers Seek From Job-Seekers." Quintessential Careers. Retrieved June 23, 2008 (www.quintcareers.com/job_skills_values.html).

Hanser, Robert D. 2002. "Labeling Theory as a Paradigm for the Etiology of Prison Rape: Implications for Understanding and Intervention." *Professional Issues in Counseling On-line Journal,* April. Retrieved July 21, 2009 (http://www.shsu.edu/~piic/summer2002/Hanser.htm).

Hardin, Garrett. 1968. "The Tragedy of the Commons." *Science* 162(3859):1243–48.

Hargreaves, D. H. 1978. "The Two Curricula and the Community." *Westminster Studies in Education* 1:31–41.

Harlingen Economic Development Corporation. 2011. "Maquiladora Advantages to U.S. and Mexico Economies." Retrieved April 9, 2011 (www.harlingenedc.com/InternationalOpportunity/Maquiladoras).

Harrington, Charlene and Carroll L. Estes. 2004. *Health Policy: Crisis and Reform in the U.S. Health Care Delivery System,* 2nd ed. Sudbury, MA: Jones and Bartlett.

Harris, Judith Rich. 2009. *The Nurture Assumption: Why Children Turn Out the Way They Do,* Revised edition. New York: The Free Press.

Harris, Marvin. 1989. *Cows, Pigs, War, and Witches: The Riddles of Culture.* New York: Random House.

Harrison, P. M. and J. C. Karberg. 2003. *Prison and Jail Inmates at Midyear 2002.* Washington, DC: U.S. Bureau of Justice Statistics.

Hart, Betty, and Todd R. Risley. 2003. "The Early Catastrophe: The 30 Million Word Gap by Age 3." *American Educator* 27(1):4–9.

Harvey, David. 1973. *Social Justice and the City.* London: Edward Arnold.

Haskins, Ron. 2006. *Work Over Welfare: The Inside Story of the 1996 Welfare Reform Law.* Washington, DC: Brookings Institution Press.

Haupt, Arthur and Thomas T. Kane. 2003. *Population Reference Bureau Report, 2003.* Washington, DC: U.S. Population Bureau.

Hayden, Tom. 2006. "Seeking a New Capitalism in Chiapas." Pp. 348–54 in *Globalization: The Transformation of Social Worlds,* edited by D. Stanley Eitzen and Maxine Baca Zinn. Belmont, CA: Wadsworth.

Hayes, Bernadette C. 1995. "The Impact of Religious Identification on Political Attitudes: An International Comparison." *Sociology of Religion* 56(2):177–94.

Haynie, Dana L. 2001. "Delinquent Peers Revisited: Does Network Structure Matter?" *American Journal of Sociology* 106(4):1013–57.

Headlam, Bruce. 2000. "Barbie PC: Fashion Over Logic." *The New York Times,* January 20, G4.

Healey, Joseph F. 2006. *Race, Ethnicity, Gender and Class: The Sociology of Group Conflict and Change,* 4th ed. Thousand Oaks, CA: Pine Forge Press.

Health Confidence Survey. 2001. "2001 Health Confidence Survey: Summary of Findings," pp. 1–9. Washington, DC: Health Confidence Survey/Employee Benefit Research.

HealthReform.gov. n.d. "Coverage Denied: How the Current Health Insurance System Leaves Millions Behind." Retrieved January 10, 2010 (*www.healthreform.gov/reports/denied_coverage/index.html*).

Heaton, Tim B. 1990. "Religious Group Characteristics, Endogamy, and Interfaith Marriages." *Sociological Analysis* 51(4):363–76.

Heilbroner, Robert L. and William Milberg. 2007. *The Making of Economic Society,* 12th ed. Englewood Cliffs, NJ: Prentice Hall.

Heilbroner, Robert L. and Lester C. Thurow. 1981. *The Economic Problem,* 6th ed. New Jersey: Prentice Hall.

Heilman, Madeline E. and Julie J. Chen. 2003. "Entrepreneurship as a Solution: The Allure of Self-Employment for Women and Minorities." *Human Resource Management Review* 13(2):347–64.

Heilman, Samuel. 2000. *Defenders of the Faith: Inside Ultra-Orthodox Jewry.* Berkeley: University of California Press.

Helfand, Duke. 2008. "Presbyterian Leaders OK Gay Clergy." *Los Angeles Times,* June 28. Retrieved August 18, 2008 (http://articles.latimes.com/2008/jun/28/local/me-ordain28).

Helgesen, Sally. 1995. *The Female Advantage: Women's Ways of Leadership.* New York: Doubleday.

Helium. 2011. "Is Overpopulation a World Threat?" Retrieved August 1, 2011 (www.helium.com/debates/166036-is-overpopulation-a-world-threat/side_by_side).

Hendry, Joy. 1987. *Becoming Japanese: The World of the Preschool Child.* Honolulu: University of Hawaii Press.

———. 1995. *Understanding Japanese Society,* 2nd ed. New York: Routledge.

Henley, Nancy, Mykol Hamilton, and Barrie Thorne. 2000. "Womanspeak and Manspeak: Sex Differences in Communication, Verbal and Nonverbal." Pp. 111–15 in *Sociological Footprints,* edited by Leonard Cargan and Jeanne Ballantine. Belmont, CA. Wadsworth.

Hensley, Christopher, M. Koscheski, and Richard Tewksbury. 2005. "Examining the Characteristics of Male Sexual Assault Targets in a Southern Maximum-Security Prison." *Journal of Interpersonal Violence* 20(6):667–79.

Henslin, James M. 2005. *Sociology: A Down-to-Earth Approach,* 8th ed. Boston: Allyn & Bacon.

Heritage House '76. 1998. "Adoption or Abortion: Decision of a Lifetime." Retrieved April 3, 2008 (www.abortionfacts.com/literature/literature_9338aa.asp).

Herman, Edward S. and David Peterson. 2006. "The Threat of Global State Terrorism: Retail vs. Wholesale Terror." Pp. 252–57 in *Globalization: The Transformation of Social Worlds,* edited by D. Stanley Eitzen and Maxine Baca Zinn. Belmont, CA: Wadsworth.

Hertsgaard, Mark. 2003. *The Eagle's Shadow: Why America Fascinates and Infuriates the World.* New York: Picador.

Hesse-Biber, Sharlene Nagy. 2007. *The Cult of Thinness,* 2nd ed. New York: Oxford University Press.

Hesse-Biber, Sharlene Nagy and Patricia Lina Leavey. 2007. *Feminist Research Practice: A Primer.* Thousand Oaks, CA: Sage.

Hewitt, John P. 2007. *Self and Society: A Symbolic Interactionism Social Psychology,* 10th ed. Boston: Allyn & Bacon.

Hiebert-White, Jane. 2009. "52 Million Uninsured Americans by 2010." *Health Affairs,* June 2. Retrieved July 26, 2011 (http://healthaffairs.org/blog/2009/06/02/52-million-uninsured-americans-by-2010/).

Hill, Christopher. 1991. *The World Turned Upside Down: Radical Ideas During the English Revolution.* New York: Penguin.

Himmelstein, David U. and Steffie Woodhandler. 1992. *National Health Program Chartbook.* Cambridge, MA: Center for National Health Program Studies.

Hinton, Christopher. 2010."Global Military Spending to Outpace GDP Growth in 2010." Market Watch. June 18. Retrieved May 10, 2011 (*www.marketwatch.com/story/worlds-militaries-see-another-budget-busting-year-2010–06–18*).

Hirschi, Travis. [1969] 2002. *Causes of Delinquency.* New Brunswick, NJ: Transaction Publishers.

Hispanic-Americans.com. 2011. "Mexico Leader Likens Drug Battle to Fight Against French." May 6. Retrieved May 9, 2011 (http://hispanic-americans.com/blog/mexiconews/mexican-leader-likens-drug-battle-to-fight-against-french.aspx).

Hitler, Adolf. 1939. *Mein Kampf.* New York: Reynal and Hitchcock.

Hochschild, Arlie. 1989. *The Second Shift: Working Parents and the Revolution at Home.* New York: Viking.

———. 1997. *The Time Bind: When Work Becomes Home and Home Becomes Work.* New York: Metropolitan Books.

Hoffer, Thomas, Andrew M. Greeley, and James S. Coleman. 1987. "Catholic High School Effects on Achievement Growth." Pp. 67–88 in *Comparing Public and Private Schools, Vol. 2, Student Achievement,* edited by Edward H. Haertel, Thomas James, and Henry M. Levin. New York: Falmer.

Hogeland, Lisa Maria. 2004. "Fear of Feminism: Why Young Women Get the Willies." Pp. 565–68 in *Women's Voices, Feminist Visions,* 2nd ed., edited by Susan M. Shaw and Janet Lee. Boston: McGraw-Hill.

Holloway, Susan. 2001. "Mothers of Japanese Preschoolers." *GSE Term Paper,* Vol. VIII, No. 1. University of California, Berkeley. Retrieved October 21, 2009 (http://gse.berkeley.edu/admin/publications/term-paper/fa1101/holloway.html).

Homans, George C. 1974. *Social Behavior: Its Elementary Forms.* New York: Harcourt, Brace Jovanovich.

Hood, Roger. 2002. *The Death Penalty: A World-wide Perspective,* 3rd ed. Oxford, UK: Clarendon Press.

Horrigan, John B. and Aaron Smith. 2007. "Home Broadband Adoption 2007 Report." Pew Internet and American Life Project. Retrieved July 21, 2009 (http://www.pewinternet.org/Reports/2007/Home-Broadband-Adoption-2007.aspx).

Hossfeld, Karen J. 2006. "Gender, Race, and Class in Silicon Valley." Pp. 264–70 in *Beyond Borders: Thinking Critically About Global Issues,* edited by Paula S. Rothenberg. New York: Worth.

Hostetler, John A. 1993. *Amish Society,* 4th ed. Baltimore: John Hopkins University Press.

Houlihan, G. Thomas. 2005. "The Importance of International Benchmarking for U.S. Educational Leaders." *Phi Delta Kappan* 87(3):217–18.

House, James S. 1994. "Social Structure and Personality: Past, Present, and Future." Pp. 77–102 in *Sociological Perspectives on Social Psychology,* edited by Karen Cook, Gary Fine, and James S. House. Boston: Allyn & Bacon.

Housing and Urban Development. 2008. "HUD Reports Drop in the Number of Chronically Homeless Persons" (News Release No. 08–113). Retrieved July 21, 2009 (http://www.hud.gov/news/release.cfm?content=pr08–113.cfm).

Howard, Adam. 2007. *Learning Privilege: Lessons of Power and Identity in Affluent Schooling.* New York: Taylor and Francis.

Howard, Philip N. and Steve Jones, eds. 2004. *Society On-Line: The Internet in Context.* Thousand Oaks, CA: Sage.

Hozien, Muhammad. n.d. "Ibn Khaldun: His Life and Work." Retrieved May 7, 2009 (www.muslimphilosophy.com/ik/klf.htm).

Hu, Winnie. 1999. "Woman Fearing Mutilation Savors Freedom." *The New York Times,* August 20, A21.

Huddy, Leonie and Stanley Feldman. 2006. "Worlds Apart: Blacks and Whites React to Hurricane Katrina." *Du Bois Review* 3(1):97–113. Retrieved July 7, 2011 (http://journals.cambridge.org/action/display Abstract?fromPage=online&aid=462978).

Huebler, Friedrich. 2008. "International Education Statistics: Adult Literacy in 2007." June 15. Retrieved July 11, 2009 (http://huebler.blogspot.com/2008/06/adult-literacy.html).

Hughes, David and Lesley Griffiths. 1999. "On Penalties and the Patient's Charter: Centralism v. De-centralised Governance in the NHS." *Sociology of Health and Illness* 21(1):71–94.

Hughes, Melanie M. 2004. *Armed Conflict, International Linkages, and Women's Parliamentary Representation in Developing Nations*. Master's Thesis, Department of Sociology, The Ohio State University.

Huizinga, David, Rolf Loeber, and Terence P. Thornberry. 1994. "Urban Delinquency and Substance Abuse: Initial Findings." *OJJDP Research Summary*. Washington, DC: Government Printing Office.

Human Development Report. 2009. "Old Age Dependency Ratio." Retrieved May 13, 2011 (http://hdrstats.undp.org/en/indicators/147.html).

Human Rights Campaign. 2003. *Answers to Questions About Marriage Equality*. Washington, DC: Human Rights Campaign, Family Net Project.

———. 2008. "Questions About Same Sex Marriage." Retrieved February 20, 2008 (http://www.hrc.org/issues/5517.htm).

———. 2011. "An Overview of Federal Rights and Protections Granted to Married Couples." Retrieved April 12, 2011 (www.hrc.org/issues/5585.htm).

Humes, Karen R., Nicholas A. Jones, and Roberto R. Ramirez. 2011. "2010 Census Briefs: Overview of Race and Hispanic Origins: 2010." Retrieved April 7, 2011 (http://www.census.gov/prod/cen2010/briefs/c2010br-02.pdf).

Hunsberger, B. 1995. "The Role of Religious Fundamentalism, Quest, and Right-Wing Authoritarianism." *Journal of Social Issues* 51(2):113–29.

Hunter College Women's Studies Collective. 2005. *Women's Realities, Women's Choices: An Introduction to Women's Studies,* 3rd ed. New York: Oxford University Press.

Hunter, James Davidson. 1983. *American Evangelicalism*. New Brunswick, NJ: Rutgers University Press.

Hurst, Charles E. 2006. *Social Inequality: Forms, Causes and Consequences,* 6th ed. Boston: Allyn & Bacon.

Iannaccone, Laurence R. 1994. "Why Strict Churches Are Strong." *American Journal of Sociology* 59(5):1180–211.

———. 1995. "Voodoo Economics? Reviewing the Rational Choice Approach to Religion." *Journal for the Scientific Study of Religion* 34(1):76–88.

Identity Theft Resource Center. 2011. "Workplace Facts and Statistics." Retrieved July 7, 2011 (http://www.idtheftcenter.org/workplace_facts.html).

Immigration Policy Center. 2010. "Immigrant Women in the United States: A Portrait of Demographic Diversity." Retrieved May 15, 2011 (http://www.immigrationpolicy.org/just-facts/immigrant-women-united-states-portrait-demographic-diversity).

Independence Hall Association. 2010. "Betsy Ross and the American Flag." Retrieved July 22, 2011 (http://www.ushistory.org/betsy/flag-tale.html).

Information Please Database. 2007. "Population Density per Square Mile of Countries." Retrieved January 6, 2010 (www.infoplease.com/ipa/A0934666.html).

———. 2010. "Significant Ongoing Armed Conflicts." Retrieved August 5, 2011 (www.infoplease.com/ipa/A0904550.html).

———. 2008. "National Voter Turnout in Federal Elections: 1960–2008." Retrieved March 17, 2009 (www.infoplease.com/ipa/A0781453.html/).

Ingersoll, Richard M. 2004. "The Status of Teaching as a Profession." Pp. 102–18 in *Schools and Society: A Sociological Approach to Education,* 2nd ed., edited by Jeanne H. Ballantine and Joan Z. Spade. Belmont, CA: Wadsworth.

———. 2005. "The Problem of Underqualified Teachers: A Sociological Perspective." *Sociology of Education* 78(2):175–79.

Ingersoll, Richard M. and Elizabeth Merrill. 2012. "The Status of Teaching as a Profession." Pp. 185–98 in *Schools and Society: A Sociological Approach to Education,* edited by Jeanne H. Ballantine and Joan Z. Spade. Thousand Oaks, CA: Pine Forge Press.

Inglehart, Ronald. 1997. *Modernization and Postmodernization: Cultural, Economic, and Political Change in 43 Societies*. Princeton, NJ: Princeton University Press.

Inglehart, Ronald and Wayne E. Baker. 2001. "Modernization's Challenge to Traditional Values: Who's Afraid of Ronald McDonald?" *The Futurist* 35(2):16–22.

Inhofe, James M. 2008. "Senate Republicans Introduce Package of Immigration Enforcement Bills." Retrieved August 5, 2011 (http://inhofe.senate.gov/public/index.cfm?FuseAction=PressRoom.PressReleases&ContentRecord_id=80247da6–802a-23ad-422d-6aba9ec1292f&Region_id=4ac9f730-b2c1-a0af-012d-a56a166ffe75&Issue_id=4e74117e-802a-23ad-40cc-02fa02764c40).

Inniss, Janis Prince. 2010. "A Closer Look at Interracial Marriage Statistics." *Everyday Sociology,* August 2. Retrieved April 10, 2011 (http://nortonbooks.typepad.com/everydaysociology/2010/08/a-closer-look-at-interracial-marriage-statistics.html).

Inside Higher Ed. 2007. "More Students and Higher Scores for ACT." August 15. Retrieved April 2, 2008 (www.insidehighered.com/news/2007/08/15/act).

———. 2008. "The SAT's Growing Gaps." August 27. Retrieved October 14, 2008 (www.insidehighered.com/news/2008/08/27/sat).

Institute for Research on Poverty. 2008. "What Are Poverty Thresholds and Poverty Guidelines?" Retrieved March 12, 2009 (www.irp.wisc.edu/faqs/faq1.htm).

Institute for Statistics. 2005. *Fact Sheet: International Literacy Day 2005*. New York: United Nations.

———. 2006a. "Global Education Digest 2006: Comparing Education Statistics Across the World." New York: UNESCO. Retrieved July 28, 2006 (www.uis.unesco.org/TEMPLATE/pdf/ged/2006/GED2006.pdf).

———. 2006b. "The State of the World's Children 2007." Retrieved August 23, 2008 (http://www.unicef.org/sowc07/report/report.php).

Intergovernmental Panel on Climate Change. 2007. *Climate Change 2007: Synthesis Report*. Retrieved August 29, 2008 (www.ipcc.ch/pdf/assessment-report/ar4/syr/ar4_syr_spm.pdf).

International Association for the Evaluation of Educational Achievement (IEA). 2007. "Trends in International Math and Science Study." Retrieved August 23, 2008 (http://www.iea.nl/timss2007.html).

International Centre for Prison Studies. 2011. "Entire World: Prison Population Rates per 100,000 of the National Population." King's College, London. Retrieved August 6, 2011 (http://www.kcl.ac.uk/depsta/law/research/icps/worldbrief/wpb_stats.php?area=all&category=wb_poprate).

International Encyclopedia of the Social Sciences. 2008. "G8 Countries." Retrieved December 21, 2009 (http://www.encyclopedia.com/doc/1G2-3045300884.html).

International Forum on Globalization. 2006. "A Critical Look at Measures of Economic Progress." Pp. 346–55 in *Beyond Borders: Thinking Critically About Global Issues,* edited by Paula S. Rothenberg. New York: Worth.

International Humanist and Ethical Union. 2009. "Untouchability in Japan: Discrimination Against Burakumin." August 21. Retrieved December 4, 2009 (www.iheu.org/untouchability-japan-discrimination-against-burakumin).

International Institute for Democracy and Electoral Assistance. 2005. "Turnout in the World—Country by Country Performance." March 7. Retrieved July 27, 2009 (http://www.idea.int/vt/survey/voter_turnout_pop2–2.cfm).

International Monetary Fund. 2003. "Proposal for a Sovereign Debt Restructuring Mechanism: A Factsheet." Retrieved September 29, 2006 (www.imf.org/external/np/exr/facts/sdrm.htm).

———. 2011. "Debt Relief Under the Heavily Indebted Poor Countries (HIPC) Initiative." March 31. Retrieved July 28, 2011 (http://www.imf.org/external/np/exr/facts/hipc.htm).

Internet World Statistics. 2009. "U.S. Internet Usage and Broadband Usage Report." Retrieved October 21, 2009 (http://www.internetworldstats.com/am/us.htm).

Inter-Parliamentary Union. 2011. "Women in National Parliaments." World Classification, May 31. Retrieved July 9, 2011 (http://www.ipu.org/wmn-e/classif.htm).

Interpol. 2007. "Interpol News." Retrieved July 27, 2009 (http://www.interpol.int/Public/News/news2009.asp).

Irvine, Leslie. 2004. *If You Tame Me: Understanding Our Connection With Animals.* Philadelphia: Temple University Press.

Irwin, John. 1985. *The Jail: Managing the Underclass in American Society.* Berkeley: University of California Press.

———. 2005. *The Warehouse Prison: Disposal of the New Dangerous Class.* Los Angeles: Roxbury.

Isaacs, John. 2011. "Current U.S. and Russian Nuclear Weapons Stockpiles." The Center for Arms Control and Non-Proliferation. Retrieved July 22, 2011 (http://www.armscontrolcenter.org/policy/nuclearweapons/article/031009_current_nuclear_weapons_stockpiles/).

Jackman, Robert W. 1993. *Power Without Force: The Political Capacity of Nation-States.* Ann Arbor: University of Michigan Press.

Jackson, Philip W. 1968. *Life in Classrooms.* New York: Holt, Rinehart, and Winston.

Jacobs, Jane. 1961. *Life and Death of Great American Cities.* New York: Random House.

———. 1969. *Economy of Cities.* New York: Random House.

Jacoby, Jeff. 2009. "A Bad Sign Illegal Immigrants Are Leaving." Center for Immigration Studies. September 6. Retrieved July 26, 2011 (http://pqasb.pqarchiver.com/boston/access/1853977161.html?FMT=ABS&date=Sep+6%2C+2009).

James, William. [1890] 1934. *The Principles of Psychology.* Mineola, NY: Dover.

Jao, Jui-Chang and Matthew McKeever. 2006. "Ethnic Inequalities and Educational Attainment in Taiwan." *Sociology of Education* 79(2):131–52.

Japan Institute for Labour Policy and Training. 2009. "The Gender Gap in the Japanese Labor Market." February 26. Retrieved November 8, 2009 (www.ikjeld.com/en/news/91/the-gender-gap-in-the-labor -market).

"Japanese Whaling Suspension From Sea Shepherd Harassment Not Good Enough for Conservationists." 2011. *The Huffington Post,* February 17. Retrieved March 8, 2011 (http://www.huffingtonpost.com/2011/02/17/japans-whaling-suspension_n_824437.html).

"Japanese Women Face Difficulties Balancing Work and Family Life." 2007. Japan for Sustainability, October 1. Retrieved March 11, 2011 (http://www.japanfs.org/en/pages/026799.html).

Jarrett, R. L., P. J. Sullivan, and N. D. Watkins. 2005. "Developing Social Capital Through Participation in Organized Youth Programs: Qualitative Insights From Three Programs." *Journal of Community Psychology* 33(1):41–55.

Jayson, Sharon. 2005a. "Cohabitation Is Replacing Dating." *USA Today,* July 17. Retrieved April 3, 2008 (www.usatoday.com/life/lifestyle/2005–07–17-cohabitation_x.htm).

———. 2005b. "Divorce Declining, but So Is Marriage." *USA Today,* July 18, 3A.

Jellinek, E. M. 1960. *The Disease Concept of Alcoholism.* New Haven, CT: Hillhouse.

Jencks, Christopher. 1972. *Inequality: A Reassessment of the Effects of Family and Schooling in America.* New York: Basic Books.

———, ed. 1979. *Who Gets Ahead? The Determinants of Economic Success in America.* New York: Harper & Row.

———. 1992. *Rethinking Social Policy: Race, Poverty, and the Underclass.* Cambridge, MA: Harvard University Press.

Jenkins, Patricia H. 1995. "School Delinquency and School Commitment." *Sociology of Education* 68(July):221–39.

Jenness, Valerie and Kendal Broad. 1997. *Hate Crimes.* New York: Aldine de Gruyter.

Jennings, Jerry T. 1992. "Voting and Registration in the Election of November 1992." *Current Population Reports* (P-20–466). Washington, DC: U.S. Department of Commerce, Economics and Statistics Administration, U.S. Bureau of the Census.

Jensen, Robert and Emily Oster. 2007. *The Power of TV: Cable Television and Women's Status in India* (NBER Working Paper No. 13305). Cambridge, MA: National Bureau of Economic Research.

Jing, Tang. 2007. "The Popularizing of China's Higher Education and Its Influence on University Mathematics Education." *Educational Studies in Mathematics* 66(1):77–82.

Johnson, David W. and Frank P. Johnson. 2006. *Joining Together: Group Theory and Group Skills,* 9th ed. Boston: Allyn & Bacon.

Johnson, Jeff. 2007. "U.S. Presses for New Nuclear Weapons." *Chemical and Engineering News* 85(12):34–37.

Johnson, Kenneth M. and Glenn V. Fuguitt. 2000. "Continuity and Change in Rural Migration Patterns, 1950–1995." *Rural Sociology* 65(1):27–49.

Johnson, Paul. 2005. "Majority of Americans Believe Homosexuality Should Not Be Illegal, Support Partner Rights: Gallup Poll." Retrieved August 22, 2006 (www.sodomylaws.org/usa/usnews141.htm).

Johnson, Ramon. 2008. "Where Is Gay Adoption Legal?" *Gay Life.* Retrieved January 2, 2010 (http://gaylife.about.com/od/gayparentingadoption/a/gaycoupleadopt.htm).

Johnson, Robert. 2002. *Hard Time: Understanding and Reforming the Prison.* Belmont, CA: Wadsworth/Thompson Learning.

Johnson, Steven. 2009. "How Twitter Will Change the Way We Live (in 140 Characters or Less)." *Time,* June 15, 32–37.

Jones, Jacqueline. 1992. *The Dispossessed: America's Underclasses From the Civil War to the Present.* New York: Basic Books.

Jones, Sydney, and Susannah Fox. 2009. "Generations Online in 2009." Pew Internet and American Life Project, January 28. Retrieved June 25, 2011 (http://www.pewinternet.org/Reports/2009/Generations-Online-in-2009.aspx).

Jordan-Bychkov, Terry G. and Mono Domoch. 1998. *The Human Mosaic,* 8th ed. New York: W. H. Freeman.

Kagan, Sharon Lynn and Vivien Stewart. 2005. "A New World View: Education in a Global Era." *Phi Delta Kappan* 87(3):185–87.

Kain, Edward L. 1990. *The Myth of Family Decline: Understanding Families in a World of Rapid Social Change.* Lanham, MD: Lexington.

———. 2005. *Family Change and the Life Course: Cohorts and Social Change in the Year 2050.* Paper presented at the annual meetings of the Southern Sociological Society, Charlotte, NC, April.

Kaiser Family Foundation. 2005. "U.S. Teen Sexual Activity." Retrieved August 7, 2008 (http://www.kff.org/youthhivstds/upload/U-S-Teen-Sexual-Activity-Fact-Sheet.pdf).

Kalleberg, Ragnvold. 2005. "What Is 'Public Sociology'? Why and How Should It Be Made Stronger?" *The British Journal of Sociology* 56(3):387–93.

Kalmijn, Matthijs. 1991. "Shifting Boundaries: Trends in Religious and Educational Homogamy." *American Sociological Review* 56(December):786–800.

Kalmijn, Matthijs and Gerbert Kraaykamp. 1996. "Race, Cultural Capital, and Schooling: An Analysis of Trends in the United States." *Sociology of Education* 69(1):22–34.

Kamens, David H. and Aaron Benavot. 1991. "Elite Knowledge for the Masses: The Origins and Spread of Mathematics and Science Education in National Curricula." *American Journal of Education* 99(2):137–80.

Kanter, Rosabeth Moss. 1977. *Men and Women of the Corporation.* New York: Basic Books.

———. 1983. *The Change Masters: Innovation for Productivity in the American Corporation.* New York: Simon and Schuster.

———. 2001a. "Creating the Culture for Innovation." In *Leading for Innovation: Managing for Results,* edited by Frances Hesselbein, Marshall Goldsmith, and Iain Somerville. San Francisco: Jossey-Bass.

———. 2001b. "From Spare Change to Real Change: The Social Sector as Beta Site for Business Innovation." *Harvard Business Review on Innovation* 77(3). Retrieved July 27, 2009 (http://hbswk.hbs.edu/item/2974.html).

———. 2005. *Commitment and Community.* Cambridge, MA: Harvard University Press.

Kantrowitz, Barbara and Pat Wingert. 2003. "What's at Stake?" *Newsweek,* January 27, 30–37.

Kao, Grace. 2004. "Social Capital and Its Relevance to Minority and Immigrant Populations." *Sociology of Education* 77:172–76.

Kaplan, Howard B. and Robert J. Johnson. 1991. "Negative Social Sanctions and Juvenile Delinquency: Effects of Labeling in a Model of Deviant Behavior." *Social Science Quarterly* 72(1):117.

Karen, David. 2005. "No Child Left Behind? Sociology Ignored!" *Sociology of Education* 78(2):165–82.

Karraker, Katherine Hildebrant. 1995. "Parent's Gender-Stereotyped Perceptions of Newborns: The Eye of the Beholder Revisited." *Sex Roles* 33(9/10):687–701.

Karthikeyan, D. 2011. "Dalits Pay the Price for Their Political Assertion." *The Hindu: Tamil Nadu*, February 18, 1.

Kates, Brian. 2002. "Black Market in Transplant Organs." *New York Daily News,* August 25. Retrieved July 21, 2009 (http://www.vachss.com/help_text/archive/black_market.html).

Katz, Jackson. 2006. *The Macho Paradox: Why Some Men Hurt Women and How All Men Can Help.* Naperville, IL: Sourcebooks.

Kaufman, Bel. 1964. *Up the Down Staircase.* Englewood Cliffs, NJ: Prentice Hall.

KBYU-TV, producer. 2005. *Small Fortunes: Microcredit and the Future of Poverty.* Provo, UT: Author.

Keeter, Scott, Juliana Horowitz, and Alec Tyson. 2008. "Young Voters in the 2008 Election." Pew Research Center for the People & the Press, November 12. Retrieved July 11, 2009 (http://pewresearch.org/pubs/1031/young-voters-in-the-2008-election).

Keigher, Ashley and Freddie Cross. 2010. "Teacher Attrition and Mobility." School and Staffing Survey. Retrieved April 17, 2011 (http://nces.ed.gov/pubs2010/2010353.pdf).

Keller, Mark. 1991. "On Defining Alcoholism." *Alcohol Health and Research World* 15(4):253–59.

Kellerhals, Merle David. 2009. "Terrorist Attacks Fell 18% in 2008." April 30. Retrieved December 17, 2009 (www.america.gov/ . . . /20090430160651dmslahrellek0.6059229.html).

———. 2010. "Terrorist Attacks, Fatalities from Attacks Decline in 2009." Bureau of International Information Programs, U.S. Department of State, August 5. Retrieved December 2, 2010 (www.america.gov).

Kemmis, S. and R. McTaggart. 2000. "Participatory Action Research." Pp. 567–606 in *Handbook of Qualitative Research,* 2nd ed., edited by Norman K. Denzin and Yvonna S. Lincoln. Thousand Oaks, CA: Sage.

Kendall, Diane. 2004. *Sociology in Our Times: The Essentials,* 4th ed. Belmont, CA: Wadsworth.

Kerbo, Harold R. 2008. *Social Stratification and Inequality,* 7th ed. Boston: McGraw-Hill.

Kerckhoff, Alan C. 2001. "Education and Social Stratification Processes in Comparative Perspective." *Sociology of Education* 74(Extra Issue: Currents of Thought: Sociology of Education at the Dawn of the 21st Century):3–18.

Kerckhoff, Alan C. and Lorraine Bell. 1998. "Hidden Capital: Vocational Credentials and Attainment in the United States." *Sociology of Education* 71(2):152–74.

Kettl, Donald F. 1993. *Sharing Power: Public Governance and Private Markets.* Washington, DC: Brookings.

Khazaleh, Lorenz. 2009. "Internet Fatwas Cautiously Support Divorce Among Women." *CULCOM,* October 29. Retrieved January 4, 2010 (http://www.culcom.uio.no/english/news/2009/bogstad.html).

Kilbourne, Jean. 1999. *Deadly Persuasion: Why Women and Girls Must Fight the Addictive Power of Advertising.* New York: The Free Press.

———. 2000. *Killing Us Softly, III* (Video). Northampton, MA: Media Education Foundation.

Kile, Shannon N., Vitaly Fedchenko, and Hans M. Kristensen. 2009. "World Nuclear Forces." *SIPRI Yearbook 2009: Armaments, Disarmament and International Security.* Retrieved July 26. 2011 (www.sipri.org/yearbook/2009/files/SIPRIYB0908.pdf).

Kilgore, Sally B. 1991. "The Organizational Context of Tracking in Schools." *American Sociological Review* 56(2):201–2.

Killian, Caitlin. 2006. *North African Women in France: Gender, Culture, and Identity.* Palo Alto, CA: Stanford University Press.

Kim, Min-Sun, Katsuya Tasaki, In-Duk Kim, and Hye-ryeon Lee. 2007. "The Influence of Social Status on Communication Predispositions Focusing on Independent and Interdependent Self-Construals." *Journal of Asian Pacific Communication* 17(2):303–29.

Kimmel, Michael S. 2003. *The Gendered Society,* 2nd ed. New York: Oxford University Press.

Kimmel, Michael S. and Michael A. Messner. 2009. *Men's Lives,* 8th ed. Boston: Allyn & Bacon.

King, Edith W. 2006. *Meeting the Challenges of Teaching in an Era of Terrorism.* Belmont, CA: Thomson.

King, Jacqueline E. 2000. *Gender Equity in Higher Education.* Washington, DC: American Council on Education, Center for Policy Analysis.

Kinsey, Alfred E., Wardell B. Pomeroy, and Clyde E. Martin. 1948. *Sexual Behavior in the Human Male.* Philadelphia: Saunders.

Kinsey, Alfred E., Wardell B. Pomeroy, Clyde E. Martin, and H. Gephard. 1953. *Sexual Behavior in the Human Female.* Philadelphia: Saunders.

Kirk, Gwyn and Margo Okazawa-Rey. 2007. *Women's Lives: Multicultural Perspectives,* 4th ed. Boston: McGraw-Hill.

Kitano, Harry H. L. 2005. *Asian Americans: Emerging Minorities,* 4th ed. Englewood Cliffs, NJ: Prentice Hall.

Kitano, Harry H., Pauline Aqbayani, and Diane de Anda. 2005. *Race Relations,* 6th edition. Englewood Cliffs, NJ: Prentice Hall.

Klaus, Patsy. 2007. *Crime and the Nation's Households, 2005* (National Crime Victimization Survey, NCJ 217198). Washington, DC: U.S. Department of Justice, Bureau of Justice Statistics.

Klein, Barbara and Steve Ember. 2008. "Technology Increases Income, Reduces Poverty in Developing Countries." Voice of America, July 1. Retrieved August 6, 2011 (http://www.voanews.com/learningenglish/home/a-23–2008–07–01-voa1–83136752.html).

Knapp, Mark L. and Judith A. Hall. 1997. *Nonverbal Communication in Human Interaction,* 4th ed. Fort Worth: Harcourt Brace College.

Knickmeyer, Ellen. 2008. "Turkey's Gul Signs Head Scarf Measure." February 23. Retrieved July 21, 2009 (http://www.washingtonpost.com/wp-dyn/content/article/2008/02/22/AR2008022202988.html).

Knox, Paul L. and Peter J. Taylor, eds. 1995. *World Cities in a World-System.* Cambridge, UK: Cambridge University Press.

Koch, Jerome R. and Evans W. Curry. 2000. "Social Context and the Presbyterian Gay/Lesbian Debate: Testing Open Systems Theory." *Review of Religious Research* 42(2):206–14.

Kodish, Bruce I. 2003. "What We Do With Language—What It Does With Us." *ETC: A Review of General Semantics* 60:383–95.

Kohlberg, Lawrence. 1971. "From Is to Ought." Pp. 151–284 in *Cognitive Development and Epistemology,* edited by T. Mischel. New York: Academic Press.

Kohn, Melvin. 1989. *Class and Conformity: A Study of Values,* 2nd ed. Chicago: University of Chicago Press.

Kollock, Peter, Philip Blumstein, and Pepper Schwartz. 1985. "Sex and Power in Interaction: Conversational Privilege and Duties." *American Sociological Review* 50(February):34–66.

Koos, Earl. 1954. *The Health of Regionville.* New York: Columbia University Press.

Korte, Charles and Stanley Milgram. 1970. "Acquaintance Networks Between Racial Groups." *Journal of Personality and Social Psychology* 15:101–08.

Kosmin, Barry A. and Seymour P. Lackman. 1993. *One Nation Under God: Religion in Contemporary American Society.* New York: Harmony Books.

Kottak, Conrad Phillip. 2010. *Prime-Time Society: An Anthropological Analysis of Television and Culture,* Updated ed. Walnut Creek, CA: Left Coast Press.

Kozol, Jonathan. 1991. *Savage Inequalities: Children in America's Schools.* New York: Crown.

———. 2005. "Confections of Apartheid: A Stick-and-Carrot Pedagogy for the Children of Our Inner-City Poor." *Phi Delta Kappan* 87(4):265–75.

———. 2006. *The Shame of the Nation: The Restoration of Apartheid Schooling in America.* New York: Crown.

Kramer, Laura. 2010. *The Sociology of Gender: A Brief Introduction,* 3rd ed. New York: Oxford University Press.

Kreider, Rose M. 2010a. "Increase In Opposite-Sex Cohabiting Couples From 2009 to 2010 in the Annual Social and Economic Supplement to the Current Population Survey." U.S. Bureau of the Census, September 15. Retrieved July 11, 2011 (http://www.census.gov/population/www/socdemo/Inc-Opp-sex-2009-to-2010.pdf).

———. 2010b. "Working Paper on the Change in Cohabiting Couples from 2009 to 2010." Retrieved July 14, 2011 (http://www.census.gov/population/www/socdemo/Inc-Opp-sex-2009-to-2010.pdf).

Krieger, Nancy. 1994. "Epidemiology and the Web of Causation: Has Anyone Seen the Spider?" *Social Science and Medicine* 39(7):887–903.

Kristof, Nicholas D. 1995. "Japan's Invisible Minority: Better Off Than in Past, but Still Outcasts." *The New York Times* International, November 30, A18.

———. 2009. "Where Sweatshops Are a Dream." *The New York Times,* January 14. Retrieved May 9, 2011 (www.nytimes.com/2009/01/15/opinion/15kristof.html).

Kristoff, Nicholas D. and Sheryl WuDunn. 2000. *Thunder From the East.* New York: Vintage Books (Random House).

Krug, E., L. Dahlberg, J. Mercy, A. Zwi, and R. Lozano. 2002. *World Report on Violence and Health.* Geneva, Switzerland: World Health Organization.

Kübler-Ross, Elizabeth. 1997. *Death, the Final Stage of Growth,* Rev. ed. New York: Scribner.

Kuhn, Manford. 1964. "Major Trends in Symbolic Interaction Theory in the Past Twenty-Five Years." *Sociological Quarterly* 5:61–84.

Kuhn, Thomas. 1970. *The Structure of Scientific Revolutions,* 2nd ed. Chicago: University of Chicago Press.

Kumlin, Johanna. 2006. "The Sex Wage Gap in Japan and Sweden: The Role of Human Capital, Workplace Sex Composition, and Family Responsibility." *European Sociological Review,* December 18. Retrieved November 8, 2009 (http://esr.oxfordjournals.org/content/23/2/203.abstract).

Kurtz, Richard A. and H. Paul Chalfant. 1991. *The Sociology of Medicine and Illness.* Boston: Allyn & Bacon.

Kyle, David and Rey Koslowski, eds. 2001. *Global Human Smuggling: Comparative Perspectives.* Baltimore, MD: Johns Hopkins University Press.

Lacayo, Richard. 1993. "Cult of Death." *Time,* March 15, 36.

Laconte, Joe. 1998. "I'll Stand Bayou: Louisiana Couples Choose a More Muscular Marriage Contract." *Policy Review* 89:30–35.

Ladson-Billings, Gloria. 2006. "Once Upon a Time When Patriotism Was What You Did." *Phi Delta Kappan* 87(8):585–88.

Laird, J. S. Lew, M. DeBell, and C. Chapman. 2006. *Dropout Rates in the United States: 2002 and 2003* (NCES 2006–062). Washington, DC: U.S. Department of Education, National Center for Education Statistics.

Lake, Robert. 1990. "An Indian Father's Plea." *Teacher Magazine* 2(September):48–53.

Lamanna, Mary Ann and Agnes Riedmann. 2008. *Marriages and Families: Making Choices Throughout the Life Cycle,* 10th ed. Belmont, CA: Wadsworth.

Lambert, Lisa. 2006. "Half of Teachers Quit in Five Years: Working Conditions, Low Salaries Cited." *The Washington Post,* May 9, A7.

Lambert, Yves. 2000. "Religion in Modernity as a New Axial Age: Secularization or New Religious Forms?" Pp. 95–125 in *The Secularization Debate,* edited by William H. Swatos, Jr., and Daniel V. A. Olson. Lanham, MD: Rowman and Littlefield.

Lamy, Philip. 1996. *Millennium Rage.* New York: Plenum.

Landau, Elizabeth. 2009. "Life Expectancy Could Be Topic in Health Care Debate." Retrieved October 20, 2009 (http://articles.cnn.com/2009–06–11/health/life.expectancy.health.care_1_life-expectancy-health-care-payer?_s=PM:HEALTH).

Landler, Mark, and Brian Stelter. 2009. "Washington Taps into a Potent New Force in Diplomacy." *New York Times,* June 17, A12. Retrieved December 22, 2009 (www.nytimes.com/2009/06/17/world/middleeast/17media.html).

Langworth, Richard, ed. 2009. *Churchill by Himself: The Definitive Collection of Quotations.* Jackson, TN: Public Affairs.

Lareau, Annette. 2003. *Unequal Childhoods: Class, Race, and Family Life.* Berkeley: University of California Press.

Lareau, Annette and Erin McNamara Horvat. 1999. "Moments of Social Inclusion and Exclusion: Race, Class, and Cultural Capital in Family-School Relationships." *Sociology of Education* 72(1):37–53.

Larimer, Tim. 1999. "The Japan Syndrome." *Time,* October 11, 50–51.

Lassey, Marie L., William R. Lassey, and Martin J. Jinks. 1997. *Health Care Systems Around the World.* Englewood Cliffs, NJ: Prentice Hall.

Lawson, Carol. 1991. "Sheltering Children of the Vietnam War." *The New York Times,* April 18. Retrieved July 21, 2009 (http://www.nytimes.com/1991/04/18/garden/sheltering-children-of-the-vietnam-war.html?scp=1&sq=Sheltering%20Children%20of%20the%20Vietnam%20War&st=cse).

Laythe, Brian, Deborah Finkel, and Lee A. Kirkpatrick. 2001. "Predicting Prejudice From the Religious Fundamentalism and Right-Wing Authoritarianism." *Journal for the Scientific Study of Religion* 41(1):1–10.

Lazare, Aaron. 2004. *On Apology.* New York: Oxford University Press.

Leach, Edmund R. 1979. "Ritualization in Man in Relation to Conceptual and Social Development." Pp. 333–37 in *Reader in Comparative Religion,* 3rd ed., edited by William A. Lessa and Evon Z. Vogt. New York: Harper & Row.

Leavitt, Jacqueline and Susan Saegert. 1990. From *Abandonment to Hope: Community-Households in Harlem* (Columbia History of Urban Life). New York: Columbia University Press.

LeBon, Gustave. [1895] 1960. *The Crowd: A Study of the Popular Mind.* New York: Viking.

LeBow, Bob. 2004. *Health Care Meltdown: Confronting the Myths and Fixing Our Failing System.* Chambersburg, PA: Alan C. Hood.

Lechner, Frank J. and John Boli. 2005. *World Culture: Origins and Consequences.* Malden, MA: Blackwell.

Lee, B. C. and J. L. Werth. 2000. *Observations on the First Year of Oregon's Death With Dignity Act.* Washington, DC: American Psychological Association.

Lee, Gang, Ronald L. Akers, and Marian J. Borg. 2004. "Social Learning and Structural Factors in Adolescent Substance Use." *Western Criminology Review* 5(1):17.

Lee, Jennifer and Frank D. Bean. 2004. "America's Changing Color Lines: Immigration, Race/Ethnicity, and Multiracial Identification." *Annual Review of Sociology* 30(August):222–42.

———. 2007. "Redrawing the Color Line?" *City and Community* 6(1):49–62.

Lee, Richard B. 1984. *The Dobe!Kung.* New York: Holt, Rinehart, and Winston.

Lee, Richard Wayne. 1992. "Christianity and the Other Religions: Interreligious Relations in a Shrinking World." *Sociological Analysis* 53(2):125–39.

Lee, Seh-Ahn. 1991. "Family Structure Effects on Student Outcomes." In *Resources and Actions: Parents, Their Children, and Schools.* Washington, DC: National Science Foundation and National Center for Educational Statistics.

Lee, Sharon M. and Barry Edmonston. 2005. "New Marriages, New Families: U.S. Racial and Hispanic Intermarriage." *Population Bulletin* 60(2). Retrieved August 7, 2008 (http://www.prb.org/pdf05/60.2NewMarriages.pdf).

Leeder, Elaine J. 2004. *The Family in Global Perspective: A Gendered Journey.* Thousand Oaks, CA: Sage.

Legters, Nettie E. 2001. "Teachers as Workers in the World System." Pp. 417–26 in *Schools and Society: A Sociological Approach to Education,* edited by Jeanne H. Ballantine and Joan Z. Spade. Belmont, CA: Wadsworth.

Lehman, Edward C., Jr. 1980. "Patterns of Lay Resistance to Women in Ministry." *Sociological Analysis* 41(4):317–38.

———. 1981. "Organizational Resistance to Women in Ministry." *Sociological Analysis* 42(2):101–18.

———. 1985. *Women Clergy: Breaking Through Gender Barriers.* New Brunswick, NJ: Transaction.

Lehne, Gregory K. 1995. "Homophobia Among Men: Supporting and Defining the Male Role." Pp. 325–36 in *Men's Lives,* 3rd ed., edited by Michael Kimmel and Michael Messner. Boston: Allyn & Bacon.

Leit, Richard A., James J. Gray, and Harrison G. Pope. 2002. "The Media's Representation of the Ideal Male Body: A Cause for Muscle Dysmorphia?" *International Journal of Eating Disorders* 31(3):334–38.

LeMay, Michael L. 2000. *The Perennial Struggle: Race, Ethnicity, and Minority Group Politics in the United States.* Upper Saddle River, NJ: Prentice Hall.

Lemert, Edwin M. 1951. *Social Pathology.* New York: McGraw-Hill.

———. 1972. *Human Deviance, Social Problems, and Social Control,* 2nd ed. Englewood Cliffs, NJ: Prentice Hall.

Lemkau, Jeanne Parr. 2006. *Fences, Volcanoes, and Embargoes.* Unpublished manuscript.

Lemkin, Raphael. 1933. "Akte der Barbarei und des Vandalismus als *delicta juris gentium*" (Acts of Barbarism and Vandalism Under the Law of Nations). *Anwaltsblatt Internationales* 19(6):117–19.

Lengermann, Patricia M. and Jill Niebrugge-Brantley. 1990. "Feminist Sociological Theory: The Near-Future Prospects." Pp. 316–44 in *Frontiers of Social Theory: The New Synthesis,* edited by George Ritzer. New York: Columbia University Press.

Lenski, Gerhard E. 1966. *Human Societies.* New York: McGraw-Hill.

Lerner, Richard M. 1992. "Sociobiology and Human Development: Arguments and Evidence." *Human Development* 35(1):12–51.

Lesko, Nancy. 2001. *Act Your Age.* London: Routledge/Taylor and Francis.

Leslie, Gerald R. and Sheila K. Korman. 1989. *The Family in Social Context,* 7th ed. New York: Oxford University Press.

Leung, K., S. Lau, and W. L. Lam. 1998. *Parenting Styles and Academic Achievement: A Cross-Cultural Study.* Detroit, MI: Wayne State University Press.

Levin, Jack and Jack McDevitt. 2003. *Hate Crimes Revisited: America's War on Those Who Are Different.* Boulder, CO: Westview.

Levine, Leslie. 2004. *If You Tame Me: Understanding Our Connection With Animals.* Philadelphia: Temple University Press.

Levine, Michael H. 2005. "Take a Giant Step: Investing in Preschool Education in Emerging Nations." *Phi Delta Kappan* 87(3):196–200.

Levinson, Arnold R. 1993. "Outcomes Assessment." *National Law Journal* 16(6):S8.

Levinson, Stephen C. 2000. "Yeli Dnye and the Theory of Basic Color Terms." *Journal of Linguistic Anthropology* 1:3–55.

Levitt, Peggy. 2001. *The Transnational Villagers.* Berkeley: University of California Press.

———. 2007. *God Needs No Passport: Immigrants and the Changing American Religious Landscape.* New York: New Press.

Levitt, Peggy and Mary Waters, eds. 2006. *The Changing Face of Home: The Transnational Lives of the Second Generation.* Russell Sage Foundation. Retrieved June 27, 2011 (http://www.russellsage.org/publications/changing-face-home).

Levy, Audrey. 2008. "No to Forced Marriages!" *France Diplomatie.* Retrieved July 9, 2011 (http://www.diplomatie.gouv.fr/en/france_159/label-france_2554/themes_3713/society-environment_3901/society_3909/human-rights-and-fights-against-discrimination_3917/no-to-forced-marriages_6066.html).

Levy, Howard S. 1990. *Chinese Footbinding: The History of a Curious Erotic Custom.* Taipei, Taiwan: SMC.

Levy, Howard S., Arthur Waley, and Wolfram Eberhard. 1992. *The Lotus Lovers: The Complete History of the Curious Erotic Custom of Footbinding in China.* New York: Prometheus.

Lewin, Tamar. 2006. "At Colleges, Women Are Leaving Men in the Dust." *The New York Times,* July 9. Retrieved July 21, 2009 (http://www.nytimes.com/2006/07/09/education/09college.html?pagewanted=1&_r=1).

Lewis, Maureen A. and Marlaine E. Lockheed. 2006. *Inexcusable Absence: Why 60 Million Girls Still Aren't in School and What to Do About It.* Washington, DC: Center for Global Development.

Lewis, Oscar. 1961. *The Children of Sánchez: Autobiography of a Mexican Family.* New York: Random House.

———. 1986. *La Vida: A Puerto Rican Family in the Culture of Poverty.* New York: Irvington.

Lichter, Daniel T., Richard N. Turner, and Sharon Sassler. 2010. "National Estimates of the Rise in Serial Cohabitation." *Social Science Research* 39(5):754–65. Retrieved April 11, 2011 (doi:10.1016/j.ssresearch.2009.11.002).

Lieberson, Stanley. 1980. *A Piece of the Pie: Blacks and White Immigrants Since 1880.* Berkeley: University of California Press.

Lieberson, Stanley, Susan Dumais, and Shyon Bauman. 2000. "The Instability of Androgynous Names: The Symbolic Maintenance of Gender Boundaries." *American Journal of Sociology* 105(5):1249–87.

Lin, BoQi, Richard Peto, Zheng-Ming Chen, Jillian Boreham, Ya-Ping Wu, Jun-Yao Li, T. Colin Campbell, and Jun-Shi Chen. 1998. "Emerging Tobacco Hazards in China: Retrospective Proportional Mortality Study of One Million Deaths." *British Medical Journal* 317(7170):1411–22.

Lincoln, Erik and Laurence Mamiya. 1990. *The Black Church in the African American Experience.* Durham, NC: Duke University Press.

Lindberg, Richard and Vesna Markovic. n.d. "Organized Crime Outlook in the New Russia: Russia Is Paying the Price of a Market Economy in Blood." Retrieved January 4, 2001 (www.search-international.com/Articles/crime/russiacrime.htm).

Lindsay, James M. 2006. "Global Warming Heats Up." Pp. 307–313 in *Globalization: The Transformation of Social Worlds,* edited by D. Stanley Eitzen and Maxine Baca Zinn. Belmont, CA: Wadsworth.

Lindsey, Linda L. 2008. *Gender Roles: A Sociological Perspective,* 4th ed. Englewood Cliffs, NJ: Prentice Hall.

———. 2011. *Gender Roles: A Sociological Perspective,* 5th ed. Englewood Cliffs, NJ: Prentice Hall.

Linton, Ralph. 1937. *The Study of Man.* New York: D. Appleton-Century.

Lips, Hilary M. 1991. *Women, Men, and Power.* Mountain View, CA: Mayfield.

———. 2007. *Sex and Gender: An Introduction,* 6th ed. Boston: McGraw-Hill.

Lipset, Seymour Martin. 1959. *Political Man.* New York: Doubleday.

———. 1963. "The Value Patterns of Democracy: A Case Study on Comparative Analysis." *American Sociological Review* 28(August):515–31.

———. 1993. "Racial and Ethnic Conflicts: A Global Perspective." *Political Science Quarterly* 107:585–606.

Lipset, Seymour Martin and Stein Rokkan. 1967. *Party Systems and Voter Alignments: Cross-National Perspectives.* New York: Free Press.

Liptak, Adam. 2008. "1 in 100 U.S. Adults Behind Bars, New Study Says." *The New York Times,* February 28. Retrieved December 20, 2009 (www.nytimes.com/2008/02/28/us/28cnd-prison.html).

Liska, Allen E. 1999. *Perspectives in Deviance,* 3rd ed. Englewood Cliffs, NJ: Prentice Hall.

Lofgren, Orvar. 1999. *On Holiday: A History of Vacationing.* Berkeley: University of California Press.

———. 2010. "The Global Beach." Pp. 37–55 in *Tourists and Tourism,* 2nd ed., edited by Sharon Bohn Gmelch. Long Grove, IL: Waveland Press.

Lofland, John and Norman Skonovd. 1981. "Conversion Motifs." *Journal for the Scientific Study of Religion* 20(4):373–85.

Logan, John and Harvey Molotch. 1987. *Urban Fortunes: The Political Economy of Place.* Berkeley: University of California Press.

Logan, John R. and Glenna D. Spitze. 1994. "Family Neighbors." *American Journal of Sociology* 100(2):453–76.

Longman, Timothy. 2005. "Rwanda: Achieving Equality or Serving an Authoritarian State?" In *Women in African Parliaments,* edited by Gretchen Bauer and Hannah Britton. Boulder, CO: Lynne Reinner.

Lopez, Ian F. Haney. 1996. *White by Law.* New York: New York University Press.

Lopez-Garza, Marta. 2002. "Convergence of the Public and Private Spheres: Women in the Informal Economy." *Race, Gender, and Class* 9(3):175–92.

Lopreato, Joseph. 2001. "Sociobiological Theorizing: Evolutionary Sociology." Pp. 405–33 in *Handbook of Sociological Theory,* edited by Jonathan H. Turner. Secaucus, NJ: Springer.

Lorber, Judith. 1998. "Reinventing the Sexes: The Biomedical Construction of Femininity and Masculinity." *Contemporary Society* 27(5):498–99.

Lorber, Judith and Lisa Jean Moore. 2007. *Gendered Bodies.* Los Angeles: Roxbury.

Lubeck, Sally. 1985. *Sandbox Society: Early Education in Black and White America.* London: Falmer.

Luby, Dianne. 2007. "Getting Real With Sex Education." *Boston Globe,* December 27. Retrieved March 17, 2008 (www.boston.com/boston-globe/editorial_opinion/oped/articles/2007/12/27/getting_real_with_sex_education/).

Lucal, Betsy. 1999a. "Building Boxes and Policing Boundaries: (De)con-structing Intersexuality, Transgender and Bisexuality." *Sociology Compass.* Retrieved April 8, 2011 (www.blackwell-compass.com/subject/sociology/article_view?article_id=soco_tr_bp1100).

———. 1999b. "What It Means to Be Gendered Me." *Gender and Society* 13(6):781–97.

Lucas, Samuel R. 1999. *Tracking Inequality: Stratification and Mobility in American High Schools.* New York: Teachers College Press.

Lucas, Samuel R. and Mark Berends. 2002. "Sociodemographic Diversity, Correlated Achievement, and De Facto Tracking." *Sociology of Education* 75(4):328–48.

Luhman, Reid and Stuart Gilman. 1980. *Race and Ethnic Relations: The Social and Political Experience of Minority Groups.* Belmont, CA: Wadsworth.

Lumsden, Charles J. and Edward O. Wilson. 1981. *Genes, Mind, and Culture: The Coevolutionary Process.* Cambridge, MA: Harvard University Press.

Lundt, John C. 2004. "Learning for Ourselves: A New Paradigm for Education." *The Futurist,* November 1. Retrieved July 23, 2009 (http://www.allbusiness.com/professional-scientific/scientific-research/238485-1.html).

Lupton, Deborah. 2001. "Medicine and Health Care in Australia." Pp. 429–40 in *The Blackwell Companion to Medical Sociology,* edited by William C. Cockerham. Malden, MA: Blackwell.

Lyons, Linda. 2002. "Church Reform: Women in the Clergy." Gallup, May 7. Retrieved August 8, 2011 (http://www.gallup.com/poll/indicators/indreligion.asp).

Macaulay, David. 1978. *Underground.* Boston: Houghton Mifflin.

MacDougal, Gary. 2005. *Make a Difference: A Spectacular Breakthrough in the Fight Against Poverty.* New York: St. Martin's Press.

MacFarlane, Ann G. 1994. "Racial Education Values." *America* 17(9):10–12.

Machalek, Richard and Michael W. Martin. 2010. "Evolution, Biology, and Society: A Conversation for the 21st Century Classroom." *Teaching Sociology* 38(1):35–45.

Macionis, John. 2010. *Sociology,* 13th ed. Upper Saddle River, NJ: Prentice Hall.

Mackey, Richard A. and Bernard A. O'Brien. 1995. *Lasting Marriages: Men and Women Growing Together.* Westport, CT: Praeger.

Mackie, Paul. 2003. "Coalition of Twelve Major U.S. Corporations and WRI Announce Largest Corporate Green Power Purchases in U.S." World Resources Institute, September 17. Retrieved July 27, 2009 (http://archive.wri.org/news.cfm?id=260).

MacLeod, Jay. 2008. *Ain't No Makin' It: Aspirations and Attainment in a Low-Income Neighborhood,* 3rd ed. Boulder, CO: Westview.

Madden, Mary and Amanda Lenhart. 2006. "Online Dating." Pew Internet and American Life Project. Retrieved January 3, 2010 (http://www.pewinternet.org/Reports/2006/Online-Dating.aspx).

Made in Mexico, Inc. 2005. "What Are Maquiladoras? The Maquiladora Industry." Retrieved March 7, 2008 (www.madeinmexicoinc.com/maquiladoras_industry.htm).

Makhmalbaf, Mohsen, producer. 2003. *Kandahar: The Sun Behind the Moon* (Film).

"Male Dominance Causes Rape." 2008. *Journal of Feminist Insight,* December 8. Retrieved April 8, 2011 (http://journaloffeministinsight.blogspot.com/2008/12/male-dominance-causes-rape.html).

Malthus, Thomas R. [1798] 1926. *An Essay on the Principle of Population.* London: Macmillan.

Mann, Michael. 1993. *Sources of Social Power,* Vol. II. Cambridge, UK: Cambridge University Press.

Manning, Jennifer E. 2011. "Membership of the 112th Congress: A Profile." Congressional Research Service Report to Congress. Retrieved July 8, 2011 (http://www.senate.gov/reference/resources/pdf/R41647.pdf).

Manning, Wendy D. and Pamela J. Smock. 1995. "Why Marry? Race and the Transition to Marriage Among Cohabitors." *Demography* 32(4):509–20.

Manza, Jeff, Michael Hout, and Clem Brooks. 1995. "Class Voting in Democratic Capitalist Societies Since World War II: Dealignment, Realignment, or Trendless Fluctuation?" *Annual Review of Sociology* 21:137–63.

Marcus, Richard R. 2008. "Kenya's Conflict Isn't 'Tribal.'" *Los Angles Times,* January 24. Retrieved June 3, 2008 (www.latimes.com/news/opinion/la-oew-marcus24jan24,0,4231203.story).

Marger, Martin N. 2012. *Race and Ethnic Relations: American and Global Perspectives,* 9th ed. Belmont, CA: Wadsworth.

Marks, Alexandra. 2008. "For Election 08, Youth Voter Turnout Swells." *Christian Science Monitor,* January 16, 1.

Marmor, Theodore R. 1999. *The Politics of Medicare,* 2nd ed. Hawthorne, NY: Aldine de Gruyter.

Marsden, Peter V. 1987. "Core Discussion Networks of Americans." *American Sociological Review* 52(1):122–31.

Marti, Gerardo. 2005. *A Mosaic of Believers: Diversity and Innovation in a Multiethnic Church.* Bloomington: Indiana University Press.

Martin, Michel and Gwen Thompkins. 2009. "Proposed Uganda Law: If You See a Homosexual, Call the Police." *National Public Radio: Tell Me More,* December 18. Retrieved April 12, 2011 (www.npr.org/templates/story/story.php?storyId=121605525).

Martin, Patricia Yancey and Robert A. Hummer. 1989. "Fraternities and Rape on Campus." *Gender and Society* 3(4):457–73.

Martineau, Harriet. [1837] (1966). *Society in America: Vol. 1.* New York: AMS Press Inc.

———. 1838. *How to Observe. Morals and Manners.* New York: Harper & Brothers.

Marty, Martin E. 1983. "Religion in America Since Mid-Century." Pp. 273–87 in *Religion and America,* edited by Mary Douglas and Stephen Tipton. Boston: Beacon.

———. 2001. "The Logic of Fundamentalism." *Boston College Magazine,* Fall, 44–46.

Marty, Martin E. and R. Scott Appleby, eds. 1991. *Fundamentalism Observed.* Chicago: University of Chicago Press.

———. 1992. *The Glory and the Power: The Fundamentalist Challenge to the Modern Age.* Boston: Beacon.

———. 2004. *Accounting for Fundamentalism: The Dynamic Character of Movements.* Chicago: University of Chicago Press.

Marx, Karl. [1844] 1963. "Contribution to the Critique of Hegel's Philosophy of Right." Pp. 43–59 in *Karl Marx: Early Writings,* trans-lated and edited by T. B. Bottomore. New York: McGraw-Hill.

———. [1844] 1964. *The Economic and Philosophical Manuscripts of 1844.* New York: International Publishers.

———. 1963. *Early Writings,* translated by Thomas B. Bottomore. New York: McGraw-Hill.

Marx, Karl and Friedrich Engels. [1848] 1969. *The Communist Manifesto.* Baltimore: Penguin.

————. 1955. *Selected Works in Two Volumes.* Moscow: Foreign Language Publishing House.

Mashberg, Tom. 2002. "Med Examiner's Office Has Secret Body-Parts Deal." *Boston Herald,* May 20, 1.

Mason-Schrock, Douglas. 1996. "Transsexual's Narrative Construction of the 'True Self.'" *Social Psychology Quarterly* 59(3):176–92.

Massey, Douglas S. 2007. *Categorically Unequal: The American Stratification System.* New York: Russell Sage Foundation.

Massey, Douglas S. and Nancy A. Denton. 1998. *American Apartheid: Segregation and the Making of the Underclass.* Cambridge, MA: Harvard University Press.

Masters, William H. and Virginia Johnson. 1966. *Human Sexual Response.* Boston: Little, Brown.

————. 1970. *Human Sexual Inadequacy.* Boston: Little, Brown.

Mather, George and Alvin J. Schmidt. 2006. *Encyclopedic Dictionary of Cults, Sects, and World Religions,* revised and updated edition. Grand Rapids, MI: Zondervan.

Mauksch, Hans. 1993. "Teaching of Applied Sociology: Opportunities and Obstacles." Pp. 1–7 in *Teaching Applied Sociology: A Resource Book,* edited by C. Howery. Washington, DC: American Sociological Association Teaching Resources Center.

Mauss, Armand. 1975. *Social Problems as Social Movements.* Philadelphia: Lippincott.

Mayo Clinic. 2011. "Domestic Violence Against Women: Recognize Patterns, Seek Help." Retrieved July 11, 2011 (http://www.mayoclinic.com/health/domestic-violence/WO00044).

Mayoux, Linda. 2002. "Women's Empowerment or Feminisation of Debt? Towards a New Agenda in African Microfinance." Report at the One World Action Conference in London, March 21–22.

————, ed. 2008. *Sustainable Learning for Women's Empowerment: Ways Forward in Micro-Finance.* Warwickshire, UK: ITDG Publishing.

McAdam, Doug. 1999. *Political Process and the Development of Black Insurgency, 1930–1970,* 2nd ed. Chicago: University of Chicago Press.

McCaghy, Charles H., Timothy A. Capron, J. D. Jamieson, and Sandra Harley Carey. 2006. *Deviant Behavior: Crime, Conflict, and Interest Groups,* 7th ed. Boston: Allyn & Bacon.

McCarthy, Ellen. 2009. "Study Links Cohabitation Before Marriage to Greater Potential for Divorce." *The Washington Post,* August 16. Retrieved July 19, 2011 (http://www.washingtonpost.com/wp-dyn/content/article/2009/08/13/AR2009081304118.html).

McCarthy, John D. and Mayer N. Zald. 1977. "Resource Mobilization and Social Movements: A Partial Theory." *American Journal of Sociology* 82(6):1212–41.

McConnell, E. D. and E. Delgado-Romero. 2004. "Latinos, Panethnicity, and Census 2000: Reality or Methodological Construction." Retrieved August 6, 2011 (http://citation.allacademic.com/meta/p_mla_apa_research_citation/1/0/9/9/9/pages109991/p109991–1.php).

McCormick, John. 1992. "A Housing Program That Actually Works." *Newsweek,* June 22, 61.

McCrone, David. 1998. *The Sociology of Nationalism.* London: Routledge.

McCrummen, Stephanie. 2008. "Women Run the Show in a Recovering Rwanda." *The Washington Post,* October 27. Retrieved May 6, 2011 (www.washingtonpost.com/wp-dyn/content/article/2008/10/26/AR2008102602197.html).

McDonald, Michael. 2009. "2008 General Election Turnout Rates." Retrieved March 17, 2009 (http://elections.gmu.edu/Turnout_2008G.html).

McEneaney, Elizabeth H. and John W. Meyer. 2000. "The Content of the Curriculum: An Institutionalist Perspective." Pp. 189–211 in *Handbook of the Sociology of Education,* edited by Maureen T. Hallinan. New York: Kluwer Academic/Plenum.

McEvoy, Alan W. 1990. "Confronting Gangs." *School Intervention Report* (Learning Publications) 3/4(February/March):1–20.

McEvoy, Alan W. and Jeff B. Brookings. 2008. *If She Is Raped: A Guidebook for the Men in Her Life,* 4th ed., abridged. Tampa, FL: Teal Ribbon Books.

McGuire, Meredith. 2002. *Religion: The Social Context,* 5th ed. Belmont, CA: Wadsworth.

McIntosh, Peggy. 1992. "White Privilege and Male Privilege: A Personal Account of Coming to See Correspondences Through Work in Women's Studies." Pp. 70–81 in *Race, Class, and Gender: An Anthology,* edited by Margaret L. Anderson and Patricia Hill Collins. Belmont, CA: Wadsworth.

————. 2002. "White Privilege: Unpacking the Invisible Knapsack." Pp. 97–101 in *White Privilege: Essential Readings on the Other Side of Racism,* edited by Paula S. Rothenberg. New York: Worth.

McKelvie, Samuel R. 1926. "What the Movies Meant to the Farmer." *Annals of the American Academy of Political and Social Science* 128(November):131.

McKinlay, John B. and L. D. Marceau. 1998. "The End of the Golden Age of Doctoring." Presented at the American Public Health Association, November, Washington, DC.

McKnight, Matthew. 2011. "False Choice: How Private School Vouchers Might Harm Minority Students." *The New Republic,* April 15. Retrieved April 17, 2011 (www.tnr.com/article/politics/86710/school-vouchers-education-republicans).

McLemore, S. Dale, Harriett Romo, and Susan Gonzalez Baker. 2001. *Racial and Ethnic Relations in America,* 6th ed. Boston: Allyn & Bacon.

McLeod, Jay. 2004. *Ain't No Makin' It: Aspirations and Attainment in a Low-Income Neighborhood,* 2nd ed. Boulder, CO: Westview.

————. 2008. *Ain't No Makin' It: Aspirations and Attainment in a Low-Income Neighborhood,* 3rd ed. Boulder, CO: Westview.

McLuhan, Marshall and Bruce R. Powers. 1992. *The Global Village: Transformations in World Life and Media in the 21st Century.* New York: Oxford University Press.

McManus, Patricia A. and Thomas A DiPrete. 2001. "Losers and Winners: The Financial Consequences of Separation and Divorce for Men." *American Sociological Review* 66(April):246–68.

McNall, Scott G. and Sally Allen McNall. 1983. *Plains Families: Exploring Sociology Through Social History.* New York: St. Martin's.

McNeal, Ralph B. 1995. "Extracurricular Activities and High School Dropouts." *Sociology of Education* 68(January):62–81.

McNeil, John J. 1993. *The Church and the Homosexual,* 4th ed. Boston: Beacon.

McTaggart, R. 1997. *Participatory Action Research: International Contexts and Consequences.* Albany: State University of New York Press.

Mead, George Herbert. [1934] 1962. *Mind, Self, and Society.* Chicago: University of Chicago Press.

Mead, Margaret. [1935] 1963. *Sex and Temperament in Three Primitive Societies.* New York: William Morrow.

————. 1973. *Coming of Age in Samoa.* New York: Morrow Quill.

Medical News Today. 2009. "World Should Contribute $23B to Increase Women's Access to Contraception, UNFPA Says." September 4. Retrieved July 23, 2011 (www.medicalnewstoday.com/articles/162997.php).

Mehan, Hugh. 1992. "Understanding Inequality in Schools: The Contribution of Interpretive Studies." *Sociology of Education* 65(1):1–20.

Mehra, Bharat, Cecelia Merkel, and Ann P. Bishop. 2004. "The Internet for Empowerment of Minority and Marginalized Users." *New Media and Society* 6:781–802.

Melton, J. Gordon. 1992. *Encyclopedic Handbook of Cults in America,* Revised and updated edition. London: Routledge.

Meltzer, B. 1978. "Mead's Social Psychology." Pp. 15–27 in *Symbolic Interactionism: A Reader in Social Psychology,* 3rd ed., edited by J. Manis and B. Meltzer. Boston: Allyn & Bacon.

Meltzer, Bernard N., John W. Petras, and Larry T. Reynolds. 1975. *Symbolic Interactionism: Genesis, Varieties and Criticism.* London: Routledge and Kegan Paul.

Meltzer, Jack. 1999. *Metropolis to Metroplex: The Social and Spatial Planning of Cities.* Baltimore: Johns Hopkins University Press.

Memorial Institute for the Prevention of Terrorism. 2007. "Number of International Terrorist Attacks." MIPT Terrorism Knowledge

Database. Retrieved April 14, 2009 (www.publicagenda.org/charts/number-international-terrorist-attacks).

Mental Health America. 2011. "Bullying in Schools: Harassment Puts Gay Youth at Risk." Retrieved April 17, 2011 (www.nmha.org/index.cfm?objectid=CA866DCF-1372–4D20-C8EB26EEB30B9982).

Merton, Robert K. 1938. "Social Structure and Anomie." *American Sociological Review* 3(October):672–82.

———. [1942] 1973. *The Sociology of Science: Theoretical and Empirical Investigations.* Chicago: University of Chicago Press.

———. 1949. "Discrimination and the American Creed." Pp. 99–126 in *Discrimination and National Welfare,* edited by Robert M. MacIver. New York: Harper.

———. 1968a. "Social Structure and Anomie." *American Sociological Review* 3(October):672–82.

———. 1968b. *Social Theory and Social Structure,* 2nd ed. New York: The Free Press.

Metropolitan Community Churches. 2010. "Our Churches." Retrieved July 19, 2011 (http://mcchurch.org/ourchurches/).

Metz, Mary Haywood. 1986. *Different by Design: The Context and Character of Three Magnet Schools.* New York: Routledge & Kegan Paul.

Meyer, John W., David Kamens, Aaron Benavot, Yun-Kyung Cha, and Suk-Ying Wong. 1992. *School Knowledge for the Masses: World Models and National Primary Curriculum Categories in the Twentieth Century.* London: Falmer.

Meyer, Madonna H. and Eliza K. Pavalko. 1996. "Family, Work, and Access to Health Insurance Among Mature Women." *Journal of Health and Social Behavior* 37(4):311–25.

Meyer, R. 1996. "The Disease Called Addiction: Emerging Evidence in a Two Hundred Year Debate." *The Lancet* 347(8995):162–66.

Michels, Robert. [1911] 1967. *Political Parties.* New York: The Free Press.

Mickelson, Roslyn Arlin. 1990. "The Attitude-Achievement Paradox Among Black Adolescents." *Sociology of Education* 63(January):44–61.

Migration Information Source. 2011. "United States." Retrieved May 15, 2011 (www.migrationinformation.org/Resources/unitedstates.cfm).

Mike, Valerie. 2003. "Evidence and the Future of Medicine." *Evaluation and the Health Professions* 26(2):127–52.

Mikkelson, Barbara and David P. Mikkelson. 2011. "Urban Legends Reference Pages: Frequently Asked Questions." Retrieved August 4, 2011 (http://www.snopes.com/info/faq.asp).

Milgram, Stanley. 1967. "The Small World Problem." *Psychology Today* 1:61–67.

Milkie, Melissa. 1999. "Social Comparisons, Reflected Appraisals, and Mass Media: The Impact of Pervasive Beauty Images on Black and White Girls' Self-Concepts." *Social Psychology Quarterly* 62(2):190–210.

Millennium Project. 2006. "The 0.7% Target: An In-Depth Look." Retrieved January 18, 2011 (www.unmillenniumproject.org/press/07.htm).

Miller, Donald E. 1997. *Reinventing American Protestantism.* Berkeley: University of California Press.

Miller, Greg. 2009. "Global Economic Crisis Called Biggest U.S. Security Threat." *Los Angeles Times.* Retrieved February 13, 2009 (http://articles.latimes.com/2009/feb/13/nation/na-security-threat13).

Miller, Matthew and Luisa Kroll. 2010. "World's Billionaires in 2010." *Forbes,* March 10. Retrieved March 18, 2011 (http://finance.yahoo.com/career-work/article/109029/worlds-billionaires-2010).

Mills, C. Wright. 1956. *The Power Elite.* New York: Oxford University Press.

———. 1959. *The Sociological Imagination.* New York: Oxford University Press.

Mills, Robert J. 2001. "Health Insurance Coverage 2000." *Current Population Reports.* Washington, DC: U.S. Census Bureau.

Mills, Theodore M. 1984. *The Sociology of Small Groups,* 2nd ed. Englewood Cliffs, NJ: Prentice Hall.

Mincy, Ronald B. 2006. *Black Males Left Behind.* Washington, DC: The Urban Institute Press.

Minkoff, Debra C. 1995. *Organizing for Equality.* New Brunswick, NJ: Rutgers University Press.

Mintz, Morton and Harry Cohen. 1971. *America, Inc.* New York: Dial Presses.

Misztal, Bronislaw and Anson D. Shupe. 1998. *Fundamentalism and Globalization: Fundamentalist Movements at the Twilight of the Twentieth Century.* Westport, CT: Praeger.

Molotch, Harvey. 2003. *Where Stuff Comes From: How Toasters, Toilets, Cars, Computers, and Many Other Things Come to Be as They Are.* London: Routledge.

Moore, Laurence. 1995. *Selling God: American Religion in the Marketplace of Culture.* New York: Oxford University Press.

Moore, Valerie Ann. 2001. "Doing Racialized and Gendered Age to Organize Peer Relations: Observing Kids in Summer Camp." *Gender and Society* 15(6):835–58.

Morbidity and Mortality Weekly Report. 1999. "Ten Great Public Health Achievements—United States, 1900–1999." *Journal of the American Medical Association* 281(16):1481–84.

———. 2006. "Tobacco Use Among Adults—United States, 2005." October 27, 55(42):1145–48. Retrieved January 23, 2010 (www.cdc.gov/mmwr/preview/mmwrhtml/mm5542a1.htm).

Morello, Carol, and Dan Keating. 2009. "Number of Foreign-Born U.S. Residents Drops." September 22. Retrieved January 5, 2010 (www.washingtonpost.com/wp-dyn/content/article/2009/09/21/AR2009092103251.html).

Morgan, Kenneth O. 1987. *Rebirth of a Nation: Wales, 1880–1980.* Oxford, UK: University of Oxford Press.

Morgan, Stephen L. 2001. "Counterfactuals, Causal Effect Heterogeneity, and the Catholic School Effect on Learning." *Sociology of Education* 74(4):341–73.

Moritz, Owen. 2001. "Trafficking in Humans Is Thriving Business in Africa." *New York Daily News,* April 17, 4.

Morris, Edward W. 2005. "From 'Middle Class' to 'Trailer Trash': Teachers' Perceptions of White Students in a Predominately Minority School." *Sociology of Education* 78(2):99–121.

Morris, Joan M. and Michael D. Grimes. 1997. *Caught in the Middle: Contradictions in the Lives of Sociologists From Working Class Backgrounds.* Westport, CT: Praeger.

Morris, Lloyd R. 1949. *Not So Long Ago.* New York: Random House.

Morrison, J. Ian. 2000. *Health Care in the New Millennium.* San Francisco: Jossey-Bass.

Mother Jones. 2007. "Aftermath: Long-Term Thinking." Retrieved July 21, 2009 (http://www.motherjones.com/politics/2007/03/iraq-101-aftermath-long-term-thinking).

Mulkey, Lynn M., Robert L. Crain, and Alexander J. C. Harrington. 1992. "One-Parent Households and Achievement: Economic and Behavioral Explanations of a Small Effect." *Sociology of Education* 65(January):48–65.

Mulligan, Susan. 2008 "Young People Make Voices Heard in Iowa." *The Boston Globe,* January 5, A9.

Mumford, Lewis. 1961. *The City in History: Its Origins, Transformations, and Prospects.* New York: Harcourt, Brace, and World.

Murdock, George Peter. 1967. *Ethnographic Atlas.* Pittsburgh, PA: University of Pittsburgh Press.

Murphy, Cullen. 1993. "Women and the Bible." *Atlantic Monthly* 272(2):39–64.

Murstein, Bernard I. 1987. "A Clarification and Extension of the SVR Theory of Dyadic Pairing." *Journal of Marriage and the Family* 49(November):929–33.

Mutamba, John. 2005. "Strategies for Increasing Women's Participation in Government." Expert Group Meeting on Democratic Governance in Africa, Nairobi, Kenya, December 6–8.

Mydans, Seth. 2002. "In Pakistan, Rape Victims Are the 'Criminals.'" *The New York Times,* May 17, A3.

———. 2009. "For Khmer Rouge Guard, It Was Kill or Be Killed." March 1. Retrieved July 11, 2009 (www.nytimes.com/2009/03/01/world/asia/01iht-guard.1.20501994.html).

Myers, John P. 2003. *Dominant-Minority Relations in America: Linking Personal History With the Convergence in the New World.* Boston: Allyn & Bacon.

Myrdal, Gunnar. 1964. *An American Dilemma.* New York: McGraw-Hill.

Nagel, Joane. 1994. "Constructing Ethnicity: Creating and Recreating Ethnic Identity and Culture." *Social Problems* 41(1):152–76.

Nakamura, Lisa. 2004. "Interrogating the Digital Divide: The Political Economy of Race and Commerce in the New Media." Pp. 71–83 in *Society On-Line: The Internet in Context,* edited by Philip N. Howard and Steve Jones. Thousand Oaks, CA: Sage.

Nanda, Serena and Richard L. Warms. 2010. *Cultural Anthropology,* 10th ed. Belmont, CA: Wadsworth.

Nardi, Peter M., ed. 1992. *Men's Friendships.* Newbury Park, CA: Sage.

National Archives and Records Administration. 2008. "What Is the Electoral College?" Retrieved March 21, 2008 (http://www.archives .gov/federal-register/electoral-college/about.html).

National Assessment of Educational Progress. 2006. "Comparing Private Schools and Public Schools Using Hierarchical Linear Modeling." NCES, U.S. Department of Education. NCES 2006–461. Retrieved July 10, 2009 (www. nytimes.com/packages/pdf/national/2006715report.pdf).

National Association for the Advancement of Colored People. 2008. "Why Are 1 in 9 Young Black Men in Prison?" March 27. Retrieved April 17, 2011 (http://naacpoc.org/2008/03/why-are-1-in-9-young-black-men-in-prison/).

National Center for Education Statistics. 2000. *Distribution of Undergraduate Enrollment Among Students Aged 24 or Younger, by Race/Ethnicity, Gender, and Income.* Washington, DC: U.S. Department of Education. Retrieved August 26, 2006 (http://nces.ed.gov/fastfacts/dailyarchive. asp?StatCat).

———. 2002. *Digest of Education Statistics* 2002. Table 207. Washington, DC: U.S. Department of Education.

———. 2004. "Trends in Educational Equity of Girls and Women: 2004." November. Retrieved August 25, 2006 (http://nces.ed.gov/pubsearch/pubsinfo.asp?pubid=2005016).

———. 2006. "Revenues and Expenditures for Public Elementary and Secondary Education, Table 1." National Center for Education Statistics. Retrieved April 17, 2011 (http://nces.ed.gov/ccd/pubs/npefs03/tables.asp).

———. 2007. "Status and Trends in the Education of Racial and Ethnic Minorities." Retrieved July 23, 2009 (http://nces.ed.gov/pubs2007/minoritytrends).

———. 2008a. "Degrees Conferred by Degree-Granting Institutions, by Level of Degree and Sex of Student" (Table 268). Retrieved January 6, 2010 (http://nces.ed.gov/programs/digest/d08/tables/dt08_268.asp).

———. 2008b. "Distance Education at Degree-Granting Postsecondary Institutions: 2006–2007." December. Retrieved April 17, 2011 (http://nces.ed.gov/pubsearch/pubsinfo.asp?pubid=2009044).

———. n.d. "Glossary." Retrieved July 10, 2009 (http://nces.ed.gov/ipeds/glossary/?charindex=D).

National Conference of State Legislatures. 2007. "2007 Enacted State Legislation Related to Immigrants and Immigration." Retrieved July 21, 2009 (http://www.ncsl.org/default.aspx?tabid=13134).

National Institutes of Health. 2007. "The Use of Complementary and Alternative Medicine in the United States." Retrieved August 6, 2011 (http://nccam.nih.gov/news/camstats/costs/costdatafs.htm).

National Labor Committee. 2002. "School Uniforms Made in El Salvador: The Case of Elder Manufacturing." Retrieved April 4, 2008 (www .nlcnet.org/campaigns/uniforms/elsalvador/report.shtml).

National Opinion Research Center. 2010. "More Than 60% of Marriages Break Up Because of Adultery." Retrieved February 28, 2011 (http://www.norc.uchicago.edu/).

National Postsecondary Education Cooperative (NPEC). 2004. *How Does Technology Affect Access in Postsecondary Education? What Do We Really Know?* (NPEC 2004–831), prepared by Ronald A. Phipps for NPEC Working Group on Access-Technology. Washington, DC: Author.

National Public Radio. 2008. "Poppy Growing in Afghanistan." *Morning Edition,* February 9.

National Science Foundation. 2005. "Children, TV, Computers and More Media: New Research Shows Pluses, Minuses." Retrieved August 6, 2011 (http://www.nsf.gov/news/news_summ.jsp?org=NSF&cntn_id=102813&preview=false).

National Security Network Ambassador. 2011. "Where Girls Are Educated 'The Standard of Living Goes Up.'" January 31. Retrieved April 17, 2011 (www.nsnetwork.org/node/1850).

National Urban League. 2006. *The State of Black America.* New York: Author.

NationMaster. 2010. "Crime Stats: China vs. Cuba." Retrieved July 5, 2011 (www.nationmaster.com/compare/China/Cuba/Crime).

———. 2011. "United States Population Pyramids." Retrieved May 19, 2011 (www.nationmaster.com/country/us/Age_distribution).

Navarro, Vicente. 1996. "Why Congress Did Not Enact Health Reform." *Journal of Health Politics, Policy and Law* 20:455–62.

Neuman, Michelle J. 2005. "Global Early Care and Education: Challenges, Responses, and Lessons." *Phi Delta Kappan* 87(3):188–92.

"New Federal Report: Sexual Abuse Plagues U.S. Prisons and Jails." 2010. *Just Detention,* August 26. Retrieved July 1, 2011 (http://www.businesswire .com/news/home/20100826005772/en/Federal-Report-Sexual-Abuse-Plagues-U.S.-Prisons).

Newman, Andy. 2008. "What Women Want (Maybe)." *The New York Times,* June 12, E6.

Newman, David. 2009. *Families: A Sociological Perspective.* New York: McGraw-Hill.

Newman, David M. and Liz Grauerholz. 2002. *Sociology of Families,* 2nd ed. Thousand Oaks, CA: Pine Forge Press.

Newport, Frank. 2004. "A Look at Americans and Religion Today." Retrieved August 6, 2011 (http://www.gallup.com/poll/11089/look-americans-religion-today.aspx).

———. 2006. "Mormons, Evangelical Protestants, Baptists Top Church Attendance List" (Gallup Poll). Retrieved August 6, 2011 (http://www.gallup.com/poll/22414/mormons-evangelical-protestants-baptists-top-church-attendance-list.aspx).

———. 2009. "This Christmas, 78% of Americans Identify as Christian." Gallup, December 24. Retrieved August 7, 2011 (http://www.gallup.com/poll/124793/this-christmas-78-americans-identify-christian.aspx).

———. 2010. "Near-Record High See Religion Losing Influence in America." *Gallup Poll Update,* December 29. Retrieved May 15, 2011 (www.gallup.com/poll/145409/near-record-high-religion-losing-influence-america.aspx).

———. 2011. "More Than 9 in 10 Americans Continue to Believe in God." Gallup, June 3. Retrieved July 14, 2011 (http://www.gallup.com/poll/147887/americans-continue-believe-god.aspx).

Newport, Frank and Elizabeth Mendes. 2009. "About One in Six United States Adults Are Without Health Insurance." Gallup Poll, July 22. Retrieved March 13, 2011 (www.gallup.com/poll/121820/one-six-adults-without-health-insurance.aspx).

News of Future. 2007. "Future News in Technology." Retrieved January 14, 2008 (www.newsoffuture.com/technology).

The New York Times. 2008. "Primary Calendar: Democratic Nominating Contests." Retrieved March 30, 2008 (http://politics.nytimes.com/election-guide/2008/primaries/democraticprimaries/index.html).

The New York Times Almanac. 2011. "Major Languages of the World." New York: Penguin Reference, p. 505.

Nguyen, Nathalie Huynh Chau. 2005. "Eurasian/Amerasian Perspectives: Kim Lefevre's *Metisse blanche,* and Kien Nguyen's *The Unwanted.*" *Asian Studies Review* 29(June):107–22.

Niebuhr, H. Richard. 1929. *The Social Sources of Denominationalism.* New York: Henry Holt.

———. 1932. *Moral Man and Immoral Society.* New York: Scribner.

Nightingale Alliance. 2007. "Fast Facts." Retrieved January 20, 2008 (www.nightingalealliance.org/cgi-bin/home.pl?section=3).

Nock, Steven L. 1999. "The Problem With Marriage." *Society* 36(5):20–27.

Nock, Steven L., James D. Wright, and Laura Sanchez. 1999. "America's Divorce Problem." *Society* 36(4):43–52.

Noel, Donald. 1968. "A Theory of the Origin of Ethnic Stratification." *Social Problems* 16(Fall):157–72.

Noguera, Pedro A. 1996. "Preventing and Producing Violence: A Critical Analysis of Responses to School Violence." *Sociology of Education* 56(3):189–212.

———. 2004. "Social Capital and the Education of Immigrant Students: Categories and Generalizations." *Sociology of Education* 77(2):180–84.

Noguera, Pedro and Robby Cohen. 2006. "Patriotism and Accountability: The Role of Educators in the War on Terrorism." *Phi Delta Kappan* 87(8):573–78.

Nolan, Cathal J. 2002a. "Terrorism." In *The Greenwood Encyclopedia of International Relations,* p. 1648–49. London: Greenwood.

———. 2002b. "War." In *The Greenwood Encyclopedia of International Relations,* p. 1803. London: Greenwood.

Nolan, Patrick and Gerhard Lenski. 2008. *Human Societies.* Boulder, CO: Paradigm.

Nolan, Patrick D., Jennifer Triplett, and Shannon McDonough. 2010. "Sociology's Suicide: A Forensic Autopsy." *The American Sociologist* 41:292–305.

Noss, David S. and John B. Noss. 1990. *A History of the World's Religions.* New York: Macmillan.

The Nuclear Information Project. 2008. "Status of World Nuclear Forces." Retrieved April 16, 2008 (www.nukestrat.com/nukestatus.htm).

Nyden, Philip W. 1984. *Steelworkers Rank and File: The Political Economy of a Union Reform Movement.* New York: Praeger.

Oakes, Jeannie. 1985. *Keeping Track: How High Schools Structure Inequality.* New Haven, CT: Yale University Press.

———. 1990. *Multiplying Inequalities: The Effects of Race, Social Class, and Tracking on Opportunities to Learn Mathematics and Science.* Santa Monica, CA: The Rand Corporation.

Oakes, Jeannie, Amy Stuart Wells, Makeba Jones, and Amanda Datnow. 1997. "Detracking: The Social Construction of Ability, Cultural Politics, and Resistance to Reform." *Teacher's College Record* 98(3):482–510.

Oberlander, Jonathan. 2002. "The U.S. Health Care System: On a Road to Nowhere?" *Canadian Medical Association Journal* 167(2):163–69.

O'Brien, Denise. 1977. "Female Husbands in Southern Bantu Societies." In *Sexual Stratification: A Cross-Cultural View,* edited by Alice Schlegel. New York: Columbia University Press.

O'Brien, Elizabeth and John-Henry Westen. 2007. "Gallup Poll Shows Highest-Ever Acceptance of Homosexuality in America." LifeSiteNews.com, May 30. Retrieved April 12, 2011 (www.lifesite news.com/news/archive/ldn/2007/may/07053003).

O'Brien, Jane, 2006. "UNICEF 'Report Card' on Child Nutrition Reveals Millions Dying, Calls for Action." May 2. Retrieved July 9, 2011 (http://www.unicef.org/nutrition/index_33721.html).

O'Brien, Jody. 2006. *The Production of Reality,* 4th ed. Thousand Oaks, CA: Pine Forge Press.

O'Brien, Patrick K., ed. 1999. *Atlas of World History.* New York: Oxford University Press.

"O Come All Ye Faithful: Special Report on Religion and Public Life." 2007. *The Economist,* November 3, 6.

O'Connel, Sanjida. 1993. "Meet My Two Husbands." *Guardian,* March 4 (Sec. 2), 12.

Office of National Statistics [UK]. 2007. "Marriages: Age at Marriage By Sex and Previous Marital Status, 1991, 2001, and 2003–2005." *Population Trends* 127. Retrieved January 6, 2010 (www.statistics. gov.uk/STATBASE/ssdataset.asp?vlnk=9599).

Ogbu, John U. 1998. "Understanding Cultural Diversity and Learning." *Educational Researcher* 21(8):5–14.

Ogburn, William F. [1922] 1938. *Social Change, With Respect to Culture and Original Nature.* New York: Viking.

———. 1933. *Recent Social Trends.* New York: McGraw-Hill.

———. 1950. *Social Change.* New York: Viking.

———. 1961. "The Hypothesis of Cultural Lag." Pp. 1270–73 in *Theories of Society: Foundations of Modern Sociological Theory,* Vol. 2, edited by Talcott Parsons, Edward Shils, Kaspar D. Naegele, and Jesse R. Pitts. New York: The Free Press.

———. 1964. *On Culture and Social Change: Selected Papers,* edited by Otis Dudley Duncan. Chicago: University of Chicago Press.

Olin, Susan Moller. 1996. "Families and Feminist Theory: Some Past and Present Issues." Pp. 13–26 in *Feminism and Families,* edited by Hilde Lindemann Nelson. New York: Routledge.

Oller, Kimbrough. 2006. "Development and Evolution in Human Vocal Communication." *Biological Theory* 1(4):349–51.

Olsen, Marvin E. 1968. *The Process of Social Organization.* New York: Holt, Rinehart, and Winston.

———. 1970. "Power as a Social Process." Pp. 2–10 in *Power in Societies,* edited by Marvin Olsen. New York: Macmillan.

Olson, C. L., H. D. Schumaker, and B. P. Yawn. 1994. "Overweight Women Delay Medical Care." *Archives of Family Medicine* 3(10):888–92.

Olsson, Sven E. 1990. *Social Policy and the Welfare State in Sweden.* Stockholm: Swedish Institute for Social Research.

Omi, Michael and Howard Winant. 1989. *Racial Formation in the United States: From the 1960s to the 1980s.* New York: Routledge.

One in Four, Inc. n.d. Retrieved July 9, 2011 (www.oneinfourusa.org).

One Laptop per Child. 2011. "Vision." Retrieved March 18, 2011 (http://laptop.org/en/vision/).

Ontario Consultants on Religious Tolerance. 2004. "'Fundamentalism' in Christianity and Islam." Retrieved July 5, 2008 (www.religious tolerance.org/reac_ter9.htm).

———. 2005. "Homosexuality and Bisexuality." Retrieved August 23, 2008 (http://www.religioustolerance.org/hom_mar6.htm).

———. 2008. "Prohibiting Same-Sex Marriages in the U.S.: Federal and State 'Doma' Legislation." Retrieved August 7, 2008 (http://www .religioustolerance.org/hom_mar6.htm).

Orfield, Gary, Mark D. Brachmeier, David R. James, and Tamela Eitle. 1997. "Deepening Segregation in American Public Schools." *Equity and Excellence in Education* 30(2):5–24.

Organisation for Economic Co-operation and Development. 2007. "Average Class Size" (Table 5.d, pp. 138–39). Retrieved July 14, 2011 (oei.es.quipu/WEI2007report.pdf).

———. 2008. "Education at a Glance 2008" (Table D2.1). Retrieved July 14, 2011 (http://dx.doi.org/10.1787/402267680060).

"Organized Crime." 2011. Retrieved August 6, 2011 (http://law.jrank.org/pages/11951/Organized-Crime.html).

Orobello, Natala. 2008. "What Our Students Need." *Academic Leadership* 6(2). Retrieved August 6, 2011 (http://www.academicleadership.org/article/What_Our_Students_Need/orobello).

Orum, Anthony M. 2001. *Introduction to Political Sociology,* 4th ed. Upper Saddle River, NJ: Prentice Hall.

Oswald, Ramona Faith. 2000. "A Member of the Wedding? Heterosexism and Family Ritual." *Journal of Social and Personal Relationships* 17(June):349–68.

———. 2001. "Religion, Family, and Ritual: The Production of Gay, Lesbian, and Transgendered Outsiders-Within." *Review of Religious Research* 43(December):39–50.

O'Toole, Tara and Donald A. Henderson. 2006. "A Clear and Present Danger: Confronting the Threat of Bioterrorism." Pp. 239–45 in *Globalization: The Transformation of Social Worlds,* edited by D. Stanley Eitzen and Maxine Baca Zinn. Belmont, CA: Wadsworth.

Ottenheimer, Martin. 1996. *Forbidden Relatives: The American Myth of Cousin Marriages.* Urbana: University of Illinois Press.

"Overseas Population Spending Threatened." 2006. *Popline* 29(March/April):1.

Padden, Brian. 2009. "Nigeria Still Fighting False Rumors About Polio Vaccine." February 17. Retrieved August 6, 2011 (http://www.voanews .com/english/news/a-13-2009-02-17-voa48-68672337.html).

Page, Ann L. and Donald A. Clelland. 1978. "The Kanawha County Textbook Controversy: A Study of the Politics of Life Style Concern." *Social Forces* 57(September):265–81.

"A Painful Tradition." 1999. *Newsweek,* July 5, 32–33.

Pan American Health Organization. 2000. *Physicians per 10,000 Inhabitants Ratio for Year 1999.* Washington, DC: Division of Health Systems and Services Development, Human Resources Development Program.

Papalia, Diane E., Sally Wendkos Olds, and Ruth Duskin Feldman. 2006. *A Child's World: Infancy Through Adolescence,* 10th ed. Boston: McGraw-Hill.

Pareto, Vilfredo. [1911] 1955. "Mathematical Economics." In *Encyclopedie des Sciences Mathematique.* New York: Macmillan.

Parish, Thomas S. and Joycelyn G. Parish. 1991. "The Effect of Family Configuration and Support System Failures During Childhood and Adolescence on College Students' Self-Concepts and Social Skills." *Adolescence* 26(102):441–47.

Park, Ken. 2006. *World Almanac and Book of Facts.* Mahwah, NJ: World Almanac.

Park, Robert Ezra, Ernest W. Burgess, and Roderick D. McKenzie. [1925] 1967. *The City.* Chicago: University of Chicago Press.

Parkinson, C. Northcote. 1957. *Parkinson's Law.* Boston: Houghton Mifflin.

Parrillo, Vincent N. 2005. *Understanding Race and Ethnic Relations,* 2nd ed. Boston: Pearson Allyn & Bacon.

Parsons, Talcott. 1951a. *The Social System.* New York: The Free Press.

———. 1951b. *Toward a General Theory of Action.* New York: Harper & Row.

———. 1975. "The Sick Role and Role of the Physician Reconsidered." *Milbank Memorial Fund Quarterly* 53(3):257–78.

Parsons, Talcott and Robert F. Bales. 1953. *Family, Socialization, and Interaction Process.* Glencoe, IL: The Free Press.

Passeron, Jean Claude and Pierre Bourdieu. 1990. *Reproduction in Education, Society, and Culture,* 2nd ed. Newbury Park, CA: Sage.

Paulhus, D. L., P. D. Trapnell, and D. Chen. 1999. *Birth Order Effects on Personality and Achievement Within Families.* Malden, MA: Blackwell.

Paxton, Pamela and Melanie M. Hughes. 2007. *Women, Politics, and Power: A Global Perspective.* Thousand Oaks, CA: Pine Forge Press.

Peace Over Violence. 2007. "Rape Prevention and Education Campaign." April 25. Retrieved July 22, 2009 (http://www.denimdayinla.org/actionkit/flyer_dd_stats.pdf).

Pearce, Fred. 2010. "As Longevity Grows, the World Might Become a Better Place." *The Washington Post.* May 25. Retrieved July 28, 2011 (www.washingtonpost.com/wp-dyn/content/article/2010/05/24/AR2010052402607.html).

Pearce, Tess. 2009. "How Does Social Class Affect Socialisation Within The Family?" April 24. IBSS/Exploring Social Science Research, April 24. Retrieved March 1, 2011 (http://ibssblog.wordpress.com/2009/04/24/how-does-social-class-affect-socialisation-within-the-family/).

Pellow, David Naguib. 2002. *Garbage Wars: The Struggle for Environmental Justice in Chicago.* Cambridge, MA: MIT Press.

Pennington, Bill. 2006. "Small Colleges, Short of Men, Embrace Football." *The New York Times,* July 12, A1.

Pentikänen, Juha. 1999. *"Silent as Waters We Live": Old Believers in Russia and Abroad.* Helsinki, Finland: Finnish Literature Society.

Perez, Miguel. 2011. "The Latino Backlash Is Coming Sooner Than Republicans Expected." April 6. Retrieved April 7, 2011 (http://www.paragoulddailypress.com/articles/2011/04/06/opinion/doc4d9ba4c-6c2ef4312670875.txt).

Perrucci, Robert and Earl Wysong. 1999. *The New Class Society.* Lanham, MD: Rowman & Littlefield.

Persell, Caroline Hodges. 2005. "Race, Education, and Inequality." Pp. 286–24 in *Blackwell Companion to Social Inequalities,* edited by M. Romero and E. Margolis. Oxford, UK: Blackwell.

Persell, Caroline Hodges and Peter W. Cookson, Jr. 1985. "Chartering and Bartering: Elite Education and Social Reproduction." *Social Problems* 33(2):114–29.

Pescosolido, Bernice A. 1992. "Beyond Rational Choice: The Social Dynamics of How People Seek Help." *American Journal of Sociology* 97(4):1113.

Pescosolido, Bernice A. and Carol A. Boyer. 2001. "The American Health Care System: Entering the Twenty-first Century With High Risk, Major Challenges, and Great Opportunities." Pp. 180–98 in *The Blackwell Companion to Medical Sociology,* edited by William C. Cockerham. Malden, MA: Blackwell.

Pescosolido, Bernice A. and Sharon Georgianna. 1989. "Durkheim, Suicide, and Religion: Toward a Network Theory of Suicide." *American Sociological Review* 54(February):33–48.

Peters, William. [1972] 1987. *A Class Divided.* New York: Doubleday.

Petersen, Larry R. and Gregory V. Donnenwerth. 1998. "Religion and Declining Support for Religious Beliefs About Gender Roles and Homosexual Rights." *Sociology of Religion* 59(4):353–71.

Peterson, Richard R. 1996. "A Re-evaluation of the Economic Consequences of Divorce." *American Sociological Review* 61(June):528–36.

Pew Forum on Religion and Public Life. 2008. "U.S. Religious Landscape Survey, 2008." Retrieved March 2, 2008 (http://religions.pewforum.org/pdf/report-religious-landscape-study-full.pdf).

———. 2009. "The Stronger Sex—Spiritually Speaking." Retrieved December 22, 2009 (http://pewforum.org/docs/?DocID=403).

———. 2010. "U.S. Religious Landscape Survey." Retrieved January 1, 2010 (http://religions.pewforum.org/reports).

Pew Hispanic Center. 2011. "Hispanics Account for 56% of Nation's Growth in Past Decade." Retrieved April 7, 2011 (http://pewhispanic.org/).

Pew Internet and American Life Project. 2008. "Computer Usage in the U.S." Retrieved January 9, 2008 (http://www.infoplease.com/ipa/A0921872.html).

———. 2010. "Change in Internet Access by Age Group, 2000–2009." Retrieved February 27, 2011 (http://www.pewinternet.org/Infographics/2010/Internet-acess-by-age-group-over-time.aspx).

Pharr, Suzanne. 1997. *Homophobia: A Weapon of Sexism,* Expanded ed. Berkeley, CA: Chardon.

Phi Delta Kappa/Gallup Poll. 2005. "The 37th Annual Phi Delta Kappa/Gallup Poll of the Public's Attitudes Toward the Public Schools." *Phi Delta Kappan* 87(1):41–57.

Phipher, Mary. 1994. *Reviving Ophelia: Saving the Selves of Adolescent Girls.* New York: Ballantine Books.

Physicians' Working Group for Single-Payer National Health Insurance. 2003. "Proposal of the Physicians' Working Group for Single-Payer National Health Insurance." *Journal of the American Medical Association* 290(13):798–805.

Piaget, Jean. 1989. *The Child's Conception of the World.* Savage, MD: Littlefield, Adams Quality Paperbacks.

Piaget, Jean and Barbel Inhelder. [1955] 1999. *Growth of Logical Thinking.* London: Routledge and Kegan Paul.

Pickard, Ruth and Daryl Poole. 2007. *The Study of Society and the Practice of Sociology.* Previously unpublished essay.

Pieterse, Jan Nederveen. 2004. *Globalization and Culture.* Lanham, MD: Rowman and Littlefield.

Pinker, Steven. 2002. *The Blank Slate: The Denial of Human Nature.* New York: Viking.

Piven, Frances Fox and Richard Cloward. 1988. *Why Americans Don't Vote.* New York: Pantheon Books.

Plato. [circa 350 BCE] 1960. *The Laws.* New York: Dutton.

Plouffe, David. 2009. *The Audactiy to Win: The Inside Story and Lessons of Barack Obama's Historic Victory.* New York: Viking.

Polgreen, Lydia. 2008. "Scorched-Earth Strategy Returns to Darfur." *The New York Times,* March 2. Retrieved July 26, 2009 (http://www.nytimes.com/2008/03/02/world/africa/02darfur.html).

Pollack, William. 1999. *Real Boys: Rescuing Our Sons From the Myths of Boyhood.* New York: Owl Books.

Pope, Liston. [1942] 1965. *Millhands and Preachers.* New Haven, CT: Yale University Press.

Popenoe, David, Jean Bethke Elshtain, and David Blankenhorn. 1996. *Promises to Keep: Decline and Renewal of Marriage in America.* Lanham, MD: Rowman and Littlefield.

Population Action International. 1993. *Closing the Gender Gap: Educating Girls: Report on Progress Toward World Population Stabilization.* Washington, DC: Population Action International.

Population Reference Bureau. 2005. "World Population Data Sheet." Retrieved September 29, 2006 (www.prb.org/pdf05/05WorldDataSheet_Eng.pdf).

———. 2007a. "World Population Data Sheet." Retrieved July 22, 2009 (http://www.prb.org/Publications/Datasheets/2007/2007WorldPopulationDataSheet.aspx).

———. 2007b. "World Population Highlights." *Population Bulletin* 62(3):2.

———. 2008. "World Population Data Sheet." Retrieved August 23, 2008 (http://www.prb.org/Publications/Datasheets/2008/2008wpds.aspx).

———. 2009a. "2009 World Population Clock." Retrieved January 5, 2010 (www.prb.org/Articles/2009/worldpopulationclock2009.aspx).

———. 2009b. "2009 World Population Data Sheet." Retrieved January 8, 2010 (www.prb.org/pdf09/09wpds_eng.pdf).

———. 2010a. "2010 World Population Clock." Retrieved May 13, 2011 (www.prb.org/Articles/2010/worldpopulationclock2010.aspx).

———. 2010b. "2010 World Population Data Sheet." Retrieved July 27, 2011 (http://www.prb.org/pdf10/10wpds_eng.pdf).

———. 2011a. "Human Population: Women." Retrieved July 23, 2011 (www.prb.org/Educators/TeachersGuides/HumanPopulation/Women.aspx).

———. 2011b. "World Population Growth, 1950–2050." Retrieved July 23, 2011 (www.prb.org/Educators/TeachersGuides/HumanPopulation/PopulationGrowth.aspx?p=1).

Porter, Eduardo. 2007. "More Than Ever, It Pays to Be the Top Executive." *The New York Times,* May 25. Retrieved July 7, 2008 (http://www.nytimes.com/2007/05/25/business/25execs.html?scp=3&sq=Porter%20%22more%20than%20ever%22&st=cse).

Portes, Alejandro and Ruben G. Rumbaut. 2001. *Legacies: The Story of the Immigrant Second Generation.* Berkeley: University of California Press.

Povik, Fili. 1994. *Zlata's Diary: A Child's Life in Sarajevo.* New York: Viking.

Powell, Andrea D. and Arnold S. Kahn. 1995. "Racial Differences in Women's Desires to Be Thin." *International Journal of Eating Disorders* 17(2):191–95.

Powell, Brian, Catherine Bolzendahl, Claudia Geist, and Lala Carr Steelman. 2010. *Counted Out: Same-Sex Relations and Americans' Definitions of Family.* New York: Russell Sage Foundation.

Powell, Michael. 2008a. "Campaigns Try to Lure the Facebook Generation." *International Herald Tribune,* January 9, 5.

———. 2008b. "Tidal Wave of Youth Buoys 2 Campaigns." *International Herald Tribune,* January 9, 2.

Powell, Walter W. 1990. "Neither Market nor Hierarchy: Network Forms of Organization." *Research in Organizational Behavior* 12:295–336.

Powley, Elizabeth. 2003. "Strengthening Governance: The Role of Women in Rwanda's Transition." In *Women Waging Peace,* edited by S. N. Anderlini. Washington, DC: Hunt Alternatives Fund.

Preston, David L. 1988. *The Social Organization of Zen Practice: Constructing Transcultural Reality.* Cambridge, UK: Cambridge University Press.

Project Vote Smart. 2007. "State Presidential Primary and Caucus Dates." Retrieved July 26, 2011 (http://www.votesmart.org/election_president_state_primary_dates.php?sort=date).

Public Citizen's Global Trade Watch. 2003. "The Ten Year Track Record of the North American Free Trade Agreement: U.S. Workers' Jobs, Wages and Economic Security." Retrieved September 29, 2006 (http://www.citizen.org/documents/NAFTA_10_jobs.pdf).

Purcell, Piper and Lara Stewart. 1990. "Dick and Jane in 1989." *Sex Roles* 22(3/4):177–85.

Putnam, Robert D. 1995. "Bowling Alone, Revisited." *The Responsive Community* (Spring):18–33.

———. 2001. *Bowling Alone: The Collapse and Revival of American Community.* New York: Simon and Schuster.

Pyle, Ralph E. 2006. "Trends in Religious Stratification: Have Religious Group Socioeconomic Distinctions Declined in Recent Decades?" *Sociology of Religion* 67(Spring):61–79.

Quinley, Harold. 1974. *The Prophetic Clergy.* New York: John Wiley.

Quinn, Jane Bryant. 2006. "Health Care's New Lottery." *Newsweek,* February 27, 47.

Quinney, Richard. 2002. *Critique of Legal Order: Crime Control in Capitalist Society.* New Brunswick, NJ: Transaction Publishers.

Radcliffe-Brown, A. R. 1935. "On the Concept of Functional in Social Science." *American Anthropologist* 37(3):394–402.

Rama Rao, Saumya and Raji Mohanam. 2003. "The Quality of Family Planning Programs: Concepts, Measurements, Interventions, and Effects." *Studies in Family Planning* 34(4):227–48.

Rampell, Catherine. 2008. "Why Obama Rocks the Vote." *The Washington Post,* March 30, B7.

Randerson, James. 2008. "Cutting TV Time Makes Children Healthier, Says U.S. Study." *The Guardian,* March 4. Retrieved April 10, 2009 (http://www.guardian.co.uk/science/2008/mar/04/medicalresearch.health).

Rankin, Bruce H. and Isık A. Aytaç. 2006. "Gender Inequality in Schooling: The Case of Turkey." *Sociology of Education* 79(1):25–43.

Read, Kristen. 2011. "Number of Internet Users Worldwide Climbs to 2 Billion." UN International Telecommunications Union, January 28. Retrieved March 18, 2011 (http://graphicartsmag.com/news/2011/01/number-of-internet-users-worldwide-climbs-to-2-billion).

Regnerus, Mark D. 2007. *Forbidden Fruit: Sex and Religion in the Lives of American Teenagers.* New York: Oxford University Press.

Reid, Scott A. and Sik Hung Ng. 2006. "The Dynamics of Intragroup Differentiation in an Intergroup Social Context." *Human Communication Research* 32:504–25.

Reiman, Jeffrey. 1998. *The Rich Get Richer and the Poor Get Prison,* 5th ed. Boston: Allyn & Bacon.

Reiman, Jeffrey and Paul Leighton. 2010a. *The Rich Get Richer and the Poor Get Prison,* 9th ed. Boston: Allyn & Bacon.

———. 2010b. *The Rich Get Richer and the Poor Get Prison: A Reader.* Boston: Allyn and Bacon.

Relethford, John H. 2008. *The Human Species: An Introduction to Biological Anthropology,* 7th ed. Boston: McGraw-Hill.

Religious Tolerance. 2009a. "U.S. Divorce Rates for Various Faith Groups, Age Groups, and Geographic Areas." Retrieved November 30, 2009 (http://www.religioustolerance.org/chr_dira.htm).

———. 2009b. "Women as Clergy." Retrieved December 24, 2009 (http://www.religioustolerance.org/femclrgy.htm).

Religious Worlds. 2007. "New Religious Movements." Retrieved August 14, 2008 (www.religiousworlds.com/newreligions.html).

Remmert, Consuelo. 2003. "Women in Reconstruction: Rwanda Promotes Women Decision-makers." *UN Chronicle* 4(3):25.

Renzetti, Claire. 2003. "Urban Violence Against Women." Speech, September 30, Dayton, Ohio, Wright State University.

Renzetti, Claire M. and Daniel J. Curran. 2003. *Women, Men, and Society,* 5th ed. Boston: Allyn & Bacon.

Report of the National Advisory Commission on Civil Disorders. 1968. New York: Bantam.

Residents of Hull House. [circa 1895] 1970. *Hull House Maps and Papers.* New York: Arno.

Richardson, James T. 1985. "Active Versus Passive Converts: Paradigm Conflict in Conversion/Recruitment Research." *Journal for the Scientific Study of Religion* 24(2):163–79.

Rideout, Victoria J., Ulla G. Foehr, and Donald F. Roberts. 2010. "Generation M2: Media in the Lives of 8- to 18-Year-Olds." A Kaiser Family Foundation Study, January. Retrieved March 2, 2011 (http://www.kff.org/entmedia/upload/8010.pdf).

Rideout, Victoria J., Ulla G. Foehr, Donald F. Roberts, and Mollyann Brodie. 1999. "Kids and Media at the New Millennium." Kaiser Family Foundation, November. Retrieved July 27, 2009 (http://www.kff.org/entmedia/upload/Kids-Media-The-New-Millennium-Executive-Summary.pdf).

Riehl, Carolyn. 2001. "Bridges to the Future: Contributions of Qualitative Research to the Sociology of Education." *Sociology of Education* (Extra Issue):115–34.

Ries, L. A. G., M. P. Eisner, C. L. Kossary, B. F. Hankey, B. A. Miller, and B. K. Edwards, eds. 2000. *SEER Cancer Statistics Review, 1973–1997.* Bethesda, MD: National Cancer Institute.

Riesman, David. 1961. *The Lonely Crowd: A Study of the Changing American Character.* New Haven, CT: Yale University Press.

Riley, Glenda. 1997. *Divorce: An American Tradition.* Lincoln: University of Nebraska Press.

Riordan, Cornelius. 2004. *Equality and Achievement: An Introduction to the Sociology of Education.* Upper Saddle River, NJ: Prentice Hall.

Ritzer, George. 1998. *The McDonaldization Thesis: Explorations and Extensions.* Thousand Oaks, CA: Pine Forge.

———. 2004. *The Globalization of Nothing.* Thousand Oaks, CA: Pine Forge.

———. 2010. *The McDonaldization of Society,* 6th ed. Thousand Oaks, CA: Pine Forge.

Ritzer, George and Douglas J. Goodman. 2004. *Sociological Theory,* 6th ed. New York: McGraw-Hill.

Roach, Ronald. 2004. "Survey Reveals 10 Biggest Trends in Internet Use." *Black Issues in Higher Education,* October 21:42.

Robbins, Mandy. 1998. "A Different Voice: A Different View." *Review of Religious Research* 40(1):75–80.

Robbins, Richard H. 2011. *Global Problems and the Culture of Capitalism,* 5th ed. Upper Saddle River, NJ: Prentice Hall.

Robbins, Thomas. 1988. *Cults, Converts, and Charisma: The Sociology of New Religious Movements.* London and Newbury Park, CA: Sage.

Roberts, Dorothy. 1997. *Killing the Black Body: Race, Reproduction, and the Meaning of Liberty.* New York: Pantheon.

———. 2002. *Shattered Bonds: The Color of Child Welfare.* New York: Basic Books.

Roberts, Keith A. 2003. *Interviews With Ohio Residents in Communities Considering Tax Levies.* Unpublished.

Roberts, Keith A. and Karen A. Donahue. 2000. "Professing Professionalism: Bureaucratization and Deprofessionalization in the Academy." *Sociological Focus* 33(4):365–83.

Roberts, Keith A. and David Yamane. 2012. *Religion in Sociological Perspective,* 5th ed. Thousand Oaks, CA: Pine Forge Press.

Robertson, Roland. 1989. "Globalization, Politics, and Religion." Pp. 1–9 in *The Changing Face of Religion,* edited by James A. Beckford and Thomas Luckmann. London: Sage.

———. 1992. *Globalization: Social Theory and Global Culture.* London: Sage.

———. 1997. "Social Theory, Cultural Relativity and the Problem of Globality." Pp. 69–90 in *Culture, Globalization and the World System,* edited by Anthony King. Minneapolis: University of Minnesota Press.

Robertson, Roland and William R. Garrett, eds. 1991. *Religion and Global Order.* New York: Paragon.

Robertson, Roland and H. H. Khondker. 1998. "Comparative Sociology, Global Sociology and Social Theory." *Sociology* 13(1):25–40.

Robson, Roy R. 1995. *Old Believers in Modern Russia.* DeKalb: Northern Illinois University Press.

Rodgers, Bill. 2009. "Developing Countries Employ 'Leapfrog Technology' With Cell Phones." The Cutting Edge, March 9. Retrieved April 28, 2009 (www.thecuttingedgenews.com/index.php?article=11174&pageid=28&pagename=Sci-Tech).

Rodriguez, Eric M. and Suzanne C. Ouellette. 2000. "Gay and Lesbian Christians: Homosexual and Religious Identity Integration in the Members and Participants of a Gay-Positive Church." *Journal for the Scientific Study of Religion* 39(3):333–47.

Roethlisberger, Fritz J. and William J. Dickson. 1939. *Management and the Worker.* Cambridge, MA: Harvard University Press.

Romano, Lois. 2008. "Generation Y: Ready to Rock the 2008 Election." *The Washington Post,* January 10, C1.

Roof, Wade Clark. 1999. *Spiritual Marketplace: Baby Boomers and the Remaking of American Religion.* Princeton, NJ: Princeton University Press.

Rose, Lowell C. and Alec M. Gallup. 2001. "The 33rd Annual Phi Delta Kappa/Gallup Poll of the Public's Attitudes Toward the Public Schools." *Phi Delta Kappan* 83(1):41–48.

Rose, Stephen. 2000. *Social Stratification in the United States.* New York: The New Press.

Rosen, Karen. 2007. "Women Still Lag in College Sports." June 6. Retrieved March 7, 2008 (www.oxfordpress.com/sports/content/shared/sports/stories/2007/WOMEN_SPORTS_0606_COX.html).

Rosenbaum, James E. 1999. "If Tracking Is Bad, Is Detracking Better? A Study of a Detracked High School." *American Schools* (Winter):24–47.

Rosenthal, Robert and Lenore Jacobson. 1968. *Pygmalion in the Classroom.* New York: Holt, Rinehart, and Winston.

Rossi, Alice S. 1984. "Gender and Parenthood." *American Sociological Review* 49(February):1–19.

Rossides, Daniel W. 1997. *Social Stratification: The Interplay of Class, Race, and Gender.* Englewood Cliffs, NJ: Prentice Hall.

Rothbard, Nancy P. and Jeanne M. Brett. 2000. "Promote Equal Opportunity by Recognizing Gender Differences in the Experience of Work and Family." In *The Blackwell Handbook of Principles of Organizational Behavior,* edited by Edwin A. Locke. Malden, MA: Blackwell.

Rothenberg, Paula S. 2006a. *Beyond Borders: Thinking Critically About Global Issues.* New York: Worth.

———. 2006b. "Transnational Institutions and the Global Economy." Pp. 411–18 in *Beyond Borders: Thinking Critically About Global Issues,* edited by Paula S. Rothenberg. New York: Worth.

———. 2007. *Race, Class, and Gender in the United States: An Integrated Study,* 7th ed. New York: Worth.

———. 2011. *White Privilege: Essential Readings on the Other Side of Racism,* 4th ed. New York: Worth.

Rothman, Robert A. 2005. *Inequality and Stratification: Race, Class, and Gender,* 5th ed. Englewood Cliffs, NJ: Prentice Hall.

Rothman, Stanley and Amy E. Black. 1998. "Who Rules Now? American Elites in the 1990s." *Society* 35(6):17–20.

Rubenstein, Grace. 2007. "Computers for Peace: The $100 Laptop." Retrieved April 21, 2009 (www.edutopia.org/computers-peace).

Rubin, Jeffrey Z. 1974. "The Eye of the Beholder: Parents' Views on Sex of Newborns." *American Journal of Orthopsychiatry* 44(4):512–19.

Rugh, Andrea B. 1997. *Within the Circle: Parents and Children in an Arab Village.* New York: Columbia University Press.

Rumbaut, Ruben G. and Alejandro Portes. 2001. *Ethnicities: Children of Immigrants in America.* Los Angeles: University of California Press.

Runningen, Roger and Brian Faler. 2008. "Bush Boosts Defense Spending in $3.1 Trillion Budget." February 4. Retrieved August 5, 2011 (http://www.bloomberg.com/apps/news?pid=newsarchive&sid=azGB.29vVkzA&refer=us).

Rusere, Patience. 2009. "Rainy Season Brings New Cholera Outbreaks in Zimbabwe, Five Deaths Reported." *Voice of America News.* November 16. Retrieved December 20, 2009 (http://www.voanews.com/zimbabwe/news/a-13–56–74–2009–11–16-voa48–70422597.html).

Rydgren, Jens. 2004. "Mechanisms of Exclusion: Ethnic Discrimination in the Swedish Labour Market." *Journal of Ethnic and Migration Studies* 30(4):687–716.

Sabol, William J. 2007. *Prisoners in 2006.* Washington, DC: U.S. Department of Justice, Bureau of Justice Statistics.

Sabol, William J., Heather C. West, and Matthew Cooper. 2009. "Prisoners in 2008." *Bureau of Justice Statistics Bulletin,* U.S. Department of Justice, December. Retrieved December 20, 2009 (http://bjs.ojp.usdoj.gov/content/pub/pdf/p08.pdf).

Sachoff, Mike. 2008. "18% of U.S. Households Have No Internet Access." Retrieved February 27, 2011 (www.WebProNews.com).

Sack, Kevin. 1996. "Burnings of Dozens of Black Churches Across the South Are Investigated." *The New York Times,* May 21, A6.

Sadker, Myra and David Sadker. 1995. *Failing at Fairness: How Our Schools Cheat Girls.* New York: Touchstone.

————. 2005. *Teachers, Schools, and Society,* 7th ed. New York: McGraw-Hill.

Sado, Stephanie and Angela Bayer. 2001. "Executive Summary: The Changing American Family." *The Changing Family* 2(12):169.

Sadovnik, Alan R. 2007. *Sociology of Education: A Critical Reader.* New York: Routledge.

Saenz, Rogelio. 2007. "The Growing Color Divide in U.S. Infant Mortality." Retrieved July 26, 2011 (www.prb.org/Articles/2007/ColorDivideinInfantMortality.aspx).

Sage, George H. 2005. "Racial Inequality and Sport." Pp. 266–75 in *Sport in Contemporary Society,* 7th ed., edited by Stanley D. Eitzen. Boulder, CO: Paradigm.

Sager, Ira, Ben Elgin, Peter Elstrom, Faith Keenan, and Pallavi Gogoi. 2006. "The Underground Web." Pp. 261–70 in *Globalization: The Transformation of Social Worlds,* edited by D. Stanley Eitzen and Maxine Baca Zinn. Belmont, CA: Wadsworth.

Saha, Lawrence and A. Gary Dworkin. 2006. "Educational Attainment and Job Status: The Role of Status Inconsistency on Occupational Burnout." Paper presented at the International Sociological Association, July 23–29, Durban, South Africa.

Saharan Vibe. 2007. "Wodaabe Beauty Ceremony." February 19. Retrieved November 6, 2009 (http://www.saharanvibe.blogspot.com/2007/02/wodaabe-beauty-ceremony.html).

Salzman, Marian and Ira Matathia. 2000. "Lifestyles of the Next Millennium: 65 Forecasts." Pp. 466–71 in *Sociological Footprints,* edited by Leonard Cargan and Jeanne Ballantine. Belmont, CA: Wadsworth.

Salzman, Michael B. 2008. "Globalization, Religious Fundamentalism and the Need for Meaning." *International Journal of Intercultural Relations* 32(4):318–27.

Samovar, Larry A. and Richard E. Porter. 2003. *Intercultural Communication.* Belmont, CA: Wadsworth.

Sanday, Peggy Reeves. 1981. "The Socio-Cultural Context of Rape: A Cross-Cultural Study." *Journal of Social Issues* 37(4):5–27.

————. 1996. *A Woman Scorned: Acquaintance Rape on Trial.* Berkeley: University of California Press.

————. 2007. *Fraternity Gang Rape: Sex, Brotherhood, and Privilege on Campus.* New York: New York University Press.

Sanday, Peggy and Ruth Gallagher Goodenough, eds. 1990. *Beyond the Second Sex.* Philadelphia: University of Pennsylvania Press.

Sander, William. 1993. "Catholicism and Intermarriage in the United States." *Journal of Marriage and the Family* 55(November):1037–41.

Sapir, Edward. 1929. "The Status of Linguistics as a Science." *Language* 5:207–14.

————. 1949. *Selected Writings of Edward Sapir in Language, Culture, and Personality,* edited by David G. Mandelbaum. Berkeley: University of California Press.

Sapiro, Virginia. 2003. *Women in American Society: An Introduction to Women's Studies,* 5th ed. Mountain View, CA: Mayfield.

Sarat, Austin. 2001. *When the State Kills: Capital Punishment and the American Condition.* Princeton, NJ: Princeton University Press.

Sargeant, Kimon Howland. 2000. *Seeker Churches: Promoting Traditional Religion in a Nontraditional Way.* New Brunswick, NJ: Rutgers University Press.

Sarno, David. 2008. "Looking for the Youth Vote? It's Online." *Los Angeles Times,* January 13, E42.

Sassler, Sharon and Amanda Miller. 2011. "Waiting to Be Asked: Gender, Power, and Relationship Progression Among Cohabiting Couples." *Journal of Family Issues* 32:482–506.

Sax, Leonard. 2005. *Why Gender Matters: What Parents and Teachers Need to Know About the Emerging Science of Sex Differences.* New York: Doubleday.

Scarce, Rik. 1999. "Good Faith, Bad Ethics: When Scholars Go the Distance and Scholarly Associations Do Not." *Law and Social Inquiry* 24(4):977–86.

Schaefer, Brett and Anthony Kim. 2010. "David Cameron and Barack Obama Must Advance Economic Freedom—Not More Foreign Aid." July 19. Retrieved July 28, 2011 (http://www.heritage.org/research/reports/2010/07/david-cameron-and-barack-obama-must-advance-economic-freedom-not-more-foreign-aid).

Schaefer, Richard T. 2012. *Racial and Ethnic Groups,* 13th ed. Upper Saddle River, NJ: Prentice Hall.

Schaefer, Richard T. and Jenifer Kunz. 2008. *Racial and Ethnic Groups,* 11th ed. Upper Saddle River, NJ: Prentice Hall.

Schaeffer, Robert K. 2003. *Understanding Globalization: The Social Consequences of Political, Economic, and Environmental Change,* 2nd ed. Lanham, MD: Rowman and Littlefield.

Schapiro, Mark. 2006. "Big Tobacco." Pp. 271–84 in *Globalization: The Transformation of Social Worlds,* edited by D. Stanley Eitzen and Maxine Baca Zinn. Belmont, CA: Wadsworth.

Schemo, Diana Jean. 2006. "Public Schools Close to Private in U.S. Study." *The New York Times,* July 15, A1, A10.

Schieber, George and Akiko Maeda. 1999. "Health Care Financing and Delivery in Developing Countries." *Health Affairs* 18(3):193–206.

Schienberg, Jonathan. 2006. "Firefighters, Doctors and Nurses Considered Most Prestigious Jobs." *CNNMoney.com,* July 31. Retrieved July 7, 2008 (http://money.cnn.com/2006/07/26/news/economy/prestigious_professions/index.htm).

Schmalleger, Frank. 2006. *Criminology Today: An Integrative Introduction,* 4th ed. Upper Saddle River, NJ: Pearson Prentice Hall.

————. 2009. *Criminology Today: An Integrative Introduction,* 5th ed. Upper Saddle River, NJ: Pearson Prentice Hall.

Schneider, Barbara and James S. Coleman. 1993. *Parents, Their Children, and Schools.* Boulder, CO: Westview.

Schneider, Barbara and David Stevenson. 1999. *The Ambitious Generation: Imagining the Future.* New Haven, CT: Yale University Press.

Schneider, Barbara and Linda J. Waite, eds. 2005. *Being Together, Working Apart: Dual-Career Families and the Work-Life Balance.* New York: Cambridge University Press.

Schneider, Linda and Arnold Silverman. 2006. *Global Sociology,* 4th ed. Boston: McGraw-Hill.

Schoeman, Ferdinand. 1991. "Book Review: Heavy Drinking: The Myth of Alcoholism as a Disease." *The Philosophical Review* 100(3):493–98.

Schoen, Robert and Robin M. Weinick. 1993. "Partner Choice in Marriages and Cohabitations." *Journal of Marriage and the Family* 55(May):408–14.

Schweinhart, Lawrence J. 1997. "Child-Initiated Learning Activities for Young Children Living in Poverty." *ERIC Digest,* October. Washington, DC: Office of Educational Research and Improvement.

Scott, Ellen K., Andrew S. London, and Nancy A. Myers. 2002. "Dangerous Dependencies: The Intersection of Welfare Reform and Domestic Violence." *Gender and Society* 16(6):878–97.

Scott, Mark. 2009. "Politicians Urge for Global Stimulus." *BusinessWeek,* March 9. Retrieved July 22, 2009 (http://www.businessweek.com/executivesummary/archives/2009/03/politicians_urg.html).

Scott-Montagu, John. 1904. "Automobile Legislation: A Criticism and Review." *North American Review* 179(573):168–77.

Scribd. 2011. "Chipko Movement." Retrieved August 6, 2011 (www.scribd.com/doc/27513230/Chipko-Movement).

Sells, Heather. 2008. "Hunger in America a Growing Reality." CBNnews.com. Retrieved April 20, 2009 (www.cbn.com/cbnnews/485723.aspx).

Senate Democratic Policy Committee Hearing. 2004. "America's Uninsured: Myths, Realities and Solutions." Retrieved July 22, 2009 (http://dpc.senate.gov/dpchearing.cfm?h=hearing9).

The Sentencing Project. 2006. "International Rates of Incarceration." Retrieved July 24, 2009 (http://www.nccd-crc.org/nccd/pubs/2006nov_factsheet_incarceration.pdf).

Sernau, Scott. 2010. *Social Inequality in a Global Age.* Thousand Oaks, CA: Pine Forge Press.

Shade, Leslie Regan. 2004. "Bending Gender Into the Net: Feminizing Content, Corporate Interests, and Research Strategy." Pp. 57–70 in *Society On-Line: The Internet in Context,* edited by Philip N. Howard and Steve Jones. Thousand Oaks, CA: Sage.

Shaffer, David Williamson, Kurt R. Squire, Richard Haverson, and James P. Gee. 2005. "Video Games and the Future of Learning." *Phi Delta Kappan* 87(2):95–112.

Shah, Anup. 2008. "Consumption and Consumerism." Retrieved December 20, 2009 (www.globalissues.org/issue/235/consumption-and-consumerism).

———. 2010. "Today, Over 25,000 Children Died Around the World." September 7. Retrieved July 18, 2011 (www.globalissues.org/article/715/today-over-25000-children-died-around-the-world).

———. 2011. "World Military Spending." Retrieved July 22, 2011 (http://www.globalissues.org/article/75/world-military-spending).

Sharp, Henry S. 1991. "Memory, Meaning, and Imaginary Time: The Construction of Knowledge in White and Chipewyan Cultures." *Ethnohistory* 38(2):149–73.

Sharp, Lauriston. 1990. "Steel Axes for Stone-Age Australians." Pp. 410–24 in *Conformity and Conflict,* 7th ed., edited by James P. Spradley and David W. McCurdy. Glenview, IL: Scott Foresman.

Sharp, Travis. 2008. "U.S. Military Spending vs. the World in 2008." Center for Arms Control and Non-Proliferation, February 20. Retrieved September 3, 2008 (www.armscontrolcenter.org/policy/securityspending/articles/us_vs_world.gif).

Shaw, Clifford R. and Henry D. McKay. 1929. *Delinquency Areas.* Chicago: University of Chicago Press.

Shaw, Susan M. and Janet Lee. 2005. *Women's Voices, Feminist Visions: Classic and Contemporary Readings,* 3rd ed. Boston: McGraw-Hill.

Shepherd, William R. 1964. *Historical Atlas,* 9th ed. New York: Barnes & Noble.

Sherif, Muzafer and Carolyn Sherif. 1953. *Groups in Harmony and Tension.* New York: Harper & Row.

Sherkat, Darren E. and Christopher G. Ellison. 1999. "Recent Developments and Current Controversies in the Sociology of Religion." *Annual Review of Sociology* 25:363–94.

Sherman, Karen J. 2005. *Complementary and Alternative Medicine in the United States.* Washington, DC: National Academies Press.

Shore, Cris. 2008. "Corruption Scandals in America and Europe: Enron and EU Fraud in Comparative Perspective." Pp. 191–98 in *Deviance Across Cultures,* edited by Robert Heiner. New York: Oxford University Press.

Short, Katherine. 2009. "Voter Participation Rate, 2008." Retrieved March 17, 2009 (www.askquestions.org/details.php?id=21094).

Shorter, Edward. 1991. *Women's Bodies: A Social History of Women's Encounter With Health, Ill-Health and Medicine.* New Brunswick, NJ: Transaction Publishers.

Shriver, Eunice Kennedy. 2007. *Add Health Study.* Washington, DC: National Institute of Child Health and Human Development. Retrieved July 10, 2009 (www.nichd.nih.gov/health/topics/add_health_study.cfm).

Shu, Xiaoling. 2004. "Education and Gender Egalitarianism: The Case of China." *Sociology of Education* 77(4):311–36.

Shupe, Anson and Jeffrey K. Hadden. 1989. "Is There Such a Thing as Global Fundamentalism?" Pp. 109–22 in *Secularization and Fundamentalism Reconsidered,* edited by Jeffrey K. Hadden and Anson Shupe. New York: Paragon.

Siegel, Dina and Hans Nelen, eds. 2008. *Organized Crime: Culture, Markets, and Policies.* Series: Studies in Organized Crime, Vol. 7. New York: Springer.

Siegel, Larry J. 2000. *Criminology,* 7th ed. St. Paul, MN: West.

———. 2009. *Criminology: Theories, Patterns, and Typologies,* 10th ed. Belmont, CA: Thomson/Wadsworth.

Simmel, Georg. [1902–17] 1950. *The Sociology of Georg Simmel,* translated by Kurt Wolff. Glencoe, IL: The Free Press.

———. 1955. *Conflict and the Web of Group Affiliation,* translated by Kurt H. Wolff. New York: The Free Press.

Simmons, Rachel. 2002. *Odd Girl Out: The Hidden Culture of Aggression in Girls.* Orlando, FL: Harcourt.

Simon, David R. 2006. *Elite Deviance,* 8th ed. Boston: Allyn & Bacon.

Sine, Richard. 2002. "Garment Workers Say Gap Aided in Cambodian Strife." Retrieved September 2, 2008 (http://laborcenter.berkeley.edu/press/sfchronicle_dec02.shtml).

Sizer, Theodore R. 1984. *Horace's Compromise: The Dilemma of the American High School.* Boston: Houghton Mifflin.

Sjoberg, Gideon. 1960. *The Preindustrial City: Past and Present.* Glencoe, IL: The Free Press.

———. 1965. "The Origin and Evolution of Cities." *Scientific American* 213(September):56–57.

Skocpol, Theda. 1979. *States and Social Revolutions: A Comparative Analysis of France, Russia, and China.* Cambridge, UK: Cambridge University Press.

Smarick, Andy. 2009. "Saving Catholic Schools." *The Baltimore Sun,* July 17. Retrieved April 17, 2011 (www.nje3.org/?p=3362).

Smelser, Neil J. 1963. *Theory of Collective Behavior.* New York: The Free Press.

———. 1988. "Social Structure." Pp. 103–29 in *Handbook of Sociology,* edited by Neil J. Smelser. Newbury Park, CA: Sage.

———. 1992. "The Rational Choice Perspective: A Theoretical Assessment." *Rationality and Society* 4:381–410.

Smerdon, Becky A. 2002. "Students' Perceptions of Membership in Their High Schools." *Sociology of Education* 75(4):287–305.

Smith, Barbara Ellen. 1999. "The Social Relations of Southern Women." Pp. 13–31 in *Neither Separate Nor Equal,* edited by B. E. Smith. Philadelphia: Temple University Press.

Smith, Brent L. 1994. *Terrorism in America: Pipe Bombs and Pipe Dreams.* New York: State University of New York Press.

Smith, Christian and Robert Faris. 2005. "Socioeconomic Inequality in the American Religious System: An Update and Assessment." *Journal for the Scientific Study of Religion* 44(1):95–104.

Smith, Daniel R. and David F. Ayers. 2006. "Culturally Responsive Pedagogy and Online Learning: Implications for the Globalized Community College." *Community College Journal of Research and Practice* 30:401–415.

Smith, David M. and Gary J. Gates. 2001. *Gay and Lesbian Families in the United States: Same-Sex Unmarried Partner Households: A Preliminary Analysis of 2000 United States Census Data.* Washington, DC: Human Rights Campaign.

Smith, Heather. 2008. "Galvanizing Young Voters." *The Washington Post,* April 5, A13.

Smith, Mark K. 2008. "Howard Gardner and Multiple Intelligences." *The Encyclopedia of Informal Education.* Retrieved April 17, 2011 (http://www.infed.org/thinkers/gardner.htm).

Smith, Phillip. 2009. "Prohibition: UN Drug Chief Says Black Market Drug Profits Propped Up Global Banking System Last Year." *Drug War Chronicle,* January 30. Retrieved June 30, 2011 (http://stopthedrugwar.org/chronicle/570/costa_UNODC_drug_trade_banks).

Smith, Russell. 2005. "How Many Have Died in Darfur?" *BBC News,* February 16. Retrieved February 16, 2005 (http://news.bbc.co.uk/2/hi/africa/4268733.stm).

Smith, Scott. 2009. "Obama Touts Stimulus in Elkhart." *Kokomo Tribune,* February 9. Retrieved February 24, 2009 (http://www.kokomotribune.com/archivesearch/local_story_040174018.html).

Smock, Pamela J. 2004. "The Wax and Wane of Marriage: Prospects for Marriage in the 21st Century." *Journal of Marriage and Family* 66(November):966–73.

Snarr, Michael T. and D. Neil Snarr, eds. 2008. *Introducing Global Issues,* 4th ed. Boulder, CO: Lynne Rienner.

Snider, Mike. 2009. "Mattell Gives Barbie Online Dream House." *USA Today,* January 27. Retrieved July 9, 2011 (http://www.usatoday.com/life/lifestyle/2009–01–26-mattel-website-main_N.htm).

Snow, David A. and L. Anderson 1993. *Down on Their Luck: A Study of Homeless Street People.* Berkeley: University of California Press.

Snow, David A., Louis A. Zurcher, Jr., and Sheldon Ekland-Olson. 1980. "Social Networks and Social Movements: A Microstructural Approach to Differential Recruitment." *American Sociological Review* 45(5):787–801.

Snyder, Benson R. 1970. *The Hidden Curriculum.* New York: Alfred A. Knopf.

Sobolewski, Juliana M. and Paul R. Amato. 2007. "Parents' Discord and Divorce, Parent-Child Relationships, and Subjective Well-Being In Early Adulthood: Is Feeling Close to Two Parents Always Better Than Feeling Close to One?" *Social Forces* 85(3):1105–124.

Social Security Online. 2010. "Isabella Reigns as New Queen of Baby Names—Takes Top Spot on Social Security's Most Popular Baby Names List." May 7. Retrieved April 7, 2011 (http://www.ssa.gov/pressoffice/pr/baby-names2009-pr.htm).

Sociologists without Borders. 2008. "Who We Are." Retrieved July 3, 2009 (http://www.sociologistswithoutborders.org/index.html).

Solow, Robert M. 2000. *Growth Theory: An Exposition,* 2nd ed. New York: Oxford University Press.

Sommerville, C. John. 2002. "Stark's Age of Faith Argument and the Secularization of Things: A Commentary." *Review of Religious Research* (Fall):361–72.

Son, Young-ho. 1992. "Korean Response to the 'Yellow Peril' and Search for Racial Accommodation in the United States." *Korean Journal* 32(2):58–74.

Song, Jason. 2008. "Return of the Youth Vote." *Los Angeles Times,* February 5, B1.

Sons of Union Veterans of the Civil War. 2010. "U.S. Flag Code (4 US Code 1)." Retrieved July 22, 2011 (http://suvcw.org/flag.htm).

Sorkin, Andrew Ross, ed. 2009. "Pay Czar Said to Plan to Disclose Top Salaries." *DealBook,* September 17. Retrieved July 18, 2011 (http://dealbook.nytimes.com/2009/09/17/pay-czar-seen-disclosing-top-executive-salaries/?scp=1&sq=Pay%20Czar%20said%20to%20Plan%20to%20Disclose%20Top%20Salaries&st=cse)

Southerland, Anne. 1986. *Gypsies: The Hidden Americans.* Prospect Heights, IL: Waveland.

Southern Poverty Law Center. 2010. "Active U.S. Hate Groups" (2010). Retrieved July 8, 2011 (http://www.splcenter.org/get-informed/hate-map).

Sowell, Thomas. 1994. *Race and Culture: A World View.* New York: Basic Books.

Spade, Joan Z. 2004. "Gender in Education in the United States." Pp. 287–95 in *Schools and Society: A Sociological Approach to Education,* 2nd ed., edited by Jeanne H. Ballantine and Joan Z. Spade. Belmont, CA: Wadsworth.

Spade, Joan Z., Lynn Columba, and Beth E. Vanfossen. 1997. "Tracking in Mathematics and Science: Courses and Course Selection Procedures." *Sociology of Education* 70(2):108–27.

Spalter-Roth, Roberta, and Nicole Van Vooren. 2008. "What Are They Doing with a Bachelor's Degree in Sociology? Data Brief on Current Jobs." January. Washington, DC: American Sociological Association: Department of Research and Development.

Spalter-Roth, Roberta, Nicole Van Vooren, and Mary S. Senter. 2009. *Decreasing the Leak from the Sociology Pipeline: Social and Cultural Capital to Enhance the Post-Baccalaureate Sociology Career.* Washington, DC: American Sociological Association, Research and Development Department. Retrieved June 25, 2011 (http://www.asanet.org/images/research/docs/pdf/Decreasing%20the%20Leak%20from%20Soc%20Pipeline.pdf).

Sperling, Gene B. 2005. "The Case for Universal Basic Education for the World's Poorest Boys and Girls." *Phi Delta Kappan* 87(3):213–16.

————. 2006. "What Works in Girls' Education." PBS Wide Angle. Retrieved July 11, 2009 (http://www.pbs.org/wnet/wideangle/episodes/time-for-school-series/essay-what-works-in-girls-education/274/).

Sperling, Gene and Barbara Herz. 2004. *What Works in Girls' Education: Evidence and Policies From the Developing World.* Washington, DC: Council on Foreign Relations.

Spero News. 2009. "China: Good news from Beijing, the Number of Billionaires Is Rising, So Is the Economy." October 13. Retrieved November 11, 2009 (www.speronews.com/a/20830/china—-good-news-from-beijing-the-number-of-billionaires-is-rising-so-is-the-economy)

Stack, Carol B. 1974. *All Our Kin: Strategies for Survival in a Black Community.* New York: Harper & Row.

Stallings, Jane A. 1995. "Ensuring Teaching and Learning in the 21st Century." *Educational Researcher* 24(6):4.

Stanback, Thomas M. 1991. *The New Suburbanization: Challenge to the Central City.* Boulder, CO: Westview.

Stanley, Scott M. and Galena K. Rhoades. 2009. "Marriages at Risk: Relationship Formation and Opportunities for Relationship Education." Pp. 21–44 in *What Works in Relationship Education,* edited by Harry Benson and Samantha Callan. Doha, Qatar: Doha International Institute for Family Studies and Development.

Staples, Brent. 1992. "Black Men and Public Space." Pp. 29–32 in *Life Studies,* edited by D. Cavitch. Boston: Bedford.

Staples, Robert. 1999. *The Black Family: Essays and Studies,* 6th ed. Belmont, CA: Wadsworth.

Stark, Rodney. 1985. "Church and Sect." Pp. 139–49 in *The Sacred in a Secular Age,* edited by P. Hammond. Berkeley: University of California Press.

————. 2000. "Secularization, R.P.I." Pp. 41–66 in *The Secularization Debate,* edited by William H. Swatos, Jr., and Daniel V. A. Olson. Lanham, MD: Rowman and Littlefield.

Stark, Rodney and William Sims Bainbridge. 1985. *The Future of Religion: Secularization, Revival, and Cult Formation.* Berkeley: University of California Press.

Stark, Rodney, Daniel P. Doyle, and Jesse Lynn Rushing. 1983. "Beyond Durkheim: Religion and Suicide." *Journal for the Scientific Study of Religion* 22(March):120–31.

Starnes, Bobby Ann. 2006. "What We Don't Know Can Hurt Them: White Teachers, Indian Children." *Phi Delta Kappan* 87(5):384–92.

Starr, Paul D. 1982. *The Social Transformation of American Medicine.* New York: Basic Books.

State of Alaska. 2006. "Workplace Alaska: How to Apply." Retrieved July 22, 2009 (http://notes4.state.ak.us/wa/mainentry.nsf/WebData/HTMLHow+to+Apply/?open).

State of Delaware. 2008. "Presidential Primary Election." Retrieved March 21, 2008 (http://elections.delaware.gov/information/elections/presidential_2008.shtml).

Statistical Handbook of Japan. 2010. "Population." Retrieved May 13, 2011 (www.stat.go.jp/english/data/handbook/c02cont.htm#cha2_2).

Steele, Stephen F. and Jammie Price. 2004. *Applied Sociology: Terms, Topics, Tools, and Tasks.* Belmont, CA: Wadsworth.

Steele, Tracey L. 2005. *Sex, Self, and Society: The Social Context of Sexuality.* Belmont, CA: Thomson Wadsworth.

Steele, Tracey and Norma Wilcox. 2003. "A View From the Inside: The Role of Redemption, Deterrence, and Masculinity on Inmate Support for the Death Penalty." *Crime and Delinquency* 49(2):285–313.

Stein, Nicholas. 2006. "No Way Out." Pp. 293–99 in *Globalization: The Transformation of Social Worlds,* edited by D. Stanley Eitzen and Maxine Baca Zinn. Belmont, CA: Wadsworth.

Stelter, Brian. 2009. "In Coverage of Iran, Amateurs Take the Lead" [Mediadecoder weblog]. *New York Times,* June 17. Retrieved December 22, 2009 (*http://mediadecoder.blogs.nytimes.com*).

Stephen Roth Institute for the Study of Contemporary Antisemitism and Racism. 2006. *Antisemitism Worldwide 2006: General Analysis.* Israel: Tel Aviv University.

Stern, Jessica. 2003. *Terror in the Name of God: Why Religious Militants Kill.* New York: HarperCollins.

Stevens, Daphne, Gary Kiger, and Pamela J. Riley. 2001. "Working Hard and Hardly Working: Domestic Labor and Marital Satisfaction Among Dual-Earner Families." *Journal of Marriage and the Family* 63(May):514–26.

Stevenson, Mark. 2003. "Mexico Finishes Repaying Restructuring Debt." *Laredo Morning Times,* June 13, A14.

Stewart, Susan D. 2007. *Brave New Stepfamilies: Diverse Paths Toward Stepfamily Living.* Thousand Oaks, CA: Sage/Pine Forge Press.

Stewart, Vivien. 2005. "A World Transformed: How Other Countries Are Preparing Students for the Interconnected World of the 21st Century." *Phi Delta Kappan* 87(3):229–32.

Stillwell, Robert. 2010. "Public School Graduates and Dropouts from the Common Core of Date: School Year 2007–08." Table 2, p. 15, "Public High School Number of Graduates and Average Freshmen Graduate Rate, by Race/Ethnicity and State or Jurisdiction: School Year 2007–08."

Stockholm International Peace Research Institute. 2010. "World Military Expenditures Increase Despite Financial Crisis." June 2. Retrieved May 10, 2011 (www.sipri.org/media/pressreleases/2010/100602ye arbooklaunch).

Stoessinger, John. 1993. *Why Nations Go to War.* New York: St. Martin's Press.

Stoltenberg, John. 1993. *The End of Manhood: A Book for Men of Conscience.* New York: Dutton.

Stolzenberg, Ross M., Mary Blair-Loy, and Linda J. Waite. 1995. "Religious Participation in Early Adulthood: Age and Family Life Cycle Effects on Church Membership." *American Sociological Review* 60(February):84–103.

Stone, Brad and Noam Cohen. 2009. "Social Networks Spread Defiance Online." *The New York Times.* June 15. Retrieved October 23, 2009 (http://www.nytimes.com/2009/06/16/world/middleeast/16media.html?_r=1&ref=world).

Stone, Norman, ed. 1991. *The Times Atlas of World History,* 3rd ed. Maplewood, NJ: Hammond.

Stout, David. 2009. "Violent Crime Fell in 2008, F.B.I. Report Says." *The New York Times,* September 14. Retrieved November 4, 2009 (http://www.nytimes.com/2009/09/15/us/15crime.html).

Straus, Murray and Richard J. Gelles. 1990. *Physical Violence in American Families.* New Brunswick, NJ: Transaction Publishers.

Straus, Murray A., Richard J. Gelles, and Suzanne K. Steinmetz. 2006. *Behind Closed Doors: Violence in the American Family.* New Brunswick, NJ: Transaction Publishers.

Stringer, Donna M. 2006. "Let Me Count the Ways: African American/European American Marriages." Pp. 170–76 in *Intercultural Communication: A Reader,* edited by Larry A. Samovar, Richard E. Porter, and Edwin R. McDaniel. Belmont, CA: Wadsworth.

Struck, Doug. 2007. "Warming Will Exacerbate Global Water Conflicts." *The Washington Post,* August 20. Retrieved August 21, 2008 (www.washingtonpost.com/wp-dyn/content/article/2007/08/19/AR2007081900967.html).

Stryker, Sheldon. 1980. *Symbolic Interactionism: A Social Structural Version.* Menlo Park, CA: Benjamin Cummings.

———. 2000. "Identity Competition: Key to Differential Social Involvement." Pp. 21–40 in *Identity, Self, and Social Movements,* edited by Sheldon Styker, Timothy Owens, and Robert White. Minneapolis: University of Minnesota Press.

Stryker, Sheldon and Anne Stratham. 1985. "Symbolic Interaction and Role Theory." Pp. 311–78 in *Handbook of Social Psychology,* edited by Gardiner Lindsey and Eliot Aronson. New York: Random House.

Stutz, Fredrick P. and Barney Warf. 2005. *The World Economy.* Upper Saddle River, NJ: Prentice Hall.

Suarez-Orozco, Marcelo M. 2005. "Rethinking Education in the Global Era." *Phi Delta Kappan* 87(3):209–12.

Sun, Yongmin. 2001. "Family Environment and Adolescents' Well-Being Before and After Parents' Disruption: A Longitudinal Analysis." *Journal of Marriage and the Family* 63(August):697–713.

Sunshine, Rebecca. 2009. "One State Looks to Cut the Death Penalty, Put Money Elsewhere." Retrieved July 1, 2009 (www.ktiv.com/Global/story.asp?S=9790524).

Sutherland, Edwin H., Donald R. Cressey, and David Luckenbil. 1992. *Criminology.* Dix Hills, NY: General Hall.

Swatos, William H., Jr., and Luftur Reimur Gissurarson. 1997. *Icelandic Spiritualism: Mediumship and Modernity in Iceland.* New Brunswick, NJ: Transaction.

Sway, Marlene. 1988. *Familiar Strangers: Gypsy Life in America.* Urbana: University of Illinois Press.

Sweet, Stephen. 2001. *College and Society: An Introduction to the Sociological Imagination.* Boston: Allyn & Bacon.

Szymanski, Linda A., Ann Sloan Devlin, Joan C. Chrisler, and Stuart A. Vyse. 1993. "Gender Role and Attitude Toward Rape in Male and Female College Students." *Sex Roles* 29:37–55.

Talbot, Margaret. 2008. "Red Sex, Blue Sex." *The New Yorker,* November 3. Retrieved July 10, 2009 (http://www.newyorker.com/reporting/2008/11/03/081103fa_fact_talbot).

Tamney, Joseph B. 1992. *The Resilience of Christianity in the Modern World.* Albany: State University of New York Press.

Tapscott, Don. 1998. *Growing Up Digital: The Rise of the Net Generation.* New York: McGraw-Hill.

Taub, Diane E. and Penelope A. McLorg. 2010. "Influences of Gender Socialization and Athletic Involvement on the Occurrence of Eating Disorders." Pp. 73–82 in *Sociological Footprints: Introductory Readings in Sociology,* 11th ed., edited by Leonard Cargan and Jeanne H. Ballantine. Belmont, CA: Wadsworth Cengage Learning.

Tavris, Carol and Carol Wade. 1984. *The Longest War,* 2nd ed. San Diego: Harcourt Brace Jovanovich.

Taylor, Humphrey. 2002. "Scientists, Doctors, Teachers, and Military Officers Top the List of Most Prestigious Occupations." *The Harris Poll,* October 16. Retrieved July 22, 2009 (http://harrisinteractive.com/harris_poll/index.asp?PID=333).

TechWench. 2010. "Physical Books to Be Overshadowed by E-Books within 5 Years." October 19. Retrieved April 17, 2011 (www.techwench.com/physical-books-to-be-overshadowed-by-e-books-within-5-years/).

Terkel, Studs. 1974. *Working.* New York: Pantheon Books.

Therborn, Goran. 1976. "What Does the Ruling Class Do When It Rules?" *Insurgent Sociologist* 6:3–16.

Thiagaraj, Henry. 2006. *Minority and Human Rights from the Dalit Perspective.* Chennai, India: Oneworld Educational Trust.

———. 2007. *Human Rights From the Dalits' Perspective.* New Delhi, India: Gyan Publications.

Thio, Alex. 2007. *Deviant Behavior,* 9th ed. Boston: Allyn & Bacon.

Thoits, Peggy A. and Lyndi N. Hewitt. 2001. "Volunteer Work and Well-Being." *Journal of Health and Social Behavior* 42(June):115–31.

Thomas, Charles B. 1985. "Clergy in Racial Controversy: A Replication of the Campbell and Pettigrew Study." *Review of Religious Research* (June):379–90.

Thomas, V. J. and F. D. Rose. 1991. "Ethnic Differences in the Experience of Pain." *Social Science and Medicine* 32(9):1063–66.

Thompson, A. C. 2009. "Katrina's Hidden Race War." *The Nation,* January 5. Retrieved July 7, 2011 (http://www.thenation.com/article/katrinas-hidden-race-war).

Thorne, Barrie. 1993. *Gender Play: Girls and Boys in School.* New Brunswick, NJ: Rutgers University Press.

Thumma, Scott and Dave Travis. 2007. *Beyond Megachurch Myths: What We Can Learn From America's Largest Churches.* Hoboken, NJ: Jossey-Bass.

Tichenor, Veronica Jaris. 1999. "Status and Income as Gendered Resources: The Case of Marital Power." *Journal of Marriage and the Family* 61(August):638–50.

Tilley, Michael. 2009. "Power Shift? Proponents Again Push for Natural Gas-Powered Vehicles." Retrieved January 17, 2011 (www.thecity-wire.com/index.php?q=node/5479).

Tipton, Steven M. 1990. "The Social Organization of Zen Practice: Constructing Transcultural Reality." *American Journal of Sociology* 96(2):488–90.

Tjaden, Patricia and Nancy Thoennes. 2000. "Full Report of the Prevalence, Incidence and Consequences of Violence Against Women: Findings From the National Violence Against Women Survey." Washington, DC: National Institute of Justice.

Toffler, Alvin and Heidi Toffler. 1980. *The Third Wave.* New York: Morrow.

Tolbert, Pamela S. and Richard H. Hall. 2009. *Organizations: Structures, Processes, and Outcomes,* 10th ed. Upper Saddle River, NJ: Prentice Hall.

Tomsen, Peter. 1994. "Cambodia Recent Developments." *U.S. Department of State Dispatch,* May 23, 343–344.

Tonnies, Ferdinand. [1887] 1963. *Community and Society.* New York: Harper & Row.

Topix.com. 2010. "Fired for Being Gay? It's Legal in 29 States." Retrieved July 10, 2011 (http://www.topix.com/forum/state/de/TP9N5JVK30FRS3NLO).

Torche, Florencia. 2005. "Privatization Reform and Inequality of Educational Opportunity: The Case of Chile." *Sociology of Education* 78(4):316–43.

Tough, Paul. 2008. *Whatever It Takes: Geoffrey Canada's Quest to Change Harlem and America.* Boston: Houghton Mifflin Harcourt.

Toynbee, Arnold and D. C. Somervell. [1934–61] 1988. *The Study of History.* Oxford, UK: Oxford University Press.

Transparency International. 2009. "Global Corruption Barometer." Retrieved July 3, 2009 (http://media.transparency.org/fbooks/reports/gcb_2009/).

Travers, Jeffrey and Stanley Milgram. 1969. "An Experimental Study of the Small World Problem." *Sociometry* 32:425–43.

Tuchman, Gaye. 1996. "Women's Depiction by the Mass Media." Pp. 11–15 in *Turning It On: A Reader on Women and Media,* edited by Helen Baehr and Ann Gray. London: Arnold.

Tumin, Melvin M. 1953. "Some Principles of Social Stratification: A Critical Analysis." *American Sociological Review* 18(August):387–94.

Turnbull, Colin M. 1962. *The Forest People.* New York: Simon & Schuster.

Turner, Bryan S. 1991a. "Politics and Culture in Islamic Globalism." Pp. 161–81 in *Religion and Global Order,* edited by Roland Robertson and William R. Garrett. New York: Paragon.

———. 1991b. *Religion and Social Theory.* London: Sage.

Turner, Jonathan H. 2003. *The Structure of Sociological Theory,* 7th ed. Belmont, CA: Wadsworth.

Turner, Ralph H. and Lewis M. Killian. 1993. "The Field of Collective Behavior." Pp. 5–20 in *Collective Behavior and Social Movements,* edited by Russell L. Curtis, Jr., and Benigno E. Aguirre. Boston: Allyn & Bacon.

Twigg, Krassimira. 2009. "Twitterers Defy China's Firewall." BBC News. Retrieved May 15, 2011 (http://news.bbc.co.uk/2/hi/asia-pacific/8091411.stm).

Tyler, Sir Edward B. [1871] 1958. *Primitive Culture: Researches Into the Development of Mythology, Philosophy, Religion, Art and Custom,* Vol. 1. London: John Murray.

UNAIDS–Joint United Nations Programme on HIV/AIDS. 2006. "Fact Sheet: Sub-Saharan Africa." Retrieved July 22, 2009 (http://data.unaids.org/pub/EpiReport/2006/20061121_epi_fs_ssa_en.pdf).

UNESCO. 2005. *Education for All Global Monitoring Report 2005: The Quality Imperative.* Paris: Author.

———. 2006. *Education for All Global Monitoring Report 2006: The Quality Imperative.* Paris: Author.

———. 2011. "Worldwide Shortage of Teachers." Retrieved July 27, 2011 (http://www.teachersforefa.unesco.org/).

UNESCO GMR. 2007. "Highlights of the EFA Report 2007." Retrieved July 10, 2009 (www.unesco.org/education/GMR/2007/highlights.pdf).

UNESCO Institute for Statistics. 2008. "Percentage of Pupils in Single-Grade or Multiple-Grade Classes, by School Location," p. 123.

———. 2010. "World Science Report 2010." November 17. Retrieved December 2, 2010 (www.uis.unesco.org/ev_en.php?ID=8167_201&ID2=DO_TOPIC).

UNHCR. 2006a. "2005 Global Refugee Trends: Statistical Overview of Populations of Refugees, Asylum-Seekers, Internally Displaced Persons, Stateless Persons, and Other Persons of Concern to UNHCR." Retrieved July 22, 2009 (http://reliefweb.int/rw/lib.nsf/db900SID/HMYT-6QLLQX?OpenDocument).

———. 2006b. *The State of the World's Refugees.* New York: United Nations. Retrieved July 22, 2009 (http://www.unhcr.org/cgi-bin/texis/vtx/search?page=search&docid=4444afc70&query=The%20State%20of%20the%20World's%20Refugees).

UNHCR Statistical Yearbook. 2006. "Asylum and Refugee Status Determination" (Chapter 5). Retrieved July 22, 2009 (http://www.unhcr.org/478ce2bd2.html).

UNICEF. 2004. *Progress for Children: A Child Survival Report Card, No. 1.* New York: Author.

UNIFEM. 2002. "Report of the Learning Oriented Assessment of Gender Mainstreaming and Women's Empowerment Strategies in Rwanda." Retrieved April 7, 2008 (http://www.unifem.org/attachments/products/rwanda_assessment_report_eng.pdf).

United Nations. 2000. *The World's Women: Trends and Statistics.* New York: United Nations.

———. 2002. *World Urbanization Prospects: The 2001 Revision.* New York: United Nations Population Division. Retrieved July 22, 2009 (http://www.un.org/esa/population/publications/wup2001/WUP2001report.htm).

———. 2003. "Demographic Yearbook: 2003." Retrieved August 6, 2011 (http://unstats.un.org/unsd/demographic/products/dyb/2000_round.htm).

———. 2006a. "Demographic Yearbook: 2006." Retrieved August 6, 2011 (http://unstats.un.org/unsd/demographic/products/dyb/2000_round.htm).

———. 2006b. "Member States of the United Nations." Retrieved July 22, 2011 (www.un.org/members/list.shtml).

———. 2008a. "Demographic Yearbook: 2008." Retrieved August 23, 2008 (http://unstats.un.org/unsd/Demographic/products/dyb/dyb2006.htm).

United Nations Climate Change Conference. 2009. Retrieved December 22, 2009 (http://en.cop15.dk/about+cop15).

United Nations, Department of Economic and Social Affairs, Population Division. 2008. "World Urbanization Prospects: The 2007 Revision." Retrieved August 1, 2011 (www.un.org/esa/population/publications/wup2007/2007WUP_Highlights_web.pdf).

———. 2010. "World Urbanization Prospects: The 2009 Revision." Retrieved July 26, 2011 (http://esa.un.org/unpd/wup/index.htm).

———. 2011. "World Population Prospects: The 2010 Revision." Retrieved July 9, 2011 (http://esa.un.org/unpd/wpp/Excel-Data/mortality.htm).

United Nations Development Programme. 2007. *Human Development Report,* November 27, p. 25. New York: Author.

———. 2007/2008. *Human Development Report: Country Fact Sheets.* New York: Author.

United Nations High Commissioner for Refugees. 2009. "UNHCR Annual Report Shows 42 Million People Uprooted Worldwide." Retrieved January 5, 2010 (www.unhcr.org/4a2fd52412d.html).

United Nations, International Telecommunication Union. 2004. "Internet Indicators by Country for 2004." Retrieved August 9, 2006 (http://www.itu.int/ITU-D/ict/statistics/at_glance/Internet 04.pdf).

United Nations Millennium Project. 2007. "Investing in Development." Retrieved April 26, 2009 (http://unmillenniumproject.org/reports/).

United Nations News Center. 2010. "Senior UN Official Cites Evidence of Growing Support for Abolishing Death Penalty." February 24. Retrieved March 8, 2011 (www.un.org/apps/news/story.asp?NewsID=33877&Cr=death+penalty&Cr1=).

United Nations, Office on Drugs and Crime. 2003. *Global Illicit Drug Trends 2003.* New York: Author.

————. 2005. *World Drug Report 2005*. Retrieved July 22, 2009 (http://www.unodc.org/unodc/en/data-and-analysis/WDR-2005.html).

————. 2006a. "Annual Reports Questionnaire Data/DELTA." Retrieved September 20, 2006 (www.unodc.org/pdf/WDR_2006/wdr2006_ex_summary.pdf).

————. 2006b. "World Drug Report 2006, Volume 1." Retrieved September 20, 2006 (http://www.unodc.org/pdf/WDR_2006/wdr2006_volume1.pdf).

United Nations Office of the High Commissioner for Human Rights. 2011. "Migration and Human Rights." Retrieved August 1, 2011 (http://www.ohchr.org/EN/Issues/Migration/Pages/MigrationAndHumanRightsIndex.aspx).

————. 2008, February. *Economic & World Urbanization Prospects: The 2007 Revision*. New York: United Nations Department of Economic and Social Affairs, Population Division. Retrieved April 24, 2009 (www.un.org/esa/population/publications/wup2007/2007WUP_Highlights_web.pdf).

United Nations Population Fund. 2005. "State of the World Population 2005." Retrieved September 29, 2006 (www.unfpa.org/swp/2005/pdf/en_swp05.pdf).

————. n.d. "Giving Birth Should Not Be a Matter of Life and Death." Retrieved May 13, 2011 (www.unfpa.org/webdav/site/global/shared/safemotherhood/docs/maternalhealth_factsheet_en.pdf).

United Nations Statistics Division. 2010. *Demographic Yearbook*. Retrieved August 1, 2011 (http://unstats.un.org/unsd/demographic/products/dyb/2000_round.htm).

United North America. 2008. "Similarities and Differences Between Canada & United States." Retrieved August 26, 2008 (www.unitednorthamerica.org/simdiff.htm).

University of Michigan. 2006. "Demographic Transition: An Historical Sociological Perspective." Retrieved July 22, 2009 (http://www.globalchange.umich.edu/globalchange2/current/lectures/pop_socio/pop_socio.html).

University of Pennsylvania. 2010. "Body Modification." Retrieved November 18, 2010 (penn.museum/sites/body_modification/bodmodpierce.shtml).

UPI. 2008. "Annan Says Kenyan Conflict 'Evolving.'" January 27. Retrieved June 3, 2008 (www.upi.com/NewsTrack/Top_News/2008/01/27/annan_says_kenyan_conflict_evolving/1912/).

The Urban Institute. 2004. "Family Support, Substance Abuse Help, and Work Release Programs Are Essential as Ex-Prisoners Restart Lives in Baltimore." Retrieved September 9, 2008 (http://www.urban.org/publications/900688.html).

U.S. Bureau of Justice Statistics. 2001. *National Crime Victimization Survey Report*. Washington, DC: U.S. Department of Justice.

————. 2002, 2003, 2004, 2005, 2006. *Sourcebook of Criminal Justice Statistics*. Washington, DC: U.S. Department of Justice.

U.S. Bureau of Labor Statistics. 2003. "American Time Use Survey." Retrieved July 27, 2009 (http://www.bls.gov/news.release/archives/atus_09142004.pdf).

————. 2007. "American Time Use Survey." Retrieved July 27, 2009 (http://www.bls.gov/news.release/archives/atus_06252008.pdf).

————. 2008–09. "Occupational Outlook Handbook, 2008–09 Edition." Retrieved April 9, 2008 (www.bls.gov/oco/ocos285.htm).

————. 2011. "American Time Use Survey Summary" (2010 Results). Retrieved July 11, 2011 (http://www.bls.gov/news.release/atus.nr0.htm).

U.S. Census Bureau. 2000. *American FactFinder: Geographic Comparison Table*. Retrieved July 4, 2009 (http://factfinder.census.gov/servlet/GCTTable?_bm=y&-geo_id=01000US&-_box_head_nbr=GCT-P6&-ds_name=DEC_2000_SF1_U&-format=US-9).

————. 2001. *Current Population Survey, 2001*. Washington, DC: U.S. Government Printing Office.

————. 2003. "2001 Annual Report: The Status of Equal Employment Opportunity and Affirmative Action in Alaska State Government." Retrieved September 29, 2006 (http://factfinder.census.gov/servlet/ACSSAFFFacts?_event=&geo_id=04000US02&_geoContext=0100   0US%7C04000US02&_street=&_county=&_cityTown=&_state=04000US02&_zip=&_lang=en&_sse=on&ActiveGeoDiv=&_useEV=&pctxt=fph&pgsl=040).

————. 2006a. "America's Families and Living Arrangements: 2005." Retrieved February 20, 2008 (http://www.census.gov/population/www/socdemo/hh-fam/cps2005.html).

————. 2006b. "Geographical Mobility Between 2004 and 2005." *Population Profile of the United States*. Washington, DC: Author.

————. 2006c. "Movers by Type of Move and Reason for Moving 2006." *Statistical Abstracts 2008*, Table 0031. Retrieved August 23, 2008 (infochimps.org/dataset/statab2008_0031_moversbytypeofmoveandreasonformovin).

————. 2008a. "Mean Earnings by Highest Degree Earned: 2005." *Statistical Abstracts of the United States, 2008* (Table 220). Washington, DC: Author.

————. 2008b. "Public Education Finances: 2006." Retrieved January 7, 2010 (www2.census.gov/govs/school/06f33pub.pdf).

————. 2009a. "American FactFinder." Retrieved July 1, 2011 (http://factfinder.census.gov/).

————. 2009b. "Fact Sheet: Alaska." Retrieved August 8, 2011 (http://factfinder.census.gov/servlet/ACSSAFFFacts?_event=Search&_state=04000US02).

————. 2009c. "Historical Income Tables—Households." Retrieved April 26, 2009 (http://www.census.gov/hhes/www/income/data/historical/household/index.html).

————. 2009d. "Income, Poverty, and Health Insurance Coverage in the U.S.: 2009." Report P-60, n. 238, Table B-2, pp. 62–7.

————. 2009e. "Nearly Half of Parents Get Full Amount of Child Support, Census Bureau Reports." Retrieved July 19, 2011 (www.census.gov/newsroom/releases/archives/children/cb09–170.html).

————. 2010. "Current Population Survey (CPS)—Definitions and Explanations." Retrieved July 11, 2011 (http://www.census.gov/population/www/cps/cpsdef.html).

————. 2011a. "College Enrollment of Recent High School Graduates" (Table 272). Retrieved May 14, 2011 (www.census.gov/compendia/statab/2011/tables/11s0272.pdf).

————. 2011b. "Educational Attainment by Race and Hispanic Origin: 1970 to 2009" (Table 225). Retrieved August 9, 2011 (http://www.census.gov/compendia/statab/2011/tables/11s0225.pdf).

————. 2011c. "Income." Retrieved July 6, 2011 (www.census.gov/hhes/www/income/income.html).

————. 2011d. "International Statistics: Population, Households." *The 2011 Statistical Abstract*. Retrieved July 11, 2011 (http://www.census.gov/compendia/statab/cats/international_statistics/population_households.html).

————. 2011e. "Marriage and Divorce Rates by Country: 1980–2008" (Table 1335). Retrieved July 11, 2011 (http://www.census.gov/compendia/statab/2011/tables/11s1335.pdf).

————. 2011f. "Mean Earnings by Highest Degree Earned: 2008" (Table 228). Retrieved April 19, 2011 (http://www.census.gov/compendia/statab/2011/tables/11s0228.pdf).

————. 2011g. "Movers by Type of Move and Reason for Moving: 2009." Retrieved May 19, 2011 (www.census.gov/compendia/statab/2011/tables/11s0031.pdf).

————. 2011h. "State & County QuickFacts: Alaska." Retrieved April 7, 2011 (http://quickfacts.census.gov/qfd/states/02000.html).

U.S. Census Bureau, Population Division, International Programs Center. 2011. "International Data Base." Retrieved August 6, 2011 (www.census.gov/ipc/www/idbnew.html).

U.S. Department of Agriculture. 2002. *Economic Research Service: Data set of United States Farm and Farm-Related Employment, 2002*. Washington, DC: Author.

———. 2005a. *Characteristics of Food Stamp Households: Fiscal Year 2004 Summary.* Washington, DC: Author

———. 2005b. "Household Food Security in the United States, 2004." *Economic Report Vol. 11.* Washington, DC: Author.

———. 2009. "USDA Releases Annual Study Which Notes That Child Born in 2008 Will Cost $221,190 to Raise." Release No. 0365.09, August 4. Retrieved July 23, 2011 (http://www.usda.gov/wps/portal/usda/usdahome?contentidonly=true&contentid=2009/08/0365.xml).

U.S. Department of Education. 2004–2005. "Integrated Postsecondary Education Data System: Glossary." Retrieved July 10, 2009 (http://nces.ed.gov/ipeds/glossary/?charindex=D).

U.S. Department of Health and Human Services. 2011. "NHE Fact Sheet." Retrieved April 7, 2011 (www.cms.gov/NationalHealthExpendData/25_NHE_Fact_Sheet.asp)

U.S. Department of Interior Office of Education. 1930. *Availability of Public School Education in Rural Communities* (Bulletin No. 34, edited by Walter H. Gaummitz). Washington, DC: Government Printing Office.

U.S. Department of Justice. 2006. *National Crime Victimization Survey.* Washington, DC: Bureau of Justice Statistics.

———. 2010a. "Crime Clock Statistics." *Crime in the United States, 2009.* Retrieved July 7, 2011 (http://www2.fbi.gov/ucr/cius2009/about/crime_clock.html).

———. 2010b. "Hate Crime Statistics." *Crime in the United States, 2009.* Retrieved July 7, 2011 (http://www2.fbi.gov/ucr/hc2009/index.html).

U.S. Department of Labor. 2005. *Current Population Survey.* Washington, DC: Bureau of Labor Statistics.

U.S. Department of State. 2001. "Patterns of Global Terrorism 2001 Report." Retrieved September 29, 2006 (http://www.state.gov/s/ct/rls/crt/2001/).

———. 2002. "Curbing Violence Against Political Activists in Cambodia" (Press release). Retrieved August 8, 2011 (http://usinfo.org/wf-archive/2002/021216/epf107.htm).

———. 2005. *Patterns of Global Terrorism 1985–2005.* Washington, DC: Berkshire Publishing Group.

———. 2006. "NCTC Statistical Annex Supplement on Terrorism Deaths, Injuries, Kidnappings of Private U.S. Citizens" [Section 2656f (d)(2)]. Retrieved June 30, 2009 (www.state.gov/s/ct/rls/crt/2005/65353.htm).

U.S. Election Project. 2010. "General Election Turnout Rates." October 6. Retrieved May 6, 2011 (http://elections.gmu.edu/Turnout_2008G.html).

U.S. Environmental Protection Agency. 2009. "Ag101: Demographics." Retrieved March 30, 2010 (http://www.epa.gov/oecaagct/ag101/demographics/html).

U.S. Executive Office of the President. 2001. *Budget of the U.S. Government.* Washington, DC: Author.

U.S. General Accounting Office. 2004. "Defense of Marriage Act—Update to Prior Report." Retrieved July 11, 2011 (http://www.gao.gov/new.items/d04353r.pdf).

U.S. Senate. 2011. "Ethnic Diversity in the Senate." Retrieved April 22, 2011 (www.senate.gov/artandhistory/history/common/briefing/minority_senators.htm).

U.S. Trade Representative. 2011. "North American Free Trade Agreement." Retrieved August 6, 2011 (http://www.ustr.gov/trade-agreements/free-trade-agreements/north-american-free-trade-agreement-nafta).

Vandell, Deborah Lowe and Sheri E. Hembree. 1994. "Peer Social Status and Friendship: Independent Contributors to Children's Social and Academic Adjustment." *Merrill-Palmer Quarterly* 40(4):461–75.

Vandenburgh, Henry. 2001. "Emerging Trends in the Provision and Consumption of Health Services." *Sociological Spectrum* 21(3):279–92.

Veblen, Thorstein. 1902. *The Theory of the Leisure Class: An Economic Study of Institutions.* New York: Macmillan.

Verbeek, Stijn and Rinus Penninx. 2009. "Employment Equity Policies in Work Organisations." Pp. 69–94 in *Equal Opportunity and Ethnic Inequality in European Labour Markets: Discrimination, Gender, and Policies of Diversity,* edited by Karen Kraal, Judith Roosblad, and John Wrench. Amsterdam: University of Amsterdam Press.

Victor, Barbara. 2003. *Army of Roses: Inside the World of Palestinian Women Suicide Bombers.* Emmaus, PA: Rodale Books.

Visual Economics. 2010. "How Countries Spend Their Money." Retrieved July 26, 2011 (www.visualeconomics.com/how-countries-spend-their-money/).

Voting and Democracy Research Center. 2008. "Primaries: Open and Closed." Retrieved March 21, 2008 (http://www.fairvote.org/?page=1801).

Wagmiller Robert, Jr., Li Kuang, J. Lawrence Aber, Mary Clare Lennon, and Philip Alberti. 2006. "The Dynamics of Economic Disadvantage and Children's Life Chances." *American Sociological Review* 71(5):847–66.

Waite, Linda J. and Maggie Gallagher. 2000. *The Case for Marriage: Why Married People Are Happier, Healthier, and Better Off Financially.* New York: Doubleday.

Waitzkin, Howard. 2000. "Changing Patient-Physician Relationships in the Changing Health-Policy Environment." Pp. 271–83 in *Handbook of Medical Sociology,* 5th ed., edited by C. Bird, P. Conrad, and A. Fremont. Englewood Cliffs, NJ: Prentice Hall.

Walby, Sylvia. 1990. *The Historical Roots of Materialist Feminism.* Paper presented at the International Sociological Association, Madrid, Spain.

Wallenstein, Peter. 2002. *Tell the Court I Love My Wife: Race, Marriage, and Law—An American History.* New York: Macmillan.

Waller, Willard. [1932] 1965. *Sociology of Teaching.* New York: Russell & Russell.

Wallerstein, Immanuel. 1974. *The Modern World System.* New York: Academic Press.

———. 1979. *The Capitalist World Economy.* London: Cambridge University Press.

———. 1991. *Geopolitics and Geoculture: Essays on the Changing World-system.* Cambridge, MA: Cambridge University Press.

———. 2004. *World Systems Analysis: An Introduction.* Durham, NC: Duke University Press.

———. 2005. "Render Unto Caesar? The Dilemmas of a Multicultural World." *Sociology of Religion* 66(2):121–33.

Wallerstein, Judith. 1996. *Surviving the Breakup: How Children and Parents Cope with Divorce.* New York: Basic Books.

Wallerstein, Judith S. and Sarah Blakeslee. 1996. *The Good Marriage: How and Why Love Lasts.* New York: Warner Books.

———. 2004. *Second Chances: Men, Women and Children a Decade After Divorce,* 15th ed. Boston: Houghton Mifflin.

The Wall Street Journal. 2009. "The Madoff Case: A Timeline." March 12. Retrieved November 5, 2009 (http://online.wsj.com/article/SB112966954231272304.html?mod=googlenews.wsj).

Walsh, Edward J. and Rex H. Warland. 1983. "Social Movement Involvement in the Wake of a Nuclear Accident: Activists and Free Riders in the IMI Area." *American Sociological Review* 48(December):764–80.

Walsh, Mark. 1991. "Students at Private Schools for Blacks Post Above-Average Scores, Study Finds." *Education Week,* October 16. Retrieved July 22, 2009 (http://www.edweek.org/ew/articles/1991/10/16/07black.h11.html?tkn=MWMFIUdaY9D8cb4L0KNvtA%2BDSE42IZxfzOKq).

Walum, Laurel Richardson. 1974. "The Changing Door Ceremony: Some Notes on the Operation of Sex Roles in Everyday Life." *Urban Life and Culture* 2(4):506–15.

Ward, Martha C. 1996. *A World Full of Women.* Boston: Allyn & Bacon.

Ward, Martha C. and Monica Edelstein. 2009. *A World Full of Women,* 5th ed. Boston: Allyn & Bacon.

Ware, Alyn. 2008. "Nuclear Stockpiles." Project of the Nuclear Age Peace Foundation. Retrieved September 3, 2008 (www.nuclearfiles.org/menu/key-issues/nuclear-weapons/basics/nuclear-stockpiles.htm).

Warner, R. Stephen. 1993. "Work in Progress Toward a New Paradigm for the Sociological Study of Religion in the United States." *American Journal of Sociology* 98(5):1044–93.

Warren, Jennifer. 1990. "Schoolbook Furor Rends Rural Town." *Los Angeles Times,* August 20, A3.

War Resisters League. 2008. "Where Your Income Tax Money Really Goes." Retrieved April 16, 2008 (www.warresisters.org/pages/piechart.htm).

WaterAid. 2004. "No Water, No School." *Oasis,* Spring/Summer. Retrieved February 25, 2008 (http://www.wateraid.org/international/about_us/oasis/springsummer_04/1465.asp).

———. 2008 "WaterAid's Key Facts and Statistics." Retrieved February 25, 2008 (http://www.wateraid.org/international/what_we_do/statistics/default.asp).

Waters, Melissa S., Will Carrington Heath, and John Keith Watson. 1995. "A Positive Model of the Determination of Religious Affiliation." *Social Science Quarterly* 76(1):105–23.

Weatherly, Leslie A. 2004. "The Rising Cost of Health Care: Strategic and Societal Considerations for Employers." *HR Magazine,* September. Retrieved April 4, 2008 (http://findarticles.com/p/articles/mi_m3495/is_9_49/ai_n6206615).

Weaver, Janelle. 2010 "Social Life Starts in the Womb." *ScienceShot,* October 12. [Retrieved October 19, 2010 (http://news.sciencemag.org/sciencenow/2010/10/scienceshot-social-life-starts-in.html?rss=1&utm_source=twitterfeed&utm_medium=twitter).

Weber, B. J. and L. M. Omotani. 1994. "The Power of Believing." *The Executive Educator* 19(September):35–38.

Weber, David. 2003. "25 Health Care Trends: What's Hot, What's Not, and What Does the Future Hold." *Physician Executive* 29(1):6–14.

Weber, Max. [1904–1905] 1958. *The Protestant Ethic and the Spirit of Capitalism,* translated by Talcott Parsons. New York: Scribner.

———. 1946. *From Max Weber: Essays in Sociology,* translated and edited by Hans H. Gerth and C. Wright Mills. New York: Oxford University Press.

———. 1947. *The Theory of Social and Economic Organization,* translated and edited by A. M. Henderson and Talcott Parsons. New York: Oxford University Press.

———. 1958. "The Three Types of Legitimate Rule," translated by Hans Gerth. *Berkeley Publications in Society and Institutions* 4(1):1–11.

Web Site Optimization. 2010. "U.S. Broadband Penetration Jumps to 45.2%; Internet Access Nearly 75%." Retrieved June 29, 2011 (http://www.websiteoptimization.com/bw/0403/).

Webster's Encyclopedic Unabridged Dictionary of the English Language. 1989. Avenel, NJ: Gramercy Books/dilithium Press Ltd.

Weeks, John R. 2011. *Population: An Introduction to Concepts and Issues,* 11th ed. Belmont, CA: Wadsworth.

Weidenbaum, Murray. 2006. "Globalization: Wonderland or Waste Land?" Pp. 53–60 in *Globalization: The Transformation of Social Worlds,* edited by D. Stanley Eitzen and Maxine Baca Zinn. Belmont, CA: Wadsworth.

Weil, Elizabeth. 2008. "Teaching to the Testosterone." *The New York Times Magazine,* March 2, 38–45, 84–87.

Weisenberg, Faye and Elizabeth Stacey. 2005. "Reflections on Teaching and Learning Online: Quality Program Design, Delivery, and Support Issues From a Cross-Global Perspective." *Distance Education,* 26(3):385–404.

Weiss, Gregory L. and Lynne E. Lonnquist. 2008. *The Sociology of Health, Healing, and Illness,* 6th ed. Englewood Cliffs, NJ: Prentice Hall.

Weitz, Rose. 1995. "What Price Independence? Social Reactions to Lesbians, Spinsters, Widows and Nuns." Pp. 448–57 in *Women: A Feminist Perspective,* edited by Jo Freeman. Mountain View, CA: Mayfield.

Weitzman, Lenore J., Deborah Eifler, Elizabeth Hokada, and Catherine Ross. 1972. "Sex-Role Socialization in Picture Books for Preschool Children." *American Journal of Sociology* 77(May):1125–50.

Welch, Michael. 1996. *Corrections: A Critical Approach.* St. Louis, MO: McGraw-Hill.

Weller, Christian E. and Adam Hersh. 2006. "Free Markets and Poverty." Pp. 69–73 in *Globalization: The Transformation of Social Worlds,* edited by D. Stanley Eitzen and Maxine Baca Zinn. Belmont, CA: Wadsworth.

Wells, Amy Stuart and Jeannie Oakes. 1996. "Potential Pitfalls of Systemic Reform: Early Lessons From Research on Detracking." *Sociology of Education* 69(Extra Issue):135–43.

Wenglinsky, Harold. 1997. "How Money Matters: The Effect of School District Spending on Academic Achievement." *Sociology of Education* 70(3):221–37.

Werner International Management Consultants. 2005. *Primary Textiles Labor Cost Comparisons.* Herndon, VA: Author.

Werth, James L., Dean Blevins, Karine L. Toussaint, and Martha R. Durham. 2002. "The Influence of Cultural Diversity on End-of-Life Care and Decisions." *American Behavioral Scientist* 46(2):204–19.

Wessinger, Catherine. 2000. *How the Millennium Comes Violently: From Jonestown to Heaven's Gate.* New York: Seven Bridges.

West, Candace and Don H. Zimmerman. 1987. "Doing Gender." *Gender and Society* 1(2):125–51.

West, Heather C., William J. Sabol, and Sarah J. Greenman. 2009. "Prisoners in 2009." U.S. Department of Justice, Bureau of Justice Statistics. Retrieved July 7, 2011 (http://bjs.ojp.usdoj.gov/content/pub/pdf/p09.pdf).

Westermann, Ted D. and James W. Burfeind. 1991. *Crime and Justice in Two Societies: Japan and the United States.* Pacific Grove, CA: Brooks/Cole.

Westheimer, Joel. 2006. "Patriotism and Education: An Introduction." *Phi Delta Kappan* 87(8):569–72.

Wheeler, David L. 1995. "A Growing Number of Scientists Reject the Concept of Race." *The Chronicle of Higher Education,* February 17, A15.

White, Kevin. 2002. *An Introduction to the Sociology of Health and Illness.* London: Sage.

White, Merry I. 1987. *The Japanese Educational Challenge: A Commitment to Children.* New York: The Free Press.

WhiteHouse.gov. 2010. "Annual Report on U.S. Contributions to the United Nations." Retrieved January 18, 2011 (www.whitehouse.gov/sites/default/files/omb/assets/legislative_reports/us_contributions_to_the_un_06112010.pdf).

Whorf, Benjamin Lee. 1956. *Language, Thought, and Reality.* New York: John Wiley.

"Why Comprehensive Health System Reform Failed" (Editorial). 1994. *American Family Physician* 50(5):919–20.

Whyte, William H. 1956. *The Organization Man.* New York: Simon and Schuster.

Wilcox, Norma and Tracey Steele. 2003. "Just the Facts: A Descriptive Analysis of Inmate Attitudes Toward Capital Punishment." *The Prison Journal* 83(4):464–82.

Williams, Brian K., Stacey C. Sawyer, and Carl M. Wahlstrom. 2009. *Marriages, Families, and Intimate Relationships,* 2nd ed. Boston: Allyn & Bacon.

Williams, Catrina. 2000. *Internet Access in U.S. Public Schools and Classrooms 1994–1999.* Washington, DC: National Center for Education Statistics.

Williams, Christine L. 1992. "The Glass Escalator: Hidden Advantages for Men in the 'Female' Professions." *Social Problems* 39(3):253–66.

Williams, D. R. and C. Collins. 1995. "U.S. Socioeconomic and Racial Differences in Health: Patterns and Explanations." *Annual Review of Sociology* 21:349–86.

Williams, Gregory H. 1996. *Life on the Color Line: The True Story of a White Boy Who Discovered He Was Black.* New York: Dutton.

Williams, Robin Murphy, Jr. 1970. *American Society: A Sociological Interpretation,* 3rd ed. New York: Alfred Knopf.

Williams, Terry and William Kornblum. 1985. *Growing Up Poor.* Lexington, MA: Lexington.

Willie, Charles Vert. 2003. *A New Look at Black Families,* 5th ed. Walnut Creek, CA: AltaMira.

Willis, Paul. 1979. *Learning to Labor: How Working Class Kids Get Working Class Jobs.* Aldershot, Hampshire, England: Saxon House.

Wilmore, Gayraud S. 1973. *Black Religion and Black Radicalism*. Garden City, NY: Doubleday.

Wilson, Bryan. 1982. *Religion in Sociological Perspective*. Oxford, UK: Oxford University Press.

Wilson, Edward O. 1980. *Sociobiology*. Cambridge, MA: Belknap.

———. 1987. *The Coevolution of Biology and Culture*. Cambridge, MA: Harvard University Press.

Wilson, Edward O., Michael S. Gregory, Anita Silvers, and Diane Sutch. 1978. "What Is Sociobiology?" *Society* 15(6):1–12.

Wilson, K. 1993. *Dialectics of Consciousness: Problems of Development, the Indian Reality*. Madras: Oneworld Educational Trust.

Wilson, Mary E. 2006. "Infectious Concerns: Modern Factors in the Spread of Disease." Pp. 313–19 in *Globalization: The Transformation of Social Worlds,* edited by D. Stanley Eitzen and Maxine Baca Zinn. Belmont, CA: Wadsworth.

Wilson, Warren H. 1924. "What the Automobile Has Done to and for the Country Church." *Annals of the American Academy of Political and Social Science* 116(November):85–86.

Wilson, William Julius. 1978. *The Declining Significance of Race: Blacks and Changing American Institutions*. Chicago: University of Chicago Press.

———. 1984. "The Black Underclass." *The Wilson Quarterly* (Spring):88–89.

———. 1987. *The Truly Disadvantaged: The Inner City, the Underclass, and Public Policy*. Chicago: University of Chicago Press.

———. 1993a. *The Ghetto Underclass: Social Science Perspectives*. Newbury Park, CA: Sage.

———. 1993b. "The New Urban Poverty and the Problem of Race." *The Tanner Lecture on Human Values,* October 22 (printed in *Michigan Quarterly Review* 33:247–73).

———. 1996. *When Work Disappears*. New York: Alfred A. Knopf.

Winders, Bill. 2004. *Changing Racial Inequality: The Rise and Fall of Systems of Racial Inequality in the U.S.* Paper presented at the Annual Meeting of the American Sociological Association, San Francisco.

Wingfield, Adia Harvey. 2009. "Racializing the Glass Escalator: Reconsidering Men's Experiences With Women's Work." *Gender and Society* 23(5):5–26.

Winkler, Karen J. 1991. "Revisiting the Nature vs. Nurture Debate: Historian Looks Anew at Influence of Biology on Behavior." *Chronicle of Higher Education,* May 22, A5, A8.

Winnick, Terri A. 2006. "Medical Doctors and Complementary and Alternative Medicine: The Context of Holistic Practice." *Health: An Interdisciplinary Journal for the Social Study of Health, Illness and Medicine* 10(2):149–73.

Winslow, Robert W. and Sheldon X. Zhang. 2008. *Criminology: A Global Perspective*. Upper Saddle River, NJ: Pearson Prentice Hall.

Wirth, Louis. 1964. "Urbanism as a Way of Life." *American Journal of Sociology* 44(1):1–24.

Witte, John F. and Christopher A. Thorn. 1996. "Who Chooses? Voucher and Interdistrict Choice Programs in Milwaukee." *American Journal of Education* 104(May):186–217.

Witzig, Ritchie. 1996. "The Medicalization of Race: Scientific Legitimation of a Flawed Social Construct." *Annals of Internal Medicine: American College of Physicians* 125(8):675–76.

Wolf, Richard. 2010. "Number of Uninsured Americans Rises to 50.7 Million." *USA Today,* September 17. Retrieved April 7, 2011 (www.usatoday.com/news/nation/2010–09–17-uninsured17_ST_N.htm).

Women's Sports Foundation. 2002. "Women's Sports: Title IX Q and A." May. Retrieved August 8, 2011 (http://lobby.la.psu.edu/_107th/135_Title%20IX/Organizational_Statements/Womens_Sports_Fdn/Womens_Sports_Fdn_Title_IX_Q_&_A_041902.htm).

Wong, Sandra L. 1991. "Evaluating the Content of Textbooks: Public Interests and Professional Authority." *Sociology of Education* 64(January):11–18.

Wood, Julia T. 2008. *Gendered Lives: Communication, Gender, and Culture,* 8th ed. Belmont, CA: Wadsworth.

Wood, Julia T. and Nina M. Reich. 2006. "Gendered Communication Styles." Pp. 177–86 in *Intercultural Communication,* 11th ed., edited by Larry A. Samovar, Richard E. Porter, and Edwin R. McDaniel. Belmont, CA: Wadsworth.

Woodberry, Robert D. and Christian S. Smith. 1998. "Fundamentalism et al.: Conservative Protestants in America." *Annual Review of Sociology* 24:25–26.

Woolhandler, Steffie, Terry Campbell, and David U. Himmelstein. 2004. "Health Care Administration in the United States and Canada: Micromanagement, Macro Costs." *International Journal of Health Services* 34(1):65–78.

World Almanac Education Group. 2001. *The World Almanac and Book of Facts 2001*. Mahwah, NJ: World Almanac Books.

World Bank. 2006. "Per Capital Income-World." World Bank Development Indicators. Retrieved February 26, 2008 (www.finfacts.com/biz10).

———. 2011. "Girls' Education: A World Bank Priority." Retrieved July 21, 2011 (www.worldbank.org/education/girls).

World Factbook. 2009. "World's 50 Most Populous Countries: 2009." Central Intelligence Agency. Retrieved July 12, 2009 (www.infoplease.com/world/statistics/most-populous-countries.html).

World Factbook. 2010a. "Country Comparison: GDP—per Capita." Central Intelligence Agency. Retrieved January 26, 2010 (*www.cia.gov/library/publications/the-world-factbook/rankorder/2004rank.html*).

World Factbook. 2010b. "Country Comparison: Infant Mortality Rate." Central Intelligence Agency. Retrieved January 23, 2010 (*www.cia.gov/library/publications/the-world-factbook/rankorder/2091rank.html*).

World Factbook. 2010c. "Country Comparison: Life Expectancy at Birth." Central Intelligence Agency. Retrieved January 23, 2010 (*www.cia.gov/library/publications/the-world-factbook/rankorder/2102rank.html*).

World Factbook. 2010d. "Country Comparison: Population." Central Intelligence Agency. Retrieved February 1, 2010 (https://www.cia.gov/library/publications/the-world-factbook/rankorder/2119rank.html).

World Factbook. 2011. "Field Listing: Suffrage." Retrieved May 6, 2011 (https://www.cia.gov/library/publications/the-world-factbook/fields/2123.html).

World Famine Timeline. 2011. "World Disasters: Famine Timeline in 21st Century." Retrieved May 13, 2011 (www.mapreport.com/subtopics/d/0.html#2010).

World Health Organization. 2002a. "The Top 10 Causes of Death: Fact Sheet." Retrieved July 18, 2011 (http://www.who.int/whr/2002/en/whr02_en.pdf).

———. 2002b. "Reducing Risks, Promoting Healthy Life." *The World Health Report 2002*. Retrieved September 29, 2006 (www.who.int/whr/2002/en/whr02_en.pdf).

———. 2007. "Fact Sheet: The Top Ten Causes of Death." February. Retrieved July 13, 2009 (http://www.who.int/mediacentre/factsheets/fs310.pdf).

———. 2011. "The Top 10 Causes of Death." Retrieved August 1, 2011 (http://www.who.int/mediacentre/factsheets/fs310/en/index.html).

World Hunger Education Service. 2011. "2011 World Hunger and Poverty Facts and Statistics." Retrieved April 12, 2011 (www.worldhunger.org/articles/Learn/world%20hunger%20facts%202002.htm#).

World Resources Institute. 2007. "Ask EarthTrends: How Much of the World's Resource Consumption Occurs in Rich Countries?" *EarthTrends*. Retrieved April 14, 2008 (http://earthtrends.wri.org/updates/node/236).

WorldWideLearn. 2007. "Guide to College Majors in Sociology." Retrieved June 23, 2008 (www.worldwidelearn.com/online-education-guide/social-science/sociology-major.htm).

Wright, Erik Olin. 2000. *Class Counts: Comparative Studies in Class Analysis,* Student ed. Cambridge, MA: Cambridge University Press.

Wright, John W., ed. 2007. *The New York Times Almanac 2008*. New York: Penguin Reference.

Wright, Stuart A. 1995. *Armageddon in Waco: Critical Perspectives on the Branch Davidian Conflict*. Chicago: University of Chicago Press.

Wuthnow, Robert, ed. 1994. *"I Come Away Stronger": How Small Groups Are Shaping American Religion*. Grand Rapids, MI: William B. Eerdmans.

Yablonski, Lewis. 1959. "The Gang as a Near-Group." *Social Problems* 7(Fall):108–17.

Yamane, David. 1997. "Secularization on Trial: In Defense of a Neosecularization Paradigm." *Journal for the Scientific Study of Religion* 36(1):109–22.

Yinger, J. Milton. 1960. "Contraculture and Subculture." *American Sociological Review* 25(October):625–35.

———. 1970. *The Scientific Study of Religion*. New York: Macmillan.

Yoon, Mi Yung. 2001. "Democratization and Women's Legislative Representation in Sub-Saharan Africa." *Democratization* 8(2):169–90.

———. 2004. "Explaining Women's Legislative Representation in Sub-Saharan Africa." *Legislative Studies Quarterly* 29(3):447–468.

———. 2008. "Special Seats for Women in the National Legislature: The Case of Tanzania." *Africa Today* 55(1):61–85.

Younger, Stephen M. 2000. "Nuclear Weapons in the Twenty-first Century." Los Alamos National Laboratory, June 27. Retrieved September 2, 2008 (www.fas.org/nuke/guide/usa/doctrine/doe/younger.htm).

youthxchange. 2007. "Women's Literacy." Retrieved July 27, 2009 (http://www.youthxchange.net).

Yung, Judith. 1995. *Unbound Feet: From China to San Francisco's Chinatown*. Berkeley: University of California Press.

Yunus, Muhammad and Alan Jolis. 1999. *Banker to the Poor: Micro-Lending and the Battle Against World Poverty*. New York: Public Affairs.

Zalman, Amy. 2009. "About.com: Terrorism Issues." *The New York Times Company*. Retrieved November 5, 2009 (http://terrorism.about.com/od/whatisterroris1/ss/DefineTerrorism_2.htm).

Zborowski, Mark. 1952. "Cultural Components in Response to Pain." *Journal of Social Issues* 8(4):16–30.

Zehr , Mary Ann. 2009. "The Problem of Tracking in Middle Schools." *Education Week,* June 4. Retrieved April 17, 2011 (http://blogs.edweek.org/edweek/curriculum/2009/06/the_problem_of_tracking_in_mid.html).

Zeleny, Jeff. 2009. "Obama Vows, 'We Will Rebuild' and 'Recover.'" *The New York Times,* February 25. Retrieved August 6, 2011 (http://www.nytimes.com/2009/02/25/us/politics/25obama.html?scp=1&sq=Obama%20Vows,%20%91We%20Will%20Rebuild%92%20and%20%91Recover&st=cse).

Zelizer, Viviana A. 1985. *Pricing the Priceless Child: The Changing Social Value of Children*. New York: Basic Books.

———. 1994. *The Social Meaning of Money*. New York: Basic Books.

Zhao, Yong. 2005. "Increasing Math and Science Achievement: The Best and Worst of the East and West." *Phi Delta Kappan* 87(3):219–22.

Zigler, Edward, Sally J. Styfco, and Elizabeth Gilman. 1993. *Headstart and Beyond*. New Haven, CT: Yale University Press.

Zimbardo, Philip C. 2004. "Power Turns Good Soldiers Into 'Bad Apples.'" *Boston Globe,* May 9. Retrieved July 5, 2008 (http://www.boston.com/news/globe/editorial_opinion/oped/articles/2004/05/09/power_turns_good_soldiers_into_bad_apples/).

———. 2010. "You Can't Be a Sweet Cucumber in a Vinegar Barrel." Pp. 122–25 in *Sociological Footprints,* 11th ed., edited by Leonard Cargan and Jeanne Ballantine. Belmont, CA: Wadsworth.

Zimbardo, Philip C., Craig Haney, Curtis Banks, and David Jaffe. 1973. "The Mind Is a Formidable Jailer: A Pirandellian Prison." *The New York Times,* April 8, 36–60.

Zimmer, Hans. 1992. *Millennium: Tribal Wisdom and the Modern World*. Milwaukee, WI: Narada Productions.

Zimolzak, Chester E. and Charles A. Stansfield, Jr. 1983. *Human Landscape,* 2nd ed. Columbus, OH: Merrill.

Credits

CO Photo, page 116 (middle left).
© iStockphoto.com
CO Photo, page 116 (top right).
© iStockphoto.com/Aaron Kohr
CO Photo, page 116 (bottom).
© iStockphoto.com/Sean Locke
Photo 5.1, page 118. © Keith Roberts
Photo 5.2, page 122. © iStockphoto.com / Jacob Wackerhausen
Photo 5.3, page 123. © iStockphoto.com / Mark Coffey
Photo 5.4, page 123. © Michael Westhoff/ iStockphoto.com
Photo 5.5, page 123. © iStockphoto.com/ Julie Deshaies
Photo 5.6, page 123. © iStockphoto.com
Photo 5.7, page 123. © istockphoto.com/ Sean Locke
Photo 5.8, page 124. © Pete Souza/White House/Handout/The White House/Corbis
Photo 5.9, page 126. © Jeanne Ballantine
Photo 5.10, page 126. © Jeff Hutchens/ Getty Images
Photo 5.11, page 126. © Jeanne Ballantine
Photo 5.12, page 126. © Kate Ballantine
Photo 5.13, page 128. © iStockphoto.com/ Nancy Louie
Photo 5.14, page 129. © Elise Roberts
Photo 5.15, page 130. © Bettmann/CORBIS
Photo 5.16, page 131. Brand X Pictures/ thinkstock.com
Photo 5.17, page 132. © Steve Starr/Corbis
Photo 5.18, page 134. N/A
Photo 5.19, page 137. © USAID
Photo 5.20, page 139. © Elise Roberts
Photo 5.21, page 140. © Clay Ballantine

Chapter 6

CO Photo, page 144 (background). © Peter Chen/istockphoto
CO Photo, page 144 (top left). © Mark Goddard/iStockphoto.com
CO Photo, page 144 (top middle). Ryan McVay
CO Photo, page 144 (top right). © Stephanie Phillips/iStockphoto.com
CO Photo, page 144 (bottom). Creatas
Photo 6.1a, page 148. © iStockphoto.com/ Jennifer Matthews
Photo 6.1b, page 148. © iStockphoto.com/ Eva Serrabassa
Photo 6.1c, page 148. © Keith Roberts
Photo 6.2a, page 149. © iStockphoto.com/ Frances Twitty
Photo 6.2b, page 149. © iStockphoto.com/ Jerry Koch
Photo 6.3, page 150. © Kim Kulish/Corbis
Photo 6.4, page 151. © iStockphoto.com/ Kathye Killer
Photo 6.5, page 152. © Keith Roberts
Photo 6.6, page 153. © www.istockphoto.com
Photo 6.7, page 154. © Simon Marcus/Corbis
Photo 6.8, page 157. © Peter Turnley/Corbis

Photo 6.9, page 158. © MICHAEL REYNOLDS/epa/Corbis
Photo 6.10, page 159. ©Warrick Page/Getty
Photo 6.11, page 161. © Jeffrey L. Rotman/ Corbis
Photo 6.12, page 162. © Ed Kashi/CORBIS
Photo 6.13, page 162. © Adam Mastoon/ CORBIS
Photo 6.14, page 165. © Ron Sachs/CNP/Corbis
Photo 6.15, page 169. © iStockphoto.com/ Brandon Laufenberg
Photo 6.16, page 171. © istockphoto.com/ Slobo Mitic

Chapter 7

CO Photo, page 178 (background). © Skip O'Donnell/istockphoto.com
CO Photo, page 178 (top left). © Elise Roberts
CO Photo, page 178 (top middle). © Jeanne Ballantine
CO Photo, page 178 (top row, 3rd photo). © Juan Collado/istockphoto.com
CO Photo, page 178 (top right). © Jeanne Ballantine
CO Photo, page 178 (bottom). © Keith Roberts
Photo 7.1, page 180. © Keith Roberts
Photo 7.2, page 181. © USAID
Photo 7.3, page 182. © iStockphoto.com/ Thania Navarro
Photo 7.4, page 182. © Dave Bartruff/CORBIS
Photo 7.5, page 184. © istockphoto.com/ Tomaz Levstek
Photo 7.6, page 185. © Jeanne Ballantine
Photo 7.7, page 187. ©Kevork Djansezian/ Getty Images
Photo 7.8, page 188. © Paul Almasy/CORBIS
Photo 7.9, page 193. © USAID
Photo 7.10, page 198. © Wendy Stone/ CORBIS
Photo 7.11, page 199. © Elise Roberts
Photo 7.12, page 204. © Keith Roberts
Photo 7.13, page 205. © istockphoto.com
Photo 7.14, page 208. © Emma Rian/zefa/Corbis
Photo 7.15, page XX. © Jeanne Ballantine
Photo 7.16, page XX. Provided by authors

Chapter 8

CO Photo, page 212 (background). Hemera Technologies
CO Photo, page 212 (top left). © Jiang Dao Hua/iStockphoto.com
CO Photo, page 212 (top middle). © Elise Roberts
CO Photo, page 212 (top right). © Jeanne Ballantine
CO Photo, page 212 (bottom). © Jeanne Ballantine
Photo 8.1, page 214. © Zana Briski and Kids with Cameras. Used with permission from "Kids with Cameras."

Photo 8.2, page 215. © Sophie Elbaz/ Sygma/Corbis
Photo 8.3, page 215. N/A
Photo 8.4, page 219. © Bettmann/CORBIS
Photo 8.5, page 220. © Elise Roberts
Photo 8.6, page 223. © Bettmann/Corbis
Photo 8.7, page 225. © Andrew Holbrooke/ Corbis
Photo 8.8, page 227. © Phil Schermeister/ CORBIS
Photo 8.9, page 231. © miafarrow.org
Photo 8.10, page 232. © Keith Roberts
Photo 8.11, page 233. © Josef Scaylea/ CORBIS
Photo 8.12, page 235. © William Campbell/ Sygma/Corbis
Photo 8.13, page 236. © Bettmann/CORBIS
Photo 8.14, page 239. N/A
Photo 8.15, page 242. © TOUHIG SION/ CORBIS SYGMA

Chapter 9

CO Photo, page 246 (background). Stockbyte
CO Photo, page 246 (top left). © Zsolt Nyulaszi/iStockphoto.com
CO Photo, page 246 (top middle). © Jo Ann Snover/iStockphoto.com
CO Photo, page 246 (top right). © Jeanne Ballantine
CO Photo, page 246 (bottom). Jupiterimages
Photo 9.1, page 249. © Atlantide Phototravel/ Corbis
Photo 9.2, page 249. © Thomas Strange/ iStockphoto.com
Photo 9.3, page 252. © epa/Corbis
Photo 9.4, page 256. © Jostein Hauge/ iStockphoto.com
Photo 9.5, page 258. NA
Photo 9.6, page 259. © Hubert Boesl/dpa/Corbis
Photo 9.7, page 260. Photo by Joe Robbins; courtesy of Hanover College
Photo 9.8, page 260. © G Newman Lowrance/Getty
Photo 9.9, page 264. © iStockphoto.com/ Pathathai Chungyam
Photo 9.10, page 265. © USAID
Photo 9.11, page 266. © Jupiterimages
Photo 9.12, page 267. © Felipe Dupouy/ thinkstock.com
Photo 9.13, page 268. © Jeanne Ballantine
Photo 9.14, page 268. © Nadeem Khawer/ epa/Corbis
Photo 9.15, page 273. © goodshot/thinkstock
Photo 9.16, page 275. © UNESCO, photo by Dominique Rogers.

Chapter 10

CO Photo, page 284 (background).
© Nancy Louie/iStockphoto.com
CO Photo, page 284 (top left).
© Sandra Gligorijevic/iStockphoto.com

Glossary/Index

Migration. In terms of demographic processes, refers to the movement of people from one place to another, 415
    deterrents, 427
    internal, 428–429
    international, 426–428
    push-pull model, and, 425
Milgram, Stanley, 119
Military spending, U.S. versus the world, 397 (figure)
Mills, C. Wright, 10, 374
Mind, Self, and Society (Mead), 46
Mini-max strategy, 453
Minority groups. Groups in a population that differ from others in some characteristics and are therefore subject to less power, fewer privileges, and discrimination, 216
    characteristics of, 216
    relations with dominant groups, policies governing, 237–242
Mixed economies, 387
Mobility, intragenerational, 191. *See also* Migration; Social mobility
Mobilization for action, 454
Mobs. Emotional crowds that engage in violence against a specific target; lynchings, killings, and hate crimes are examples, 454
    riots and, differences, 455 (figure)
Monogamy. The most familiar form of marriage in industrial and postindustrial societies; refers to marriage of two individuals, 298
Moon, Sun Myung, 351
Moonies, 351
Moore, Wilbert, 183
More, Thomas, 43
Mores, 76
Mormons, 82, 345, 353
Morrison, Toni, 326
Mortality. In terms of demographic processes, refers to the death rate, 415
    diseases and plagues, from the spread of, 421–422
    rates, 419–425
Mortification, 168
Mother Earth, 75
Mothers Against Drunk Driving (MADD), 401
Mother Teresa, 180
Mugabe, Robert, 369, 444
Muhammad, 341
Multilevel analysis, 52. *See also* Macro-level analysis; Meso-level analysis; Micro-level analysis
Muñoz, Carlos, 329
Murdock, George, 299
Muslim girl, 249 (photo)
Muslim prayer, 346 (photo)
Myrdal, Gunnar, 75
Myths. Stories, true or not, embodying ideas about the world and transmitting values, 345

NAFTA (North American Free Trade Association), 443
National Association for the Advancement of Colored People (NAACP), 50, 237
National culture. A culture of common values and beliefs that tie a nation's citizens together; there may be subcultures within the national culture, 84
National Incident-Based Reporting System (NIBRS), 160

National crime, 164–165
National level
    change at the, 443–445
    crime at the, 164–165
    See also Macro-level analysis
National Organization for Women (NOW), 276
National society. Includes a population of people usually living within a specified geographic area, who are connected by common ideas, are subject to a particular political authority, and cooperate for the attainment of common goals, 19, 84
Nation-state. A political, geographical, and cultural unit with recognizable boundaries and a system of government, 387
Native American cultures, and applied sociology, 238
Natural disaster, 23 (map), 444 (photo)
Nature versus nurture, or working together, 95
Nazi Holocaust, 242
Negotiation, and avoiding war, 397–398
Neolithic Revolution, 63
Nerds, 320
Networks
    micro, meso, and macro levels, 119–121
    national and global, macro level, 138–139
    power and, 268 (photo)
    social, 119
Newport, RI, 180, 180 (photo)
New religious movements (NRMs). Innovative religious groups that may become established new religions if they survive for several generations, 351
New York Stock Exchange, 381 (photo)
Ng, Wendy, 108
Nichols, Terry, 399
Nickel and Dimed (Ehrenreich), 197
9/11 terrorist attacks, consequences, 442 (photo)
No Child Left Behind, 336, 444
Nongovernmental organizations (NGOs), 373, 457
Nonmaterial culture. The thoughts, language, feelings, beliefs, values, and attitudes that make up much of our culture, 72
Nonparticipant observation, 38
Nonverbal communication. Interactions without words using facial expressions, the head, eye contact, body posture, gestures, touch, walk, status symbols, and personal space, 122
Nonverbal language, 79
Nonviolent resistance, 236–237
Norms. Rules of behavior shared by members of a society and rooted in the value system, 75
    violations of, 77 (table)
Norris, Michael, 251
North American Free Trade Association (NAFTA), 443
NRMs. *See* New religious movements
Nuclear family. Family consisting of mother, father, and children-or any two of the three, 299
Nuremberg Trials, 242

Obama, Barack, 124 (photo), 158 (photo), 220, 228, 336, 368 (photo), 379–380, 387, 417
Obama, Michelle, 52 (photo), 99

Objectivity. Steps taken to use methods that do not contaminate one's findings and that limit the impact of the researcher's opinions or biases on the study being planned, data collection, and analysis of evidence about the social world, 33
Observation studies (field methods). Involve systematic, planned observation and recording of behavior or interaction in natural settings, 38
Occupation, outflow from father to son/daughter, 191 (table)
Occupational categories, for sociology graduates, 16 (figure)
Occupational crime, 164
Of Mice and Men (Steinbeck), 326
Ogburn, William F., 447, 462
Old Believer communities, 350, 350 (photo)
Old Order Amish, 82
Olsen, Ashley, 259 (photo)
Olsen, Mary-Kate, 259 (photo)
One Laptop per Child foundation, 209
Open class system. A societal system that allows movement between classes, 191
Open system models, 457, 457 (figure)
Operationalization, 36
Operation PUSH, 237
Organic solidarity. Émile Durkheim's term for social coherence (glue) based on division of labor, with each member playing a highly specialized role in the society and each person being dependent on others due to interrelated, interdependent tasks, 62, 267
Organizational change, models for planning, 456–457
Organizations
    bureaucracies and, meso-macro connection, 133–138
    change in, 456–457
    crime and, 162–166
    evolution of modern, 133–134
    formal, 133
    modern life, and, 134–135
    See also Meso-level analysis
Organized crime. Ongoing criminal enterprises by an organized group whose ultimate purpose is economic gain through illegitimate means, 163
Orr, Amy J., 340
Outcaste group, 198–199
Out-group. A group to which an individual does not belong and that is often in competition or in opposition to an in-group, 132. *See also* Groups
Overpopulation, 414 (photo)

Palestinian schoolchildren, 395 (photo)
Palin, Sarah, 353, 379
Panethnicity, 220
Panic. Occurs when a large number of individuals become fearful or try to flee threatening situations that are beyond their control, sometimes putting their lives in danger, 455
Pareto, 374
Parsons, Talcott, 48
Participant observation, 38
Participation, power, and individuals: micro-level analysis, 374–376
Passing, 235–236

Past-in-present discrimination. Practices from the past that may no longer be allowed but that continue to affect people today, 228, 264
Patil, Pratibha, 252 (photo)
Patriarchy, 268
Patriot Act, 443
Payos, 82
Peace, and religion, 359–361
Peace Corps, 210
Pearl Harbor, 223
Peer groups. A group of people who are roughly equal in some status within the society, such as age or occupation, 97
Pellerin, Robert M., 328
People, social nature of, 6–7
PerfectMatch, 296
Personal distance, 123
Personal space, 122–123
Perspectives, cooperative versus competitive, 53 (figure)
Petty bourgeoisie, 185
Physical anthropology, 13
Piaget, Jean, 217
Planned change, process of, 457. See also Change
Planned or centralized system. Economic systems that attempt to limit private ownership of property and have the government do planning of production and distribution, 385
Plato, 411
Play stage. The stage in developing the self when a child develops the ability to take a role from the perspective of one person at a time; simple role-taking or play-acting, 101
Pluralism. Occurs when each ethnic or racial group in a country maintains its own culture and separate set of institutions, 231
Pluralist theory, 373, 373 (photo)
Political behavior, 190
Political institutions, purposes served, 378–381
Political protests, and social networking, 390
Political science, 13
Political systems, types of
  authoritarian political systems, 382
  democratic systems, 382–385
  totalitarian, 383
Politics
  economics and, 366–401
  levels of participation in, 376
Pollack, William, 256
Polyandry, 299, 299 (photo)
Polygamy. Marriage of one person to more than one partner at the same time, 299
  Mormons and, 82
Ponzi, Charles, 164
Ponzi scheme, 164
Popeneo, David, 305
Popular culture, 68
Population(s). Permanent societies, states, communities, adherents of a common religious faith, racial or ethnic groups, kinship or clan groups, professions, and other identifiable categories of people, 407
  centers, ten largest, 429 (table)
  change, institutional influences on, 415–429

clock, 406 (table)
health and, 404–436
momentum, 407
overcrowding, 409 (photo)
percent of by race and Hispanic origin, 222 (figure)
trends, and social mobility, 193–195
Population patterns
  conflict theory and, 414
  demographic transition, 413–414
  economic development and, 415
  Malthus' theory of population, 411–413
  micro-level, 432–435
Population pyramids. Pyramidal diagrams that illustrate sex ratios and dependency ratios, 410
  affluence, by levels of, 411 (figure)
  Japan's population over age 65, 410 (figure)
  predicting community needs and services, 434
  U.S. population, 2010 and 2050, 432 (figure)
Population transfer. The removal, often forced, of a minority group from a region or country, 230
Postindustrial societies. Societies that have moved from human labor and manufacturing to automated production and service jobs, largely processing information, 65
Pot, Pol, 383
Poverty, 188 (photo)
  feminization of, 204–205
  functions of, 206
  multilevel determinants and social policy, 203–205
  relative 204
Power. Ability of a person or group to realize its own will in groups, even against the resistance of others, 201
  arenas, 369
  elite theory, 373–374
  factor, in achieved status, 201
  illegitimate, 370
  individuals, 372–373
  legitimacy of, 370–374
  national, 376–387
  national and global systems of, 387–399
  nation-state, 387–388
  pluralism theorists, 201
  pluralist theory, 373
  social constructions of, 370–372
  systems of, 381–387
  theoretical perspectives on, 370–374
  Weber's formula regarding, 372 (figure)
  what is, 369
Power elite. Power held by top leaders in corporations, politics, and military; these interlocking elites make major decisions guiding the nation, 201
Precipitating factor, 454
Predatory crimes, 161
Predestination, 281
Preference policies, 240, 241
Prejudice. Attitudes that prejudge a group, usually negatively and not based on facts, 222
  effects of, 234–237
  explanations of, 223–225
  micro-level analysis, 222–225
  minority reactions to, 235–237
  nature of, 222–223

Prestige. The esteem and recognition one receives, based on wealth, position, or accomplishments, 201
  factor, in achieved status, 201
  rankings, professions and occupations, 201 (table)
Primary deviance. A violation of a norm that may be an isolated act or an initial act of rule breaking, 154. See also Deviance
Primary groups. Groups characterized by close, intimate, long-term contacts, cooperation, and relationships, 130. See also Groups
Prince Charles, 372 (photo)
Prisons, 168–170
Privilege, in the social world, 369–370. See also Power
Process (social dynamics), 44. See also Social process
Progress, 441
Proletariat. Karl Marx's term for exploited "have-not" workers who do not own means of production and sell their labor to survive, 49, 385
Promise Academy, 337
Pronatalist policies, 416
Property factor, in achieved status, 200–201
Protestant Ethic and the Spirit of Capitalism, The (Weber), 52
Prussian model, for education, 325
Psychology, 13
Public distance, 123
Public order crimes. Acts committed by or between individual consenting adults, 161. See also Victimless crimes
Public sociology, 53–54
Push-pull model, 425

Questionnaires, 37
Quota system, 240
Quran, 261

Race. Physical characteristics that allow individuals or groups to be singled out, often for differential treatment, 216
  class and, 218, 270
  concept of, 216–219
  ethnic group stratification, 212–244
  gender and, 270
  percent of population, 222 (figure)
  social construction of, 217–218
Racism. Any institutional arrangement that favors one racial group over another, 226
  costs of, 234–235
  forms of, 225
  micro-level analysis, 222–225
  minority reactions to, 235–237
  religion and, 355–357
Rainbow PUSH Coalition, 237
Random sample, 40
Rape, 269
Rational choice theory. A theory that focuses on humans as fundamentally concerned with self-interests, making rational decisions based on weighing costs and rewards of the projected outcome, 47
  collective behavior and, 453
  deviance and, 152
  educational settings and, 319–320
  family and, 290
  interaction and, 124–125
  males and, 51
  religion and, 347–348

OUR SOCIAL WORLD
Condensed Version | SECOND EDITION

*The introductory text authored by two recipients of the
ASA's Distinguished Contributions to Teaching Award!*

The **Second Edition** of **Our Social World: Condensed Version** inspires students to see how larger social structures and global trends affect their personal lives, to develop their sociological imaginations, and to view both world events and their personal experiences from a sociological perspective. In each chapter, the authors provide a balanced organizing theme to help students see relationships between various levels of the social system.

New to the Second Edition

- Draws in students with new, **interactive "Engaging Sociology" features** that help them discover the appeal of analyzing events through a sociological lens

- Presents a **new section opener on institutions** that more clearly explains the concept for increased student comprehension

- Offers **new examples, topics, and data throughout**, as well as **30 new boxed features, 315 new references**, and **120 new photos**

Key Features

- *Offers a strong global focus*: A global perspective is integrated into each chapter to encourage students to think of global society as a logical extension of their own micro world.

- *Illustrates the practical side of sociology*: Boxes highlight careers and volunteer opportunities for those with a background in sociology as well as policy issues that sociologists influence.

- *Provides a connection between concepts and real-life situations:* The authors demonstrate the relation between sociological concepts and students' day-to-day lives.

A companion website at www.sagepub.com/oswcondensed2e includes

- A password-protected **Instructor Teaching Site** featuring test banks, PowerPoint presentations, sample syllabi, video links, audio resources, class assignments, web resources, and more.

- An open-access **Student Study Site** featuring web quizzes, flashcards, video and audio resources, web exercises, SAGE journals articles, recommended readings, and more.

FSC
www.fsc.org
MIX
Paper from
responsible sources
FSC® C011825

Cover Image: Tatiana Cardeal Furlaneto

ISBN 978-1-4129-8727-1

90000

9 781412 987271

SAGE
www.sagepublications.com
Los Angeles • London • New Delhi • Singapore • Washington DC